## The Duxbury Series in Statistics and Decision Sciences

*Applications, Basics, and Computing of Exploratory Data Analysis,* Velleman and Hoa
*Applied Regression Analysis and Other Multivariable Methods,* Second Edition, Kleinbaum, Kupper, and Muller
*Classical and Modern Regression with Applications,* Myers
*A Course in Business Statistics,* Second Edition, Mendenhall
*Elementary Statistics for Business,* Second Edition, Johnson and Siskin
*Elementary Statistics,* Fifth Edition, Johnson
*Elementary Survey Sampling,* Third Edition, Scheaffer, Mendenhall, and Ott
*Essential Business Statistics: A Minitab Framework,* Bond and Scott
*Fundamental Statistics for the Behavioral Sciences,* Second Edition, Howell
*Fundamentals of Biostatistics,* Second Edition, Rosner
*Fundamentals of Statistics in the Biological, Medical, and Health Sciences,* Runyon
*Introduction to Contemporary Statistical Methods,* Second Edition, Koopmans
*Introduction to Probability and Mathematical Statistics,* Bain and Engelhardt
*Introduction to Probability and Statistics,* Seventh Edition, Mendenhall
*An Introduction to Statistical Methods and Data Analysis,* Third Edition, Ott
*Introductory Statistical Methods: An Integrated Approach Using Minitab,* Groenevelc
*Introductory Statistics for Management and Economics,* Third Edition, Kenkel
*Linear Statistical Models: An Applied Approach,* Bowerman, O'Connell, and Dickey
*Mathematical Statistics with Applications,* Third Edition, Mendenhall, Scheaffer, and Wackerly
*Minitab Handbook,* Second Edition, Ryan, Joiner, and Ryan
*Minitab Handbook for Business and Economics,* Miller
*Operations Research: Applications and Algorithms,* Winston
*Probability Modeling and Computer Simulation,* Matloff
*Probability and Statistics for Engineers,* Second Edition, Scheaffer and McClave
*Probability and Statistics for Modern Engineering,* Lapin
*Quantitative Forecasting Methods,* Farnum and Stanton
*Quantitative Models for Management,* Second Edition, Davis and McKeown
*Statistical Experiments Using BASIC,* Dowdy
*Statistical Methods for Psychology,* Second Edition, Howell
*Statistical Thinking for Behavioral Scientists,* Hildebrand
*Statistical Thinking for Managers,* Second Edition, Hildebrand and Ott
*Statistics for Business and Economics,* Bechtold and Johnson
*Statistics for Management and Economics,* Sixth Edition, Mendenhall, Reinmuth, and Beaver
*Statistics: A Tool for the Social Sciences,* Fourth Edition, Ott, Larson, and Mendenhall
*Time Series Analysis,* Cryer
*Time Series Forecasting: Unified Concepts and Computer Implementation,* Second Edition, Bowerman and O'Connell
*Understanding Statistics,* Fourth Edition, Ott and Mendenhall

# QUANTITATIVE FORECASTING METHODS

# QUANTITATIVE FORECASTING METHODS

Nicholas R. Farnum
California State University-Fullerton

LaVerne W. Stanton
California State University-Fullerton

PWS-KENT Publishing Company
Boston

PWS-KENT
Publishing Company

20 Park Plaza
Boston, Massachusetts 02116

*To my parents*
N.R.F.

*To Nan, and Herman too.*
L.W.S.



PWS-KENT Publishing Company is a division of Wadsworth, Inc.

**Library of Congress Cataloging-in-Publication Data**
Farnum, Nicholas R.
Quantitative forecasting methods/by Nicholas R. Farnum and LaVerne W. Stanton.
p. cm.
Includes bibliographies and index.
ISBN 0-534-91686-4
1. Prediction theory. 2. Regression analysis. 3. Time-series analysis. I. Stanton, LaVerne W. II. Title.
QA279.2.F37 1989 88-21078
519.5′4—dc 19 CIP

Printed in the United States of America

89 90 91 92 93—10 9 8 7 6 5 4 3 2 1

Editor: Michael R. Payne
Production Editor: S. London
Production: Sara Hunsaker/Ex Libris
Interior Designer: Jean Coulombre
Cover Designer and Illustrator: Steve Snider
Interior Illustrator: Carl Brown
Manufacturing Coordinator: Ellen J. Glisker
Typesetting: Arcata Graphics, Kingsport
Cover Printing: John P. Pow Company
Print/Bind: R. R. Donnelley & Sons

# Preface

*Quantitative Forecasting Methods* is an applications-oriented statistical forecasting book. It was designed to serve both as a quick reference source for forecasting practitioners and as a college-level text in introductory statistical forecasting courses, often taught in business schools at the upper division and graduate level. Knowledge of algebra and basic statistics are the only prerequisites. While every effort has been made to assure that the material is mathematically correct, mathematical proofs are not emphasized. Instead, the text focuses on in-depth descriptions of the rationale underlying the various forecasting techniques, guidelines for choosing among the techniques, and detailed procedures for applying each technique.

The chapters have been organized around the classical time series notions of trend, seasonal, cyclical, and irregular components. Starting with an overview of these four components and of basic time series graphs and accuracy measures in Chapter 1, succeeding chapters cover each of the four components in a progressive manner: Chapter 2 deals with stationary series. Chapter 3 covers series exhibiting trend and irregular movements. Chapter 5 handles series with trend, seasonal, and irregular components. Chapter 6 discusses series having all four components. Classifying the topics in this manner provides a simple way of organizing the study of forecasting literature, which can otherwise resemble a collection of *ad hoc* procedures having little relation to one another. The organization also allows the material to proceed from the simplest to the more complex models as the chapters progress.

Not all forecasting topics fit into the simple framework of time series components. In particular, regression analysis and the Box-Jenkins procedures are each assigned chapters of their own. These chapters are placed roughly in the order of their relative complexity.

Regression analysis, which is used in many forecasting methods, is introduced in Chapter 4. This chapter starts with a self-contained development of simple regression and multiple regression, then proceeds to specific time series applications of regression. Readers already familiar with regression analysis may wish to skip immediately to the time series section of this chapter.

The Box-Jenkins methodology, which consists of an organized set of procedures for analyzing any time series and which is the most mathematically complex of the techniques presented, appears in Chapter 7. Every effort has been expended to keep the mathematics accessible to the text's intended audience.

The final chapter, Chapter 8, presents some of the currently available techniques for building fast adaptivity into forecasting models. While these schemes are not the book's most complex, they nonetheless are placed at the end of the text because their main application is to existing forecasting schemes.

Within chapters, material is organized as follows: (1) introduction; (2) description of the particular type of model to be discussed; (3) examples intended to give some intuitive feel for the model type; (4) statistical and other techniques for determining whether the model type is appropriate for the time series being studied; (5) model-fitting techniques; and, finally, (6) diagnostic checking procedures.

There is plenty of material for a two-semester forecasting course. A one-semester course can be constructed by picking and choosing among the topics in Chapters 3–8, but care should be taken not to skip over the material in Chapters 1 and 2, since the basic topics developed there (time series components, measures of accuracy, differencing, the autocorrelation function, least squares, etc.) are essential to developments in the other chapters.

Finally, a word about the computer. The authors recognize that meaningful development and use of forecasting models requires the use of computers. Computer printouts from various programs are integrated into the text's discussions. However, because there is such a wide variety of software available, the text is not tied to any particular package. Any of the techniques dealt with in the text can be handled by the more popular statistical software, such as Minitab, SAS, BMDP, and SPSS, among others.

The authors wish to thank the professors that reviewed our manuscript, among them were: J. Scott Armstrong, The Wharton School, University of Pennsylvania; Joseph Glaz, University of Connecticut; Stanley R. Schulz, Cleveland State University; David Stoffer, University of Pittsburgh; Paula Texter, University of South Carolina; Mary Sue Younger, University of Tennessee; and Louis H. Zincone, Jr., East Carolina University.

*Nicholas R. Farnum*
*LaVerne W. Stanton*
Fullerton, California

# Contents

# Forecasting: An Introductory Discussion

**Objective:** *To become familiar with the fundamental concepts and problems associated with the forecasting process and study criteria for choosing and evaluating models.*

## 1.1 Introduction

Forecasting is a universal endeavor. At one time or another everyone perceives a need to forecast. The practice of basing a current decision upon a forecast of the future is so common, it is hard to imagine any individual or organization that doesn't do so at one time or another.

While there is little question as to the need for forecasting, there are many questions concerning the actual process. Every time a forecast is obtained the following issues must necessarily be resolved:

- Exactly what is to be forecast?
- Which forecasting method will be used?

Sometimes these decisions are very easy to make; sometimes they require great care. For example, if we set out to predict cash flow for an upcoming month, the correct approach seems straightforward: Divide the various sources of income and expenditures into categories, estimate the dollar values of each category for the month, then combine the estimates. Exactly what to forecast seems self-evident as well (expenses, income), although there is probably some small amount of decision making involved. For example, do we make "entertainment" and "miscellaneous" separate categories or not? The method of categorizing, estimating, and combining estimates seems so reliable and easy that we don't feel any need to consider other methods. But notice how our awareness of these decisions changes if we consider making a cash flow forecast for a large business. There are many more categories of revenue and expenditures, so the right choices will be harder to make. More importantly, though, the best method of forecasting is no longer at all obvious. As before, we would probably combine estimates of the categories, but the method of obtaining these individual estimates is less clear. How will such variables as sales revenues, inventory-related costs, personnel changes, and interest rates be estimated? There will be many methods from which to select, each with its own virtues and drawbacks.

There are other questions that are also basic to any forecasting process:

- How accurate is a particular method?
- How much time and money should be devoted to obtaining the forecast?
- How is the forecast to be obtained?

The amount of effort spent answering these questions will depend, as before, on the situation (think about them in the context of the previous cash flow example). Also, notice how interrelated the questions are. Accuracy depends on the method used; the method used depends on available time and money, as well as what will be forecast and how the forecast will be used. More money might be made available, for example, if better accuracy could be achieved, and so forth.

Underlying all of these considerations is a working definition of the word "forecast." The definition depends upon where our interest lies. Do we want to predict whether an event (corporate merger, earthquake, war, sports event outcome) will or will not occur? Or are we interested in the future values of numerically measured quantities (quarterly sales, expenses, prices)? In this text we will be concerned with approaches using analysis of historical data to predict or estimate (forecast) future values of numerical quantities (variables). While we will not focus on predicting events, it should be recognized that the occurrence of some events (e.g., "the stock market will rise") are inextricably tied to numerical quantities (e.g., the Dow Jones Index), so these two approaches to prediction are not necessarily mutually exclusive.

Sections 1.4 and 1.5 discuss, in detail, the questions, definitions, notation, and terminology common to all quantitative forecasting techniques. In Section 1.6 a decomposition method for analyzing time series data is introduced. The characteristics (or components) of a time series that are identified by this decomposition method will provide a convenient structure on which to base our study of forecasting practice.

## 1.2 The Nature of Forecasting

### 1.2.1 Terminology

As was mentioned, this text deals with those methods involving the analysis of *data* as a means of producing forecasts, that is, statistical or quantitative forecasting methods. The basic premise of these methods is that discernible time-dependent patterns and relationships exist in data and are stable enough to allow extrapolation into the future.

**DEFINITION:** A **statistical forecasting technique** is one that generates forecasts by extrapolating patterns in historical data.

Hereafter, for the sake of convenience, we will use the words "forecasting," "statistical forecasting," and "quantitative forecasting" interchangeably.

The data referred to in the above definition usually consist of measurements on one or more **numerical variables** collected over some **time period.** Such data are said to form a **time series.** A little later on we will give a more precise definition of time series and discuss some approaches to analyzing them.

### Nomenclature

Many different forecasting techniques have been developed and studied over the years, and many names are used when referring to them. Having so many names around can be confusing. However, you will find that most of these names fall into one of two categories:

- Names describing a *specific item* or *area* of forecast interest (sales forecasting, financial forecasting, technology forecasting, etc.)
- Names describing a *particular* forecasting *scheme* (exponential smoothing, autoregression, decomposition, etc.)

A few of the names can justifiably be put into either group (econometric methods), a few don't seem to fall into either (Box-Jenkins methods), but for the majority, this classification works well. Even when a forecasting technique is named in honor of its inventor, a secondary name from one of these groups is usually added (Holt's two-parameter *exponential smoothing*). Thinking about nomenclature in this way will help lend relevance to the various terms, make them less intimidating, and serve to focus attention on the more important questions about what is being forecast and what method will be used.

## 1.2.2 The Need for Forecasting

A widespread need for forecasts as inputs to decision making has already been mentioned, but there are other aspects to consider.

### Perceived Value

Resources spent on forecasting usually vary with time, depending chiefly on the attitudes of management. For example, in times of stable economic growth the level of effort devoted to obtaining forecasts could decline. Feeling that corporate growth will naturally follow the upward economic trend, management may divert resources towards other concerns, deciding that further or more precise forecasts are unnecessary. In light of today's sensitive world economy, such an attitude may seem short-sighted, but it does illustrate the fact that there is constant concern about what the value forecasts will add to the decision-making process. The impact of this concern on the choice of a forecasting technique will be more completely addressed later.

#### Use with Economic Series

While forecasting methods can be used for any time series, economic time series is one area of broad application. Examples include not only time series measured in monetary terms, such as sales, GNP, expenditures, etc., but also nonmonetary series, such as unemployment, interest rates, market share, etc., which have economic relevance even though not measured directly in dollars. Concentration of effort in this area is not surprising considering the diversity of measures that have been created to provide visibility into economic processes.

### 1.2.3 Forecasting in Practice

Many applications arise in fields where forecasts are needed on a regular basis. As you would expect, much attention has been devoted to developing forecasting models specific to these fields. Some of the areas in which forecasting is an ongoing concern include:

- sales forecasting
- strategic planning
- financial forecasting
- technology forecasting
- energy forecasting
- market forecasting
- econometric modeling
- inventory control
- material requirements planning
- service product forecasting
- stock market forecasting

#### Planning

Planning and forecasting are closely related functions in any decision-making process. When forecasting is a regular activity in a business environment, it is usually because planning is viewed as critical to the success of the company. Planning involves several stages. First, current forecasts are used as inputs in developing an interim plan. Then, as more information and the effects of planning are observed, the forecasts are revised, and the plan is updated (Armstrong, 1978, pp. 5–7). This process is repeated until a suitable plan is finalized. The influence which forecasts are allowed to have as the plan develops depends on their reliability and on the costs of over- and underforecasting. Questions about what is being forecast and how the forecast is to be generated should be kept in mind when studying a particular application area. Answers to these questions will provide a good overview of the practice of forecasting in that area. In the next few paragraphs this approach is used to examine some specific areas where forecasting is applied.

### Sales Forecasting

The forecasting of sales is probably near the top of any list of important applications. Here, the answers to our question of "what is to be forecast?" are numerous. There is much more than just the sales of a particular product, or even an entire company's sales. Market potential, which is the total dollar volume or number of units sold by a company and all its competitors, would certainly be of interest as well. Estimates of a company's or competitor's share of the market and competitive pricing strategies are also needed. Many different methods are used to obtain these forecasts. One can simply predict that sales will grow by a fixed percentage from period to period. Or one can use forecasting models specific to a particular product. It is likely that the quantitative and qualitative inputs provided by sales representatives will also be used, either as an aid in generating forecasts or as a way of validating the forecasts. Adding new product lines or dropping old ones, planning purchases, setting sales quotas, and making personnel, advertising, or financial decisions are just a few of the areas impacted by sales forecasts. As a matter of fact, few company functions, if any, go unaffected.

### Economic Forecasting

Economic (or econometric) forecasting should also be included on any list of ongoing forecasting activities. As was pointed out when discussing nomenclature, economic forecasting actually refers either to a certain type of methodology used to obtain forecasts or to the practice of forecasting key indicators of the economy. We will focus on the latter meaning here and defer discussion of the methodology of econometric forecasting to a later section.

Estimation of economic indicators is so great a concern that there are entire companies formed expressly for this purpose. Many universities and large financial institutions also develop and maintain their own forecasting services.

The approach used by most organizations undertaking this formidable task is to use a combination of three or four forecasting techniques: econometric modeling, time series analysis, expert opinion, and other types of data analysis. When the forecasts generated by these methods are eventually combined, the econometric model is given the majority of the weight. The econometric model employed is designed by the organization and is so fundamental to the firm's forecasts that the model's name is often synonymous with that of the firm. Well-known examples include the Chase model, the Wharton model, the DRI (Data Resources, Inc.) model, and the Kent model.

The scale of such projects is large. Forecasts for between several hundred and several thousand economic variables are generated, usually on a monthly basis, and the results are stored in a computer along with vast historical data banks. To effectively use such a huge information resource, clients must be given a way to interface with it. The organization provides this interface in the form of consultants and various subscription packages. Consultants help clients clarify their information needs (i.e., What do they want to forecast?)

and suggest suitable formats for the presentation of data. The subscription services vary in format from simple data bank access privileges to individually tailored forecasting packages.

Although initially used as a governmental planning tool, econometric forecasts have been increasingly used by private organizations, and recent indications point to a continuation of this trend.

### Manufacturing Forecasting

Production and inventory control, which comprise a major portion of any manufacturing activity, also make wide use of forecasting techniques. Because of the costs associated with manufacturing, it is not only important to have reasonable forecasts of the demand for finished goods, but also to have a mechanism for efficiently reacting to changes. Changes in the forecasts for demand of end-items will have a bearing on decisions about safety stock levels. Similarly, the effect of a cancelled order must be assessed quickly so that material received for use on that order can be efficiently rerouted. Since products are usually assembled in a step-by-step fashion, scheduling is required, which further complicates matters. Materials should be on hand when needed, but should not arrive too early—otherwise, storage space will be tied up. More importantly, such material represents capital that cannot immediately be used or invested.

How demand is forecast depends upon where an item is to be used in the assembly process. If the item is a finished good or cannot be considered as a component within some larger assembly, it is said to experience *independent* demand; otherwise its demand is *dependent* (Orlicky, 1975). Most of the standard techniques from sales and econometric forecasting can be used to forecast the need for independent demand pieces; however, the dependent demand items require a different approach. In one sense, dependent demand items should be easy to forecast once a reliable forecast for the finished goods had been made. If we forecast a demand for X automobiles, then our forecast for the number of wheels needed should be 4X (if each car requires 4 wheels). But now, what is the best way to handle scheduling? Should we order the exact number of components and schedule their arrival to coincide with the exact time and place of installation, or should we maintain a reasonable inventory of each component at convenient locations along the assembly line, replenishing at regular intervals? Which of these two policies is chosen will dictate the type of forecasting tool to be used. For example, if we schedule arrival when needed, the task is best handled by computer, often through sophisticated materials requirements planning (MRP) software. On the other hand, maintaining inventory at various points along the assembly line usually involves some type of economic order quantity (EOQ) model that signals when to place new orders. We will take a closer look at these approaches to manufacturing forecasting in Section 1.3.

### Stock Market Forecasting

Forecasting, by every imaginable means, is done daily in the stock market. This is quite an exciting arena in which to examine forecasting methodology, since most people have strong opinions about what to forecast, which methods are best, and which methods are useless. In addition, the effects of good or bad forecasts can be measured in very real terms.

Almost without exception, the diverse approaches to predicting changes in security investments fall into one of two broad groups: technical analysis or fundamental analysis (Lorie and Hamilton, 1973). **Technical analysis** is based on the premise that there are identifiable trends in the past data on any security and that, to some extent, this information should be valuable in guiding one's investment strategies. Technical analysts, often called "chartists," search stock market records for indicators of turning points, business cycles, and trends. The graphs and indices they construct are then used as signals of the appropriate times to buy and sell securities.

**Fundamental analysis,** on the other hand, uses information from annual reports, financial news sources, and mathematical models to estimate the "intrinsic value" of a given security. These estimates are then compared to the current market price for the security and over- or undervalued securities are identified.

There is a fairly clean split between the two schools of thought. That is, most security analysts identify themselves with only one of the two approaches. It is probably safe to say, however, that about 90 percent of the analysts are "fundamentalists" rather than "chartists." The belief that a security's value can be determined from such factors as a company's earnings per share or dividend payments, among other things, is very widely held.

Deciding whether to use fundamental or technical analysis is complicated by some results in the field of stock market research. These results are summarized in the **efficient market hypothesis.** In its simplest form, this hypothesis says that security prices adjust so quickly to new information that, in the long run, all attempts to predict future price changes will be of no value. Stated differently, the various forms of the efficient market hypothesis imply that, in the long run, neither fundamental analysis, technical analysis, nor even the possession of privileged information will allow selection of a portfolio of stocks that will produce consistently superior investment results. Since it seems to say that market movements are not predictable, this hypothesis is often called the "random walk theory." Needless to say, the implications of these statements are enormous and the controversy surrounding the subject has been correspondingly heated. A reader interested in a more thorough discussion of the problem will have no trouble finding many extensive references on the matter (e.g., Malkiel, 1985).

## 1.3 Examples

### 1.3.1 Sales Forecasting

To illustrate some of the questions that arise in sales forecasting, consider the following scenario: A company making computer peripherals (terminals, printers, microfilm output devices, etc.) has noticed recent declines in sales of one of its established products. Such a decline naturally evokes important questions about the future market for that product:

- Where is the product in its life cycle? Do recent patterns arise from changes in the market for the product, or do they only reflect movements in the general economy?
- If the product is in the declining phase of its life cycle, should it be rejuvenated or dropped?
- How is the competition affecting the market share for this product?
- If the product is to be continued, what advertising, distribution, promotion, selling, and service resources are necessary to maintain a desirable sales level?
- How should sales efforts be distributed over this product and others in the same product family?

The maturity of the product in question should ensure that a good deal of historical sales data are available. Management, too, can be expected to have valuable insights concerning competition, new market trends, and other relevant economic variables. With so much input, quantitative forecasting techniques should be readily applicable. Either a "top-down" or "bottom-up" approach can be adopted (Shlifer and Wolff, 1979).

The **"top-down"** approach is one of *disaggregation.* It starts with an estimate of sales of the entire peripherals industry. Such estimates can be obtained from the company's own econometric models or, perhaps, by subscribing to an industry-specific forecasting service. The company then estimates its percentage share of total industry sales and multiplies by the industry sales figure to give a forecast of company sales. This process is then repeated using the product's share of the total company sales to yield the sales forecast for the product. In practice, breaking down a "top" forecast into successively smaller ones may require many steps, depending on the choice of subcategories. For example, we could continue by breaking total product sales into sales by territory, by distributor, by salesperson, and so on.

The **"bottom-up"** method works by *aggregating* data. It starts by looking at sales data that is more detailed than at product level. For example, the forecasts of sales by each salesperson in a particular territory may be summed to give a forecast of total sales for that territory. Territory forecasts can then be aggregated to yield a forecast of total product sales. As in the "top-down" approach, there may be many steps in the process.

### 1.3.2 Manufacturing Forecasting

Manufacturing is a complex process that is sensitive to errors in demand forecasts or changes in the economic climate. Large plants, expensive machinery, large stocks of parts and materials, sizeable labor forces, and complex planning represent a serious commitment of resources. The complexity creates diverse forecasting requirements with little tolerance for bad forecasts (Orlicky, 1975, ch. 10). Strangely enough, this very complexity also allows for the construction of procedures for reacting to forecast errors. Forecasts of overall demand can be quickly and accurately transformed into the required forecasts of materials, labor, parts, and so forth.

Figure 1.1 Bill of Materials (BOM) Diagrams

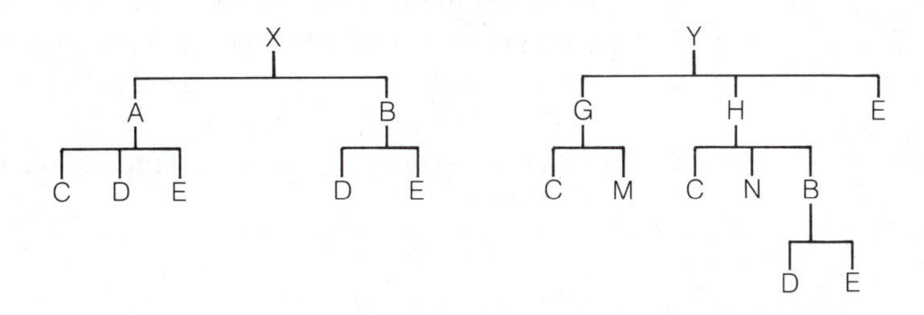


As an illustration of the way in which forecasting and manufacturing processes interact, assume that a particular manufacturer makes only two products, X and Y. These products, along with their component parts, are shown in Figure 1.1. Interpreting the diagram for product X, we see it has two subassemblies, A and B, each made up of smaller parts. Assembly A consists of parts C, D, and E, while assembly B requires two parts, D and E. The diagram for product Y is interpreted similarly. Defining a product in terms of its subcomponents is referred to as specifying the **bill of material (BOM)** of the product.

If we have initial forecasts for the number of X's and Y's demanded, the requirements for all the subcomponents can be calculated. For example, an anticipated demand for 3 X's translates into a need for 3 C's, 6 D's, and 6 E's, while a demand for 2 Y's will require 4 C's, 2 D's, 4 E's, 2 M's and 2 N's. In the same way, demand for the intermediate assemblies A, B, G, and H can be calculated. This process of calculating the gross requirements for parts, based upon forecast demand, is called "exploding" the bill of material.

Theoretically, then, the forecasting process will involve obtaining a demand forecast for X and Y, "exploding" this into forecasts for required parts or material, and scheduling the ordering and delivery of these components. Complications arise, however, when cancelled orders, new orders, and delays in deliveries of crucial parts begin to show up. Suppose the demand for X's decreases by 2 units and the demand for Y's simultaneously increases by 1 unit. What will be the resulting effect on the number of C's required? Using the BOMs of X and Y, we can see that the net change in the number of C's will be 0, so that the changed demand for X and Y has had no impact on the future gross requirements for part C. Had the demands been reversed, the X's increasing by 1 and the Y's decreasing by 2, the net change for C's would be $-3$. That is, 3 fewer C's would be needed. The effects on demand for the other parts may be examined similarly. Even when there is no net change in the demand for a part, scheduling will be affected because of the changes in the relative numbers of end-items (i.e., X's and Y's) being made.

Assembly lines cannot grind to a halt while such analyses are carried out. Instead, the ability to rapidly react to forecast errors must be incorporated into the manufacturing system itself. The EOQ and MRP systems discussed in Section 1.2 are two of the basic mechanisms used for such purposes.

After assessing the effects of demand changes, other basic questions can be addressed, such as:

- Will there be enough plant capacity to accommodate the new demand?
- What are the future prices of parts and materials likely to be?
- How are procurement lead times and job scheduling affected?
- How much work is already in progress? And how will it be impacted by the changes?
- Would a revised production plan still meet the company's objective for return on investment?

### 1.3.3 Econometric Analysis and Forecasting

In the search for appropriate forecasting methods it may be possible to identify a set of explanatory variables that affect, and are perhaps affected by, the quantity being forecast. It is natural to try to express the interrelationships between these variables mathematically so that knowledge of the "explanatory" variables, once available, can be used to compute forecasts. This approach to forecasting is termed "econometric forecasting" or "econometric analysis." More specifically, the steps in **econometric forecasting** are:

- *Identification* of explanatory variables
- *Construction and testing* of a mathematical model relating the explanatory variables to the variable(s) being forecast
- Use of the model either to *explain* the behavior of the original time series, to *forecast* the series, or to *evaluate* the effects of policy decisions on the series

The following example illustrates the econometric modeling process.

Suppose one wants to forecast "price" and "quantity sold" of a widely used product (e.g., a certain make of automobile or a certain agricultural good). One approach is to consider the "quantity demanded" separately from the "quantity supplied." It may be that "quantity demanded" depends chiefly on two things: (1) the "current price" and (2) the "quantity demanded in the *previous* time period." Similarly, the "quantity supplied may depend on the "current price" and the "current cost of labor." To tie these variables together, economists then assume that, eventually, an equilibrium is reached where "quantity demanded" equals "quantity supplied." One mathematical expression of these relationships might be:

$$\text{(current quantity demanded)} = A_1 \cdot \text{(current price)} + B_1 \cdot \text{(previous quantity demanded)} + C_1$$

$$\text{(current quantity supplied)} = A_2 \cdot \text{(current price)} + B_2 \cdot \text{(cost of labor)} + C_2$$

along with the equilibrium condition

$$\text{(current quantity demanded)} = \text{(current quantity supplied)}$$

$C_1$ and $C_2$ are included as *random error,* or simply *error* terms, to take into account the fact that the equations do not hold perfectly. That is, $A_1 \cdot$ (current price) + $B_1 \cdot$ (previous quantity demanded) may in fact be a good estimator of, but not exactly equal to, the current quantity demanded. The amount of error, however small, is then included so that the equals sign can legitimately be put in the equation. Without going into the actual mechanics of solving these equations (the interested reader may consult any econometrics text, e.g., Judge, et al., 1980; Intriligator, 1978; Wonnacott and Wonnacott, 1979), this simple model can be used to exhibit some of the basic concerns in econometric analysis.

Examine the variables in the model. There is usually great latitude in the selection of explanatory variables. In particular, any previous (*lagged*) value of any variable can be used as an explanatory variable. "Quantity demanded in previous period" was used as a lagged explanatory variable here. Also, in the equation for "quantity supplied," "price in the previous period" could certainly have been included. It was excluded to keep the model simple.

Economists classify the variables in a model as endogenous or exogenous. **Endogenous** variables are those whose values the model is built to explain, while the **exogenous** variables are those that are not determined by the model but nonetheless impact on it. In the example, "price" and "quantity demanded" and "quantity supplied" would most likely be thought of as endogenous and "cost of labor" as exogenous. The next step in an analysis would be to solve the equations for the endogenous variables as functions of the exogenous ones. It is worth mentioning that the specific form of the model is a matter of choice and experimentation. Simple linear equations were used here, but more mathematically complex models could certainly have been chosen.

Eventually, the constants (called **parameters**) $A_1$, $A_2$, $B_1$, and $B_2$ used in the model will have to be estimated. Estimation of these constants, just one part of the process of solving the model equations, is achieved by using historical data. Finally, when the solution for endogenous variables in terms of exogenous ones has been found, the model can be used for the econometric activities mentioned above (explanation of behavior, forecasting, and policy evaluation).

Good econometric modeling cannot be undertaken without a reasonable knowledge of economics. Without some economics background, an analyst probably would not have separated "quantity demanded" and "quantity supplied" in the above example, and would, therefore, have omitted the equilibrium condition relating them. Additionally, knowing how economic variables interact is a great aid in choosing the model's mathematical form.

## 1.4 Basic Notation and Concepts

### 1.4.1 Terminology, Notation, and Models

#### Terminology

Every measurement, whether taken for forecasting purposes or not, is taken at some point in time. In many analyses, this time aspect of the data may be harmlessly suppressed or ignored. Forecasting methods, on the other hand, since they deal with estimating future behavior, necessarily exploit the data-time relationship. To handle this emphasis on time, some new terminology is needed.

Numerical measurements on one or more variables are called **data.** Data can be collected on a one-time basis for the purpose of solving a particular problem or answering a specific question. Frequently, though, the collection of data is part of some ongoing record-keeping process which makes the observations time dependent. This time dependence raises some important questions. For example, should the numbers be recorded *continuously,* as they arise, or should "collection periods" of fixed length be established, with the data being summarized and reported at the *end of each period?* If the latter option is chosen, should the length of the collection periods be fixed, or allowed to depend on the data (such as reporting data only when certain predetermined criteria are satisfied)?

Because so much of human activity is tied to the calendar, with its equal time periods (weeks, years), fixed collection periods of equal length are a convenient choice. In the case of sales data, for instance, it is customary to report sales figures on a monthly, quarterly, or yearly basis (shorter collection periods are possible too). Stock market indexes are recorded hourly, daily, and weekly. Collection periods of varying length, although used to a lesser extent, can arise quite naturally. Inventory control yields a case in point. In a commonly used inventory policy, parts are reordered whenever stock drops below a safety threshold. Recording the number of parts ordered each time would yield data at unequal time intervals because reorder points occur at irregular intervals.

Sometimes, data that appear to be reported for equal intervals of time actually are not, as would be the case if a business bases its records on an "accounting" calendar. The "13-week," or "4–5–4," calendar in Figure 1.2 forces each of its "months" to contain exactly four or five weeks.

Regardless of the decisions about how to report time-dependent data, the sequence of measurements that results is called a time series.

---

**DEFINITION:** A **time series** is a sequence of measurements of some numerical quantity made at or during successive periods of time.

---

A city's daily temperatures recorded at noon each day or closing stock market prices for several consecutive days would form a time series. These measurements are made at specific points in time, and their number is called the **length** of the series. Sales data, on the other hand, might be reported quarterly. This type of data would be measured and summarized, or aggregated, during each quarter, and two years' worth of such data would form a time series eight periods in length. The distinction between data measured *at* specific points in time and data measured *over* intervals of time is important for understanding basic time series notation.

**Figure 1.2 The 13-Week, or "4–5–4," Calendar (for 1988)**

JANUARY

| Sat | Sun | Mon | Tue | Wed | Thu | Fri |
|---|---|---|---|---|---|---|
| | | | | | | 1 |
| 2 | 3 | 4 | 5 | 6 | 7 | 8 |
| 9 | 10 | 11 | 12 | 13 | 14 | 15 |
| 16 | 17 | 18 | 19 | 20 | 21 | 22 |
| 23 | 24 | 25 | 26 | 27 | 28 | 29 |

FEBRARY

| Sat | Sun | Mon | Tue | Wed | Thu | Fri |
|---|---|---|---|---|---|---|
| 30 | 31 | 1 | 2 | 3 | 4 | 5 |
| 6 | 7 | 8 | 9 | 10 | 11 | 12 |
| 13 | 14 | 15 | 16 | 17 | 18 | 19 |
| 20 | 21 | 22 | 23 | 24 | 25 | 26 |

MARCH

| Sat | Sun | Mon | Tue | Wed | Thu | Fri |
|---|---|---|---|---|---|---|
| 27 | 28 | 29 | 1 | 2 | 3 | 4 |
| 5 | 6 | 7 | 8 | 9 | 10 | 11 |
| 12 | 13 | 14 | 15 | 16 | 17 | 18 |
| 19 | 20 | 21 | 22 | 23 | 24 | 25 |

APRIL

| Sat | Sun | Mon | Tue | Wed | Thu | Fri |
|---|---|---|---|---|---|---|
| 26 | 27 | 28 | 29 | 30 | 31 | 1 |
| 2 | 3 | 4 | 5 | 6 | 7 | 8 |
| 9 | 10 | 11 | 12 | 13 | 14 | 15 |
| 16 | 17 | 18 | 19 | 20 | 21 | 22 |
| 23 | 24 | 25 | 26 | 27 | 28 | 29 |

MAY

| Sat | Sun | Mon | Tue | Wed | Thu | Fri |
|---|---|---|---|---|---|---|
| 30 | 1 | 2 | 3 | 4 | 5 | 6 |
| 7 | 8 | 9 | 10 | 11 | 12 | 13 |
| 14 | 15 | 16 | 17 | 18 | 19 | 20 |
| 21 | 22 | 23 | 24 | 25 | 26 | 27 |

JUNE

| Sat | Sun | Mon | Tue | Wed | Thu | Fri |
|---|---|---|---|---|---|---|
| 28 | 29 | 30 | 31 | 1 | 2 | 3 |
| 4 | 5 | 6 | 7 | 8 | 9 | 10 |
| 11 | 12 | 13 | 14 | 15 | 16 | 17 |
| 18 | 19 | 20 | 21 | 22 | 23 | 24 |

JULY

| Sat | Sun | Mon | Tue | Wed | Thu | Fri |
|---|---|---|---|---|---|---|
| 25 | 26 | 27 | 28 | 29 | 30 | 1 |
| 2 | 3 | 4 | 5 | 6 | 7 | 8 |
| 9 | 10 | 11 | 12 | 13 | 14 | 15 |
| 16 | 17 | 18 | 19 | 20 | 21 | 22 |
| 23 | 24 | 25 | 26 | 27 | 28 | 29 |

AUGUST

| Sat | Sun | Mon | Tue | Wed | Thu | Fri |
|---|---|---|---|---|---|---|
| 30 | 31 | 1 | 2 | 3 | 4 | 5 |
| 6 | 7 | 8 | 9 | 10 | 11 | 12 |
| 13 | 14 | 15 | 16 | 17 | 18 | 19 |
| 20 | 21 | 22 | 23 | 24 | 25 | 26 |

SEPTEMBER

| Sat | Sun | Mon | Tue | Wed | Thu | Fri |
|---|---|---|---|---|---|---|
| 27 | 28 | 29 | 30 | 31 | 1 | 2 |
| 3 | 4 | 5 | 6 | 7 | 8 | 9 |
| 10 | 11 | 12 | 13 | 14 | 15 | 16 |
| 17 | 18 | 19 | 20 | 21 | 22 | 23 |
| 24 | 25 | 26 | 27 | 28 | 29 | 30 |

OCTOBER

| Sat | Sun | Mon | Tue | Wed | Thu | Fri |
|---|---|---|---|---|---|---|
| 1 | 2 | 3 | 4 | 5 | 6 | 7 |
| 8 | 9 | 10 | 11 | 12 | 13 | 14 |
| 15 | 16 | 17 | 18 | 19 | 20 | 21 |
| 22 | 23 | 24 | 25 | 26 | 27 | 28 |

NOVEMBER

| Sat | Sun | Mon | Tue | Wed | Thu | Fri |
|---|---|---|---|---|---|---|
| 29 | 30 | 31 | 1 | 2 | 3 | 4 |
| 5 | 6 | 7 | 8 | 9 | 10 | 11 |
| 12 | 13 | 14 | 15 | 16 | 17 | 18 |
| 19 | 20 | 21 | 22 | 23 | 24 | 25 |

DECEMBER

| Sat | Sun | Mon | Tue | Wed | Thu | Fri |
|---|---|---|---|---|---|---|
| 26 | 27 | 28 | 29 | 30 | 1 | 2 |
| 3 | 4 | 5 | 6 | 7 | 8 | 9 |
| 10 | 11 | 12 | 13 | 14 | 15 | 16 |
| 17 | 18 | 19 | 20 | 21 | 22 | 23 |
| 24 | 25 | 26 | 27 | 28 | 29 | 30 |
| 31 | | | | | | |

## Notation

Quantitative forecasting, as the name implies, involves mathematical analysis of time series data. With time series, mathematical notation offers the usual convenience of reducing complex statements and ideas to a short series of symbols. In forecasting, this notation must distinguish between a time series and *forecasts* of that series. It must also unambiguously refer to the underlying *time periods.*

It is conventional to use a lowercase letter to denote the quantity being measured in a time series. For uniformity, the letter $y$ will always be used for the time series variable unless there is more than one time series involved, in which case additional letters (usually $x$'s) become necessary. The particular time period associated with an observation is then included as a subscript. Thus, $y_t$ means "the value of the series at or during time period $t$." We might use $y_1, y_2, \ldots, y_{24}$ to denote a two-year series of monthly sales data. It is inherent in this scheme of subscripting that the time series start somewhere. In this example, since we choose to focus on a specific two-year period, the first month in this two-year span is where the series begins, i.e., at time $t = 1$. Subsequent months are then numbered sequentially without reference to the particular year within which they occur. To avoid having to write out the whole sequence of terms in a time series, we will often use the notation $y_t$, $t = 1, 2, \ldots$ or $y_1, y_2, \ldots$ to represent the entire series. Occasionally, if the context makes it clear that time series are being discussed, we will even suppress the $t$ in $y_t$ when doing so simplifies the presentation.

Notation is also needed for distinguishing between an *actual* value of a time series (at time $t$) and a *forecast* value. We will use $\hat{y}_t$ in this context. That is, $\hat{y}_t$ is an estimate of $y_t$ generated by some model. (In general, a "^" placed above a quantity will indicate that the quantity is being estimated.) Since $\hat{y}_t$ is a forecast, its value must be determined *before* time $t$ and, in practice, is often calculated just one period before, at time $(t - 1)$.

The usefulness of a forecasting method is usually judged by comparing the original series, $y_1, y_2, \ldots$, to the series of forecasts, $\hat{y}_1, \hat{y}_2, \ldots$. Later in this section some of the widely accepted ways of assessing forecast accuracy will be presented. These methods each depend, in some manner, on the series of "forecast errors" created by using $\hat{y}_t$ to predict $y_t$. The **forecast error** series, often referred to as the **series of residuals,** is denoted $e_t$ and is calculated by simply subtracting the series $\hat{y}_t$, term-for-term, from the series $y_t$, giving $y_t - \hat{y}_t = e_t$. Note that a negative value of $e_t$ implies an overforecast, and a positive one an underforecast.

Essentially, the $e_t$'s measure how much our forecasts missed the actual values. Intuitively, it seems that the behavior of the residual series $e_t$ will reflect much about the effectiveness of the forecasting model used. The very best we could hope for is an error series all of whose terms are 0, that is, only perfect forecasts. In practice, of course, this ideal is rarely, if ever, achieved, and we settle for a series of uniformly small $e_t$'s. The occurrence of a few unusually large values, or of noticeably oscillating patterns, in the residual series may be taken as evidence of model inadequacy, and signals the need for corrective action. Although this may mean searching for an entirely different model, more often it entails only some adjustments or fine-tuning of the model already in use. Throughout this text, analysis of the residual series is emphasized as the *key* diagnostic tool available to a forecaster.

### Basic Forecasting Notation

$y_t$ = the **value** of a time series at or during period $t$

$\hat{y}_t$ = the **forecast** of $y_t$

$e_t = y_t - \hat{y}_t$ = the **error,** or **residual,** created by using $\hat{y}_t$ to predict $y_t$

There is more forecasting notation, but much of it tends to be specific either to a particular type of analysis or to a particular problem area. This other notation will be introduced in the context of the model or analysis to which it applies.

## Models

Throughout this chapter we have repeatedly referred to forecasting "models." Until now, only the vague notion of a model as a process or system for generating forecasts has been sufficient. We will now describe the modeling process in more detail.

In statistical forecasting, the following definitions apply:

**DEFINITIONS:** A **model** is a mathematical description of the process which generates a given time series.

A forecasting **procedure** or **scheme** as derived from the model is a procedure whose input is historical data and whose output is a forecast.

Good models need not necessarily be complex. In fact, one of the main goals of modeling is to provide a simplified but acceptably accurate view of a more complicated situation. A later section discusses simplicity and accuracy as they affect model choice. For the moment, however, let us look at some concrete examples of forecasting models.

There are two kinds of models, often chosen because they are simple and intuitively appealing. The first, the **"no-change"** model, assumes that the values of the time series are relatively unchanging from period to period. The current value of the series is then used as the forecast for the next period. That is,

$$\hat{y}_{t+1} = y_t$$

In a situation where sales for the current month total $5 million, for example, the "no-change" model would forecast sales of $5 million for the next month. If, on the other hand, sales have been tending to increase by about 1 percent each month, the **"%-change"** model would be used. It forecasts next month by increasing current sales by a fixed percentage (1% in this example). That is,

$$\hat{y}_{t+1} = (1 + c)y_t$$

where $c$ is the rate of increase (or decrease if $c$ is negative). For the current example,

$$\hat{y}_{t+1} = 1.01 \cdot y_t$$

so that sales of $5.0 million this month would produce a forecast of $5.05 million for next month. Both the "no-change" and "%-change" models are based on simple mathematical expressions, and both base their calculations on historical values (the most recent value) of the time series. Many texts refer to them as "naive" models because of their obvious simplicity.

To demonstrate the application of these models, we use them to forecast

**Table 1.1 Forecasting with the "No-Change" and "%-Change" Models**
$y_t$ = sales ($ million)

| Quarter | | "No Change" | | "3% Change" | |
|---|---|---|---|---|---|
| $t$ | $y_t$ | $\hat{y}_t$ | $(e_t)$ | $\hat{y}_t$ | $(e_t)$ |
| 1 | 5.10 | — | (—) | — | (—) |
| 2 | 5.40 | 5.10 | (+.30) | 5.25 | (+.15) |
| 3 | 5.30 | 5.40 | (−.10) | 5.56 | (−.26) |
| 4 | 5.70 | 5.30 | (+.40) | 5.46 | (+.24) |
| 5 | 5.72 | 5.70 | (+.02) | 5.87 | (−.15) |
| 6 | 6.08 | 5.72 | (+.36) | 5.89 | (+.19) |
| 7 | 6.05 | 6.08 | (−.03) | 6.26 | (−.21) |
| 8 | 6.42 | 6.05 | (+.37) | 6.23 | (+.19) |
| 9 | ? | 6.42 | (?) | 6.61 | (?) |

a quarterly series. The series, which consists of two years' worth of quarterly sales data (in $ million), appears in Table 1.1. The "no-change" and "%-change" forecasts and their associated residual series are also included there. For purposes of illustration, we have assumed that sales are growing by about 12% per year (5.72/5.10 = 1.12), which translates into about a 3% quarterly growth rate. Neither method can be used to forecast Quarter 1 because no prior data are available.

Examining the error series, we find that there are problems with both models. The "no-change" forecasts are slightly erratic, sometimes giving very good results (periods 5 and 7) and sometimes missing badly (periods 4 and 8). The "3%-change" model never gives any really bad predictions, but never gives particularly good ones either. Further analysis of the residual series might reveal much more, but even the simple discoveries just made demonstrate a lot about the general process of modeling. Analyzing the residual series not only yields information about a particular model, but also allows comparisons of models. In addition, since neither model seems entirely acceptable, the search for an appropriate model has really only just begun. Thus, you can probably already see some simple adjustments which would make the "no-change" and "3%-change" models more accurate. Modeling is typically an *iterative* process, in which an initial model is chosen, tested, adjusted, retested, and so forth.

Before turning to a discussion of time series graphs, we note that some problems may arise in the modeling process if a series is not measured at equally spaced intervals of time, that is, if the data are accumulated for intervals of unequal length. Most forecasting models assume equal-length intervals, and this is reflected in the various coefficients or parameters of the model. For example, in the "3%-change" model ($\hat{y}_{t+1} = 1.03y_t$), the coefficient 1.03 depends upon equal spacing. If the data were collected for unequal periods, we wouldn't forecast 3% growth for every period; instead, the 3% figure would be adjusted according to the amount of time in the period. The adjustment procedure would be fairly simple but is nonetheless necessary.

## 1.4.2 Graphing a Time Series

Analysis of time series involves searching for patterns in historical data and attempting to explain their underlying causes. One of the most powerful, yet simple, methods of doing this analysis is to create a visual display of a series. Patterns that are not at all apparent in tabulated data often emerge readily from graphical presentation (Tufte, 1983).

The use of visual displays in time series analysis has a long history. The first graph of an economic series appeared in 1786 in *The Commercial and Political Atlas* by William Playfair. Several approaches have been proposed over the years. One method, which has become standard, is applicable to any time series and, at its simplest, consists of plotting the series (on the vertical axis) versus time (on the horizontal axis).

### Standard Graphs

To illustrate some of the frequently used types of time series graphs we will use the data in Table 1.2. The series consists of quarterly estimates of the total purchases of goods and services made by state and local governments from 1978 through 1982. The figures are *annualized,* that is, the actual quarterly expenditures are multiplied by 4 to convert them to yearly rates. (The same principle is involved when one multiplies a monthly interest rate by 12 to get the nominal yearly rate.)

All of the graphs to be presented are obtained by plotting a particular function of the original series, $y_t$, against $t$. The most commonly used functions are:

**Table 1.2 State and Local Government Purchases, Quarterly, 1978 thru 1982**
$y_t$ = "annualized rate ($ billion)"

| Year | Quarter | $t$ | $y_t$ | Year | Quarter | $t$ | $y_t$ |
|---|---|---|---|---|---|---|---|
| 1978 | I | 1 | 266.2 | 1980 | III | 11 | 345.2 |
| | II | 2 | 276.0 | | IV | 12 | 352.8 |
| | III | 3 | 284.2 | 1981 | I | 13 | 361.1 |
| | IV | 4 | 290.6 | | II | 14 | 365.0 |
| 1979 | I | 5 | 293.4 | | III | 15 | 370.1 |
| | II | 6 | 301.6 | | IV | 16 | 375.7 |
| | III | 7 | 310.4 | 1982 | I | 17 | 380.4 |
| | IV | 8 | 318.3 | | II | 18 | 386.6 |
| 1980 | I | 9 | 329.6 | | III | 19 | 392.7 |
| | II | 10 | 337.2 | | IV | 20 | 398.9 |

*Source: Business Conditions Digest* (1978–1982).

$y_t$ versus $t$ → the original series

$\log y_t$ versus $t$ → logarithms of the original series

$y_t/y_{t-1}$ versus $t$ → the "link relative" series

$y_t - y_{t-1}$ versus $t$ → the series of "first differences"

All four plots are shown in Figure 1.3. These plots can result in very different pictures of the same data because each transformation brings a specific aspect of the series into focus.

## The Plot of $y_t$ Versus $t$

Every analysis should include this simple plot. It provides a history of the data, is undistorted by transformations of the data, and aids in searching for trends and patterns. Since the vertical axis is scaled in the same units of measurement

**Figure 1.3** Standard Graphs of State and Local Government Purchase Data

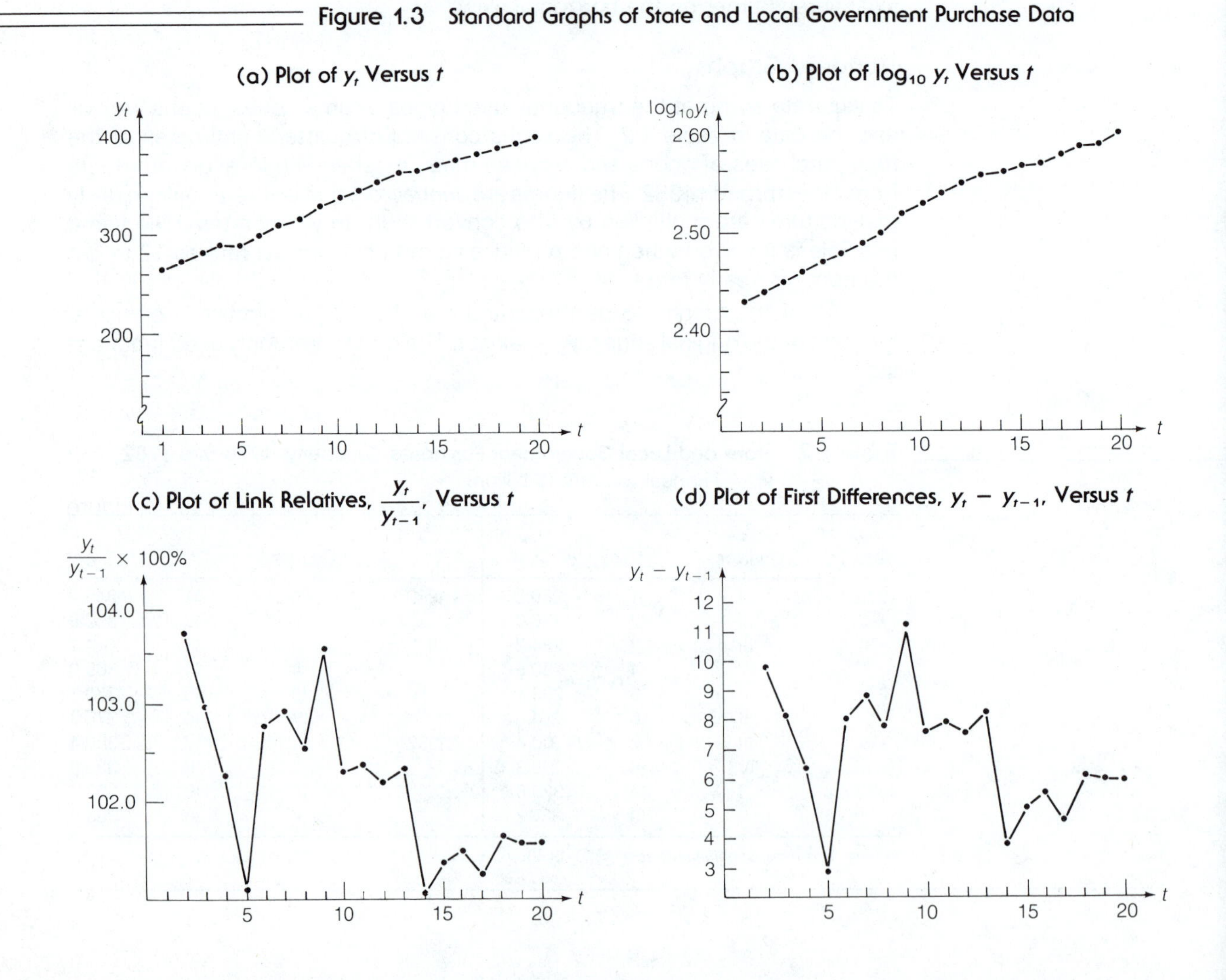

as $y_t$, the difference in heights of the graph at any two points in time reflects the actual change in the series between the two times. To emphasize this property, the plot is said to be made on an **arithmetic scale.**

### Plot of log $y_t$ Versus $t$

In this plot, the vertical scale is not in the same units as the original series. Attention is drawn to this fact by saying that the plot was done on a **ratio scale** or **log scale.** The difference in heights at two points is proportional to the corresponding percent change in the original series (cf. Exercise 1.18). Figure 1.3(b) shows a fairly stable percentage growth in the series from year to year, with a slight decline during 1981 and 1982. Because many economic series tend to grow or decline at relatively constant percentage rates, log scale plots are often used to highlight changes in these rates.

The log scale can also be used to compare two time series. When both series are changing by about the same percentage each period, the fact will be revealed by the similarity of their log scale plots. Their arithmetic scale plots, however, may be quite dissimilar (cf. Exercise 1.19).

### Plot of $y_t/y_{t-1}$ Versus $t$

A point on this graph indicates how much the current value of the series has increased or decreased relative to the previous value. Since each term connects successive values of $y_t$, this series of ratios is called the **link relative series.** It is customary to multiply the ratios by 100 when constructing the vertical scale for this graph. This allows the point to be interpreted as period-to-period percent changes. For example, in Figure 1.3(c), the link relative for period 5 was about 101, indicating that the series grew about 1% from period 4 to period 5. Overall, since all the link relatives are greater than 100% and tending downward over time, the figure shows that the series is constantly growing, but at a decreasing rate. More important, the actual magnitudes of growth (roughly between 1% and 3%) are readily apparent in the link relative plot. The same information would be harder to obtain from the log plot in Figure 1.3(b).

### Plot of $y_t - y_{t-1}$ Versus $t$

The plot of the series of *first differences* versus *time* exhibits the actual changes in magnitude between successive time periods. In Figure 1.3(d), for example, we see that the period-to-period changes in purchases have been in the range of around \$4 billion to \$9 billion, annualized, or about \$1 billion to \$2.25 billion per quarter.

The differencing process can be extended by differencing the time series two times, that is, by differencing the series of first differences to obtain the series of second differences. Third differences and beyond are obtained in a similar manner. Series derived in this way play an important role in many time series applications discussed later.

### 1.4.3 Diagnostic Graphs

Diagnostic checks of the accuracy of a model can be made either by plotting the actual series against the predicted series or by applying some of the standard graphical methods above to the series of residuals.

#### Plot of $y_t$ Versus $\hat{y}_t$

When plotting the *actual* against the *predicted* series, the same scale is often used for both axes, so that the 45° line through the origin becomes the basis for comparison. That is, if the model forecasts perfectly ($y_t = \hat{y}_t$), then the plotted points will fall exactly along the 45° line. Otherwise, the more imperfect the forecasts, the farther the plotted points will depart from the 45° line. Points falling above the line represent underforecasts, those below are overforecasts.

This kind of plot is used in Figure 1.4 to examine the "no-change" and "+3%-change" models of Table 1.1. The plots confirm our previous conclusions and make it quite apparent that the "no-change" model is frequently underforecasting.

#### Plot of $e_t$ Versus $t$

Plotting the **residual series** $e_t$ versus *time* can reveal patterns of variability which the model has failed to explain. The presence of noticeable patterns in such a plot indicates shortcomings or defects in the model. With a good model, the residual plot should appear to be a *random scattering* of points.

The residual plots for the "no-change" and "3%-change" models of Table 1.1 are shown in Figure 1.5. For the "no-change" model the points mostly lie above 0, indicating once again that this model underforecasts. The noticeable sawtooth patterns in Figure 1.5 imply that both models miss some of the information in the original series. These patterns may be used to select different

**Figure 1.4 Diagnostic Graphs for the "No-Change" and "3%-Change" Models for Sales Data in Table 1.1** ($y_t$ versus $\hat{y}_t$)

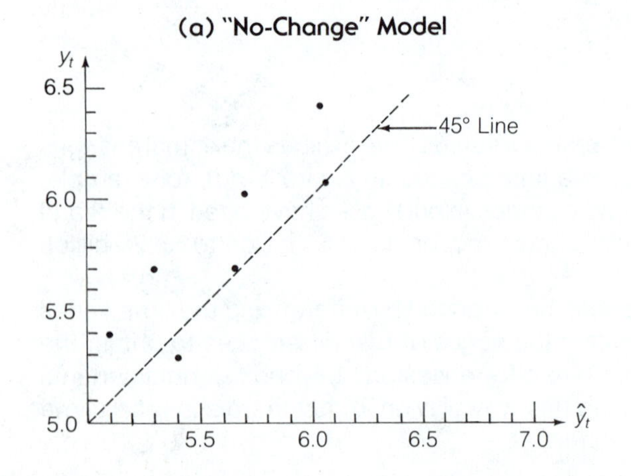


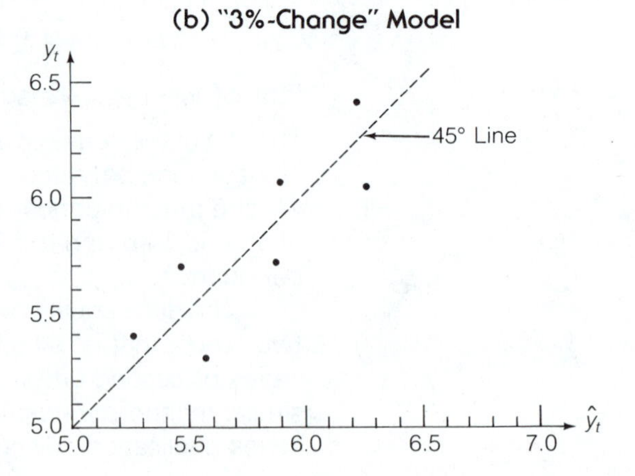

Figure 1.5 Residual Plots for the "No-Change" and "3%-Change" Models

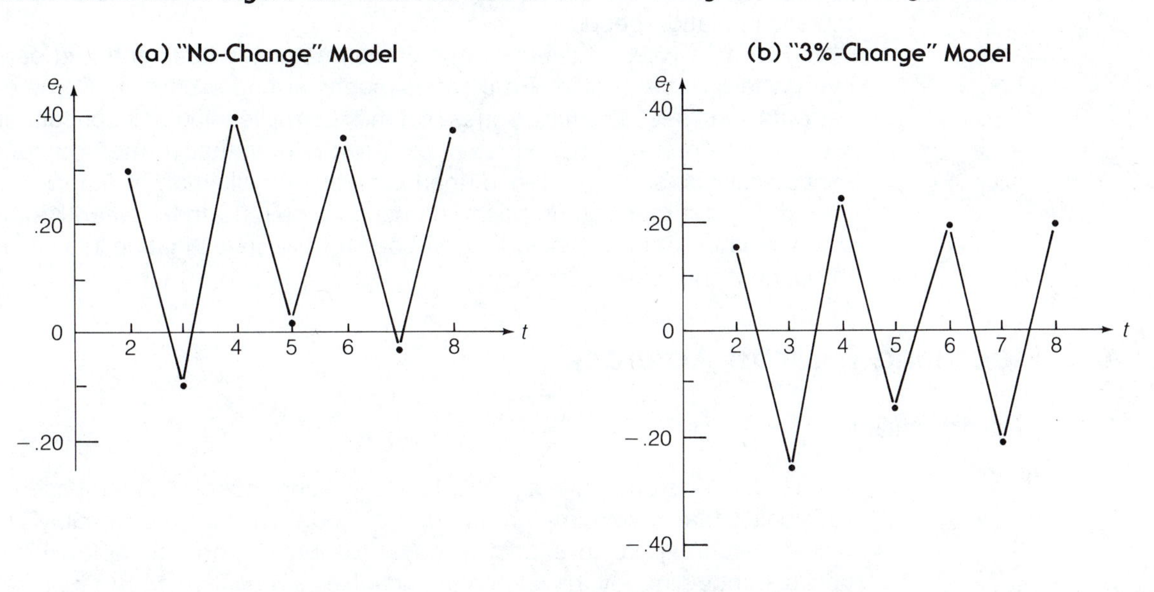


models or make adjustments to the current ones. In later chapters we discuss how to incorporate the results of residual analysis into the modeling process.

### 1.4.4 General Considerations When Plotting Data

In graphing we do more than just plot points at various time periods. The points are also joined by line segments, a practice so familiar that it may well escape notice. The purpose of the line segments is twofold:

- They make the visual display easier to read.
- They reinforce the feeling that a continuous time scale exists between the plotted points.

The convention we have adopted is to draw the line segments strictly *between* successive points, leaving a small gap, so as not to disguise the exact location of the plotted points (which can be a problem if you simply "connect the dots"). Caution should be exercised when viewing these graphs, since using line segments implies that the series changes in a linear fashion between successive time periods, which may not be the case. When the time periods are short, problems do not usually arise from interpreting the lines in this way. But for longer periods, as with a yearly series, it could be misleading.

Some conventions must be established concerning exactly where to plot successive values of a given series. Since time is continuous, we can imagine a continuous scale on the horizontal axis on which any instant in time may be represented. Then, if the time series measures "point," or **instantaneous,** data like "temperature at time *t*," we simply plot the temperature directly above the point *t*. For "accumulated," or **aggregated,** data, like "February sales," we adopt the convention of plotting the sales above the *midpoint* of the accumulation period.

Finally, a word of caution: It is human nature to seek and find patterns everywhere in everything. Time series graphs are no exception. Some of the patterns you "see" in a time series plot may be misleading or inconsequential. Much of the rest of this text is devoted to the quantitative methods of time series analysis, so dependence upon subjective evaluation of graphs is minimized. A good rule of thumb is to demand that a pattern be patently obvious or undeniable before concluding without further investigation that it exists throughout the series.

## 1.5 Measuring Forecast Accuracy

### 1.5.1 Introduction

Part of the decision to use a particular forecasting model must be based upon your belief that, when implemented, the model will work reasonably well. It would be unrealistic to expect a model to predict perfectly all of the time, because modeling, by its very nature, involves *simplification.* But it *is* realistic to expect to find a model that produces relatively small forecast errors. The performance of a model, then, can be judged by looking at the sizes and signs of the residuals. To judge a model quantitatively, we need both measures which summarize this information and ways of interpreting these measures.

Measures of forecast accuracy are used to:

- Provide a single, easily interpreted *measure of* a model's *usefulness, or reliability.*
- *Compare* the *accuracy* of two different models.
- *Search for* an *optimal model.*
- *Monitor* a model's *performance.*

The first three items on this list will be illustrated shortly. The subject of monitoring a forecasting process's accuracy will be taken up in Chapter 8.

### 1.5.2 Descriptive Measures of Accuracy

A model's performance can be evaluated over any span of time. The selection of the time span is up to the forecaster's judgment, although we would generally expect a model's recent performance to be the most reliable barometer of future performance. In what follows we will assume a time span, $t = 1, 2, \ldots, n$, has been chosen and that the forecasts over this interval can be obtained. The resulting residual series will be denoted by $e_1, e_2, \ldots, e_n$, where $n$ is the number of periods over which we are measuring the model's accuracy (*which may or may not be the same as the number of observations in the original series*).



### 1.5.3 Mean Absolute Deviation (MAD)

The **mean absolute deviation (MAD)** me
averaging the *magnitudes* of the forecast error
of the $e_t$'s. (Just averaging the $e_t$'s themselve
measure, since a residual series most often co
and negative terms which, when added, tend to
for the mean absolute deviation of the error se

$$\text{MAD} = \frac{\Sigma\,|\text{forecast error}|}{\text{number of forecasts}} = \frac{\Sigma\,|y_t - \hat{y}_t|}{n} = \frac{\Sigma\,|e_t|}{n}$$

***Note:*** To avoid confusion later, we point out here that it
that forecasts are available for all $n$ periods and th
is $n$. In later applications the number of forecasts
case the denominator should be changed accordin

The calculation of the MAD for the "no-c
els of Table 1.1 would look as follows:

$$\text{MAD("no-change")} = \frac{.30 + .10 + .40 +}{7}$$

$$\text{MAD("3\%-change")} = \frac{.15 + .26 + .24 +}{7}$$

Recall that since each model reaches back
gives forecasts for Quarter 1, so we are act
(Quarters 2 through 8). This explains why w

A useful feature of the MAD is that it i
the original series and can be easily interpre
instance, was in error by about $226,000 pe
model produced an average error of $199,00
However, the MAD does not reveal that the
been underforecasting by a significant amo
a graph of the residuals, Figure 1.5(a), along
lighted this deficiency in the "no-change" mo
fact: *The MAD, like any measure of foreca
measure. It should be used in conjunction*

### 1.5.4 Mean Square Error (MSE) and Root Mean Square Error (RMSE)

The descriptive **mean square error (MS**
attempts to average the sizes of the forec
of positive and negative terms, however, by

1.5.5 M

converted by expressing each $e_t$ as a percentage of the corresponding $y_t$. For the data we have been analyzing (Table 1.1), the first forecast for the "3%-change" model is 5.25, which is in error by .15, or 2.8% of the actual value, 5.40. Repeating this operation on all the terms in the residual series and then averaging the magnitudes of the resulting percentages, we arrive at a measure called the **mean absolute percent error (MAPE)** (notice that if the MAPE is to be used, none of the $y_t$ can be zero):

$$\text{MAPE} = \frac{\Sigma\,|\text{forecast error/actual value}|}{\text{number of forecasts}}$$

$$= \frac{\Sigma\,|e_t/y_t|}{n} \cdot 100\% \qquad (y_t \neq 0)$$

Thus, in Table 1.1,

$$\text{MAPE("no-change")} = \frac{.056 + .019 + \cdots + .058}{7} \cdot 100\% = 3.9\%$$

$$\text{MAPE("3\%-change")} = \frac{0.28 + .049 + \cdots + .030}{7} \cdot 100\% = 3.4\%$$

Like the MAD, the MAPE doesn't distinguish much between these two models, but it does give an indication of how large the forecast errors are in comparison to the actual values of the series.

Because it is dimensionless, or unit-free, the MAPE can be used to compare the accuracy of the same or different models on two entirely different series. The efficiency of a given forecasting procedure is often studied by using it to forecast a wide variety of different series, calculating the MAPE in each case, and then averaging the MAPEs for all the series. Such a procedure would not be possible with the MAD, MSE, or RMSE if there were any differences in the units of measurement among the series.

### 1.5.6 Correlation Coefficient ($r$)

When the diagnostic plot of $y_t$ versus $\hat{y}_t$ is used, the criterion for a good model is that the plotted points fall close to the 45° line. Evaluating a particular model or comparing several distinct models can be done by inspecting their $y_t$ versus $\hat{y}_t$ plot(s), as in Figure 1.4. For many reasons it is also desirable to have a numerical measure of the degree of fit. For one thing, it is easier, less subjective, and more precise to communicate the degree of fit by a number rather than an adjective (an "excellent" fit, a "good" fit, a "poor" fit, etc.). Sometimes, too, it is difficult to make a visual distinction between plots, whereas a comparison of their associated measures would be easy, immediate, and not dependent upon human judgment.

A measure often used in conjunction with the plot of $y_t$ versus $\hat{y}_t$ is the *correlation coefficient.* There are many types of correlation coefficients around, each based on a different approach to measuring association between two quantities. We will use **Pearson's coefficient of correlation,** traditionally denoted by the letter $r$. (In case you are unfamiliar with $r$, its calculation is not really necessary for the present discussion. For completeness, however, formulas for this calculation are given in Appendix 1A. More complete details of its use and interpretation will be given in later chapters.)

When evaluating a diagnostic plot, we suggest using $r^2$, the **coefficient of determination,** rather than $r$ itself. The correlation coefficient ranges between $-1$ and $+1$, so $r^2$ will lie between 0 and 1. The closer it is to 1, the closer the plotted points come to falling on a straight line. Keep in mind, though, that being close to 1 only means close to *some* straight line, not necessarily the perfect forecast (45°) line. Using $r^2$ to compare the "no-change" and "3%-change" models, we find that $r^2 = .716$ in *both* cases. The plots of Figure 1.4 are repeated in Figure 1.6 but with lines added that fit the data best. (*Note:* These lines are actually the "least squares" lines and are related to $r$. Their determination, like that of $r$, will be treated in more detail later.) The line along which the points tend to fall for the "no-change" model (line 1) deviates from the perfect forecast line, while the points for the "3%-change" model fall around a line (line 2) which is closer to the perfect forecast line. The coefficient of determination is about the same in each case because both models exhibit the same degree of packing around their respective lines.

To make proper use of $r^2$, first make sure that the line in the diagnostic plot is not materially different from the 45° line; then $r^2$ will be a valid measure of forecast accuracy. Failure of the plot to fall along the 45° line indicates some systematic bias in the model.

### 1.5.7 Theil's Inequality Coefficient ($U$)

In a sense, the "no-change" model is the simplest we can use. It is practically free of costs (time or money), and its use requires little skill. It would be difficult to justify using a more costly or complex model whose forecasts were no more accurate.

A model can be compared to the "no-change" model by using a measure introduced by econometrician Henri Theil (Theil, 1966). The measure is referred

**Figure 1.6 Diagnostic Graphs with "Best Fitting" Line Added to Figure 1.4**

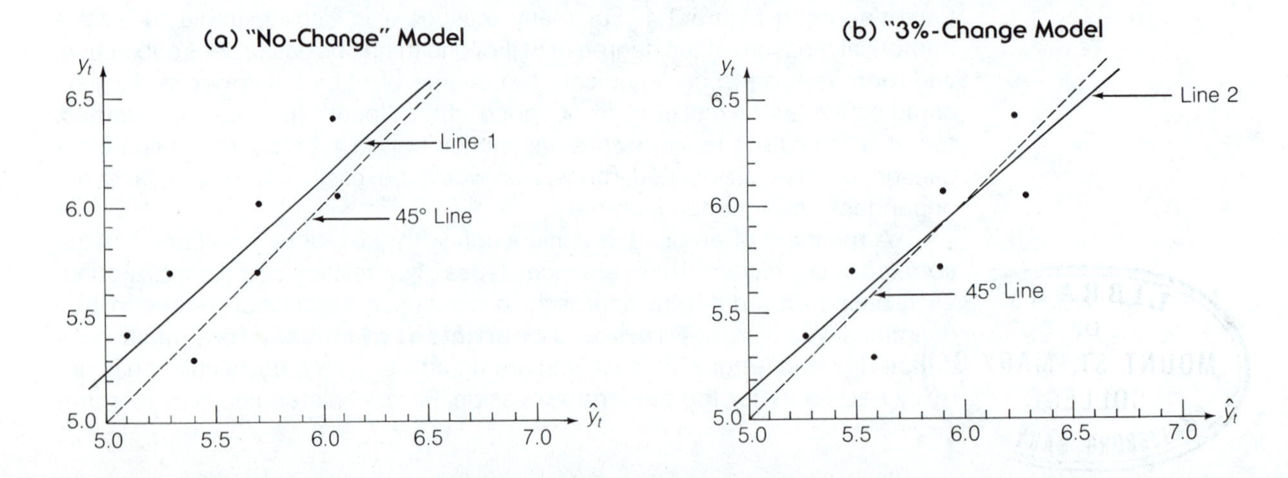

to as **Theil's *U*** or **Theil's inequality coefficient.** The formula for $U$ can be written three different ways, each useful in different situations:

*Formula 1:* $$U = \frac{\sqrt{\Sigma\, e_t^2}}{\sqrt{\Sigma\, (y_t - y_{t-1})^2}}$$

*Formula 2:* $$U = \frac{\sqrt{\text{MSE(model)}}}{\sqrt{\text{MSE(``no-change'' model)}}}$$

*Formula 3:* $$U = \frac{\sqrt{\Sigma\, (A_t - P_t)^2}}{\sqrt{\Sigma\, A_t^2}}$$

where

$\hat{y}_t$ = forecast using the model being compared

$e_t = y_t - \hat{y}_t$ = the forecast error for the model being compared

$P_t = \hat{y}_t - y_{t-1}$ = the predicted change in period $t$

$A_t = y_t - y_{t-1}$ = the actual change in period $t$

Formula 1 is definitional and is used for calculating $U$ from raw data. Formula 2 clearly shows how the comparison to the "no-change" model is to be made. It is useful when the MSE calculations have already been done. Formula 3 is most relevant when the objective is to forecast changes in a series, often done in economic modeling.

$U$ is most easily interpreted using the second of the three formulas. If a model predicts perfectly, its MSE will be 0 and hence $U$ will be 0. If the model does exactly as well (in terms of MSE) as the "no-change" model, then MSE(model) = MSE("no-change") and $U$ will be 1.0. If the model is worse than a "no-change" model, then MSE(model) will be greater than MSE("no-change") and $U$ will be greater than 1.0. Studies have shown that, in practice, values of around .55 or less are very good (Lindberg, 1982; McNees, 1979).

Continuing our analysis of Table 1.1, we find

$$U(\text{``no-change''}) = 1 \qquad \text{(no calculations necessary)}$$

$$U(\text{``3\%-change''}) = \frac{\sqrt{\text{MSE(``3\%-change'')}}}{\sqrt{\text{MSE(``no-change'')}}} = \frac{\sqrt{.041}}{\sqrt{.075}} = .739$$

The fact that $U$("no-change") = 1 follows immediately from Formula 2. Having previously calculated the MSE for each model, Formula 2 is also the easiest to use for the "3%-change" model. The net result, in agreement with earlier analyses, is that, in terms of the inequality coefficient, the "3%-change" model outperforms the "no-change" model.

**Interpretation of Thiel's *U***

| | | |
|---|---|---|
| $U = 0$ | → | The model forecasts perfectly. |
| $U < 1$ | → | The model performs better than the "no-change" model. |
| $U = 1$ | → | The model performs about as well as the "no-change" model. |
| $U > 1$ | → | The model performs worse than the "no-change" model. |

Various modifications to $U$ have been suggested. These suggestions have generally involved shifting the base of comparison to standards other than the "no-change" model. The recommendations have ranged from simple trend extrapolation to models based on sophisticated Box-Jenkins methodology (Chapter 7). None has found universal acceptance, but perhaps the most practical suggestion has been to use the current model as the base. In other words, start by using the "no-change" model as a standard, use it to search for a better model, and if one is found, use the new one as the standard for future comparisons. This process could be repeated until competing models can no longer add significantly to prediction accuracy. Using the inequality coefficient like this is really a quantitative implementation of the idea that a model should be used "until something better comes along."

## 1.5.8 Theil's Decomposition of the MSE

Together with the inequality coefficient, Theil developed a simple decomposition of the MSE into three components, each addressing a different aspect of forecast accuracy (Theil, 1966). That decomposition is:

$$\text{MSE} = \underbrace{(\bar{\hat{y}} - \bar{y})^2}_{\text{component 1}} + \underbrace{(s'_{\hat{y}} - r s'_y)^2}_{\text{component 2}} + \underbrace{(1 - r^2) s'^2_y}_{\text{component 3}}$$

where

$y_t$ = the actual value of the series

$\hat{y}_t$ = the value predicted by the model

$\bar{\hat{y}}$ = the average of the predicted values

$\bar{y}$ = the average of the actual series

$$s'_y = \sqrt{\frac{\Sigma (y_t - \bar{y})^2}{n}}$$

$$s'_{\hat{y}} = \sqrt{\frac{\Sigma (\hat{y}_t - \bar{\hat{y}})^2}{n}}$$

$r$ = Pearson's coefficient of correlation between the actual and predicted series

$$\text{MSE} = \frac{\Sigma (y_t - \hat{y}_t)^2}{n}$$

For the "no-change" model of Table 1.1 the decomposition is illustrated in Table 1.3.

Theil suggested that the decomposition could best be interpreted if both sides of the identity were first divided by the MSE:

$$1 = \underbrace{\left[\frac{(\bar{\hat{y}} - \bar{y})^2}{\text{MSE}}\right]}_{U_M} + \underbrace{\left[\frac{(s'_{\hat{y}} - r s'_y)^2}{\text{MSE}}\right]}_{U_R} + \underbrace{\left[\frac{(1 - r^2) s'^2_y}{\text{MSE}}\right]}_{U_D}$$

Each of the three components, called $U_M$, $U_R$, and $U_D$ by Theil, can now be interpreted as a proportion or percentage of the MSE. The decomposition in Table 1.3 becomes

**Table 1.3 Decomposition of MSE for the "No-Change" Model for the Series in Table 1.1**

| Quarter $t$ | Actual Series $y_t$ | "No-Change" Forecasts $\hat{y}_t$ |
|---|---|---|
| 2 | 5.40 | 5.10 |
| 3 | 5.30 | 5.40 |
| 4 | 5.70 | 5.30 |
| 5 | 5.72 | 5.70 |
| 6 | 6.08 | 5.72 |
| 7 | 6.05 | 6.08 |
| 8 | 6.42 | 6.05 |

$\bar{y} = 5.810$ $\quad\bar{\hat{y}} = 5.621$

$s'_y = .3686$ $\quad s'_{\hat{y}} = .3452$

**Correlation Coefficient:** $r = 0.846$

**Slope of "Regression" Line:** $b = .9034$

**"No-Change" Model Mean Square Error:** $\text{MSE} = .0754$

**Decomposition:**

$$\text{MSE} = \underbrace{(5.621 - 5.810)^2}_{\text{component 1}} + \underbrace{(.3452 - .846 \times .3686)^2}_{\text{component 2}} + \underbrace{(1 - .846^2) \times .3686^2}_{\text{component 3}}$$

$$= .0357 + .0011 + .0386$$

$$= .0754$$

$$1 = .473 + .015 + .512$$

or, as percentages,

$$100\% = 47.3\% + 1.5\% + 51.2\%$$

$U_M$ ($M$ stands for the difference in the "means") measures the proportion of the MSE that is caused by *bias* in the forecasting model. In our example, $U_M = 47.3\%$, which indicates that quite a bit of the forecasting error is due to some sort of bias. It was already clear from the residual plot (Figure 1.5) that the "no-change" model tended to underforecast. $U_M$ now gives a numerical indicator of this bias.

$U_R$ ($R$ stands for "regression," cf. Chapter 4) is related to the diagnostic plot of $y_t$ versus $\hat{y}_t$. There is a line which "best fits" the plotted points (like lines 1 and 2 of Figure 1.6). This line is called the "regression line." $U_R$ measures how much of the MSE is due to this line's *deviating from the 45° line.* Denoting the slope of the "regression" line by $b$ (again, calculations are not necessary for the present discussion, but formulas are given in Appendix 1A), it can be shown that

$$U_R = \frac{(1 - b)^2 s_{\hat{y}}'^2}{\text{MSE}}$$

Remembering that the slope of the 45° line is 1.0, we see that the closer the "regression" line is to the perfect line, the closer $b$ is to 1.0 and the closer $U_R$ will be to 0. $U_R$ has a value of 1.5% for the "no-change" model, which implies that a very small amount of the forecasting error is due to the tendency of the points on the diagnostic plot to fall in a direction different from 45°.

$U_D$ ($D$ stands for "disturbance"), which = 51.2% for the "no-change" model, is a measure of the imperfect correlation between the actual and the predicted series. The larger it is, the worse the correlation between the two series. $U_D$ is the component that responds to *random "disturbances"* in the series and thus measures the amount of error over which we have no control. Of the three terms in the decomposition, $U_D$ is the one that, ideally, should be large (i.e., close to 1.0). While this may seem confusing at first, remember that all three coefficients are interpreted *relative to the MSE.* A certain amount of MSE is inevitable, since models do not forecast perfectly. But, hopefully, the imperfection is not caused by systematic deficiency, which would be the case if either $U_M$ or $U_R$ were large. The error that does exist, large or small, should be due mainly to the model's nonsystematic deviation from perfect forecasts.

In summary, then, the most desirable decomposition is $U_M = 0$, $U_R = 0$, and $U_D = 1$. In practice, we only expect $U_M$ and $U_R$ to be close to 0 and $U_D$ to be close to 1. Comparing the "no-change" and "3%-change" models via Theil's decomposition we get the following table:

| Model | $U_M$ | $U_R$ | $U_D$ |
|---|---|---|---|
| "no-change" | 47.3% | 1.5% | 51.2% |
| "3%-change" | 1.1% | 4.6% | 94.3% |

The decomposition for the "3%-change" model is close to ideal for these data, which doesn't necessarily mean that the model is the best possible or even that it is free of other serious problems. It merely means that the particular types of systematic errors measured by $U_M$ and $U_R$ are of negligible importance. As previously mentioned, when the "no-change" model is applied to the same data, the decomposition reveals serious bias. The "no-change" model, though it may work with other time series, does not work well with this one.

### 1.5.9 Applying Measures of Accuracy

As we have seen, most measures of forecast accuracy involve averaging some function (e.g., absolute value, square) of the residuals. The MSE is prominent among these measures because its theoretical behavior has been extensively studied. Consequently, it is often used in deriving other measures of accuracy (e.g., RMSE, $U$, Theil's decomposition). Non-MSE-based measures may actually be preferable in various situations, but generally they occupy a smaller role in the study of forecasting models.

Measures of accuracy are often employed when searching for an optimal model. The procedure is similar to that outlined at the end of the discussion of Theil's *U:* Select a measure, calculate its value for a variety of models, and choose that model which minimizes the measure. In terms of the MSE, this means searching for a model that minimizes the MSE for a fixed set of data. This procedure will guide many of our model selection decisions in the remainder of the text.

For evaluating the adequacy of a model the following guidelines are suggested:

- Use more than one measure of accuracy; especially try the MSE and the MAD.
- Use the diagnostic graphs of $y_t$ versus $\hat{y}_t$ and $e_t$ versus $t$.
- Consolidate the information gathered above when judging the model's behavior, since each group or measure only focuses on one or two aspects of the data.

## 1.6 General Approaches to Model Building

### 1.6.1 Introduction

When a forecasting project begins, an essential decision must be made about the overall approach to the modeling process. The choice will strongly influence model selection and, ultimately, affect the quality of our forecasts. This initial decision involves the selection of variables from which the model will be built. We may decide to use a time series model, based solely on time ($t$) and/or previous values of the series itself ($y_{t-1}, y_{t-2}, \ldots$). Such an approach reflects a belief that the underlying patterns in the series are fairly stable and can be reliably extrapolated. On the other hand, if we believe *other* time series can be useful in explaining the behavior of $y_t$, and attempt to identify key variables which "cause" $y_t$ to behave as it does, the result is called a causal model.

#### Time Series Models

**DEFINITION:** A **time series model** for predicting $y_t$ is any mathematical model having only time and past values of $y_t$ as its inputs.

The "no-change" and "3%-change" models of Table 1.1 are both examples of time series models, since they base their forecasts solely on simple functions of $y_t$. Some other models qualifying as time series models would be:

$\hat{y}_t = \beta_0 + \beta_1 t$ → Time, $t$, is the only input.

$\hat{y}_t = \beta_1 y_{t-1} + \beta_2 y_{t-2} + \beta_3 y_{t-3}$ → Previous values of $y$ are inputs.

$\hat{y}_t = \alpha y_{t-1} + (1 - \alpha)\hat{y}_{t-1}$ → Past forecasts and past values of $y$ are inputs.

### Causal Models

**DEFINITION:** A **causal model** for predicting $y_t$ is any mathematical model having any number of time series *other* than $y_t$ as its inputs.

Most econometric models would be considered causal models. Other examples include:

$\hat{y}_t = \beta_1 x_t + \beta_2 w_t$ → Concurrent values of other series are inputs.

$\hat{y}_t = \beta_0 x_t z_t u_t$ → A multiplicative model.

$\hat{y}_t = \beta_1 x_t + \beta_2 x_{t-1} + \beta_3 w_t$ → Lagged values of other series are inputs.

### Mixed Models

**DEFINITION:** A **mixed time series–causal model** for predicting $y_t$ is one having time, past values of $y_t$, and other time series as its inputs.

Essentially, mixed models can contain any combination of time series and explanatory variables. An example of a mixed model would be:

$$\hat{y}_t = \beta_0 + \beta_1 t + \beta_2 y_{t-1} + \beta_3 x_t$$

Mathematical methods for building causal models are generally called "econometric methods." A comprehensive treatment of these methods could, in itself, fill a book and will not be undertaken here. Our discussion of causal models will be limited to the regression models in Chapter 4.

## 1.6.2 Decomposition of Time Series

An immediate temptation in the analysis of time series data, whether tabulated or graphically displayed, is to try to "explain" or account for the behavior of the series. Why did each turning point occur? Does there appear to be an overall upward or downward trend? What causes the series to be unusually large each December? Did a certain economic event have an effect on the series, and, if so, exactly when was the effect felt? To avoid wasted effort, what is needed is a systematic approach to analyzing the series.

### Components of a Time Series

Historically, the search for this systematic method began with efforts to *decompose* a time series into *components* (Nerlove, Grether, and Carvalho, 1979). Each component was defined to be a major factor or force (e.g., long-term trends, climatic influences, cyclical movements) that could affect any time series. Since each component had a specific interpretation, the behavior of the original series could be explained by combining the results of the separate analyses for each component.

There is a problem with this approach, however, in that the components are *unobservable.* We can't simply go to some other source, record the values of each component, and then combine the results to get the original series. Instead, the components must be estimated from the original data. This process is called "decomposing" the time series.

## Classical Decomposition

There is a particular decomposition that has stood the test of time. This method identifies four components, each having a simple interpretation relative to the time series. The four components are most easily understood by examining the graph of a particular time series, such as the one in Figure 1.7. This series consists of 20 quarterly measurements. It has been decomposed into trend ($T$), seasonal ($S$), cycle ($C$), and irregular ($I$) factors.

### Trend ($T$)

**Trend,** often called **secular trend,** refers to the long-term ("secular") tendency of a series to rise or fall. This may be thought of as the underlying growth ("upward" or "positive" trend) or decline ("downward" or "negative" trend) component in the series. In Figure 1.7, an upward trend is noticeable.

### Seasonality ($S$)

The **seasonal** component is concerned with the periodic fluctuation in the series *within each year.* Such fluctuations form a relatively fixed pattern that tends to repeat year after year. In Figure 1.7, the seasonal component is pronounced: Starting at the first quarter of *any* year, the series decreases from Quarter 1 to Quarter 2; decreases from Quarter 2 to Quarter 3; increases from Quarter 3 to Quarter 4; and finally increases from Quarter 4 to Quarter 1 of the next year. Seasonal fluctuations, or "movements," are most often attributable to social custom, weather changes, or various institutional arrangements such as school calendars, tax filing schedules, and the like. The causes of seasonal movements are explored more fully in Chapter 5.

**Figure 1.7** Twenty Quarterly Measurements of a Time Series

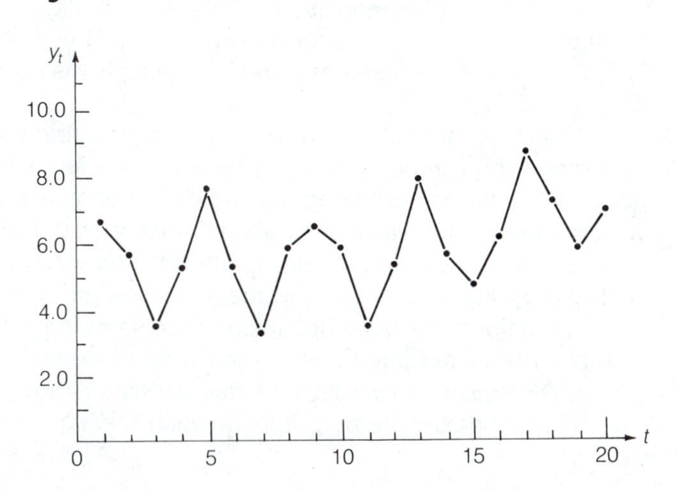

### Cycles ($C$)

**Cyclical behavior** in a time series is similar to seasonality in that it is evidenced by a wavelike pattern of ups and downs. But it differs from seasonality in two important ways: First, cycles are viewed as broad contractions and expansions that take place over a period of years, *not within each year.* Second, the length of time between the successive peaks (or troughs) of a cycle is *not* necessarily *fixed,* as it is with seasonal movement. The familiar expansion-recession-recovery-expansion of the "business-cycle" is a good example of cyclical movement. In Figure 1.7, the cyclical character of the series is not easy to discern, but it does appear that the graph initially turns down before eventually climbing.

### Irregular Movement ($I$ or $\epsilon$)

Trend, seasonal, and cyclical movements may explain much of the behavior of a series. The movement left over after accounting for these factors is called the **irregular** component. Borrowing from engineering terminology, the irregular component is often called the **noise** in the series, while the other three components comprise the **signal.**

### Additive and Multiplicative Decomposition

Actual estimation of trend, seasonal, cyclical, and irregular components is an involved process, and is discussed later in the book. The discussion that follows illustrates, instead, the process of combining the components *assuming they have already been estimated,* i.e., that $\hat{T}, \hat{S}, \hat{C},$ and $\hat{I}$ have been obtained.

The principal methods of combining these factors involve simply adding or multiplying the four factors. The **additive model** expresses a series as the sum of the components:

$$y_t = \hat{T}_t + \hat{S}_t + \hat{C}_t + \hat{I}_t$$

while the multiplicative model uses the product:

$$y_t = \hat{T}_t \cdot \hat{S}_t \cdot \hat{C}_t \cdot \hat{I}_t$$

An additive decomposition of the series in Figure 1.7 is shown in Table 1.4. One feature of this decomposition is that the units of measurement of $\hat{T}, \hat{S}, \hat{C},$ and $\hat{I}$ are the same as those of $y$ (e.g., if $y$ is measured in dollars, then so are $\hat{T}, \hat{S}, \hat{C},$ and $\hat{I}$).

Since the four estimated components form four time series, the additive decomposition can be shown graphically as in Figure 1.8. Graphing the components facilitates our understanding of the original series, $y_t$. For example, we can see that the value of $y_9$ is lower than would have been expected from looking only at trend and seasonal patterns. The value is lower because of the declining cyclical factor, $\hat{C}_9$, and the slightly negative irregular movement, $\hat{I}_9$.

A multiplicative decomposition of the same time series might appear as in Table 1.5. Here, only trend is measured in the same units as the original series. The seasonal, cyclical, and irregular factors are "indexes" and are unit-free. These indexes are proportions which measure how far a component is

**Table 1.4 Additive Decomposition of the Series in Figure 1.7**

| $t$ | $y_t$ | = | $\hat{T}_t$ | + | $\hat{S}_t$ | + | $\hat{C}_t$ | + | $\hat{I}_t$ |
|---|---|---|---|---|---|---|---|---|---|
| 1 | 6.750 | | 5.000 | | 1.5 | | .000 | | .25 |
| 2 | 5.75 | | 5.125 | | 0.0 | | .125 | | .50 |
| 3 | 3.500 | | 5.250 | | −1.5 | | .250 | | −.50 |
| 4 | 5.375 | | 5.375 | | 0.0 | | .250 | | −.25 |
| 5 | 7.625 | | 5.500 | | 1.5 | | .125 | | .50 |
| 6 | 5.375 | | 5.625 | | 0.0 | | .000 | | −.25 |
| 7 | 3.375 | | 5.750 | | −1.5 | | −.125 | | −.75 |
| 8 | 5.875 | | 5.875 | | 0.0 | | −.250 | | .25 |
| 9 | 6.625 | | 6.000 | | 1.5 | | −.375 | | −.50 |
| 10 | 5.875 | | 6.125 | | 0.0 | | −.500 | | .25 |
| 11 | 3.500 | | 6.250 | | −1.5 | | −.500 | | −.75 |
| 12 | 5.375 | | 6.375 | | 0.0 | | −.500 | | −.50 |
| 13 | 8.000 | | 6.500 | | 1.5 | | −.500 | | .50 |
| 14 | 5.625 | | 6.625 | | 0.0 | | −.500 | | −.50 |
| 15 | 4.750 | | 6.750 | | −1.5 | | −.500 | | .00 |
| 16 | 6.125 | | 6.875 | | 0.0 | | −.500 | | −.25 |
| 17 | 8.875 | | 7.000 | | 1.5 | | −.375 | | .75 |
| 18 | 7.125 | | 7.125 | | 0.0 | | −.250 | | .25 |
| 19 | 5.875 | | 7.250 | | −1.5 | | −.125 | | .25 |
| 20 | 6.875 | | 7.375 | | 0.0 | | .000 | | −.50 |

above or below the average value. As proportions, indexes are expressed with a base of 1. It is common practice, however, to report seasonal, cyclical, and irregular indexes as percentages, in which case 100% becomes the base value.

In Table 1.5, the seasonal index for period 1 ($\hat{S}_1 = 1.330$, or 133%) would be interpreted as follows: If there were no seasonal influences, $\hat{S}_1$ would be 1 (or 100%). But seasonality is present, and it has been responsible for a 33% rise in the series during the first quarter. Similarly, the seasonal index for period 3 ($\hat{S}_3 = 0.699$, or 69.9%) indicates that seasonal factors have depressed the series by 30.1% during that quarter.

In Chapters 5 and 6, we discuss how to choose between the additive and multiplicative or some other decomposition. It turns out that the multiplicative model is appropriate for many economic series. It also serves as the foundation for the Census II and X-II computer programs of the U.S. Bureau of the Census. These programs are widely used for the analysis and deseasonalization (seasonal adjustment) of time series data.

### 1.6.3 The Structure of This Text

If no attempt is made at organization, forecasting methodology can appear to consist of a large body of unrelated, ad hoc formulas and procedures. The subject matter then seems disconnected, and it becomes difficult to choose an appropriate technique. In this book, we provide structure by centering the study of forecasting methods around the four classical time series components.

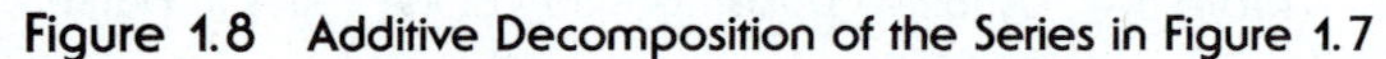

**Figure 1.8 Additive Decomposition of the Series in Figure 1.7**

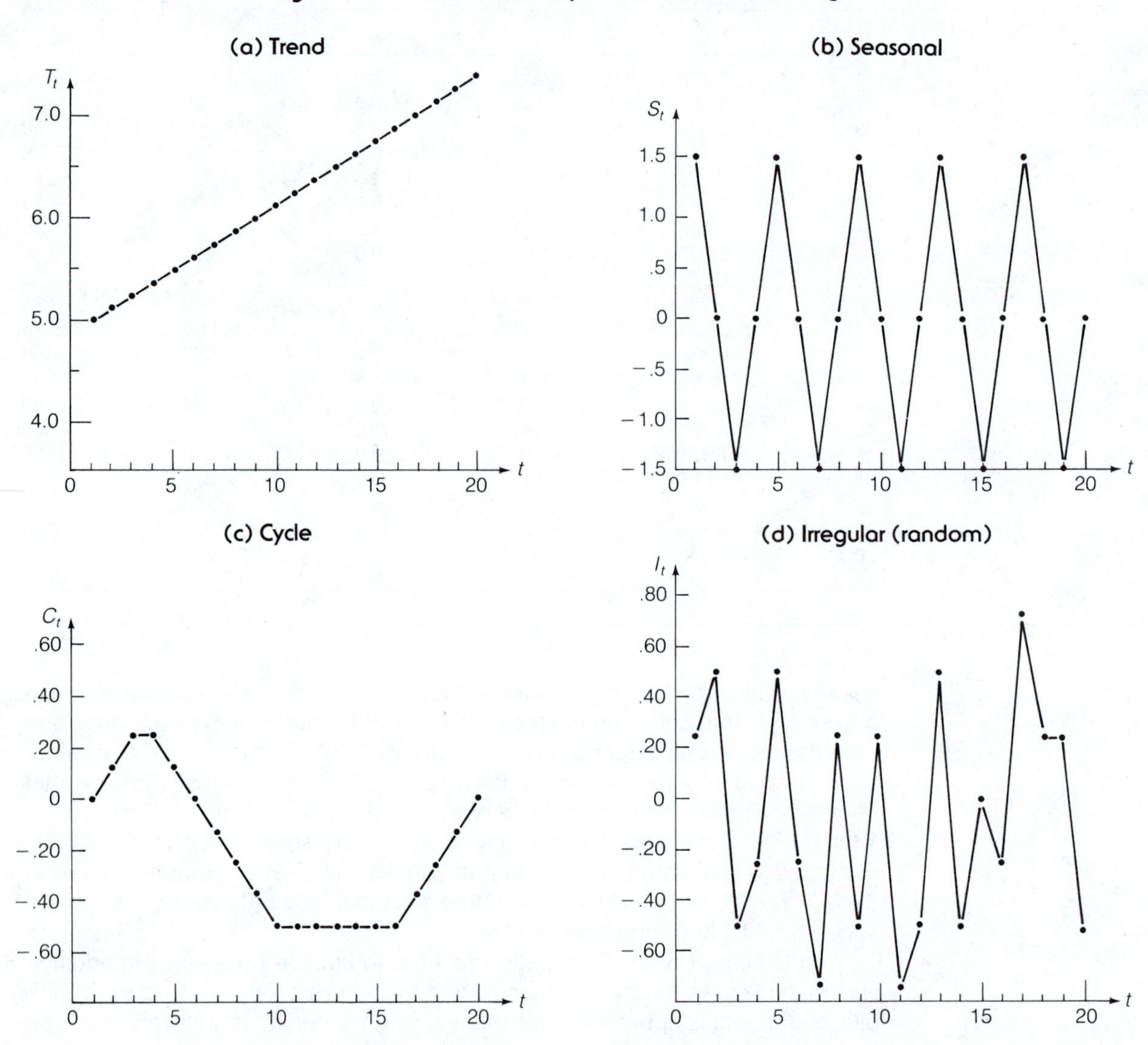


Attention is first isolated to models designed for handling series exhibiting neither trend, seasonality, nor cyclical movement (Chapter 2). Complexity is then added in stages by successively considering models for trend (Chapter 3), trend and seasonality (Chapter 5), and finally, trend, seasonality, and cyclical movement together (Chapter 6). Special topics such as regression analysis (Chapter 4), Box-Jenkins methodology (Chapter 7), and adaptive methods (Chapter 8) are extensive enough to each warrant a chapter and are presented as needed. But Chapters 2, 3, 5, and 6 represent the core of the text.

# 1.7 Choosing a Forecasting Technique

## 1.7.1 Introduction

The overview of forecasting in Section 1.1 included a short list of questions to be asked at the outset of a forecasting project. The questions concerned issues such as what to forecast and various cost-benefit decisions (Mohn and Reid, 1977; Brown, 1982). We may now take a more detailed look at these questions.

**Table 1.5** Multiplicative Decomposition of the Series in Figure 1.7

| $t$ | $y_t$ | = | $\hat{T}_t$ | · | $\hat{S}_t$ | · | $\hat{C}_t$ | · | $\hat{I}_t$ |
|---|---|---|---|---|---|---|---|---|---|
| 1 | 6.750 | | 5.18 | | 1.330 | | .951 | | 1.030 |
| 2 | 5.750 | | 5.26 | | 1.006 | | .970 | | 1.120 |
| 3 | 3.500 | | 5.34 | | .699 | | 1.002 | | .978 |
| 4 | 5.375 | | 5.42 | | .995 | | 1.028 | | .970 |
| 5 | 7.625 | | 5.50 | | 1.330 | | .996 | | 1.047 |
| 6 | 5.375 | | 5.58 | | 1.006 | | .977 | | .980 |
| 7 | 3.375 | | 5.66 | | .699 | | .986 | | .905 |
| 8 | 5.875 | | 5.74 | | .995 | | .929 | | 1.107 |
| 9 | 6.625 | | 5.82 | | 1.330 | | .938 | | .912 |
| 10 | 5.875 | | 5.90 | | 1.006 | | .931 | | 1.063 |
| 11 | 3.500 | | 5.98 | | .699 | | .898 | | .974 |
| 12 | 5.375 | | 6.06 | | .995 | | .943 | | .945 |
| 13 | 8.000 | | 6.14 | | 1.330 | | .921 | | 1.064 |
| 14 | 5.625 | | 6.22 | | 1.006 | | .960 | | .936 |
| 15 | 4.750 | | 6.30 | | .699 | | .979 | | 1.151 |
| 16 | 6.125 | | 6.38 | | .995 | | 1.001 | | .964 |
| 17 | 8.875 | | 6.46 | | 1.330 | | 1.048 | | .986 |
| 18 | 7.125 | | 6.54 | | 1.006 | | 1.078 | | 1.005 |
| 19 | 5.875 | | 6.62 | | .699 | | 1.094 | | 1.213 |
| 20 | 6.875 | | 6.70 | | .995 | | 1.110 | | .929 |

## 1.7.2 What to Forecast

Specifying the series to be forecast is not as simple as it might first appear. The basic problem is that there are usually many ways to quantify what we want to forecast. In earlier sections, for example, it was easy to create a list of measurable attributes of "sales," i.e., "units sold," "dollar volume," "market share," "cost of goods sold," "profit margin," "sales by territory," and so on.

The list of candidate variables can be trimmed either by using common sense or by exploiting any relationships among the variables. Some of the relationships among the sales variables above are:

$$\text{(dollar volume)} = \text{(units sold)} \cdot \text{(price per unit)}$$

$$\text{(units sold)} = \text{(industry sales)} \cdot \text{(market share)}$$

$$\text{(company sales)} = \text{(sum of territory sales)}$$

Since some variables (such as dollar volume) can be determined from a knowledge of other ones (units sold), it is not necessary to *separately* generate forecasts of all the variables. Identifying these interrelationships is an effective way of determining which variables to forecast.

### 1.7.3 Cost/Benefit Considerations

The costs involved in a forecasting project fall into two categories: money costs and time costs. A basic checklist of cost-related items includes:

- **Data acquisition** Some data can be obtained cheaply from company records, some must be purchased from outside sources.
- **Data storage** Since most likely a computer is used, there are data entry and storage costs.
- **Computer time** Some forecasting methods are very expensive of computer time, others are not.
- **Model** This always involves implementation and maintenance costs.
- **Dissemination** Summarization and communication of the results is necessary.

#### Forecast Accuracy

The fundamental issues are that the chosen model yield forecasts that are adequate for planning and that the model be at least as accurate as the one currently being used. Things to consider:

- **Measuring accuracy** Select an appropriate measure, probably from the list in Section 1.4.
- **Level of accuracy** Establish desired and/or acceptable levels, taking into account the costs of over- versus underforecasting.
- **Current model** If some forecasting procedure is already in use, determine its level of accuracy for comparison.
- **Feedback law** The forecasts you generate may themselves impact the series being forecast, e.g., when sales forecasts become sales goals (Smith, 1964).

#### Time Horizon

Forecasts may be needed for short-term planning (a few periods ahead) or long-term planning (several periods ahead). As a rule, long-term forecasts are less accurate than short-term ones. Matters of concern include:

- **Time or forecast horizon** This means the time span over which planning will have an impact, i.e., the number of periods ahead for which the forecasts are required.
- **Time units** The units of time at which data are collected (day, week, month, quarter, half-year, year, decade, etc.) are usually dictated by the time horizon (smaller units are used for the short term, larger units for the long term).
- **Data availability** Data must become available soon enough to be used in the forecasting model; using lagged variables can sometimes overcome this problem.

## 1.7.4 How to Use This Book

As already mentioned, Chapters 2, 3, 5, and 6 form the core of this text. Except for some specialized procedures (Chapters 4, 7, and 8), then, the majority of the models we present are found in those four chapters. In addition, each chapter is itself structured around an algorithmic approach to model selection. The first sections of a chapter give examples and graphs of some series that are typical of the ones handled in the chapter. The next few sections contain statistical and other methods for verifying the presence of the specific component (trend, seasonal, etc.) under discussion. Next come sections presenting the relevant forecasting models and illustrating their application. The final section, where appropriate, discusses various factors (diagnostics) to be considered when checking the adequacy of the chosen model.

## 1.7.5 Model Selection Flowchart

The following flowchart summarizes how this book can be used to find an appropriate model, test it, and make any necessary adjustments.

**Model Selection Flowchart**

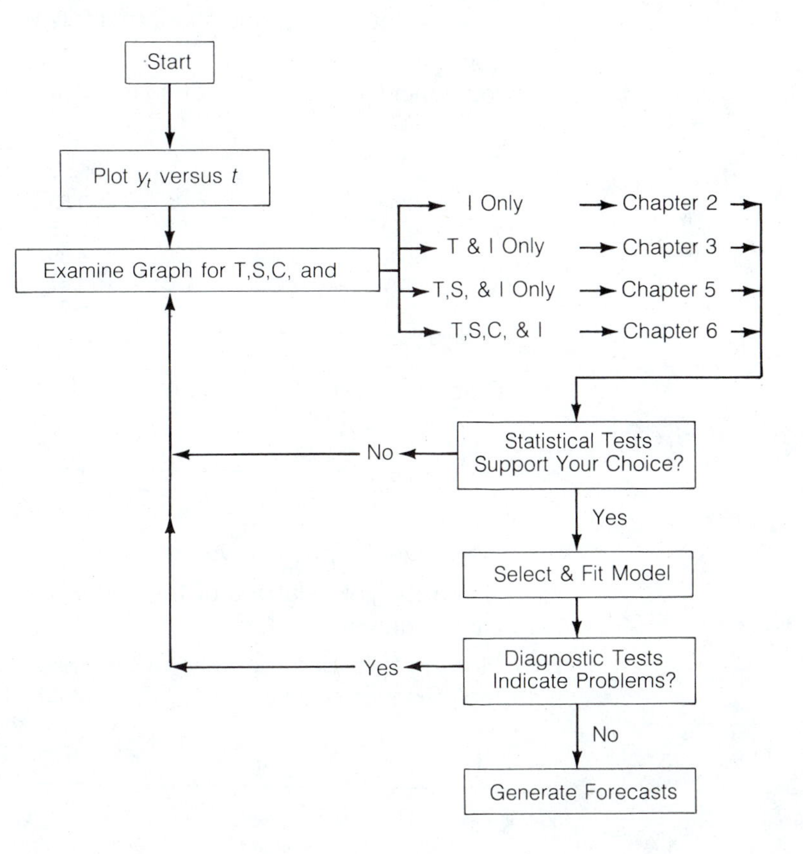

## 1.8 Exercises

**1.1** What is the distinction between a time series model and a forecasting model?

**1.2** Which measure of forecast accuracy, MSE or MAD, penalizes you most for large forecast errors? Why?

**1.3** What purpose is served by drawing lines between successive points on a time series plot? What cautions should be exercised when interpreting these plots?

**1.4** In an additive decomposition, what units of measure (with respect to those of $y_t$) are used for $T$, $S$, $C$, and $I$? What units of measure would be used for these components in a multiplicative decomposition?

**1.5** What forecasting model serves as the base of comparison when using Theil's uncertainty coefficient $U$? Why?

**1.6** For which of the following series would the "%-change" model be inappropriate? Justify your answers.

**a.** $y_t$ = sales of jewelry in the U.S., yearly ($)
**b.** $y_t$ = sales of jewelry in the U.S., monthly ($)
**c.** $y_t$ = total U.S. population, yearly
**d.** $y_t$ = total U.S. population, monthly

**1.7** Compute Theil's $U$ for the following time series and forecasts. What do you conclude about the adequacy of the forecasting model used?

| $t$ | $y_t$ | $\hat{y}_t$ | $t$ | $y_t$ | $\hat{y}_t$ |
|---|---|---|---|---|---|
| 1 | 5.63 | 7.25 | 6 | 9.63 | 7.25 |
| 2 | 9.41 | 7.25 | 7 | 5.35 | 7.25 |
| 3 | 4.95 | 7.25 | 8 | 8.95 | 7.25 |
| 4 | 8.78 | 7.25 | 9 | 4.87 | 7.25 |
| 5 | 5.20 | 7.25 | 10 | 9.21 | 7.25 |

**1.8** Graph the first differences of the following time series:

| $t$: | 1 | 2 | 3 | 4 | 5 | 6 | 7 | 8 | 9 | 10 |
|---|---|---|---|---|---|---|---|---|---|---|
| $y_t$: | 1.21 | 1.72 | 2.23 | 2.73 | 3.21 | 3.72 | 4.24 | 4.72 | 5.23 | 5.71 |

What conclusions can you draw from this graph?

**1.9** Graph the link relatives of the series in Problem 1.8. What conclusions can be drawn?

**1.10** Plot the following monthly time series. Does a visual inspection reveal the presence of seasonality? If so, what is the length of the seasonality?

| $t$: | 1 | 2 | 3 | 4 | 5 | 6 | 7 | 8 | 9 | 10 | 11 | 12 |
|---|---|---|---|---|---|---|---|---|---|---|---|---|
| $y_t$: | 2.3 | 2.5 | 6.1 | 2.1 | 1.9 | 6.2 | 2.4 | 2.0 | 6.1 | 2.3 | 2.3 | 5.4 |

**1.11** In Table 1.1, examine the error series for the "3%-change" model more carefully. What pattern do you observe? Use this result and your intuition to develop a forecasting model that you believe will outperform the "3%-change" model.

**1.12** Graph $y_t$ versus $\hat{y}_t$ for the following data. What do you conclude about the adequacy of the forecasting model that was used to generate the $\hat{y}_t$'s?

| $t$ | $y_t$ | $\hat{y}_t$ |
|---|---|---|
| 1 | 1.0 | 0.8 |
| 2 | 1.6 | 1.3 |
| 3 | 2.1 | 2.0 |
| 4 | 1.8 | 1.7 |
| 5 | 0.9 | 0.8 |
| 6 | 1.4 | 1.2 |
| 7 | 1.9 | 1.6 |
| 8 | 1.7 | 1.5 |

**1.13** Calculate the components $U_M$, $U_R$, and $U_D$ in Theil's decomposition for the data in Problem 1.12. Do your results agree with your answer to 1.12? (You will need to know that $r = .981$ for the above data.)

**1.14** Show that the following two series are the same.

**a.** The first differences of the log $y_t$ series.

**b.** The series formed by the logarithms of the link relative series.

**1.15** For the naive model, $\hat{y}_{t+1} = Cy_t$, find the value of $C$ that minimizes Theil's $U$ coefficient. (Calculus will help here.)

**1.16** Suppose the link relative series is very stable and averages about 104%. Would a "%-change" model be appropriate in this case? Why/why not? What if the average level of the link relatives shows an upward trend?

**1.17** If the formula for the first differences of the series $y_t$ is $y_t - y_{t-1}$, what is the formula for the second differences? the third differences? Do you see a pattern here?

**1.18** At periods 9 and 10 of a certain series, suppose the difference between their heights on the ratio scale plot (natural logarithms) is .02. What was the *exact* percentage change in the original series from period 9 to period 10? (*Note:* For small percentage changes, the difference log $y_t$ − log $y_{t-1}$ itself is a good *approximation* to the percentage change in the series.)

**1.19** Each of the following series grows by exactly 7% each period. Plot both of the series on the same graph using an arithmetic scale, then plot both series using a ratio (log) scale. What conclusions can you draw?

| $t$: | 1 | 2 | 3 | 4 | 5 | 6 | 7 | 8 | 9 | 10 |
|---|---|---|---|---|---|---|---|---|---|---|
| $y_t$: | 107.0 | 114.5 | 122.5 | 131.1 | 140.3 | 150.1 | 160.6 | 171.8 | 183.8 | 196.7 |
| $x_t$: | 10.7 | 11.4 | 12.3 | 13.1 | 14.0 | 15.0 | 16.1 | 17.2 | 18.4 | 19.7 |

**1.20** One method for estimating $C$ in the "%-change" model $\hat{y}_{t+1} = Cy_t$ is to:

(i) Plot $y_t$ versus $y_{t-1}$.

(ii) Fit a straight line through this plot (by eye or by some other method of your choosing).

(iii) Use the slope of this straight line as an estimate of $C$.

**a.** Try this method on the following quarterly sales data (units: $1,000) by first estimating $C$ and then computing the forecasts and forecast errors for the resulting model.

**b.** How does the MAD for this model compare to that for a model based on estimating $C$ as the average link relative of the series? How about the MSE for each of the two models?

| Qtr $t$ | Sales $y_t$ | Qtr $t$ | Sales $y_t$ |
|---|---|---|---|
| 1 | 10.1 | 9 | 14.4 |
| 2 | 10.1 | 10 | 14.7 |
| 3 | 10.6 | 11 | 17.0 |
| 4 | 10.9 | 12 | 18.0 |
| 5 | 11.8 | 13 | 18.0 |
| 6 | 13.3 | 14 | 18.6 |
| 7 | 13.6 | 15 | 19.5 |
| 8 | 14.9 | | |

**1.21** Suppose a certain time series $y_t$ is growing (approximately) by a fixed percentage each time period. Discuss how you could use the *ratio-scale* plot to estimate the period-to-period growth.

**1.22** Consider the following time series: A coin is tossed successively letting $x = +1$ if heads occurs and $x = -1$ if tails occurs; for each successive toss of the coin, let $y_0 = 0$ and $y_t - y_{t-1} = x$. In this way, $y_t$ will keep track of the difference between the number of heads and tails tossed.

**a.** In 50 tosses of the coin, how many times would you expect $y_t$ to become zero? (a) never, (b) infrequently, (c) 20–30 times, (d) frequently .

**b.** Perform an experiment by tossing the coin 50 times and plotting the resulting time series ($y_t$ versus $t$). How many times did your graph cross the $t$-axis (become zero)? (This series is an example of a *random walk* (cf. Chapter 4.)

**1.23** For the residuals from *any* forecasting model, the MAD is never larger than the RMSE. Verify this inequality by taking any ten numbers and, letting them represent ten forecast errors, calculate the MAD and the RMSE. (*Note:* MAD = RMSE only when all the forecast errors happen to have the same magnitude.)

**1.24** Under what circumstances would one prefer the MAPE to the MSE?

**1.25** Describe two shortcomings of the MSE as a measure of forecast accuracy?

**1.26** If a series has both trend and a strong seasonal component but, for some reason, a forecaster only models the trend, what would you expect the graphs of $e_t$ vs. $t$ and $y_t$ vs. $t$ to look like?

**1.27** Calculate the MAD, MSE, RMSE, and MAPE for the following series of forecast errors:

| $t$: | 1 | 2 | 3 | 4 | 5 | 6 | 7 | 8 | 9 | 10 |
|---|---|---|---|---|---|---|---|---|---|---|
| $e_t$: | 4.4 | 3.4 | 5.3 | 5.0 | 5.1 | 4.4 | 3.7 | 5.9 | 5.0 | 3.8 |

**1.28** For the following series, find $U_M$, $U_R$, and $U_D$ in Theil's decomposition of the MSE:

| $t$: | 1 | 2 | 3 | 4 | 5 | 6 | 7 |
|---|---|---|---|---|---|---|---|
| $y_t$: | 10.2 | 10.4 | 10.6 | 10.8 | 11.0 | 11.2 | 11.4 |
| $\hat{y}_t$: | 10.0 | 5.0 | 2.0 | 0.0 | 2.0 | 5.0 | 10.0 |

# APPENDIX 1A

## Computation Formulas for Theil's Decomposition of the Mean Square Error

Letting

$y$ = actual value of the time series

$\hat{y}$ = predicted value of the time series

$n$ = the number of forecasts

and

$$\bar{y} = \frac{\Sigma\, y}{n}$$

$$\bar{\hat{y}} = \frac{\Sigma\, \hat{y}}{n}$$

$$SS_{yy} = \Sigma\,(y - \bar{y})^2 = \Sigma\, y^2 - \frac{(\Sigma\, y)^2}{n}$$

$$SS_{\hat{y}\hat{y}} = \Sigma\,(\hat{y} - \bar{\hat{y}})^2 = \Sigma\, \hat{y}^2 - \frac{(\Sigma\, \hat{y})^2}{n}$$

$$SS_{y\hat{y}} = \Sigma\,(y - \bar{y})(\hat{y} - \bar{\hat{y}}) = \Sigma\, y \cdot \hat{y} - \frac{(\Sigma\, y)(\Sigma\, \hat{y})}{n}$$

then

$$s'_y = \sqrt{\frac{SS_{yy}}{n}}$$

$$s'_{\hat{y}} = \sqrt{\frac{SS_{\hat{y}\hat{y}}}{n}}$$

$$b = \frac{SS_{y\hat{y}}}{SS_{\hat{y}\hat{y}}} \qquad \text{and} \qquad r = \frac{SS_{y\hat{y}}}{\sqrt{SS_{yy}SS_{\hat{y}\hat{y}}}}$$

and

$$\text{MSE} = (\bar{\hat{y}} - \bar{y})^2 + (s'_{\hat{y}} - rs'_y)^2 + (1 - r^2)s'^2_y$$

## References

Armstrong, J. S. (1978). *Long-Range Forecasting.* New York: Wiley.

Armstrong, J. S. (1986). "The Ombudsman: Research on Forecasting: "A Quarter-Century Review, 1960–1984." *Interfaces* 16(1): 89–109.

Brown, R. G. (1982). "The Balance of Effort in Forecasting." *J. Forecasting* 1: 49–53.

Gilchrist, W. G. (1976). *Statistical Forecasting.* New York: Wiley.

Intriligator, M. D. (1978). *Econometric Models, Techniques, and Applications.* Englewood Cliffs, N. J.: Prentice-Hall.

Judge, G. G., Griffiths, W. E., Hill, R. C., and Lee, T. (1980). *The Theory and Practice of Economics.* New York: Wiley.

Lindberg, B. C. (1982). "International Comparison of Growth in Demand for a New Durable Consumer Product." *J. Market. Res.* 19: 364–371.

Lorie, J. H., and Hamilton, M. T. (1973). *The Stock Market, Theories and Evidence.* Homewood, Ill.: Richard D. Irwin.

Makridakis, S., and Hibon, M. (1979). "Accuracy of Forecasting: An Empirical Investigation (with Discussion)." *J. Roy. Statist. Soc.* A(2), 142: 97–145.

Makridakis, S., Wheelwright, S. C., and McGee, V. E. (1983). *Forecasting: Methods and Applications,* 2nd Ed. New York: Wiley.

Malkiel, B. G. (1985). *A Random Walk Down Wall Street.* 4th Ed. New York: Norton.

McNees, M. K. (1979). "The Forecasting Record for the 1970's." *New England Econ. Rev.* Sept–Oct.

Mohn, N. C., and Reid, J. C. (1977). "Some Practical Guidelines for the Corporate Forecaster." *Interfaces* 7(3): 70–75.

Nerlove, N., Grether, D. M., and Carvalho, J. L. (1979). *Analysis of Economic Time Series.* New York: Academic Press.

Orlicky, J. (1975). *Material Requirements Planning.* New York: McGraw-Hill.

Schmid, C. F. (1986). "Whatever Has Happened to the Semilogarithmic Chart?" *Amer. Statistician* 40(3): 238–244.

Shlifer, E., and Wolff, R. W. (1979). "Aggregation and Proration in Forecasting." *Management Sci.* 25(6): 594–603.

Smith, G. C. (1964). "The Law of Forecast Feedback." *Amer. Statistician* 18(4): 11–14.

Theil, H. (1966). *Applied Economic Forecasting.* Amsterdam: North-Holland.

Tufte, E. R. (1983). *The Visual Display of Quantitative Information.* Chesire, Conn.: Graphics Press.

Wonnacott, R. J., and Wonnacott, T. H. (1979). *Econometrics,* 2nd Ed. New York: Wiley.

# CHAPTER 2 Forecasting Series with No Trend

**Objective:** *To present procedures for forecasting time series that, except for random variation, remain essentially constant over time.*

## 2.1 Introduction

To a great degree, forecasting may be thought of as constructing an "expected" future value for a time series. It is reasonable, then, that the simpler forecasting situations involve series whose "expected" or average value remains relatively constant over the period for which the forecasts are being constructed. Such situations arise when the behavioral or demand patterns influencing the time series become relatively stable. It may also be possible to stabilize a nonconstant series by changing measurement units. For example, one can use "per capita" or "per unit" figures to correct for trends due solely to population change. Other stabilizing transformations may also be applied, as when period-to-period changes (first differences) are used in lieu of the original series. These kinds of stable series are often referred to as "stationary," "constant mean," or "horizontal" series, the latter term reflecting the fact that the graph of such a series will appear roughly parallel to the horizontal (time) axis. In the language of classical time series analysis discussed in Chapter 1, these series do not exhibit the general upward or downward movement called trend. Thus, we may also refer to them as "no-trend" series. Although some of the procedures developed in this chapter will handle seasonal or cyclical movements to a greater or lesser degree, they are not specifically designed to do so, and should not be expected to be particularly efficient in that regard.

In its simplest form, forecasting a no-trend series involves using the available history of the series to estimate its average value. This estimate then becomes the forecast for future values. More complicated schemes involve "updating" the estimate as new information becomes available. These schemes are useful when initial estimates are unreliable or when the stability of the average is in question. In addition, updating procedures inherently provide some degree of responsiveness to changes in the underlying structure of the series.

These relatively simple models will be studied at some length, not only because they arise often in practice, but also because they provide a convenient framework around which to build concepts fundamental to approaching forecasting problems of all kinds. It is also true that the residuals (forecast errors) from a well-constructed forecasting model should form a stationary series with an average value of zero, an important consideration when performing diagnostic checks for model adequacy.

## 2.2 The Nature of Stationary Series

### 2.2.1 Introduction

The term "stationary" is commonly used by the more mathematically inclined when referring to horizontal time series. Although the term has a rigorous mathematical definition, for our purposes it is sufficient to think of a *stationary* series as one which appears about the same, *on the average,* no matter when it is observed. (Mathematicians would refer to this type of stationarity more precisely as stationarity *in the mean.* See Jenkins and Watts, 1968, pp. 149–151.) A slightly more formal definition is the following:

**DEFINITION:** A time series $y_t$; $t = 1, 2, \ldots$ is said to be **stationary** if the expected value of $y_t$ is the same for any time $t$, i.e., $E(y_t) = \beta_0$ for $t = 1, 2, \ldots$

### 2.2.2 Characteristics of Stationary Series

A model for a stationary time series may be expressed as follows:

$y_t$ = the actual value of the time series at time $t$

$\beta_0$ = the unchanging average level of the series

$\epsilon_t$ = a random variable representing irregular fluctuations around the average level at time $t$

**No-Trend Model**

$$y_t = \beta_0 + \epsilon_t \qquad t = 1, 2, \ldots$$

Notice that we have chosen to use $\epsilon_t$ for the random element, rather than the $I_t$ mentioned in Chapter 1. This notation is more modern and more to our liking.

The nature of these irregular movements, or "error terms," depends on the specific application, although the mean value of $\epsilon_t$ is generally assumed to be zero. The $\epsilon_t$'s are also quite often assumed to be independent. The series produced when the error terms are independent varies around the average level, $\beta_0$, with no discernible pattern (Figure 2.1). Dependent ("autocorrelated") error terms, on the other hand, produce either a series which tends to alternate rapidly above and below the average (Figure 2.2) or one which drifts above and below for longer periods of time (Figure 2.3). The term "autocorrelation" is used to describe correlation between successive observations in a time series.

Forecasting stationary series involves estimating $\beta_0$, unless its value is known from other sources. The nature of the $\epsilon_t$'s, revealed from an analysis of historical data, determines how $\beta_0$ is estimated and affects the assessment of forecast accuracy.

**Figure 2.1** Independent Error Terms

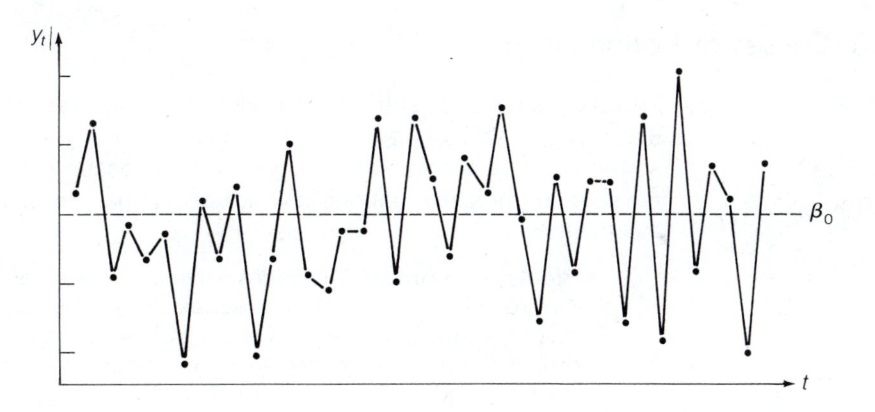


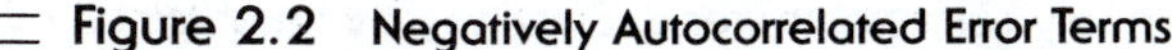

**Figure 2.2** Negatively Autocorrelated Error Terms

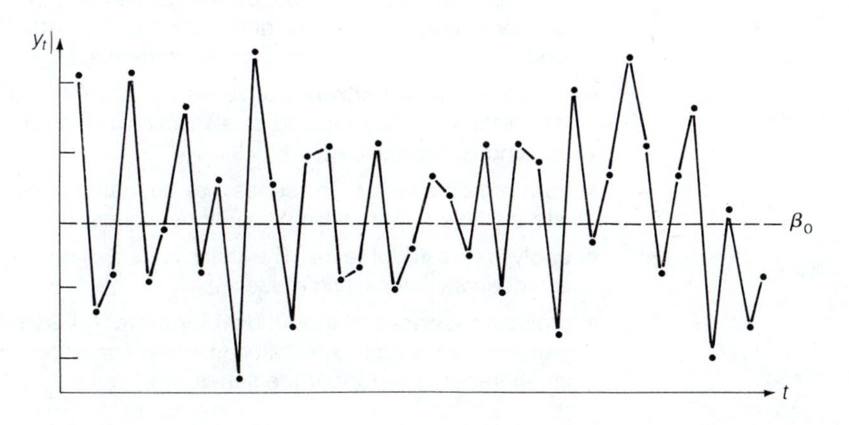

**Figure 2.3 Positively Autocorrelated Error Terms**

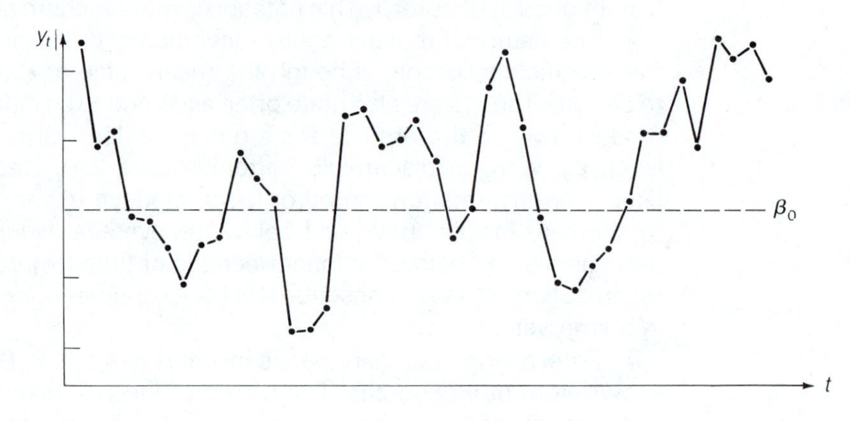


## 2.2.3 Causes of Stationarity

The simplicity of a stationary model belies its usefulness somewhat. The underlying stability needed to produce a stationary series may, at first, seem so restrictive a condition that it would rarely occur in practice. Actually, there are several kinds of situations in which a stationary model may prove adequate, including:

- **stable environment** The forces generating the series have stabilized, and the environment in which the series exists is relatively unchanging. Examples might be: the unit sales of a product or service in the mature stage of its life cycle; the number of assembly line breakdowns per week in a facility with uniform production rate; the number of appointments per week requested of a doctor or lawyer whose patient or client population is fixed; or the number of sales resulting from a constant level and quality of salesperson effort.
- **easily correctable trend** Stability may be obtained by making simple corrections for factors such as population growth and inflation. Changing demand to "per capita" demand or dollar sales to "constant dollar" (deflated) sales would be examples of such simple corrections.
- **short forecasting horizon** The series has a trend, but the period of time over which forecasts are needed is relatively short, so the amount of change due to trend is negligible.
- **transformable series** The series may be mathematically altered (transformed) into a stable one by taking logarithms, square roots, differences, etc.
- **analysis of residual series** The series is a set of residuals (forecast errors) from an adequate forecasting model.
- **preliminary stages of model development** A very simple model is required, perhaps for ease of explanation or implementation, or because of the lack of an appreciable history of the series.

## 2.3 Examples

### 2.3.1 Introduction

It is important to remember that no-trend models occur when economic, behavioral, or other forces affecting the value of the series are relatively stable. Factors such as population, standard of living, and economic health either change little over the forecasting horizon or do not appreciably affect the average value of the series.

### 2.3.2 Example 2.1—Mainline Plumbing Co.

As an example, consider a plumbing repair service that maintains a fleet of small vans. The vans are dispatched on service calls as needed. The gasoline purchases for this fleet over the past several weeks are summarized in Table 2.1. Visual inspection of a graph of these data (Figure 2.4) reveals no apparent pattern or trend in the series, but, rather, a jagged line varying around an average level of about 350–450 gallons.

Stationarity could be anticipated in this situation for several reasons:

- The number of vans in the fleet remains constant over the period being recorded, stabilizing the total gasoline consumption.
- The metropolitan area in which the service operates has sufficient demand for the company's services so that it operates at essentially full capacity; relatively independent of other plumbing service companies in the area, of population change in the city as a whole, and of small changes in the economic climate.
- The period over which the data have been collected is a relatively short one so that any change in average consumption due to trend, if any, can probably be neglected.
- No changes in advertising or other promotional activities have taken place.

**Table 2.1 Gasoline Purchases for Mainline Plumbing Co. Fleet**

| Week $t$ | Gasoline Purchased $y_t$ (gal) | Week $t$ | Gasoline Purchased $y_t$ (gal) |
|---|---|---|---|
| 1 | 409 | 11 | 380 |
| 2 | 289 | 12 | 598 |
| 3 | 509 | 13 | 418 |
| 4 | 364 | 14 | 359 |
| 5 | 404 | 15 | 432 |
| 6 | 445 | 16 | 252 |
| 7 | 310 | 17 | 446 |
| 8 | 372 | 18 | 473 |
| 9 | 440 | 19 | 337 |
| 10 | 414 | 20 | 478 |

**Figure 2.4** Gasoline Purchases For Mainline Plumbing Co. Fleet

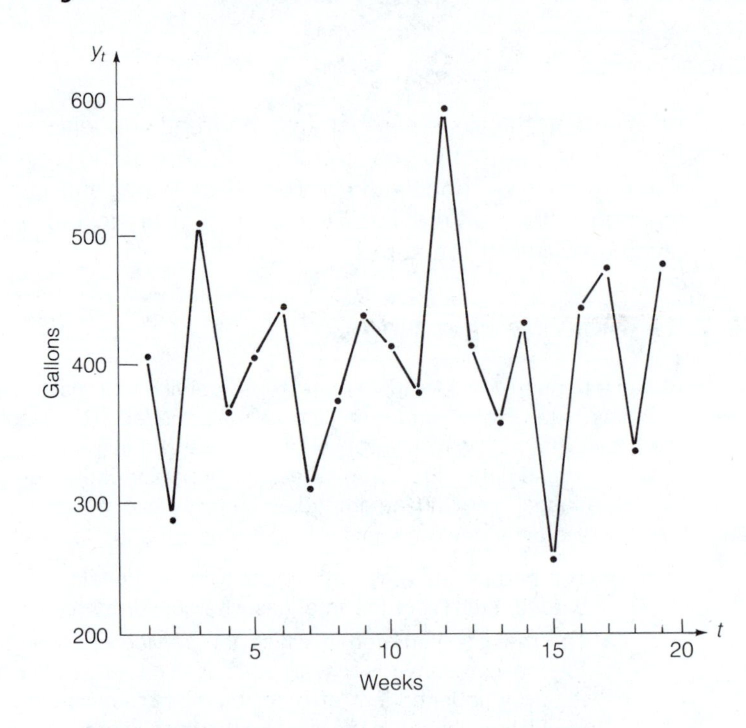


- No change in the basic method of delivering service has occurred (such as more efficient scheduling/dispatching of the vans, better maintenance, or driver training).
- The actual amount of gasoline consumed by a van on a given day is determined almost solely by the random locations around the city of the various service points and by the random number of services that may be completed within the normal work day. ●

### 2.3.3 Example 2.2—Rogers Electronics Specialties, Inc.

Many no-trend series exhibit a sort of undulation above and below the average level because the conditions that affect the time series at one point in time tend not to disappear immediately. These conditions have varying degrees of effect on other observations near them in time. Influences that cause a movement above the average in, say, June may not have entirely disappeared by July and would then influence the July value of the series. Other newly occurring influences may move the series in either direction (above or below $\beta_0$), but some influence from June's condition remains. As a matter of fact, June's conditions may intensify in July and for several months thereafter (when "fads" develop), then decline for several months, producing long swings above, then below the average demand. Consider, for example, a small manufacturing concern, call it Rogers Electronics Specialties, Inc., with a work force of around 200. Its personnel records are examined and the number of absentees among assembly line personnel is aggregated for each week over a period of weeks. The resulting number of absentee days each week is given in Table 2.2. (Note that one employee absent for, say, 3 days would count as 3 absentee days that week.)

If the work force is relatively fixed, the average number of absentee days each

**Table 2.2 Rogers Electronics Specialties, Inc.**
Absenteeism

| Week $t$ | Absentee Days $y_t$ | Week $t$ | Absentee Days $y_t$ |
|---|---|---|---|
| 1 | 32 | 9 | 59 |
| 2 | 40 | 10 | 25 |
| 3 | 26 | 11 | 30 |
| 4 | 37 | 12 | 28 |
| 5 | 48 | 13 | 30 |
| 6 | 66 | 14 | 40 |
| 7 | 73 | 15 | 36 |
| 8 | 62 | | |

week should remain nearly constant. There are factors, however, which tend to introduce interdependence into the error terms. For one thing, psychological interaction among employees could generate "rashes" of absenteeism. Employees who see others out sick develop a feeling that it would be OK if they were sick too. For another thing, there are varying degrees of contagion present in many types of illness, causing a cycle of illness. Finally, illnesses whose course is long will produce two, three, or even a week or more of absentee days for each infected employee. The overall result is that when a particular week shows a higher-than-average number of absentee days, we can expect higher-than-average values in nearby subsequent weeks, with an eventual return to normal conditions or, perhaps, a swing below normal as employees feel they have used up their "OK" sick days. Examining a graph of the data in Figure 2.5,

**Figure 2.5 Rogers Electronics Specialties, Inc.: Absenteeism**

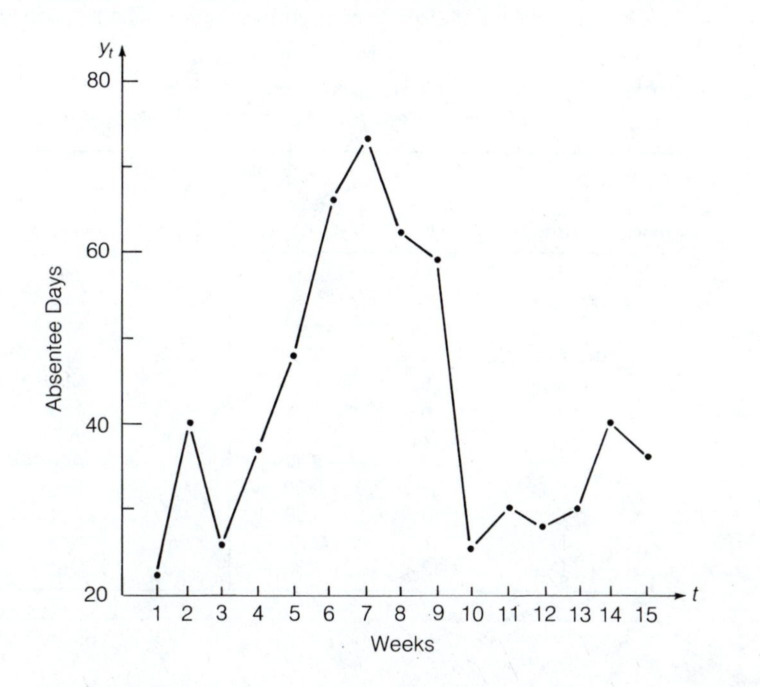

we see a "normal" absentee rate of around 30 or so per week, with a decided swing above the average over weeks 5 through 9 and a slight swing above average in the last couple of weeks, perhaps presaging another cycle of increased absenteeism. (An interesting question here might be: What would one forecast for the number of absentee days in weeks 16 through 20?) Notice, in Figure 2.5, that at week 9, if we were trying to forecast this series, we might well have come to the conclusion that a trend of some kind was present. One should be very careful about drawing such conclusions solely on the basis of a small set of data. •

### 2.3.4 Example 2.3—Superior Ink Products Corp.

Another phenomenon which may be present in no-trend series is a tendency to oscillate rapidly above and below the normal level. One might expect such patterns in a series representing orders for some product or service from a fixed clientele or for which there is steady demand. If an unusually large number of orders is received during one period, we may expect, perhaps, fewer-than-normal orders in the next period. Such conditions will tend to produce negatively autocorrelated error terms and give rise to the sort of data in Table 2.3 which contains figures on the Region III monthly orders for a certain style of replacement ribbon cartridge from Superior Ink Products Corp. during 1982 and 1983.

A graph of these data in Figure 2.6 shows a series with no apparent trend. It has the jagged appearance characteristic of the presence of strong negative first-order (lag 1) autocorrelation. On further examination, the average monthly order appears to be around 250 boxes, with a variation of around 100 boxes above and below the average. In addition, a large swing above the average in one period is quite often followed in the next period by a rather large compensating swing below. The apparent lack of trend may be thought of as a consequence of a stable clientele, that is, a relatively constant number of retail outlets generating the ribbon orders. The up-down swing in demand could result from the ordering policies of the retailers involved. One such policy might involve the use of a naive forecasting model, as mentioned in Section 1.4, to generate order sizes from month to month. The retailers' monthly orders would

**Table 2.3 Superior Ink Products Corp.**
Monthly Orders for Replacement Ribbon #140964

| Year | Month | Time $t$ | Total Orders $y_t$ (boxes) | Year | Month | Time $t$ | Total Orders $y_t$ (boxes) |
|---|---|---|---|---|---|---|---|
| 1982 | Jan | 1 | 1.41 | 1983 | Jan | 13 | 3.43 |
| | Feb | 2 | 3.26 | | Feb | 14 | 2.74 |
| | Mar | 3 | 1.91 | | Mar | 15 | 3.87 |
| | Apr | 4 | 2.68 | | Apr | 16 | 1.16 |
| | May | 5 | 2.50 | | May | 17 | 2.63 |
| | Jun | 6 | 3.99 | | Jun | 18 | 2.76 |
| | Jul | 7 | 1.52 | | Jul | 19 | 1.54 |
| | Aug | 8 | 2.12 | | Aug | 20 | 3.24 |
| | Sep | 9 | 3.80 | | Sep | 21 | 2.52 |
| | Oct | 10 | 1.58 | | Oct | 22 | 3.95 |
| | Nov | 11 | 2.85 | | Nov | 23 | 1.64 |
| | Dec | 12 | 1.86 | | Dec | 24 | 2.52 |

**Figure 2.6** Superior Ink Products Corp.: Monthly Orders for Replacement Ribbon #140964, January 1982–December 1983

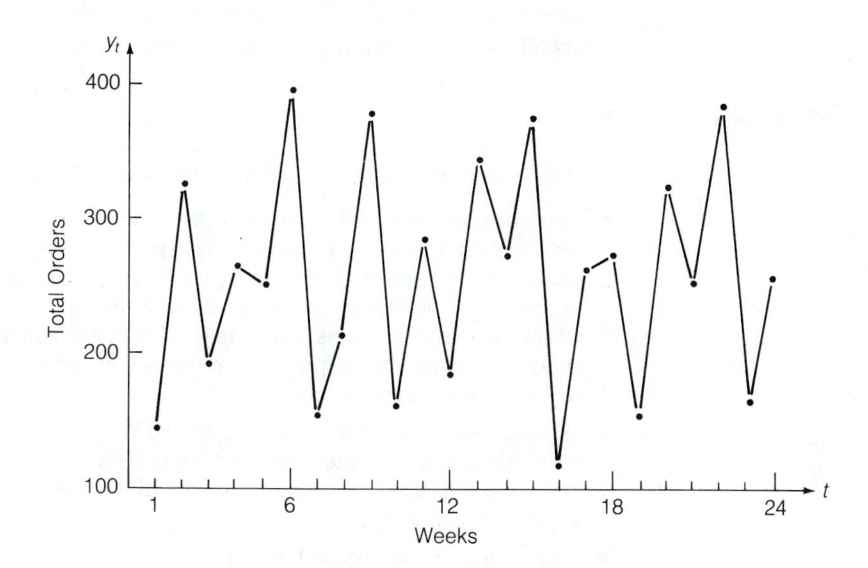


then be chosen as last month's demand, plus, perhaps, a safety margin to guard against unanticipated demand, and less the stock on hand. The net effect of such an ordering policy in the face of constant average demand is to generate a highly variable order quantity to the supplier. If an unusually large demand at the retail level occurs during one period, stocks will be depleted and large orders generated for the next period. If the demand at the retail level then falls to near or below normal during the next period, overstock occurs and exaggeratedly small orders are generated. The net result is a rapid up-down movement in Superior's orders. ●

These examples illustrate situations where no-trend models could be directly applied. From them one can begin to develop a general feel for the types of conditions that might call for the use of such models. There are many other valuable applications areas for horizontal models, including, most importantly, cases where relatively simple alterations to trended series will produce untrended ones. Such models are also used in the analysis of forecast errors.

## 2.4 The Decision to Use No-Trend Models

### 2.4.1 Introduction

When using statistical forecasting techniques, sooner or later—usually sooner—we face the problem of deciding which model would work best, or even work well, for our data. The general factors affecting this decision (such as cost, availability of data, and desired accuracy) were discussed in Section 1.4. More specific items may be divided into three categories:

- Items that should be considered whichever model (no-trend or any other) is being examined.
- Criteria that apply specifically to no-trend models.
- Statistical tests and other mathematical techniques.

## 2.4.2 General Practices

The following are important considerations in any model fitting context.

- **familiarity with the series** Become as knowledgeable as time and cost will permit about the series you are trying to predict, by learning about the process generating the series. If you are trying to forecast sales of tomato sauce, talk to the people who sell it, those who buy it, those who make it, and to others who may affect its sales (e.g., marketing staff). Anything you can find out about the series may help you avoid serious mistakes in model selection and/or avoid embarrassingly large errors.
- **organization of the data** All data should be organized into easily examined form. A neatly constructed table or array of data with perhaps some preliminary statistical measures computed should suffice. There are computer programs around to help with this.
- **graphical presentation of the data** A carefully constructed graph of the data is a simple device, and the effort involved in constructing it pays substantial dividends. Standard graphs you may use have been described in Section 1.4. Thoughtful examination of various of these graphs will develop familiarity with the data that can be obtained in no other way.

## 2.4.3 Specific Criteria for No-Trend Models

After the above preliminary steps, and keeping in mind the stability conditions discussed earlier, a horizontal model should be useful—provided, that is, that the following are true.

- Your knowledge of the series supports, or at least does not contradict, the hypothesis that the series is stable.
- Inspection of the graph of the series reveals no trend, so the general appearance of the series is horizontal.
- The first differences $y_t - y_{t-1}$ of the series *themselves* form a horizontal series. The average value of that series should be zero. A handy way to check is to compute and graph the first differences.
- The link relatives $y_t/y_{t-1}$ of the series also form a horizontal series, with an average value of 100%.
- If the series drifts above and below the average level, there is no regularity to this drift. If seasonality is present, it may be handled using methods discussed in Chapter 5.

## 2.4.4 Statistical Tests for Stationarity and Randomness

Often, trend in the series will be apparent from the graphical analysis. In such cases, no statistical testing is really needed. Sometimes, however, even if the series generally seems to conform to these criteria, one may feel uncertain about stationarity. Usually the "randomness," or "noise," in the series causes this insecurity.

Hypothesis testing may help resolve the question of stationarity. The format of such hypothesis tests is:

$H_0$: The series is stationary/random.

$H_a$: The series is not stationary/random.

and the conclusions that may be drawn are:

Do not reject $H_0$, meaning there is insufficient evidence at the $\alpha \cdot 100\%$ level of significance that the series is not stationary/random.

Reject $H_0$, meaning there is sufficient evidence at the $\alpha \cdot 100\%$ level that the series is not stationary/random.

As a result, the tests given in the next sections will not *prove* whether a model is stationary. However, "Do not reject $H_0$" decisions may be considered *supportive* of stationarity.

## 2.5 Nonparametric Tests for Stationarity and Randomness

### 2.5.1 Introduction

We begin our discussion of statistical tests with those that require no assumptions about the nature of the probability distribution of the errors $\epsilon_t$. Such tests are usually referred to as **nonparametric** or **distribution-free.**

### 2.5.2 Runs Test: A Simple Test for Randomness

The **runs test** is based on the premise that any observation from a horizontal time series with independent error terms is equally likely to be above or below the median of the series. To run the test we first compute the median, $\tilde{y}$, of the series, then assign a plus sign to observations above the median and a minus sign to observations below it. Finally we list the pluses and minuses in chronological order and count the number of "runs," or blocks of pluses and minuses (Brownlee, 1965, sec. 6.5).

The number of pluses and minuses depends on whether the series has an even or an odd number of observations. If $n$ is odd, the median, which is itself an observation, is ignored and we get $(n - 1)/2$ pluses and $(n - 1)/2$ minuses. Otherwise we get $n/2$ of each. Suppose we let:

$$
\begin{aligned}
m &= \text{the number of } +\text{'s} = \text{the number of } -\text{'s} \\
&= \frac{n-1}{2} \qquad n \text{ odd} \\
&= \frac{n}{2} \qquad n \text{ even}
\end{aligned}
$$

Then a random no-trend series should produce a random string of pluses and minuses, that is, a string with neither too few nor too many runs. A series with trend, or one with large positive autocorrelation, will tend to have few runs, and a series with negative autocorrelation will tend to have many. Figure 2.7(a), for example, shows the effect of trend on runs. The early observations in the series tend to be below the median (minuses), while later ones tend to be above it (pluses). A very pronounced upward trend can lead to a sequence with only two runs, a sequence of minuses followed by a sequence of pluses (vice versa for a downward trend). In Figure 2.7(b), the effect of positive autocorrelation can be seen: a relatively small number of runs as the series drifts above and below the median. Finally, a random series, such as that in Figure 2.7(c), will produce a moderate number of smaller runs of pluses and minuses.

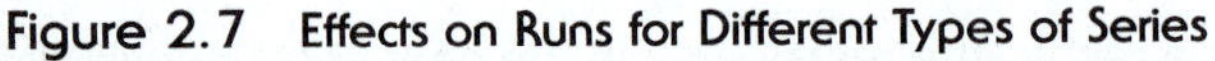

**Figure 2.7** Effects on Runs for Different Types of Series

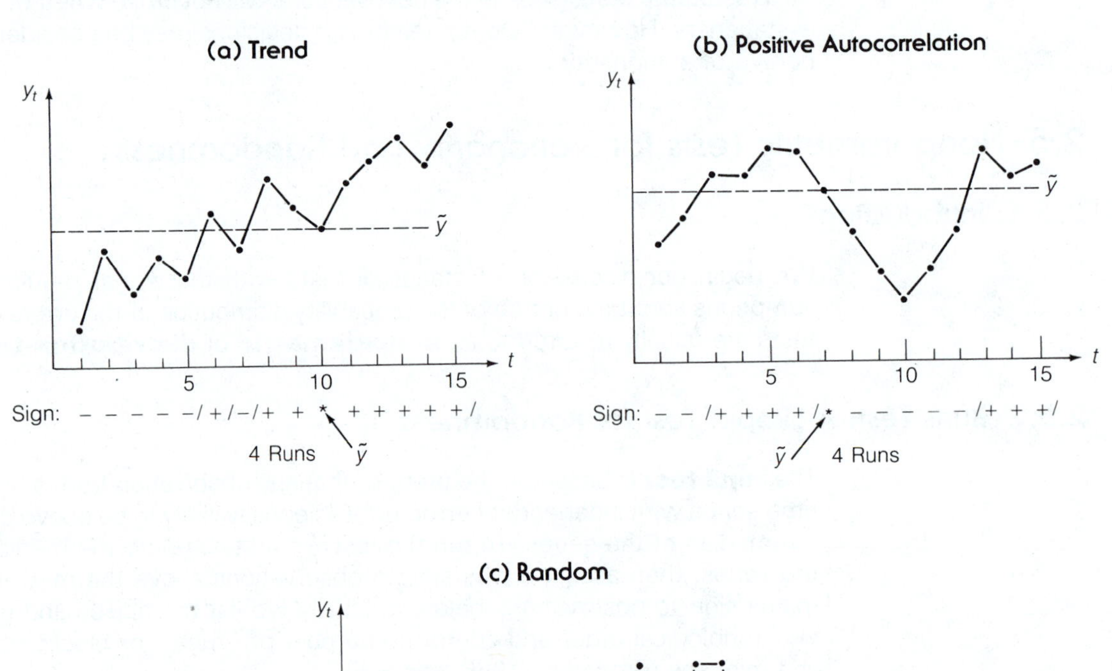

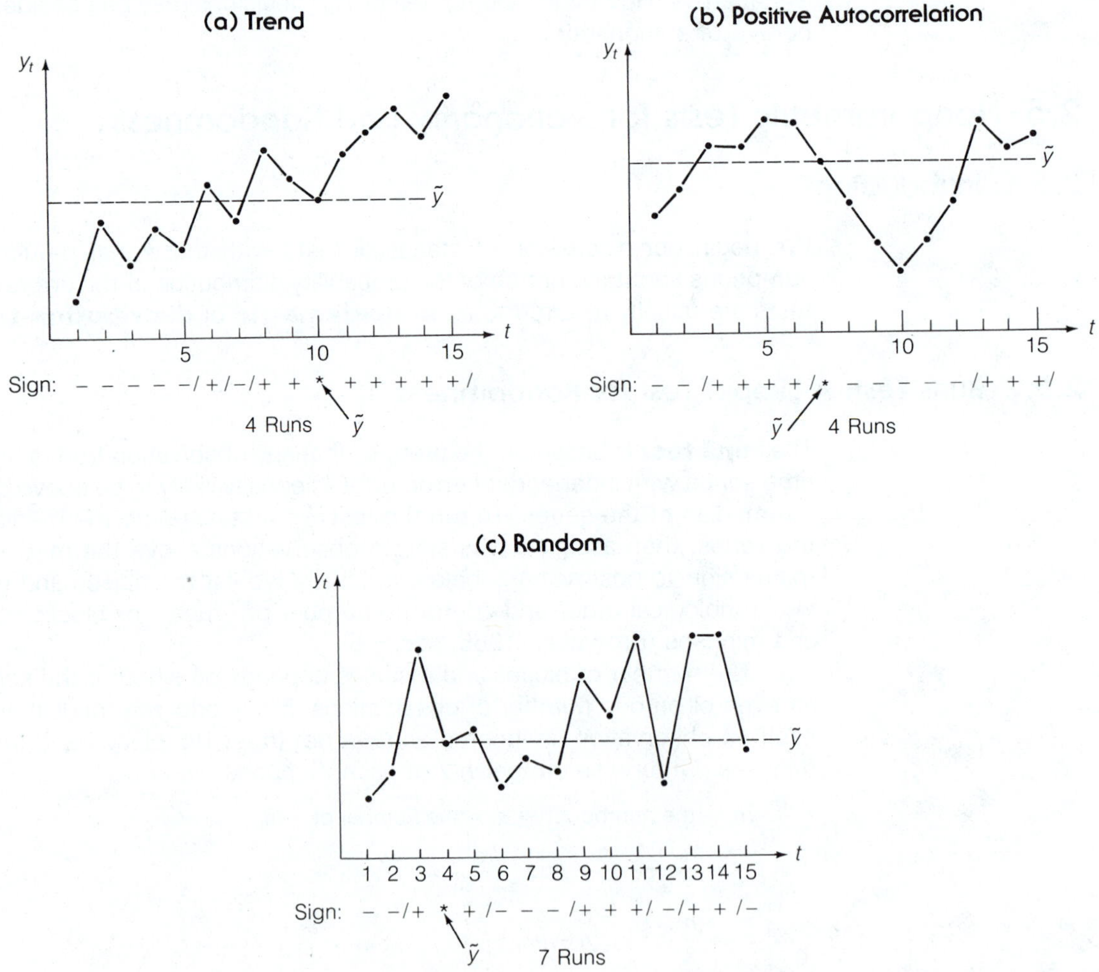

**(c) Random**

$y_t$ $\tilde{y}$ t

1 2 3 4 5 6 7 8 9 10 11 12 13 14 15

Sign: – –/+ * + /– – – /+ + +/ –/+ + /–

$\tilde{y}$ 7 Runs

The statistic $R$ = the number of runs in a *random* sequence of $m$ pluses and $m$ minuses can be shown (with arguments the details of which are not necessary for our discussion here) to have the following mean and standard deviation:

$$\mu_R = \text{expected number of runs} = m + 1$$

$$\sigma_R = \text{standard deviation of the number of runs} = \sqrt{\frac{m(m-1)}{2m-1}}$$

For small values of $m$ ($m \leq 20$), the upper and lower critical values of $R$, call them $R_U$ and $R_L$, for one-tailed tests with 5% significance levels have

---

**Test Procedure:** *Runs Test*

$H_0$: The series is a no-trend series with independent errors.

$H_a$: The series has trend and/or autocorrelated errors.

or, more formally:

$H_0$: $y_t = \beta_0 + \epsilon_t$  $\beta_0$ constant and $\epsilon_t$ independent

$H_a$: $y_t = \beta_0 + \epsilon_t$  $\beta_0$ not constant or $\epsilon_t$ not independent

**Test Statistic:** $R$ = the number of runs above and below the median

**Decision Rule (Small Sample):** ($m \leq 20$, $\alpha = 10\%$)

Reject $H_0$ if $R \geq R_U$ or $R \leq R_L$. **(Appendix AA)**

Otherwise do not reject $H_0$.

**Decision Rule (Large Sample):** ($m > 20$, $\alpha$)

Reject $H_0$ if $|z| > z_{\alpha/2}$.

Otherwise do not reject $H_0$.

$$z = \frac{|R - \mu_R|}{\sigma_R}$$

$$\mu_R = m + 1 \qquad \sigma_R = \sqrt{\frac{m(m-1)}{2m-1}} \qquad \text{and} \qquad \begin{cases} m = n/2 & \text{for } n \text{ even} \\ m = (n-1)/2 & \text{for } n \text{ odd} \end{cases}$$

$z_{\alpha/2}$ = the upper ($\alpha/2 \times 100\%$ point of the standard normal distribution **(Appendix BB)**

**Conclusion:** If $H_0$ is rejected, the test has shown with $(1 - \alpha) \times 100\%$ confidence that a no-trend independent model is inappropriate.

If $H_0$ is not rejected, the test has shown some support for using a no-trend model.

---

been tabulated (Swed and Eisenhart, 1943). These critical values will give two-tailed tests with approximately 10% significance levels (see Appendix AA). For $m > 20$, $R$ may be treated as a normal random variable with mean $\mu_R$ and standard deviation $\sigma_R$, as given above (Brownlee, 1965). Critical values may be found from a table of the standard normal distribution (see Appendix BB).

**EXAMPLE 2.4** *Burglary Statistics*

Many series generated by the demands, customs, or other activities of a growing population tend to have an upward trend. Take, for example, the crime statistics of a community located on the outskirts of a large metropolitan area. The number of burglaries reported to the city police during the year for a 20-year period are given in Table 2.4 and graphed in Figure 2.8. (These are actual data coded at the request of the source personnel.)

Inspection of the graph should make it clear that there is a trend in the number of burglaries, so a horizontal model would be a poor choice. However, for purposes of illustration, we show how a runs test would proceed.

$H_0$: The number of burglaries per year follows a no-trend model with independent error terms.

$H_a$: The number of burglaries per year has trend and/or autocorrelated error terms.

**Test Statistic:** $R$ = the number of runs above and below the sample median

**Decision Rule:** (Small Sample, $m = n/2 = 10$, $\alpha = 10\%$)

Reject $H_0$ if $R \geq 16$ or $R \leq 6$. **(Appendix AA)**

Otherwise do not reject $H_0$.

To calculate $R$:

- Compute the median. Since $n = 20$, the median, $\tilde{y}$, is the midpoint between

**Table 2.4 Number of Burglaries Reported to City Police**

| Year | Time $t$ | Burglaries $y_t$ | Sign | Year | Time $t$ | Burglaries $y_t$ | Sign |
|---|---|---|---|---|---|---|---|
| 1964 | 1 | 22 | − | 1974 | 11 | 114 | + |
| 1965 | 2 | 50 | − | 1975 | 12 | 131 | + |
| 1966 | 3 | 34 | − | 1976 | 13 | 91 | + |
| 1967 | 4 | 58 | − | 1977 | 14 | 127 | + |
| 1968 | 5 | 59 | − | 1978 | 15 | 122 | + |
| 1969 | 6 | 53 | − | 1979 | 16 | 158 | + |
| 1970 | 7 | 90 | − | 1980 | 17 | 134 | + |
| 1971 | 8 | 85 | − | 1981 | 18 | 237 | + |
| 1972 | 9 | 73 | − | 1982 | 19 | 270 | + |
| 1973 | 10 | 52 | − | 1983 | 20 | 261 | + |

the 10th and 11th observations (arranged in numerical, *not time,* order). Since the 10th observation = 90 and the 11th observation = 91,

$$\tilde{y} = y_{(10.5)} = \frac{(90 + 91)}{2} = 90.5$$

***Note:*** By $y_{(j)}$ is meant the $j$th largest observation, called the *jth order statistic* (see Appendix 2A). Thus, $y_{(1)}$ is the smallest observation, $y_{(2)}$ the next smallest, and so forth. Also, $y_{(j+.5)}$ means the midpoint between $y_{(j)}$ and $y_{(j+1)}$.

- Assign plus signs to observations above 90.5 and minus signs to those below, as in Table 2.4.
- Determine the number of runs. In this case, $R = 2$, a run of 10 minuses followed by a run of 10 pluses.

Since $R$ here is less than 6, we reject the null hypothesis and conclude, at the 10% level of significance, that a no-trend model is inappropriate. In particular, from an inspection of the graph in Figure 2.8 and from the fact that $R$ is small, we can conclude that the average number of burglaries per year is increasing.

Although $m = 10$ is a rather small value for which to use the large-sample decision rule, the calculations below illustrate how the test would be run. The criterion for rejecting $H_0$ at the 10% significance level becomes:

**Decision Rule:** Reject $H_0$ if $|z| > z_{.05} = 1.645$

$$\mu_R = m + 1 = 10 + 1 = 11 \qquad \sigma_R = \sqrt{\frac{m(m-1)}{(2m-1)}}$$

$$R = 2 \qquad\qquad = \sqrt{\frac{10(10-1)}{(2 \times 10 - 1)}} = 2.18$$

$$z = \frac{R - \mu_R}{\sigma_R} = \frac{2 - 11}{2.18} = -4.13$$

**Figure 2.8 Yearly Number of Burglaries Reported to City Police, 1964–1983**

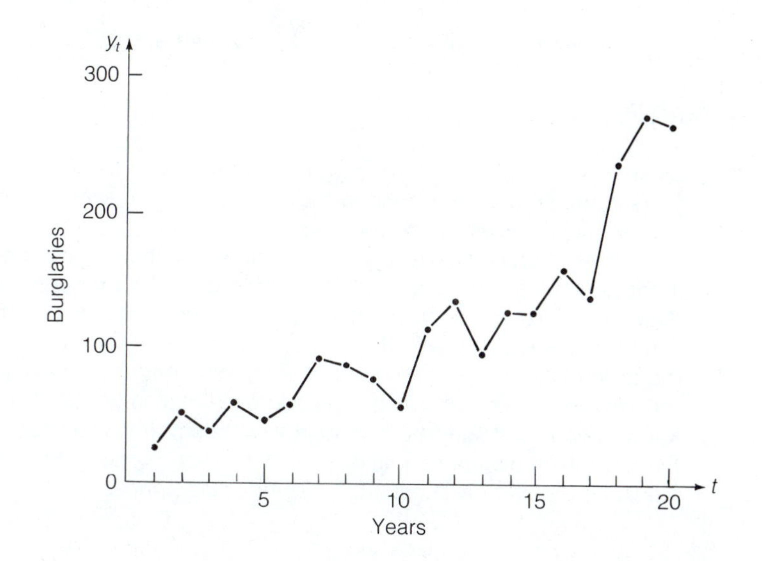

**Table 2.5 Burglary Rates from City Police Reports**

$x_t$ = population estimated by the Bureau of Vital Statistics

$y_t$ = number of burglaries per 10,000 residents

| Time $t$ | Population $x_t$ | Rate $y_t$ | Sign | Time $t$ | Population $x_t$ | Rate $y_t$ | Sign |
|---|---|---|---|---|---|---|---|
| 1 | 16,046 | 1.37 | − | 11 | 28,756 | 3.96 | + |
| 2 | 16,880 | 2.96 | − | 12 | 31,258 | 4.19 | + |
| 3 | 17,809 | 1.91 | − | 13 | 33,540 | 2.71 | − |
| 4 | 18,699 | 3.10 | + | 14 | 37,095 | 3.42 | + |
| 5 | 19,709 | 2.08 | − | 15 | 40,434 | 3.02 | + |
| 6 | 20,852 | 2.54 | − | 16 | 44,639 | 3.54 | + |
| 7 | 22,124 | 4.07 | + | 17 | 50,353 | 2.66 | − |
| 8 | 23,496 | 3.62 | + | 18 | 57,654 | 4.11 | + |
| 9 | 25,118 | 2.91 | − | 19 | 63,534 | 4.25 | + |
| 10 | 26,750 | 1.94 | − | 20 | 69,507 | 3.76 | + |

Since $|-4.13| = 4.13 > 1.645$, we would again reject the null hypothesis and conclude that a no-trend model is inappropriate.

It is quite possible that this trend in burglaries is caused by increasing population. If so, by expressing the data as "burglary rates" or "per capita burglaries," a no-trend model may yet be useful. To accomplish this modification, population figures for the years in question must be available. Estimates of the city's population for 1964–1983, as determined by the local government, are given in Table 2.5, along with the burglary rates computed from these figures. Having altered the data in this way, a runs test may again be performed. The median burglary rate is 3.00, giving the signs in Table 2.5 ("−" indicates $y_t < 3.00$, "+" that $y_t > 3.00$).

$H_0$: The burglary rates follow a no-trend model with independent error terms.

$H_a$: The burglary rates have trend and/or autocorrelated errors.

**Test Statistic:** $R$ = the number of runs above and below the median.

**Decision Rule:** Reject $H_0$ if $R \geq 16$ or $R \leq 6$. **(Appendix AA)**

Since $R = 10$ (Table 2.5), we do not reject the null hypothesis, concluding that, while there is very probably a trend in the *number of burglaries,* there is some support for a horizontal model for *burglary rates.*

In the burglary rates graphed in Figure 2.9, there appears to be an upward trend in the series that the runs test has failed to detect. Further statistical evaluation of the data seems in order. Its failure to detect this trend points out one of the major problems with the runs test—a lack of statistical "power." Since the test uses only a portion of the information present in the data (only whether an observation is above or below the median), it may well mistakenly fail to reject a no-trend model. It should be pointed out however, that the test is simple to conduct. It is also free of assumptions about the underlying distribution, and, should it reject $H_0$, no further, more complex testing would be required to substantiate lack of randomness. •

Figure 2.9 Yearly Burglary Rates per Thousand (from Police Reports), 1964–1983

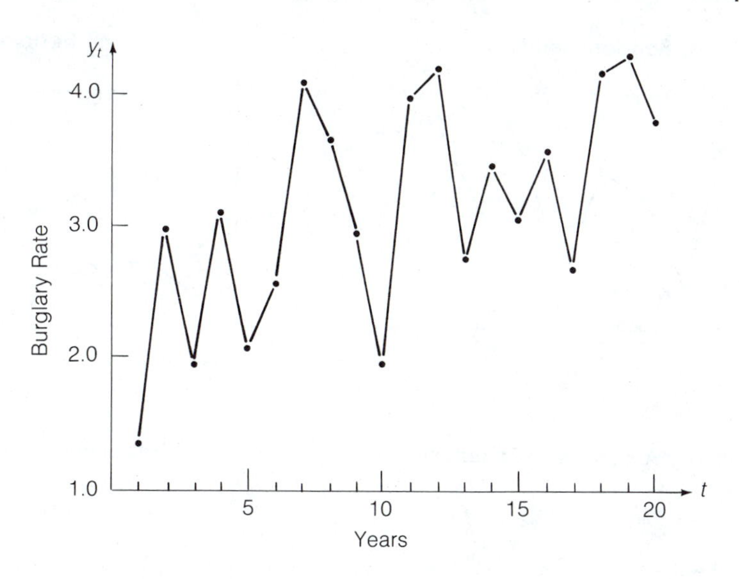


## 2.5.3 Turning Points Test: Randomness

Another test used to check no-trend models is called the **turning points test** (Gilchrist, 1976; Kendall, 1973). This test uses more of the information in the data than does the runs test and, therefore, is more likely to detect nonstationarity (i.e., it is more powerful).

A *turning point* in a time series is a point where the series changes directions. Each such point represents either a local "peak" or a local "trough" in the series. A trended or positively autocorrelated series should have fewer turning points than a random one. A negatively autocorrelated series should have more. To clarify, four illustrative series are given in Figures 2.10(a) through 2.10(d).

A convenient way to compute the number of turning points is to assign a plus or minus sign to a period depending on whether its first difference $y_t - y_{t-1}$ is positive or negative. A plus sign thus indicates that the series went up in that time period, and a minus sign, that it went down. A turning point is a time period whose sign is different from that of the *next* period.

If the series is actually a random no-trend series, the sampling distribution of the number of turning points $U$ is approximately normal for even moderately small values of $n$. A test for stationarity may, therefore, be constructed using percentage points of the normal distribution.

Returning to the data of Example 2.4, the turning points in yearly burglary rates are shown in Table 2.6. A turning points test for this data set is performed as follows:

$H_0$: Burglary rate is a random no-trend series.

$H_a$: Burglary rate has trend and/or autocorrelated error terms.

**Figure 2.10 Turning Points for Various Types of Series**

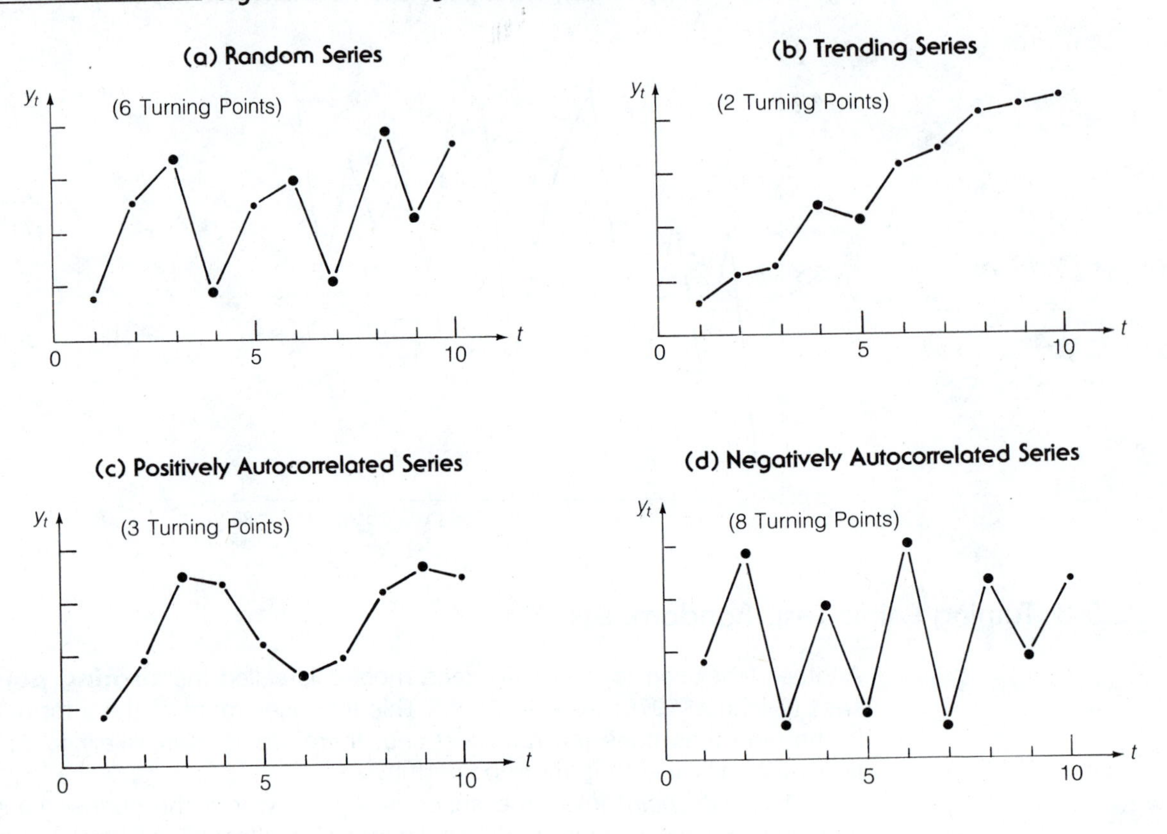


**Table 2.6 Determination of Turning Points for Burglary Rate Data**

| Time $t$ | Burglary Rate $y_t$ | Sign of $y_t - y_{t-1}$ | Time $t$ | Burglary Rate $y_t$ | Sign of $y_t - y_{t-1}$ |
|---|---|---|---|---|---|
| 1 | 1.37 | — | 11 | 3.96 | + |
| 2 | 2.96 | + [TP] | 12 | 4.19 | + [TP] |
| 3 | 1.91 | − [TP] | 13 | 2.71 | − [TP] |
| 4 | 3.10 | + [TP] | 14 | 3.42 | + [TP] |
| 5 | 2.08 | − [TP] | 15 | 3.02 | − [TP] |
| 6 | 2.54 | + | 16 | 3.54 | + [TP] |
| 7 | 4.07 | + [TP] | 17 | 2.66 | − [TP] |
| 8 | 3.62 | − | 18 | 4.11 | + |
| 9 | 2.91 | − | 19 | 4.25 | + [TP] |
| 10 | 1.94 | − [TP] | 20 | 3.96 | − |

*Note:* [TP] = turning point.

---

**Test Procedure:** *Turning Points Test*

$H_0$: The series is a random no-trend series.

$H_a$: The series is either trended or has autocorrelated errors.

**Test Statistic:** $U$ = the number of turning points in a series of $n$ observations

**Decision Rule:** (Moderate to Large Sample, $n \geq 10$)

Reject $H_0$ if $|z| > z_{\alpha/2}$.

Otherwise do not reject $H_0$

$$z = \left| \frac{U - \mu_U}{\sigma_U} \right|$$

$$\mu_U = \frac{2(n-2)}{3} \qquad \sigma_U = \sqrt{\frac{16n - 29}{90}}$$

$z_{\alpha/2}$ = the upper $(\alpha/2) \times 100\%$ point of the standard normal distribution **(Appendix BB)**

**Conclusion:** If $H_0$ is rejected, the test has shown with $(1 - \alpha) \times 100\%$ confidence that a no-trend independent model is inappropriate. (If, in addition, $z$ is positive, we conclude there is trend or positive autocorrelation; or, if $z$ is negative, we conclude there is negative autocorrelation.

If $H_0$ is not rejected, the test has shown some support for a no-trend independent model.

***Note:*** This test is equivalent to what in nonparametric statistics is often termed a "test for runs up and down" or a "test for runs of signs of first differences" (see Edgington, 1961; Bradley, 1968).

---

**Test Statistic:** $U$ = the number of turning points = 13

**Decision Rule:** ($\alpha$ = 5%)

Reject $H_0$ if $|z| > z_{.025} = 1.960$.

$$n = 20 \qquad \mu_U = \frac{2(20-2)}{3} = 12 \qquad \sigma_U = \sqrt{\frac{16 \times 20 - 29}{90}} = 1.80$$

$$|z| = \frac{|13 - 12|}{1.80} = 0.56 \rightarrow \text{Do not reject } H_0.$$

**Conclusion:** There is some support for a no-trend independent model.

The turning points test, even though more powerful than the runs test, has also failed to detect the apparent trend in the burglary rate data. In fact, the test is designed more to respond to autocorrelative errors. It is not too surprising, then, that it missed the trend in this case.

## 2.5.4 Sign Test: A Simple Test for Trend

Once the signs of the first differences have been determined for the turning points test, another nonparametric procedure, the sign test, may be carried out with little additional calculation. This test is designed specifically to detect trend (Conover, 1980).

As we remarked earlier, the first differences of a stationary series should themselves be stationary, with an average value of zero. If the average value is zero it is reasonable to assume that the median is close to zero. In that case, each first difference should be as likely to be above zero (+) as below (−). In other words, the probability of a positive first difference $p$ is $\frac{1}{2}$. To run the sign test, we compute the number of positive first differences and use a $z$-test to decide whether the hypothesis that $p = \frac{1}{2}$ is defensible.

---

**Test Procedure:** ***Sign Test for Trend***

$H_0$: The series has no trend and is as likely to increase as decrease in any period.

$H_a$: The series has trend, either upward or downward.

Or, more formally,

$H_0$: $p = \frac{1}{2}$

$H_a$: $p \neq \frac{1}{2}$

**Test Statistic:** $V$ = the number of positive first differences in the series

**Decision Rule:** (Large Sample, $n \geq 20$, level $\alpha$)

Reject $H_0$ if $|z| > z_{\alpha/2}$.

Otherwise do not reject $H_0$.

$n$ = the number of nonzero first differences

$$\mu_V = \frac{n}{2} \qquad \sigma_V = \sqrt{\frac{n}{4}} \qquad z = \frac{V - \mu_V}{\sigma_V}$$

$z_{\alpha/2}$ = the upper $(\alpha/2) \times 100\%$ point of the standard normal distribution **(Appendix BB)**

**Conclusion:** If $H_0$ is rejected, we conclude with $(1 - \alpha) \times 100\%$ confidence that a trend is present in the series, so a horizontal model is inappropriate. If, in addition, $z$ is positive, the trend is upward, or, if $z$ is negative, the trend is downward.

If $H_0$ is not rejected, we have some support for a horizontal model.

When applied to the burglary rate data, the test takes the following form.

$H_0$: The series has no trend ($p = \frac{1}{2}$)

$H_a$: The series has trend ($p \neq \frac{1}{2}$).

**Test Statistic:**

$$V = 9$$

$$n = 20 \qquad \mu_V = \frac{20}{2} = 10 \qquad \sigma_V = \sqrt{\frac{20}{4}} = 2.24$$

$$|z| = \frac{|9 - 10|}{2.24} = -0.45$$

**Decision Rule:** ($\alpha = .05$)

Reject $H_0$ if $|z| > z_{.025} = 1.960$

**Conclusion:** Since $|z| = 0.45 < 1.960$, we do not reject $H_0$ and we conclude that this test provides some support for a horizontal model.

## 2.5.5 Daniels' Test: Trend

More powerful nonparametric tests for stationarity can be constructed using either Spearman's correlation coefficient ($\rho$ = rho) or Kendall's correlation coefficient ($\tau$ = tau). These tests are designed specifically for trend and will not detect autocorrelation.

The test based on Spearman's coefficient has become known as **Daniels' test** for trend (Daniels, 1950). To accomplish the test, Spearman's rho is computed for the $n$ pairs $(t, y_t)$ and tested for significance using critical values given in Appendix DD for small samples or a normal approximation for larger samples. (See Appendices 2A and 2C for discussions of ranks and the details of computation of Spearman's rho.)

As an illustration of Daniels' test we return once more to the burglary rate data of Example 2.4. Computation of $r_s$ is shown in Table 2.7. The test is completed as follows.

$H_0$: There is no trend in the burglary rate series.

$H_a$: There is trend in the burglary rate series, either upward or downward.

**Test Statistic:** (Small Sample, $n = 20$)

$$r_s = 1 - \frac{6 \times 562}{20(400 - 1)} = +.5774$$

**Decision Rule:** ($\alpha = 5\%$)

Reject $H_0$ if $|r_s| > r_{.025} = .4451$ **(Appendix DD)**

**Conclusion:** Since $r_s = .5774 > .4451$, we reject $H_0$ and conclude with 95% confidence that there is upward trend in the data ($r_s$ is positive).

***Note:*** Although $n = 20$ allows use of Appendix DD, had we chosen to use the normal approximation we would have gotten the result shown in the test procedure.

**Test Procedure:** ***Daniels' Test for Trend***

$H_0$: The series has no trend.

$H_a$: The series has trend (upward or downward).

**Test Statistic:** (Small Sample, $n \leq 30$)

$$r_s = 1 - \frac{6\Sigma\, d_t^2}{n(n^2 - 1)}$$

$$d_t = t - \text{rank of } y_t = t - R(y_t)$$

(Large Sample, $n > 30$)

$$z = \frac{r_s - \mu_{rs}}{\sigma_{rs}} \qquad \mu_{rs} = 0 \qquad \sigma_{rs} = \frac{1}{\sqrt{n - 1}}$$

**Decision Rule:** (Level $\alpha$)
(Small Sample, $n \leq 30$)

Reject $H_0$ if $|r_s| > r_{\alpha/2}$. **(Appendix DD)**

(Large Sample, $n > 30$)

Reject $H_0$ if $|z| > z_{\alpha/2}$. **(Appendix BB)**

**Conclusion:** If $H_0$ is rejected, we conclude with $(1 - \alpha) \times 100\%$ confidence that the series has trend. If, in addition, $r_s$ or $z$ is positive, we conclude the trend is upward; if $r_s$ or $z$ is negative, we conclude the trend is downward.

If $H_0$ is not rejected, we have some support for a no-trend model.

---

**Decision Rule:** ($\alpha = 5\%$)

Reject $H_0$ if $|z| > z_{.025} = 1.960$.

$$\mu_{rs} = 0 \qquad \sigma_{rs} = \frac{1}{\sqrt{19}} = .2294$$

$$z = \frac{.5774 - 0}{.2294} = 2.517$$

**Conclusion:** Since $z = 2.517 > 1.960$, we reject $H_0$ and conclude with 95% confidence that upward trend is present in the burglary rate data ($z$ is positive).

**Table 2.7** Computation of Spearman's rho for Daniels' Test for Burglary Rate Data

| Time $t$ | Burglary Rate $y_t$ | Rank of $y_t$ $R(y_t)$ | $d_t = t - R(y_t)$ | $d_t^2$ |
|---|---|---|---|---|
| 1 | 1.37 | 1 | 0 | 0 |
| 2 | 2.96 | 9 | −7 | 49 |
| 3 | 1.91 | 2 | 1 | 1 |
| 4 | 3.10 | 11 | −7 | 49 |
| 5 | 2.08 | 4 | 1 | 1 |
| 6 | 2.54 | 5 | 1 | 1 |
| 7 | 4.07 | 17 | −10 | 100 |
| 8 | 3.62 | 14 | −6 | 36 |
| 9 | 2.91 | 8 | 1 | 1 |
| 10 | 1.94 | 3 | 7 | 49 |
| 11 | 3.96 | 16 | −5 | 25 |
| 12 | 4.19 | 19 | −7 | 49 |
| 13 | 2.71 | 7 | 4 | 16 |
| 14 | 3.42 | 12 | 2 | 4 |
| 15 | 3.02 | 10 | 5 | 25 |
| 16 | 3.54 | 13 | 3 | 9 |
| 17 | 2.66 | 6 | 11 | 121 |
| 18 | 4.11 | 18 | 0 | 0 |
| 19 | 4.25 | 20 | −1 | 1 |
| 20 | 3.76 | 15 | 5 | 25 |
| | | | | $\Sigma d_t^2 = 562$ |

Our efforts have finally led us to the conclusion that there is trend in the burglary rate series as well as in the original data on the number of burglaries. The increased power of Daniels' procedure over the runs, turning points, and sign tests enabled us to recognize statistically what seemed apparent visually in Figure 2.9. We hasten to point out, however, that, although less powerful, these latter tests are generally quicker and easier, and in some cases are powerful enough.

### 2.5.6 Test Based on Kendall's tau: Trend

A test nearly equivalent to Daniels' may be derived from Kendall's correlation coefficient tau (Gilchrist, 1976, p. 185). To run this test we examine the $n(n - 1)/2$ possible pairs that can be constructed from the observations in the series. An observation from an upward-trending series will tend to be larger than earlier observations and smaller than later ones (and vice versa for a downward-trending series). Suppose we describe an "upward pair" as a pair in which the later observation is larger than the earlier one and a "downward pair" as a pair in which the earlier observation is the greater of the two. Letting

$N_u$ = the number of "upward" pairs and $N_d$ = the number of "downward" pairs, we would expect the following:

- For a no-trend series there would be about the same number of upward and downward pairs, that is $N_u - N_d \doteq 0$.
- For an upward-trending series, the number of upward pairs would be greater than the number of downward pairs, that is $N_u - N_d > 0$.
- For a downward-trending series, the number of upward pairs would be smaller than the number of downward pairs, that is $N_u - N_d < 0$.

There are $n(n - 1)/2$ pairs altogether, and each is either upward or downward, so $N_u + N_d = n(n - 1)/2$. Kendall's correlation coefficient can be computed from $N_u$ and $N_d$ in any of three equivalent ways:

$$\tau = \frac{N_u - N_d}{n(n - 1)/2} = 1 - \frac{4N_d}{n(n - 1)} = \frac{4N_u}{n(n - 1)} - 1$$

For small sample sizes ($n \leq 10$), critical values of $\tau$ for two-tailed tests (upward or downward trend) with a significance level of $\alpha$ are given in Appendix CC. For larger samples ($n > 10$), a normal approximation provides adequate accuracy.

**Table 2.8 Computation of Kendall's Coefficient for Burglary Rate Data**

| Time $t$ | Burglary Rate $y_t$ | No. of Later Observations Greater than $y_t$ | Time $t$ | Burglary Rate $y_t$ | No. of Later Observations Greater than $y_t$ |
|---|---|---|---|---|---|
| 1 | 1.37 | 19 | 11 | 3.96 | 3 |
| 2 | 2.96 | 11 | 12 | 4.19 | 1 |
| 3 | 1.91 | 17 | 13 | 2.71 | 6 |
| 4 | 3.10 | 9 | 14 | 3.42 | 4 |
| 5 | 2.08 | 14 | 15 | 3.02 | 4 |
| 6 | 2.54 | 13 | 16 | 3.54 | 3 |
| 7 | 4.07 | 3 | 17 | 2.66 | 3 |
| 8 | 3.62 | 5 | 18 | 4.11 | 1 |
| 9 | 2.91 | 8 | 19 | 4.25 | 0 |
| 10 | 1.94 | 10 | 20 | 3.76 | 0 |
| | | | | | Total = 134 = $N_u$ |

**Test Procedure:** *Kendall's Correlation Coefficient*

$H_0$: The series is a random no-trend series.

$H_a$: The series has trend, either upward or downward.

**Test Statistic:** (Small Sample, $n \leq 10$)

$$\tau = \frac{N_u - N_d}{n(n-1)/2}$$

(Large Sample, $n > 10$)

$$z_\tau = \frac{\tau - \mu_\tau}{\sigma_\tau}$$

where

$$\mu_\tau = 0 \qquad \text{and} \qquad \sigma_\tau = \sqrt{\frac{2(2n+5)}{9n(n-1)}}$$

**Decision Rule:** (Two-tailed test, Level $\alpha$)
(Small Sample, $n \leq 10$)

Reject $H_0$ if $|\tau| > \tau_{\alpha/2}$. **(Appendix CC)**

(Large Sample, $n > 10$)

Reject $H_0$ if $|z_\tau| > z_{\alpha/2}$. **(Appendix BB)**

Otherwise do not reject $H_0$.

**Conclusion:** If $H_0$ is rejected, we conclude with $(1 - \alpha) \times 100\%$ confidence that the series is trended. If, in addition, $\tau$ or $z_\tau$ is positive, we conclude that the trend is upward; or if $\tau$ or $z_\tau$ is negative, the trend is downward.

If $H_0$ is not rejected, we have some support for a no-trend model.

---

Returning to our investigation of the burglary rate data, a simple algorithm is used in Table 2.8 to determine the value of $N_u$ needed to compute Kendall's tau. To use the algorithm, for each observation $y_t$ we merely count the number of chronologically later observations greater than $y_t$. This is the number that would form upward pairs with $y_t$. We can then find $N_u$ by adding the counts for all time periods.

$H_0$: The burglary rate data have no trend.

$H_a$: The burglary rate data have trend, either upward or downward.

**Test Statistic:** (Large Sample, $n = 20$)

$$N_u = 134 \qquad \text{(Table 2.8)}$$

$$\frac{n(n-1)}{2} = \frac{20 \times 19}{2} = 190$$

$$N_d = \frac{n(n-1)}{2} - N_u = 190 - 134 = 56$$

$$\tau = \frac{N_u - N_d}{n(n-1)/2} = \frac{134 - 56}{190} = +.4105$$

$$\mu_\tau = 0 \qquad \sigma_\tau = \sqrt{\frac{2(2n+5)}{9n(n-1)}} = \sqrt{\frac{2(2 \times 20 + 5)}{9 \times 20 \times (20 - 1)}} = .1622$$

$$z = \frac{.4105 - 0}{.1622} = 2.53$$

**Decision Rule:** ($\alpha = 5\%$)

Reject $H_0$ if $|z| > z_{.025} = 1.960$ **(Appendix BB)**

**Conclusion:** Since $|z| = 2.53 > 1.960$, we reject $H_0$ and conclude again, as we should have anticipated in light of the results for Daniels' test (see remarks below), that upward trend is present in the burglary rates ($\tau$ is positive).

This test and Daniels' nearly always give the same results. They are nearly equal in statistical power as well. While there is little to recommend one over the other, a slight preference may derive from the following:

- Kendall's coefficient is somewhat simpler to compute and has, perhaps, more intuitive appeal in a time series context.
- Kendall's coefficient tends to normality somewhat faster than Spearman's and may thus be preferable for smaller sample sizes if the normal approximation is used.
- Spearman's coefficient may be more familiar, since it is a direct nonparametric analog of Pearson's product moment correlation coefficient, which is widely used in parametric correlation and regression analysis. Because of this, many standard computer programs will handle Spearman's but not Kendall's coefficient.
- Spearman's coefficient is usually larger in magnitude than Kendall's.

## 2.6 Parametric Tests for Stationarity and Randomness

### 2.6.1 Introduction

The main difference between nonparametric tests and the tests we will discuss now is that the error terms $\epsilon_t$ and, by implication, the series $y_t$ are required to be normally distributed. The results of these tests may differ from those of the nonparametric ones. In the presence of normality, they are more powerful, especially for smaller sample sizes, than are the nonparametric procedures. However, the requisite normality may be difficult to verify.

### 2.6.2 Test for $\rho = 0$: Trend

A straightforward test for stationarity can be based on the usual procedure for testing the correlation coefficient in simple linear correlation analysis. If the series is stationary, the test should not reject the hypothesis that the population correlation coefficient $\rho$ between $t$ and $y_t$ is zero. If it does reject, the series is nonstationary and trend is present. Refer to the discussion in Section 3.4 for the details of running this test and interpreting the results.

### 2.6.3 Mean Square Successive Difference Test (von Neumann's ratio): Randomness

Another way to test the stationarity (or randomness) of the series is to use a procedure developed by von Neumann (1941). The test uses the statistic $M$ = the ratio of the average squared first difference to the sample variance. Critical values of the test statistic were tabulated by Hart (1942) and refined by Nelson (1980). We have modified Nelson's results to suit our purposes and present them in Appendix EE. In practice, the statistic $M$ may be computed by taking the ratio of the sum of squared first differences

$$SS_{\Delta y} = \sum_{2}^{n} (y_t - y_{t-1})^2 = \sum_{2}^{n} (\Delta y_t)^2$$

to the sum of squared deviations from the mean

$$SS_{yy} = \sum_{1}^{n} (y_t - \bar{y})^2$$

so that

$$M = \frac{SS_{\Delta y}}{SS_{yy}}$$

$SS_{yy}$ may be computed in the usual way:

$$SS_{yy} = \Sigma\, y^2 - \frac{(\Sigma\, y)^2}{n}$$

It can be shown that $M$ always lies between 0 and 4. Within this interval the value of $M$ is affected by trend and autocorrelation, as summarized in the following table.

**Effects of Trend and Autocorrelation on $M$**

| If the Series Is: | The Value of $M$ Tends to Be: |
|---|---|
| Stationary with independent errors | Around 2.0 |
| Trended | Smaller than 2.0 |
| Stationary with error terms positively autocorrelated | Smaller than 2.0 |
| Stationary with error terms negatively autocorrelated | Larger than 2.0 |

---

**Test Procedure:** ***Mean Square Successive Difference Test (von Neumann's Ratio)***

$H_0$: The series is random and stationary.

$H_a$: The series is either nonstationary or has autocorrelated errors.

**Test Statistic:**

$$M = \frac{SS_{\Delta y}}{SS_{yy}} = \frac{\sum_{2}^{n} (\Delta y)^2}{\sum_{1}^{n} (y - \bar{y})^2}$$

$$SS_{yy} = \sum_{1}^{n} y^2 - \frac{\left(\sum_{1}^{n} y\right)^2}{n} \qquad SS_{\Delta y} = \sum_{2}^{n} (y_t - y_{t-1})^2$$

**Decision Rule:** (Level $\alpha$, $n$)

Reject $H_0$ if $M < M_{1-\alpha/2}$ or if $M > M_{\alpha/2} = 4 - M_{1-\alpha/2}$ **(Appendix EE)**

**Conclusion:** If $H_0$ is rejected with $M < M_{1-\alpha/2}$, we conclude that the series is nonstationary. Graphing the series will reveal whether trend or positive autocorrelation caused the rejection.

If $H_0$ is rejected with $M > M_{\alpha/2}$, we conclude that the series has negatively autocorrelated error terms.

If $H_0$ is not rejected, we have some support for a random no-trend model.

---

Values of $M_{.90}$, $M_{.95}$, and $M_{.99}$ may be obtained from Appendix EE (also, for an approximation, see Bingham and Nelson, 1981). From these one can construct two-tailed tests with 20%, 10%,and 2% significance levels, respectively. Note that larger values of $\alpha$ tend to strengthen "Do not reject $H_0$" decisions.

A one-tailed test may also be constructed from the critical values of Appendix EE by using only the appropriate upper or lower critical value. The one-sided test, which is used most often, tests the following hypotheses.

$H_0$: The series is random and has no trend.

$H_a$: The series is trended or positively autocorrelated.

**Decision Rule:** (Level $\alpha$)

Reject $H_0$ if $M < M_{1-\alpha/2}$. **(Appendix EE)**

Otherwise do not reject $H_0$.

**Conclusion:** Reject $H_0$, meaning the series is either nonstationary or positively autocorrelated.

Do not reject $H_0$, meaning there is some support for a random no-trend model.

**EXAMPLE 2.5** ***Return Rates for Questionnaires***

In order to determine subscribers' experience with various products, a consumer research agency makes annual mailings of questionnaires. As part of a study of the return rate for these questionnaires, the data in Table 2.9 were collected for the years 1967–1986. A preliminary graph of the data (Figure 2.11) suggested a stationary model.

One of the executives in the company questions the stationarity, feeling that increasing subscriber acceptance of the questionnaire should have resulted in a gradual increase in the return rate. He also feels that psychological forces could cause oscillations in the rate that might mask this trend. The mean square successive difference test may be used to check the feasibility of a random stationary model.

$H_0$: Return rate is a random no-trend series.

$H_a$: Return rate is trended and/or has autocorrelated errors.

**Table 2.9 Return Rates of Annual Questionnaire Mailings, 1967–1986**

| Year | $t$ | Return Rate (%) $y_t$ | $\Delta y_t$ | Year | $t$ | Return Rate (%) $y_t$ | $\Delta y_t$ |
|---|---|---|---|---|---|---|---|
| 1967 | 1 | 9.4 | — | 1977 | 11 | 14.0 | +0.7 |
| 1968 | 2 | 14.9 | +5.5 | 1978 | 12 | 10.0 | −4.0 |
| 1969 | 3 | 11.1 | −3.8 | 1979 | 13 | 12.8 | +2.8 |
| 1970 | 4 | 10.6 | −0.5 | 1980 | 14 | 10.9 | −1.9 |
| 1971 | 5 | 15.1 | +4.5 | 1981 | 15 | 9.2 | −1.7 |
| 1972 | 6 | 8.9 | −6.2 | 1982 | 16 | 11.1 | +1.9 |
| 1973 | 7 | 12.6 | +3.7 | 1983 | 17 | 12.5 | +1.4 |
| 1974 | 8 | 13.0 | +0.4 | 1984 | 18 | 12.6 | +0.1 |
| 1975 | 9 | 17.3 | +4.3 | 1985 | 19 | 13.9 | +1.3 |
| 1976 | 10 | 13.3 | −4.0 | 1986 | 20 | 10.4 | −3.5 |

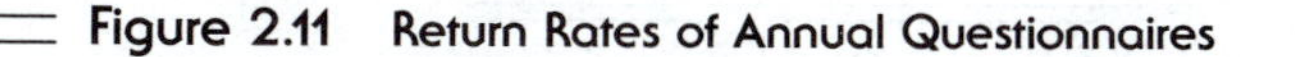
**Figure 2.11 Return Rates of Annual Questionnaires**

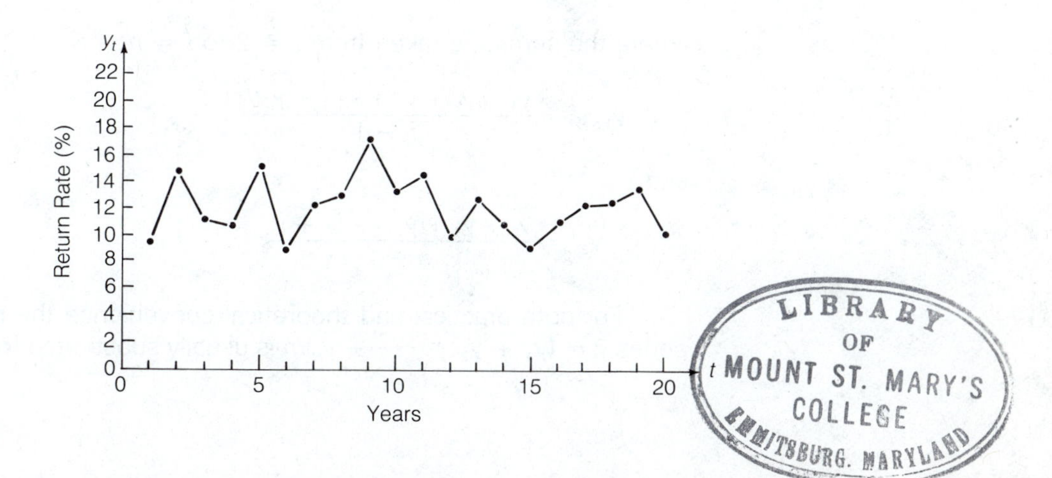

**Test Statistic:**

$$M = \frac{SS_{\Delta y}}{SS_{yy}} = \frac{202.32}{92.932} = 2.177$$

$$\Sigma\,(\Delta y)^2 = SS_{\Delta y} = 5.5^2 + (-3.8)^2 + \cdots + (-3.5)^2 = 202.32$$

$$\Sigma\,y = 9.4 + 14.9 + \cdots + 10.4 = 243.6$$

$$\Sigma\,y^2 = 9.4^2 + 14.9^2 + \cdots + 10.4^2 = 3059.98$$

$$SS_{yy} = \Sigma\,y^2 - \frac{(\Sigma\,y)^2}{n} = 3059.98 - \frac{(243.6)^2}{20} = 92.932$$

**Decision Rule:** ($\alpha = 10\%,\ n = 20$)

Reject $H_0$ if $M < M_{.95} = 1.300$ or if $M > M_{.05} = 4 - 1.300 = 2.700$

**(Appendix EE)**

**Conclusion:** Since $M = 2.177$ is between 1.300 and 2.700, we do not reject $H_0$, concluding that there is some support for a random stationary model; i.e., the data do not support the executive's contention that trend is present. ●

## 2.6.4 The Autocorrelation Function (acf)

When concerned about independence in time series it is natural to consider computing the correlation coefficient between successive observations in the series. We could treat the $(n - 1)$ pairs

$$(y_1, y_2), (y_2, y_3), (y_3, y_4), \ldots, (y_{n-1}, y_n)$$

as $(x, y)$ pairs are treated in correlation analysis. The usual formula for the correlation coefficient, i.e.,

$$r = \frac{\Sigma\,(x - \bar{x})\,(y - \bar{y})}{\sqrt{\Sigma\,(x - \bar{x})^2 \Sigma\,(y - \bar{y})^2}}$$

would then become

$$r = \frac{\Sigma\,(y_{i-1} - \bar{y}_{1,n-1})\,(y_i - \bar{y}_{2,n})}{\sqrt{\Sigma\,(y_{i-1} - \bar{y}_{1,n-1})^2 \Sigma\,(y_n - \bar{y}_{2,n})^2}}$$

where the sums are taken from $i = 2$ to $i = n$:

$$\bar{y}_{1,n-1} = \frac{y_1 + y_2 + y_3 + \cdots + y_{n-1}}{n - 1}$$

and

$$\bar{y}_{2,n} = \frac{y_2 + y_3 + y_4 + \cdots + y_n}{n - 1}$$

For both practical and theoretical convenience the mean of the entire series $\bar{y} = (y_1 + y_2 + \cdots + y_n)/n$ is usually substituted for the partial means

$\bar{y}_{2,n}$ and $\bar{y}_{1,n-1}$, and the sum of squares of the entire series is substituted for the partial sums of squares. Once this substitution is made, the result is called the **sample autocorrelation coefficient of lag 1, $r_1$,** and is computed from the equation

$$r_1 = \frac{\Sigma\,(y_{i-1} - \bar{y})\,(y_i - \bar{y})}{\Sigma\,(y_i - \bar{y})^2}$$

(It is interesting to note that $r_1$ here is related to von Neumann's ratio $M$ in that $r_1 \approx 1 - M/2$. See the Exercises.)

Similarly, the sample autocorrelation of lag $k$ is computed for the $(n - k)$ pairs:

$$\begin{array}{l} (y_1, y_{k+1}) \\ (y_2, y_{k+2}) \\ (y_3, y_{k+3}) \\ \quad\cdot \\ \quad\cdot \\ \quad\cdot \\ (y_{n-k}, y_n) \end{array}$$

and

$$r_k = \frac{\Sigma\,(y_{i-k} - \bar{y})\,(y_i - \bar{y})}{\Sigma\,(y_i - \bar{y})^2} = \frac{\Sigma\,(y_{i-k} - \bar{y})\,(y_i - \bar{y})}{SS_{yy}}$$

As long as the number of observations in the series is large compared to $k$, the substitution of total means and sums of squares for partial ones does not greatly change the value of $r$. However, one should be careful when the available history of the series is short, in which case such substitutions *do* have a material effect and may give values of $r$ greater than 1 (see also Stigler, 1986).

When the sample autocorrelation coefficients are evaluated for lag 1, lag 2, lag 3, . . . and graphed versus $k$, the result is usually called the **sample autocorrelation function (acf).** This graph is very useful both when examining stationarity or autocorrelation and when selecting from among various nonstationary models. It is one of the major tools, for example, in the Box-Jenkins modeling process discussed in Chapter 7. For the present, we will deal only with the acf's use as an indicator of either nonstationarity or autocorrelated error terms.

In this regard, a summary of the general characteristics of the theoretical acf's of different types of series is given in Table 2.10. Notice that, for stationary series without seasonality, either the autocorrelations are all zero (if the error terms are random) or they differ from zero only for short lags. Remember, too, that because there are sampling errors involved when the observed series is finite, the sample autocorrelations will differ somewhat from theoretical values.

**Table 2.10 Autocorrelation Functions (Theoretical) for Different Types of Time Series**

| Series type | Autocorrelation function |
|---|---|
| **Random No-Trend Series:** | Autocorrelations for all lags are zero. |
| **Trended Series:** | Autocorrelations are large and positive for short lags, decreasing slowly as the lag increases. The acf itself appears to have a downward trend as lag length increases. |
| **Untrended Series with Seasonal Movement:** | Autocorrelations are largest for lags 12, 24, . . . for monthly data (4, 8, . . . for quarterly data), becoming smaller as the lag length increases. Autocorrelations for other lags are dependent on the nature of the other movement in the series. |
| **Trended Series with Seasonal Movement:** | Autocorrelations are similar to trended series, but they oscillate, with peaks at lags 12, 24, . . . for monthly data (4, 8, . . . for quarterly data). |
| **Stationary Series with Positively Autocorrelated Error Terms:** | Low-order autocorrelations are large but damp out rapidly as lag length increases. |

## 2.6.5 "Rule of Thumb" Procedure

A "rule of thumb" procedure based on a large-sample normal approximation is often used to decide whether a particular sample autocorrelation is within sampling error of zero (Bartlett, 1946). If $\rho_k$, the population autocorrelation of lag $k$, is zero for $k = 1, 2, \ldots$, then, for fairly large $n$, $r_k$ will be approximately normally distributed, with a mean of zero and a standard deviation equal to $1/\sqrt{n}$. Since roughly 95% of a normal population is within 2 standard deviations of the mean, a test that rejects the hypothesis that $\rho_k = 0$ when $r_k$ is greater than $2/\sqrt{n}$ has a significance level of approximately 5%. The test is summarized as follows.

**Test Procedure:** ***"Rule of Thumb" Procedure***

$H_0$: $\rho_k = 0$

$H_a$: $\rho_k \neq 0$

**Test Statistic:** $r_k$ = sample autocorrelation of lag $k$

**Decision Rule:** ($\alpha \simeq 5\%$, $n$ large)

Reject $H_0$ if $|r_k| > 2/\sqrt{n}$

Otherwise do not reject $H_0$.

**Conclusion:** If $H_0$ is rejected, we conclude with approximately 95% confidence that autocorrelation of lag $k$ is present.

If $H_0$ is not rejected, we have some support for a model with no lag $k$ autocorrelation.

The "rule of thumb" procedure may be conducted for any or all values of $k$, but a reasonable strategy to adopt is to run the test both for lag 1 and for whichever lag corresponds to seasonal movement ($k = 12$ for monthly, $k = 4$ for quarterly, etc.). If the former test rejects, the acf should be examined in more detail. If the latter rejects, seasonality is likely to be present. Failure to reject $H_0$ in both cases supports a random no-trend model.

A final note of caution is in order if you decide to test several $r_k$'s. Although the nominal value of $\alpha$ is approximately 5% for each $r_k$ tested, the actual probability of incorrectly deciding that at least one of the $\rho_k$'s is nonzero may be considerably larger. (If we test 20 coefficients from a random no-trend series, for example, the probability of obtaining at least one "false rejection" is about 64%.) The net impact is that finding one or two significant $r_k$'s when testing several individually does not necessarily imply that the series is nonrandom. However, if any of the $r_k$'s are large enough to cause a rejection, some further investigation of the randomness of the series is probably in order.

**EXAMPLE 2.6** *Reconsideration of Examples 2.1, 2.2, and 2.3*

To illustrate the use of the sample acf we recall data from this chapter's previous examples.

## Mainline Plumbing Data

The gasoline purchase data of Example 2.1 represents a random no-trend series. A graph of these data together with a graph of the sample acf for lags 1 through 10 are given in Figure 2.12. The autocorrelation coefficients are computed as demonstrated in Table 2.11, which uses $r_5$ as an example (the computation is usually done by computer). Note that to determine $\bar{y}$ and $SS_{yy}$, the sums are taken over the entire 20 terms of the series. The remaining terms of the acf are computed in a similar way; the results are given in Table 2.12.

Since data are collected weekly and cover only a 20-week period, we cannot test for seasonal movement. A test for $r_1$ follows.

$H_0$: $\rho_1 = 0$

$H_a$: $\rho_1 \neq 0$

**Test Statistic:** $r_1$ = the sample autocorrelation coefficient of lag 1 $= -.333$ (see Table 2.11)

**Decision Rule:** Reject $H_0$ if $|r_1| > \dfrac{2}{\sqrt{20}} = .447$.

**Conclusion:** Since $|r_1| = .333 < .447$, we do not reject $H_0$.

Our decision supports a random no-trend model. Note that if one tests each of the ten $r_k$'s, all are less than the critical value (.447) except $r_4 = -.523$. For the following reasons, a random no-trend model (or at least a no-trend model) may well be appropriate in spite of the size of $r_4$:

- Most important, there is nothing about the nature of gasoline purchases that would lead one to believe that a movement during one particular week should be followed four weeks later by a movement in the opposite direction.

**Table 2.11** **Computation of Sample Autocorrelation Coefficient of Lag 5 for Mainline Plumbing Data**

| $t$ | $y_{t-5}$ | $y_t$ | $(y_{t-5} - \bar{y})$ | $(y_t - \bar{y})$ | $(y_{t-5} - \bar{y}) \cdot (y_t - \bar{y})$ |
|---|---|---|---|---|---|
| 6 | 409 | 445 | 2.55 | 38.55 | 98.3 |
| 7 | 289 | 310 | −117.45 | −96.45 | 11,328.1 |
| 8 | 509 | 372 | 102.55 | −34.45 | −3,532.8 |
| 9 | 364 | 440 | −42.45 | 33.55 | −1,424.2 |
| 10 | 404 | 414 | −2.45 | 7.55 | −18.5 |
| 11 | 445 | 380 | 38.55 | −26.45 | −1,019.6 |
| 12 | 310 | 598 | −96.45 | 191.55 | −18,475.0 |
| 13 | 372 | 418 | −34.45 | 11.55 | −397.9 |
| 14 | 440 | 359 | 33.55 | −47.45 | −1,591.9 |
| 15 | 414 | 432 | 7.55 | 25.55 | 192.9 |
| 16 | 380 | 252 | −26.45 | −154.45 | 4,085.2 |
| 17 | 598 | 446 | 191.55 | 39.55 | 7,575.8 |
| 18 | 418 | 473 | 11.55 | 66.55 | 768.7 |
| 19 | 359 | 337 | −47.45 | −69.45 | 3,295.4 |
| 20 | 432 | 478 | 25.55 | 71.55 | 1,828.1 |
| | | | | | 2,712.4 |

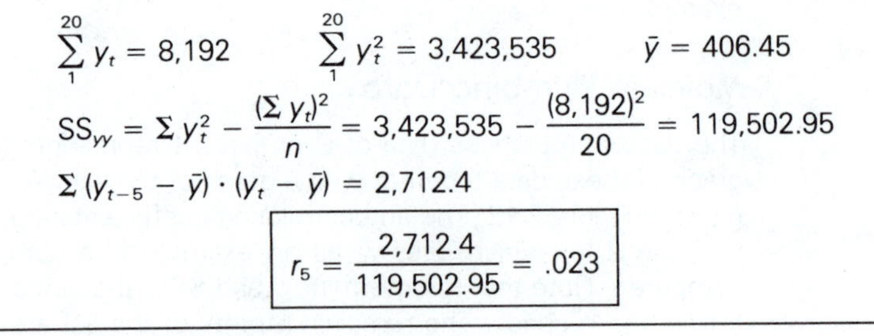

$$\sum_{1}^{20} y_t = 8{,}192 \qquad \sum_{1}^{20} y_t^2 = 3{,}423{,}535 \qquad \bar{y} = 406.45$$

$$SS_{yy} = \Sigma y_t^2 - \frac{(\Sigma y_t)^2}{n} = 3{,}423{,}535 - \frac{(8{,}192)^2}{20} = 119{,}502.95$$

$$\Sigma (y_{t-5} - \bar{y}) \cdot (y_t - \bar{y}) = 2{,}712.4$$

$$r_5 = \frac{2{,}712.4}{119{,}502.95} = .023$$

**Table 2.12** **Autocorrelation Coefficients for Gasoline Purchases of Mainline Plumbing Co.**

| Lag $k$ | $r_k$ | Lag $k$ | $r_k$ |
|---|---|---|---|
| 1 | −.333 | 6 | .185 |
| 2 | −.081 | 7 | −.147 |
| 3 | .285 | 8 | .056 |
| 4 | −.523 | 9 | .307 |
| 5 | .023 | 10 | −.278 |

- Ten coefficients were tested and only one was significant. With a nominal $\alpha$ of 5% for each, it is possible that a "false rejection" has occurred.
- Lag 4 autocorrelation alone, even if present, would imply only a certain amount of nonrandomness, not trend.

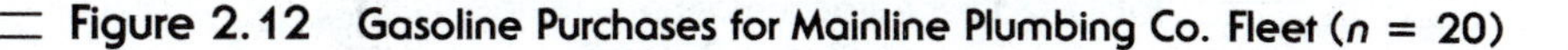

**Figure 2.12 Gasoline Purchases for Mainline Plumbing Co. Fleet ($n$ = 20)**

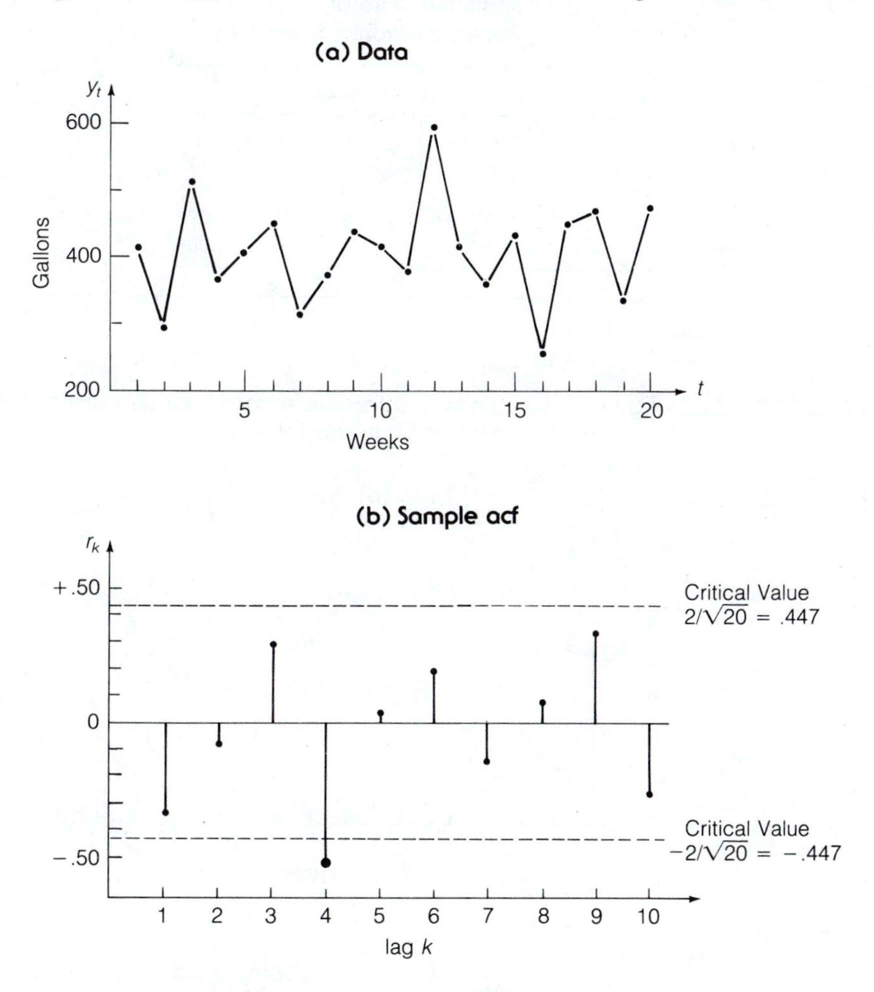


## Rogers Electronics Specialties Data

The absentee data from Rogers Electronics Specialties, Inc., display the up and down "drift" produced by positively autocorrelated errors. The sample autocorrelations for these data are given in Table 2.13 and a graph of the data along with the acf are shown in Figure 2.13. The reason the acf is positive for shorter lags is that once the series has begun to rise it tends to keep rising for a few periods, and once it turns down it goes down for a few periods. As the lag length increases, the acf decreases and eventually becomes negative because the lag length overlaps the "cycles" in the series. The values of the lags for which the acf changes signs give some idea of the "cycle length" of the series.

The significance of $r_1$ may be tested using the "rule of thumb" as follows.

$H_0$: $\rho_1 = 0$

$H_a$: $\rho_1 \neq 0$

**Table 2.13** Autocorrelation Coefficients for Absentee Data of Rogers Electronics Specialties

| Lag $k$ | $r_k$ | Lag $k$ | $r_k$ |
|---|---|---|---|
| 1 | +.646 | 6 | −.424 |
| 2 | +.313 | 7 | −.133 |
| 3 | −.194 | 8 | −.051 |
| 4 | −.511 | 9 | +.096 |
| 5 | −.495 | 10 | +.068 |

**Figure 2.13** Data and Sample Autocorrelation Function for Rogers Electronic Specialties Company ($n = 15$)

(a) Data

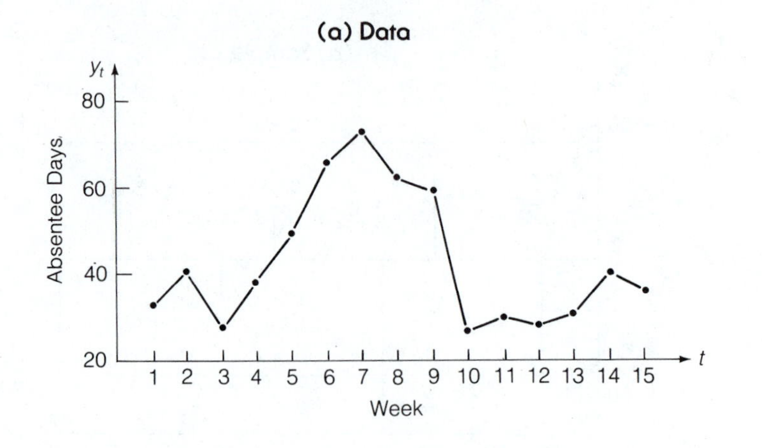


(b) Sample acf

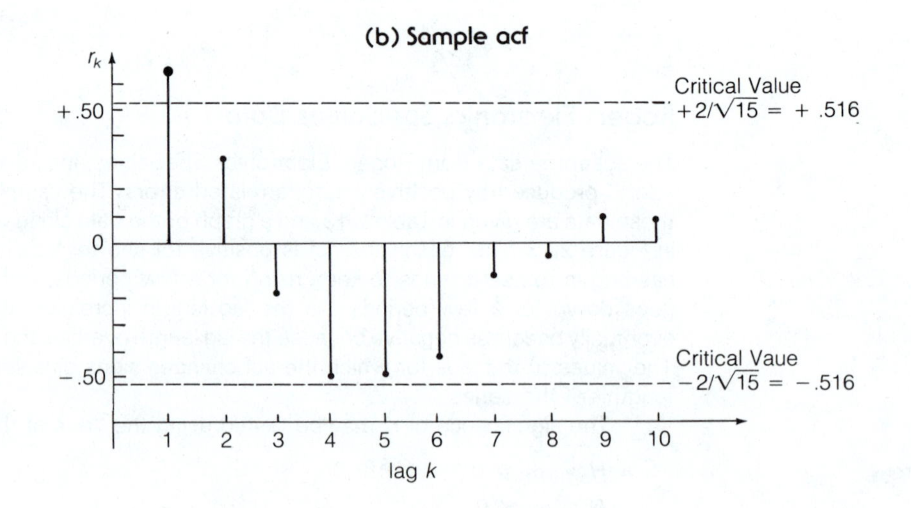

**Test Statistic:** $r_1$ = the sample autocorrelation coefficient of lag 1 = +.646

**Decision Rule:** Reject $H_0$ if $|r_1| > \frac{2}{\sqrt{15}} = .516$.

**Conclusion:** Since $|r_1| = +.646 > .516$, we reject $H_0$.

We can then conclude with approximately 95% confidence that positive lag 1 autocorrelation is present in the series. Notice that none of the other coefficients are greater than .516 in magnitude and are therefore not significant. This fact together with an inspection of the acf in Figure 2.13 would lead us to the conclusion that we have a horizontal series with cyclical drift of about 5–8 weeks' duration (the acf becomes negative at about lag 3 and reaches a negative peak at about lag 4).

## Burglary Data

The acf's of the data in Example 2.4 for the number of burglaries per week and for the burglary rates are given in Table 2.14 and graphed in Figure 2.14. The acf for the number of burglaries is typical of that for a series with a pronounced trend. The autocorrelations are large and positive for short lags. For longer lags they slowly decrease and eventually become negative. This pattern would be similar to that for a series with a long cycle length, since a trend, in a sense, resembles a long cycle. Notice from Figure 2.14 that, of the $r_k$'s, only $r_1$ and $r_2$ are significant.

**Table 2.14 Autocorrelation Coefficients for Number of Burglaries and Burglary Rates**

| Lag $k$ | Number $r_k$ | Rates $r_k$ | Lag $k$ | Number $r_k$ | Rates $r_k$ |
|---|---|---|---|---|---|
| 1 | +.726 | +.121 | 6 | +.068 | −.236 |
| 2 | +.503 | +.044 | 7 | +.059 | +.158 |
| 3 | +.306 | −.003 | 8 | −.005 | +.033 |
| 4 | +.271 | +.272 | 9 | −.122 | −.033 |
| 5 | +.145 | +.118 | 10 | −.227 | −.162 |

When the data are changed to rates, the pattern disappears. None of the first 10 coefficients test significant, and consequently we have support for a horizontal random model. However, the graph of the data shows an apparent trend. This trend was confirmed earlier using Spearman's rank correlation test. The failure of the current procedure to detect the trend may be attributed to the large amount of "noise," or irregular movement, in the series. Because of the complex interrelationships of the correlation coefficients for different lags, the sample acf is somewhat sensitive to this noise, and tests based on it tend to lack power when too much noise is present. In addition, the acf is primarily a device for studying cyclical phenomena, rather than trend, in time series. Tests specifically designed to detect trend, such as Spearman's procedure or the regression procedures of Chapter 3 and 4 would be better choices for deciding whether trend is present. ●

**Figure 2.14** Sample Autocorrelation Function

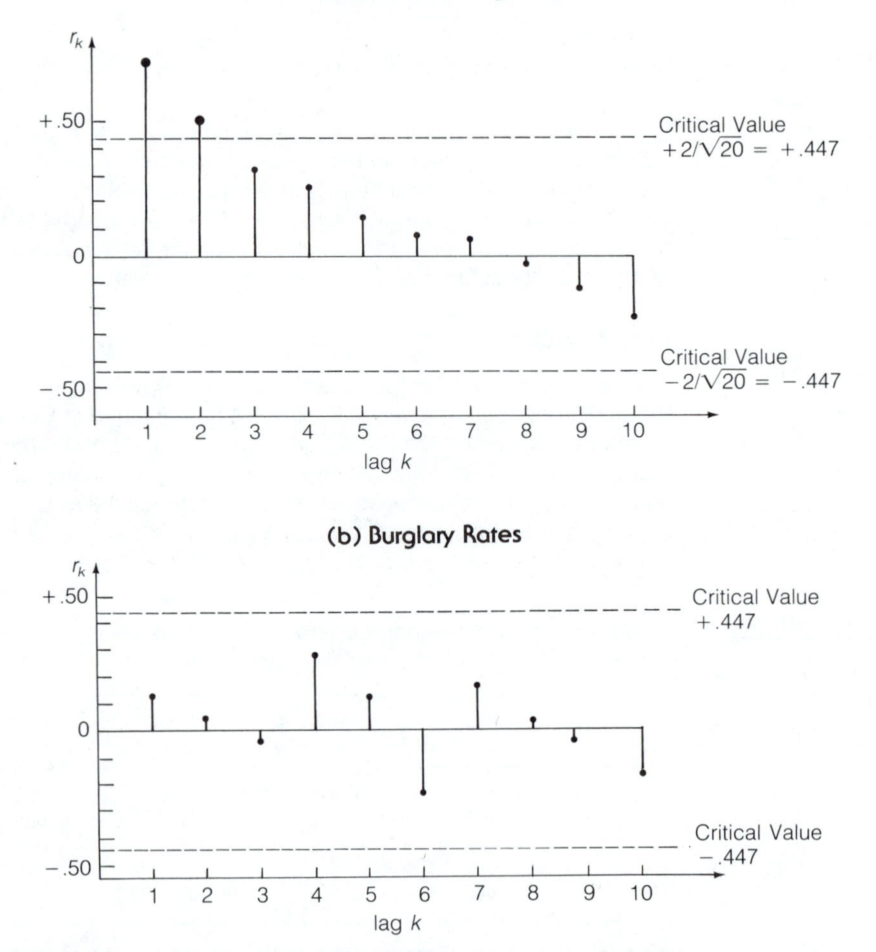


### 2.6.6 Modified Box-Pierce $Q$: A Portmanteau Test

Recall that one difficulty with the "rule of thumb" procedure is that, to test autocorrelations for several different lags, several separate tests must be run. The significance level of the resulting combination of tests—that is, the probability of getting at least one mistaken rejection—may be quite different from 5%. However, a set of $m$ sample autocorrelations may be tested all at once, thereby controlling the significance level, by using a test developed by Box and Pierce (1970) and modified by Ljung and Box (1978).

If the history of the series is large, the test statistic, known as the **modified Box-Pierce** (or Ljung-Box-Pierce) **Q statistic,** has an approximate chi-square distribution with degrees of freedom equal to the number of sample autocorrelations $m$ being tested. (A slight modification of the number of de-

**Test Procedure:** *Modified Box-Pierce Q*

$H_0$: $\rho_k = 0$ for all $k \leq m$

$H_a$: $\rho_k \neq 0$ for some value of $k \leq m$

**Test Statistic:**

$$Q_m = n(n+2)\sum_{k=1}^{m}\frac{r_k^2}{n-k}$$

$m$ = the number of coefficients being tested

$n$ = the number of observations in the series

**Decision Rule:** (Level $\alpha$, $n$, $m$)

Reject $H_0$ if $Q_m > \chi^2_\alpha(m)$. **(Appendix FF)**

Otherwise do not reject $H_0$.

($\chi^2_\alpha(m)$ is the upper $\alpha \times 100\%$ point of the chi-square distribution with $m$ degrees of freedom.)

**Conclusion:** If $H_0$ is rejected, we conclude with approximately $(1 - \alpha) \times 100\%$ confidence that the series is nonrandom (i.e, it is either trended or autocorrelated). A graph of the series in conjunction with the acf may be used to determine the nature of the autocorrelation or trend.

If $H_0$ is not rejected, we have some support for a random no-trend model.

---

grees of freedom is necessary when the test is used in the Box-Jenkins methodology, as described in Chapter 7.) Percentage points necessary to run the test are given in Appendix FF.

Although the value of $m$ is arbitrary, it should not be too large, usually no larger than around one-quarter of the sample size.

Reconsidering the Mainline Plumbing gasoline purchase data for a moment, we may test, say, the first three coefficients in Table 2.12 (lags 1, 2, and 3) with a single test:

$H_0$: $\rho_k = 0$ for all $k \leq 3$

$H_a$: $\rho_k \neq 0$ for some $k \leq 3$

or

$H_0$: The series has no autocorrelation at lags of 3 or less.

$H_a$: The series has autocorrelation at some lag of 3 or less.

**Test Statistic:**

$$Q_3 = 20(20+2)\left[\frac{(-.333)^2}{19} + \frac{(-.081)^2}{18} + \frac{(.285)^2}{17}\right] = 4.86$$

**Decision Rule:** ($n = 20$, $m = 3$, $\alpha = 5\%$)

Reject $H_0$ if $Q_3 > \chi^2_{.05}(3) = 7.81$. **(Appendix FF)**

**Conclusion:** Since $Q_3 = 4.86 < 7.81$, we do not reject $H_0$ and thus have some support for a random no-trend model.

It should be noted that both the ''rule of thumb'' and modified Box-Pierce procedures have been shown to be somewhat lacking in statistical power (Granger and Newbold, 1986, p. 100; Davies and Newbold, 1979). Their failure to reject the null hypothesis should be considered supportive of, but not sole proof for, stationarity and/or randomness.

## 2.7 Other Considerations

### 2.7.1 Other Tests

We have discussed a variety of the more widely used procedures for testing stationarity and randomness. In most situations where a test is required, one of these will suffice. However, there are many other tests that could be useful under special circumstances but that are somewhat beyond the scope of this book. In fact, any procedure designed to test the hypothesis that the $y_t$'s are independent and identically distributed (against some specific alternative) might serve.

### 2.7.2 Transformations

Sometimes when a time series is trended, relatively simple corrective measures may be taken to ''detrend'' or ''stationarize'' it. One may then forecast the detrended series with stationary models and avoid the necessity of fitting the more complicated trend models of Chapter 3. The most commonly used methods are:

- **conversion to ''rates''** The data are divided by the values of a concurrent time series, such as population, to give ''per capita,'' ''per occurrence,'' or other types of rate figures. This method was used when attempting to detrend the burglary data of Example 2.4. The main advantage of the method is that it is easy to understand and interpret. Its main disadvantages are that it requires knowledge of a concurrent series and that it may well require forecasting both the concurrent series and the series under study. Obtaining burglary ''rates,'' for instance, required gathering additional data on population. Forecasting the number of burglaries for a future period would require forecasting both the burglary rate and the population for that period and multiplying the two.
- **differencing** If the series has a straight-line trend, the first differences ($\Delta y_t$) of the series will have no trend. If the trend is quadratic (second degree), taking second differences ($\Delta^2 y_t$) will detrend it, and so forth. (Remember, from Section 1.4, that second differences are first differences of the first differences, i.e., $\Delta^2 y_t = \Delta y_t - \Delta y_{t-1}$. Third- and higher-order differences are defined and denoted analogously.) To illustrate, recall that in the burglary data of Example

**Table 2.15** First Differences for Burglary Rate Data

| $t$ | $y_t$ | $y_{t-1}$ | $\Delta y_t$ | $t$ | $y_t$ | $y_{t-1}$ | $\Delta y_t$ |
|---|---|---|---|---|---|---|---|
| 1 | 1.37 | — | — | 11 | 3.96 | 1.94 | +2.02 |
| 2 | 2.96 | 1.37 | +1.59 | 12 | 4.19 | 3.96 | +0.23 |
| 3 | 1.91 | 2.96 | −1.05 | 13 | 2.71 | 4.19 | −1.48 |
| 4 | 3.10 | 1.91 | +1.19 | 14 | 3.42 | 2.71 | +0.71 |
| 5 | 2.08 | 3.10 | −1.02 | 15 | 3.02 | 3.42 | −0.40 |
| 6 | 2.54 | 2.08 | +0.46 | 16 | 3.54 | 3.02 | +0.52 |
| 7 | 4.07 | 2.54 | +1.53 | 17 | 2.66 | 3.54 | −0.88 |
| 8 | 3.62 | 4.07 | −0.45 | 18 | 4.11 | 2.66 | +1.45 |
| 9 | 2.91 | 3.62 | −0.71 | 19 | 4.25 | 4.11 | +0.14 |
| 10 | 1.94 | 2.91 | −0.97 | 20 | 3.76 | 4.25 | −0.49 |

2.4, converting to rates did not completely detrend the data. The first differences for the rate data are given in Table 2.15 and plotted in Figure 2.15. Kendall's tau for these first differences is $\tau = -.041$, giving $z = -.245$, which is not significant. The trend that was relatively obvious in Figure 2.9 is no longer present in Figure 2.15.

Differencing is a simple and straightforward way to stationarize a time series. However, it has two main disadvantages: First, each time differencing

**Figure 2.15** Graph of First Differences for Burglary Rate Data

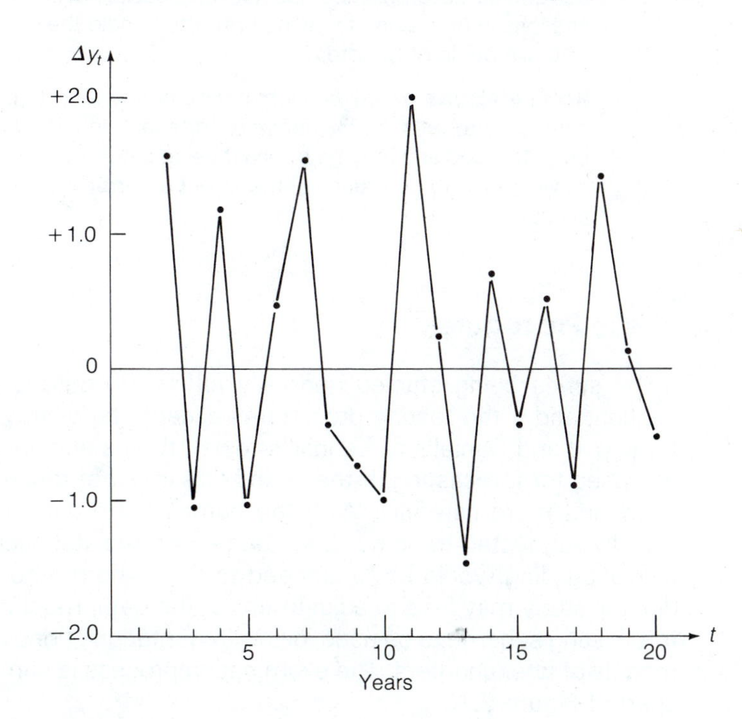

is done an observation is lost (for the burglary rate data, 20 observations yield 19 first differences, which yield 18 second differences, and so on). Second, and more important, differencing tends to increase the error variation, thus reducing estimation and forecasting precision. For any series, $y_t = T_t + \epsilon_t$, where $T_t$ is the trend, and $\epsilon_t$ are independent, and the standard deviation of $y_t$ is $\sigma_y$.

$$\sigma_{\Delta y} = \sqrt{2} \cdot \sigma_y = 1.41\sigma_y$$

$$\sigma_{\Delta^2 y} = \sqrt{6} \cdot \sigma_y = 2.45\sigma_y$$

$$\sigma_{\Delta^3 y} = \sqrt{20} \cdot \sigma_y = 4.47\sigma_y$$

$$\vdots$$

$$\sigma_{\Delta^n y} = \left[\frac{(2n)!}{n!n!}\right]^{1/2} \cdot \sigma_y$$

***Note:*** $n! = n(n-1)(n-2)\ldots 1$

If the errors are autocorrelated, the effect is more complicated. Additionally, because $\Delta y_t = y_t - y_{t-1}$ and $\Delta y_{t+1} = y_{t+1} - y_t$ have $y_t$ in common in their calculation, successive first differences are not independent. In any case, when using differencing it is important to weigh the advantages of any detrending accomplished against the possible resulting increase in variation and loss in precision. It may be better to handle the trend in the data with one of the models of Chapter 3.

- **link relatives** Many economic time series tend to grow at a constant percentage rate ($p \times 100\%$ per year, for example). The link relatives $y_t/y_{t-1}$ will be untrended and may be forecast using a no-trend model. With this method, as well as with differencing, the effect on error variation should be taken into account.

### 2.7.3 How to Use These Procedures

If the series being studied generally fulfills the criteria given earlier in this section, and if the testing procedures accept stationarity (especially the test for $\rho = 0$ and Daniel's or Kendall's tests), then a stationary model may be fit and used for forecasting. If the series fails in either respect, then it should be examined more carefully. With the burglary rate data, for example, once a visually suggested trend has been demonstrated statistically, the next step in model building would be to proceed to the trend methods of Chapter 3. Additional study may lead to adjustment of the data, modification of the type of forecasting scheme to be used, or implementation of one of the more complex models of later chapters. The examination process is summarized in the flowchart of Figure 2.16.

**Figure 2.16 Flowchart for Decision to Use No-Trend Models**

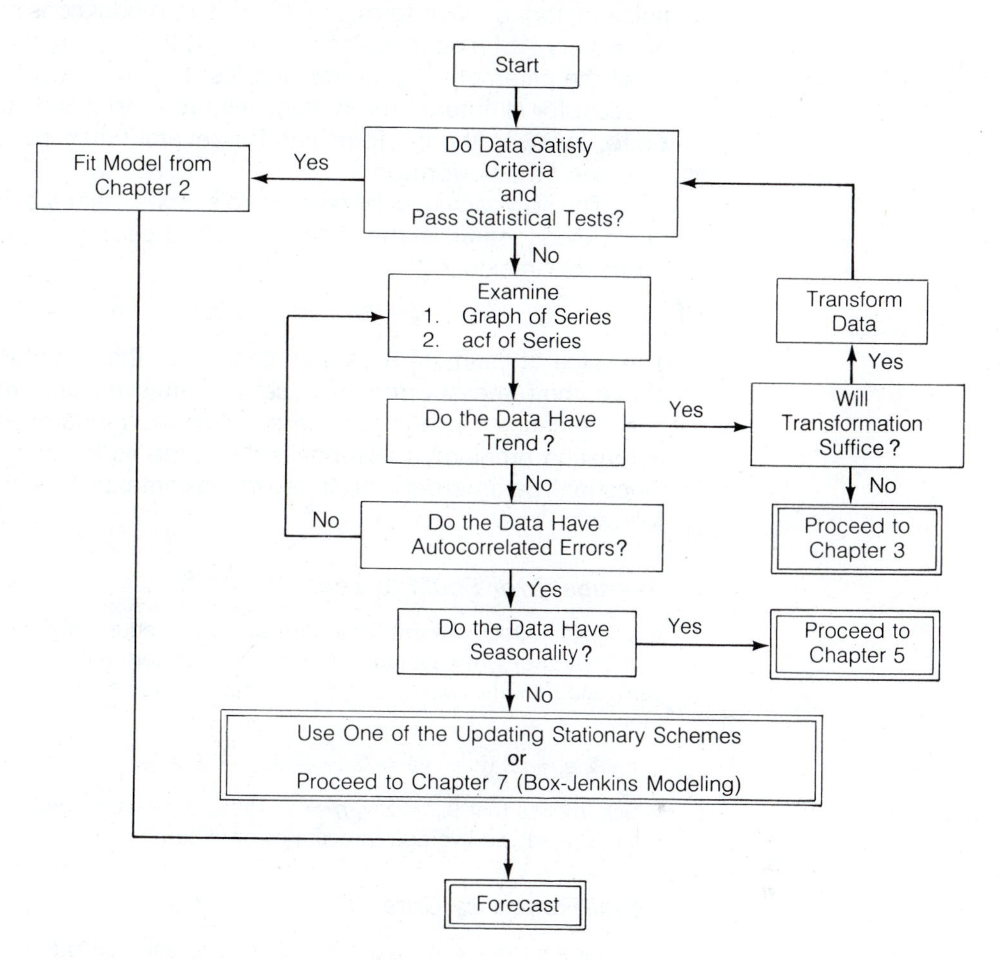


## 2.8 Single-Forecast Models

### 2.8.1 Introduction

Once it has been decided that a no-trend model might be appropriate, the next step is to select a model/procedure and "fit" it to the data.

### 2.8.2 The Single-Forecast Case

As noted earlier, the basic model for a random no-trend series is:

**No-Trend Model**

$$y_t = \beta_0 + \epsilon_t$$

where $\beta_0$ is the average level of the series and the $\epsilon_t$ represents unpredictable noise factors or error terms. "Fitting" this model consists of using a sample, or "history," of $n$ observations, $y_t$; $t = 1, 2, \ldots, n$, to calculate an estimate $\hat{\beta}_0$ of the parameter $\beta_0$. In the simplest type of forecasting, $\beta_0$ becomes the forecast for all future time periods. We refer to this as the **single-forecast case,** for even though forecasts for several future periods may be needed, *only one fit* is performed.

The approach usually taken is to suppose that a certain value of $\beta_0$ were being used to forecast the observations in the sample, giving a set of $n$ forecast errors, or "residuals."

$$e_t = y_t - \hat{y}_t = y_t - \beta_0 \qquad t = 1, 2, \ldots, n$$

The value of estimate $\hat{\beta}_0$ is then chosen so that when it is substituted for $\beta_0$ above, the forecast errors have certain intuitively or mathematically desirable characteristics. Because of the stability and randomness of the series, this estimation problem is essentially the same as for one-sample estimation of "location" parameters in nonforecasting contexts. Five criteria of "desirability" follow.

## I *Average Error Equal to Zero*

A reasonable requirement for a forecasting procedure might be that it overforecast as much (by the same average amount) as it underforecasts, or, equivalently, that the sum (and thus the average) of the forecast errors be zero. In the single-forecast case, then,

$$\Sigma\, e_t = \Sigma\, (y_t - \hat{y}_t) = \Sigma\, (y_t - \hat{\beta}_0) = \Sigma\, y_t - n\hat{\beta}_0 = 0$$

which implies that $\hat{\beta}_0 = \Sigma\, y_t/n = \bar{y}$. Therefore, when used as a forecast, the sample mean $\bar{y}$ yields an average forecast error of zero.

## II *Equal-Frequency Criterion*

Another approach that might be taken is to ask that the forecast be as likely to be above as below the actual value, or that it overforecast as often as it underforecasts. This condition is satisfied for a sample when $\hat{\beta}_0$ is chosen to be the sample median:

$$\hat{\beta}_0 = \tilde{y} = y_{\left(\frac{n+1}{2}\right)} = \left(\frac{n+1}{2}\right)\text{th-order statistic} = \text{sample median}$$

***Note:*** In the notation of order statistics (see Appendix 2A), $y_{(1)}$ = smallest observation = the first order statistic, $y_{(2)}$ = the second smallest observation = the second order statistic and so on, and by $y_{(6.5)}$, say, is meant the average of the 6th and 7th observations.

## III *Minimax Criterion*

In some situations, consistent moderate forecast errors are acceptable; but occasional large forecast errors are disastrous. For example, a greatly overforecast demand might lead to a company's bankruptcy. Under such circumstances it would be better to choose a value for $\hat{\beta}_0$ that would yield the smallest maximum error, that is,

$$\min_{\beta_0} \max_{t} |e_t| = \min_{\beta_0} \max_{t} |y_t - \beta_0|$$

The value that fulfills this condition is the midpoint between the largest and smallest $y_t$. It is called the **sample midrange** and is computed by averaging the 1st- and $n$th-order statistics, $y_{(1)}$ and $y_{(n)}$ or, if you prefer, $y_{min}$ and $y_{max}$, of the sample:

$$\hat{\beta}_0 = \frac{y_{(1)} + y_{(n)}}{2} = \text{sample midrange}$$

The sample midrange is often called the "minimax estimate of $\beta_0$" for the single-forecast case.

## IV *Min MAD Criterion*

As mentioned in Section 1.5, positive and negative forecast errors tend to cancel in the averaging process so that their size may not be accurately portrayed by the average error. As a result, we usually prefer minimizing the average magnitude, that is, by using the MAD or Mean Absolute Deviation, rather than requiring that the average signed forecast error be zero (Criterion I). We then select $\hat{\beta}_0$ as the value that achieves this minimum, that is,

$$\min_{\beta_0} \frac{\Sigma |e_t|}{n}$$

In the single-forecast case, the **sample median** has this property (Arbel, 1985), so to satisfy the min MAD criterion we choose

$$\hat{\beta}_0 = \tilde{y} = \text{sample median}$$

as we did for Criterion II.

## V *Least-Squares Criterion*

The most widely used statistical criterion for goodness of model fit is "least squares." The reasons for preferring this criterion were explained in Section 1.5 and will not be repeated here.

The method selects as $\hat{\beta}_0$ the value of $\beta_0$ that minimizes the sum of squared forecast errors,

$$Q = \Sigma (y - \hat{y})^2 = \Sigma (y - \beta_0)^2$$

Notice that this value of $\beta_0$ also minimizes the mean squared error (MSE) and the root mean squared error (RMSE). If we differentiate $Q$ with respect to $\beta_0$, and equate the results to zero we have

$$\frac{dQ}{d\beta_0} = \Sigma[-2(y - \beta_0)] \rightarrow -2 \cdot \Sigma (y - \hat{\beta}_0) = 0$$

$$\rightarrow \Sigma y - n\hat{\beta}_0 = 0 \; \hat{\beta}_0 = \frac{\Sigma y}{n}$$

and the estimate is

$$\hat{\beta}_0 = \bar{y} = \text{sample mean}$$

**Table 2.16 Single Forecasts for a Random No-Trend Model**

**Given:** A sample of $n$ observations, $y_1, y_2, \ldots, y_n$

**Model Equation:** $y_t = \beta_0 + \epsilon_t$

**Criterion:**
I Zero Average Error
II Equal Frequency
III Minimax
IV Min MAD
V Least Squares

**Estimates:**

I: $\hat{\beta}_0 = \bar{y} = \text{sample mean} = \Sigma\, y/n$

II: $\hat{\beta}_0 = \tilde{y} = \text{sample median} = y_{\left(\frac{n+1}{2}\right)}$

III: $\hat{\beta}_0 = y^* = \text{sample midrange} = (y_{\min} + y_{\max})/2$

IV: $\hat{\beta}_0 = \tilde{y} = \text{sample median} = y_{\left(\frac{n+1}{2}\right)}$

V: $\hat{\beta}_0 = \bar{y} = \text{sample mean} = \Sigma\, y/n$

**Forecast:** (At time $t$ for $p$-steps ahead)

$\hat{y}_{t+p} = \hat{\beta}_0 \qquad p \geq 1$

which, incidentally, satisfies Criterion I as well. The sample mean is then referred to as the "least squares estimate of $\beta_0$." The details for the various criteria for the single-forecast case are summarized in Table 2.16.

**EXAMPLE 2.7** *Single Forecast for Mainline Plumbing Data*

The gasoline purchase data of Mainline Plumbing Company in Example 2.1 are reproduced in Table 2.17. These data pass the visual and statistical checks for trend, so a no-trend forecasting model seems a reasonable choice. The forecasts for each of the criteria and several measures of goodness of model fit are given in the table. Notice that each of the estimates has the optimal value for its specific criteria. ●

Of the mean, median, and midrange, the mean is by far the most widely used, mainly because of its intuitive appeal and mathematical tractability. The median is also commonly used because it is relatively insensitive to aberrant or "outlying" observations. The midrange's main advantage lies in its ease of computation; however, it has the rather undesirable feature that its standard error $\sigma_{y^*}$ does not decrease as sample size increases. The use of the midrange is generally confined to situations in which samples are small, ease of computation is paramount, and large errors carry inordinately high risks.

## 2.8.3 Single Forecast for Normal Errors

When experience or other factors lead us to believe that the error terms $\epsilon_t$ are independently normally distributed, we may use standard statistical methods for a more detailed analysis of the data. If the error terms are normally

**Table 2.17** Single Forecasts, Gasoline Purchase Data of Mainline Plumbing Fleet

| $t$ | $y_t$ (gal) | $t$ | $y_t$ (gal) |
|---|---|---|---|
| 1 | 409 | 11 | 380 |
| 2 | 289 | 12 | 598 |
| 3 | 509 | 13 | 418 |
| 4 | 364 | 14 | 359 |
| 5 | 404 | 15 | 432 |
| 6 | 445 | 16 | 252 |
| 7 | 310 | 17 | 446 |
| 8 | 372 | 18 | 473 |
| 9 | 440 | 19 | 337 |
| 10 | 414 | 20 | 478 |

**Computation of Forecasts:**

Criteria I, V: $\Sigma y = 8{,}129 \quad \rightarrow \hat{\beta}_0 = \bar{y} = 406.45$

Criteria II, IV: $y_{(10)} = 409$

$$y_{(11)} = 414 \quad \rightarrow \hat{\beta}_0 = \tilde{y} = \frac{409 + 414}{2} = 411.5$$

Criteria III: $y_{(1)} = 252$

$$y_{(20)} = 598 \quad \rightarrow \hat{\beta}_0 = y^* = \frac{252 + 598}{2} = 425.0$$

**Measures of Goodness of Model Fit:**

| Criterion | Forecast | MAD | RMSE | MAPE | $\max\lvert e_t\rvert$ |
|---|---|---|---|---|---|
| I, V | 406.45 | 59.1 | 77.3 | 15.6% | 191.6 |
| II, IV | 411.50 | 58.8 | 77.5 | 15.8% | 186.5 |
| III | 425.00 | 60.6 | 79.1 | 16.7% | 173.0 |

distributed with a mean of zero and a standard deviation of $\sigma_\epsilon$, which we denote, $\epsilon_t \sim N(0, \sigma_\epsilon)$, then $y_t \sim N(\beta_0, \sigma_\epsilon)$. The nature of the forecasts depends on whether the values of $\beta_0$ and $\sigma_\epsilon$ are known or not.

If *both* $\beta_0$ and $\sigma_\epsilon$ are known, then the forecast for future periods will be $\beta_0$, and a prediction interval may be constructed from percentage points of the standard normal distribution using $\sigma_\epsilon = \sigma_{y-\beta_0} = \sigma_y = \sigma_\epsilon$. If *neither* $\beta_0$ nor $\sigma_e$ is known, both must be estimated. A prediction interval may be constructed from percentage points of Student's $t$-distribution (Whitmore, 1986), noting that, since $\bar{y}$ is independent of future values of $y$,

$$\sigma_e^2 = \sigma_{y-\bar{y}}^2 = \sigma_y^2 + \sigma_{\bar{y}}^2 = \sigma_\epsilon^2 + \frac{\sigma_\epsilon^2}{n} = \sigma_\epsilon^2\left(1 + \frac{1}{n}\right)$$

The formulas necessary for the various possibilities are given in Table 2.18.

**EXAMPLE 2.8** *Forecasts of Medical Insurance Claims*

A provider of medical insurance for a moderate-size manufacturing firm has found from past experience that the number of claims filed by employees is reasonably steady

**Table 2.18 Single Forecast with $\epsilon_t \sim N(0, \sigma_\epsilon)$**

I. $\beta_0$ and $\sigma_\epsilon$ Are Known

**Forecast:** $\hat{y}_t = \hat{\beta}_0$ for all $t$

**$(1 - \alpha) \times 100\%$ Prediction Interval:**

$$\hat{y}_t \pm B = \hat{\beta}_0 \pm B$$
$$B = z_{\alpha/2}\sigma_\epsilon$$

where $z_{\alpha/2}$ is the upper $(\alpha/2) \times 100\%$ point of the standard normal distribution and $B$ is a $(1 - \alpha) \times 100\%$ bound on the forecast error $e_t = y_t - \hat{y}_t$.

II. $\beta_0$ and $\sigma_\epsilon$ Are Unknown

**Given:** A sample of $n$ observations, $y_1, y_2, \ldots, y_n$

**Estimates:** $\hat{\beta}_0 = \bar{y}$ $\quad \hat{\sigma}_\epsilon = \hat{\sigma}_y = \dfrac{SS_{yy}}{n - 1}$

where $SS_{yy} = \Sigma (y - \bar{y})^2 = \Sigma y^2 - (\Sigma y)^2/n$

**Forecast:** $\hat{y}_t = \hat{\beta}_0 = \bar{y}$ for all $t$

**$(1 - \alpha) \times 100\%$ Prediction Interval:**

$$\hat{y}_t \pm B = \bar{y} \pm B$$
$$B = t_{\alpha/2}\hat{\sigma}_e$$
$$\hat{\sigma}_e = \hat{\sigma}_\epsilon \sqrt{1 + \frac{1}{n}}$$

**(Appendix GG)**

where $t_{\alpha/2}$ is the upper $(\alpha/2) \times 100\%$ point of Student's $t$-distribution, with $(n - 1)$ degrees of freedom, and $B$ is a $(1 - \alpha) \times 100\%$ bound on the forecast error.

year-round and, for that size company, is normally distributed, with an average of about 95 per month and a standard deviation of 30 per month. In December, the insurer wishes to:

1. Forecast the number of claims per month for the following January through June using a random no-trend model.
2. Construct 95% and 98% prediction intervals for January through June.
3. Forecast and construct a 95% prediction interval for the total number of claims for the first half of the year.

The model is

$$y_t = 95 + \epsilon_t \qquad \epsilon_t \sim N(0, 30)$$

so that the probability distribution of $y_t$ is

$$y_t \sim N(95, 30)$$

This population is pictured in Figure 2.17.

1. *Forecasts:* $\hat{y}_t = \beta_0 = 95$ for any value of $t$, so

$$\hat{y}_{Jan} = \hat{y}_{Feb} = \hat{y}_{Mar} = \cdots = \hat{y}_{Jun} = 95$$

**Figure 2.17** Distribution of $y_t$ = the number of claims in a month

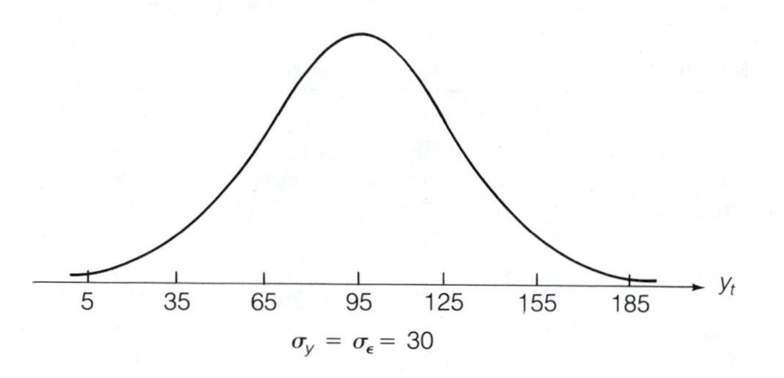


2. *95% Prediction Interval:*

$$\alpha = .05 \qquad z_{\alpha/2} = z_{.025} = 1.960$$
$$B = z_{\alpha/2}\sigma_\epsilon = 1.960 \times 30 = 58.8$$
$$\hat{y} = B = 95 \pm 58.8 = (36.2, 153.8)$$
$$\doteq (36, 154) \qquad \text{for any } t$$

*98% Prediction Interval:*

$$\alpha = .02 \qquad z_{\alpha/2} = z_{.01} = 2.326$$
$$B = z_{\alpha/2}\sigma_e = 2.326 \times 30 = 69.8$$
$$\hat{y} \pm B = 95 \pm 69.8 = (25.2, 164.8)$$
$$\doteq (25, 165) \qquad \text{for any } t$$

So, we can be 95% confident that there will be between 36 and 154 claims each month and 98% confident that there will be between 25 and 165 claims each month.

3. The total number of claims in the first half of the year is the sum of 6 months of values of $y$ and

$$\mu_{\Sigma y} = n\beta_0 = 6 \times 95 = 570$$
$$\sigma_{\Sigma y} = \sqrt{n}\sigma_y = \sqrt{n}\sigma_\epsilon = \sqrt{6} \cdot 30 = 73.5$$

*Forecast of* $\Sigma\, y$: $\Sigma\, \hat{y} = 570$

*95% Prediction Interval for* $\Sigma\, y$:

$$\alpha = .05 \qquad z_{\alpha/2} = z_{.025} = 1.960$$
$$B = z_{.025}\sigma_{\Sigma y} = 1.960 \times 73.5 = 144.1$$
$$\Sigma\, \hat{y} \pm B = 570 \pm 144.1 = (425.9, 714.1)$$
$$\doteq (426, 714) \qquad \text{for Jan through Jun}$$

The actual numbers of claims in January through October are given in Table 2.19. Because so many of the forecast errors in Table 2.19 are negative, the insurer believes that the model suggested by experience last December was a poor one and that the proposed mean and standard deviation may well have been in error. We can test whether the data contradict the hypothesis that the mean number of claims is 95 by

**Table 2.19 Number of Insurance Claims**

| Month | $t$ | $y_t$ | $\hat{y}_t$ | $e_t = y_t - \hat{y}_t$ |
|---|---|---|---|---|
| Jan | 1 | 87 | 95 | −8 |
| Feb | 2 | 82 | 95 | −13 |
| Mar | 3 | 69 | 95 | −26 |
| Apr | 4 | 54 | 95 | −41 |
| May | 5 | 84 | 95 | −11 |
| Jun | 6 | 104 | 95 | +9 |
| Jul | 7 | 47 | 95 | −48 |
| Aug | 8 | 39 | 95 | −56 |
| Sep | 9 | 103 | 95 | +8 |
| Oct | 10 | 83 | 95 | −12 |

using a $t$-test (Appendix 2D). Because we are no longer sure that $\sigma_\epsilon$ is 30, it must be estimated from the data. The test proceeds as follows:

$$H_0: \quad \mu = \beta_0 = 95$$

$$H_a: \quad \mu = \beta_0 \neq 95$$

**Test Statistic:** $t = \dfrac{\bar{y} - 95}{\hat{\sigma}_{\bar{y}}}$

**Decision Rule:** ($\alpha = .05$, $n = 10$, df $= 9$)

Reject $H_0$ if $|t| > t_{.025} = 2.262$ **(Appendix GG)**

Calculations:

$$\Sigma y = 752 \qquad \Sigma y^2 = 61{,}070$$

$$\bar{y} = \frac{752}{10} = 75.2 \qquad SS_{yy} = 61{,}070 - \frac{(752)^2}{10} = 4{,}519.6$$

$$\hat{\sigma}_\epsilon = \hat{\sigma}_y = \sqrt{\frac{SS_{yy}}{n-1}} = \sqrt{\frac{4519.6}{9}} = 22.4$$

$$\hat{\sigma}_{\bar{y}} = \frac{\hat{\sigma}_y}{\sqrt{n}} = \frac{22.4}{\sqrt{10}} = 7.08$$

$$t = \frac{75.2 - 95}{7.08} = -2.80$$

**Conclusion:** Since $|t| = 2.80 > 2.262$, we reject $H_0$; and since $t$ is negative, we conclude with 95% confidence that the true average number of claims per month is less than 95.

Our initial suspicions were, therefore, justified. Further forecasts should be constructed using the formulas that are appropriate for unknown $\beta_0$ and $\sigma_\epsilon$. Thus, to prepare forecasts and, say, 95% prediction intervals for November and December:

*Forecast:* $\hat{y} = \bar{y} = 75.2$ for any $t$

*95% Prediction Intervals:*

$$B = t_{\alpha/2}\hat{\sigma}_\epsilon \sqrt{1 + \frac{1}{n}} = 2.262 \times 22.4 \times \sqrt{1 + \frac{1}{10}} = 53.1$$

$$\hat{y} \pm B = 75.2 \pm 53.1 = (22.1, 128.3) \doteq (22, 128) \qquad \text{for } t = 11, 12$$

●

## 2.9 Updating Procedures

### 2.9.1 Introduction

When forecasting is an ongoing activity we might object to the single-forecast model as being overly restrictive and unrealistic, because once data have been collected and a model has been fit, that *single* model becomes the basis for *all* future forecasts. Since model fitting has inherent sampling error, it is natural to want to "update" the model now and then as more observations become available. In addition, if forecast errors develop a pattern as time goes by, we may wish to "refit" the model, perhaps discounting or discarding older observations.

### 2.9.2 Updating Single Forecasts

A straightforward way to do this updating is to refit the model as each new observation comes in, using the single-forecast criteria at each step. In this way a **one-step-ahead forecast** is obtained from the new single forecast each time. For Criteria I or V, we would recompute the mean each time, for II and IV, the median, and for III, the midrange. More precisely, the procedure goes like this: First $y_1$ becomes available, then a forecast for $y_2$ is made, after which $y_2$ becomes available, then a forecast for $y_3$ is made, and so on. The forecasts would then take the form given in Table 2.20.

**Table 2.20 Updated Forecasts Using Single-Forecast Procedures**

| Time $t$ | Observations Available | One-Step-Ahead Forecast: Criteria I, V | Criteria II, IV | Criterion III |
|---|---|---|---|---|
| 1 | $y_1$ | $y_1$ | $y_1$ | $y_1$ |
| 2 | $y_1, y_2$ | $\frac{y_1 + y_2}{2}$ | $\frac{y_{(1)} + y_{(2)}}{2}$ | $\frac{y_{(1)} + y_{(2)}}{2}$ |
| 3 | $y_1, y_2, y_3$ | $\frac{y_1 + y_2 + y_3}{3}$ | $y_{(2)}$ | $\frac{y_{(1)} + y_{(3)}}{2}$ |
| 4 | $y_1, y_2, y_3, y_4$ | $\frac{y_1 + y_2 + y_3 + y_4}{4}$ | $\frac{y_{(2)} + y_{(3)}}{2}$ | $\frac{y_{(1)} + y_{(4)}}{2}$ |
| ⋮ | | | ⋮ | |
| $t$ | $y_1, y_2, \cdots, y_t$ | $\bar{y}_t = \frac{\Sigma y}{t}$ | $\tilde{y}_t = y_{\left(\frac{t+1}{2}\right)}$ | $y_t^* = \frac{y_{(1)} + y_{(t)}}{2}$ |

Before proceeding with the discussion of updating schemes it should be pointed out that, in line with the comments about the MAD in Chapter 1, the denominators for the MAD, MSE, and MAPE will differ for the various updating schemes. For example, with the updated mean, median, or midrange, only $n - 1$ forecasts are generated, so one divides by $n - 1$ rather than $n$. The guideline to remember is that, since these measures are averages, the denominator is always the number of forecasts entered into the calculation of the sum.

**EXAMPLE 2.9** *One-Step-Ahead Forecasts for Mainline Plumbing*

Using the Mainline Plumbing Co. gasoline purchase data again, the forecasts for Criteria I–V, along with their corresponding forecast errors, are shown in Table 2.21. It should

**Table 2.21 Updating Forecasts for Gasoline Purchase Data**

| Time $t$ | Gasoline Purchases $y_t$ (gal) | Criteria I, V ($\bar{y}$) $\hat{y}_t$ | Criteria I, V ($\bar{y}$) $e_t$ | Criteria II, IV ($\tilde{y}$) $\hat{y}_t$ | Criteria II, IV ($\tilde{y}$) $e_t$ | Criterion III ($y^*$) $\hat{y}_t$ | Criterion III ($y^*$) $e_t$ |
|---|---|---|---|---|---|---|---|
| 1 | 409 | — | — | — | — | — | — |
| 2 | 289 | 409.0 | −120.0 | 409.0 | −120.0 | 409.0 | −120.0 |
| 3 | 509 | 349.0 | 160.0 | 349.0 | 160.0 | 349.0 | 160.0 |
| 4 | 364 | 402.3 | −38.3 | 409.0 | −49.0 | 399.0 | −45.0 |
| 5 | 404 | 392.8 | 11.2 | 386.5 | 17.5 | 399.0 | 5.0 |
| 6 | 445 | 395.0 | 50.0 | 404.0 | 41.0 | 399.0 | 46.0 |
| 7 | 310 | 403.3 | −93.3 | 406.5 | −96.5 | 399.0 | −89.0 |
| 8 | 372 | 390.0 | −18.0 | 404.0 | −32.0 | 399.0 | −27.0 |
| 9 | 440 | 387.8 | 52.2 | 388.0 | 52.0 | 399.0 | 41.0 |
| 10 | 414 | 393.6 | 20.4 | 404.0 | 10.0 | 399.0 | 15.0 |
| 11 | 380 | 395.6 | −15.6 | 406.5 | −26.5 | 399.0 | −19.0 |
| 12 | 598 | 394.2 | 203.8 | 404.0 | 194.0 | 399.0 | 199.0 |
| 13 | 418 | 411.2 | 6.8 | 406.5 | 11.5 | 399.0 | −25.5 |
| 14 | 359 | 411.7 | −52.7 | 409.0 | −50.0 | 399.0 | 84.5 |
| 15 | 432 | 407.9 | 24.1 | 406.5 | 25.5 | 443.5 | −11.5 |
| 16 | 252 | 409.5 | −157.5 | 409.0 | −157.0 | 443.5 | −191.5 |
| 17 | 446 | 399.7 | 46.3 | 406.5 | 39.5 | 443.5 | 2.5 |
| 18 | 473 | 402.4 | 70.6 | 409.0 | 64.0 | 443.5 | 29.5 |
| 19 | 337 | 406.3 | −69.3 | 411.5 | −74.5 | 425.0 | −88.0 |
| 20 | 478 | 402.7 | 75.3 | 409.0 | 69.0 | 425.0 | 53.0 |
| 21 | ? | 406.0 | ? | 411.5 | ? | 425.0 | ? |

**Measures of Goodness of Model Fit**

| Criteria | $\bar{e}$ | MAD | RMSE | max $\lvert e_t \rvert$ | MAPE |
|---|---|---|---|---|---|
| I, V | +8.21 | 67.7 | 87.0 | 203.8 | 45.9% |
| II, IV | +4.34 | 67.7 | 85.7 | 194.0 | 44.5% |
| III | −5.42 | 67.3 | 89.4 | 199.0 | 48.4% |

be noted that, while $\bar{y}$, $\tilde{y}$, and $y^*$ optimize their respective criteria *at each step,* they do not necessarily optimize these same measures for the whole process. In Table 2.21, for example, we see that the median gives a smaller process RMSE than the mean. By the same token, the midrange gives a slightly smaller MAD than either the mean or the median. In addition, the median gives a smaller max $|e_t|$ than the midrange. Finally, none of these methods gives an average forecast error of zero. •

## 2.9.3 Updating Equations

There are many ways updating can be accomplished. The basic principle is to compute a new estimate of $\beta_0$ each time a new observation is made. If we let

$$\hat{\beta}_0(t) = \text{the estimate of } \beta_0 \text{ based on information available at or before time } t$$

then the sequence of forecasts

$$\hat{y}_{t+1} = \hat{\beta}_0(t) \qquad t = 1, 2, \ldots$$

is the general form of an updating scheme. Because of the stationarity of the model, $p$-step-ahead and one-step-ahead forecasts are the same, so

$$\hat{y}_{t+p} = \hat{y}_{t+1} \qquad p = 2, 3, \ldots$$

For the schemes discussed previously, the expressions become

*Criteria I, V:*

$$\hat{y}_{t+1} = \hat{\beta}_0(t) = \frac{y_1 + y_2 + \cdots + y_t}{t} = \bar{y}_t$$

*Criteria II, IV:*

$$\hat{y}_{t+1} = \hat{\beta}_0(t) = y_{\left(\frac{t+1}{2}\right)} = \tilde{y}_t$$

*Criterion III:*

$$\hat{y}_{t+1} = \hat{\beta}_0(t) = \frac{y_{(1)} + y_{(t)}}{2} = y_t^*$$

It is often convenient to rewrite forecasting equations in a form that allows for easier calculation (and computer implementation). In their rewritten form, they may also provide substantial insight into the nature of the effect of each new observation on the forecast. Such expressions are usually referred to as **updating equations.** The least-squares forecast $\hat{y}_{t+1} = \bar{y}_t$, for example, may be rewritten as

$$\begin{aligned}\hat{y}_{t+1} &= \frac{y_1 + y_2 + \cdots + y_t}{t} = \frac{y_1 + y_2 + \cdots + y_{t-1}}{t} + \frac{y_t}{t} \\ &= \left(\frac{t-1}{t}\right) \times \frac{y_1 + y_2 + \cdots + y_{t-1}}{t-1} + \left(\frac{1}{t}\right)y_t \\ &= \left(\frac{t-1}{t}\right)\bar{y}_{t-1} + \left(\frac{1}{t}\right)y_t \\ &= \left(\frac{t-1}{t}\right)\hat{y}_t + \left(\frac{1}{t}\right)y_t\end{aligned}$$

When written in this way, we see that the new, "updated" forecast $\hat{y}_{t+1}$ is a weighted average of the old (previous) forecast $\hat{y}_t$ and the newly obtained observation, with weight $(t - 1)/t$ given to the former and weight $1/t$ given to the latter. That is,

$$\text{(new forecast)} = \left(\frac{t-1}{t}\right) \times \text{(old forecast)} + \frac{1}{t} \times \text{(new observation)}$$

It is easy to see that, as time passes, each new observation has less and less impact on the new forecast ($1/t$ becomes smaller and smaller). As a result, for a long series, least-squares "updated" forecasts ultimately become the same as "single" forecasts.

Another way to write the updating equation is

$$\hat{y}_{t+1} = \left(\frac{t-1}{t}\right)\hat{y}_t + \left(\frac{1}{t}\right)y_t = \hat{y}_t + \left(\frac{1}{t}\right)(y_t - \hat{y}_t)$$

or

$$\text{(new forecast)} = \text{(old forecast)} + \frac{1}{t} \times \text{(old forecast error)}$$

which means that the new forecast (at time $t$) is just the old forecast modified by one-$t$th of the just-observed error. At the tenth step, then, if we had underforecast by 20 units, we would modify the old forecast upward by $\frac{1}{10}$ of 20, or 2 units. Here, too, the amount of adjustment tends to become smaller as time passes, so the impact of each new piece of data diminishes with time.

The updating equations for the median and midrange are not so straightforward, since they involve the order statistics from the sample. For the midrange, the new forecast is the same as the old one unless the new observation is bigger than (or smaller than) all the rest. Avoiding complex notation, the relationship

$$\hat{y}_{t+1} = y_t^* = \text{the midrange at time } t$$

becomes

$$\hat{y}_{t+1} = \hat{y}_t = y_{t-1}^* \qquad \text{if } y_t \text{ lies between the largest and smallest of the previous } t-1 \text{ observations}$$

$$= \frac{y_{(1)} + y_t}{2} \qquad \text{if } y_t \text{ exceeds all } t-1 \text{ previous observations (i.e., } y_t = y_{(t)} \text{ of the new sample of } t \text{ observations)}$$

$$= \frac{y_t + y_{(t)}}{2} \qquad \text{if } y_t \text{ exceeds none of the } t-1 \text{ previous observations (i.e., } y_t = y_{(1)} \text{ of the new sample of } t \text{ observations)}$$

From these relationships we can conclude for the midrange that, as time passes, it becomes less and less likely that the new observation will change the forecast.

Updating forms of forecasting equations also can tell us what information is needed to perform the update. Examining the updating equation for the mean, we can see that updating requires the time $t$, the old forecast $\hat{y}_t$, and the new observation $y_t$. With the midrange, the largest and smallest of the previous observations and the new observation are needed.

Note that there is no simplified way to update the median. As each new observation comes in we have to see where that observation ranks among all previous observations, for only then can we tell what the median of the updated sample will be. Therefore all previous observations must be available when the updating is done. This trait is often undesirable, especially when large numbers of time series are being forecast, since a large amount of data must be kept in storage, ready for immediate access.

### EXAMPLE 2.10 *Updating Forecasts of Medical Insurance Claims*

To demonstrate the use of these updating formulas, let us suppose that the insurer of Example 2.8 has forecast 95 claims for January based on the previous 12 months' observations, so January is $t = 13$ and $\hat{y}_{13} = 95$.

To update the mean:

***At t = 13:*** $y_{13} = 87$ becomes available and so

$$\hat{y}_{14} = \left(\frac{12}{13}\right) \times 95 + \left(\frac{1}{13}\right) \times 87 = 94.4$$

or, in the alternate form,

$$\hat{y}_{14} = 95 + \left(\frac{1}{13}\right) \times (87 - 95) = 94.4$$

***At t = 14:*** $y_{14}$ becomes available and so

$$\hat{y}_{15} = \left(\frac{13}{14}\right) \times 94.4 + \left(\frac{1}{14}\right) \times 82 = 93.5$$

or, in the alternate form,

$$\hat{y}_{15} = 94.4 + \left(\frac{1}{14}\right) \times (82 - 94.4) = 93.5$$

and so forth, giving the forecasts in Table 2.22.

**Table 2.22 Updated Mean Forecasts of Insurance Claims**

| Month | $t$ | $y_t$ | $\hat{y}_{t+1}$ |
|---|---|---|---|
| Dec | 12 | — | 95 ($\hat{y}_{13}$) |
| Jan | 13 | 87 | 94.4 |
| Feb | 14 | 82 | 93.5 |
| Mar | 15 | 69 | 91.9 |
| Apr | 16 | 54 | 89.5 |
| May | 17 | 84 | 89.2 |
| Jun | 18 | 104 | 90.0 |
| Jul | 19 | 47 | 87.7 |
| Aug | 20 | 39 | 85.3 |
| Sep | 21 | 103 | 86.1 |
| Oct | 22 | 83 | 86.0 ($\hat{y}_{23}$) |
| Nov | 23 | ? | ? |

For the midrange, the largest and smallest of the first 12 observations are needed. Suppose they were 127 and 63, respectively. To update the midrange:

*At* t = *13:* $y_{13} = 87$, and 87 is between 63 and 127, so

$$\hat{y}_{14} = \hat{y}_{13} = 95$$

*At* t = *14:* $y_{14} = 82$, and 82 is between 63 and 127, so

$$\hat{y}_{15} = \hat{y}_{14} = 95$$

*At* t = *15:* $y_{15} = 69$, and 69 is between 63 and 127, so

$$\hat{y}_{16} = \hat{y}_{15} = 95$$

*At* t = *16:* $y_{16} = 54$, and 54 is less than 63, so $y_{(1)}$ becomes 54 and

$$\hat{y}_{17} = \frac{y_{16} + y_{(16)}}{2} = \frac{54 + 127}{2} = 90.5$$

*At* t = *17:* $y_{17} = 84$, and 84 is between 54 and 127, so

$$\hat{y}_{18} = \hat{y}_{17} = 90.5$$

and so forth, giving the forecasts in Table 2.23.

**Table 2.23 Updated Midrange Forecasts of Insurance Claims**

| Month | $t$ | $y_t$ | $\hat{y}_{t+1}$ | $y_{(1)}$ | $y_{(t)}$ |
|---|---|---|---|---|---|
| Dec | 12 | — | 95.0 | 63 | 127 |
| Jan | 13 | 87 | 95.0 | 63 | 127 |
| Feb | 14 | 82 | 95.0 | 63 | 127 |
| Mar | 15 | 69 | 95.0 | 63 | 127 |
| Apr | 16 | 54 | 90.5 | 54 | 127 |
| May | 17 | 84 | 90.5 | 54 | 127 |
| Jun | 18 | 104 | 90.5 | 54 | 127 |
| Jul | 19 | 47 | 87.0 | 47 | 127 |
| Aug | 20 | 39 | 83.0 | 39 | 127 |
| Sep | 21 | 103 | 83.0 | 39 | 127 |
| Oct | 22 | 83 | 83.0 | 39 | 127 |
| Nov | 23 | ? | ? | ? | ? |

## 2.9.4 Other Updating Schemes

Updating also provides a means of discounting or even discarding older observations. One might feel that observations more than a year old should not have the same influence as more recent ones, or that the influence of each observation should decrease as it grows older. In cases where autocorrelation is present, or where we are not sure whether the series is (or will remain) stationary, such procedures provide:

- A way to track drift due to autocorrelation.
- A built-in means of reacting to or tracking model changes, such as changes in the value of $\beta_0$.

Desirable as these two characteristics might be, if the series is stationary and remains that way, discounting procedures will not generally forecast as well as single-forecast or updated single-forecast procedures.

### 2.9.5 Smoothing and Tracking

While discounting could be achieved by computing the median or the midrange of, say, only the most recent 5 or 10 observations, it is often done by computing some sort of weighted average. The averaging involved is often referred to as *smoothing,* in the sense that it averages, or "smooths out," error movements. With regard to this smoothing concept, a useful principle to keep in mind is that the more smoothing done (i.e., the more terms averaged over), the less the forecasts track model changes and/or autocorrelative drift. Smoothing causes the forecasts to lag the actual series. Too much tracking tends to cause the forecasts to overreact to random errors. It is usually best to maintain some middle ground between smoothing and tracking, with perhaps the greater emphasis being given to smoothing to avoid undue fluctuations in the forecasts.

### 2.9.6 Moving Average (MA)

To compute the weighted average of a set of data, first each observation is multiplied by a weight representing its relative importance, then the products are summed, and finally this sum is divided by the sum of the weights. We may avoid division as the last step by dividing each weight by the sum of the weights *before* multiplying. The resulting weights then sum to 1.0 and are called **normalized weights.** These normalized weights, often expressed as percentages, give quite a clear picture of the proportionate influence of each observation on the value of the average. In forecasting formulas it is convenient to use the following notation:

$w_0$ = the weight given to the most recent observation

$w_1$ = the weight given to the one-period-old observation

$w_2$ = the weight given to the two-period-old observation

$\vdots$

$w_i$ = the weight given to the $i$-period-old observation

In smoothing schemes with normalized weights, then, the one-step-ahead forecast is

$$\begin{aligned} \hat{y}_{t+1} = \hat{\beta}_0(t) &= w_0 y_t + w_1 y_{t-1} + \cdots + w_{k-1} y_{t-k+1} \qquad k \leq t \\ &= \sum_{0}^{k-1} w_i y_{t-i} \end{aligned}$$

and the $p$-step-ahead forecast is

$$\hat{y}_{t+p} = \hat{y}_{t+1} \qquad p \geq 1$$

Because the weights "move" forward one step each time a new observation is obtained, an average computed as above is called a **moving average.**

Using the updating procedure for the mean discussed earlier in this section, the weighting scheme changes at each step, i.e., the weights at time $t$ are $1/t, 1/t, 1/t, \ldots, 1/t$ and at time $t + 1$ become $1/(t + 1), 1/(t + 1), \ldots, 1/(t + 1)$. For convenience, we will continue to refer to this method as **updating the mean.** Since all previous observations enter equally into the average, updating the mean provides a maximal amount of smoothing and, thus, a minimal amount of tracking.

Another concept useful in interpreting moving averages is the **center of the average.** We can think of the center as that point in time which the average most represents. The center is not necessarily the middle of the interval; it depends on the weighting scheme. However, the average used in updating the mean is centered at $(t + 1)/2$, the middle of the interval from 1 to $t$, because of the symmetry of the weights. The average of 4 observations, $y_1$, $y_2$, $y_3$, and $y_4$, would then be centered at time $(4 + 1)/2 = 2.5$, as seen below:

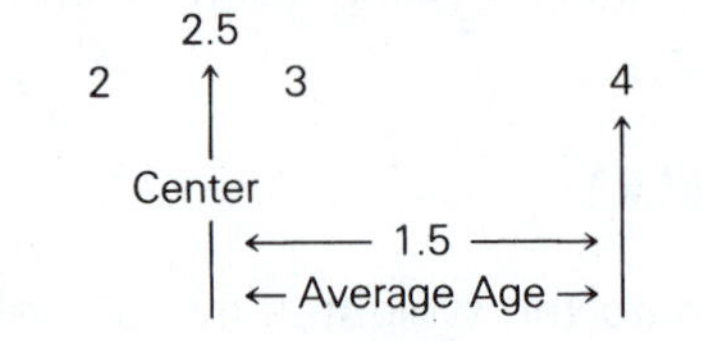


This center moves farther and farther into the past as time goes by and the mean is updated, so the average age of the forecast gets greater and greater. (The "average age" is the distance between the center of the average and $t$. For an updated mean forecast the average age would be $t - (t + 1)/2 = (t - 1)/2$.)

**Updating the Mean**

**Forecast:** $\hat{y}_{t+p} = \dfrac{\Sigma y_t}{t} \qquad p \geq 1$

**Updating Equation:** $\hat{y}_{t+1} = \left(\dfrac{t-1}{t}\right)\hat{y}_t + \left(\dfrac{1}{t}\right)y_t$

Weighting Scheme (at time $t$)

| | | | | | |
|---|---|---|---|---|---|
| **Observation:** | $y_1$ | $y_2$ | $y_3$ | $\cdots$ | $y_t$ |
| **Weight:** | $\frac{1}{t}$ | $\frac{1}{t}$ | $\frac{1}{t}$ | $\cdots$ | $\frac{1}{t}$ |

**Center:** $\dfrac{t+1}{2}$

**Average Age:** $\dfrac{t-1}{2}$

### 2.9.7 Naive Forecasting

The "no-change" model mentioned in Chapter 1 provides the opposite extreme in the smoothing/tracking tradeoff. Use of this model is often termed **naive forecasting** because of the simplicity of the ideas involved. You will recall that the forecast for each period is the immediately preceding observation. Thus 100% of the weight is given to the current value of the series. Because it discards all other values, this scheme tracks very rapidly ("no-change" is pretty much of a misnomer because, in a sense, the scheme is built to anticipate changes at all times). In fact, its outstanding fault is that random (noise) fluctuations are tracked as faithfully as model (signal) changes.

**Naive Forecasting**

**Forecast:** $\hat{y}_{t+p} = y_t \qquad p \geq 1$

**Updating Equation:** $\hat{y}_{t+1} = y_t$

Weighting Scheme (at time $t$)

| | | | | | |
|---|---|---|---|---|---|
| **Observation:** | $y_1$ | $y_2$ | $y_3$ | $\cdots$ | $y_t$ |
| **Weight:** | 0 | 0 | 0 | $\cdots$ | 1 |

**Center:** $t$

**Average Age:** 0

If a series is really random and stationary, using naive forecasting, where forecast errors are actually just first differences, will increase the mean squared forecast error by about 41% as compared to updating the mean. An argument may still be made for its use, however, if one expects precipitous changes in the model or strong autocorrelative movements (cycles).

Figures 2.18(a)–(c) show how updating the mean and naive forecasting differ in their responses to different series. In series (a), there is no noise and a step change in $\beta_0$ of about 100% at time $t = 4$. Since this type of series should be tracked quickly, naive forecasting produces better forecasts. Series (b) has normally distributed noise and an average value of $\beta_0 = 15$. Under these conditions, naive forecasting merely tracks the noise and tends to yield larger errors than do the much smoother updated mean forecasts. Finally, series (c) has strong autocorrelation, producing oscillatory movement. Naive forecasting tracks these oscillations well, while the updated mean reacts only slowly to them, reacting even more slowly as time passes.

### 2.9.8 Moving Average of Length *k*

Varying amounts of smoothing (and tracking) can be obtained by altering the weighting scheme. A common way to reduce the extreme smoothing of updated mean forecasts is to compute the mean of only the $k$ most recent observations. The result is called a **moving average of length *k*.** Varying

**Figure 2.18** Responses of Updating the Mean and Naive Forecasting to Different Series

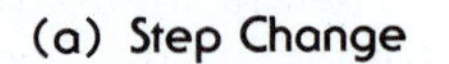

**(a) Step Change**

| $t$ | $y_t$ | $\hat{y}_t = \bar{y}_{t-1}$ | $\hat{y}_t = y_{t-1}$ |
|---|---|---|---|
| 1 | 10 | – | – |
| 2 | 10 | 10 | 10 |
| 3 | 10 | 10 | 10 |
| 4 | 10 | 10 | 10 |
| 5 | 20 | 10 | 10 |
| 6 | 20 | 12 | 20 |
| 7 | 20 | 13.3 | 20 |
| 8 | 20 | 14.3 | 20 |
| 9 | 20 | 15.0 | 20 |
| 10 | 20 | 15.6 | 20 |
| 11 | ? | 16.0 | 20 |

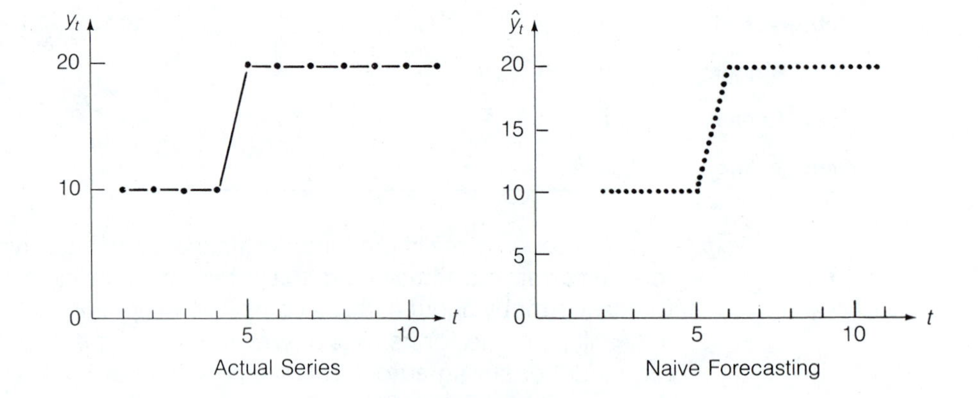


**(b) Noise Only, $\beta_0 = 15$**

| $t$ | $y_t$ | $\hat{y}_t = \bar{y}_{t-1}$ | $\hat{y}_t = y_{t-1}$ |
|---|---|---|---|
| 1 | 13 | – | – |
| 2 | 16 | 13 | 13 |
| 3 | 19 | 14.5 | 16 |
| 4 | 15 | 16.0 | 19 |
| 5 | 17 | 15.8 | 15 |
| 6 | 19 | 16.0 | 17 |
| 7 | 15 | 16.5 | 19 |
| 8 | 20 | 16.3 | 15 |
| 9 | 16 | 16.8 | 20 |
| 10 | 10 | 16.7 | 16 |
| 11 | ? | 16.0 | 10 |

### (b) Noise Only, $\beta_0 = 15$ (Continued)

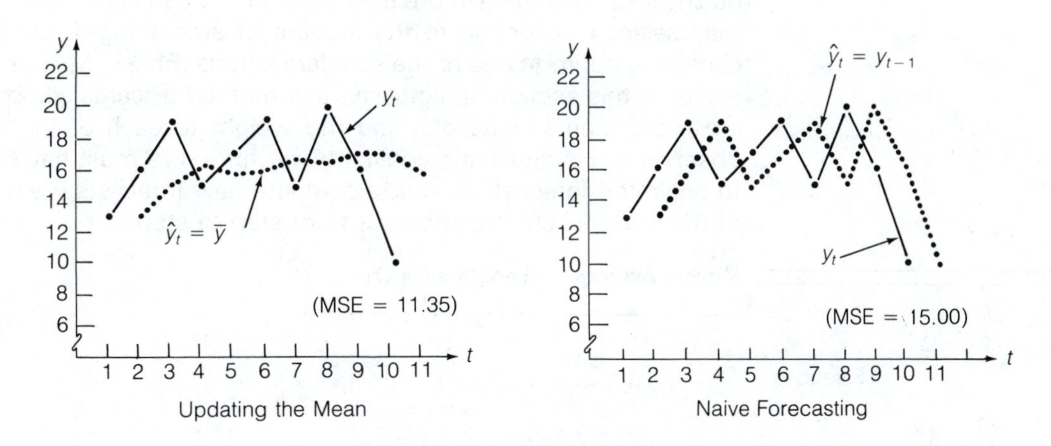


### (c) Autocorrelative Drift

| $t$ | $y_t$ | $\hat{y}_t = \bar{y}_{t-1}$ | $\hat{y}_t = y_{t-1}$ |
|---|---|---|---|
| 1 | 15 | – | – |
| 2 | 18 | 15 | 15 |
| 3 | 19 | 16.5 | 18 |
| 4 | 20 | 17.3 | 19 |
| 5 | 20 | 18.0 | 20 |
| 6 | 19 | 18.4 | 20 |
| 7 | 16 | 18.5 | 19 |
| 8 | 14 | 18.1 | 16 |
| 9 | 13 | 17.6 | 14 |
| 10 | 14 | 17.1 | 13 |
| 11 | ? | 16.8 | 14 |

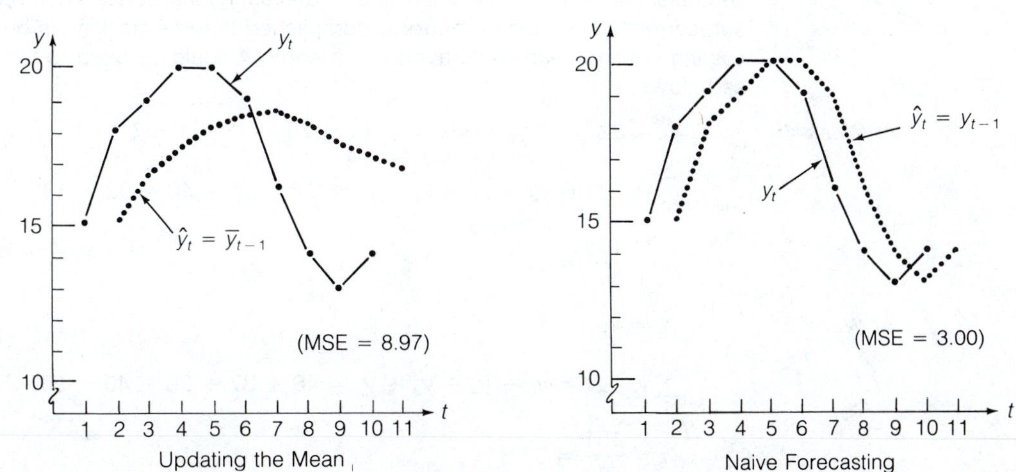

the value of $k$ varies the amount of smoothing, with smaller values of $k$ producing less smoothing (more tracking). For any particular series, the forecaster may select $k$ according to the amount of smoothing desired, or $k$ may be chosen to optimize one of the standard criteria (RMSE, MAD, etc.) mentioned earlier in this section. In actuality, this method discards all observations that are more than $k$ units old, and the weight to each of the $k$ most recent observations is equal and constant over time. We must have $k$ observations to begin the forecasting, and to compute new forecasts we must retain all $k$ of the most recent observations from step to step.

**Moving Average of Length $k$ (MA)**

**Forecast:**

$$\hat{y}_{t+p} = \frac{\sum_{1}^{k} y_{t-i+1}}{k} \qquad p \geq 1$$

$$= \frac{y_t + y_{t-1} + \cdots + y_{t-k+1}}{k}$$

**Updating Equation:**

$$\hat{y}_{t+1} = \hat{y}_t + \left(\frac{1}{k}\right)(y_t - y_{t-k})$$

Weighting Scheme (at time $t$)

| | | | | | | | | |
|---|---|---|---|---|---|---|---|---|
| **Observation:** | $y_1$ | $y_2$ | $\cdots$ | $y_{t-k}$ | | $y_{t-k+1}$ | $\cdots$ | $y_t$ |
| **Weight:** | 0 | 0 | $\cdots$ | 0 | $\cdots$ | $\frac{1}{k}$ | $\cdots$ | $\frac{1}{k}$ |

**Center:** $t - \frac{k-1}{2}$

**Average Age:** $\frac{k-1}{2}$

**EXAMPLE 2.11** *MA Forecasts for Rogers Electronics Specialties Data*

We might want to apply a moving average to the Rogers Electronics Specialties absenteeism data of Example 2.2 because forecasting that series, which contains positive autocorrelation, would be better accomplished if some tracking is built into the forecasting scheme. A moving average of length 4 could be used to forecast the series as follows:

*At Week 4:*

$$y_1 = 32 \qquad y_2 = 40 \qquad y_3 = 26 \qquad y_4 = 37$$

$$\sum_{1}^{4} y_{4-i+1} = y_4 + y_3 + y_2 + y_1 = 37 + 26 + 40 + 32 = 135$$

$$\hat{y}_5 = \frac{135}{4} = 33.8$$

*At Week 5:*

$$y_5 = 48$$

$$\sum_{1}^{4} y_{5-i+1} = y_5 + y_4 + y_3 + y_2 = 48 + 37 + 26 + 40 = 151$$

$$\hat{y}_6 = \frac{151}{4} = 37.8$$

*At Week 6:*

$y_6 = 66$

$$\sum_{1}^{4} y_{6-i+1} = y_6 + y_5 + y_4 + y_3 = 666 + 48 + 37 + 26 = 177$$

$$\hat{y}_7 = \frac{177}{4} = 44.2 \qquad \text{and so forth.}$$

The quantity $\Sigma\, y_{t-i+1}$ is usually called the **moving total.** The forecast can be updated by updating the moving total:

(new moving total) = (old moving total) + (new observation) − ($k$-period old observation)

Then $$\text{(new forecast)} = \frac{\text{(new moving total)}}{k}$$

Updating can also be done in terms of the old forecast:

*At Week 4:*

$y_1 = 32 \qquad y_2 = 40 \qquad y_3 = 26 \qquad y_4 = 37$

$$\hat{y}_5 = \frac{135}{4} = 33.8 \qquad \text{(as before)}$$

*At Week 5:*

$y_t = y_5 = 48 \qquad y_{t-k} = y_{5-4} = y_1 = 32$

$$\hat{y}_6 = \hat{y}_5 + \frac{(y_5 - y_4)}{4} = 33.8 + \frac{48 - 32}{4} = 37.8$$

*At Week 6:*

$y_t = y_6 = 66 \qquad y_{t-k} = y_{6-4} = y_2 = 40$

$$\hat{y}_7 = \hat{y}_6 + \frac{y_6 - y_2}{4} = 37.8 + \frac{66 - 40}{4} = 44.2 \qquad \text{and so forth.}$$

The entire sequence of forecasts (up to $\hat{y}_{16}$) for moving averages of length 2 and 4 are given in Table 2.24 and graphed along with the actual series in Figure 2.19. These

**Table 2.24** Moving Average Forecasts for Rogers Electronics Absenteeism Data

| | | $k = 2$ | | $k = 4$ | |
|---|---|---|---|---|---|
| Week $t$ | Absenteeism $y_t$ | Moving Total | $\hat{y}_{t+1}$ | Moving Total | $\hat{y}_{t+1}$ |
| 1 | 32 | — | — | — | — |
| 2 | 40 | 72 | 36.0 | — | — |
| 3 | 26 | 66 | 33.0 | — | — |
| 4 | 37 | 63 | 31.5 | 135 | 33.8 |
| 5 | 48 | 85 | 42.5 | 151 | 37.8 |
| 6 | 66 | 114 | 57.0 | 177 | 44.2 |
| 7 | 73 | 139 | 69.5 | 224 | 56.0 |
| 8 | 62 | 135 | 67.5 | 249 | 62.2 |
| 9 | 59 | 121 | 60.5 | 260 | 65.0 |
| 10 | 25 | 84 | 42.0 | 219 | 54.8 |
| 11 | 30 | 55 | 27.5 | 176 | 44.0 |
| 12 | 28 | 58 | 29.0 | 142 | 35.5 |
| 13 | 30 | 58 | 29.0 | 113 | 28.2 |
| 14 | 40 | 70 | 35.0 | 128 | 32.0 |
| 15 | 36 | 76 | 38.0 | 134 | 33.5 |

**Figure 2.19** Moving Averages of Lengths $k = 2, 4$ Applied to Rogers Electronics Specialties Data

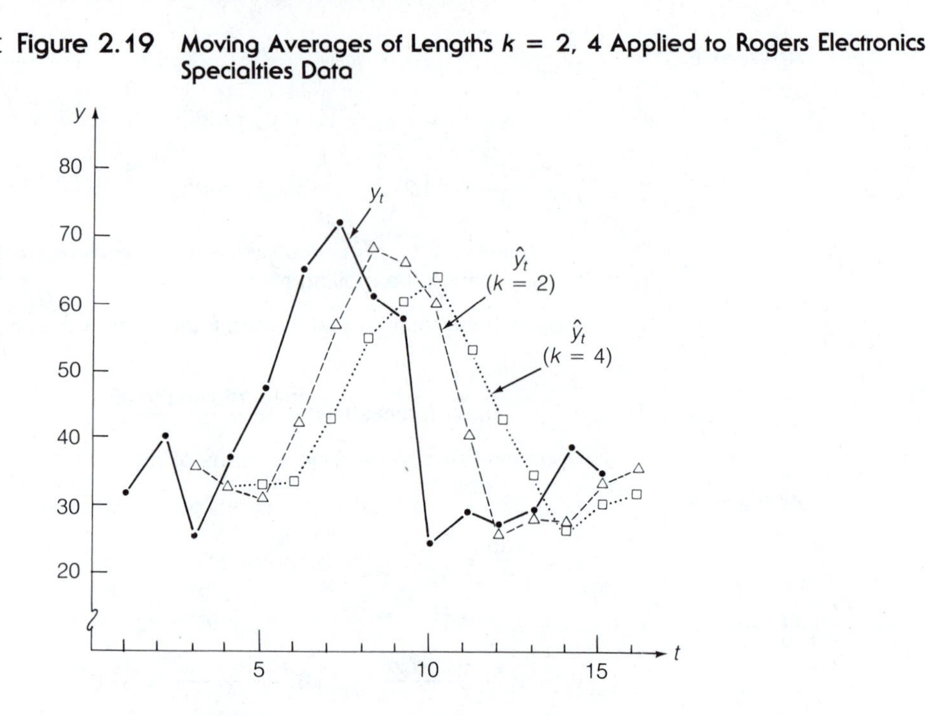

**Figure 2.20** Root Mean Square Forecast Errors For Moving Averages of Lengths 1 Through 5 Applied to Rogers Electronics Specialties Data

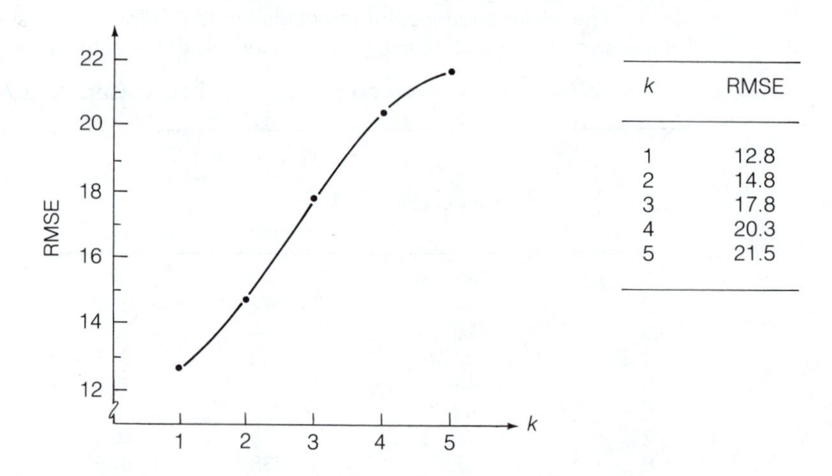


| $k$ | RMSE |
|---|---|
| 1 | 12.8 |
| 2 | 14.8 |
| 3 | 17.8 |
| 4 | 20.3 |
| 5 | 21.5 |

illustrations show that the forecast series lag somewhat behind the actual series, the lag becoming greater as the length of the moving average is increased. The RMSEs for lengths $k = 1, 2, 3, 4$, and 5 are shown in Figure 2.20. They indicate that $k = 1$ (which is actually naive forecasting) is the optimal length for the moving average.

Tracking the autocorrelative movement as quickly as possible provides better forecasts than, say, the updated mean (RMSE = 17.0) or longer-length moving averages would.

## 2.9.9 Weighted Moving Average (WMA)

It is also possible to use a **weighted moving average (WMA)** of length $k$ in which each of the observations is assigned a desired weight. For example, we might use a moving average of length 4 but give 50% of the weight in the average to the most recent observation, 30% to the next most recent, and 10% to each of the other two. Applying this to the data from Figure 2.18(c) would give Table 2.25.

**Table 2.25** Weighted Moving Average Forecasts of Data from Figure 2.18(c)

| $t$ | $y_t$ | $\hat{y}_{t+1}$ | = | $.5y_t$ | + | $.3y_{t-1}$ | + | $.1y_{t-2}$ | + | $.1y_{t-3}$ |
|---|---|---|---|---|---|---|---|---|---|---|
| 1 | 15 | — | | | | | | | | |
| 2 | 18 | — | | | | | | | | |
| 3 | 19 | — | | | | | | | | |
| 4 | 20 | 19.0 | = | $.5 \times 20$ | + | $.3 \times 19$ | + | $.1 \times 18$ | + | $.1 \times 15 = \hat{y}_5$ |
| 5 | 20 | 19.7 | = | $.5 \times 20$ | + | $.3 \times 20$ | + | $.1 \times 19$ | + | $.1 \times 18 = \hat{y}_6$ |
| 6 | 19 | 19.4 | | | | | | | | |
| 7 | 16 | 17.7 | | | | | | | | |
| 8 | 14 | 15.7 | | | | | | | | |
| 9 | 13 | 14.2 | | | | | | | | |
| 10 | 14 | 13.9 | = | $.5 \times 14$ | + | $.3 \times 13$ | + | $.1 \times 14$ | + | $.1 \times 16 = \hat{y}_{11}$ |
| 11 | ? | | | | | | | | | |

*Note:* RMSE = 2.38, based on forecasts for $t = 4$ through $t = 10$.

The average of the observations used to compute these forecasts is found by first multiplying the age of each observation in the average by its weight and then summing. The age of the most recent observation is 0, the age of the next most recent is 1, and so on, so

$$\begin{aligned}\text{(average age)} &= .5 \times 0 + .3 \times 1 + .1 \times 2 + .1 \times 3 \\ &= .8 \text{ time periods}\end{aligned}$$

which is slightly older than a simple (unweighted) moving average of length 2 (age = .5) and younger than one of length 3 (age = 1.0).

Although WMAs allow flexibility in choosing weighting schemes, updating them requires cumbersome calculations. As each new observation is

**Weighted Moving Average (WMA) of Length $k$**

**Forecast:** $$\hat{y}_{t+p} = \sum_{0}^{k-1} w_i y_{t-i} \qquad \left(\sum_{0}^{k-1} w_i = 1.0, \qquad p \geq 1\right)$$

where

$w_i$ = weight to the $(t - i)$th observation,
$i = 0, 1, \cdots, (k - 1)$

No Updating Equation

Weighting Scheme (at time $t$)

| | | | | | | | | | |
|---|---|---|---|---|---|---|---|---|---|
| **Observation:** | $y_1$ | $y_2$ | $\cdots$ | $y_{t-k+1}$ | $\cdots$ | $y_{t-2}$ | $y_{t-1}$ | $y_t$ |
| **Weight:** | 0 | 0 | $\cdots$ | $w_{k-1}$ | $\cdots$ | $w_2$ | $w_1$ | $w_0$ |

**Center:** $t - \sum_{0}^{k-1} iw_i$

**Average Age:** $\sum_{0}^{k-1} iw_i$

received, the weights move forward one time period, which necessitates multiplying all the observations used in the forecast by their new weights. For computational purposes this means that $k - 1$ old observations and $k$ weights must be stored from one step to the next and that there is no simple updating equation.

# 2.10 Exponential Smoothing

## 2.10.1 Introduction

A system of diminishing weights is more often achieved by using what is known as **exponential smoothing** or **simple exponential smoothing (SES).** (See Brown (1963) for the original work, and Gardiner (1985) for an excellent survey.) This method is quite popular for several reasons, including intuitive appeal, ease in updating, and the small amount of information (compared to WMAs) that must be carried from one period to the next.

## 2.10.2 Updating Equations

The SES procedure is most easily represented (and understood) by considering one form or other of its updating equations. In *smoothing form,* the new forecast (for time $t + 1$) may be thought of as a weighted average of the old forecast (for time $t$) and the new observation (at time $t$), with weight $\alpha$ (called the **smoothing constant**) given to the newly observed value and weight $1 - \alpha$ given to the old forecast, assuming $0 \le \alpha \le 1$. Thus

$$\text{(new forecast)} = (1 - \alpha) \times \text{(old forecast)} + \alpha \times \text{(new observation)}$$

or, more formally,

*Smoothing Form:*

$$\hat{y}_{t+1} = (1 - \alpha)\hat{y}_t + \alpha y_t$$

When $\alpha$ is large (near 1.0), great weight is given to the current value of the series at each update, so the forecast tracks quite quickly. When $\alpha$ is small (near 0.0), great weight is given to the old forecast at each update, so only relatively small changes can result from new observations. Thus, the larger $\alpha$ is, the greater is the tracking ability of the scheme. Notice, too, that choosing $\alpha = 1.0$ in reality leads to naive forecasting, while $\alpha = 0.0$ results in constant forecasts.

The *error correction form* of the updating equation is quite easily derived by rearranging the terms on the right-hand side of the smoothing form:

$$\begin{aligned}\hat{y}_{t+1} &= (1 - \alpha)\hat{y}_t + \alpha y_t = \hat{y}_t - \alpha\hat{y}_t + \alpha y_t \\ &= \hat{y}_t + \alpha(y_t - \hat{y}_t) = \hat{y}_t + \alpha e_t\end{aligned}$$

giving

*Error Correction Form:*

$$\hat{y}_{t+1} = \hat{y}_t + \alpha e_t$$

Thus we can think of the new forecast as being obtained by moving the old forecast (by a fraction, $\alpha$, of the forecast error) in the direction that would have improved the forecast. (Suppose, for example, we had forecast January sales to be $40,000 and actually observed $30,000. Using $\alpha = 20\%$, we would revise the forecast downward by 20% of $10,000, or $2,000. February's forecast would then be $40,000 − $2,000 = $38,000.)

By substituting

$$\hat{y}_t = (1 - \alpha)\hat{y}_{t-1} + \alpha y_{t-1} \tag{2.1}$$

$$\hat{y}_{t-1} = (1 - \alpha)\hat{y}_{t-2} + \alpha y_{t-2} \tag{2.2}$$

$$\hat{y}_{t-2} = (1 - \alpha)\hat{y}_{t-3} + \alpha y_{t-3} \tag{2.3}$$

$$\vdots$$

into the smoothing form of the updating equation and simplifying, we can show, from Equation 2.1, that

$$\begin{aligned}\hat{y}_{t+1} &= \alpha y_t + (1 - \alpha)\hat{y}_t \\ &= \alpha y_t + (1 - \alpha)[(1 - \alpha)\hat{y}_{t-1}\, \alpha y_{t-1}] \\ &= \alpha y_t + \alpha(1 - \alpha)y_{t-1} + (1 - \alpha)^2\hat{y}_{t-1}\end{aligned} \tag{2.4}$$

Then, from Equations 2.2 through 2.4, we can show that

$$\begin{aligned}\hat{y}_{t+1} = \alpha y_t &+ \alpha(1 - \alpha)y_{t-1} + \alpha(1 - \alpha)^2 y_{t-2} + (1 - \alpha)^3 y_{t-3} \\ &+ \cdots + \alpha(1 - \alpha)^k y_{t-k} + (1 - \alpha)^{k+1}\hat{y}_{t-k}\end{aligned}$$

Exponentially smoothed forecasts are, therefore, weighted averages of the previous observations and the oldest forecast. The weights decay exponentially (thus the name), with $1 - \alpha$ acting as a "discounting" factor, giving weight $w_i = \alpha(1 - \alpha)^i$ to observation $y_{t-i}$; $i = 0, 1, \ldots, k$ and weight $w_{k+1} = (1 - \alpha)^{k+1}$ to the $k$-period old forecast, $\hat{y}_{t-k}$. Because in practice the observed series is only of finite length, the oldest forecast is actually an "initial" forecast, which we will denote $\hat{y}_{\text{init}}$. It may be calculated in several ways and will be discussed in some detail shortly.

Forecasting begins at time $t = 1$ and continues until time $t$, when the forecast $\hat{y}_{t+1}$ for time $t + 1$ is made. It is quite popular to use $y_1$ as the initial forecast, but this gives weight $(1 - \alpha)^{t-1}$ to $y_1$, which is $1/\alpha$ times the weight it would normally receive and may be inordinate unless $t$ is large.

**EXAMPLE 2.12** *SES Forecasts for the Data of Figures 2.18*

As an illustration, let us forecast the data of Figure 2.18(c) using simple exponential smoothing with a 10% smoothing constant ($\alpha = .1$).

*At t = 1:*

$$y_1 = 15$$
$$\hat{y}_2 = y_1 = 15 \; (= \hat{y}_{\text{init}})$$

*At t = 2:*

$$y_2 = 18$$
$$e_2 = y_2 - \hat{y}_2 = 18 - 15 = +3$$
$$\hat{y}_3 = (1 - \alpha)\hat{y}_2 + \alpha y_2 \qquad \text{(Smoothing Form)}$$
$$= .9 \times 15 + .1 \times 18 = 15.3$$

or

$$\hat{y}_3 = \hat{y}_2 + \alpha e_2 = 15 + .1 \times 3 = 15.3 \qquad \text{(Error Form)}$$

*At t = 3:*

$$y_3 = 19$$
$$e_3 = y_3 - \hat{y}_3 = 19 - 15.3 = +3.7$$
$$\hat{y}_4 = .9 \cdot \hat{y}_3 + .1 \cdot y_3 \qquad \text{(Smoothing Form)}$$
$$= .9 \times 15.3 + .1 \times 19 = 15.7$$

or

$$\hat{y}_4 = \hat{y}_3 + \alpha e_3 \qquad \text{(Error Form)}$$
$$= 15.3 + .1 \times 3.7 = 15.7$$

*At t = 4:*

$$y_4 = 20$$
$$e_4 = +4.3$$
$$\hat{y}_5 = .9 \times 15.7 + .1 \times 20 = 16.1$$

or

$$\hat{y}_5 = 15.7 + .1 \times 4.3 = 16.1$$

and so forth.

The complete series of forecasts is given in Table 2.26. By graphically comparing the forecasts with the actual series in Figure 2.21, it can be seen that using $\alpha = 10\%$ smooths the data quite a lot and tracks only very slowly, lagging far behind movements in the series. Raising the smoothing constant $\alpha$ to 50% (see Table 2.26 and Figure 2.21) improves the tracking markedly and reduces the MAD and RMSE. ●

**Table 2.26 Exponential Smoothing Forecasts of Positively Autocorrelated Series from Figure 2.18**

| | | $\alpha$ = 10% | | $\alpha$ = 50% | |
|---|---|---|---|---|---|
| $t$ | $y_t$ | $\hat{y}_t$ | $e_t$ | $\hat{y}_t$ | $e_t$ |
| 1 | 15 | — | — | — | — |
| 2 | 18 | 15.0 | 3.0 | 15.0 | 3.0 |
| 3 | 19 | 15.3 | 3.7 | 16.5 | 2.5 |
| 4 | 20 | 15.7 | 4.3 | 17.8 | 2.2 |
| 5 | 20 | 16.1 | 3.9 | 18.9 | 1.1 |
| 6 | 19 | 16.5 | 2.5 | 19.4 | −.4 |
| 7 | 16 | 16.8 | −.8 | 19.2 | −3.2 |
| 8 | 14 | 16.7 | −2.9 | 17.6 | −3.6 |
| 9 | 13 | 16.4 | −3.6 | 15.8 | 2.8 |
| 10 | 14 | 16.0 | −2.0 | 14.4 | −.4 |

## Simple Exponential Smoothing (SES)

**Given:** $y_1, y_2, \ldots, y_t$

**Forecast:**

$$\hat{y}_{t+1} = \alpha y_t + \alpha(1-\alpha)y_{t-1} + \alpha(1-\alpha)^2 y_{t-2} + \cdots + \alpha(1-\alpha)^k y_{t-k} + (1-\alpha)^{k+1}\hat{y}_{\text{init}}$$

where $\hat{y}_{\text{init}}$ = initial forecast of $y_{t-k}$, $\quad k \leq t-1$

$$\hat{y}_{t+p} = \hat{y}_{t+1} \qquad p \geq 1$$

Updating Equations

$$\hat{y}_{t+1} = (1-\alpha)\hat{y}_t + \alpha y_t \qquad \text{(Smoothing Form)}$$

$$\hat{y}_{t+1} = \hat{y}_t + \alpha e_t \qquad \text{(Error Correction Form)}$$

$$e_t = y_t - \hat{y}_t$$

Weighting Scheme I: (at time $t$, $\hat{y}_{\text{init}} = \hat{y}_2 = y_1$)

Weighting Scheme II: (at time $t$, $\hat{y}_{\text{init}} = \hat{y}_{t-k}$)

| | | | | | | |
|---|---|---|---|---|---|---|
| **Observation (I):** | $y_1$ | $y_2$ | | $y_{t-2}$ | $y_{t-1}$ | $y_t$ |
| **Weight (I):** | $(1-\alpha)^{t-1}$ | $\alpha(1-\alpha)^{t-2}$ | $\cdots$ | $\alpha(1-\alpha)^2$ | $\alpha(1-\alpha)$ | $\alpha$ |
| **Observation (II):** | $\hat{y}_{\text{init}}$ | $y_{t-k}$ | $\cdots$ | $y_{t-2}$ | $y_{t-1}$ | $y_t$ |
| **Weight (II):** | $(1-\alpha)^{k+1}$ | $\alpha(1-\alpha)^k$ | $\cdots$ | $\alpha(1-\alpha)^2$ | $\alpha(1-\alpha)$ | $\alpha$ |

**Center (Approx.):** $t - \dfrac{1-\alpha}{\alpha}$

**Average Age (Approx.):** $\dfrac{1-\alpha}{\alpha}$

**Figure 2.21** Original and Forecast Series, $\alpha$ = .1, .5 from Table 2.26

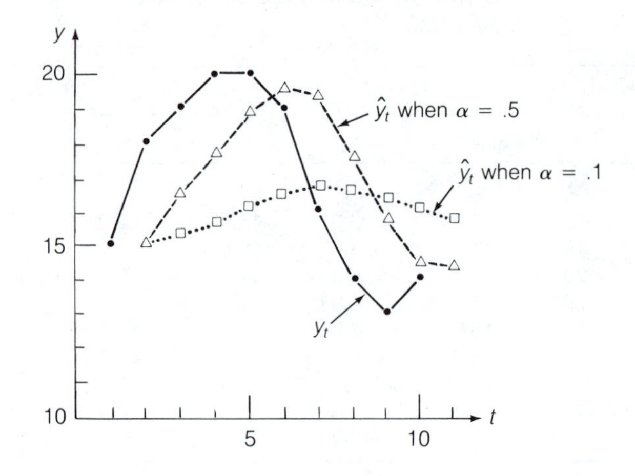


## EXAMPLE 2.13 *Forecon Associates, Inc.*

Forecon Associates, Inc., a consulting concern in the Midwest, would like, as part of its cash flow control efforts, to forecast each of several line items in its monthly budget. The telephone toll charges for the last 22 months are given in Table 2.27. A graph of the data (Figure 2.22) indicates the series has trend. Since the first differences, shown in Table 2.28 and Figure 2.23, appear to be untrended, it is decided to forecast them rather than the actual series. A horizontal model and exponential smoothing will be used to construct these forecasts.

**Table 2.27** Monthly Telephone Toll Charges, Forecon Associates, Inc., March 1986 thru December 1987

| Month | $t$ | Charge ($) $y_t$ | Month | $t$ | Charge ($) $y_t$ |
|---|---|---|---|---|---|
| Mar 86 | 1 | 326 | Feb 87 | 12 | 731 |
| Apr | 2 | 375 | Mar | 13 | 731 |
| May | 3 | 414 | Apr | 14 | 701 |
| Jun | 4 | 486 | May | 15 | 618 |
| Jul | 5 | 527 | Jun | 16 | 539 |
| Aug | 6 | 572 | Jul | 17 | 558 |
| Sep | 7 | 657 | Aug | 18 | 625 |
| Oct | 8 | 764 | Sep | 19 | 630 |
| Nov | 9 | 722 | Oct | 20 | 604 |
| Dec | 10 | 769 | Nov | 21 | 561 |
| Jan 87 | 11 | 785 | Dec | 22 | 581 |

**Figure 2.22** Forecon Toll Charge Data

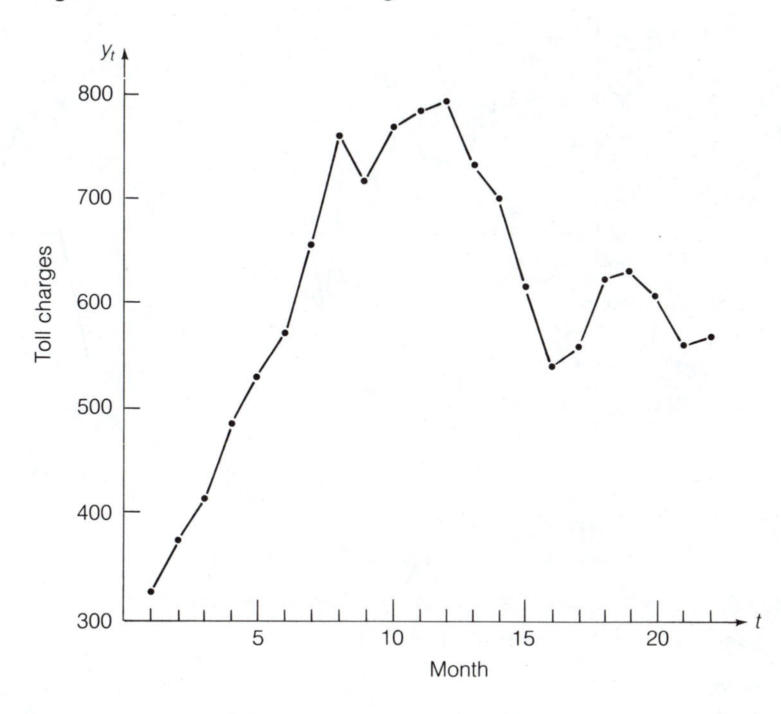


**Table 2.28** First Differences and Forecasts ($\alpha = 30\%$) of Telephone Toll Charges for Forecon Associates, Inc.

| Month $t$ | First Diff. $\Delta y_t$ | Forecast $\Delta \hat{y}_t$ | Month $t$ | First Diff. $\Delta y_t$ | *Forecast* $\Delta \hat{y}_t$ |
|---|---|---|---|---|---|
| 2 (Apr 86) | 49 | — | 13 (Mar 87) | −60 | 25.5 |
| 3 | 39 | 49.0 | 14 | −30 | −0.2 |
| 4 | 72 | 46.0 | 15 | −83 | −9.1 |
| 5 | 41 | 53.8 | 16 | −79 | −31.3 |
| 6 | 45 | 50.0 | 17 | 19 | −45.4 |
| 7 | 85 | 48.5 | 18 | 67 | −26.2 |
| 8 | 107 | 59.5 | 19 | 5 | 1.8 |
| 9 | −42 | 73.8 | 20 | −26 | 2.8 |
| 10 | 47 | 39.1 | 21 | −43 | −5.8 |
| 11 | 16 | 41.5 | 22 | 20 | −17.0 |
| 12 | 6 | 33.9 | 23 (Jan 88) | ? | −5.9 |

Notice that the first differences for $t = 1$ (March 1986) would be $y_{\text{Mar}} - y_{\text{Feb}}$. This value cannot be computed, since data for February 1986 are not available. A smoothing constant of 30% ($\alpha = .3$) is used to forecast $\Delta y_{23}$, starting with $\Delta \hat{y}_3 = \Delta \hat{y}_{\text{init}} = \Delta y_2$:

**Figure 2.23** First Differences of Forecon Toll Charge Data

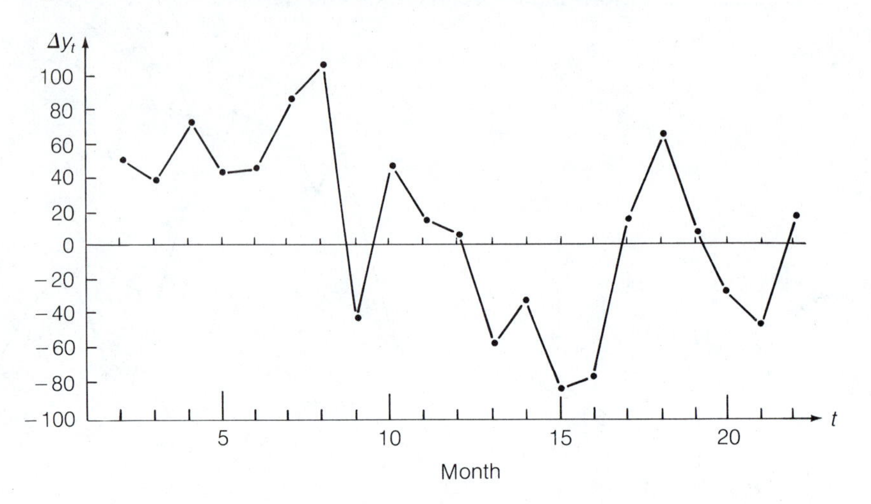


*At* t = *2 (Apr 86):*

$$\Delta y_2 = 49.0$$
$$\Delta\hat{y}_3 = \Delta\hat{y}_{\text{init}} = \Delta y_2 = 49.0$$

*At* t = *3 (May 86):*

$$\Delta y_3 = 39 \qquad e_3 = 39 - 49 = -10.0$$
$$\Delta\hat{y}_4 = 49 + .3(-10.0) = 46.0$$

*At* t = *4 (Jun 86):*

$$\Delta y_4 = 72 \qquad e_4 = 72 - 46.0 = +26.0$$
$$\Delta\hat{y}_5 = 46.0 + .3 \times 26.0 = 53.8$$

continuing until

*At* t = *22 (Dec 87):*

$$\Delta y_{22} = 20 \qquad e_{22} = 20 - (-17.0) = +37.0$$
$$\Delta\hat{y}_{23} = -17.0 + .3 \times 37.0 = -5.9$$

Note that the −5.9 above is a forecast of the *first difference* of the toll charges for January 1988, not of the toll charges themselves. This means we have forecast January 1988 toll charges to be \$5.90 less than December 1987 charges. The forecast for January 1988 charges themselves would then be $y_{22} + \Delta\hat{y}_{23} = \$581 - \$5.90 = \$575.10$.

Strictly speaking, since a horizontal model for first differences is being used, all the forecasts of future first differences would be the same, i.e., $\Delta\hat{y}_{23} = \Delta\hat{y}_{24} = \cdots$, and thus forecasts for February, March, and April 1988 would be

| | |
|---|---|
| February | \$575.10 − \$5.90 = \$569.20 |
| March | \$569.20 − \$5.90 = \$563.30 |
| April | \$563.30 − \$5.90 = \$557.40 |

## 2.10.3 Choosing the Initial Forecast

The choice of $\hat{y}_{\text{init}}$ always affects the value of subsequent forecasts (Abraham and Ledolter, 1983). Choosing $y_1$ for an initial forecast, for example, gives $y_1$ too much weight in later forecasts. Generally speaking, though, the effect of

**Table 2.29 Weight Given to First Observation for an Observed Sample of Length $t$ When $y_1$ Is Used as the Initial Forecast**

| | Actual Weight Given to First Observation* | | | | | |
|---|---|---|---|---|---|---|
| Length $t$ | $1/\alpha = 10$<br>$\alpha = 0.1$ | 5.0<br>.2 | 3.5<br>.3 | 2.0<br>.5 | 1.43<br>.7 | 1.11<br>.9 |
| 1 | 100% | 100% | 100% | 100% | 100% | 100% |
| 2 | 90 | 80 | 70 | 50 | 30 | 10 |
| 3 | 81 | 64 | 49 | 25 | 9 | 1 |
| 4 | 73 | 51 | 34 | 13 | 3 | <1% |
| 6 | 59 | 33 | 17 | 3 | <1% | |
| 8 | 48 | 21 | 8 | 1 | | |
| 10 | 39 | 13 | 4 | <1% | | |
| 15 | 23 | 4 | 1 | | | |
| 20 | 14 | 1 | <1% | | | |
| 30 | 5 | <1% | | | | |
| 50 | <1% | | | | | |

* The weight that $y_1$ would normally receive is $\alpha$ times the tabulated weight, or the tabulated weight is $1/\alpha$ times the normal weight.

$\hat{y}_{\text{init}}$ on $\hat{y}_{t+1}$ diminishes as $t$ increases. That is, the influence of the initial forecast "washes out" as forecasting progresses. Table 2.29 demonstrates this phenomenon in the case where $\hat{y}_{\text{init}} = y_1$.

## The Average of the First Several Observations

The average of the first $m$ observations may also be used for $\hat{y}_{\text{init}}$. Smoothing then begins with

$$\hat{y}_{\text{init}} = \hat{y}_{m+1} = \bar{y}_m$$

and continues until $\hat{y}_{t+1}$ is obtained.

**EXAMPLE 2.14** *SES Forecasts of Forecon Data Using the Average of the First Ten First Differences as the Initial Forecast*

We will use the Forecon toll charge data as an example, changing the notation a bit to simplify the expressions. Let $t = 1$ in April 1986, since that is the first period for which $\Delta y_t$ is available, and for simplicity let $d_t = y_t - y_{t-1} = \Delta y_t$ (Table 2.30).

Using $m = 10$, the first forecast, $\hat{d}_{11} = \hat{d}_{\text{init}} = \bar{d}_{10}$, would be

$$\hat{d}_{11} = \bar{d}_{10} = \frac{d_1 + d_2 + \cdots + d_{10}}{10} = \frac{49 + 39 + \cdots + 16}{10} = 45.9$$

Then, using $\alpha = .3$, the forecasts are:

*At t = 11:* $d_{11} = 6 \qquad e_{11} = 6 - 45.9 = -39.9$

$\hat{d}_{12} = 45.9 + .3(-39.9) = +33.3$

*At t = 12:* $d_{12} = -60 \qquad e_{12} = -60 - 33.3 = -93.3$

$\hat{d}_{13} = 33.3 + .3(-93.3) = +5.3$

**Table 2.30 First Differences of Forecon Toll Charge Data (Revised Notation)**

| $t$ | $d_t$ | $t$ | $d_t$ |
|---|---|---|---|
| 1 (Apr 86) | 49 | 12 (Apr 87) | −60 |
| 2 | 39 | 13 | −30 |
| 3 | 72 | 14 | −83 |
| 4 | 41 | 15 | −79 |
| 5 | 45 | 16 | 19 |
| 6 | 85 | 17 | 67 |
| 7 | 107 | 18 | 5 |
| 8 | −42 | 19 | −26 |
| 9 | 47 | 20 | −43 |
| 10 | 16 | 21 | 20 |
| 11 | 6 | 22 (Jan 88) | ? |

**Table 2.31 Forecasts Using the Average of the First Ten First Differences as the Initial Forecast of First Differences for Forecon Toll Charge Data**

| $t$ | $d_t$ | $\hat{d}_t$ | $e_t$ | $t$ | $d_t$ | $\hat{d}_t$ | $e_t$ |
|---|---|---|---|---|---|---|---|
| 11 | 6 | 45.9 | −39.9 | 17 | 67 | −24.9 | 91.9 |
| 12 | −60 | 33.3 | −93.3 | 18 | 5 | 2.7 | 2.3 |
| 13 | −30 | 5.3 | −35.3 | 19 | −26 | 3.4 | −29.4 |
| 14 | −83 | −5.3 | −77.7 | 20 | −43 | −5.4 | −37.6 |
| 15 | −79 | −28.6 | −50.4 | 21 | 20 | −16.7 | 36.7 |
| 16 | 19 | −43.7 | 62.7 | 22 | ? | −5.7 | ? |

and so on, giving the results in Table 2.31. The forecast for January 1988 toll charges would then be $y_{\text{Dec87}} + \hat{d}_{22} = \$58 - \$5.70 = \$575.30$, which is very nearly that obtained using $\hat{d}_{\text{init}} = d_1$ ($\Delta\hat{y}_{\text{init}} = \Delta y_1$) because of the length of the observed series (21 observations). ●

## The "Backcast" for Time Zero

An initial forecast may also be constructed by computing a **"backcast" for time zero.** To do this we act as though the series were reversed, i.e., as though $y_t$ were the earliest observation, $y_{t-1}$ the next, $y_{t-2}$ the next, and so forth. The forecast (which is actually a "backcast") for $y_0$ is then used as the initial forecast $\hat{y}_{\text{init}}$. The first true forecast then is $\hat{y}_1 = \hat{y}_{\text{init}} = \hat{y}_0$, and $\hat{y}_{t+1}$ is computed in the normal manner (Abraham and Ledolter, 1983). A brief example, using $\alpha = 20\%$ and the following time series, should clarify the process. (If a computer program is being used for the smoothing, one could merely enter the data in reverse order.)

| Year $t$ | Sales ($1,000) $y_t$ | Year $t$ | Sales ($1,000) $y_t$ |
|---|---|---|---|
| 1 | 20.6 | 4 | 31.4 |
| 2 | 16.4 | 5 | 28.2 |
| 3 | 38.3 | 6 | 19.4 |

The "backwards" updating equations are

$$\hat{y}_{t-1} = (1 - \alpha)\hat{y}_t + \alpha y_t \qquad \text{(Smoothing Form)}$$
$$= \hat{y}_t + \alpha e_t \qquad \text{(Error Form)}$$

and the latest (most recent) observation is used as the initial "backcast."

*At Year 6:* $y_6 = 19.4$

$\hat{y}_5 = 19.4$

*At Year 5:* $y_5 = 28.2 \qquad e_5 = 28.2 - 19.4 = 8.8$

$\hat{y}_4 = y_5 + .2e_5 = 28.2 + .2 \times 8.8 = 30.0$

*At Year 4:* $y_4 = 31.4 \qquad e_4 = 31.4 - 30.0 = 1.4$

$\hat{y}_3 = \hat{y}_4 + .2e_4 = 30.0 + .2 \times 1.4 = 30.3$

*At Year 3:* $y_3 = 38.3 \qquad e_3 = 38.3 - 30.3 = 8.0$

$\hat{y}_2 = 30.3 + .2 \times 8.0 = 31.9$

*At Year 2:* $y_2 = 16.4 \qquad e_2 = 16.4 - 31.9 = -15.5$

$\hat{y}_1 = 31.9 + .2(-15.5) = 28.8$

*At Year 1:* $y_1 = 20.6 \qquad e_1 = 20.6 - 28.8 = -8.2$

$\hat{y}_0 = 28.8 + .2(-8.2) = 27.2 = \hat{y}_{\text{init}}$

This value, $\hat{y}_{\text{init}} = 27.2$, is then used as $\hat{y}_1$, and forecasts are computed as usual, i.e., with $\alpha = .2$ again, $y_1 = 20.6$, $e_1 = 20.6 - 27.2 = -6.6$, $\hat{y}_2 = 27.2 + .2(-6.6) = 25.9, \ldots$, giving the following forecast through Year 7:

| $t$ | $y_t$ | $\hat{y}_t$ | $e_t$ |
|---|---|---|---|
| 1 | 20.6 | 27.2 | −6.6 |
| 2 | 16.4 | 25.9 | −9.5 |
| 3 | 38.3 | 24.0 | 14.3 |
| 4 | 31.4 | 26.9 | 4.5 |
| 5 | 28.2 | 27.8 | 0.4 |
| 6 | 19.4 | 27.9 | −8.5 |
| 7 | ? | 26.2 | ? |

*Note:* MSE = 77.48, RMSE = 8.80, based on $t = 2$ through $t = 6$.

The common method, using $\hat{y}_{\text{init}} = \hat{y}_2 = y_1$, would have given

| $t$ | $y_t$ | $\hat{y}_t$ | $e_t$ |
|---|---|---|---|
| 1 | 20.6 | — | — |
| 2 | 16.4 | 20.6 | −4.2 |
| 3 | 38.3 | 19.8 | 18.5 |
| 4 | 31.4 | 23.5 | 7.9 |
| 5 | 28.2 | 25.1 | 3.1 |
| 6 | 19.4 | 25.7 | −6.3 |
| 7 | ? | 24.4 | ? |

*Note:* MSE = 94.32, RMSE = 9.71, based on $t$ = 2 through $t$ = 6.

with a final forecast of 24.4 as opposed to 26.2 obtained using the "backcast" initial forecast. It is interesting to note that, if the "backcast" method is used with small values of $\alpha$ and $t$, the $\hat{y}_{\text{init}}$ obtained is about the same as if one had used as the initial forecast the overall series average $\bar{y}_t$.

Finally, if no data are available, other models or subjective evaluation may be used to get an initial forecast. Under these conditions one may wish to start with a large value for $\alpha$ to get the forecast tracking the series, then decrease $\alpha$ to the desired value after the first few updates.

## 2.10.4 Selection of $\alpha$

The other major problem in initiating exponential smoothing schemes is that of selecting an appropriate smoothing constant (Brown, 1963). The main consideration in this choice is the tradeoff between smoothing and tracking provided by different $\alpha$ values. More specifically, a large smoothing constant, while providing quick reaction to changes in $\beta_0$, will also effectively track random noise in the series and result in larger forecast errors if $\beta_0$ remains constant. On the other hand, a small smoothing constant, which lessens the tracking of noise, will cause the forecast to lag considerably behind any changes in $\beta_0$ that may occur.

### Equivalent-Length Moving Average

One of the reasons exponential smoothing is so popular is that it provides a way of achieving a moving average forecast without requiring that a lot of observations be carried from one time period to the next. Because of the weights that result from the exponential discounting, small values of $\alpha$ give rough equivalents of long-length moving averages and large values of $\alpha$ give short-length ones. A relationship between $\alpha$ and $k$, the length of a roughly equivalent unweighted moving average, may be derived by looking at the average age of the two types of forecasts.

For exponential smoothing with a given $\alpha$, the average age is $0 \times \alpha + 1\alpha(1-\alpha) + 2\alpha(1-\alpha)^2 + 3\alpha(1-\alpha)^3 + \cdots$, which is equal to $(1-\alpha)/\alpha$ if the series is extended to infinity and nearly that if $t$ is large. The actual weights and corresponding average for $\alpha = 0.1, 0.2, \ldots, 1.0$ are given in Table 2.32.

**Table 2.32** Weight Given to $y_{t-k} \rightarrow \alpha(1-\alpha)^k$ and Average Age of an Exponentially Weighted Moving Average for Given $\alpha$

| $\alpha$ | Weight Given to: $y_t$ | $y_{t-1}$ | $y_{t-2}$ | $y_{t-3}$ | $y_{t-4}$ | $y_{t-5}$ | $y_{t-6}$ | $y_{t-7}$ | $y_{t-8}$ | $y_{t-9}$ | Average Age $(1-\alpha)/\alpha$ |
|---|---|---|---|---|---|---|---|---|---|---|---|
| .1 | 10% | 9% | 8.1% | 7.3% | 6.6% | 5.9% | 5.3% | 4.8% | 4.3% | 3.8% | 9.00 |
| .2 | 20 | 16 | 12.8 | 10.2 | 8.2 | 6.6 | 5.2 | 4.2 | 3.3 | 2.7 | 4.00 |
| .3 | 30 | 21 | 14.7 | 10.3 | 7.2 | 5.0 | 3.5 | 2.5 | 1.7 | 1.2 | 2.33 |
| .4 | 40 | 24 | 14.4 | 8.6 | 5.2 | 3.1 | 1.9 | 1.1 | .7 | .4 | 1.50 |
| .5 | 50 | 25 | 12.5 | 6.2 | 3.1 | 1.6 | .8 | .4 | .2 | .1 | 1.00 |
| .6 | 60 | 24 | 9.6 | 3.8 | 1.5 | .6 | .2 | .1 | — | — | .67 |
| .7 | 70 | 21 | 6.3 | 1.9 | .6 | .2 | .1 | — | — | — | .43 |
| .8 | 80 | 16 | 3.2 | .6 | .1 | — | — | — | — | — | .25 |
| .9 | 90 | 9 | .9 | .1 | — | — | — | — | — | — | .11 |
| 1.0 | 100 | — | — | — | — | — | — | — | — | — | .00 |

**Table 2.33** Roughly Equivalent Values of $\alpha$ in SES and of $k$ in a Simple Moving Average

| $k$ | $\alpha = \frac{2}{k+1}$ | $\alpha$ | $k = \frac{2-\alpha}{\alpha}$ |
|---|---|---|---|
| 1 | 1.00 | .1 | 19.0 |
| 2 | .67 | .2 | 9.0 |
| 3 | .50 | .3 | 5.7 |
| 4 | .40 | .4 | 4.0 |
| 5 | .33 | .5 | 3.0 |
| 6 | .29 | .6 | 2.3 |
| 8 | .22 | .7 | 1.9 |
| 12 | .15 | .8 | 1.5 |
| 16 | .12 | .9 | 1.2 |
| 20 | .09 | 1.0 | 1.0 |
| 24 | .08 | | |
| 50 | .04 | | |
| 100 | .01 | | |

For a moving average of length $k$, the average age is $(k-1)/2$. If we equate this to $(1-\alpha)/\alpha$, we get

$$\alpha = \frac{2}{k+1} \quad \text{or} \quad k = \frac{2-\alpha}{\alpha}$$

Exponential smoothing with this value of $\alpha$ is roughly equivalent to a simple moving average of length $k$. Some equivalent values are given, for reference, in Table 2.33.

## Pragmatic Choice

It has also become quite common to choose $\alpha = 10\%$ as a kind of standard if other considerations do not call for a different value. This choice yields the equivalent of about an 18-period moving average. It smooths quite heavily but gives some tracking ability while avoiding wildly fluctuating forecasts, which may be difficult to defend/implement. If a jump in $\beta_0$ is anticipated, however, be aware that using $\alpha = 10\%$ will give forecasts that "home in" only very slowly on the new value.

### Minimum RMSE

If the requisite time and computing power are available, one may select the smoothing constant to optimize some performance measure. Usually the RMSE (or, equivalently, the MSE or SSE) is chosen. This gives a "least squares" value of $\alpha$. Although there is no explicit expression for computing this optimal $\alpha$, it can be approximated by selecting several different values and using the one that gives the best performance. The RMSE values for $\alpha$ between 0.1 and 1.0 (using $\hat{y}_{\text{init}} = y_1$) are given and graphed in Figure 2.24 for the Forecon toll charge data of Example 2.13. These computations indicate that a smoothing constant of around 35% gives the smallest RMSE. As is often the case, the RMSE function is relatively flat in the vicinity of the optimal value, so knowing the "exact" optimal value is not critical.

**Figure 2.24** **Root Mean Square Errors Given by Different Values of $\alpha$ for Forecon Toll Charges**

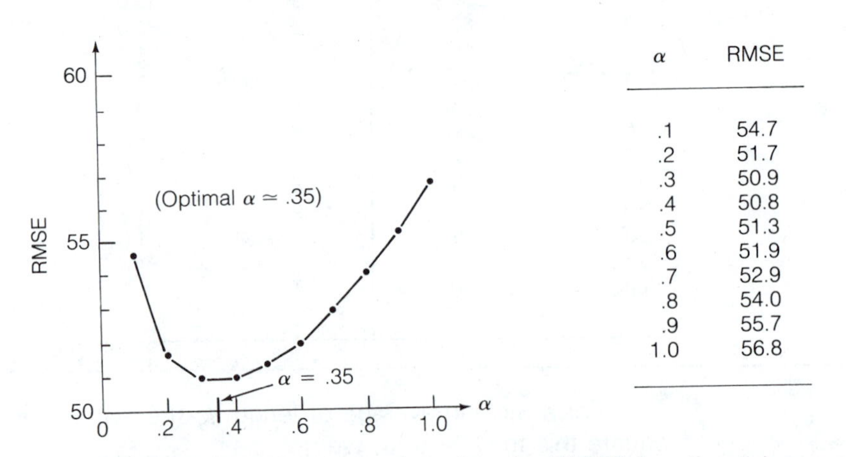


| $\alpha$ | RMSE |
|---|---|
| .1 | 54.7 |
| .2 | 51.7 |
| .3 | 50.9 |
| .4 | 50.8 |
| .5 | 51.3 |
| .6 | 51.9 |
| .7 | 52.9 |
| .8 | 54.0 |
| .9 | 55.7 |
| 1.0 | 56.8 |

It also happens for some data that the RMSE function takes one of the forms shown in Figure 2.25. A graph such as that in part (a) would indicate that the series is mostly "noise." Tracking would be needed only if or when $\beta_0$ undergoes a change. A trended series, on the other hand, or one that has very pronounced autocorrelation, will give a graph similar to Figure 2.25(b). If trend is present, it should be detected in the preliminary analysis of the data and would be better handled using one of the models of Chapter 3. Oscillatory movement with no trend resembles frequent step changes in $\beta_0$. Since tracking ability would then be of primary importance, large values of $\alpha$ would be desirable.

Another factor that should be considered when using this method is that

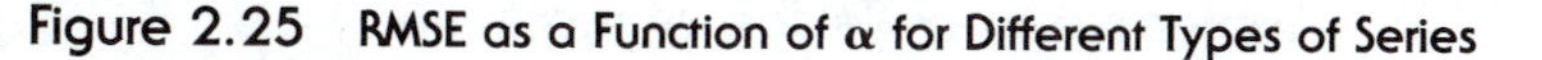

Figure 2.25 RMSE as a Function of α for Different Types of Series

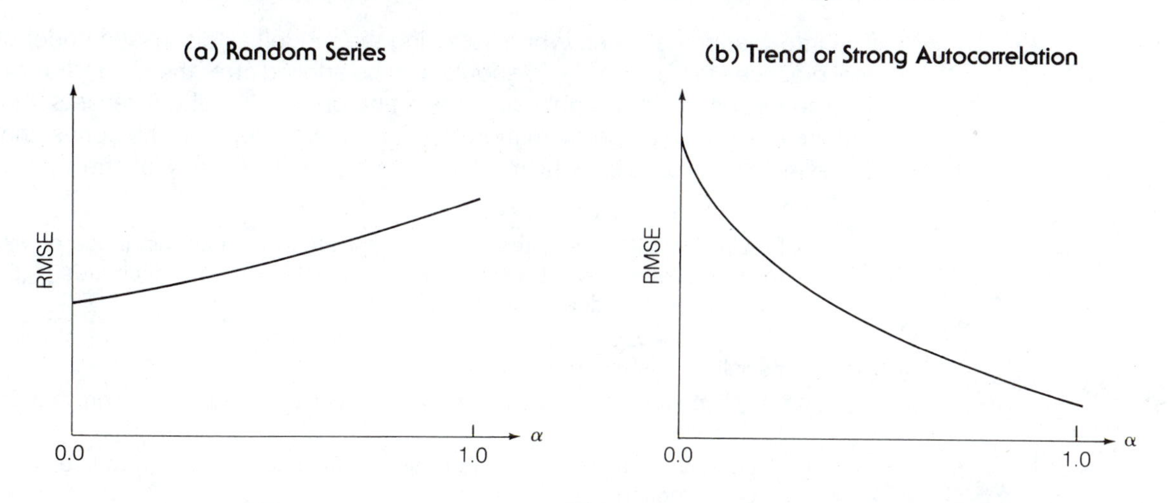


the initial forecast will affect the choice of $\alpha$. Because of the fast tracking needed early in the series to correct it, a poor initial forecast will tend to lead to a larger smoothing constant than might otherwise be chosen. Large forecast errors early in the process, which moderate later on, would indicate that a better initial forecast should be sought or $\alpha$ chosen somewhat less than the "least squares" value. As mentioned before, however, if the sample size is reasonably large, this problem should be a minor one.

## 2.11 Diagnostics

### 2.11.1 Purpose

This section includes some advice about models described earlier in the chapter and some suggested procedures for checking model adequacy. Because *any* model may be fit to *any* data, it is important that the results of such fitting be examined with a critical eye toward judging whether the model will satisfy the demands of the forecaster. Is the model *appropriate* for the specific context in which it is being applied? For example, in the case of horizontal models, is it reasonable to suppose that the average level of the series being modeled is stable? How *accurate* will the forecasts be? Will the model forecast with *sufficient* accuracy? Could the accuracy be easily *improved?*

### 2.11.2 Tools for Model Appraisal

There are two main sources of information for model appraisal: (a) subjective knowledge of the model and series, and (b) empirical knowledge of the scheme's previous and ongoing accuracy.

## Subjective Knowledge of the Series

The information gathered when selecting the model, as discussed under general practices in Section 2.4.2, should also be applied after the model has been fit. The values of the model parameters and behavior of the forecasts should reflect a reasonable approximation to your understanding of the series and its behavior. For example, your model might be faulty for any of the following reasons.

- Previous experience with the series indicates that a certain range of values for the average level $\beta_0$ would be appropriate, yet your estimates $\hat{\beta}_0(t)$ are falling near the edge or outside of that range.
- The optimal value of $k$ for a moving average scheme seems out of line, especially if it is too low.
- Very large (or very small) values of the exponential smoothing constant seem to be called for.
- Successive estimates of $\beta_0$ from an updating scheme seem to keep drifting in one direction.

Your model might also be faulty if any other behavior or characteristics of the forecasts seem to contradict your understanding of either the model or the series being forecast.

## Empirical Knowledge of the Series

The history of the series $y_1, y_2, \ldots, y_n$ that was used for model fitting also contains very important information for model appraisal. The general idea is to see how the forecasting scheme would have performed had it been used to forecast all or part of the history. To do this, a forecast $\hat{y}_t$ for each historical value is generated and the corresponding forecast errors $e_t = y_t - \hat{y}_t$ determined. These forecast errors then provide a quantitative measure of how well the scheme forecasts. In examining the historical forecast errors keep in mind one thing that may be easily overlooked: Some parameter of the underlying model may have been estimated directly from the same data for which forecast errors are now being computed. Updated mean, median, and midrange forecasts do not present this problem, nor do moving average or exponential-smoothing schemes, provided the data were not used to determine an optimal length $k$ or smoothing constant $\alpha$. However, single-forecast schemes, moving averages where $k$ is optimized for the data, and exponential-smoothing schemes where a "least squares" value of $\alpha$ is used are subject to this problem. In the latter cases, the model has been especially designed to optimize some criterion that is a function of these same forecast errors. In a sense, the model is "custom-fit" to the sample history, so judging its performance based on historical $e_t$'s tends to give an unduly rosy picture of how the scheme will perform in the future.

As a matter of fact, some analysts reserve a portion of the history, usually the most recent portion (often called a "test set"), to be used for later appraisal (Makridakis, Wheelwright, and McGee, 1983, p. 122). The parameters are then estimated solely from the unreserved part of the history, and forecast errors are computed only for the reserved portion. The shortcoming of this technique is that an appreciable portion of the sample must be excluded from influencing the estimates of the parameters. This problem is quite acute when only a small history is available to begin with. A related difficulty is that most analysts, when initiating a forecasting procedure, would feel more secure if the most recent observations *did* influence parameter estimates. It would even make good sense to give the most recent observations more weight than older ones. The more common practice is to include all the data in estimation and appraisal procedures.

### 2.11.3 Initial Step

After the forecast errors have been computed, at minimum a diagnostic graph of $e_t$ versus $t$ (a residual plot) should be drawn. Since the $e_t$'s for a well-functioning scheme should form a horizontal random series with a mean of zero, this graph should not show any clearly discernible pattern. It should display the characteristics mentioned in Section 2.2.2. In particular, there should be no obvious trend or autocorrelation (oscillations). The presence of either is an indication that the scheme could be improved. An attempt at improvement may simply involve refitting a parameter or selecting another horizontal scheme (e.g., changing from updated-mean to moving average forecasts), or it may require using a more complex model. If you are satisfied that there are no clear patterns in the residual plot, you may wish to do no further analysis.

### 2.11.4 Further Steps

If there is any doubt, or if, for any reason, a more objective and detailed analysis is desired, the following three additional steps are highly recommended.

1. Construct a diagnostic graph of $y_t$ versus $\hat{y}_t$.
2. Compute the acf of the residuals.
3. Compute Thiel's $U$ and, for updating schemes, Theil's decomposition of the MSE ($U_M$, $U_R$, and $U_D$).

The interpretation of the graph of $y_t$ versus $\hat{y}_t$ and of Theil's $U$ and decomposition components were discussed in Section 1.5. The acf was covered in Section 2.6.4. The precise effect of fitting a horizontal model to a trended, seasonal, or autocorrelated series depends upon the forecasting scheme. It also depends on the nature of the trend, seasonality, or autocorrelation, and the amount of random variation in the series. However, a few general statements may be made.

- For single forecasts, an upward trend will show up in the graph of $y_t$ versus $\hat{y}_t$ as a tendency to overforecast smaller values of $y$ (early observations) and underforecast larger values of $y$ (later observations), vice versa for a downward trend. $U_M$ may well be zero (if $\hat{y}_t = \bar{y}$) or near zero, but $U_R$ will reflect the forecasting inaccuracies. The acf may resemble that of a trended series (recall the characteristics of the acf in Table 2.10).

- For updating schemes, trend will usually result in a tendency to consistently underforecast (or overforecast) $y_t$. This will show up in the graph of $y_t$ versus $\hat{y}_t$ and in Theil's $U_M$ component. The acf, however, may be quite similar to that of random series.
- Seasonal variation and positive autocorrelation will show up in the graph of $y_t$ versus $\hat{y}_t$ as a tendency to overforecast small values of $y$ and underforecast large ones. $U_R$ will generally be inflated. $U_M$ may or may not be near zero. The acf may resemble that of a seasonal or autocorrelated series (Table 2.10). The more tracking the scheme provides, the less marked will be these effects.
- A value of around 1.00 or more for Theil's $U$ probably indicates the presence of trend, seasonality, or autocorrelation, because it means your model is not tracking quickly enough to compare favorably with naive forecasting.

### 2.11.5 Accuracy

To measure model accuracy, either the MSE, MAD, MAPE, or max $|e_t|$ may be used, according to the analyst's preference. Time and computer resources permitting, though, it is good practice to compute all four and compare them, since each contains slightly different information about forecast accuracy. It is also helpful to know the value of $\bar{e}$, which is an overall measure of forecast bias.

Finally, any of the other techniques of measuring model performance discussed in Section 1.5 should be considered, since each may be of particular interest under specific sets of circumstances.

### 2.11.6 Yule-Slutsky Effect

In connection with interpreting residual graphs and autocorrelation functions, you should be aware that updating schemes usually introduce autocorrelation into the forecasts even if none were present in the original time series. This autocorrelation can cause drift in the forecasts and may lead one to erroneously conclude that drift exists in the original data.

In particular, running a moving average through purely random data will produce forecasts that are autocorrelated for lags up to the length of the moving average. The phenomenon is called the **Yule-Slutsky effect,** after the two statisticians who studied it (Slutsky, 1937). If the actual series is random and $\hat{y}_t$ is computed from a moving average of length $k$, the theoretical autocorrelation coefficient between forecasts that are $j$ time periods apart is (Gilchrist, 1976, p. 119)

$$\begin{aligned} \rho_j &= \frac{k-j}{k} && \text{for } j < k \\ &= 0 && \text{for } j \geq k \end{aligned}$$

Intuitively, the reason for this correlation is that successive updates are influenced to a greater or lesser degree by common observations. If we used a moving average of length 3 and started forecasting at time $t = 4$, for example, we would have

$$\hat{y}_4 = \frac{y_1 + y_2 + y_3}{3}$$

$$\hat{y}_5 = \frac{y_2 + y_3 + y_4}{3}$$

$$\hat{y}_6 = \frac{y_3 + y_4 + y_5}{3}$$

$$\hat{y}_7 = \frac{y_4 + y_5 + y_6}{3}$$

$$\vdots$$

In this computation, $\hat{y}_4$ has two observations ($y_2$ and $y_3$) in common with $\hat{y}_5$, one observation ($y_3$) in common with $\hat{y}_6$, and no observations in common with $y_7, y_8, \ldots$. Thus $\hat{y}_4$ will be most highly correlated to $\hat{y}_5$, next most highly to $\hat{y}_6$, and uncorrelated with $\hat{y}_7, \hat{y}_8, \ldots$ ($\rho_1 = \frac{2}{3}$, $\rho_2 = \frac{1}{3}$, $\rho_3 = 0, \ldots$). In general, adjacent observations have $k - 1$ observations in common, forecasts two periods apart have $k - 2$ in common, and so forth until, finally, observations $k$ periods apart have none in common and are uncorrelated. Although the effect is usually associated with moving averages, it is present to some degree whenever the computation of successive updates involves the use of common observations.

If the forecasts are correlated, the residuals most likely will be as well (since $e = y - \hat{y}$). For instance, when forecasting with a moving average of length $k$, the residuals from a random series will have autocorrelations

$$\begin{aligned} \rho_j &= \text{theoretical autocorrelation between } j\text{-step-apart residuals } e_t \text{ and } e_{t-j} \\ &= \frac{-j}{k(k+1)} \qquad \text{for } 1 \le j < k \\ &= 0 \qquad \text{for } j \ge k \end{aligned}$$

Although the magnitude of the autocorrelations will generally be reduced for residuals, you may well see its effect in the acf and, less obviously, in the residual plot. The phenomenon is mentioned here because it illustrates that care must be taken when interpreting diagnostic tools.

### 2.11.7 The Problem of Lagging

The forecasting schemes resulting from horizontal models all suffer to a certain extent from the problem of lagging behind autocorrelative drift or model parameter changes (changes in $\beta_0$). Even schemes that do no smoothing, such as moving averages of length 1, or exponentially smoothed averages with $\alpha = 1$ (naive forecasts), although they track very quickly, still lag one time period behind. Users of horizontal models should expect to encounter this problem and its associated effects (systematic under- or overforecasts) whenever autocorrelation or parameter changes occur. If it is imperative to minimize lag, more complex models (with their associated difficulties) described in later chapters are generally required.

## 2.11.8 An Example

The example below is an illustration of a diagnostic analysis. While it is impossible to include all eventualities in this example, it gives a general idea of the process.

**EXAMPLE 2.15** *Scott Enterprises, Inc.*

Scott Enterprises, Inc., is owned by an entrepreneur who conducts a small mail-order business through magazine advertisements and direct mail. The company markets a line of customized rubber stamps. The history of its orders for a particular type of small, self-inking stamp is given in Table 2.34. Since actual dollar sales of the stamp were affected by seasonal influences and had autocorrelative drift due to the appearance of advertisements and mailings, the variable was expressed as a percentage of gross dollar sales (of the entire product line); therefore a horizontal model would be more

**Table 2.34** **Scott Enterprises Sales ($) of Self-Inking Rubber Stamp #S1009A as a Percentage of Gross Sales**

| Month $t$ | Percent of Sales $y_t$ | *Month* $t$ | Percent of Sales $y_t$ |
|---|---|---|---|
| 1 | 3.28% | 10 | 2.61 |
| 2 | 2.71 | 11 | 2.48 |
| 3 | 1.87 | 12 | 3.49 |
| 4 | 3.69 | 13 | 3.80 |
| 5 | 1.90 | 14 | 3.59 |
| 6 | 3.32 | 15 | 5.22 |
| 7 | 4.19 | 16 | 3.92 |
| 8 | 2.52 | 17 | 3.51 |
| 9 | 3.77 | 18 | 2.84 |

**Figure 2.26** **Scott Enterprises Sales ($) of Self-Inking Rubber Stamp #S1009A as a Percentage of Gross Sales**

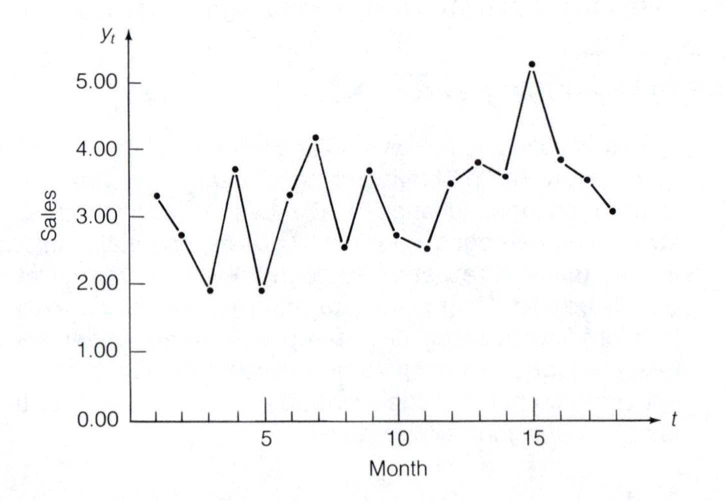

appropriate. A graph of the $y_t$'s (Figure 2.26) and several statistical procedures failed to reveal any trend or other nonstationarity in the data, so it was decided that a horizontal model would probably be adequate for forecasting purposes.

A variety of schemes was tested and appraised, including moving averages of length 1, 3, and 5; the updated mean; and exponential smoothing with $\alpha = 10\%$. Forecast errors for each of these methods were computed and graphed in Figure 2.27(a)–(e). These residual plots do not show any obvious patterns, but the errors for

**Figure 2.27** Residual Plots for Scott Enterprises Sales Data

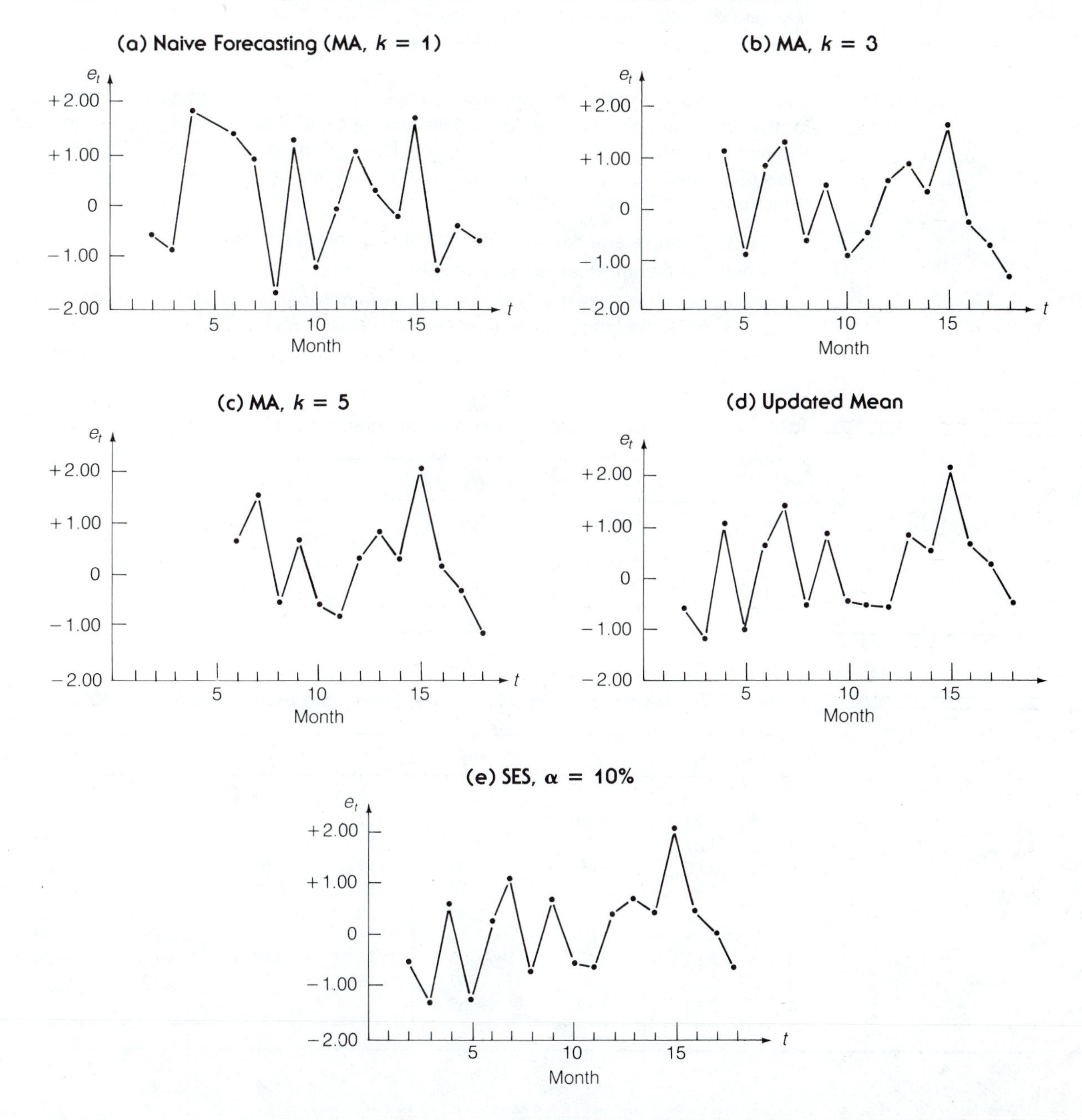

**Table 2.35** Evaluative Measures for Different Forecasting Schemes Used with Scott Enterprises Sales Data

| Scheme | RMSE | MAD | max $\|e_t\|$ | $\bar{e}$ | $U$ |
|---|---|---|---|---|---|
| MA, $k = 1$ | 1.142 | 1.004 | 1.82 | **−.026** | 1.000 |
| MA, $k = 3$ | .893 | .813 | **1.59** | .114 | **.567** |
| MA, $k = 5$ | .916 | .772 | 2.03 | .239 | .747 |
| Updated Mean | .906 | .790 | 2.13 | .257 | .629 |
| SES, $\alpha = 10\%$ | **.861** | **.738** | 2.07 | .072 | .569 |

*Note:* Boldface value is the minimum for that measure.

naive forecasting and for a moving average of length 3 do seem somewhat larger than for the other methods. A more detailed analysis helps to choose among the schemes. Values of the RMSE, MAD, max $|e_t|$, $\bar{e}$, and Theil's $U$ are given in Table 2.35. On the basis of these data, SES with $\alpha = 10\%$ was chosen as the probable best forecasting scheme, for the following reasons:

- It is simple and has minimal data storage requirements.
- It has the smallest RMSE and MAD.
- Although its forecast bias $\bar{e}$ is slightly larger than for naive forecasting (MA, $k = 1$), the latter has an unacceptably large RMSE and MAD.
- Theil's $U$ is only very slightly larger than for the moving average of length 3.

**Table 2.36** Autocorrelation Function, Scott Enterprises

| lag $j$ | $r_j$ | lag $j$ | $r_j$ |
|---|---|---|---|
| 1 | .066 | 5 | −.034 |
| 2 | .161 | 6 | .086 |
| 3 | .138 | 7 | −.053 |
| 4 | −.384 | 8 | .214 |

**Table 2.37** Residuals for Scott Enterprises Sales Data Using SES with $\alpha = 10\%$

| $t$ | $y_t$ | $\hat{y}_t$ | $e_t$ | $t$ | $y_t$ | $\hat{y}_t$ | $e_t$ |
|---|---|---|---|---|---|---|---|
| 1 | 3.28 | — | — | 11 | 2.48 | 3.11 | −.63 |
| 2 | 2.71 | 3.28 | −.57 | 12 | 3.49 | 3.05 | .44 |
| 3 | 1.87 | 3.22 | −1.35 | 13 | 3.80 | 3.09 | .71 |
| 4 | 3.69 | 3.09 | .60 | 14 | 3.59 | 3.16 | .43 |
| 5 | 1.90 | 3.15 | −1.25 | 15 | 5.22 | 3.21 | 2.07 |
| 6 | 3.32 | 3.02 | .30 | 16 | 3.92 | 3.41 | .51 |
| 7 | 4.19 | 3.05 | 1.14 | 17 | 3.51 | 3.46 | .05 |
| 8 | 2.52 | 3.17 | −.65 | 18 | 2.84 | 3.46 | −.62 |
| 9 | 3.77 | 3.10 | .67 | 19 | ? | 3.40 | ? |
| 10 | 2.61 | 3.17 | −.56 | | | | |

Since the MA, $k = 3$ scheme was still considered a possible alternative if significant autocorrelation were present in the original series, the acf for $y_t$ was computed and is given in Table 2.36. None of the sample autocorrelations tested significant, so the need for a quick-tracking model was obviated and SES with $\alpha = 10\%$ was chosen. As a final check, the acf of the residuals from the SES model were determined (Tables 2.37 and 2.38) and Theil's decomposition components $U_M$, $U_R$, and $U_D$ were computed (from the data in Table 2.37) as follows:

$$\Sigma \hat{y} = 54.20 \qquad \Sigma \hat{y}^2 = 173.11 \qquad \bar{\hat{y}} = 3.188$$

$$SS_{\hat{y}\hat{y}} = \Sigma \hat{y}^2 - \frac{(\Sigma \hat{y})^2}{n} = 173.11 - \frac{(54.20)^2}{17} = .308$$

$$s_{\hat{y}}'^2 = \frac{SS_{\hat{y}\hat{y}}}{n} = \frac{.308}{17} = .0181 \qquad s_{\hat{y}}' = \sqrt{.0181} = .135$$

$$\Sigma y = 55.43 \qquad \Sigma y^2 = 192.68 \qquad \bar{y} = 3.261$$

$$SS_{yy} = \Sigma y^2 - \frac{(\Sigma y)^2}{n} = 192.68 - \frac{(55.43)^2}{17} = 11.946$$

$$s_y'^2 = \frac{SS_{yy}}{n} = \frac{11.946}{17} = .703 \qquad s_y' = \sqrt{.703} = .838$$

$r_{y\hat{y}} = -.069$ (This is the coefficient of correlation between $y$ and $\hat{y}$, which can be computed by substituting $\hat{y}$ for $x$ in the formula given in Section 2.6 for the coefficient of correlation between $x$ and $y$.)

$$\Sigma e_t^2 = 12.611 \qquad MSE = \frac{\Sigma e_t^2}{n} = \frac{12.611}{17} = .742$$

$$U_M = \frac{(\bar{\hat{y}} - \bar{y})^2 \times 100\%}{MSE} = \frac{(3.188 - 3.261)^2 \times 100\%}{.742} = 0.72\%$$

$$U_R = \frac{(s_{\hat{y}}' - r_{y\hat{y}} \cdot s_y')^2 \times 100\%}{MSE} = \frac{[.135 - (-.069) \times .838]^2 \times 100\%}{.742} = 5.01\%$$

$$U_D = \frac{(1 - r_{y\hat{y}}^2)\, S_y'^2 \times 100\%}{MSE} = \frac{[1 - (-.069)^2] \times .703 \times 100\%}{.742} = 94.29\%$$

**Table 2.38 Residual Autocorrelation Function, Scott Enterprises Sales: SES, $\alpha = 10\%$**

| lag $k$ | $r_k$ | lag $k$ | $r_k$ |
|---|---|---|---|
| 1 | .002 | 5 | −.040 |
| 2 | .100 | 6 | .108 |
| 3 | .097 | 7 | −.036 |
| 4 | −.423 | 8 | .241 |

*Note:* Critical Value: $2/\sqrt{n} = .485$. None significant.

Since $U_M = 0.7\%$, $U_R = 5.0\%$, and $U_D = 94.3\%$, most of the squared forecast error can be attributed to random disturbances. Very little is due to bias (0.7%) or regression-type inaccuracies (5.0%). This conclusion is borne out by both the $y$ versus $\hat{y}$ diagnostic graph (Figure 2.28), which shows random scatter around the $y = \hat{y}$ line, and the residual acf (Table 2.38), which contains no significant sample autocorrelations.

**Figure 2.28** Graph of Actual Versus Forecasts for Scott Enterprises Sales Data Using SES with $\alpha = 10\%$

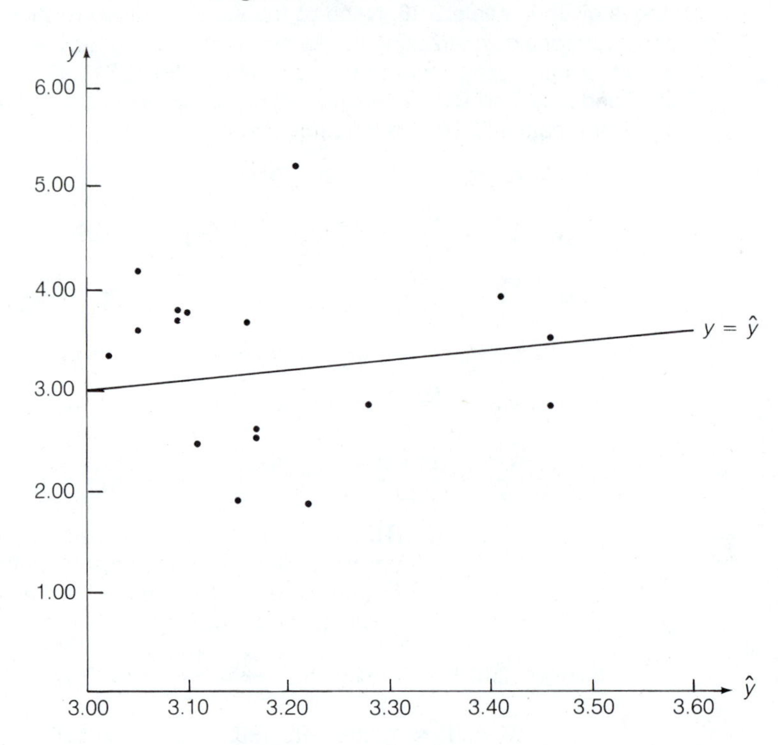


The final conclusion, then, is that, based on the available history, an SES scheme with $\alpha = 10\%$ is a reasonable one and should provide relatively unbiased forecasts. The RMSE should be about .86, and little in the way of data storage and computation is required to generate each new forecast. ●

## 2.12 Exercises

**2.1** In nonmathematical terms, what is the single most distinctive characteristic of a stationary series?

**2.2** What are three important reasons for studying horizontal models?

**2.3** In the context of $y_t$ = the number of "wrong numbers" received at a particular telephone in a week, explain the meaning of $\beta_0$ and $\epsilon_t$ in the following model:

$$y_t = \beta_0 + \epsilon_t \qquad t = 1, 2, \ldots$$

**2.4** Classify each of the following time series as (1) probably stationary, (2) either stationary or nonstationary, or (3) probably nonstationary. Explain why you believe each would fit into that particular category.

**a.** $y_t$ = yearly number of summer days when the temperature exceeds 90°F (1900–1990)

b. $y_t$ = per capita yearly unit sales of eyeglasses in the U.S. (1970–1990)
c. $y_t$ = per capita yearly unit sales of microcomputers in the U.S. (1975–1995)
d. $y_t$ = per capita dollar expenditures for life insurance in the U.S. (1970–1990)
e. $y_t$= the number of errors in the unemployment rate forecasts of the U.S. Bureau of Labor Statistics (1970–1990)
f. $y_t$ = the proportion of a large firm's yearly operating cost arising from the purchase and maintenance of janitorial equipment (1970–1990)
g. $y_t$ = the daily discrepancies between cash register receipts and cash drawer contents at a large supermarket (January 1985–January 1986)
h. $y_t$ = per capita consumption of chewing gum in the U.S. (1970–1980)

**2.5** What is the difference between parametric and nonparametric tests for stationarity?

**2.6** What is the difference between nonstationarity and nonrandomness?

**2.7** Suppose 50 observations are taken from a random series and $R$, the number of runs above and below the median, is determined.
a. How many runs are expected?
b. What is the standard deviation of the number of runs?
c. Suppose $R$ = 18 runs are observed. Test the hypothesis that the series is stationary/random at the 1% level of significance. What is your conclusion?
d. Suppose $R$ = 18 runs are observed. Test the hypothesis that the series is stationary/random at the 5% level of significance. What is your conclusion?

**2.8** Suppose 19 observations are taken from a random series and $R$, the number of runs above and below the median, is determined.
a. How many runs do we expect?
b. What is the standard deviation of the number of runs?
c. Suppose $R$ = 7 runs are observed. Test the hypothesis that the series is random at the 10% level of significance. What is your conclusion about whether a horizontal/random model is appropriate or not?

**2.9** Suppose the time series below were observed:

| $t$ | $y_t$ | $t$ | $y_t$ | $t$ | $y_t$ |
|---|---|---|---|---|---|
| 1 | 72.1 | 9 | 92.5 | 17 | 78.3 |
| 2 | 63.9 | 10 | 83.0 | 18 | 69.3 |
| 3 | 61.6 | 11 | 65.6 | 19 | 87.9 |
| 4 | 82.8 | 12 | 59.1 | 20 | 94.9 |
| 5 | 84.2 | 13 | 76.4 | 21 | 79.3 |
| 6 | 69.1 | 14 | 85.0 | 22 | 74.3 |
| 7 | 76.3 | 15 | 68.2 | 23 | 87.4 |
| 8 | 81.5 | 16 | 78.4 | 24 | 91.3 |

a. Graph this time series.
b. Does a runs test ($\alpha$ = 5%) provide support for use of a horizontal model?

**c.** Does a turning points test ($\alpha = 5\%$) provide support for use of a horizontal model?

**d.** Determine the signs of the first differences, and test the hypothesis that the series is random using a sign test for trend ($\alpha = 5\%$). Does this test provide support for use of a horizontal model?

**e.** Does the test based on Kendall's tau provide support for a horizontal model ($\alpha = 5\%$)?

**f.** Does Daniels' test for trend provide support for a horizontal model ($\alpha = 5\%$)?

**g.** Considering the results of all the above tests, what would you think about using a horizontal model?

**2.10** A department store forecasts its quarterly sales ($\$10^5$) using the following forecasting model:

$$\hat{y}_{t+1} = y_t + 4$$

Its sales over the last few quarters are:

| Year | Quarter | Sales ($\$10^5$) |
|---|---|---|
| 1985 | II | 37.4 |
| | III | 39.3 |
| | IV | 49.3 |
| 1986 | I | 48.2 |
| | II | 54.6 |

**a.** Compute Theil's $U$ for the department store's model.

**b.** Compute Theil's decomposition for the department store's model.

**2.11** A runs test has been conducted and the null hypothesis rejected with $R < R_L$. Explain how you would decide whether the rejection was caused by trend, autocorrelation, or both.

**2.12** Why might one be reluctant to use a no-trend model based solely on the fact that the runs test does not reject stationarity or randomness?

**2.13** In a stationary random time series with 50 observations, how likely are we to observe 10 or fewer runs above and below the median?

**2.14** A retail store wishes to forecast the revenues from one of its departments. The following data have been collected:

| Year | Revenues ($100,000) | Year | Revenues ($100,000) |
|---|---|---|---|
| 1978 | 10 | 1982 | 22 |
| 1979 | 14 | 1983 | 30 |
| 1980 | 18 | 1984 | 29 |
| 1981 | 19 | | |

**a.** On the basis of Spearman's rho, would one reject a stationary model?

**b.** On the basis of a runs test, would one reject a stationary random model?

c. On the basis of Pearson's correlation coefficient, would you reject a stationary model? ($r_{ty} = +.974$)

**2.15** A statistician is constructing a forecasting model for the following time series:

| Month $t$ | $y_t$ | $y_{(t)}$ |
|---|---|---|
| 1 | 10.3 | 9.7 |
| 2 | 12.8 | 9.9 |
| 3 | 9.7 | 10.0 |
| 4 | 10.0 | 10.3 |
| 5 | 11.7 | 11.7 |
| 6 | 12.1 | 12.1 |
| 7 | 9.9 | 12.8 |

Suppose the statistician wishes to use a single-forecast technique. What would the forecast for Month 8 be under each of the following conditions?

a. The cumulative forecast error should be as small as possible.
b. Underforecasts should be no more likely than overforecasts.
c. Large forecast errors should be avoided at all costs.
d. The cumulative magnitude of the forecast errors should be as small as possible.

**2.16** Suppose the statistician in Exercise 2.15 were all along forecasting $y_t$ using updating schemes.

a. What would be the forecast for Month 8 as of Month 4 using the updated mean?
b. What would be the forecast for Month 6 as of Month 5 using a simple moving average of length 3?
c. What would be the forecast for Month 9 as of Month 7 using a weighted moving average of length 3 with $w_1 = 2w_2 = 3w_3$?
d. If $y_7$ had been forecast by a previous (experienced) employee to be $\hat{y}_7 = 12.9$, what would be the forecast for week 9 using SES with a 20% smoothing constant?

**2.17** Calculate the autocorrelation coefficient of lag 2 for the following time series:

| $t$: | 1 | 2 | 3 | 4 | 5 | 6 | 7 | 8 | 9 | 10 |
|---|---|---|---|---|---|---|---|---|---|---|
| $y_t$: | −1 | −3 | −4 | 2 | 0 | 1 | 2 | 1 | −1 | 3 |

**2.18** Suppose $y_1, y_2, \ldots, y_n$ are observations from a random no-trend series such that the $y_t$'s are independent of one another. Does the series of second differences also form a random no-trend series, or are the second differences autocorrelated? Explain your answer.

**2.19** Show algebraically that, for no-trend series, the autocorrelation coefficient of lag 1 ($r_1$) is related to von Neumann's ratio $M$ in the following way:

$$r_1 \simeq 1 - \frac{M}{2}$$

**2.20** A time series of 20 observations produces the following acf:

```
           -0.8 -0.6 -0.4 -0.2  0.0  0.2  0.4  0.6  0.8
 j    r_j     +----+----+----+----+----+----+----+----+
                                  +
 1    0.237                       XXXXXXX
 2    0.020                       XX
 3    0.001                       X
 4   -0.092                     XXX
 5   -0.183                  XXXXXX
 6   -0.164                   XXXXX
 7   -0.243                 XXXXXXX
 8   -0.172                   XXXXX
 9    0.128                       XXXX
10    0.079                       XXX
```

**a.** What is the approximate value of von Neumann's ratio for this series?
**b.** Using a 10% significance level, test the series for randomness using von Neumann's ratio.

**2.21** Given the following time series:

| $t$ | $y_t$ | $t$ | $y_t$ |
|---|---|---|---|
| 1 | 16.3 | 7 | 19.6 |
| 2 | 14.1 | 8 | 17.4 |
| 3 | 14.0 | 9 | 17.0 |
| 4 | 15.7 | 10 | 16.7 |
| 5 | 17.8 | 11 | 14.9 |
| 6 | 18.3 | 12 | 14.1 |

**a.** Graph this series.
**b.** Using von Neumann's ratio ($\alpha = 10\%$), test the hypothesis that the series is random and stationary.
**c.** Compute the sample autocorrelation coefficients of lag 1, lag 2, and lag 3.
**d.** Test the hypothesis that $\rho_1 = 0$ using the "rule of thumb."
**e.** Test the hypothesis that $\rho_1 = \rho_2 = \rho_3 = 0$ at the 5% level of significance.
**f.** What is your conclusion about whether autocorrelation is present in the series?
**g.** What would you say about whether the series is horizontal?

**2.22** For the following time series data, find the backcast for time 0 using exponential smoothing. Use the last value in the series to initiate the backcasting procedure and use a value of .10 for the smoothing constant.

| $t$: | 1 | 2 | 3 | 4 | 5 | 6 | 7 | 8 |
|---|---|---|---|---|---|---|---|---|
| $y_t$: | 1 | 8 | 9 | 4 | 0 | 8 | 4 | 7 |

**2.23** A small, overnight gas station/minimarket has kept records of its average gross sales ($) on Sunday mornings between 12:00 midnight and 6:00 A.M. for the past 24-month period. These records are given below. They would like to forecast sales for each of the next 12 months along with total gross sales for the year.

| Month | Average Sales ($) | Month | Average Sales ($) | Month | Average Sales ($) |
|---|---|---|---|---|---|
| 1 | 690 | 9 | 833 | 17 | 815 |
| 2 | 590 | 10 | 722 | 18 | 748 |
| 3 | 745 | 11 | 821 | 19 | 781 |
| 4 | 635 | 12 | 757 | 20 | 889 |
| 5 | 725 | 13 | 652 | 21 | 846 |
| 6 | 738 | 14 | 812 | 22 | 863 |
| 7 | 632 | 15 | 723 | 23 | 734 |
| 8 | 765 | 16 | 708 | 24 | 788 |

**a.** Graph $y_t$ and $\Delta y_t$ versus $t$. Do these graphs seem to indicate that a no-trend model would be appropriate?
**b.** Does there seem to be autocorrelation in the series? Base your conclusion on statistical evidence.
**c.** Does there seem to be seasonality in the series?
**d.** Construct single forecasts for next month using the following criteria:
  **(i)** average error equals zero
  **(ii)** equal frequency of under- and overforecasts
  **(iii)** minimized maximum absolute error
  **(iv)** minimum average absolute error
  **(v)** minimum mean square error
**e.** Construct forecasts for next month using the following criteria:
  **(i)** updated mean
  **(ii)** updated median
  **(iii)** updated midrange
  **(iv)** moving average of length 12
  **(v)** SES, $\alpha = 10\%$, $\hat{y}_{\text{init}} = y_1$
  **(vi)** SES, $\alpha = 20\%$, $\hat{y}_{\text{init}} = y_1$

**2.24** Compute the RMSE, MAD, max $|e_t|$, and $\bar{e}$ for each estimate of $\beta_0$ in Exercise 2.23(d).

**2.25** Compute the RMSE, MAD, max $|e_t|$, $\bar{e}$, and Theil's $U$ for each of the forecasting schemes in Exercise 2.23(e).

**2.26** Why is it not entirely appropriate to compare the results of Exercise 2.24 and Exercise 2.25?

**2.27** Suppose you were deciding only between using the updated mean and SES with a 10% smoothing constant. Would you have any preference based on the results of Exercises 2.21 through Exercise 2.24?

**2.28** Given the following monthly history of the demand for a particular brand of soft drink, with $y_t$ = number of cases (1000's), for the last 13 months:

| Month | No. of Cases (1000's) | Month | No. of Cases (1000's) |
|---|---|---|---|
| Jan 1984 | 1.07 | Aug | 1.33 |
| Feb | 1.15 | Sep | 1.09 |
| Mar | 1.30 | Oct | 0.84 |
| Apr | 1.01 | Nov | 1.19 |
| May | 1.34 | Dec | 1.23 |
| Jun | 1.11 | Jan 1985 | 1.37 |
| Jul | 1.35 | | |

**a.** Graph the series $y_t$ versus $t$, to begin to familiarize yourself with the data.

**b.** Graph the series $\Delta y_t$ vs $t$.

**c.** Based on your feelings at this point, what sort of forecasting scheme might work well?

**d.** Using SES with $\hat{y}_{init} = y_1 = \hat{y}_2$, what would the optimal value of $\alpha$ be, to the nearest .01?

**e.** What would be the optimal length MA using RMSE as your optimizing criterion? How about if you use the MAPE?

**f.** Using SES with $\alpha = .15$ and "backcasting" $y_0$ to get an initial forecast, give the forecast for $y_{Feb85}$.

**g.** Using SES with $\alpha = .15$ and the average of the first 6 observations as your initial forecast, give the forecast for $y_{Feb85}$.

**h.** Using SES with $\alpha = .15$ and $y_1$ as your initial forecast, give the forecast for $y_{Feb85}$.

**2.29** The conventional way to initialize an SES scheme is to use the first observation as the initial forecast. What is the main drawback of this conventional method?

**2.30** When using a moving average of length $k$ to forecast a no-trend series, how does one choose $k$, the length of the moving average?

**2.31** What would be the general result of using SES to forecast a series that contained an upward linear trend component?

**2.32** As part of a production scheduling model, a simple exponential-smoothing scheme with a 20% smoothing constant is used to forecast demand for a certain product type. A forecast for December 1981 is to be made. During November 1981, in preparation for making December's forecast, the following data were stored in the company's data base:

| | |
|---|---|
| November 1981 Demand, Forecast: | 22.3K units |
| November 1981 Demand Actual: | 29.4K units |

**a.** Calculate and give the forecast for December 1981.

**b.** May the November data now be purged from the data base as far as constructing the January 1982 forecast is concerned? Justify your answer.

**c.** Using this scheme, what weight would be given to the actual value of the demand for September 1981 in the forecast for December 1981?

**d.** What smoothing constant should be used if it is desired that the current observation get 3 times as much weight in the forecasts as the 2-month-old observation?

**2.33** A series of 100 observations begins with $y_1 = 5$ and ends with $y_{100} = 75$. What is the average of the first differences for this series?

**2.34** For the following data, neither $y_t$ nor $\log y_t$ is stationary, but the first differences of $\log y_t$ do appear to be stationary:

| $t$ | $y_t$ | $\log y_t$ | $\Delta\log y_t$ |
|---|---|---|---|
| 1 | 5 | 1.609 | — |
| 2 | 5 | 1.609 | 0.000 |
| 3 | 4 | 1.386 | −0.223 |
| 4 | 10 | 2.303 | 0.917 |
| 5 | 9 | 2.197 | −0.106 |
| 6 | 11 | 2.398 | 0.201 |
| 7 | 16 | 2.773 | 0.375 |
| 8 | 29 | 3.367 | 0.594 |
| 9 | 23 | 3.135 | −0.232 |
| 10 | 39 | 3.664 | 0.529 |

Simple exponential smoothing would not be appropriate for $y_t$ or $\log y_t$ but would be for $\Delta\log y_t$. Using SES with a 30% smoothing constant and the first value ($\Delta\log y_2$) to initiate the smoothing, forecast $y_{11}$.

# APPENDIX 2A

## Ranking and Order Statistics

Suppose one has a time series, i.e., a set of chronologically ordered observations,

$$y_1, y_2, \ldots, y_n$$

Then the set of **order statistics** is denoted

$$y_{(1)}, y_{(2)}, \ldots, y_{(n)}$$

where

$$y_{(1)} \leq y_{(2)} \leq \cdots \leq y_{(n)}$$

So parenthetical subscripts denote numerically ordered observations, with

$y_{(1)}$ = the smallest observation

$y_{(2)}$ = the second smallest observation

.

.

.

$y_{(n)}$ = the largest observation

Fractional subscripts are used to denote points between successive numerically ordered observations, i.e.,

$$y_{(j \cdot p)} = y_{(j)} + p \cdot (y_{(j+1)} \cdot p - y_{(j)})$$

and, in particular,

$$y_{(j \cdot 5)} = \frac{y_{(j)} + y_{(j+1)}}{2}$$

For example, suppose we have

$y_1 = 24$ $y_2 = 18$ $y_3 = 29$ $y_4 = 36$

$y_5 = 47$ $y_6 = 29$ $y_7 = 19$

Then

$y_{(1)} = 18$ $y_{(2)} = 19$ $y_{(3)} = 24$ $y_{(4)} = y_{(5)} = 29$

$y_{(6)} = 36$ $y_{(7)} = 47$

and

$$y_{(3.5)} = \frac{y_{(3)} + y_{(4)}}{2} = \frac{24 + 29}{2} = 26.5$$

$$y_{(4.5)} = \frac{y_{(4)} + y_{(5)}}{2} = \frac{29 + 29}{2} = 29$$

$$y_{(2.8)} = y_{(2)} + .8 \times (y_{(3)} - y_{(2)}) = 19 + .8 \times (24 - 19) = 2.3$$

Finally, by **rank of $y_t$,** or **$R(y_t)$,** is meant the parenthetical subscript of $y_t$, that is, the position of $y_t$ in the numerically ordered time series. For example, rank of $y_7 = R(y_7) = 2$. When several observations have the same numerical value, the observations are said to have tied ranks and the value of the rank is usually considered to be the average of all the parenthetical subscripts of the tied observations, i.e.,

$$R(y_3) = R(y_6) = \frac{4 + 5}{2} = 4.5$$

# APPENDIX 2B

## Pearson's Product Moment Correlation Coefficient $r$

The **correlation coefficient (Pearson's) $r$** is a measure of the linear association between two variables. If we have a sample of pairs of measurements, $(x_1, y_1), (x_2, y_2), \ldots, (x_n, y_n)$, then $r$ measures how closely the pairs come, when graphed, to falling on a straight line. It is calculated from

$$r = \frac{\Sigma (x - \bar{x})(y - \bar{y})}{\sqrt{\Sigma (x - \bar{x})^2 \cdot \Sigma (y - \bar{y})^2}}$$

Or, letting,

$$SS_{xx} = \Sigma (x - \bar{x})^2 = \Sigma x^2 - \frac{(\Sigma x)^2}{n}$$

$$SS_{yy} = \Sigma (y - \bar{y})^2 = \Sigma y^2 - \frac{(\Sigma y)^2}{n}$$

$$SS_{xy} = \Sigma (x - \bar{x})(y - \bar{y}) = \Sigma xy - \frac{(\Sigma x)(\Sigma y)}{n}$$

then

$$r = \frac{SS_{xy}}{\sqrt{SS_{xx} \cdot SS_{yy}}}$$

The main properties of the correlation coefficient are:

1. $-1 \leq r \leq 1$
2. $|r| = 1$ implies that the sample of $(x, y)$ pairs falls on a perfectly straight line (meaning, in most cases, that they are deterministically related).
3. $r = 0$ implies that there is no *linear* association between $x$ and $y$ in the sample (meaning, in most cases, that they are not related).

## Example

Suppose we are given the following set of observations:

$$x_1 = 8 \quad x_2 = 10 \quad x_3 = 13 \quad x_4 = 16$$
$$y_1 = 17 \quad y_2 = 25 \quad y_3 = 24 \quad y_4 = 29$$

Then $r$ is calculated as below:

| $x$ | $y$ | $xy$ | $x^2$ | $y^2$ |
|---|---|---|---|---|
| 8 | 17 | 136 | 64 | 289 |
| 10 | 25 | 250 | 100 | 625 |
| 13 | 24 | 312 | 169 | 576 |
| 16 | 29 | 464 | 256 | 841 |
| 47 | 95 | 1162 | 589 | 2331 |

$$\Sigma x = 47 \quad \Sigma y = 95 \quad \Sigma xy = 1162 \quad \Sigma x^2 = 589 \quad \Sigma y^2 = 2331$$

$$SS_{xx} = \Sigma x^2 - \frac{(\Sigma x)^2}{n} = 589 - \frac{47^2}{4} = 36.75$$

$$SS_{yy} = \Sigma y^2 - \frac{(\Sigma y)^2}{n} = 2331 - \frac{95^2}{4} = 74.75$$

$$SS_{xy} = \Sigma xy - \frac{(\Sigma x)(\Sigma y)}{n} = 1162 - \frac{47 \times 95}{4} = 45.75$$

$$r = \frac{45.75}{\sqrt{36.75 \times 74.75}} = +.873$$

# APPENDIX 2C

## Spearman's Rank Correlation Coefficient $r_s$

**Spearman's rank correlation coefficient,** or **Spearman's rho,** is found by calculating Pearson's $r$ between the ranks of two variables. For a sample of pairs of measurements, $(x_1, y_1), (x_2, y_2), \ldots, (x_n, y_n)$, we replace $x_i$ and $y_i$ by their respective ranks $R(x_i)$ and $R(y_i)$ and then calculate $r$ for the new sets of pairs of ranks, i.e.,

$$r_s = \frac{\Sigma\,[R(x) - \overline{R(x)}]\,[R(y) - \overline{R(y)}]}{\sqrt{\Sigma\,[R(x) - \overline{R(x)}]^2 \cdot \Sigma\,[R(y) - \overline{R(y)}]^2}}$$

The calculation of $r_s$ is relatively simple because the ranks always have the values 1, 2, . . . , $n$. With some algebraic manipulation, $r_s$ may be shown to be more easily computed from

$$r_s = 1 - \frac{6\,\Sigma\,d^2}{n(n^2 - 1)}$$

where $d = R(y) - R(x)$

**Example**

Suppose we are given the following set of observations:

$x_1 = 10$ $x_2 = 8$ $x_3 = 7$ $x_4 = 12$ $x_5 = 11$

$y_1 = 45$ $y_2 = 19$ $y_3 = 31$ $y_4 = 39$ $y_5 = 16$

Then $r_s$ is calculated as below:

| $x$ | $y$ | $R(x)$ | $R(y)$ | $d$ | $d^2$ |
|---|---|---|---|---|---|
| 10 | 45 | 3 | 5 | −2 | 4 |
| 8 | 19 | 2 | 2 | 0 | 0 |
| 7 | 31 | 1 | 3 | −2 | 4 |
| 12 | 39 | 5 | 4 | 1 | 1 |
| 11 | 16 | 4 | 1 | 3 | 9 |
| | | | | $\Sigma\,d^2 =$ | 18 |

$$r_s = 1 - \frac{6 \times 18}{5(5^2 - 1)} = 1 - \frac{108}{5 \times 24} = 1 - .900 = .100$$

# APPENDIX 2D

## Student's *t*-Test

To test whether a specific hypothesized value $\mu_0$ is reasonable for the mean of a normal population in light of an observed sample $y_1, y_2, \ldots, y_n$ from that population, one can use Student's *t*-test.

**Test Procedure:** *Student's* t-*test*

$H_0$: $\mu = \mu_0$

$H_a$: $\mu \neq \mu_0$

**Test Statistic:**

$$t = \frac{\bar{y} - \mu_0}{s/\sqrt{n}}$$

where $\bar{y}$ = the sample mean = $\Sigma\, y/n$

$s$ = the sample standard deviation

$$= \sqrt{\frac{\Sigma\,(y - \bar{y})^2}{(n - 1)}} = \sqrt{\frac{SS_{yy}}{(n - 1)}}$$

$$SS_{yy} = \Sigma\,(y - \bar{y})^2 = \Sigma\, y^2 - \frac{(\Sigma\, y)^2}{n}$$

**Decision Rule:**

Reject $H_0$ if $|t| > t_{\alpha/2}$.
Otherwise do not reject $H_0$.

**(Appendix GG, df = $n - 1$)**

### Example

Suppose we wish to test ($\alpha = 5\%$) whether a population mean of 50 is reasonable for the normal population from which the following sample was drawn:

$y_1 = 72$ $\quad y_2 = 43$ $\quad y_3 = 68$ $\quad y_4 = 62$

$\Sigma\, y = 245$ $\quad \bar{y} = 61.25$ $\quad \Sigma\, y^2 = 15{,}501$

$$SS_{yy} = 15{,}501 - \frac{245^2}{4} = 494.75$$

$$s = \frac{\sqrt{494.75}}{3} = 12.84$$

## Test Procedure

The test would be:

$$H_0: \quad \mu = 50$$
$$H_a: \quad \mu \neq 50$$

**Test Statistic:** $t = \dfrac{61.5 - 50}{12.84/\sqrt{4}} = 1.752$

**Decision Rule:** ($\alpha = 5\%$, df $= 3$)
Reject $H_0$ if $|t| > t_{.025} = 3.182$.

**Conclusion:** Since $1.752 < 3.182$, we do not reject $H_0$, concluding that, at the 5% significance level, the sample does not contradict the hypothesis that the population mean is 50.

# References

Abraham, B., and Ledolter, J. (1983). *Statistical Methods for Forecasting.* New York: Wiley.

Arbel, T. (1985). "Minimizing the Sum of Absolute Deviations." *Teaching Statistics* 7: 88–89.

Bartlett, M. S. (1946). "On the Theoretical Specification of Sampling Properties of Autocorrelated Time Series." *J. Roy. Stat. Soc.* B, 8: 27–41.

Bingham, C., and Nelson, L. S. (1981). "An Approximation for the Distribution of the von Neumann Ratio." *Technometrics* 23(3): 285–288.

Box, G. E. P., and Pierce, D. A. (1970). "Distributions of Residual Autocorrelations in Autoregressive-Integrated Moving Average Models." *J. Amer. Stat. Assoc.* 65: 1509–1526.

Bradley, J. V. (1968). *Distribution-Free Statistical Tests.* Englewood Cliffs, N.J.: Prentice-Hall.

Brown, R. G. (1963). *Smoothing, Forecasting, and Prediction of Discrete Time Series.* Englewood Cliffs, N.J.: Prentice-Hall.

Brownlee, K. A. (1965). *Statistical Theory and Methodology in Science and Engineering,* 2nd Ed. New York: Wiley.

Conover, W. J. (1980). *Practical Nonparametric Statistics,* 2nd Ed. New York: Wiley.

Daniels, H. E. (1950). "Rank Correlation and Population Models." *J. Roy. Stat. Soc.* B, 12: 171–181.

Davies, N., and Newbold, P. (1979). "Some Power Studies of a Portmanteau Test of Time Series Model Specification." *Biometrika* 66: 153–155.

Edgington, E. S. (1961). "Probability Table for Number of Runs of Signs of First Differences in Ordered Series." *J. Amer. Statist. Assoc.* 56: 156–159.

Gardner, E. S. (1985). "Exponential Smoothing: The State of the Art." *J. Forecasting* 4(1): 1–28.

Gilchrist, W. (1976). *Statistical Forecasting.* New York: Wiley.

Granger, C. W. J., and Newbold, P. (1986). *Forecasting Economic Time Series,* 2nd Ed. New York: Academic Press.

Hart, B. I. (1942). "Tabulation of the Probabilities for the Ratio of the Mean Square Successive Difference to the Variance." *Ann. Math. Stat.* 13: 207–214.

Jenkins, G. M., and Watts, D. G. (1968). *Spectral Analysis and Its Applications.* San Francisco: Holden-Day.

Kendall, M. G. (1973). *Time Series.* London: Griffin.

Ljung, G. M., and Box, G. E. P. (1978). "On a Measure of Lack of Fit in Time Series Models." *Biometrika* 65: 67–72.

Makridakis, S., Wheelwright, S. C., and McGee, V. E. (1983). *Forecasting: Methods and Applications,* 2nd Ed. New York: Wiley.

Nelson, L. S. (1980). "The Mean Square Successive Difference Test." *J. Qual. Tech.* 12(3): 174–175.

Slutsky, E. (1937). "The Summation of Random Causes as the Source of Cyclic Processes." *Econometrika* 5: 105–146.

Stigler, S. M. (1986). "Estimating Serial Correlation by Visual Inspection of Diagnostic Plots." *Amer. Statistician* 40(2): 111–116.

Swed, F. S., and Eisenhart, C. (1943). "Tables for Testing Randomness of Grouping in a Sequence of Alternatives." *Ann. Math. Stat.* 14: 66–87.

von Neumann, J. (1941). "Distribution of the Mean Square Difference to the Variance." *Ann. Math. Stat.* 12: 367–395.

Whitmore, G. A. (1986). "Prediction Limits for a Univariate Normal Observation." *Amer. Statistician* 40(2): 141–143.

# Forecasting Series with Trend

**Objective:** *To study methods for forecasting time series that have long-term patterns of growth or decline.*

## 3.1 Introduction

As we have mentioned before, the expected value of a series may change as time passes because of the influence of factors such as population growth, technological advance, and other societal change. If the movement is long term and only in one direction (up or down), it is called **trend** and the resultant series is *trended.* We will now deal with the problem of detecting this trend and discuss some procedures for modeling and forecasting trended series.

With trended series, we generally assume there is some functional relationship between the expected value of the series and the time variable. This function usually involves some unknown parameters $\beta_0$, $\beta_1$, . . . that must be estimated from the history of the series. As was the case for no-trend series in Chapter 2, both single-forecast (one-time estimates) and updating schemes will be discussed.

This chapter, then, focuses on:

- Statistical tests designed to detect trend in a series.
- Different types of functional relationships (linear, polynomial, exponential growth, etc.) used to describe trend.
- Methods for estimating the parameters of trend models and for updating estimates.
- Special aspects of diagnostics when applied to trend models.

## 3.2 The Nature of Trend

### 3.2.1 Introduction

A time series whose average value changes over time is called **nonstationary.** Trend, as we use the term here, is a particular kind of nonstationarity in which the pattern of change has two main properties (Bell and Hillmer, 1984):

- It is of *long duration* compared to the forecast horizon (so that substantially the same trend model will be valid throughout the period to be forecast).
- It is predominantly in *one direction only,* either upward (growth) or downward (decline).

This type of time series movement is summarized a little more formally in the following definition.

**DEFINITION:** A time series $y_1, y_2, \ldots$ is said to have **trend** if the expected value of $y_t$ changes over time so that

$$E(y_t) = f(\beta_0, \beta_1, \ldots; t) \qquad t = 1, 2, \ldots$$

where $f(\beta_0, \beta_1, \ldots; t)$ is predominantly either an increasing or a decreasing function of $t$ over the forecasting horizon (the time for which forecasts are to be constructed).

### 3.2.2 Characteristics of Trended Series

A model for a trended time series may be stated as follows:

$$\begin{aligned} y_t &= \text{the actual value of the series at time } t \\ T_t &= \text{the expected value (trend) of the series at time } t \\ &= f(\beta_0, \beta_1, \ldots; t) \\ \epsilon_t &= \text{a random variable representing irregular fluctuations around the trend} \end{aligned}$$

where $f(\beta_0, \beta_1, \ldots; t)$ is an increasing or decreasing function describing the trend pattern.

**General Trend Model**

$$y_t = T_t + \epsilon_t \qquad t = 1, 2, \ldots$$

This model describes a consistently growing or declining series with variations from trend ($T_t$) caused by random influences ($\epsilon_t$). In most cases it is assumed that the $\epsilon_t$'s form a stationary random series with $E(\epsilon_t) = 0$ for all $t$. As a hypothetical example to show how the model works, suppose a company's profits $y_t$ are growing at an average rate of 10 percent per year, starting at \$125,000 in 1979. The trend would then be

$$T_t = 125(1.1)^t$$

where $t$ is 0 in 1979 with 1 year units and $T_t$ is in \$1,000 units. (Notice that, when writing trend equations, the origin (zero point) and units to be used for $t$ should be given as well as the units in which $T_t$ is expressed; more about this later.) Profits are affected each year by a hypothetical amount $\epsilon_t$ given in Table 3.1. In Figure 3.1, $T_t$ and $\epsilon_t$ are given separately and then combined to give $y_t$.

Sometimes there may be confusion about whether a movement that appears to be trend is, in fact, only a part of a very long cycle. However, practically speaking, a portion of a long cycle may be modeled as trend if the cycle has not changed direction during the time period of interest to the

**Table 3.1 Example of Trend Model**

| Year | $t$ | Trend $T_t$ | Error $\epsilon_t$ | Profits $y_t = T_t + \epsilon_t$ |
|---|---|---|---|---|
| 1979 | 1 | 125.0 | −2.6 | 122.4 |
| 1980 | 2 | 137.5 | 14.4 | 151.9 |
| 1981 | 3 | 151.2 | 22.0 | 173.2 |
| 1982 | 4 | 166.4 | −18.8 | 147.6 |
| 1983 | 5 | 183.0 | 6.4 | 189.4 |
| 1984 | 6 | 201.3 | −20.6 | 180.7 |
| 1985 | 7 | 221.4 | −9.2 | 212.2 |
| 1986 | 8 | 243.6 | −15.8 | 227.8 |
| 1987 | 9 | 267.9 | 16.6 | 284.5 |
| 1988 | 10 | 294.7 | −1.2 | 293.5 |

**Figure 3.1 Separate and Combined Components of Time Series for Trend Model in Table 3.1**

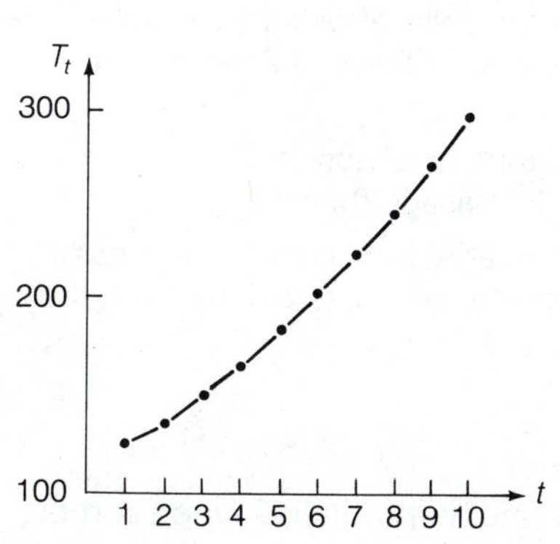


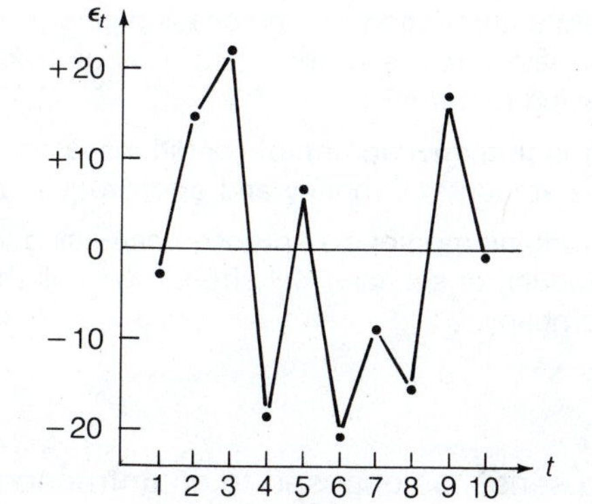


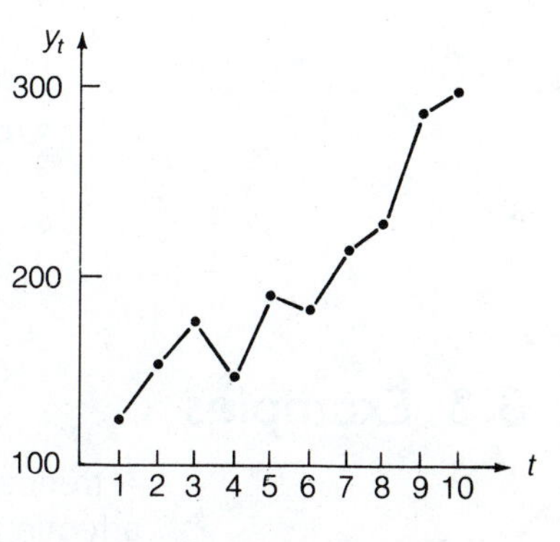

forecaster and if no such change is anticipated. Under these circumstances it is probably only academic whether the movement is "truly" a trend or a cycle. Some caution should be exercised, though, because if a cycle is modeled as trend and it happens to turn over during the forecast horizon, embarrassingly large forecast errors may result.

When considering model equations, it is convenient to classify trends into two major categories:

- **linear trend** constant incremental change, that is, the *expected* first differences $E(y_t - y_{t-1})$, or equivalently, $T_t - T_{t-1}$, are the same for all $t$.
- **curvilinear trend** the amount of change (but not the direction) varies as time passes.

Linear models are the easiest to handle. In addition, they can serve as rough approximations, especially over the short term, to curvilinear models, which are generally more difficult. Various kinds of curvilinear models are available for describing different patterns of change. Both types of models will be discussed in later sections.

### 3.2.3 Causes of Trend

It is quite common for time series, especially economic ones, to contain trend. Any motive force, such as "the economy," acting slowly and persistently over time will tend to produce a trended series. Among the major sources of trend are the following.

- **population changes** changing population causes changes in demand and in the variables associated with that demand, such as energy consumption, use of raw materials, manpower use, and sales revenues.
- **technological changes** improving technology and increased productivity lead to changes in life quality and standard of living; items that were once luxuries become necessities, necessities become outdated, and newly developed products and services go from rare to commonplace.
- **changes in social custom** slowly changing tastes, habits, and mores alter consumption and other patterns; divorce rates slowly climb, halloween trick-or-treating subsides, hours spent watching television increase, cigarette smoking declines, and so on.
- **inflation/deflation** the purchasing power of the dollar affects prices, salaries, interest rates, and such less obvious matters as labor strife and consumer saving patterns.
- **changing environmental conditions** as environmental conditions vary, so do the amounts of money and effort required to manage them.
- **changing market acceptance** increasing familiarity with a firm's brand name, product, or services will affect nearly all the time series associated with that company.

## 3.3 Examples

A trended series, as opposed to an untrended one, is produced when forces affecting the series change, causing the average value to rise or fall steadily.

Different patterns of trend occur, depending on how the forecast variable interacts with these forces.

## 3.3.1 Example 3.1—Valley Discount Liquor, Inc.

Valley Discount Liquor supplies beer and wine to retail outlets in a moderate-size western city. It has experienced steadily increasing demand for various brands and types of white wines. Data on the number of cases sold over five years are shown in Table 3.2.

The increasing demand is apparent in the graph of these data in Figure 3.2. The trend in $y_t$ could be a result of several factors:

**Table 3.2** Purchases of White Wine, Valley Discount Liquor, Incorporated

| Year | Half | Time $t$ | No. of Cases Purchased $y_t$ (1,000's) |
|---|---|---|---|
| 1981 | I | 1 | 1.34 |
| | II | 2 | 2.81 |
| 1982 | I | 3 | 2.41 |
| | II | 4 | 3.62 |
| 1983 | I | 5 | 3.64 |
| | II | 6 | 4.29 |
| 1984 | I | 7 | 4.18 |
| | II | 8 | 6.09 |
| 1985 | I | 9 | 4.95 |
| | II | 10 | 5.68 |

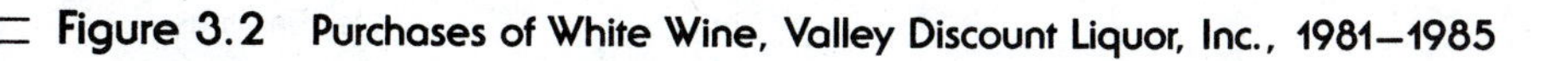

**Figure 3.2** Purchases of White Wine, Valley Discount Liquor, Inc., 1981–1985

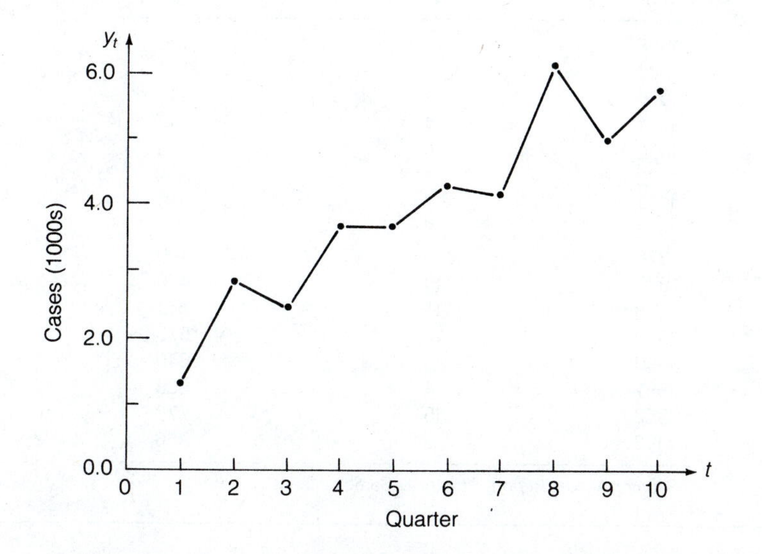

- The population in the city where Valley operates has been steadily increasing.
- Valley Discount, which started business in the '70s, has taken steps (management, marketing, public relations, etc.) to enhance its competitive position and is becoming known as a reliable, efficient, low-cost supplier.
- Consumption patterns are drifting away from "hard" distilled liquors to lighter drinks.
- A "faddish" popularity has been developing for white wine drinks. •

### 3.3.2 Example 3.2—Procter & Gamble, Inc.

Another example of a clearly trended series may be seen in Table 3.3 and Figure 3.3. The data are the earnings ($ millions) of Procter & Gamble for the years 1968 through 1983 (*Source:* Procter & Gamble Annual Report, 1984). Although a straight-line trend seems adequate in the case of Valley Discount sales, the yearly incremental growth is not constant for Procter & Gamble. This results in the upward-curving graph of Figure 3.3 and indicates a curvilinear model may be needed. The trend in earnings could be attributed to:

- Increasing population and/or disposable income producing increased demand for Procter & Gamble's various products.
- Growth in the company resulting from diversification, good management, advertising, new product acceptance, etc.
- Inflation, which will tend to cause all series measured in dollar units to increase (an effect that may be corrected by expressing $y_t$ in constant dollars using some price index such as the Consumer Price Index of the Bureau of Labor Statistics).

**Table 3.3 Net Earnings, Procter & Gamble Inc., 1968–1983**

| Year | Time $t$ | Net Earnings $y_t$ ($$10^6$) |
|---|---|---|
| 1968 | 1 | 182 |
| 1969 | 2 | 187 |
| 1970 | 3 | 211 |
| 1971 | 4 | 237 |
| 1972 | 5 | 276 |
| 1973 | 6 | 301 |
| 1974 | 7 | 316 |
| 1975 | 8 | 332 |
| 1976 | 9 | 400 |
| 1977 | 10 | 460 |
| 1978 | 11 | 510 |
| 1979 | 12 | 575 |
| 1980 | 13 | 640 |
| 1981 | 14 | 668 |
| 1982 | 15 | 777 |
| 1983 | 16 | 866 |

*Source: Annual Report,* Procter & Gamble, 1984.

Figure 3.3 Net Earnings of Procter and Gamble Inc., 1968–1983

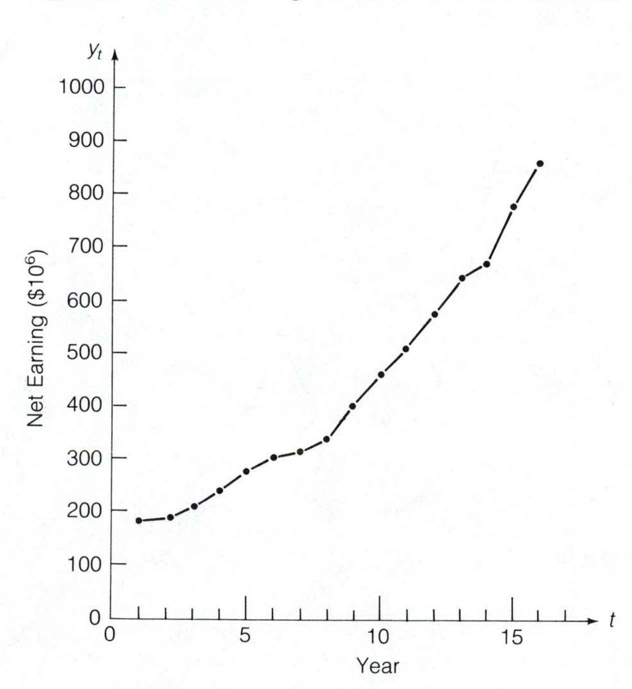


## 3.3.3 Example 3.3—United States Distilled Spirits Production

Changing social customs can produce changes in time series. For example, if one examines the production figures for distilled spirits in the United States from 1971 to 1983 (*Source: Business Conditions Digest*) (Table 3.4 and Figure 3.4), there seems to

**Table 3.4** U.S. Distilled Spirits Production, 1971–1983

| Year | Time $t$ | Production $y_t$ ($10^6$ taxable gallons) |
|---|---|---|
| 1971 | 1 | 183.3 |
| 1972 | 2 | 183.7 |
| 1973 | 3 | 183.1 |
| 1974 | 4 | 162.6 |
| 1975 | 5 | 144.2 |
| 1976 | 6 | 160.4 |
| 1977 | 7 | 159.3 |
| 1978 | 8 | 166.6 |
| 1979 | 9 | 186.7 |
| 1980 | 10 | 140.5 |
| 1981 | 11 | 152.0 |
| 1982 | 12 | 138.1 |
| 1983 | 13 | 119.4 |

*Source: Business Conditions Digest.*

Figure 3.4 U.S. Distilled Spirits Production, 1971–1983

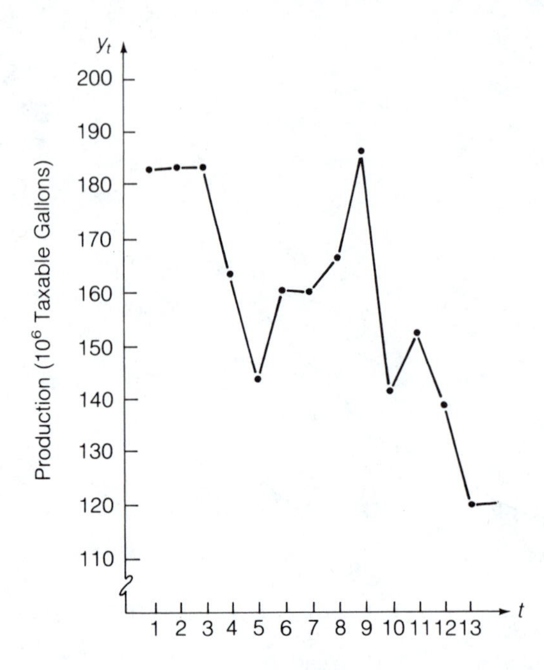


be a downward trend. This trend may be caused by a popular shift away from stronger spirits to ones with lower alcohol content, such as beer and wine. Other factors influencing the series may be:

- Increased awareness of and treatment for problems associated with heavy drinking.
- Admission by health insurance providers that alcoholism treatment is a valid medical expense.
- Increasingly heavy taxation of "hard" liquor.
- Increased importation of foreign distilled spirits, such as Scotch whiskey.
- Legalization of home production of wine and beer. ●

## 3.4 The Decision to Use Trend Models

### 3.4.1 Introduction

Trended series are much more common than untrended ones, so forecasters very often have to fit some sort of trend model. In many of these cases the presence of trend is obvious and detailed testing procedures are not needed. If there is some doubt, however, or if more evidence is desired, procedures that may be followed fall into the same three categories as those outlined in Section 2.4 for no-trend models, i.e., (a) general practices, (b) specific criteria for trend, and (c) statistical tests.

### 3.4.2 General Practices

These considerations are virtually the same for any modeling decision. They are restated here for emphasis and easy reference.

- **familiarity with the series** knowing the process that generates the series.
- **organization of the data** getting the data into orderly and examinable form.
- **graphical presentation of the data** use of standard graphing methods to study patterns in the data.

### 3.4.3 Specific Criteria for Trend Models

Having taken the above general preliminary steps, a trend model should be useful, provided the following hold true.

- Your knowledge of the series suggests that it should be growing or declining as time passes.
- A trend is discernible in the graph of the series ($y_t$ versus $t$), so the general appearance of the series is not horizontal.
- The first differences, $y_t - y_{t-1}$ of the series themselves form either a horizontal series with a nonzero average or a trended series.
- The link relatives $y_t/y_{t-1}$ of the series form either a nonhorizontal series or a horizontal series whose average is not 100% (greater than 100% for growth, less for decline).

### 3.4.4 Statistical Tests for Trend

Hypothesis tests may also be used to substantiate the presence of trend. The statistical question that must be addressed here is whether the hypothesis that the series is stationary/random can be defended against the specific alternative that trend is present. The general format of these hypothesis tests is:

$H_0$: The series is stationary/random (has no trend)

$H_a$: The series is trended

and the conclusions that may be drawn are:

Reject $H_0$, meaning there is sufficient evidence, at the $\alpha \times 100\%$ level, that the series is trended.

Do not reject $H_0$, meaning there is not sufficient evidence, at the $\alpha \times 100\%$ level, that the series is trended.

A decision to reject $H_0$, then, is a strong indication (statistical proof at the $(1 - \alpha) \times 100\%$ confidence level) that trend is present.

### 3.4.5 Nonparametric Tests for Trend

Since trend is a form of nonstationarity, its presence will tend to cause tests for stationarity to reject $H_0$. Several such tests of the nonparametric type were given in Chapter 2. Among them, Daniels' test, based on Spearman's rho, is a good overall choice.

**EXAMPLE 3.4** ***Daniels' Test Applied to U.S. Spirits Production***

As a brief review, we will apply Daniels' test to the distilled spirits production data of Example 3.3.

$H_0$: There is *no trend* in U.S. Distilled Spirits production

$H_a$: There is *trend* in U.S. Distilled Spirits production, either upward or downward.

**Test Statistic:** (Small sample, $n = 13$)

$$r_s = 1 - \frac{6 \sum d_t^2}{n(n^2 - 1)} = 1 - \frac{6 \cdot 614}{13(169 - 1)}$$

$$= -.687 \quad \text{(See Table 3.5 for computations.)}$$

$$d_t = t - \text{Rank}(y_t)$$

**Decision Rule:** ($\alpha = 5\%$)

Reject $H_0$ if $|r_s| > r_{.025} = .555$. **(Appendix AA)**

**Conclusion:** Since $|r_s| = .687 > .555$, we reject $H_0$ and conclude with 95% confidence that there is downward trend in U.S. Distilled Spirits production ($r_s$ is negative). ●

As mentioned in Chapter 2, the test based on Kendall's tau is as good as Daniels' and almost always gives the same results. The other nonparametric

**Table 3.5 Computation of Spearman's Rho for U.S. Distilled Spirits Production**

| Year | Time $t$ | *Production* $y_t$ | Rank of $y_t$ $R(y_t)$ | $d_t$ | $d_t^2$ |
|---|---|---|---|---|---|
| 1971 | 1 | 183.3 | 11 | −10 | 100 |
| 1972 | 2 | 183.7 | 12 | −10 | 100 |
| 1973 | 3 | 183.1 | 10 | −7 | 49 |
| 1974 | 4 | 162.6 | 8 | −4 | 16 |
| 1975 | 5 | 144.2 | 4 | 1 | 1 |
| 1976 | 6 | 160.4 | 7 | −1 | 1 |
| 1977 | 7 | 159.3 | 6 | 1 | 1 |
| 1978 | 8 | 166.6 | 9 | −1 | 1 |
| 1979 | 9 | 186.7 | 13 | −4 | 16 |
| 1980 | 10 | 140.5 | 3 | 7 | 49 |
| 1981 | 11 | 152.0 | 5 | 6 | 36 |
| 1982 | 12 | 138.1 | 2 | 10 | 100 |
| 1983 | 13 | 119.4 | 1 | 12 | 144 |
| | | | | | $\sum d_t^2 = 614$ |

tests of Chapter 2 can also be applied. Though somewhat less powerful, they are simpler to carry out.

### Runs Test—Trend

A series with trend produces only a few runs above and below the median. This causes the runs test to reject stationarity, with the number of runs being less than the lower critical value (or $z$ being less than $-z_{\alpha/2}$ in the large-sample case.) However, pronounced positive autocorrelation (say, a cycle occurring during the available history of the series) could produce the same sort of results. If it is not clear which phenomenon has produced the rejection, a graph will usually help clarify the matter.

### Turning Points Test—Trend

Trended series generally have fewer turning points than untrended ones, which causes the turning points test to reject $H_0$ with the test statistic ($z$) below the lower critical value ($-z_{\alpha/2}$). The turning points test suffers from the same deficiencies as the runs test. For example, neither test rejects stationarity for the distilled spirits series.

### Sign Test—Trend

For a relatively simple way to detect trend, the sign test is a better choice than the runs or turning points tests. Trend usually produces a preponderance of plus or minus signs for the first differences, leading the test to reject with a large positive value for $z$ (upward trend) or a large negative one (downward trend). However, owing to its ignoring the magnitude of the upward or downward change in $y_t$, the test is not as powerful as those based on rho or tau. It is simple to run, though, and when it rejects, indicating trend, no further testing is necessary.

## 3.4.6 Parametric Tests for Trend

### Pearson's *r*

If it is reasonable to expect that the $y_t$ are *normally distributed,* a test based on Pearson's product moment correlation coefficient is very useful for deciding whether a trend model should be used. The test is designed primarily to detect linear trend but will often reject $H_0$ even if the actual trend is curvilinear. The sample correlation coefficient between $t$ and $y_t$ is computed from

$$SS_{tt} = \Sigma\,(t - \bar{t})^2 = \Sigma\, t^2 - \frac{(\Sigma\, t)^2}{n} \qquad \left(\text{or } \frac{n(n^2 - 1)}{12}\text{; see below}\right)$$

$$SS_{yy} = \Sigma\,(y - \bar{y})^2 = \Sigma\, y^2 - \frac{(\Sigma\, y)^2}{n}$$

$$SS_{ty} = \Sigma\,(t - \bar{t})(y - \bar{y})^2 = \Sigma\, ty - \frac{(\Sigma\, t)\cdot(\Sigma\, y)}{n}$$

$$r = \frac{SS_{ty}}{\sqrt{SS_{tt}\cdot SS_{yy}}}$$

**Test Procedure:** *Test for $\rho = 0$*

$H_0$: $\rho = 0$

$H_a$: $\rho \neq 0$

**Test Statistic:**

$$t_r = \frac{r\sqrt{n-2}}{\sqrt{1-r^2}}$$

$$r = \frac{SS_{ty}}{\sqrt{SS_{tt}SS_{yy}}}$$

$$SS_{tt} = \Sigma t^2 - \frac{(\Sigma t)^2}{n} = \frac{n(n^2-1)}{12} \qquad \text{if } t = 1, 2, \ldots, n$$

$$SS_{ty} = \Sigma ty - \frac{(\Sigma t) \cdot (\Sigma y)}{n}$$

$$SS_{yy} = \Sigma y^2 - (\Sigma y)^2$$

**Decision Rule:** (Level $\alpha$)

Reject $H_0$ if $|t_r| > t_{\alpha/2}$. **(Appendix GG)**

where $t_{\alpha/2} = (\alpha/2) \times 100\%$ point of Student's *t*-distribution with $(n - 2)$ degrees of freedom.

**Conclusion:** If $H_0$ is rejected, we conclude, with $(1 - \alpha) \times 100\%$ confidence, that the series has trend. If, in addition, $t_r$ is positive, we conclude the trend is upward, or if $t_r$ is negative, that the trend is downward.

If $H_0$ is not rejected, we have some support for a no-trend model.

---

If $t$ represents times $1, 2, \ldots, n$, then $SS_{tt}$ may be more simply computed as $SS_{tt} = n(n^2 - 1)/12$.

If the population correlation coefficient $\rho$ is zero, then the series is not linearly correlated with time, i.e., not trended, and $r$ should be within sampling error of zero. The test statistic

$$t_r = \frac{r\sqrt{n-2}}{\sqrt{1-r^2}}$$

follows Student's *t*-distribution with $(n - 2)$ degrees of freedom if $\rho$ is zero and $y_t$ are normally distributed. The test for trend is based on this distribution.

**EXAMPLE 3.5** *Test for $\rho = 0$, Distilled Spirits Production*

Although Daniels' test has already indicated that there is trend in the distilled spirits series, it is illustrative to subject the data to the test for $\rho = 0$ as well. To evaluate

**Table 3.6** Computation of Pearson's Correlation Coefficient for U.S. Distilled Spirits Production

| Year | Time $t$ | Production $y_t$ | $t \cdot y_t$ | $y_t^2$ | $t^2$ |
|---|---|---|---|---|---|
| 1971 | 1 | 183.3 | 183.3 | 33,598.9 | 1 |
| 1972 | 2 | 183.7 | 367.4 | 33,745.7 | 4 |
| 1973 | 3 | 183.1 | 549.3 | 33,525.6 | 9 |
| 1974 | 4 | 162.6 | 650.4 | 26,438.8 | 16 |
| 1975 | 5 | 144.2 | 721.0 | 20,793.6 | 25 |
| 1976 | 6 | 160.4 | 962.4 | 25,728.2 | 36 |
| 1977 | 7 | 159.3 | 1115.1 | 25,376.5 | 49 |
| 1978 | 8 | 166.6 | 1332.8 | 27,755.6 | 64 |
| 1979 | 9 | 186.7 | 1680.3 | 34,856.9 | 81 |
| 1980 | 10 | 140.5 | 1405.0 | 19,740.3 | 100 |
| 1981 | 11 | 152.0 | 1672.0 | 23,104.0 | 121 |
| 1982 | 12 | 138.1 | 1657.2 | 19,071.6 | 144 |
| 1983 | 13 | 119.4 | 1552.2 | 14,256.4 | 169 |
| $\Sigma$ | 91 | 2079.9 | 13,848.4 | 337,992.1 | 819 |

the test statistic, $t_r$, we must first calculate the five values $\Sigma y$, $\Sigma y^2$, $\Sigma t$, $\Sigma t^2$, and $\Sigma ty$ as shown in Table 3.6. The results are:

$$\Sigma y = 2079.9 \qquad \Sigma y^2 = 337{,}992 \qquad \Sigma ty = 13{,}848$$
$$\Sigma t = 91 \qquad \Sigma t^2 = 819$$

From this we can determine $SS_{yy}$, $SS_{tt}$, and $SS_{ty}$ as follows:

$$SS_{yy} = \Sigma y^2 - \frac{(\Sigma y)^2}{n} = 337{,}992 - \frac{(2079.9)^2}{13} = 5224.0$$

$$SS_{ty} = \Sigma ty - \frac{(\Sigma t)(\Sigma y)}{n} = 13{,}848 - \frac{91 \cdot 2079.9}{13} = -711.3$$

$$SS_{tt} = \Sigma t^2 - \frac{(\Sigma t)^2}{n} = 819 - \frac{(91)^2}{13} = 182$$

$$\left(\text{or } SS_{tt} = \frac{n(n^2 - 1)}{12} = \frac{13 \cdot (13^2 - 1)}{12} = 182 \qquad \text{since } t = 1, \ldots, 13\right)$$

The correlation coefficient may now be calculated:

$$r = \frac{SS_{ty}}{\sqrt{SS_{tt}SS_{yy}}} = \frac{-711.3}{\sqrt{182 \cdot 5224.0}} = -.729$$

and the test for trend would be as follows:

Test for $\rho = 0$

$H_0$: $\rho = 0$ (No trend)
$H_a$: $\rho = 0$ (Trend)

**Test Statistic:** $t_r = \dfrac{r\sqrt{n-2}}{\sqrt{1-r^2}} = \dfrac{-.729\sqrt{13-2}}{\sqrt{1-(-.729)^2}} = -3.532$

**Decision Rule:** ($\alpha = 5\%$, df $= 13 - 2 = 11$)

Reject $H_0$ if $|t_r| > t_{.025} = 2.201$. **(Appendix GG)**

**Conclusion:** Since $|t_r| = 3.532 > 2.201$, we reject $H_0$ and conclude with 95% confidence that there is downward trend in U.S. Distilled Spirits production ($t_r < 0$).

The test for $\rho = 0$, then, leads us to the same decision that Daniels' test had reached previously. •

### Other Parametric Tests

The mean square successive difference test may indicate trend if it rejects the stationarity hypothesis with a *small value* of the test statistic $M$ (see Chapter 2). It has the difficulty (as do the runs and turning points tests, incidently) that positively autocorrelated errors may cause the same type of reject decision. As before, an apparent trend in the graph of the series together with such a rejection would justify a trend model.

The sample autocorrelation function may also indicate trend if it has the characteristics mentioned in Table 2.10, i.e., large positive autocorrelation for short lags, decreasing slowly as the lag length increases. If several of the shorter lag sample autocorrelations are significant using the "rule of thumb" procedure and the acf graph has the appearance mentioned above, a trend model should be considered. It would be a good idea under these conditions to perform the test for $\rho = 0$ (which will probably reject $H_0$) to provide more explicit justification for the model.

## 3.4.7 Other Considerations

The tests we have discussed in this section should be sufficient to decide the trend/no-trend question in the vast majority of cases. Daniels' (or Kendall's) test and the one based on Pearson's coefficient are the most powerful among those given. Many other tests are known or could be constructed. For the sake of brevity, they are not discussed here.

If trend is discovered in a series, there is also the possibility that it may be detrended using rates, differences, or link relatives, as noted in Chapter 2. Once the detrending is done, a no-trend model may be adequate.

Finally, with trend models there is an additional question as to whether the trend is linear or curvilinear and, if curvilinear, whether to choose a quadratic, exponential, or some other model. Matters affecting this choice will be handled in the context of model fitting in the next few sections.

## 3.5 Models Equations for Trend

### 3.5.1 Introduction

Trended models are more complicated to fit than untrended ones. Horizontal models have only one form,

$$y_t = \beta_0 + \epsilon_t$$

and fitting involves estimating only one value, $\beta_0$, in some appropriate way. On the other hand, fitting trended models,

$$y_t = T_t + \epsilon_t$$

requires first choosing the functional form of $T_t$ (i.e., linear, quadratic, exponential, etc.) and then estimating the parameters of whatever functional form is chosen. With trend equations it is also necessary to pay more attention to the scale of measurement of the time variable, $t$.

### 3.5.2 Scale and Origin of *t*

When the average level of the series does not change with time, the scale and origin (units and zero point) of the time variable are minor concerns. However, trend equations are functions involving $t$. To evaluate the trend for some point in time, we must know the numerical value to substitute for $t$. To illustrate, suppose the trend equation

$$T_t = 1300(1.03)^t$$

is to be evaluated for 1987. We must know what zero point and units were used in developing the equation in order to know what the value of $t$ is in 1987. This information should be part of the statement of the equation. Thus, a more complete form of the expression is:

$$T_t = 1300 \cdot (1.03)^t \qquad \text{Origin: } 1981 \quad \text{Units } (t)\text{: Yearly}$$

We now know that $t = 6$ in 1987 (1981 = 0, 1982 = 1, 1983 = 2, . . . , 1987 = 6), so the calculated trend value for 1987 would be

$$T_t = T_6 = 1300 \times (1.03)^6 = 1300 \times (1.194) = 1{,}552$$

Another problem: While $t$ represents a *point* in time, 1981 is an interval. When working with units, especially when converting to smaller units, we need to know what point in 1981 is zero. Conventionally, any interval of time is assumed to be associated with its center point; therefore 1981 is actually taken to mean July 1, 1981, or 7/1/81, or, if you insist on more precision, midnight between June 30 and July 1, 1981. Getting it right to the day is sufficiently precise for most purposes, and often we can get by with just the

**Table 3.7 Trend Equation Conventions for Center Points**

| Units | Center Point |
|---|---|
| Yearly | 7/1/xx |
| Monthly | 1/15/xx, 2/15/xx, . . . , 12/15/xx |
| Quarterly | 2/15/xx, 5/15/xx, 8/15/xx, 11/15/xx |
| Semiannually | 4/1/xx, 10/1/xx |

month or quarter. The amount of precision required will depend on circumstances. In our trend expressions the conventions used will be those in Table 3.7.

### 3.5.3 Changing the Scale and Origin of $t$

When working with trend equations it is often convenient to convert one time scale to another. To evaluate the trend expression discussed earlier in this section, what was done amounted to converting 1987 in everyday time reference (Origin: Birth of Christ, Units ($t$): Yearly) to 6 in the scale used in the trend equation (Origin: 1981, Units ($t$): Yearly). To do this we noted that

$$\text{trend equation } t = \text{everyday } t - 1981$$

or

$$6 = 1987 - 1981$$

The general approach to shifting the origin or changing units can be summarized in a few simple rules. In multiple changes, each step should involve only one change (units or origin). Any number of steps may be carried out successively.

- **Rule 1** To shift the origin forward $k$ units, substitute $t + k$ for $t$ in the current form of the trend equation.
- **Rule 2** To shift the origin backward $k$ units, substitute $t - k$ for $t$ in the current form of the trend equation.
- **Rule 3** To decrease units by a factor of $c$, substitute $t/c$ for $t$ in the current form of the trend equation.
- **Rule 4** To increase units by a factor of $c$, substitute $ct$ for $t$ in the current form of the trend equation.

For instance, supposing that the trend equation is currently expressed in yearly units.

- To shift the origin forward 3 years, substitute $t + 3$ for $t$.
- To shift the origin backward 5 years, substitute $t - 5$ for $t$.
- To change to quarterly units, that is, to reduce units to one-quarter what they currently are, substitute $t/4$ for $t$.

- To change to 10-year units, ten times what they currently are, substitute $10t$ for $t$.

Finally, care should be taken to center the origin in the conventional middle of the new time-unit intervals. If you are changing from quarterly units to yearly units, remember that the origin for quarterly units is centered in the middle of a quarter (2/15/xx, 5/15/xx, 8/15/xx, or 11/15/xx). In years, the center should be at 7/1/xx, so if the origin is currently in the second quarter, it will need to be shifted forward one half-quarter to get it set at 7/1/xx. Reasons for this attention to origin and scale will become more apparent as more complicated trend models are encountered.

### EXAMPLE 3.6 *Changing the Scale and Origin of a Trend Equation, Linear*

On a section of interstate highway, $y_t$ = the yearly number of accidents has been increasing according to the following trend model:

$$\hat{T}_t = 40.31 + 4.20t \qquad \text{Origin: } 1980 \quad \text{Units } (t)\text{: Yearly}$$

Suppose we wish to change to monthly units with a January 1987 origin.

- **Move the origin to 1987** Since 1987 is $t = 7$ with the current origin, this move requires a 7-year ($k = 7$) forward shift, and $t$ becomes $t + 7$:

$$\hat{T}_t = 40.31 + 4.20 \cdot (t + 7) = 40.31 + 4.20t + 29.40$$
$$= 69.71 + 4.20t \qquad \text{Origin: } 1987 \quad \text{Units } (t)\text{: Yearly}$$

- **Change to monthly units** Since this is a decrease by a factor of 12, $t$ becomes $t/12$:

$$\hat{T}_t = 69.71 + 4.20 \cdot \frac{t}{12} = 69.71 + \frac{4.20}{12} \cdot t$$
$$= 69.74 + .35t \qquad \text{Origin: } 1987\ (7/1/87) \quad \text{Units } (t)\text{: Monthly}$$

Notice that the 7/1/87 needs to be made explicit because of the unconventional origin for monthly units as the equation now stands. (Also remember that the intercept, 69.74, is still in the units of $y_t$, i.e., yearly number of accidents. For a discussion of manipulating the units of $y_t$, see the next section.)

- **Change the origin to the conventional middle** Currently the model has monthly units centered at the first of the month. We could just shift forward one-half monthly unit, giving an origin of July 1987. But suppose we prefer a January 1987 origin instead, to make the equation more convenient for our intended use. This requires shifting the origin backward 5.5 monthly units (from the beginning of July back to the middle of January), then $t$ becomes $t - 5.5$:

$$\hat{T}_t = 69.71 + .35 \cdot (t - 5.5) = 69.71 + .35t - 1.93$$
$$= 67.78 + .35t \qquad \text{Origin: Jan } 1987 \quad \text{Units } (t)\text{: Monthly}$$

which is the desired form of the model. The first and third steps above could have been combined in one step by shifting the origin backward to January 15, 1980, to begin with ($t$ becomes $t - 5.5/12 = t - .4578$), but it is less confusing if the procedure is broken into simpler steps. ●

**EXAMPLE 3.7** *Changing the Scale and Origin, Exponential*

A trade union has found that $y_t$ = its members' yearly earnings ($) is described by the following trend model:

$$\hat{T}_t = 18.760 \cdot (1.056)^t \qquad \text{Origin: } 1982 \quad \text{Units } (t)\text{: Yearly}$$

It is desired to change to half-yearly (semiannual) units with an origin in the first half of 1980.

- **Change the origin to 1980, $t$ becomes $t - 2$**

$$\hat{T}_t = 18{,}760 \cdot (1.056)^{(t-2)} = 18{,}760 \cdot (1.056)^{-2} \cdot (1.056)^t$$
$$= 16{,}823 \cdot (1.056)^t \qquad \text{Origin: } 1980 \quad \text{Units } (t)\text{: Yearly}$$

- **Change to half-yearly units, $t$ becomes $t/2$**

$$\hat{T}_t = 16{,}823 \cdot (1.056)^{t/2} = 16{,}823 \cdot (\sqrt{1.056})^t$$
$$= 16{,}823 \cdot (1.028)^t \qquad \text{Origin: } 1980 \quad \text{Units } (t)\text{: Semiannual}$$

- **Change origin to the conventional one** For semiannual units the origin is at 4/1/xx, so we can move backward $\frac{1}{4}$ of a year, which is $\frac{1}{2}$ unit in the equation's current half-year-unit scale ($t$ becomes $t - .5$):

$$\hat{T}_t = 16{,}823 \cdot (1.028)^{t-.5} = 16{,}823 \cdot (1.028)^{-.5} \cdot (1.028)^t$$
$$= 16{,}592 \cdot (1.028)^t \qquad \text{Origin: I, } 1980 \quad \text{Units } (t)\text{: Semiannual}$$

which is the desired form of the model. ●

## 3.5.4 Changing the Units of $y_t$

The trend equation in Example 3.6 was for $y_t$ = the yearly number of accidents; this is called a *yearly aggregate.* When we express the model for $y_t$ in monthly time units, we are evaluating a yearly aggregate each month (the yearly aggregate presumably being taken over the preceding and succeeding six months). The interpretation of such a mixture of units can be confusing. For example, in the accident model

$$\hat{T}_t = 6.78 + .35t$$

the monthly change in the yearly number of accidents is .35, so we would have to say that the *yearly* number of accidents was increasing at a rate of .35 accidents per *month,* i.e., a rate of .35 accidents *per year per month.* Often the model is less confusing if the aggregate units of $y_t$ are changed to monthly as well. If we assume that the aggregate is evenly spread over the period, then a monthly aggregate is $\frac{1}{12}$ of a yearly aggregate. Likewise, a yearly aggregate would be 4 times a quarterly aggregate, and so forth. The operation may be accomplished with the following rules:

- **Rule 5** To increase the aggregate units of $y_t$ by a factor of $c$, we multiply the entire right-hand side of the trend equation by $c$.
- **Rule 6** To decrease the aggregate units of $y_t$ by a factor of $c$, we divide the entire right-hand side of the trend equation by $c$.

Thus changing from $y_t$ = the yearly number of accidents to $y_t$ = the monthly number of accidents gives

$$\hat{T}_t = \frac{67.78 + .35t}{12}$$
$$= 5.65 + .029t \qquad \text{Origin: Jan 1987} \quad \text{Units } (t)\text{: Monthly}$$

In this form of the model, .029 is the monthly change in the monthly number of accidents, so our interpretation would be that the number of accidents is growing at a rate of .029 accidents *per month per month.*

If we wanted to change from yearly to half-yearly aggregate earnings in Example 3.7 we would have

$$\hat{T}_t = \frac{16{,}592 \cdot (1.028)^t}{2}$$
$$= 8{,}296 \cdot (1.028)^t \qquad \text{Origin: I, 1980} \quad \text{Units } (t)\text{: Semiannual}$$

which would indicate a rate of increase of 2.8% *per half-year per half-year.* Notice that the rate, being a percentage, did not change when the aggregate was changed.

## 3.6 Linear Trend Models

### 3.6.1 Introduction

Our discussion of functional forms begins with linear models for several reasons. They represent the smallest step up in complexity from no-trend models and are, therefore, easiest to understand. They are widely implemented, and the mathematics of their application is well developed. Finally, many curvilinear series are nearly linear over limited intervals of time and may be approximated by linear models with reasonable accuracy over the short term.

In the linear case $T_t$ takes the form

$$T_t = \beta_0 + \beta_1 t$$

so the model becomes

$y_t$ = the actual value of the time series at time $t$

$\beta_0$ = the trend value at time zero

$\beta_1$ = the change in trend per unit time

$\epsilon_t$ = irregular fluctuation away from trend at time $t$

**Linear Trend Model**

$$y_t = \beta_0 + \beta_1 t + \epsilon_t$$

The parameter $\beta_0$ is called the **intercept,** or **constant term.** It is the trend value at the origin, $t = 0$. The other parameter, $\beta_1$, is called the **slope.** It is the amount by which the trend changes between successive time periods.

## 3.6.2 Properties of Linear Trend

The single most distinguishing characteristic of linear trend is its *constant incremental change.* Let's look at the trend at times $t$ and $t + 1$:

*At Time* t: $T_t = \beta_0 + \beta_1 t$

*At Time* t + *1*:

$$T_{t+1} = \beta_0 + \beta_1(t + 1) = (\beta_0 + \beta_1 t) + \beta_1$$
$$= T_t + \beta_1$$

We see that the trend changes by a constant amount ($\beta_1$) each time period. As a result, the first differences $\Delta y_t = y_t - y_{t-1}$ are

$$y_t = \beta_0 + \beta_1 t + \epsilon_t$$
$$y_{t-1} = \beta_0 + \beta_1(t - 1) + \epsilon_{t-1} = \beta_0 + \beta_1 t - \beta_1 + \epsilon_{t-1}$$
$$\Delta y_t = y_t - y_{t-1} = \beta_0 + \beta_1 t - \beta_0 - \beta_1 t + \beta_1 + (\epsilon_t - \epsilon_{t-1})$$
$$= \beta_1 + (\epsilon_t - \epsilon_{t-1}) = \beta_1 + \epsilon_t^*$$

So the $\Delta y_t$ from a linearly trended series form a horizontal series (average level $\beta_1$) with correlated error terms $\epsilon_t^* = \epsilon_t - \epsilon_{t-1}$. When graphed, the first differences should not show trend themselves. The link relatives $y_t/y_{t-1}$, however, will show trend because the constant change $\beta_1$ will tend to be a smaller percentage of $y_t$ as $y_t$ grows,or a lesser percentage as $y_t$ declines. The effect is usually not pronounced, however. A linear model may then be considered a reasonable choice if

- the graph of $y_t$ versus $t$ is not apparently curvilinear, and
- the first differences of the series are constant except for random error, so
- the graph of first differences does not show trend.

## 3.6.3 Single Forecast

As you will recall, in the single forecast case, a history of $n$ observations $y_1$, $y_2$, . . . , $y_n$ is used to estimate the parameters. In the linear case, these parameters would be $\beta_0$ and $\beta_1$. The estimates $\hat{\beta}_0$ and $\hat{\beta}_1$ then are used to generate forecasts for all future observations. There are several different ways to arrive at these estimates. The central aim of each is to find a line that passes "close" to all the points $(t, y_t)$.

### 3.6.4 Fitting a Line Through Two Points

Before we begin the discussion of different estimation methods, it will be helpful to recall how one fits a line through two points. Suppose we wish to know the slope $\hat{\beta}_1$ and intercept $\hat{\beta}_0$ of the line passing through two points, say, $(t1, y1)$ and $(t2, y2)$.

***Note:*** For convenience we are using the nonsubscripted notation $y1$, $y2$, rather than $y_1$, $y_2$.

Since the line must fit at these points, we know that

$$y1 = \hat{\beta}_0 + \hat{\beta}_1 t1$$

and

$$y2 = \hat{\beta}_0 + \hat{\beta}_1 t2$$

These two equations may be solved simultaneously, giving, for a general solution,

$$\hat{\beta}_1 = \frac{y2 - y1}{t2 - t1} \tag{3.1}$$

$$\hat{\beta}_0 = y1 - \hat{\beta}_1 t1 \qquad \text{or} \qquad \hat{\beta}_0 = y2 - \hat{\beta}_1 t2 \tag{3.2}$$

and the forecasts would be computed from

$$\hat{y}_t = \hat{T}_t = \hat{\beta}_0 + \hat{\beta}_1 \cdot t$$

or, at time $t$ for $p$ steps ahead,

$$\hat{y}_{t+p} = \hat{T}_{t+p} = \hat{\beta}_0 + \hat{\beta}_1 \cdot (t + p) = \hat{T}_t + p \cdot \hat{\beta}_1 \tag{3.3}$$

These expressions will be useful in the discussions that follow.

### 3.6.5 Free-hand Method

The least mathematically sophisticated approach to estimating the parameters of the linear model is called the **free-hand,** or **"eyeball," method.** To accomplish it we draw a scatter diagram of the points $(1, y_1)$, $(2, y_2)$, . . . , $(n, y_n)$ and then place a transparent ruler on the graph, arranging the ruler visually so that, in the aggregate, its edge passes as close to all the points as possible. Estimates of the slope and intercept may then be determined graphically from a line drawn along the edge of the ruler. First we read two points from the graph of the line, which become $(t1, y1)$, $(t2, y2)$ in Equations (3.1) and (3.2).

**EXAMPLE 3.8** *Fitting a Linear Model, Free-hand Method*

The free-hand procedure was carried out for the Valley Discount Liquor data of Example 3.1. Figure 3.5 shows the scatter diagram and the line, which was fit by eye. The circled points on the graph in Figure 3.5 were determined to be

$$(t1, y1) = (0.0, 1.5)$$
$$(t2, y2) = (10.0, 5.8)$$

The corresponding estimates computed from Equations 3.1 and 3.2 are

$$\hat{\beta}_1 = \frac{y2 - y1}{t2 - t1} = \frac{5.8 - 1.5}{10.0 - 0.0} = .43$$

$$\hat{\beta}_0 = y1 - \hat{\beta}_1 \cdot t1 = 1.5 - .43 \times 0 = 1.50$$

and

$$\hat{T}_t = 1.50 + .43t \qquad \text{Origin: II, 1980} \quad \text{Units } (t)\text{: Semiannual}$$

Forecasts calculated at time $t$ for $p$ periods ahead can now be calculated using

$$\hat{y}_{t+p} = \hat{T}_{t+p} = \underbrace{\hat{\beta}_0 + \hat{\beta}_1 \cdot (t + p)}_{\text{Form 1}} = \underbrace{\hat{T}_t + p \cdot \hat{\beta}_1}_{\text{Form 2}}$$

$$= 1.50 + .43 \cdot (t + p) = 1.50 + .43 \cdot t + .43 \cdot p$$

Either Form 1 or Form 2 may be used to compute the forecasts (they are entirely equivalent). We have elected to include both forms (although Form 1 is the more usual one) to give some emphasis to the practical facts that forecasts are always computed

**Figure 3.5** Free-hand Line Fit to Valley Discount Liquor Data

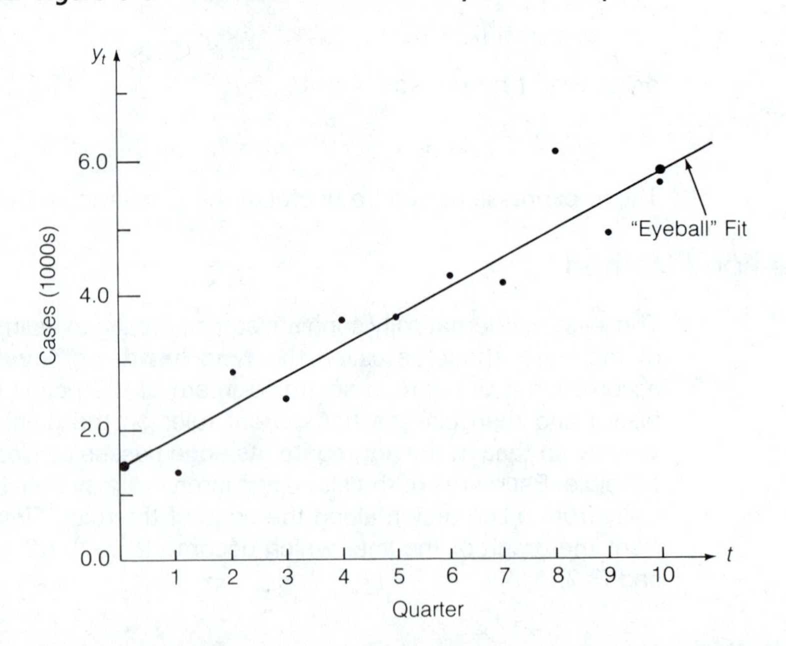

at some point in time and that, for the linear model, the forecast trend is just the estimated trend at that time plus $p$ times the slope.

Forecasts in II, 1985 ($t = 10$) for I, 1986 through II, 1987 ($t + p = 11, 12, 13, 14$ or $p = 1, 2, 3, 4$) are, then,

$$\hat{T}_t = \hat{T}_{10} = 1.50 + .43 \times 10 = 5.80$$

and

*I, 1986* ($p = 1$): $\hat{y}_{11} = 5.80 + 1 \times .43 = 6.23$ (6,230 cases)

*II, 1986* ($p = 2$): $\hat{y}_{12} = 5.80 + 2 \times .43 = 6.66$ (6,660 cases)

*I, 1987* ($p = 3$): $\hat{y}_{13} = 5.80 + 3 \times .43 = 7.09$ (7,090 cases)

*II, 1987* ($p = 4$): $\hat{y}_{14} = 5.80 + 4 \times .43 = 7.52$ (7,520 cases) •

**Single Forecasts Using Free-hand Method** Linear Model

**Given:** A sample of $n$ observations $(t, y_t)$; $t = 1, 2, \ldots, n$

**Model:** $y_t = \beta_0 + \beta_1 t + \epsilon_t$

**Estimates:** A line is fit free-hand to a scatter diagram of $(t, y_t)$ and two points $(t1, y1)$ and $(t2, y2)$ are read from the graph of the line. Then

$$\hat{\beta}_1 = \frac{y2 - y1}{t2 - t1}$$

$$\hat{\beta}_0 = y1 - \hat{\beta}_1 \cdot t1 \quad \text{or} \quad \hat{\beta}_0 = y2 - \hat{\beta}_1 \cdot t2$$

$$\hat{T}_t = \hat{\beta}_0 + \hat{\beta}_1 \cdot t$$

**Forecast:** (At time $t$ for $p$ periods ahead)

$$\hat{y}_{t+p} = \hat{\beta}_0 + \hat{\beta}_1 \cdot (t + p) = \hat{T}_t + p \cdot \hat{\beta}_1$$

The free-hand method is simple and intuitive. It also allows the analyst to interject subjective weighting into the model-fitting process. Aberrant or atypical data points (outliers) may be given diminished weight simply by partially ignoring them when adjusting the ruler for "best" fit. More current observations may be weighted more heavily by allowing them increased influence in determining the placement of the ruler.

Because of the subjectivity associated with eyeballing a fit, the free-hand method has distinct mathematical disadvantages. The sampling properties of the estimates are not mathematically treatable and the fit is not "repeatable." That is, the estimates of the slope and intercept depend on the perceptions and biases of the person doing the eyeballing. The fit line does not optimize an objective criterion (e.g., MSE, max $|e_t|$, MAD). However, from a practical standpoint, if the data are not overridden by noise (random error), different analysts should tend to produce comparable fits.

## 3.6.6 Method of Selected Points

Another way to estimate $\beta_0$ and $\beta_1$ is to select two points $(t1, y1)$ and $(t2, y2)$ from among the $n$ pairs $(1, y_1), (2, y_2), \ldots, (n, y_n)$ and fit the line through these two points. Any two points judged to be typical or representative of the trend may be used. Often the earliest and latest points in the history are chosen. They are the farthest apart in time and, therefore, represent the longest possible experience with the series. Once the points are selected, Equations 3.1 and 3.2 may be used to fit a line through them much the same as for the free-hand method. If $t1$ and $t2$ are chosen as 1 and $n$, respectively, then a slightly simplified form of the equations may be used:

$$\hat{\beta}_1 = \frac{y_n - y_1}{n - 1} \quad \text{and} \quad \hat{\beta}_0 = y_1 - \hat{\beta}_1$$

and

$$\hat{T}_t = \hat{\beta}_0 + \hat{\beta}_1 t$$

***Note:*** Choosing $t1 = 1$ and $t2 = n$ results in $\hat{\beta}_1$ being the average of the first differences $\overline{\Delta y_t} = \Sigma\, \Delta y_t/(n - 1)$.

**EXAMPLE 3.9** *Fitting a Linear Model, Selected Points Method, Valley Discount*

Applying selected points to the Valley Discount data using $t1 = 1$ and $t2 = n$, i.e., the points (1, 1.34) and (10, 5.68), gives

$$\hat{\beta}_1 = \frac{y_n - y_1}{n - 1} = \frac{5.68 - 1.34}{10 - 1} = \frac{4.34}{9} = .482$$

$$\hat{\beta}_0 = y_1 - \hat{\beta}_1 = 1.34 - .482 = .858$$

and

$$\hat{T}_t = .858 + .482t$$

Origin: II, 1980
Units ($t$): Semiannual

Using the selected-points line, forecasts in II, 1985 ($t = 10$) for 1986–1987 would be

$$\hat{T}_{10} = .858 + .482 \times 10 = 5.678$$

*I, 1986* (p = *1*): $\hat{y}_{11} = 5.678 + 1 \times .482 = 6.16$ (6,160 cases)

*II, 1986* (p = *2*): $\hat{y}_{12} = 5.678 + 2 \times .482 = 6.64$ (6,640 cases)

*I, 1987* (p = *3*): $\hat{y}_{13} = 5.678 + 3 \times .482 = 7.12$ (7,120 cases)

*II, 1987* (p = *4*): $\hat{y}_{14} = 5.678 + 4 \times .482 = 7.61$ (7,610 cases)

compared to $\hat{y}_{11} = 6.23$, $\hat{y}_{12} = 6.66$, $\hat{y}_{13} = 7.09$, and $\hat{y}_{14} = 7.52$ using the previously fit free-hand line. ●

The selected-points method is simple and straightforward, and the calculations are easy. It has intuitive appeal, since using $t1 = 1$ and $t2 = n$ is equivalent to setting the slope of the trend line equal to the average first difference $\overline{\Delta y_t}$, which is actually the average change per unit time from beginning to end of the series. A serious drawback, however, is that the results depend entirely on the two selected points and in no way on any of the others. Consequently, the estimates are very sensitive to whatever irregular movement may have taken place at $t = t1$ and $t = t2$. In addition, selecting which points are "representative" involves a certain amount of subjectivity.

**Single Forecasts Using Selected Points (First and Last Points)** Linear Model

**Given:** A sample of $n$ observations $(t, y_t)$; $t = 1, 2, \ldots, n$

**Model:** $y_t = \beta_0 + \beta_1 t + \epsilon_t$

**Estimates:** $\hat{\beta}_1 = \dfrac{y_n - y_1}{n - 1}$ $\quad \hat{\beta}_0 = y_1 - \hat{\beta}_1$ $\quad \hat{T}_t = \hat{\beta}_0 + \hat{\beta}_1 \cdot t$

**Forecasts:** (At time $t$ for $p$ periods ahead)

$$\hat{y}_{t+p} = \hat{\beta}_0 + \hat{\beta}_1 \cdot (t + p) = \hat{T}_t + p \cdot \hat{\beta}_1$$

### 3.6.7 Method of Semiaverages

The idea of estimating $\beta_0$ and $\beta_1$ by fitting a line through two typical selected points may be expanded to allow all (or nearly all) points to influence the estimates by dividing the sample into two equal parts, computing the average value of $t$ and $y_t$ in each part, and fitting the line through these two average points. In case there is an odd number of observations, several approaches are possible. A weighted average could be computed, giving, say, the middle observation a weight of $\frac{1}{2}$ in each part. Alternatively, one point could simply be discarded, perhaps the latest, or the oldest, or the middle point, or some point that seems atypical of the overall trend pattern.

We can then let

$\overline{y1}$ = the average value of $y_t$ in the early half of the series
$\overline{y2}$ = the average value of $y_t$ in the later half of the series
$\overline{t1}$ = the average time value in the early half of the series
$\overline{t2}$ = the average time value in the later half of the series

and fit the line through the two points $(\overline{t1}, \overline{y1})$ and $(\overline{t2}, \overline{y2})$, which gives us

$$\hat{\beta}_1 = \frac{\overline{y2} - \overline{y1}}{\overline{t2} - \overline{t1}} \quad \text{and} \quad \hat{\beta}_0 = y1 - \hat{\beta}_1 \overline{t1}$$

If there are $n$ observations $y_1, y_2, \ldots, y_n$, with $n$ even, then

$$\overline{t1} = \frac{1 + 2 + \cdots + n/2}{n/2} = \frac{n + 2}{4}$$

$$\overline{t2} = \frac{3n + 2}{4}$$

$$\overline{t2} - \overline{t1} = \frac{3n + 2}{4} - \frac{n + 2}{4} = \frac{n}{2}$$

So in this simplified case, the expressions for $\hat{\beta}_1$ and $\hat{\beta}_0$ are

$$\hat{\beta}_1 = \frac{2(\overline{y2} - \overline{y1})}{n} \quad \text{and} \quad \hat{\beta}_0 = \overline{y1} - \frac{n + 2}{4} \hat{\beta}_1$$

For convenience, if $n$ is odd we assume that the oldest point has been discarded and the $n - 1$ remaining historical observations have been renumbered from 1 to $n - 1$, effectively shifting the origin of the time scale one unit forward. The expressions then hold using the new value of $n$.

**EXAMPLE 3.10** *Fitting a Linear Model, Semiaverage Method*

Returning to the Valley Discount data for illustration: $n$ is 10, so there would be 5 observations in each subgroup and we would get

$$\overline{y1} = \frac{y_1 + y_2 + y_3 + y_4 + y_5}{5} = \frac{1.34 + 2.81 + 2.41 + 3.62 + 3.64}{5}$$

$$= \frac{13.82}{5} = 2.764$$

$$\overline{y2} = \frac{y_6 + y_7 + y_8 + y_9 + y_{10}}{5} = \frac{4.29 + 4.18 + 6.09 + 4.95 + 5.68}{5}$$

$$= \frac{25.19}{5} = 5.038$$

The estimates would then be

$$\hat{\beta}_1 = \frac{2(\overline{y2} - \overline{y1})}{n} = \frac{2(5.038 - 2.764)}{10} = .455$$

$$\hat{\beta}_0 = \overline{y1} - \frac{n+2}{4}\hat{\beta}_1 = 2.764 - \frac{10+2}{4} \times .455 = 1.399$$

and, finally,

$$\hat{y}_t = \hat{T}_t = 1.399 + .455t \qquad \text{Origin: II, 1980} \quad \text{Units } (t)\text{: Semiannual}$$

Forecasts in II, 1984 ($t = 10$) for 1985 and 1986 would be

$$\hat{T}_{10} = 1.399 + .455 \times 10 = 5.949$$

*For I, 1985* (p = *1*): $\hat{y}_{11} = 5.949 + 1 \times .455 = 6.40$ (6,400 cases)

*For II, 1985* (p = *2*): $\hat{y}_{12} = 5.949 + 2 \times .455 = 6.86$ (6,860 cases)

*For I, 1986* (p = *3*): $\hat{y}_{13} = 5.949 + 3 \times .455 = 7.31$ (7,310 cases)

*For II, 1986* (p = *4*): $\hat{y}_{14} = 5.949 + 4 \times .455 = 7.77$ (7,770 cases) •

## Pros and Cons

Without introducing an inordinate amount of mathematical complexity into the calculations, then, semiaveraging overcomes two of the chief shortcomings of selecting points: All of the points (or at worst, all but one) enter into the calculation of the estimates and there is little or no subjectivity involved in determining the two points through which the fit is made. The chief complaint about the method is that it may not completely optimize any of the usual criteria (such as the MSE and MAD), but a good case can be made for using semiaveraging because of its intuitive attractiveness. A reasonably good intuitive method is more likely to be implemented than a more complex optimal one that is ill-understood.

**Single Forecasts Using Semiaverages** Linear Model

**Given:** A sample of $n$ observations $(t, y_t)$; $t = 1, 2, \ldots, n$, $n$ even. (For odd sample sizes, the oldest point is discarded and the observations renumbered.)

**Model:** $y_t = \beta_0 + \beta_1 t + \epsilon_t$

**Estimates:** $\hat{\beta}_1 = \dfrac{2(\overline{y2} - \overline{y1})}{n}$ $\quad \hat{\beta}_0 = \overline{y1} - \dfrac{n+2}{4}\hat{\beta}_1$ $\quad \hat{T}_t = \hat{\beta}_0 + \hat{\beta}_1 t$

(If it is desired to return the time scale to its value prior to discarding a point for $n$ odd, then $\hat{\beta}_1$ becomes $\hat{\beta}_0 - \hat{\beta}_1$.)

**Forecasts:** (At time $t$ for $p$ steps ahead)

$$\hat{y}_{t+p} = \hat{\beta}_0 + \hat{\beta}_1 \cdot (t + p) = \hat{T}_t + p \cdot \hat{\beta}_1$$

## 3.6.8 Least Squares Method

The most popular method of estimating $\beta_0$ nd $\beta_1$ is to choose the values that would have predicted the historical data best, in the sense that the aggregate squared forecast errors would have been as small as possible, that is, the value that would have optimized the MSE criterion discussed in Chapter 2. This **least squares method** amounts to selecting the line that passes as "near" as possible to all the points in the scatter diagram of $y_t$ versus $t$, nearness being measured by squared distance between what actually happened and what the line would have predicted. The concept is similar to that involved in free-hand curve fitting when the ruler is positioned visually so as to be near all the points. The important difference is that we now have a precise measure, and the only subjectivity involved is that of choosing the squared distance as the nearness criterion.

Suppose we have a sample of $n$ observations $y_1, y_2, \ldots, y_n$ at times $t_1, t_2, \ldots, t_n$. To find least squares estimates we need to find the values $\hat{\beta}_0$ and $\hat{\beta}_1$ that minimize the sum of squared forecast errors. This sum of squared errors, SSE, often called the *error sum of squares,* is then

$$SSE = \Sigma (y_t - \hat{y}_t)^2 = \Sigma (y_t - \hat{\beta}_0 - \hat{\beta}_1 t)^2$$

The resulting estimates, called **least squares regression coefficients** (discussed at length in Chapter 4), can be shown to be

$$\hat{\beta}_1 = \frac{SS_{ty}}{SS_{tt}} \qquad \hat{\beta}_0 = \bar{y} - \hat{\beta}_1 \bar{t}$$

and the forecasting equations are

$$\hat{T}_t = \hat{\beta}_0 + \hat{\beta}_1 t$$

$$\hat{y}_{t+p} = \hat{\beta}_0 + \hat{\beta}_1 \cdot (t + p) = \hat{T}_t + p \cdot \hat{\beta}_1$$

where $S_{ty}$ and $SS_{tt}$ are as defined in Section 3.4.

If the time scale is adjusted so the sample consists of observations $(1, y_1), (2, y_2), \ldots, (n, y_n)$, these expressions can be simplified to

$$SS_{tt} = \frac{n(n^2 - 1)}{12} \qquad SS_{ty} = \Sigma\, ty - \frac{n+1}{2} \cdot \Sigma\, y$$

$$\hat{\beta}_1 = \frac{12 \cdot SS_{ty}}{n(n^2 - 1)} \qquad \hat{\beta}_0 = \bar{y} - \frac{n+1}{2}\hat{\beta}_1$$

**Single Forecasts Using Least Squares** Linear Model

**Given:** A sample of $n$ observations $y_1, y_2, \ldots y_n$ at times $t_1, t_2, \ldots, t_n$.

**Model:** $y_t = \beta_0 + \beta_1 t + \epsilon_t$

**Estimates:**

$$SS_{ty} = \Sigma\,(t - \bar{t})(y - \bar{y}) = \Sigma\, ty - \frac{(\Sigma\, t)(\Sigma\, y)}{n}$$

$$SS_{tt} = \Sigma\,(t - \bar{t})^2 = \Sigma\, t^2 - \frac{(\Sigma\, t)^2}{n}$$

$$\hat{\beta}_1 = \frac{SS_{ty}}{SS_{tt}} \qquad \hat{\beta}_0 = \bar{y} - \hat{\beta}_1\bar{t}$$

$$\hat{y}_t = \hat{T}_t = \hat{\beta}_0 + \hat{\beta}_1 t$$

In the usual case where the time scale is adjusted so that $t_1 = 1, t_2 = 2, \ldots, t_n = n$,

$$\bar{t} = \frac{n+1}{2}$$

$$SS_{tt} = \frac{n(n^2 - 1)}{12} \qquad SS_{ty} = \Sigma\, ty - \frac{n+1}{2}\Sigma\, y$$

$$\hat{\beta}_1 = \frac{12SS_{ty}}{n(n^2 - 1)} \qquad \hat{\beta}_0 = \bar{y} - \frac{n+1}{2}\hat{\beta}_1$$

**Forecasts:** (At time $t$ for $p$ steps ahead)

$$\hat{y}_{t+p} = \hat{\beta}_0 + \hat{\beta}_1\,(t + p) = \hat{T}_t + p \cdot \hat{\beta}_1$$

**EXAMPLE 3.11** ***Fitting a Linear Model, Least Squares Method, Valley Discount***

Turning once more to the Valley Discount data for illustration, the calculations necessary for determining $\hat{\beta}_0$ and $\hat{\beta}_1$ are carried out in Table 3.8. Then

$$SS_{tt} = \Sigma\, t^2 - \frac{(\Sigma\, t)^2}{n} = 385 - \frac{(55)^2}{10} = 82.5$$

$$SS_{ty} = \Sigma\, ty - (\Sigma\, t) \cdot \frac{\Sigma\, y}{n} = 251.94 - 55 \cdot \frac{(39.01)}{10} = 37.385$$

and the estimates are

$$\hat{\beta}_1 = \frac{SS_{ty}}{SS_{tt}} = \frac{37.385}{82.5} = .453$$

$$\hat{\beta}_0 = \bar{y} - \hat{\beta}_1\bar{t} = \frac{39.01}{10} - .453 \times \frac{55}{10} = 1.409$$

$$\hat{T}_t = 1.409 + .453t$$

Origin: II, 1980
Units ($t$): Semiannual
Units ($y$): $10^3$ cases

**Table 3.8 Calculations for Least Squares Fit, Valley Discount Liquor, Inc.**

| $t$ | $y_t$ | $ty_t$ | $t^2$ | $y_t^2$ |
|---|---|---|---|---|
| 1 | 1.34 | 1.34 | 1 | 1.80 |
| 2 | 2.81 | 5.62 | 4 | 7.90 |
| 3 | 2.41 | 7.23 | 9 | 5.81 |
| 4 | 3.62 | 14.48 | 16 | 13.10 |
| 5 | 3.64 | 18.20 | 25 | 13.25 |
| 6 | 4.29 | 25.74 | 36 | 18.40 |
| 7 | 4.18 | 29.26 | 49 | 17.47 |
| 8 | 6.09 | 48.72 | 64 | 37.09 |
| 9 | 4.95 | 44.55 | 81 | 24.50 |
| 10 | 5.68 | 56.80 | 100 | 32.26 |
| 55 | 39.01 | 251.94 | 385 | 171.58 |

Because the $t$ values are 1, 2, . . . , 10, the alternative calculation could be used to get

$$\hat{\beta}_1 = \frac{12 \cdot SS_{ty}}{n(n^2 - 1)} = \frac{12 \times 37.385}{10(10^2 - 1)} = .453$$

$$\hat{\beta}_0 = \bar{y} - \frac{n+1}{2}\hat{\beta}_1 = \frac{39.01}{10} - \frac{10+1}{2} \times .453 = 1.409$$

which are the same results as before.

Using the least squares model in II, 1985 ($t = 10$) to forecast 1986 and 1987 ($p = 1$, 2, 3, and 4 periods ahead), we get

$$\hat{T}_{10} = 1.409 + .453 \times 10 = 5{,}939$$

*I, 1986* ($p = 1$): $\hat{y}_{11} = 5.939 + 1 \times .453 = 6.39$ (6,390 cases)

*II, 1986* ($p = 2$): $\hat{y}_{12} = 5.939 + 2 \times .453 = 6.84$ (6,840 cases)

*I, 1987* ($p = 3$): $\hat{y}_{13} = 5.939 + 3 \times .453 = 7.30$ (7,300 cases)

*II, 1987* ($p = 4$): $\hat{y}_{14} = 5.939 + 4 \times .453 = 7.75$ (7,750 cases) •

For purposes of comparison, the forecasts yielded by all four methods (free-hand (FH), selected points (SP), semiaverages (SA), and least squares (LS)) are given in Table 3.9. You will notice that there is no appreciable difference

**Table 3.9 Forecasts in 1984 of White Wine Purchases for 1985 and 1986, Valley Discount Liquor, Inc.**

| Period | Periods Ahead | Forecast Number of Cases Purchased | | | |
|---|---|---|---|---|---|
| | | FH | SP | SA | LS |
| I, 1985 | 1 | 6,230 | 6,160 | 6,400 | 6,390 |
| II, 1985 | 2 | 6,660 | 6,640 | 6,860 | 6,840 |
| I, 1986 | 3 | 7,090 | 7,120 | 7,310 | 7,300 |
| II, 1986 | 4 | 7,520 | 7,610 | 7,770 | 7,750 |

FH = free-hand; SP = selected points; SA = semiaverages; LS = least squares.

ference between the SA and LS forecasts, which are around 200 or so cases larger than the FH and SP forecasts. The difference is due to the II, 1984 observation ($t = 8$) (see Figure 3.4), which was quite a bit higher than would have been anticipated. This observation did not enter into the SP estimation and was given less subjective weight by the analyst when positioning the transparent ruler for the FH method. However, this point was given equal weight with the other observations when computing the SA and LS lines and tended to "pull" the fit line up towards it.

### Pros and Cons

By and large, least squares is the preferred of the four methods discussed. The main reasons for this preference include the mathematical tractability of least squares methods, and their optimality under an objective criterion (MSE). The sampling behavior of the estimates is well known, provided one assumes that the irregular movements $\epsilon_t$ are independent and normally distributed. Thus it is possible to construct confidence intervals for $\beta_0$ and $\beta_1$ and prediction intervals for $y_{t+p}$.

On the negative side, least squares is the most complex of the four methods, both conceptually and computationally. It is difficult for those with little mathematical training to understand it fully and, hence, could lead to forecasts' being ignored or given little credence by those who are supposed to be using them. Other criticisms of least squares might include questioning the justification of "squared error" as a "goodness" criterion and noting the fact that the squared-error criterion gives larger inherent weight to aberrant data values.

## 3.6.9 Single Forecast, Normal Errors

As mentioned above, if we are prepared to assume that the error terms are independent and normally distributed, it is possible to construct prediction intervals for $y_{t+p}$, i.e., intervals in which we have confidence that the actual value of $y_{t+p}$ will fall. Specifically, if the $\epsilon_t$'s are normally distributed and independent, and the actual series $y_t$ consists of a linear trend, $\beta_0 + \beta_1 t$, plus the error term, then $y_t$ will follow a normal distribution with a mean of $\beta_0 + \beta_1 t$ and a standard deviation of $\sigma_\epsilon$; i.e., $y_t \sim N(\beta_0 + \beta_1 t, \sigma_\epsilon)$. Most of the time, $\beta_0$, $\beta_1$, and $\sigma_\epsilon$ are unknown and need to be estimated from the data. The least squares estimate of $\sigma_\epsilon$ is referred to variously as the "estimated standard error about the trend line," the "estimated standard deviation about the least squares line," or, simply, "the standard error of the estimate." It is computed from

$$s_\epsilon = \hat{\sigma}_\epsilon = \text{the estimated standard error about the trend line}$$

$$= \sqrt{\frac{\text{sum of squared forecast errors}}{\text{number of observations} - 2}}$$

$$= \sqrt{\frac{\Sigma (y_t - \hat{y}_t)^2}{n - 2}} = \sqrt{\frac{\text{SSE}}{n - 2}}$$

where $\text{SSE} = \Sigma (y_t - \hat{y}_t)^2$.

Note that $\hat{\sigma}_\epsilon$ is similar to, but not quite the same as, the RMSE defined in Chapter 1. The mathematics of least squares methodology require the division by $n - 2$ rather than $n$.

Usually, rather than computing each forecast error, squaring and summing, it is easier to compute the sum of squared errors SSE from the following relationship:

$$\text{SSE} = \text{SS}_{yy} - \hat{\beta}_1 \text{SS}_{ty} = \text{SS}_{yy} - \frac{\text{SS}_{ty}^2}{\text{SS}_{tt}}$$

So

$$s_\epsilon = \hat{\sigma}_\epsilon = \sqrt{\frac{\text{SS}_{yy} - \text{SS}_{ty}^2/\text{SS}_{tt}}{n - 2}}$$

Prediction intervals may now be constructed using percentage points of the $t$-distribution and the formulas given in Table 3.10. (For a more detailed discussion of these formulas, the reader is referred to any good introductory statistics text, such as Mendenhall, et al., 1986.)

**Table 3.10 Single Forecasts Using Least Squares, Normal Case** $y_t \sim N(\beta_0 + \beta_1 t, \sigma_\epsilon)$

**Given:** A sample of $n$ observations $(t_1, y_1), \ldots, (t_n, y_n)$, with $\beta_0$, $\beta_1$, and $\sigma_\epsilon$ unknown.

**Estimates:**

$$\hat{\beta}_1 = \frac{\text{SS}_{ty}}{\text{SS}_{tt}} \qquad \hat{\beta}_0 = \bar{y} - \hat{\beta}_1 \bar{t}$$

$$s_\epsilon = \hat{\sigma}_\epsilon = \sqrt{\frac{\text{SSE}}{n - 2}} \qquad \text{SSE} = \Sigma (y_t - \hat{y}_t)^2$$

$$= \text{SS}_{yy} - \frac{\text{SS}_{ty}^2}{\text{SS}_{tt}} = \text{SS}_{yy} - \hat{\beta}_1 \text{SS}_{ty}$$

If $t_1 = 1, t_2 = 2, \ldots, t_n = n$, then one may use:

$$\text{SS}_{tt} = \frac{n(n^2 - 1)}{12} \qquad \text{SS}_{ty} = \Sigma\, ty - \frac{n + 1}{2} \Sigma\, y \qquad \bar{t} = \frac{n + 1}{2}$$

$$\hat{\beta}_1 = \frac{12\text{SS}_{ty}}{n(n^2 - 1)} \qquad \hat{\beta}_0 = \bar{y} - \frac{n + 1}{2} \hat{\beta}_1$$

**Forecast:** (At time $t$ for $p$ periods ahead)

$$\hat{y}_{t+p} = \hat{\beta}_0 + \hat{\beta}_1 (t + p) = T_t + p \cdot \hat{\beta}_1$$

$(1 - \alpha) \times 100\%$ Prediction Interval:

$$\hat{y}_{t+p} \pm B = \hat{\beta}_0 + \hat{\beta}_1 (t + p) \pm B$$

$$B = t_{\alpha/2} s_\epsilon \sqrt{1 + \frac{1}{n} + \frac{(t + p - \bar{t})^2}{\text{SS}_{tt}}}$$

where $t_{\alpha/2}$ is the upper $\alpha/2 \times 100\%$ point of Student's $t$-distribution with $(n - 2)$ degrees of freedom.

**EXAMPLE 3.12** *Least Squares Prediction Interval, Valley Discount Liquor*

To construct prediction intervals for the least squares forecasts obtained for Valley Discount (which will require assuming that the random errors are independent and normally distributed), we recall that

$$n = 10 \qquad SS_{yy} = 19.402 \qquad SS_{ty} = 37.385 \qquad SS_{tt} = 82.5 \qquad \Sigma t = 55$$

$$\hat{\beta}_0 = 1.409 \qquad \hat{\beta}_1 = .453$$

and

$$\hat{T}_t = 1.409 + .453t \qquad \text{Origin: II, 1980} \quad \text{Units } (t)\text{: Semiannual}$$

Then the estimated standard error about the trend would be obtained by first calculating the error sum of squares:

$$SSE = SS_{yy} - \frac{SS_{ty}^2}{SS_{tt}} = 19.402 - \frac{37.385^2}{82.5} = 2.485$$

and then calculating

$$s_\epsilon = \sqrt{\frac{SSE}{n-2}} = \sqrt{\frac{2.485}{8}} = .5573$$

A 95% prediction interval ($\alpha = 5\%$) for *I*, 1986 ($t = 11$) as of II, 1985 ($t = 10$) would be

$$\hat{T}_{10} = 1.409 + .453 \times 10 = 5.939$$

$$(p = 1) \qquad \hat{y}_{11} = 5.939 + 1 \times .453 = 6.39$$

$$\frac{\alpha}{2} = .025$$

$$t_{.025} = 2.306 \qquad (n - 2 = 10 - 2 = 8 \text{ degrees of freedom})$$

$$B = 2.306 \times .5573\sqrt{1 + \frac{1}{10} + \frac{(10 + 1 - 5.5)^2}{82.5}} = 1.285\sqrt{1.467}$$

$$= 1.556$$

95% Prediction Interval:

$$6.39 \pm 1.56 = (4.83, 7.95)$$

Providing then that the assumptions of independence and normally distributed error terms are valid, we can be 95% confident that unit sales will be between 4,830 cases and 7,950 cases.

The bound on the forecast error becomes larger the farther into the future you wish to forecast, because the $(t + p - \bar{t})^2$ term in the formula for $B$ grows as $p$ grows. The 95% bound, which was 1.56, or 1,560 cases, for I, 1986, would be

$$B = 2.306 \times .5573\sqrt{1 + \frac{1}{10} + \frac{(10 + 4 - 5.5)^2}{82.5}} = 1.81$$

or 1,810 cases for II, 1986 ($p = 4$), and

$$B = 2.306 \times .5573\sqrt{1 + \frac{1}{10} + \frac{(10 + 8 - 5.5)^2}{82.5}} = 2.22$$

or 2,220 cases for II, 1989 ($p = 8$). •

## 3.7 Curvilinear Trend Models

### 3.7.1 Introduction

A curvilinear model is needed if the amount by which a series may be expected to grow (or decline) from period to period changes markedly over the forecast horizon. Under these circumstances, the graph of the series itself will be nonlinear and that of the first differences will not be horizontal. The first differences of the Procter & Gamble data of Example 3.2, graphed in Figure 3.6, show an upward trend, indicating that a curvilinear model is needed. The Valley Discount Liquor data, Figure 3.7, show no such trend, so a linear model will probably suffice.

**Figure 3.6** First Differences of Procter & Gamble Data

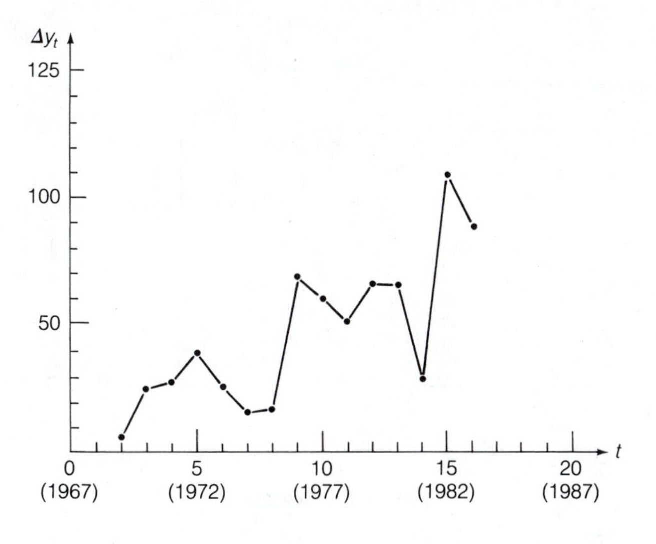

**Figure 3.7** First Differences of Valley Discount Data

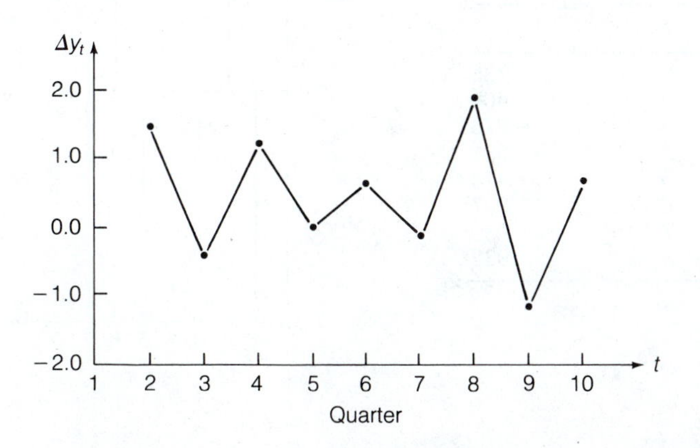

The sample autocorrelation function of the first-difference series can also provide information about curvilinearity. If the acf of this series has characteristics corresponding to trend (see Table 2.10 and Figure 2.14), a curvilinear model will be needed. The acf's of Procter & Gamble and Valley Discount in Figure 3.8 illustrate the point. Note that for Procter & Gamble the short lag autocorrelations are positive (up to lag 6) then turn negative, indicating trend. The same is not true for Valley Discount.

The type of curvilinear model to use depends on the shape of the time series, in particular on the way that the first differences are changing. Many different functional forms have been used, and individual circumstances may

**Figure 3.8 Autocorrelation Functions of First Differences for Proctor & Gamble and Valley Discount**

**(a) Proctor & Gamble**

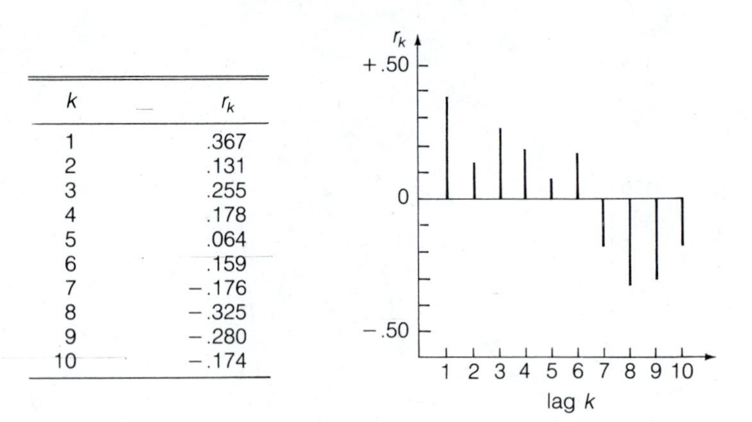

| $k$ | $r_k$ |
|---|---|
| 1 | .367 |
| 2 | .131 |
| 3 | .255 |
| 4 | .178 |
| 5 | .064 |
| 6 | .159 |
| 7 | −.176 |
| 8 | −.325 |
| 9 | −.280 |
| 10 | −.174 |



**(b) Valley Discount**

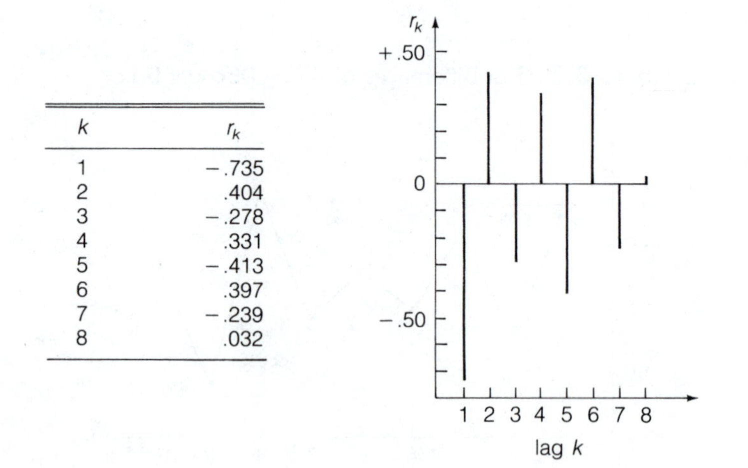

| $k$ | $r_k$ |
|---|---|
| 1 | −.735 |
| 2 | .404 |
| 3 | −.278 |
| 4 | .331 |
| 5 | −.413 |
| 6 | .397 |
| 7 | −.239 |
| 8 | .032 |

require specialized types of curves. However, there are several widely used curves that will now be discussed (Gilchrist, 1976, ch. 9; Meade, 1984).

### 3.7.2 Quadratic Models

Using the fact that most mathematical functions can be closely approximated by polynomials, one step in constructing a curvilinear model might be going from a first-degree (linear) polynomial to a second-degree **(quadratic)** polynomial. In this form, $T_t = \beta_0 + \beta_1 t + \beta_2 t^2$, and the model is as follows.

**Quadratic Trend Model**

$$y_t = \beta_0 + \beta_1 t + \beta_2 t^2 + \epsilon_t$$

A quadratic trend has linear first differences, since

$$\begin{aligned}\Delta T_t = T_t - T_{t-1} &= \beta_0 + \beta_1 t + \beta_2 t^2 - \beta_0 - \beta_1(t-1) - \beta_2(t-1)^2 \\ &= (\beta_1 + \beta_2) + 2\beta_2 t\end{aligned}$$

which is the equation of a line with intercept $(\beta_1 + \beta_2)$ and slope $2\beta_2$. The second differences are constant, however:

$$\Delta^2 T_t = \Delta T_t - \Delta T_{t-1} = (\beta_1 + \beta_2) + 2\beta_2 t - (\beta_1 + \beta_2) - 2\beta_2(t-1) = \beta_2$$

When graphed, $T_t$ will be a section of a parabola, as shown in Figure 3.9(a). One useful form of the model for time series purposes occurs when both $\beta_1$ and $\beta_2$ are positive (Figure 3.9(b)), which leads to ever-increasing first differences (positive second differences). Positive $\beta_1$ and negative $\beta_2$ (Figure 3.9(c)) will yield decreasing first differences (negative second differences). However, note that setting the derivative

$$\frac{dT_t}{dt} = \frac{d(\beta_0 + \beta_1 t + \beta_2 t^2)}{dt} = \beta_1 + 2\beta_2 t$$

equal to zero reveals that the curve turns over (reaches a maximum, since $dT_t/dt^2 = 2\beta_2 < 0$) at

$$t = \frac{-\beta_1}{2\beta_2}$$

So extrapolating the model far ahead in time may produce misleading forecasts (dashed line in Figure 3.9(c)).

It is quite possible to use polynomials of degree greater than 2, but we recommend against it. In fact, one (rather naive) philosophy of model fitting is to keep raising the power of the polynomial until no better fit can be obtained. This practice is a very bad one, both from the point of view of its lack of a rational basis and because of the poor statistical properties of the resulting forecasts. Even quadratic models suffer from these problems to a certain degree.

**Figure 3.9 Representative Quadratic Trend Curves**

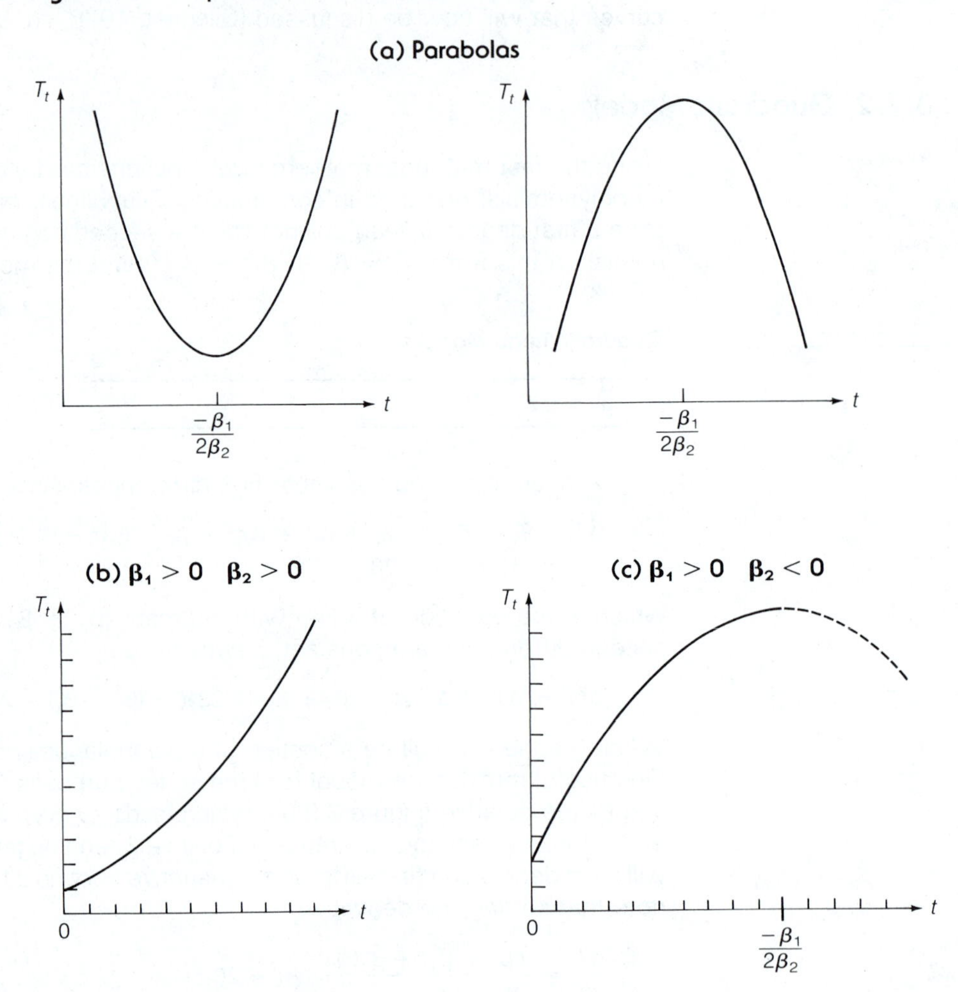


## 3.7.3 Simple Exponential Growth Models

Since many time series seem to grow (or decline) at a constant percentage rate, a trend equation with this characteristic is often useful. The **simple exponential** model, representing unbridled period-to-period percentage growth, has trend equation $T_t = \beta_0 \cdot \beta_1^t$ and is expressed as follows.

**Simple Exponential Trend Model**

$$y_t = \beta_0 \cdot \beta_1^t \cdot \epsilon_t$$

Notice that the error term in this model is a multiplicative factor rather than an additive one, making it a bit different from models previously encountered.

As a result, $\epsilon_t$ is a dimensionless variable (much like a link relative) by which the trend value is multiplied to give the actual time series value. The elements of the model may be interpreted as

$\beta_0$ = the value of the trend at time zero

$\beta_1$ = the amount by which the trend value in one period is multiplied to calculate the trend value in the next period

$\epsilon_t$ = the relative movement of the time series away from the trend at time $t$

If $\beta_1$ is written as $\beta_1 = 1 + r$, then the trend is growing (or declining) at constant rate $r \times 100\%$ per time period. Thus the equation

$$T_t = 2600(1.027)^t = 2600(1 + .027)^t$$

implies the trend value at the origin is 2600 and trend is growing at a constant rate of 2.7% per period. The series at time $t = 10$, using $\epsilon_{10} = .934$ for hypothetical example, would be

$$y_{10} = T_{10} \cdot \epsilon_{10} = 2600(1.027)^{10}(.934) = 3{,}394 \times .934 = 3{,}170$$

which means there is a movement of −6.6% (.934 = 1 − .066, or downward 6.6%) from trend (3, 394) in period 10. A trend equation such as

$$T_t = 83.2(.982)^t = 83.2(1 - .018)^t$$

would represent a decline of 1.8% per unit of time. The link relatives $T_t/T_{t-1}$ of a simple exponential trend are constant, since

$$\frac{T_t}{T_{t-1}} = \frac{\beta_0 \cdot \beta_1^t}{\beta_0 \cdot \beta_1^{t-1}} = \beta_1$$

So a time series to be modeled with such a trend should have no trend or other pattern (except random movement) in the graph of its link relatives $y_t/y_{t-1}$. Examples of a simple exponential trend for $\beta_1 > 1$ (growth) and $\beta_1 < 1$ (decline) are shown in Figure 3.10.

**Figure 3.10** **Simple Exponential Trend Curves**

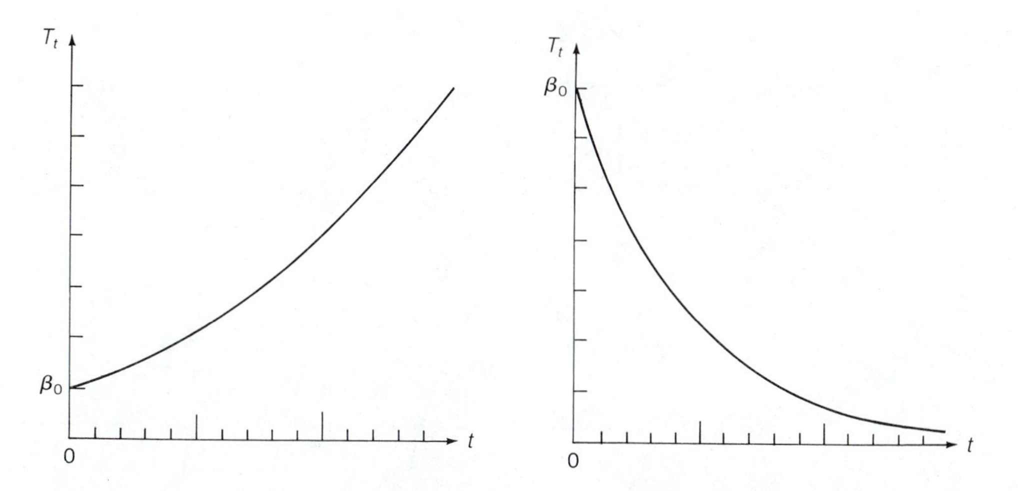

### 3.7.4 Modified Exponential Models

Use of a simple exponential model is sometimes criticized because there is no limit to its growth. It may be more realistic to use a trend curve with an upper limit on this growth. Such a limit is provided by the **modified exponential** model, whose trend equation is $T_t = \beta_0 + \beta_1 \cdot \beta_2^t$.

**Modified Exponential Trend Model**

$$y_t = \beta_0 + \beta_1 \cdot \beta_2^t \cdot \epsilon_t$$

or

$$y_t = \beta_0 + \beta_1 \cdot \beta_2^t + \epsilon_t$$

The characteristics of this model are more complex than those of the other models we have discussed. The error terms may be either additive or

**Figure 3.11** Modified Exponential Trend Curves

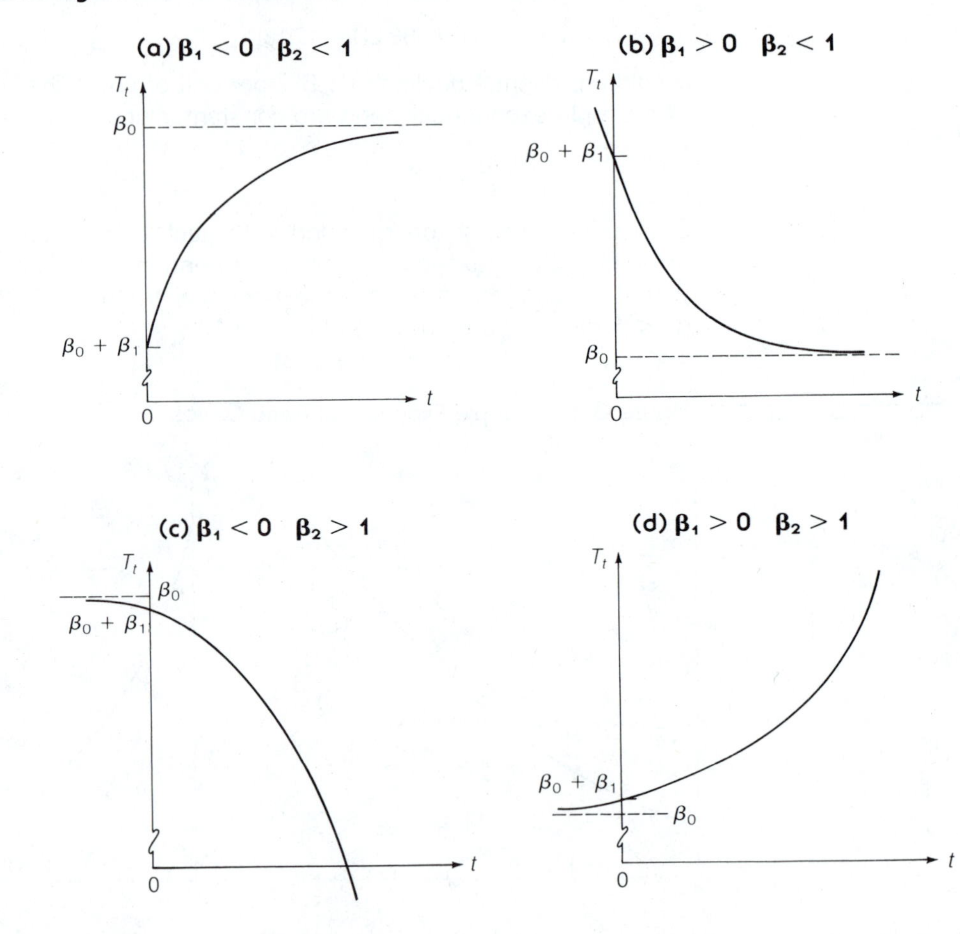

multiplicative. The ratio of successive first differences, $(y_t - y_{t-1})/(y_{t-2} - y_{t-2}) = \Delta y_t/\Delta y_{t-1} = \beta_2$, are constant. Also,

$$\beta_0 = \text{upper (or lower) limit on the trend values}$$

$$\beta_0 + \beta_1 = \text{the trend value at the origin}$$

If $\beta_1$ is less than zero, $\beta_0$ is an **upper asymptote (limit),** and if $\beta_1$ is greater than zero, $\beta_0$ is a **lower asymptote (limit).** The parameters $\beta_1$ and $\beta_2$ do not have simple intuitive interpretations, but some insight may be obtained by noting that $y_t^* = y_t - \beta_0$ follows a simple exponential trend with parameters $\beta_1$ and $\beta_2$; i.e.,

$$y_t^* = \beta_1 \cdot (\beta_2)^t \cdot \epsilon_t$$

(If $\beta_0 = 0$, the modified and simple exponential models are identical.) We can then see that $\beta_1$ is the trend value of $y_t - \beta_0$ at the origin, and $y_t - \beta_0$ is changing at a constant rate of $1 - \beta_2$ per unit of time. Graphs of the four possible modified exponential curves are shown in Figure 3.11. The most useful of these is probably that in Figure 3.11(a) with $\beta_1 < 0$ and $\beta_2 < 1$, which gives the upper limit we mentioned.

## 3.7.5 "S-shaped" Trends

Some phenomena (e.g., demand for a newly developed product) exhibit rapid growth early in their history followed by a decelerating growth rate until they reach an upper "saturation" level accompanied by practically no growth. A trend curve with an "S-shape" may be useful in these cases. Such a curve can be obtained by fitting a modified exponential model to either the natural logarithms $\log_e y_t$ or the reciprocals $1/y_t$ of the series.

### The Gompertz Model

If the natural logarithms of the trend are described by a modified exponential curve, so that

$$T_t' = \log_e T_t = \beta_0' + \beta_1' \cdot \beta_2'^t$$

then the trend curve itself is

$$T_t = \beta_0 \cdot \beta_1^{\beta_2^t} \tag{3.4}$$

where

$$\beta_0 = e^{\beta_0'} \qquad \beta_1 = e^{\beta_1'} \qquad \beta_2 = \beta_2'$$

***Note:*** "$e$" here is the base of the natural logarithms, $e = 2.71828 \ldots$ . Also, when it is not likely to cause confusion we will suppress the use of $\log_e$ and simply use log as the notation for natural logarithms. In reality it does not matter whether we use common ($\log_{10}$) or natural ($\log_e$) logarithms. We have chosen $\log_e$ here simply because it's use is somewhat more convenient in statistical applications.

Equation 3.4 is known as the **Gompertz curve.** If the parameter $\beta_1$ is between 0 and 1, then

$$\beta_0 = \text{the upper ``saturation'' level}$$

$$\beta_0 \cdot \beta_1 = \text{the value of the trend at the origin}$$

and $\beta_2$ determines how fast the rate of growth declines. For the model to be useful, $\beta_1$ should lie between 0 and 1. A representative curve is shown in Figure 3.12.

**Gompertz Trend Model**

$y_t = \beta_0 \cdot \beta_1^{\beta_2^t} \cdot \epsilon_t$

or

$y'_t = \log_e y_t = \beta'_0 + \beta'_1 \cdot \beta_2'^t + \epsilon'_t$

$\beta_0 = e^{\beta'_0} \qquad \beta_1 = e^{\beta'_1} \qquad \epsilon'_t = \log \epsilon_t$

## The Logistic (Pearl-Reed) Model

Using reciprocals rather than logs will give a very similar curve with a slightly gentler slope. Setting

$$T'_t = \frac{1}{T_t} = \beta_0 + \beta_1 \cdot \beta_2^t$$

we get

$$T_t = \frac{1}{\beta_0 + \beta_1 \cdot \beta_2^t}$$

which gives a curve similar to the one shown in Figure 3.13 (if $\beta_1 > 0$ and $\beta_2 < 1$) with

$$\frac{1}{\beta_0} = \text{the upper saturation level}$$

$$\frac{1}{\beta_0 + \beta_1} = \text{value of the trend at the origin}$$

**Figure 3.12** Gompertz Trend Curve

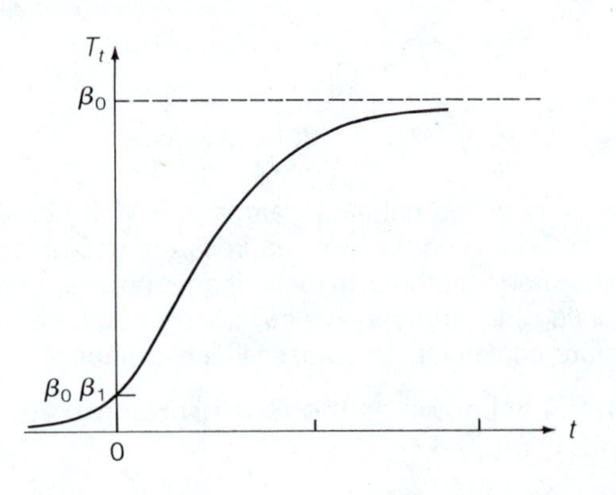

Figure 3.13 Logistic (Pearl-Reed) Trend Curve

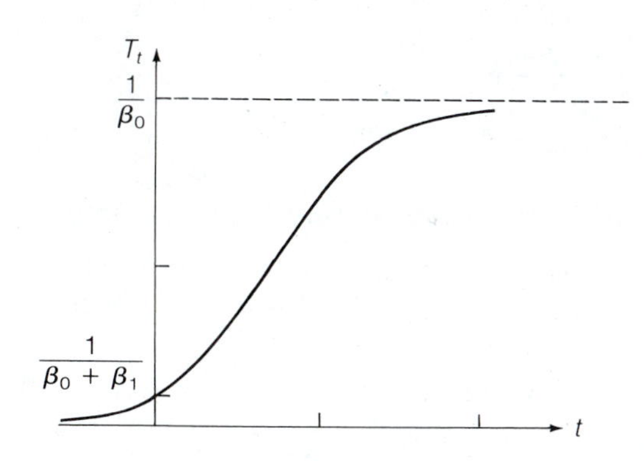


The closer $\beta_2$ is to zero, the more slowly the curve will grow toward its upper saturation level.

Logistic (Pearl-Reed) Trend Model

$$y_t = \frac{1}{\beta_0 + \beta_1 \cdot \beta_2^t} + \epsilon_t$$

## 3.8 Single Forecasts, Curvilinear Models

### 3.8.1 Introduction

Because of their mathematical complexity, curvilinear models are more difficult to fit than linear ones. While the quadratic and simple exponential models may be fit in several ways, including least squares, the models derived from the modified exponential curve are generally fit using **partial totals,** a variation of the semiaverage method. They are not easily fit using least squares.

### 3.8.2 Quadratic Model Fit, Selected Points

Because three parameters must be estimated, the selected-points method requires three points to fit a quadratic trend. Suppose these points are ($t1$, $y1$), ($t2$, $y2$), and ($t3$, $y3$). Then we get three linear equations in three unknowns:

$$y_{t1} = \hat{\beta}_0 + \hat{\beta}_1 \cdot t1 + \hat{\beta}_2 \cdot t1^2$$
$$y_{t2} = \hat{\beta}_0 + \hat{\beta}_1 \cdot t2 + \hat{\beta}_2 \cdot t2^2$$
$$y_{t3} = \hat{\beta}_0 + \hat{\beta}_1 \cdot t3 + \hat{\beta}_2 \cdot t3^2$$

which may be solved simultaneously. Let us assume that $t1 < t2 < t3$. Then the solution may be computed from

$$\hat{\beta}_2 = \frac{(y3 - y2)/(t3 - t2) - (y2 - y1)/(t2 - t1)}{t3 - t1} \tag{3.5}$$

$$\hat{\beta}_1 = \frac{y3 - y2}{t3 - t2} - \hat{\beta}_2 \cdot (t3 + t2) \tag{3.6}$$

$$\hat{\beta}_0 = y3 - \hat{\beta}_1 \cdot t3 - \hat{\beta}_2 \cdot t3^2 \tag{3.7}$$

**EXAMPLE 3.13** *Quadratic Fit to Procter & Gamble Data, Selected Points*

It is reasonable to select points near the beginning, middle, and end of the series being fit. Suppose for the Procter & Gamble data that we select 1968, 1975, and 1983 (Table 3.3). Then

$$t1 = 1 \qquad y1 = 182$$
$$t2 = 8 \qquad y2 = 332$$
$$t3 = 16 \qquad y3 = 866$$

$$\hat{\beta}_2 = \frac{(866 - 332)/(16 - 8) - (332 - 182)/(8 - 1)}{16 - 1} = 3.021$$

$$\hat{\beta}_1 = \frac{866 - 332}{16 - 8} - 3.021 \times (16 + 8) = -5.754$$

$$\hat{\beta}_0 = 866 - (-5.754) \times 16 - 3.021 \times 16^2 = 184.7$$

and the estimated model is

$$\hat{T}_t = 184.7 - 5.754t + 3.021t^2 \qquad \text{Origin: } 1967 \quad \text{Units } (t)\text{: Yearly}$$

If we wish to forecast 1984 and 1985, then

*1984* (t = *17*): $\hat{y}_{17} = \hat{T}_{17} = 184.7 - 5.754 \times 17 + 3.021 \times 17^2 = 960.0$

*1985* (t = *18*): $\hat{y}_{18} = \hat{T}_{18} = 184.7 - 5.754 \times 18 + 3.021 \times 18^2 = 1{,}059.9$ •

**Single Forecasts Using Selected Points** Quadratic Model

**Given:** Three points $(t1, y1)$, $(t2, y2)$, and $(t3, y3)$

**Model:** $y_t = \beta_0 + \beta_1 t + \beta_2 t^2 + \epsilon_t$

**Estimates:**

$$\hat{\beta}_2 = \frac{(y3 - y2)/(t3 - t2) - (y2 - y1)/(t2 - t1)}{t3 - t1}$$

$$\hat{\beta}_1 = \frac{y3 - y2}{t3 - t2} - \hat{\beta}_2 \cdot (t3 + t2)$$

$$\hat{\beta}_0 = y3 - \hat{\beta}_1 \cdot t3 - \hat{\beta}_2 \cdot t3^2$$

**Forecast:** (At time $t$ for $p$ periods ahead)

$$\hat{y}_{t+p} = \hat{\beta}_0 + \hat{\beta}_1 \cdot (t + p) + \hat{\beta}_2 \cdot (t + p)^2 \qquad p \geq 1$$

### 3.8.3 Quadratic Model Fit, Semiaverages

The semiaverage method permits much more of the data to influence the estimates. To use it, we divide the series into 3 (preferably equal) groups (omitting one or two points if desired) and average $t$ and $y_t$ in each group. That is, we let

$\overline{y1}$ = the average value of $y_t$ in Group I
$\overline{y2}$ = the average value of $y_t$ in Group II
$\overline{y3}$ = the average value of $y_t$ in Group III
$\overline{t1}$ = the average value of $t$ in Group I
$\overline{t2}$ = the average value of $t$ in Group II
$\overline{t3}$ = the average value of $t$ in Group III

These values may then be substituted for ($t1$, $y1$), ($t2$, $y2$), and ($t3$, $y3$) in the selected-points formulas (Equations 3.5–3.7), giving

$$\hat{\beta}_2 = \frac{(\overline{y3} - \overline{y2})/(\overline{t3} - \overline{t2}) - (\overline{y2} - \overline{y1})/(\overline{t2} - \overline{t1})}{\overline{t3} - \overline{t1}} \quad (3.8)$$

$$\hat{\beta}_1 = \frac{\overline{y3} - \overline{y2}}{\overline{t3} - \overline{t2}} - \hat{\beta}_2(\overline{t3} + \overline{t2}) \quad (3.9)$$

$$\hat{\beta}_0 = \overline{y3} - \hat{\beta}_1\overline{t3} - \hat{\beta}_2\overline{t3}^2 \quad (3.10)$$

If we have three equal-size groups of $m$ observations each, starting with $(1, y_1)$ and ending with $(3m, y_{3m})$, then knowing that

$$\overline{t1} = \frac{m+1}{2} \qquad \overline{t2} = \frac{3m+1}{2} \qquad \overline{t3} = \frac{5m+1}{2}$$

facilitates the computations.

**Single Forecasts Using Semiaverages** Quadratic Model

**Given:** A sample of $n$ observations, divided into (preferably equal) Groups, I, II, and III

$\overline{y1}$, $\overline{y2}$, and $\overline{y3}$ = average value of $y_t$ in Groups I, II, and III respectively
$\overline{t1}$, $\overline{t2}$, and $\overline{t3}$ = average value of $t$ in Groups I, II, and III respectively

**Model:** $y_t = \beta_0 + \beta_1 t + \beta_2 t^2 + \epsilon_t$

**Estimates:**

$$\hat{\beta}_2 = \frac{(\overline{y3} - \overline{y2})/(\overline{t3} - \overline{t2}) - (\overline{y2} - \overline{y1})/(\overline{t2} - \overline{t1})}{\overline{t3} - \overline{t1}}$$

$$\hat{\beta}_1 = \frac{\overline{y3} - \overline{y2}}{\overline{t3} - \overline{t2}} - \hat{\beta}_2 \cdot (\overline{t3} + \overline{t2})$$

$$\hat{\beta}_0 = \overline{y3} - \beta_1 \cdot \overline{t3} - \beta_2 \cdot \overline{t3}^2$$

**Forecast:** (At time $t$ for $p$ periods ahead)

$$\hat{y}_{t+p} = \hat{\beta}_0 + \hat{\beta}_1 \cdot (t + p) + \hat{\beta}_2 \cdot (t + p)^2$$

**EXAMPLE 3.14** ***Quadratic Fit to Procter & Gamble Data, Semiaverages***

Dividing the Procter & Gamble data into three equal groups (omitting 1968) we get the following:

| Group I | | Group II | | Group III | |
|---|---|---|---|---|---|
| $t$ | $y_t$ | $t$ | $y_t$ | $t$ | $y_t$ |
| 2 | 187 | 7 | 316 | 12 | 575 |
| 3 | 211 | 8 | 332 | 13 | 640 |
| 4 | 237 | 9 | 400 | 14 | 668 |
| 5 | 276 | 10 | 460 | 15 | 777 |
| 6 | 301 | 11 | 510 | 16 | 866 |
| Σ 20 | 1,212 | 45 | 2,018 | 70 | 3,526 |

$\overline{y1} = 242.4$ $\quad\overline{y2} = 403.6$ $\quad\overline{y3} = 705.2$

$\overline{t1} = 4.0$ $\quad\overline{t2} = 9.0$ $\quad\overline{t3} = 14.0$

Then substituting in Equations 3.8–3.10 we get

$$\hat{\beta}_2 = \frac{(705.2 - 403.6)/(14 - 9) - (403.6 - 242.4)/(9 - 4)}{14 - 9} = 2.808$$

$$\hat{\beta}_1 = \frac{705.2 - 403.6}{14 - 9} - 2.808 \times (14 + 9) = -4.264$$

$$\hat{\beta}_0 = 705.2 - (-4.264) \times 14 - 2.808 \times 14^2 = 214.5$$

$$\hat{T}_t = 214.5 - 4.264t + 2.808t^2$$

Origin: 1967
Units ($t$): Yearly

Thus, forecasts for 1984 and 1985 would be:

*1984* (t = *17*): $\hat{y}_{17} = \hat{T}_{17} = 214.5 - 4.264 \times 17 + 2.808 \times 17^2 = 953.5$

*1985* (t = *18*): $\hat{y}_{18} = \hat{T}_{18} = 214.5 - 4.264 \times 18 + 2.808 \times 17^2 = 1.047.5$

which are comparable to the forecasts obtained using selected points. •

## 3.8.4 Quadratic Model Fit, Least Squares

To compute least squares estimates of the quadratic-model parameters we again minimize the aggregate squared forecast errors, selecting $\hat{\beta}_0$, $\hat{\beta}_1$, and $\hat{\beta}_2$ so that

$$\Sigma (y_t - \hat{y}_t)^2 = \Sigma (y_t - \hat{\beta}_0 - \hat{\beta}_1 t - \hat{\beta}_2 t^2)^2$$

is as small as possible. Three linear equations in three unknowns must be solved for $\hat{\beta}_0$, $\hat{\beta}_1$, and $\hat{\beta}_2$. They are

$$\Sigma y = n \cdot \hat{\beta}_0 + \Sigma t \cdot \hat{\beta}_1 + \Sigma t^2 \cdot \hat{\beta}_2$$

$$\Sigma ty = \Sigma t \cdot \hat{\beta}_0 + \Sigma t^2 \cdot \hat{\beta}_1 + \Sigma t^3 \cdot \hat{\beta}_2$$

$$\Sigma t^2y = \Sigma t^2 \cdot \hat{\beta}_0 + \Sigma t^3 \cdot \hat{\beta}_1 + \Sigma t^4 \cdot \hat{\beta}_2$$

and their solution requires that $\Sigma t$, $\Sigma t^2$, $\Sigma t^3$, $\Sigma t^4$, $\Sigma y$, $\Sigma ty$, and $\Sigma t^2y$ be calculated from the data. We can adjust the scale and origin of $t$ as follows:

*n Odd:* Shift origin forward $(n + 1)/2$ units, as follows.

$$t: \quad 1 \quad 2 \cdots \frac{n+1}{2} \cdots n-1 \quad n$$

$$\text{adj. } t = t': \quad -\frac{n-1}{2} \cdots -2 \quad -1 \quad 0 \quad +1 \quad +2 \cdots +\frac{n-1}{2}$$

For example:

| $t$: | 1 | 2 | 3 | 4 | 5 | 6 | 7 | 8 | 9 |
|---|---|---|---|---|---|---|---|---|---|
| adj. $t = t'$: | −4 | −3 | −2 | −1 | 0 | +1 | +2 | +3 | +4 |

*n Even:* Shift origin forward $(n + 1)/2$ units, then decrease units by ½, e.g., yearly to semi-annual, monthly to biweekly, as follows.

$$t: \quad 1 \quad 2 \cdots \quad \frac{n}{2} \quad \frac{n}{2} + 1 \cdots \quad n$$

$$\text{adj. } t = t': \quad -(n-1) \quad -(n-3) \quad -1 \quad +1 \cdots +(n-3) \quad +(n-1)$$

For example:

| $t$: | 1 | 2 | 3 | 4 | 5 | 6 | 7 | 8 |
|---|---|---|---|---|---|---|---|---|
| adj. $t = t'$: | −7 | −5 | −3 | −1 | +1 | +3 | +5 | +7 |

With these adjusted $t$-scales, $\Sigma t' = \Sigma t'^3 = 0$, and the equations become

$$\Sigma y = n \cdot \hat{\beta}_0 + \Sigma t'^2 \cdot \hat{\beta}_2 \tag{3.11}$$

$$\Sigma ty = \Sigma t'^2 \cdot \hat{\beta}_1 \tag{3.12}$$

$$\Sigma t^2 y = \Sigma t'^2 \cdot \hat{\beta}_0 + \Sigma t'^4 \cdot \hat{\beta}_2 \tag{3.13}$$

which are more convenient to solve. After the solutions are found, the time scale and origin in the resulting estimated trend equations may be returned to their original values, if desired.

Using our conventional notation,

$$SS_{t'^2y} = \Sigma t'^2 y - \frac{(\Sigma t'^2) \cdot (\Sigma y)}{n}$$

$$SS_{t'^2t'^2} = \Sigma t'^4 - \frac{(\Sigma t'^2)^2}{n}$$

$$\bar{y} = \frac{\Sigma y}{n} \qquad \overline{t'^2} = \frac{\Sigma t'^2}{n}$$

then we can solve Equations 3.11 through 3.13, to get

$$\hat{\beta}_2 = \frac{SS_{t'^2y}}{SS_{t'^2t'^2}}$$

$$\hat{\beta}_1 = \frac{\Sigma t'y}{\Sigma t'^2}$$

$$\hat{\beta}_0 = \bar{y} - \hat{\beta}_2 \cdot \overline{t'^2}$$

which are similar in many respects to the least squares linear trend equations, but with $t$ becoming $t'^2$.

As a simple example of the use of these equations, let's fit a quadratic model to the following time series:

| Year | $t$ | $y_t$ |
|---|---|---|
| 1980 | 1 | 2.4 |
| 1981 | 2 | 3.6 |
| 1982 | 3 | 5.4 |
| 1983 | 4 | 7.8 |
| 1984 | 5 | 11.6 |
| 1985 | 6 | 17.3 |

**Computations for Least Squares Quadratic Fit**

| Year | $t$ | $t'$ | $y_t$ | $t'y_t$ | $t'^2$ | $t'^4$ | $t'^2y_t$ |
|---|---|---|---|---|---|---|---|
| 1980 | 1 | −5 | 2.4 | −12.0 | 25 | 625 | 60.0 |
| 1981 | 2 | −3 | 3.6 | −10.8 | 9 | 81 | 32.4 |
| 1982 | 3 | −1 | 5.4 | −5.4 | 1 | 1 | 5.4 |
| 1983 | 4 | 1 | 7.8 | 7.8 | 1 | 1 | 7.8 |
| 1984 | 5 | 3 | 11.6 | 34.8 | 9 | 81 | 104.4 |
| 1985 | 6 | 5 | 17.3 | 86.5 | 25 | 625 | 432.5 |
| | | Σ 0 | 48.1 | 100.9 | 70 | 1,414 | 642.5 |

The necessary calculations, shown in the Least Squares Quadratic Fit table, give:

$$\Sigma y = 48.1 \qquad \Sigma t'^2 = 70 \qquad \Sigma t'^4 = 1{,}414$$
$$\Sigma t'y = 100.9 \qquad \Sigma t'^2 y = 642.5$$

and

$$SS_{t'^2y} = 642.5 - \frac{70 \times 48.1}{6} = 81.33$$

$$SS_{t'^2t'^2} = 1{,}414 - \frac{70^2}{6} = 597.33$$

$$\hat{\beta}_1 = \frac{100.9}{70} = 1.441$$

$$\hat{\beta}_2 = \frac{81.33}{597.33} = .1362$$

$$\hat{\beta}_0 = \frac{48.1}{6} - \frac{70}{6} \times .1362 = 6.428$$

and, finally,

$$\hat{T}_t = 6.428 + 1.441 \cdot t' + .1362 \cdot t'^2 \qquad \text{Origin: } 1/1/84 \qquad \text{Units } (t')\text{: Semiannual}$$

Even with the modified $t$-scale, however, the calculations required to find the estimates (and readjust the time scale) are cumbersome enough that computers are usually preferred for fitting quadratic models. Any multiple regression program will produce the required estimates by using $t$ and $t^2$ as two independent variables. While individual programs differ, the output for the Procter & Gamble data from a widely used program called Minitab, shown in the Minitab table, should serve as a representative example.

**Minitab Output for Quadratic Fit of Procter & Gamble Data**

```
THE REGRESSION EQUATION IS
Y =      181.  + 2.72 X1 +   2.45 X2
           ↓         ↓            ↓
           β̂0        β̂1           β̂2

                                          ST. DEV.      T-RATIO =
         COLUMN      COEFFICIENT          OF COEF.      COEF S.D.
         --               181.01  →  β̂0     12.95          13.97
X1       C2 →  t           2.724  →  β̂1     3.507           0.78
X2       C3 →  t²         2.4541  →  β̂2    0.2006          12.24

THE ST. DEV. OF Y ABOUT REGRESSION LINE IS
S = 15.16
WITH (  16 - 3) = 13 DEGREES OF FREEDOM

R-SQUARED = 99.6 PERCENT
R-SQUARED = 99.5  PERCENT, ADJUSTED FOR D.F.

ANALYSIS OF VARIANCE

 DUE TO           DF         SS
REGRESSION         2   705997.1         MS = SS/DF
RESIDUAL          13     2986.7          352998.5
TOTAL             15   708983.8             229.7
```

**Single Forecasts Using Least Squares** Quadratic Model

**Given:** A sample of $n$ observations $y_1, y_2, \ldots, y_n$ at times $t_1, t_2, \ldots, t_n$

**Model:** $y_t = \beta_0 + \beta_1 t + \beta_2 t^2 + \epsilon_t$

**Estimates:** May be obtained from a multiple regression computer program with $t$ and $t^2$ as independent variables, or by adjusting the $t$ scale so $\Sigma\, t' = \Sigma\, t'^3 = 0$ and then solving:

$$\Sigma\, y = n \cdot \hat{\beta}_0 + \Sigma\, t'^2 \cdot \hat{\beta}_2$$

$$\Sigma\, yt' = \Sigma\, t'^2 \cdot \hat{\beta}_1$$

$$\Sigma\, t'^2 y = \Sigma\, t'^2 \cdot \hat{\beta}_0 + \Sigma\, t'^4 \cdot \hat{\beta}_2$$

**Forecasts:** (At time $t$ for $p$ steps ahead)

$$\hat{y}_{t+p} = \hat{\beta}_0 + \hat{\beta}_1 \cdot (t + p) + \hat{\beta}_2 \cdot (t + p)^2$$

The estimates of the parameters are shown in boldface in the printout. (The boldface is ours, not the program's.) They are

$$\hat{\beta}_0 = 181.0 \qquad \hat{\beta}_1 = 2.724 \qquad \hat{\beta}_2 = 2.454$$

So the estimated trend is

$$\hat{T}_t = 181.0 + 2.724t + 2.454t^2$$

Origin: 1967
Units ($t$): Yearly

And forecasts for 1984 and 1985 would be as follows.

*1984* (t = *17*): $\hat{y}_{17} = \hat{T}_{17} = 181.0 + 2.724 \times 17 + 2.454 \times 17^2 = 936.5$

*1985* (t = *18*): $\hat{y}_{18} = \hat{T}_{18} = 181.0 + 2.724 \times 18 + 2.454 \times 18^2 = 1{,}025.1$

### 3.8.5 Simple Exponential Model Fitting

Simple exponential models are fit by changing the model to its *log-linear* form, then applying any of the methods used to fit linear models, i.e., free-hand, selected points, semiaverages, or least squares. The log-linear form is obtained by taking logarithms of both sides of the model equation. So

$$y_t = \beta_0 \cdot \beta_1^t \cdot \epsilon_t$$

becomes

$$y_t' = \log y_t = \log \beta_0 + t \cdot \log \beta_1 + \log \epsilon_t = \beta_1' + \beta_2' t + \epsilon_t'$$

where $\beta_0' = \log \beta_0$, $\beta_1' = \log \beta_1$, and $\epsilon_t' = \log \epsilon_t$. The log-linear trend equation is then

$$T_t' = \log T_t = \beta_0' + \beta_1' t$$

After the parameters $\beta_1'$ and $\beta_2'$ are estimated, the model may be converted back to its original form with

$$\hat{\beta}_0 = \text{antilog } \hat{\beta}_0' = e^{\hat{\beta}_0'}$$
$$\hat{\beta}_1 = \text{antilog } \hat{\beta}_1' = e^{\hat{\beta}_1'}$$

Essentially, the process involves converting the time series $y_1, y_2, \ldots, y_n$ to $y_1' = \log y_1$, $y_2' = \log y_2, \ldots, y_n' = \log y_n$ and fitting a linear model to the $y_t'$ values as if they were the series of interest. To illustrate, the Procter & Gamble data are shown converted to logarithms in Table 3.11. Fitting a linear model to these logs using the four different methods of Section 3.6, we get:

- **Free-hand** (a transparent ruler was placed on the graph in Figure 3.14 and the following points read from the free-hand line)

$$y1' = 5.13 \qquad t1 = 1$$
$$y2' = 6.75 \qquad t2 = 16$$
$$\hat{\beta}_1' = \frac{6.75 - 5.13}{16 - 1} = .1080$$
$$\hat{\beta}_0' = y2' - \hat{\beta}_1' \cdot t2 = 6.75 - .108 \times 16 = 5.022$$
$$\hat{y}_t' = \hat{T}_t' = 5.022 + .108t$$

Origin: 1967
Units ($t$): Yearly
Units ($y'$): log \$$10^6$

**Table 3.11** Logarithms of Procter & Gamble Data and Graph

| Year | $t$ | $y_t$ | $y'_t = \log y_t$ | Year | $t$ | $y_t$ | $y'_t = \log y_t$ |
|---|---|---|---|---|---|---|---|
| 1968 | 1 | 182 | 5.204 | 1976 | 9 | 400 | 5.991 |
| 1969 | 2 | 187 | 5.231 | 1977 | 10 | 460 | 6.131 |
| 1970 | 3 | 211 | 5.352 | 1978 | 11 | 520 | 6.254 |
| 1971 | 4 | 237 | 5.468 | 1979 | 12 | 575 | 6.354 |
| 1972 | 5 | 276 | 5.620 | 1980 | 13 | 640 | 6.461 |
| 1973 | 6 | 301 | 5.707 | 1981 | 14 | 668 | 6.504 |
| 1974 | 7 | 316 | 5.756 | 1982 | 15 | 777 | 6.655 |
| 1975 | 8 | 332 | 5.805 | 1983 | 16 | 866 | 6.763 |

**Figure 3.14** Graph Based on Procter & Gamble Data

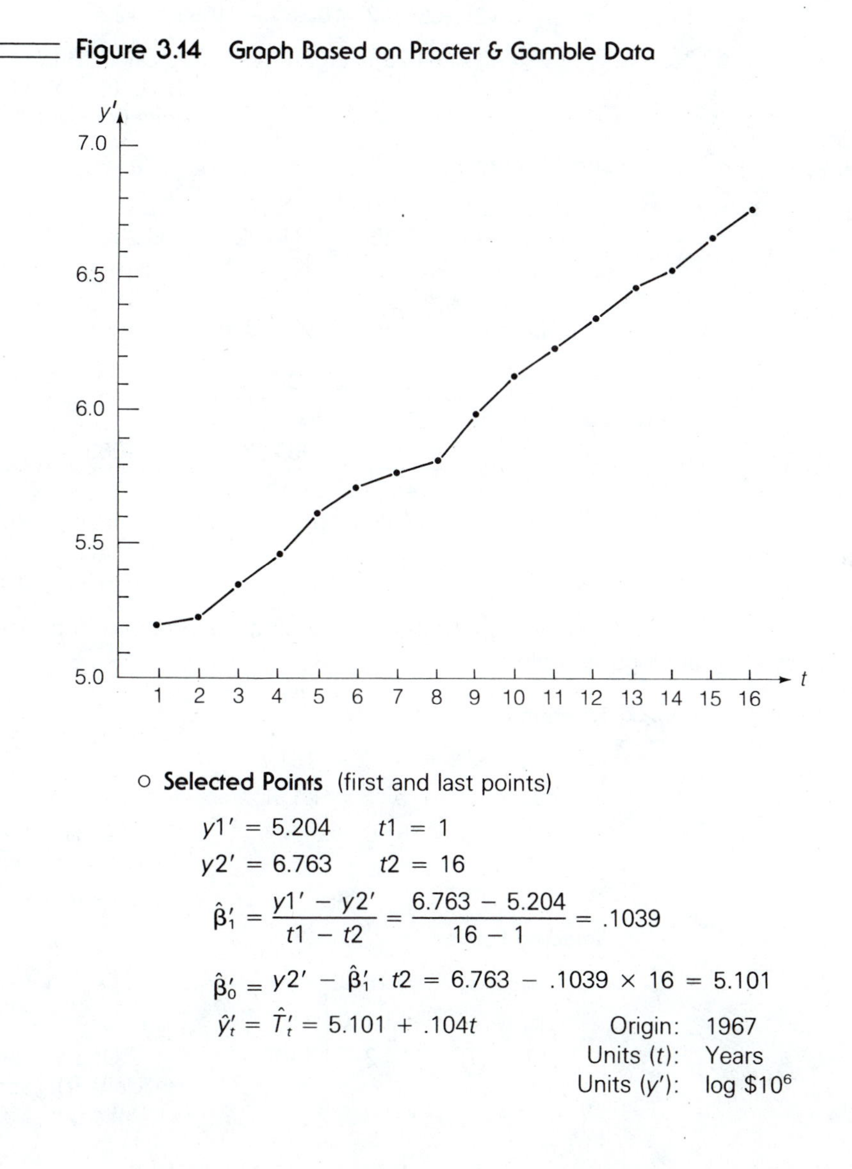


- **Selected Points** (first and last points)

$y1' = 5.204 \qquad t1 = 1$

$y2' = 6.763 \qquad t2 = 16$

$$\hat{\beta}'_1 = \frac{y1' - y2'}{t1 - t2} = \frac{6.763 - 5.204}{16 - 1} = .1039$$

$$\hat{\beta}'_0 = y2' - \hat{\beta}'_1 \cdot t2 = 6.763 - .1039 \times 16 = 5.101$$

$$\hat{y}'_t = \hat{T}'_t = 5.101 + .104t$$

Origin: 1967
Units ($t$): Years
Units ($y'$): log $\$10^6$

***Note:*** Fitting selected points to first and last points for the simple exponential is equivalent to the common method of computing the "average" percentage rate of change as $\sqrt[n]{\text{(terminal value)/(initial value)}} - 1$, which is $\hat{\beta}_1 - 1$.

- **Semiaverages**

$$\overline{y1'} = \frac{5.204 + 5.231 + \cdots + 5.805}{8} = 5.518 \qquad \overline{t1} = 4.5$$

$$\overline{y2'} = \frac{5.991 + 6.131 + \cdots + 6.763}{8} = 6.389 \qquad \overline{t2} = 12.5$$

$$\hat{\beta}_1' = \frac{6.389 - 5.518}{12.5 - 4.5} = .1089$$

$$\hat{\beta}_0' = \overline{y2'} - \hat{\beta}_1' \cdot \overline{t2} = 6.389 - .1089 \times 12.5$$

$$\hat{y}_t' = \hat{T}_t' = 5.028 + .109t$$

Origin: 1967
Units ($t$): Yearly
Units ($y'$): log $\$10^6$

- **Least Squares**

| $n$ | $\Sigma\, t$ | $\Sigma\, t^2$ | $\Sigma\, ty'$ | $\Sigma\, y'$ | $\Sigma\, y'^2$ |
|---|---|---|---|---|---|
| 16 | 136 | 1,496 | 846.021 | 95.256 | 571.012 |

$$SS_{ty'} = \Sigma\, ty' - \frac{n+1}{2}\Sigma\, y' = 846.021 - \frac{17 \times 95.256}{2} = 35.345$$

$$\hat{\beta}_1' = \frac{12 \cdot SS_{ty'}}{n^3 - n} = \frac{12 \times 36.345}{16^3 - 16} = .1069$$

$$\hat{\beta}_0 = \bar{y} - \frac{n+1}{2}\hat{\beta}_1' = \frac{95.256}{16} - \frac{17 \times .1069}{2} = 5.046$$

$$\hat{y}_t' = \hat{T}_t' = 5.046 + .107t$$

Origin: 1967
Units ($t$): Yearly
Units ($y'$): log $\$10^6$

These forecasting equations may all be expressed in net earnings by taking antilogs as follows:

- **Freehand**

$$\hat{\beta}_0 = e^{\hat{\beta}_0'} = e^{5.022} = 151.7$$

$$\hat{\beta}_1 = e^{\hat{\beta}_1'} = e^{.108} = 1.114$$

$$\hat{y}_t = \hat{T}_t = 151.6 \cdot (1.114)^t$$

Origin: 1967
Units ($t$): Yearly
Units ($y$): $\$10^6$

- **Selected Points**

$$\hat{\beta}_0 = e^{5.101} = 164.2$$

$$\hat{\beta}_1 = e^{.108} = 1.110$$

$$\hat{y}_t = \hat{T}_t = 164.2 \cdot (1.110)^t$$

Origin: 1967
Units ($t$): Yearly
Units ($y$): $\$10^6$

○ **Semiaverages**

$$\hat{\beta}_0 = e^{5.028} = 152.6$$
$$\hat{\beta}_1 = e^{.109} = 1.115$$
$$\hat{y}_t = \hat{T}_t = 155.4 \cdot (1.115)^t$$

Origin: 1967
Units ($t$): Yearly
Units ($y$): $\$10^6$

○ **Least Squares**

$$\hat{\beta}_0 = e^{5.046} = 155.4$$
$$\hat{\beta}_1 = e^{.107} = 1.113$$
$$\hat{y}_t = \hat{T}_t = 155.4 \cdot (1.113)^t$$

Origin: 1967
Units ($t$): Yearly
Units ($y$): $\$10^6$

Finally, in the least squares case, if we assume that the $\epsilon_t'$ are normally distributed (so that the $y_t'$ will be normally distributed as well), then the procedures given for least squares linear fits may be used to construct prediction intervals for $y_t'$. These may then be converted to prediction intervals for $y_t$ by taking antilogs. For instance, suppose in 1983 ($t = 16$) we wished to construct a 95% prediction interval for 1984 net earnings (one period ahead, $p = 1$). Then, since $t = 1, 2, \ldots, n$,

$$SS_{tt} = \frac{n^3 - n}{12} = \frac{16^3 - 16}{12} = 340$$

$$\bar{t} = \frac{16 + 1}{2} = 8.5$$

$$SS_{y'y'} = \Sigma y'^2 - \frac{(\Sigma y')^2}{n} = 571.012 - \frac{(95.256)^2}{2} = 3.9045$$

$$SS_{ty'} = 36.345 \qquad \text{(from previous calculations)}$$

$$SSE = SS_{y'y'} - \frac{SS_{ty'}^2}{SS_{tt}} = 3.9054 - \frac{36.345^2}{340} = .02023$$

$$\hat{\sigma}_\epsilon = \sqrt{\frac{SSE}{n-2}} = \sqrt{\frac{.02023}{14}} = .03801$$

$$t_{\alpha/2} = t_{.025} = 2.110 \qquad (n - 2 = 14 \text{ df, Appendix GG})$$

$$B = t_{\alpha/2} \cdot \hat{\sigma}_\epsilon \sqrt{1 + \frac{1}{n} + \frac{(t + p - \bar{t})^2}{SS_{tt}}}$$

$$= 2.110 \times .03801 \times \sqrt{1 + \frac{1}{16} + \frac{(17 - 8.5)^2}{340}} = .0906$$

$$\hat{y}_{17}' = \hat{y}_{16}' + \hat{\beta}_1' = 5.046 + .107 \times 16 + .107 = 6.865$$

95% prediction interval for $y_t'$:

$$\hat{y}_{17}' \pm B = 6.865 \pm .0906 = (6.774, 6.956)$$

95% prediction interval for $y_t$ (taking antilogs):

$$(e^{6.774}, e^{6.956}) = (874.8;\ 1{,}049.4)$$

### 3.8.6 Modified Exponential Model Fitting

The parameters of the modified exponential may be estimated using a method similar to semiaverages, called the **method of partial totals.** There are three parameters to fit, so three "partial totals" are needed. These totals are found by dividing the history of the series into three equal contiguous groups and summing the $y_t$ values and $t$ values in each group. If the number of data points is not an even multiple of 3, the one or two earliest points are discarded. Then, if we let

$m$ = the number of observations in each group

$s1$ = the sum of the first group of $y_t$ values

$s2$ = the sum of the second group of $y_t$ values

$s3$ = the sum of the third group of $y_t$ values

then the partial totals estimates will be

$$\hat{\beta}_2^m = \frac{s3 - s2}{s2 - s1}$$

$$\hat{\beta}_1 = \frac{(s2 - s1) \cdot (\beta_2 - 1)}{(\hat{\beta}_2^m - 1)^2}$$

$$\hat{\beta}_0 = \frac{s1}{m} - \frac{s2 - s1}{(\hat{\beta}_2^m - 1)^2}$$

These expressions give an equation that has its origin at the first observation of the first group of observations. In order to place the origin so $t = 1$ at the first observation in the first group, we use Rule 2 from Section 3.5.3 to shift backward one unit. This changes only $\hat{\beta}_1$, which becomes $\hat{\beta}_1/\hat{\beta}_2$ with the new origin.

**EXAMPLE 3.15** *Springston Population*

Let us consider Springston, a hypothetical suburb of a metropolitan area on the West Coast. It had been a small community until the expanding city made it a desirable location for commuters. Its growth was phenomenal over several years but appeared to be slowing down when available land became almost fully developed. In 1980, city planners undertook the task of forecasting the population over the next five-year period. To construct the forecasting model, yearly estimates of the population for the previous sixteen years were used as the raw data. These data are given in Table 3.12 and graphed in Figure 3.15.

The computations to fit a modified exponential trend (with 1970 omitted and only the fifteen latest observations used) are as follows:

| Group 1 (1971–75) | Group 2 (1976–80) | Group 3 (1981–85) |
|---|---|---|
| 28.1 | 68.1 | 89.1 |
| 37.7 | 74.2 | 93.8 |
| 46.3 | 77.2 | 97.1 |
| 50.9 | 82.1 | 99.8 |
| 60.3 | 84.8 | 101.2 |
| Σ 223.3 | 386.4 | 481.0 |

$$m = \frac{15}{3} = 5$$

$$s1 = 223.3 \qquad s2 = 386.4 \qquad s3 = 481.0$$

$$\hat{\beta}_2 = \left[\frac{s3 - s2}{s2 - s1}\right]^{1/m}$$

$$= \left[\frac{481.0 - 386.4}{386.4 - 223.3}\right]^{1/5} = (.5800)^{1/5} = .8968$$

$$\hat{\beta}_1 = \frac{(s2 - s1)(\hat{\beta}_2 - 1)}{(\hat{\beta}_2^5 - 1)^2}$$

$$= \frac{(386.4 - 223.3)(.8968 - 1)}{(.5800 - 1)^2} = -95.419$$

$$\hat{\beta}_0 = \frac{1}{m}\left(s1 - \frac{s2 - s1}{\hat{\beta}_2^m - 1}\right)$$

$$= \frac{1}{5}\left(223.3 - \frac{386.4 - 223.3}{.5800 - 1}\right) = 122.327$$

And the trend equation is

$$\hat{T}_t = 122.33 - 95.42(.897)^t$$

Origin: 1971
Units ($t$): Yearly
Units ($y$): 1,000's

Moving the origin to 1970 so $t = 1$ for 1971 means $\beta_1$ becomes

$$\frac{\hat{\beta}_1}{\hat{\beta}_2} = \frac{-95.419}{.8968} = -106.40$$

This gives the final estimated trend equation as

$$\hat{T}_t = 122.33 - 106.40(.897)^t$$

Origin: 1970
Units ($t$): Yearly
Units ($y$): 1,000's

The required forecasts for 1986 through 1990 ($t = 16$ through 20) can now be computed:

$$\hat{\beta}_1 \cdot (\hat{\beta}_2)^{15} = -106.4 \times (.897)^{15} = -20.837$$

*For* $t = 16$:

$$\hat{\beta}_1 \cdot (\hat{\beta}_2)^{16} = -20.837 \times .897 = -18.691$$

$$\hat{y}_{16} = \hat{T}_{16} = \hat{\beta}_0 + \hat{\beta}_1 \cdot (\hat{\beta}_2)^{16} = 122.33 - 18.690 = 103.6$$

*For* $t = 17$:

$$\hat{\beta}_1 \cdot (\hat{\beta}_2)^{17} = -18.690 \times .897 = -16.765$$

$$\hat{y}_{17} = \hat{T}_{17} = 122.33 - 16.766 = 105.6$$

*For* $t = 18$: $\hat{y}_{18} = 122.33 + (-16.765) \times .897 = 122.33 - 15.039 = 107.3$

*For* $t = 19$: $\hat{y}_{19} = 122.33 + (-15.039) \times .897 = 122.33 - 13.490 = 108.8$

*For* $t = 20$: $\hat{y}_{20} = 122.33 + (-13.490) \times .897 = 122.33 - 12.100 = 110.2$

Notice that $\hat{\beta}_0 = 122.33$ (122,330) is the estimated upper limit (asymptote) for Springston's population, $\hat{\beta}_1 + \hat{\beta}_2 = 122.33 - 106.4 = 15.93$ is the estimated trend in 1970 ($t = 0$), and .897 measures the rate at which the trend is approaching its upper limit.

**Table 3.12 Estimated Population of Springston, 1970–1985**

| Year | Population (1,000's) |
|---|---|
| 1970 | 14.8 |
| 1971 | 28.1 |
| 1972 | 37.7 |
| 1973 | 46.3 |
| 1974 | 50.9 |
| 1975 | 60.3 |
| 1976 | 68.1 |
| 1977 | 74.2 |
| 1978 | 77.2 |
| 1979 | 82.1 |
| 1980 | 84.8 |
| 1981 | 89.1 |
| 1982 | 93.8 |
| 1983 | 97.1 |
| 1984 | 99.8 |
| 1985 | 101.2 |

**Figure 3.15 Estimated Population of Springston, 1970–1985**

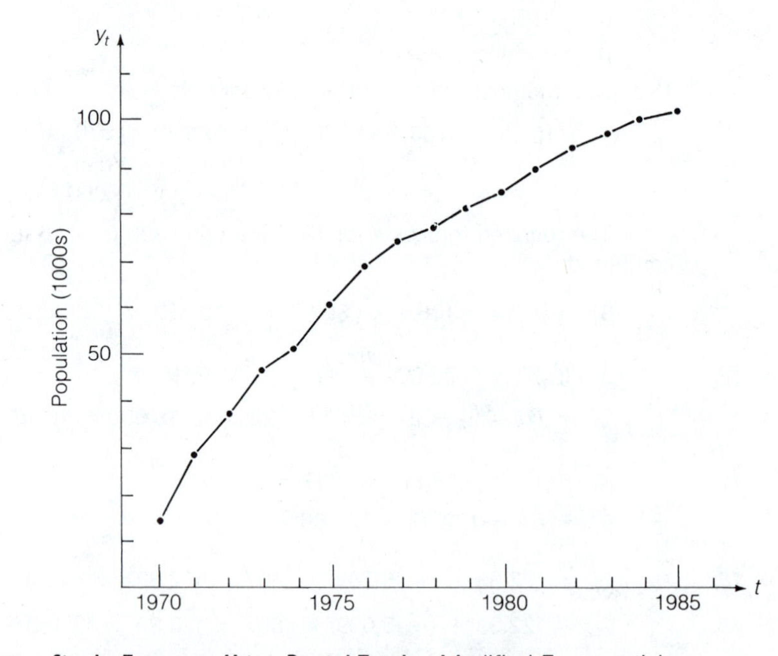

**Single Forecasts Using Partial Totals** Modified Exponential

**Given:** A sample of $n$ observations $y_1, y_2, \ldots, y_n$ with $n$ evenly divisible by 3. If $n$ is not evenly divisible by 3, the oldest observations are discarded and the observations renumbered.

**Model:** $y_t = \beta_0 + \beta_1 \cdot (\beta_2)^t \cdot \epsilon_t$ or $\beta_0 + \beta_1 \cdot (\beta_2)^t + \epsilon_t$

**Estimates:**

$m = n/3$

$s1$ = the sum of the first $m$ observations

$s2$ = the sum of the second $m$ observations

$s3$ = the sum of the third $m$ observations

$$\hat{\beta}_2 = \left[\frac{s3 - s2}{s2 - s1}\right]^{1/m}$$

$$\hat{\beta}_1 = \frac{(s2 - s1)(\hat{\beta}_2 - 1)}{(\hat{\beta}_2^m - 1)^2}$$

$$\hat{\beta}_0 = \frac{1}{m}\left(s1 - \frac{s2 - s1}{\hat{\beta}_2^m - 1}\right)$$

This gives a trend equation with $t = 0$ for $y_1$. To shift the origin so $t = 1$ for $y_1$ means $\hat{\beta}_1$ becomes $\hat{\beta}_1/\hat{\beta}_2$.

**Forecasts:**

(At time $t$ for $p$ periods ahead)

$$\hat{y}_{t+p} = \hat{\beta}_0 + \hat{\beta}_1 \cdot (\hat{\beta}_2)^t \cdot (\hat{\beta}_2)^p \qquad p \geq 1$$

## 3.8.7 Gompertz and Logistic Model Fitting

In much the same way as linear model-fitting techniques are used with simple exponential trends after changing $y_t$ to $y_t' = \log y_t$, Gompertz and logistic curves are fit using the modified exponential formulas on the logarithms and reciprocals of $y_t$; i.e.,

*Gompertz:*

$$y_t' = \log y_t$$

*Logistic:*

$$y_t' = \frac{1}{y_t}$$

In practice, when the logistic trend is being fit, the reciprocals are usually multiplied by a power of ten ($y_t' = 10^a/y_t$), where $a$ is chosen to make the numbers more convenient. The estimated trend models would be as follows.

*Gompertz:*

$$\hat{T}_t' = \log \hat{T}_t = \hat{\beta}_0' + \hat{\beta}_1' \cdot (\hat{\beta}_2')^t$$

$$\hat{T}_t = e^{\hat{\beta}_0'} \cdot [e^{\hat{\beta}_1'}]^{\hat{\beta}_2'^t} = \hat{\beta}_0 \cdot (\hat{\beta}_1)^{\hat{\beta}_2^t}$$

where

$$\hat{\beta}_1 = e^{\hat{\beta}_1'} \qquad \text{and} \qquad \hat{\beta}_2 = e^{\hat{\beta}_2'}$$

*Logistic:*

$$\hat{T}_t' = \frac{10^a}{\hat{T}_t} = \hat{\beta}_0 + \hat{\beta}_1 \cdot (\hat{\beta}_2)^t$$

$$\hat{T}_t = \frac{10^a}{\hat{\beta}_0 + \hat{\beta}_1 \cdot (\hat{\beta}_2)^t}$$

Both of these curves are used mostly for long-term forecasting, and reliable estimation of the parameters requires a reasonable amount of "unnoisy" data. In particular, estimates of the upper asymptote, or growth limit, will be untrustworthy unless you have some history of the series beyond the inflection point (the point where the S-curve turns over). The data in Table 3.13, graphed in Figure 3.16, were generated by adding computer-generated random noise to the trend curve

$$T_t = 1105(.005)^{.75^t}$$

**Table 3.13 Computer-Generated Trend Curve and Forecasts Using a Gompertz Model** $T_t = 1105(.005)^{.75^t}$

| | | | At $t$ = 15 | At $t$ = 14 |
|---|---|---|---|---|
| $t$ | $y_t$ | $y'_t = \log y_t$ | $\hat{y}_t$ | $\hat{y}_t$ |
| 1 | 79 | 4.369 | 35 | 106 |
| 2 | 29 | 3.367 | 76 | 149 |
| 3 | 162 | 5.088 | 140 | 202 |
| 4 | 247 | 5.509 | 225 | 264 |
| 5 | 299 | 5.700 | 327 | 336 |
| 6 | 621 | 6.431 | 437 | 415 |
| 7 | 510 | 6.234 | 548 | 501 |
| 8 | 701 | 6.553 | 655 | 592 |
| 9 | 734 | 6.599 | 753 | 687 |
| 10 | 608 | 6.410 | 839 | 784 |
| 11 | 993 | 6.901 | 913 | 881 |
| 12 | 1,017 | 6.925 | 976 | 978 |
| 13 | 1,006 | 6.914 | 1,028 | 1,072 |
| 14 | 1,057 | 6.963 | 1,071 | 1,164 |
| 15 | 1,009 | 6.917 | 1,105 | 1,252 |
| 16 | ? | ? | 1,133 | 1,335 |
| 17 | ? | ? | 1,155 | 1,414 |
| 18 | ? | ? | 1,173 | 1,488 |
| 19 | ? | ? | 1,187 | 1,557 |
| 20 | ? | ? | 1,198 | 1,621 |

*At* t = *14:* $\hat{\beta}_0 = 2{,}228$ $\hat{\beta}_1 = .0324$ $\hat{\beta}_2 = .888$

*At* t = *15:* $\hat{\beta}_0 = 1{,}238$ $\hat{\beta}_1 = .0104$ $\hat{\beta}_2 = .782$

**Figure 3.16 Computer-Generated Gompertz Curve**

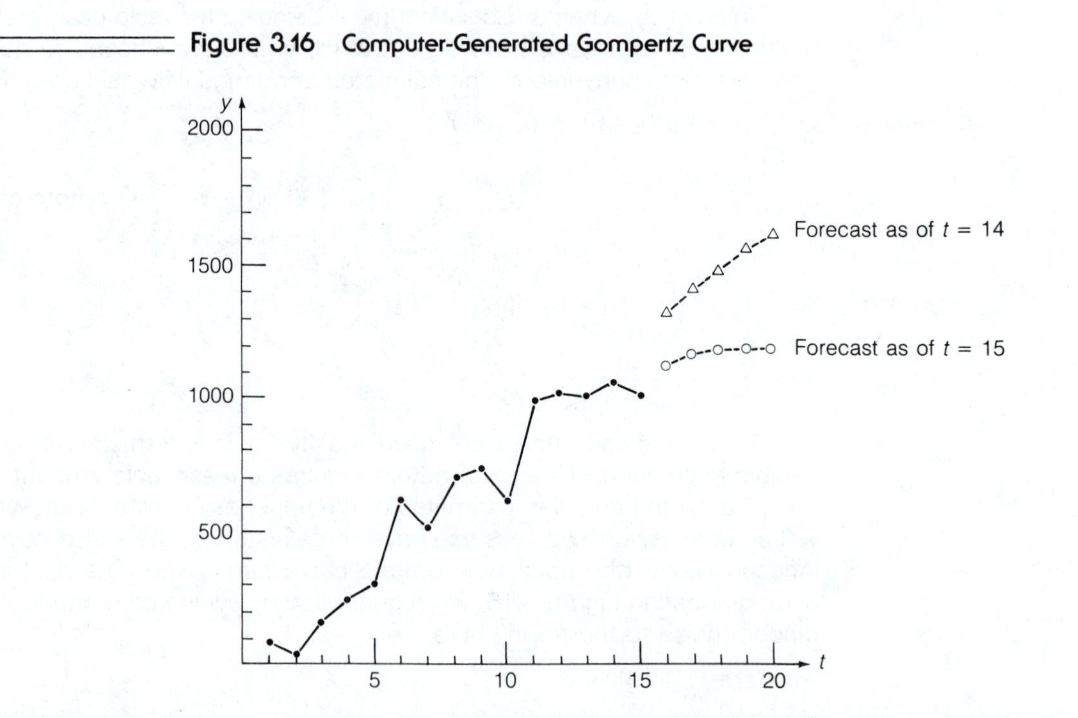

### Gompertz Fit

If one fits a Gompertz model at time $t = 15$, then

$$m = \frac{15}{3} = 5$$

$$s1 = 4.369 + 3.367 + 5.088 + 5.509 + 5.570 = 24.033$$

$$s2 = 6.431 + 6.234 + 6.553 + 6.599 + 6.410 = 32.227$$

$$s3 = 6.901 + 6.925 + 6.914 + 6.963 + 6.917 = 34.620$$

$$\hat{\beta}_2'^5 = \frac{34.620 - 32.227}{32.227 - 24.033} = \frac{2.393}{8.194} = .2920$$

$$\hat{\beta}_2' = \sqrt[5]{.29204} = .7818$$

$$\hat{\beta}_1' = \frac{8.194(.7818 - 1)}{(.2920 - 1)^2} = -3.567$$

$$\hat{\beta}_0' = \frac{1}{5}\left(24.033 - \frac{8.194}{.2920 - 1}\right) = 7.121$$

To shift the origin back 1 unit, $\hat{\beta}_1'$ becomes

$$\frac{\hat{\beta}_1'}{\hat{\beta}_2'} = \frac{-3.576}{.7818} = -4.562$$

and

$$\hat{T}_t' = \log \hat{T}_t = 7.121 - 4.562(.782)^t$$

$$\hat{T}_t = e^{7.121} \cdot (e^{-4.562})^{.782^t} = 1{,}238(.0104)^{.782^t}$$

which gives the first set of forecasts in Table 3.13, i.e.,

$$\hat{y}_1 = \hat{T}_1 = 1{,}238(.0104)^{.782^1} = 34.8 \doteq 35$$

$$\hat{y}_2 = \hat{T}_2 = 1{,}238(.0104)^{.782^2} = 75.9 \doteq 76$$

and so forth.

To show how sensitive the estimate of the upper asymptote can be to the noise in the series, forecasts were also computed as though one were forecasting at time $t = 14$ (so $y_{15}$ would not be available). Then

$$m = \frac{12}{3} = 4$$

$$s1 = 5.088 + 5.509 + 5.700 + 6.431 = 22.728$$

$$s2 = 6.234 + 6.553 + 6.599 + 6.410 = 25.796$$

$$s3 = 6.901 + 6.925 + 6.914 + 6.963 = 27.703$$

and the parameter estimates (with the same origin as the previous trend equation) would be

$$\hat{\beta}_0 = 2{,}228 \qquad \hat{\beta}_1 = .0324 \qquad \hat{\beta}_2 = .888$$

Notice that the estimate of the upper asymptote would have been considerably different at $t = 14$ than at $t = 15$. The forecasts for $t = 16$ through $t = 20$, shown in Figure 3.15, would also have been quite different (by as much as

**Single Forecasts Using Partial Totals** Gompertz and Logistic Models

**Given:** A sample of $n$ observations $y_t$; $t = 1, 2, \ldots n$ where $n$ is divisible by three (if not, delete early observations and renumber)

**Models:**

Gompertz: $y_t = \beta_0 \cdot (\beta_1)^{\beta_2^t} \cdot \epsilon_t \qquad y'_t = \log y_t = \beta'_0 + \beta'_1 \cdot (\beta'_2)^t + \epsilon'_t$

where $\beta'_0 = \log \beta_0$, $\beta'_1 = \log \beta_1$, $\beta'_2 = \beta_2$, $\epsilon'_t = \log \epsilon_t$.

Logistic: $y_t = \dfrac{10^a}{\beta'_0 + \beta'_1 \cdot (\beta'_2)^t \cdot \epsilon'_t} \qquad y'_t = \dfrac{10^a}{y_t} = \beta_0 + \beta_1 \cdot (\beta_2)^t \cdot \epsilon_t$

where $a$ is chosen for convenience, $\beta_0 = \beta'_0$, $\beta_1 = \beta'_1$, $\beta_2 = \beta'_2$, and $\epsilon_t = \epsilon'_t$.

**Estimates:** Modified Exponential parameters are fit to the series $\{y'_t;\ t = 1, 2, \ldots, n\}$, yielding $\hat{\beta}'_0$, $\hat{\beta}'_1$, and $\hat{\beta}'_2$.

Gompertz: $\hat{\beta}_0 = e^{\hat{\beta}'_0} \qquad \hat{\beta}_1 = e^{\hat{\beta}'_1} \qquad \hat{\beta}_2 = \hat{\beta}'_2$

Logistic: $\hat{\beta}_0 = \hat{\beta}'_0 \qquad \hat{\beta}_1 = \hat{\beta}'_1 \qquad \hat{\beta}_2 = \hat{\beta}'_2$

**Forecasts:**

Gompertz: $\hat{y}'_{t+p} = \hat{T}'_{t+p} = \hat{\beta}'_0 + \hat{\beta}'_1 \cdot (\hat{\beta}'_2)^t \cdot (\hat{\beta}'_2)^p$

$\hat{y}_{t+p} = e^{\hat{y}'_{t+p}}$

Logistic: $\hat{y}'_{t+p} = \hat{T}'_{t+p} = \hat{\beta}_0 + \hat{\beta}_1 \cdot (\hat{\beta}_2)^t \cdot (\hat{\beta}_2)^p$

$\hat{y}'_{t+p} = \dfrac{10^a}{\hat{y}_{t+p}}$

---

35% at $t = 20$). Thus, fitting and extrapolating a Gompertz curve for a noisy series can give very misleading forecasts (a fact some analysts have discovered to their chagrin).

## Logistic Fit

We will also use these data to demonstrate the mechanics of fitting a logistic model. To do the fitting we first note that the maximum value of $y_t$ is 1057, so $1/y_t$ would be as small as .00095. Multiplying by $10^5 = 100{,}000$ will make the values less unwieldy. At time $t = 15$,

$$\frac{m}{3} = 5$$

$$s1 = \frac{10^5}{79} + \frac{10^5}{29} + \frac{10^5}{162} + \frac{10^5}{247} + \frac{10^5}{299}$$

$$= 1266 + 3448 + 617 + 405 + 334 = 6070$$

$$s2 = 161 + 196 + 143 + 136 + 164 = 800$$

$$s3 = 101 + 98 + 99 + 95 + 99 = 492$$

and from the estimating equations for the modified exponential we get (with the same origin as the data in Table 3.13)

$$\hat{\beta}_0 = 94.6 \qquad \hat{\beta}_1 = 4{,}545 \qquad \hat{\beta}_2 = .567$$

So the estimated trend model is

$$\hat{T}_t = \frac{10^5}{94.6 + 4,545(.567)^t}$$

which approaches an upper asymptote of $10^5/94.6 = 1,057$. At time $t = 14$, this asymptote would have been $10^5/74.3$, about a 27% difference. The forecasts for, say, times $t = 16$ and $t = 17$ would be

$$\hat{\beta}_1 \cdot (\hat{\beta}_2)^{15} = .915$$

$$\hat{\beta}_1 \cdot (\hat{\beta}_2)^{16} = .915 \times .567 = .519$$

$$\hat{\beta}_0 + \hat{\beta}_1 \cdot (\hat{\beta}_2)^{16} = 94.6 + .519 = 95.119$$

$$\hat{y}_{16} = \hat{T}_{16} = \frac{10^5}{95.119} = 1,051$$

# 3.9 Updating Procedures

## 3.9.1 Introduction

Since it is possible for the parameters of a trend model to change over time, we might prefer an updating scheme similar in concept to those used for no-trend models. The most widely used such schemes assume the trend is linear. To some extent these linear updating schemes will track series whose actual trend is curvilinear or that have seasonal or cyclical components, since such movement might be loosely characterized as linear trend with a constantly changing slope. It should be remembered, however, that the updated forecasts so produced will lag the seasonal and cyclical components. A better way to handle such components (especially the seasonal) is to incorporate them explicitly into the model.

Many of the updating schemes used for no-trend models can be modified to allow for a trend component. It is possible, for example, to simply reestimate the model parameters each time a new observation becomes available much the same as with the updated mean, median, and midrange for no-trend updating. Or one could compute parameter estimates using only the most recent $k$ observations as with simple and weighted moving averages. However, because trend fitting lacks the simplicity of no-trend estimation, updating trend fits is not so straightforward. The updating equations are not so convenient, and providing new forecasts involves considerably more computation. However, some direct analogs of the moving average and exponentially smoothed forecasts have been developed for trended series and are relatively easy to use.

The main difficulty with using simple moving averages on a trended series is that the average is centered some distance in the past. If the average level of the series is unchanging, the age of the moving average is no problem. But if the level is changing, an "aged" average will lag behind the current and future levels of the series. If we know the functional form of the series' trend, we can correct the average for age to produce more accurate forecasts.

For example, the age of a moving average of length $k$ is $(k - 1)/2$ (see Section 2.9.8). If we know that the trend is linear so

$$T_t = \beta_0 + \beta_1 t$$

and we know the value of the slope, then we can simply multiply the slope by $(p + (k - 1)/2)$ and add the product to that moving average to get a $p$-step-ahead forecast. That is,

$$\hat{y}_{t+p} = (\text{moving average at time } t) + \left(\frac{k-1}{2} + p\right) \times (\text{slope})$$
$$= \text{MA}_t + \left(\frac{k-1}{2} + p\right) \cdot \beta_1$$

Since $\beta_1$ is generally *un*known, however, some scheme of estimating it from the moving average must be devised.

### 3.9.2 First Differences of a Moving Average

A very intuitive method of obtaining an estimate of the slope $\beta_1$ of a linear trend would be to note that since successive moving averages are centered one unit apart in time, they will tend to differ by an amount equal to $\beta_1$. We could then estimate $\beta_1$ at time $t$ as

$$\hat{\beta}_1(t) = \text{MA}_t - \text{MA}_{t-1}$$

If we notice that

$$\text{MA}_t - \text{MA}_{t-1} = \frac{y_t + \cdots + y_{t-k+1}}{k} - \frac{y_{t-1} + \cdots + y_{t-k}}{k} = \frac{y_t - y_{t-k}}{k}$$

then we can see that using this method would be similar to fitting a linear model by the method of selected points to the first and last of the $k$ most recent observations, refitting the model each time a new data point is obtained. It is also the same as estimating $\beta_1$ to be the average of the $k$ most recent first differences. Because the method suffers from the main drawback of selected points (i.e., sensitivity to the noise in the points), another procedure similar in concept to semiaveraging of the $k$ most recent points is generally preferred.

### 3.9.3 Double Moving Average (DMA)

Since a series $\text{MA}_t$ of moving averages of length $k$ is centered at $t - (k - 1)/2$, it may be used as an estimate of the linear trend at that time, i.e., an estimate of

$$T_{t-(k-1)/2} = \text{trend at time } \left(t - \frac{k-1}{2}\right)$$
$$= (\text{trend at time } t) - \frac{k-1}{2} \times (\text{slope})$$

or

$$T_{t-(k-1)/2} = \beta_0 + \beta_1\left(t - \frac{k-1}{2}\right) = T_t - \frac{k-1}{2}\beta_1$$

If we construct a second series of moving averages of the moving averages, both of length $k$, this second series, which we will call $MA'_t$, can be shown to be centered at $t - (k - 1)$, so it may be used to estimate the linear trend at $t - (k - 1)$, which is

$$T_{t-(k-1)} = \beta_0 + \beta_1[t - (k - 1)] = T_t - (k - 1)\beta_1$$

Then the difference between the two series, $MA_t - MA'_t$, is an estimate of the difference:

$$T_{t-(k-1)/2} - T_{t-(k-1)} = T_t - \left(\frac{k-1}{2}\right)\beta_1 - T_t + (k-1)\beta_1 = \frac{k-1}{2}\beta_1$$

So dividing the right-hand side by $(k - 1)/2$ [multiplying by $2/(k - 1)$], we get

$$\begin{aligned}\hat{\beta}_1(t) &= \text{estimate of } \beta_1 \text{ as of time } t\\ &= \left(\frac{2}{k-1}\right)(MA_t - MA'_t)\end{aligned}$$

These are commonly called **double moving averages** or **linear moving averages** (Thomopoulos, 1980). They may also be used to estimate the trend at time $t$, that is, $\beta_0 + \beta_1 t$, by noting that

$2MA_t$ is an estimate of $2\beta_0 + \beta_1(2t - k + 1)$.

$MA'_t$ is an estimate of $\beta_0 + \beta_1(t - k + 1)$.

Therefore

$2MA_t - MA'_t$ is an estimate of $\beta_0 + \beta_1 t$

That is,

$$\begin{aligned}\hat{T}_t(t) &= \text{estimated linear trend at time } t \text{ as of time } t\\ &= \text{estimate of } \beta_0 + \beta_1 t \text{ as of time } t\\ &= 2MA_t - MA'_t\end{aligned}$$

A $p$-step-ahead forecast of the series computed as of time $t$, $\hat{y}_{t+p}(t)$, could then be computed as

$$\begin{aligned}\hat{y}_{t+p}(t) &= \text{estimate of } \beta_0 + (t + p)\cdot\beta_1 \text{ as of time } t\\ &= \hat{T}_t(t) + p\cdot\hat{\beta}_1(t)\end{aligned}$$

This practice of first smoothing a series, then resmoothing the smoothed series to obtain parameter estimates is used in several applications. If both "smoothings" are done over $k$ terms, the result is often called a "$k$-by-$k$ moving average" or a "$k \times k$ MA." Note that forecasting cannot begin until the $(2k - 1)$th period, so that forecasts using a double moving average of length 4, say, cannot be generated until 7 observations have been obtained. In more

**Double Moving Average of Length $k$ ($k \times k$ Moving Average)**

**Given:** Most recent $2k - 1$ observations, $y_{t-(2k-1)+1}, \ldots, y_{t-1}, y_t$

**Estimates:**

$$\mathrm{MT}_t = \sum_{1}^{k} y_{t-i+1} \qquad \mathrm{MA}_t = \frac{\mathrm{MT}_t}{k}$$

$$\mathrm{MT}'_t = \sum_{1}^{k} \mathrm{MT}_{t-i+1} \qquad \mathrm{MA}'_t = \frac{\mathrm{MT}'_t}{k^2}$$

$$\begin{aligned} \hat{T}_t(t) &= \text{estimate of } \beta_0 + \beta_1 t \text{ as of time } t \\ &= \text{estimate of } T_t \text{ as of time } t \\ &= 2\mathrm{MA}_t - \mathrm{MA}'_t \end{aligned}$$

$$\begin{aligned} \hat{\beta}_1(t) &= \text{estimate of } \beta_1 \text{ as of time } t \\ &= \frac{2}{k-1}(\mathrm{MA}_t - \mathrm{MA}'_t) \end{aligned}$$

**Updating Equations:** (For updating using $\mathrm{MT}_t$ and $\mathrm{MT}'_t$)

$$\mathrm{MT}_t = \mathrm{MT}_{t-1} + y_t - y_{t-k}$$

$$\mathrm{MT}'_t = \mathrm{MT}'_{t-1} + \mathrm{MT}_t - \mathrm{MT}_{t-k}$$

$$\mathrm{MA}_t = \frac{\mathrm{MT}_t}{k}$$

$$\mathrm{MA}'_t = \frac{\mathrm{MT}'_t}{k^2}$$

**Forecast:** (At time $t$ for $p$ periods ahead)

$$\hat{y}_{t+p}(t) = \hat{T}_t(t) + p \cdot \hat{\beta}_1(t)$$

general applications each "smoothing" may be done over a different number of terms, say $k_2$ for the second and $k_1$ for the first (a length-$k_2$ MA of a length-$k_1$ MA), the result is called a $k_2$-by-$k_1$ moving average (denoted a $k_2 \times k_1$ MA) and requires $k_2 + k_1 - 1$ observations to begin. The actual updating is most conveniently done by updating the required moving totals, $\mathrm{MT}_t$ and $\mathrm{MT}'_t$, as follows:

$$\mathrm{MT}_t = \mathrm{MT}_{t-1} + y_t - y_{t-k_1}$$

$$\mathrm{MT}'_t = \mathrm{MT}'_{t-1} + \mathrm{MT}_t - \mathrm{MT}_{t-k_2}$$

and then computing the corresponding averages

$$\mathrm{MA}_t = \frac{\mathrm{MT}_t}{k_1}$$

$$\mathrm{MA}'_t = \frac{\mathrm{MT}'_t}{k_1 k_2}$$

## EXAMPLE 3.16 *Double Moving Averages, Metro Video Society*

Metro Video Society operates several videotape rental outlets in a moderate-size midwestern city. The company has grown rapidly over the last year or so and has been expanding its inventory to accommodate the increasing demand for its services. As an aid in planning, a forecast of the next three months' revenues is needed each month. Because the tape rental market in the area could change rapidly, an updating scheme of some type is required. Revenue data for the last 17 months are given in Table 3.14 and graphed in Figure 3.17. Since the data are obviously trended, a double moving average scheme was investigated. The forecasts that such a scheme would have

**Table 3.14** Monthly Revenues, Metro Video Society, Sept 1984–Jan 1986

| Month | $t$ | Revenues ($10) $y_t$ | Month | $t$ | Revenues ($10) $y_t$ |
|---|---|---|---|---|---|
| Sep 84 | 1 | 999 | Jun | 10 | 3,024 |
| Oct | 2 | 1,123 | Jul | 11 | 3,467 |
| Nov | 3 | 1,503 | Aug | 12 | 3,528 |
| Dec | 4 | 1,762 | Sep | 13 | 3,441 |
| Jan 85 | 5 | 2,126 | Oct | 14 | 3,558 |
| Feb | 6 | 2,315 | Nov | 15 | 3,746 |
| Mar | 7 | 2,239 | Dec | 16 | 3,628 |
| Apr | 8 | 2,655 | Jan 86 | 17 | 4,021 |
| May | 9 | 2,787 | | | |

**Figure 3.17** Monthly Revenues, Metro Video Society, Sept 1984–Feb 1986

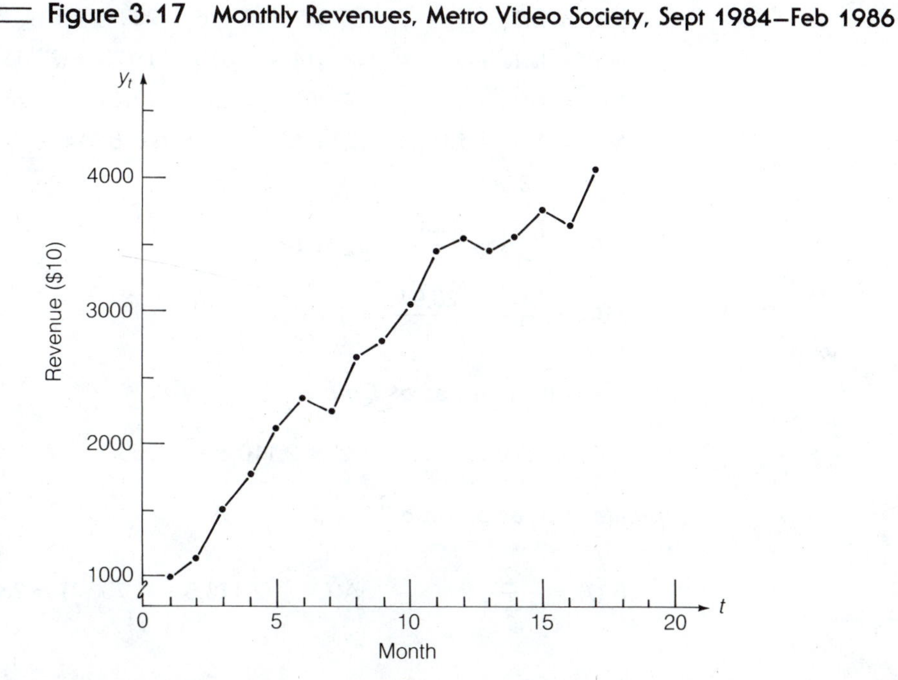

**Table 3.15 Forecasts for Metro Video Data Using a Double Moving Average of Length 4**

| $t$ | $y_t$ | $MT_t$ | $MT'_t$ | $MA_t$ | $MA'_t$ | $\hat{T}_t(t)$ | $\hat{\beta}_1(t)$ | $\hat{y}_{t+1}(t)$ |
|---|---|---|---|---|---|---|---|---|
| 1 | 999 | | | | | | | |
| 2 | 1,123 | | | | | | | |
| 3 | 1,503 | | | | | | | |
| 4 | 1,762 | 5,387 | | 1,346.8 | | | | |
| 5 | 2,126 | 6,514 | | 1,628.5 | | | | |
| 6 | 2,315 | 7,706 | | 1,926.5 | | | | |
| 7 | 2,239 | 8,442 | 28,049 | 2,110.5 | 1,753.1 | 2,467.9 | 238.3 | 2,706 |
| 8 | 2,655 | 9,335 | 31,997 | 2,333.8 | 1,999.8 | 2,667.7 | 222.6 | 2,890 |
| 9 | 2,787 | 9,996 | 35,479 | 2,499.0 | 2,217.4 | 2,780.6 | 187.7 | 2,968 |
| 10 | 3,024 | 10,705 | 38,478 | 2,676.3 | 2,404.9 | 2,947.6 | 180.9 | 3,129 |
| 11 | 3,467 | 11,933 | 41,969 | 2,983.3 | 2,623.1 | 3,343.4 | 240.1 | 3,584 |
| 12 | 3,528 | 12,806 | 45,440 | 3,201.5 | 2,840.0 | 3,563.0 | 241.0 | 3,804 |
| 13 | 3,441 | 13,460 | 48,904 | 3,365.0 | 3,056.5 | 3,673.5 | 205.7 | 3,879 |
| 14 | 3,558 | 13,994 | 52,193 | 3,498.5 | 3,262.1 | 3,734.9 | 157.6 | 3,893 |
| 15 | 3,746 | 14,273 | 54,533 | 3,568.3 | 3,408.3 | 3,728.8 | 106.6 | 3,835 |
| 16 | 3,628 | 14,373 | 56,100 | 3,593.3 | 3,506.3 | 3,680.3 | 58.0 | 3,738 |
| 17 | 4,021 | 14,953 | 57,593 | 3,738.3 | 3,599.6 | 3,876.9 | 92.5 | 3,969 |

produced for $k = 4$ are shown in Table 3.15. Forecasting could begin at time $t = 2k - 1 = 2 \times 4 - 1 = 7$ (April 1985) as follows:

*In April 1985* (t = *7*):

$$MT_4 = y_1 + y_2 + y_3 + y_4 = 999 + 1{,}123 + 1{,}503 + 1{,}762 = 5{,}387$$

$$MT_5 = MT_4 + y_5 - y_1 = 5{,}387 + 2{,}126 - 999 = 6{,}514$$

$$MT_6 = MT_5 + y_6 - y_2 = 6{,}514 + 2{,}315 - 1{,}123 = 7{,}706$$

$$MT_7 = MT_6 + y_7 - y_3 = 7{,}706 + 2{,}239 - 1{,}503 = 8{,}442$$

$$MT'_7 = MT_4 + MT_5 + MT_6 + MT_7 = 5{,}387 + 6{,}514 + 7{,}706 + 8{,}442 = 28{,}049$$

$$MA_7 = \frac{MT_7}{4} = \frac{8{,}442}{4} = 2{,}110.5$$

$$MA'_7 = \frac{MT'_7}{4^2} = \frac{28{,}049}{16} = 1{,}753.1$$

*Estimate of* $\beta_0 + 7\beta_1$ *as of Time 7*:

$$\hat{T}_7(7) = 2MA_7 - MA'_7 = 2 \times 2{,}110.5 - 1{,}753.1 = 2{,}467.9$$

*Estimate of* $\beta_1$ *as of Time 7*:

$$\hat{\beta}_1(7) = \frac{2}{4-1}(MA_7 - MA'_7) = \frac{2}{3}(2{,}110.5 - 1{,}753.1) = 238.3$$

*Forecasts for the Next 3 Months as of the Current Month, That Is, Estimates of* $\beta_0 + 8\beta_1$, $\beta_0 + 9\beta_1$, *and* $\beta_0 + 10\beta_1$ *as of Time 7*:

*May 1985* (t = *8*): $\hat{y}_8(7) = \hat{y}_{7+1}(7) = \hat{T}_7(7) + 1 \cdot \hat{\beta}_1(7) = 2{,}467.9 + 1 \times 238.3 = 2{,}706$

*June 1985* (t = *9*): $\hat{y}_9(7) = \hat{y}_{7+2}(7) = \hat{T}_7(7) + 2 \cdot \hat{\beta}_1(7) = 2{,}467.9 + 2 \times 238.3 = 2{,}944$

*July 1985* (t = *10*): $\hat{y}_{10}(7) = \hat{y}_{7+3}(7) = \hat{T}_7(7) + 3 \cdot \hat{\beta}_1(t) = 2{,}467.9 + 3 \times 238.3 = 3{,}183$

*In May 1985* (t = *8*):

$$MT_8 = MT_7 + y_8 - y_4 = 8{,}442 + 2{,}655 - 1{,}762 = 9{,}335$$

$$MT'_8 = MT'_7 + MT_8 - MT_4 = 28{,}049 + 9{,}335 - 5{,}387 = 31{,}997$$

$$MA_8 = \frac{MT_8}{4} = \frac{9{,}335}{4}$$

$$MA'_8 = \frac{MT'_8}{16} = \frac{31{,}997}{4}$$

*Estimate of* $\beta_0 + 8\beta_1$ *as of Time 8*:

$$\hat{T}_8(8) = 2MA_8 - MA'_8 = 2 \times 2{,}333.8 - 1{,}999.8 = 2{,}667.7$$

*Estimate of* $\beta_1$ *as of Time 8*:

$$\hat{\beta}_1(8) = \frac{2}{3}(2{,}333.8 - 1{,}999.8) = 222.6$$

*Forecasts for June, July, and August 1985* (t = *9, 10, 11*) *as of* t = *8*:

$$\hat{y}_9(8) = \hat{y}_{8+1}(8) = \hat{T}_8(8) + 1 \cdot \hat{\beta}_1(8) = 2{,}667.7 + 222.6 = 2{,}890$$

$$\hat{y}_{10}(8) = 2{,}667.7 + 445.2 = 3{,}113$$

$$\hat{y}_{11}(8) = 2{,}667.7 + 667.8 = 3{,}336$$

*In June 1985* (t = *9*):

$$MT_9 = 9{,}335 + 2{,}787 - 2{,}126 = 9{,}996$$

$$MT'_9 = 31{,}997 + 9{,}996 - 6{,}514 = 35{,}479$$

$$MA_9 = \frac{9{,}996}{4} = 2{,}449.0$$

$$MA'_9 = \frac{35{,}479}{16} = 2{,}217.4$$

$$\hat{T}_9(9) = 2 \times 2{,}449.0 - 2{,}217.4 = 2{,}780.6$$

$$\hat{\beta}_1(9) = \frac{2}{3}(2{,}449.0 - 2{,}217.4) = 187.7$$

$$\hat{y}_{10} = 2{,}780.6 + 1 \times 187.7 = 2{,}968$$

and so forth, giving the results in Table 3.16.

Table 3.16 RMSEs for Different-Length Double Moving Averages, Metro Video Data

| $k$ | MSE | RMSE |
|---|---|---|
| 2 | 61,556 | 248.1 |
| 3 | 51,337 | 226.6 |
| 4 | 57,009 | 238.8 |
| 5 | 72,130 | 268.6 |
| 6 | 83,213 | 288.5 |

### Choice of *k*

In the same way as for simple moving averages, the length $k$ of the average affects its tracking/smoothing tradeoff. The length may be selected according to the analyst's subjective judgment, e.g., to smooth out seasonal components or to satisfy data-handling constraints. An objective approach would be to select the value of $k$ that optimizes one of the accuracy measures, usually the MSE. In the case of Metro Video, the RMSEs for different values of $k$ are given in Table 3.16, which shows that the smallest MSE is obtained when $k = 3$. That such a small value of $k$ (short length) is optimal would indicate that fast tracking is required, which in turn may indicate that the trend is actually curvilinear or that the slope of the linear trend line has changed during the history of the series.

## 3.9.4 Generalized Double Moving Averages

Taking a moving average of a set of moving averages is a useful technique in contexts other than the linear trend updating discussed above. It may be used, for example, to get smoothed values centered at, rather than between, time periods. That is, if $k$ is even, a single MA is centered *between* time periods, but the double MA is centered *on* the time periods. This idea is used to center the moving averages used in the ratio-to-moving-average method of modeling seasonal movement to be discussed in Chapter 5.

As previously stated, in general application the DMA of length $k$ would be referred to as a $k \times k$ moving average, implying that both averages are computed over $k$ values and that a $k_1 \times k_2$ moving average is a length-$k_1$ MA of length-$k_2$ MAs. Thus, to compute a $4 \times 3$ MA we would compute length-4 MAs of length-3 MAs. The idea can be extended to any number of levels. Thus, a $3 \times 2 \times 4$ MA would be a length-3 MA of length-2 MAs of length-4 MAs. And so on.

In reality, these multiple-level moving averages are just weighted single moving averages with the weights chosen to achieve the desired results (such as estimates of $\beta_1$ or $T_t$). To see this, consider a $4 \times 4$ MA (double MA of length 4),

$$
\begin{aligned}
\mathrm{MT}_4 &= y_1 + y_2 + y_3 + y_4 \\
+\mathrm{MT}_5 &= y_2 + y_3 + y_4 + y_5 \\
+\mathrm{MT}_6 &= y_3 + y_4 + y_5 + y_6 \\
+\mathrm{MT}_7 &= y_4 + y_5 + y_6 + y_7 \\
\hline
\mathrm{MT}'_7 &= y_1 + 2y_2 + 3y_3 + 4y_4 + 3y_5 + 2y_6 + y_7 \\
\mathrm{MA}'_7 &= \frac{y_1 + 2y_2 + 3y_3 + 4y_4 + 3y_5 + 2y_6 + y_7}{16}
\end{aligned}
$$

which is a weighted MA of length 7 (as defined in Section 2.4.9) with weights

$$
w_0 = \frac{1}{16} \qquad w_1 = \frac{2}{16} \qquad w_2 = \frac{3}{16} \qquad w_3 = \frac{4}{16}
$$

$$
w_4 = \frac{3}{16} \qquad w_5 = \frac{2}{16} \qquad w_6 = \frac{1}{16}
$$

And as a matter of fact, since $\mathrm{MA}_7 = (y_4 + y_5 + y_6 + y_7)/4$, the estimate of the slope of the linear trend as of time 7 is

$$
\hat{\beta}_1(7) = \frac{2}{3}(\mathrm{MA}_7 - \mathrm{MA}'_7) = \frac{(y_5 + 2y_6 + 3y_7) - (3y_3 + 2y_2 + y_1)}{24}
$$

which is one-half the difference between the weighted averages of the three most recent points ($y_5$, $y_6$, $y_7$) and the three oldest points ($y_1$, $y_2$, $y_3$). The estimated trend at time $t$ would be

$$
\hat{T}_7(7) = 2\mathrm{MA}_7 - \mathrm{MA}'_7 = \frac{(4y_4 + 5y_5 + 6y_6 + 7y_7) - (y_1 + 2y_2 + 3y_3)}{16}
$$

which again is a weighted MA of length 7, but with weights

$$
w_0 = \frac{7}{16} \qquad w_1 = \frac{6}{16} \qquad w_2 = \frac{5}{16} \qquad w_3 = \frac{4}{16}
$$

$$
w_4 = \frac{-3}{16} \qquad w_5 = \frac{-2}{16} \qquad w_6 = \frac{-1}{16}
$$

The $4 \times 4$ MA method of updating is, therefore, similar to refitting a linear model to the latest seven points using semiaverages, except that the "semi-averaging" is done with weighted averages.

Three and higher-level moving averages may be used to update quadratic and higher-order polynomials. However, the complexity of the updating equations increases quite rapidly. More extensive treatment of the general topic may be found in Kendall (1973) and others.

### 3.9.5 Double Exponential Smoothing (DES), Brown's Method

Double moving averages have the disadvantage that $2k - 1$ points must be carried from one time period to the next in order to do the updating. If we assume, as we did with DMAs, that a linear trend model is appropriate, then using exponentially smoothed MAs can alleviate this problem as it did for simple MAs in the no-trend case. The resulting forecasting scheme is often referred to as **Brown's Method** or simply **Double Exponential Smoothing (DES)** (Brown, 1963, ch. 9).

Because the smoothed series themselves are not the forecasts, the updating equations are easier to understand if we adopt the following notation:

$A_t$ = exponentially smoothed value of $y_t$ at time $t$

$A'_t$ = double exponentially smoothed value of $y_t$ at time $t$

Then

$$\text{(new smoothed value)} = (1 - \alpha) \times \text{(old smoothed value)} + \alpha \times \text{(new observation)}$$

or

$$A_t = (1 - \alpha)A_{t-1} + \alpha y_t$$

and

$$\text{(new doubly smoothed value)} = (1 - \alpha) \times \text{(old doubly smoothed value} + \alpha \times \text{(new smoothed value)}$$

or

$$A'_t = (1 - \alpha) \cdot A'_{t-1} + \alpha \cdot A_t$$

And updated estimates of the trend and slope may be computed from

$$\begin{aligned} \hat{T}_t(t) &= \text{estimate linear trend at time } t \text{ as of time } t \\ &= 2A_t - A'_t \\ \hat{\beta}_1(t) &= \text{estimated slope of the trend line as of time } t \\ &= \frac{\alpha}{1-\alpha}(A_t - A'_t) \\ \hat{y}_{t+p}(t) &= \text{estimate of the linear trend at time } (t + p) \text{ as of time } t \\ &= p\text{-step-ahead forecast of } y_{t+p} \text{ as of time } t \\ &= \hat{T}_t(t) + p \cdot \hat{\beta}_1(t) \end{aligned}$$

The coefficient $\alpha/(1 - \alpha)$ in the expression for the estimated slope $\hat{\beta}_1(t)$ at time $t$ is necessary in the same way that $2/(k - 1)$ was needed for double moving averages. $A_t$ is centered at (approx.)

$$t - \frac{1-\alpha}{\alpha}$$

and $A'_t$ at (approx.)

$$t - \frac{2(1-\alpha)}{\alpha}$$

which is a time difference of

$$\left(t - \frac{2(1-\alpha)}{\alpha}\right) - \left(t - \frac{1-\alpha}{\alpha}\right) = \frac{1-\alpha}{\alpha}$$

Thus, the expected difference between $A_t$ and $A'_t$ is

$$\frac{1-\alpha}{\alpha}\beta_1$$

The estimated trend $\hat{T}_t(t)$ is derived along similar lines.

Updating equations for the trend and slope estimates can also be written in a kind of *error correction form* that is convenient to use and lends insight into the true nature of the forecast revision process. They are

$$\hat{T}_t(t) = \hat{y}_t(t-1) + [1-(1-\alpha)^2]e_t \tag{3.14}$$
$$= \hat{T}_{t-1}(t-1) + \hat{\beta}_1(t-1) + [1-(1-\alpha)^2]e_t$$
$$\hat{\beta}_1(t) = \hat{\beta}_1(t-1) + \alpha^2 e_t \tag{3.15}$$

where

$$e_t = y_t - \hat{y}_t(t-1)$$

is the one-step-ahead forecast error at time $t$. (Recall that $\hat{y}_t(t-1)$ is the

**Double Exponential Smoothing (DES)** Brown's Method

**Given:** The current observation of the time series $y_t$ and

**Smoothing Form:**

$A_{t-1}$ = exponentially smoothed value of $y$ at time $t-1$

$A'_{t-1}$ = double exponentially smoothed value of $y$ at time $t-1$

**Error Correction Form:**

$\hat{T}_{t-1}(t-1)$ = estimated linear trend at time $t-1$ as of time $t-1$

$\hat{\beta}_1(t-1)$ = estimated slope as of time $t-1$

So

$\hat{y}_t(t-1)$ = one-step-ahead forecast of $y_t$ as of time $t-1$

$= \hat{T}_{t-1}(t-1) + \hat{\beta}_1(t-1)$

Updating Equations:

**Smoothing Form:**

$$A_t = (1-\alpha)\cdot A_{t-1} + \alpha\cdot y_t$$
$$A'_t = (1-\alpha)\cdot A'_{t-1} + \alpha\cdot A_t$$
$$\hat{T}_t(t) = 2A_t - A'_t$$
$$\hat{\beta}_1(t) = \frac{\alpha}{1-\alpha}(A_t - A'_t)$$

**Error Correction Form:**

$$\hat{T}_t(t) = \hat{y}_t(t-1) + [1-(1-\alpha)^2]e_t$$
$$\hat{\beta}_1(t) = \hat{\beta}_1(t-1) + \alpha^2\cdot e_t$$

where $\alpha$ = smoothing constant

$e_t = y_t - \hat{y}_t(t-1)$

**Forecast:** (At time $t$ for $p$ periods ahead)

$$\hat{y}_{t+p}(t) = \hat{T}_t(t) + p\cdot\hat{\beta}_1(t)$$

forecast for $y_t$ as of time $t - 1$.) If Equation 3.14 is rewritten in terms of one-step-ahead forecasts, we have

$$\hat{T}_t(t) = (1-\alpha)^2 \cdot \underbrace{\hat{y}_t(t-1)}_{\downarrow\ \overbrace{\hat{T}_{t-1}(t-1) + \hat{\beta}_1(t-1) = \hat{T}_t(t-1)}^{}} + [1-(1-\alpha)^2]y_t \tag{3.16}$$

(estimate of the trend at time $t$ as of time $t - 1$)

or

$$\text{(new estimate of trend)} = (1-\alpha)^2 \times \text{(old estimated trend)} + [1-(1-\alpha)^2] \times \text{(new observation)}$$

which means that the trend estimates form a single exponentially smoothed series with a smoothing constant of $1 - (1 - \alpha)^2$. In addition, the new estimate of the slope (which is added to the new estimate of the trend to get the new forecast) is the old estimate corrected by a portion $\alpha^2$ of the current forecast error.

## Initial Values

As with simple exponential smoothing, there is a problem with getting the forecasting going, since to begin with there are no smoothed values to update. The problem may be handled in a number of ways, depending on how much data are available at the time smoothing begins.

- **If no history is available and forecasting is to begin for $y_1$** Some estimates of the trend and slope at $t = 0$, $\hat{T}_0(0)$ and $\hat{\beta}_1(0)$, or of the current value of the series, $\hat{y}_0(0)$, must be obtained. Any information may be used, such as a rough estimate based on your own or others' knowledge of the series. Other forecasts of the series or of similar series may do as well. Then
  - $\hat{T}_0(0)$ and $\hat{\beta}_1(0)$ available:

$$A_0 = \hat{T}_0(0) - \frac{1-\alpha}{\alpha}\hat{\beta}_1(0)$$

$$A_0' = \hat{T}_0(0) - \frac{2(1-\alpha)}{\alpha}\hat{\beta}_1(0)$$

$$\hat{y}_1(0) = \hat{T}_0(0) + \hat{\beta}_1(0)$$

  - $\hat{y}_0(0)$ available:

$$A_0 = A_0' = \hat{y}_0(0)$$

$$\hat{T}_0(0) = \hat{y}_0(0)$$

$$\hat{\beta}_1(0) = 0$$

$$\hat{y}_1(0) = \hat{T}_0(0) + \hat{\beta}_1(0)$$

  and updating is accomplished normally from $t = 1$ on.

- **If exactly one observation is available and forecasting is to begin for $y_2$**

$$A_1 = A_1' = y_1$$

or

$$\hat{T}_1(1) = y_1$$
$$\hat{\beta}_1(1) = 0$$

and updating is accomplished normally from $t = 2$ on.

If questionable initial values are used (particularly if $\beta_1(0) = 0$), it is a very good idea to use large values of $\alpha$ for the first few forecasts to get the system tracking accurately. The smoothing constant may then be lowered to the desired value after relatively stable forecasts are being obtained.

- **If a history of $n$ observations is available and forecasting is to begin for $y_{n+1}$**

1. Use $\hat{T}_1(1) = y_1$, $\hat{\beta}_1(1) = 0$ ($A_1 = A_1' = y_1$), and large $\alpha$ values for the first few forecasts and smooth the series as if you had been forecasting all along. When time $n$ is reached, a forecast of $y_{n+1}$ will be obtained.

2. Use

$$\hat{T}_1(1) = y_1 \qquad (A_1 = A_1' = y_1)$$
$$\hat{\beta}_1(1) = 0$$
$$\hat{\beta}_1(2) = y_2 - y_1 = \Delta y_2$$
$$\hat{\beta}_1(3) = \frac{y_3 - y_1}{2}$$
$$\hat{\beta}_1(4) = \frac{(y_2 - y_1) + (y_4 - y_3)}{2} = \frac{\Delta y_2 + \Delta y_4}{2}$$

and smooth in the normal manner from $t = 5$ on.

3. Use least squares or semiaverages estimates of the slope and intercept at $t = n$ to generate $\hat{y}_{n+1}(n)$, smoothing normally thereafter.

4. Use list points 1 or 2 above to backcast to time $t = 0$. Then forecast to time $t = n + 1$ using the backcast estimates of $\hat{T}_1(1)$ and $\hat{\beta}_1(1)$ to initiate the forecasting.

## Choice of $\alpha$

The double smoothing constant may be chosen in much the same manner as for single smoothing, that is,

- to optimize an accuracy criterion (very often the MSE)
- around 10%–20% to give fairly smooth forecasts with some tracking ability

It is interesting to note that $\alpha$ does not play exactly the same role in double exponential smoothing as in simple exponential smoothing. For a hor-

**Table 3.17 Equivalent Values for $\alpha_{SES}$ and $\alpha_{DES}$**

| $\alpha_{SES}$ | $\alpha_{DES} = 1 - \sqrt{1 - \alpha_{SES}}$ |
|---|---|
| .1 | .051 |
| .2 | .105 |
| .3 | .163 |
| .4 | .225 |
| .5 | .293 |
| .6 | .367 |
| .7 | .452 |
| .8 | .552 |
| .9 | .683 |

izontal model (single smoothing), $T_t = \beta_0$ and we can write the trend estimation equation as

$$\begin{aligned}\hat{T}_t(t) &= \hat{\beta}_0(t) = \hat{y}_{t+1}(t) \\ &= \hat{y}_t(t-1) + \alpha[y_t - \hat{y}_t(t-1)] \\ &= \hat{y}_t(t-1) + \alpha e_t\end{aligned}$$

and for double smoothing (Equation 3.16) as

$$\hat{T}_t(t) = \hat{y}_t(t-1) + [1 - (1-\alpha)^2]e_t$$

So the value $1 - (1-\alpha)^2$ of DES plays a role similar to $\alpha$ of SES. A value $\alpha_{DES}$ for DES that gives an amount of trend smoothing roughly equivalent to $\alpha_{SES}$ for SES may be found by setting

$$\alpha_{SES} = 1 - (1 - \alpha_{DES})^2$$

or

$$\alpha_{DES} = 1 - \sqrt{1 - \alpha_{SES}}$$

Table 3.17 shows representative equivalent values for $\alpha_{DES}$ and $\alpha_{SES}$. It is good to keep in mind, however, that the slope estimates are being smoothed with constant $\alpha^2$. So if changes in the slope are anticipated, $\alpha$ values higher than those in Table 3.17 may be desired.

**EXAMPLE 3.17** ***Brown's Method, Metro Video Society***

Applying double smoothing to the Metro Video data for one-step-ahead forecasts, suppose we start with $\hat{y}_2(1)$, choose $\alpha = .1$ so $1 - (1-\alpha)^2 = 1 - .81 = .19$, and use the simplest initialization technique, i.e., $\hat{T}_1(1) = y_1 = 999$, $\hat{\beta}_1(1) = 0$.

*At* t = *1:* $\hat{y}_2(1) = \hat{T}_1(1) + \hat{\beta}_1(1) = 999 + 0 = 999$

*At* t = *2:* $y_2 = 1{,}123 \qquad e_2 = y_2 - \hat{y}_2(1) = 1{,}123 - 999 = 124$

$$\hat{T}_2(2) = \hat{y}_2(1) + [1 - (1 - \alpha)^2]e_3 = 999 + [1 - (1 - .1)^2] \times 124$$
$$= 999 + .19 \times 124 = 1{,}023$$

$$\hat{\beta}_1(2) = \hat{\beta}_1(1) + \alpha^2 \cdot e_2 = 0 + .01 \times 124 = 1.24$$

$$\hat{y}_3(2) = \hat{T}_2(2) + \hat{\beta}_1(2) = 1{,}023 + 1.24 = 1{,}024$$

*At* t = *3:* $y_3 = 1{,}503 \qquad e_3 = 1{,}503 - 1{,}024 = 479$

$$\hat{T}_3(3) = \hat{y}_3(2) + .19 \cdot e_3 = 1{,}024 + .19 \times 479 = 1{,}115$$

$$\hat{\beta}_1(3) = \hat{\beta}_1(2) + .01 \cdot e_3 = 1.24 + .01 \times 479 = 6.03$$

$$\hat{y}_4(3) = \hat{T}_3(3) + \hat{\beta}_1(3) = 1{,}115 + 6.03 = 1{,}121$$

*At* t = *4:* $y_4 = 1{,}762 \qquad e_4 = 1{,}762 - 1{,}121 = 641$

$$\hat{T}_4(4) = 1{,}121 + .19 \times 641 = 1{,}243$$

$$\hat{\beta}_1(4) = 6.03 + .01 \times 641 = 12.44$$

$$\hat{y}_5(4) = 1{,}243 + 12.44 = 1{,}255$$

And so forth. The complete set of updates is given in Table 3.18.

**Table 3.18** Forecasts for Metro Video Data Using Brown's Double Smoothing with $\alpha = .1$

| $t$ | $y_t$ | $\hat{y}_t(t-1)$ | $e_t$ | $\hat{T}_t(t)$ | $\hat{\beta}_1(t)$ | $\hat{y}_{t+1}(t)$ |
|---|---|---|---|---|---|---|
| 1 | 999 | | | 999 | 0 | 999 |
| 2 | 1,123 | 999 | 124 | 1,023 | 1.24 | 1,024 |
| 3 | 1,503 | 1,024 | 479 | 1,115 | 6.03 | 1,121 |
| 4 | 1,762 | 1,121 | 641 | 1,243 | 12.44 | 1,255 |
| 5 | 2,126 | 1,255 | 871 | 1,420 | 21.15 | 1,441 |
| 6 | 2,315 | 1,441 | 874 | 1,607 | 29.89 | 1,637 |
| 7 | 2,239 | 1,637 | 602 | 1,751 | 35.91 | 1,787 |
| 8 | 2,655 | 1,787 | 868 | 1,952 | 44.59 | 1,997 |
| 9 | 2,787 | 1,997 | 790 | 2,147 | 52.49 | 2,199 |
| 10 | 3,024 | 2,199 | 825 | 2,356 | 60.74 | 2,417 |
| 11 | 3,467 | 2,417 | 1,050 | 2,617 | 71.24 | 2,688 |
| 12 | 3,528 | 2,688 | 840 | 2,848 | 79.64 | 2,928 |
| 13 | 3,441 | 2,928 | 513 | 3,025 | 84.77 | 3,110 |
| 14 | 3,558 | 3,110 | 448 | 3,195 | 89.25 | 3,284 |
| 15 | 3,746 | 3,284 | 462 | 3,372 | 93.87 | 3,466 |
| 16 | 3,628 | 3,466 | 162 | 3,497 | 95.49 | 3,592 |
| 17 | 4,021 | 3,592 | 429 | 3,674 | 99.78 | 3,774 |

If we use the smoothing form of the updating equations rather than the error correction form we get the following.

*At* t = *1:*

$$A_1 = A_1' = y_1 = 999$$

$$\hat{T}_1(1) = 2A_1 - A_1' = 2y_1 - y_1 = y_1 = 999$$

$$\hat{\beta}_1(1) = \frac{\alpha}{1-\alpha}(A_1 - A_1') = 0.00$$

$$\hat{y}_2(1) = 999 + 0.00 = 999$$

*At* t = *2:*

$$y_2 = 1{,}123$$

$$A_2 = (1-\alpha)\cdot A_1 + \alpha \cdot y_2 = .9 \times 999 + .1 \times 1{,}123 = 1{,}011.4$$

$$A_2' = (1-\alpha)\cdot A_1' + \alpha \cdot A_1 = .9 \times 999 + .1 \times 1{,}011.4 = 1{,}000.2$$

$$\hat{T}_2(2) = 2A_2 - A_2' = 2 \times 1{,}011.4 + 1{,}000.2 = 1{,}023$$

$$\hat{\beta}_1(2) = \frac{\alpha}{1-\alpha}(A_2 - A_2') = \frac{.1}{.9}(1{,}011.4 - 1{,}000.2) = 1.24$$

$$\hat{y}_3(2) = 1{,}023 + 1.24 = 1{,}024$$

*At* t = *3:*

$$y_3 = 1{,}503$$

$$A_3 = .9 \times 1{,}011.4 + .1 \times 1{,}503 = 1{,}060.6$$

$$A_3' = .9 \times 1{,}000.2 + .1 \times 1{,}060.6 = 1{,}006.2$$

$$\hat{T}_3(3) = 2 \times 1{,}060.6 - 1{,}006.2 = 1{,}115$$

$$\hat{\beta}_1(3) = \frac{.1}{.9}(1{,}060.6 - 1{,}006.2) = 6.03$$

$$\hat{y}_4(3) = 1{,}115 + 6.03 = 1{,}121$$

And so on. The forecasts for the next three months (February, March, April 1985) as of $t = 17$ (January 1985) would be:

$$\hat{y}_{18}(17) = \hat{T}_{17}(17) + 1 \cdot \hat{\beta}_1(17) = 3{,}674 + 1 \times 99.78 = 3{,}774$$

$$\hat{y}_{19}(17) = \hat{T}_{19}(17) + 2 \cdot \hat{\beta}_1(17) = 3{,}674 + 2 \times 99.78 = 3{,}874$$

$$\hat{y}_{20}(17) = \hat{T}_{20}(17) + 3 \cdot \hat{\beta}_1(17) = 3{,}674 + 3 \times 99.78 = 3{,}973$$

*(Example continues next page.)*

The double smoothed model as applied above with $\alpha = .1$ does not work well with the Metro Video data. The forecast errors are all positive (under-forecasts) with $\bar{e} = +623.6$ and MAPE = 23.9%. In fact, Theil's $U = 2.61$ indicates a "no-change" model would have done a much better job of forecasting. The poor initial value of the slope, $\hat{\beta}_1(1) = 0$, is the problem. Recall from the updating equations for $\hat{T}_t(t)$ and $\hat{\beta}_1(t)$ that for $\alpha = .1$ the effective smoothing constant is .19 for $\hat{T}_t(t)$ and .01 for $\hat{\beta}_1(t)$. Because of the latter value, $\hat{\beta}_1(t)$ is corrected very slowly, and the initial zero estimate is still a strong influence on the one-step-ahead forecasts even at time 17. The trend is corrected relatively quickly, but we also need a decent estimate of the slope to combine with the estimated trend to produce the forecasts. The low initial slope estimate is causing too little to be added to the trend, thus producing the underforecasts.

**EXAMPLE 3.17** *(continued)*

To attempt to correct this difficulty we will use initialization procedure 2.

*At* t = *1:*

$y_1 = 999$

$\hat{T}_1(1) = 999$

$\hat{\beta}_1(1) = 0$

$\hat{y}_2(1) = 999$ (as before)

*At* t = *2:*

$y_2 = 1{,}123 \qquad e_2 = 1{,}123 - 999 = 124$

$\hat{T}_2(2) = 999 + .19 \times 124 = 1{,}023$

$\hat{\beta}_1(2) = \Delta y_2 = y_2 - y_1 = 1{,}123 - 999 = 124$

$\hat{y}_3(2) = 1{,}023 + 124 = 1{,}147$

*At* t = *3:*

$y_3 = 1{,}503 \qquad e_3 = 1{,}503 - 1{,}147 = 356$

$\hat{T}_3(3) = 1{,}147 + .19 \times 356 = 1{,}215$

$$\hat{\beta}_1(3) = \frac{y_3 - y_1}{2} = \frac{1{,}503 - 999}{2} = 252.0$$

$\hat{y}_4(3) = 1{,}215 + 252 = 1{,}467$

*At* t = *4:*

$y_4 = 1{,}762 \qquad e_4 = 1{,}762 - 1{,}467 = 295$

$\hat{T}_4(4) = 1{,}467 + .19 \times 295 = 1{,}563$

$$\hat{\beta}_1(4) = \frac{\Delta y_2 + \Delta y_4}{2} = \frac{(y_2 - y_1) + (y_4 - y_3)}{2}$$

$$= \frac{(1{,}123 - 999) + (1{,}762 - 1{,}503)}{2} = 191.5$$

$\hat{y}_5(4) = 1{,}523 + 191.5 = 1{,}715$

*At* t = *5:*

$y_5 = 2{,}126 \qquad e_5 = 411$

$\hat{T}_5(5) = 1{,}715 + .19 \times 411 = 1{,}793$

$\hat{\beta}_1(5) = 191.5 + .01 \times 411 = 195.6$

$\hat{y}_6(5) = 1{,}793 + 195.6 = 1{,}989$

*At* t = *6:*

$y_6 = 2{,}315 \qquad e_6 = 326$

$\hat{T}_6(6) = 1{,}989 + .19 \times 326 = 2{,}051$

$\hat{\beta}_1(6) = 195.6 + .01 \times 326 = 198.9$

$\hat{y}_7(6) = 2{,}051 + 198.9 = 2{,}250$

And so forth. Using this better initialization, the mean error $\bar{e}$ drops to 64.6, the MAPE to 9.4%, and Theil's $U$ to 1.02.

To illustrate initialization method 4, we will assume that forecasting is begun at time $t = 10$ and least squares initial estimates are used to generate a one-step-ahead forecast for $t = 11$. Then DES is used to generate $\hat{y}_{12}(11), \hat{y}_{13}(12), \ldots, \hat{y}_{18}(17)$.

***At* t = *10:***

$$SS_{tt} = \frac{10^3 - 10}{12} = 82.5 \qquad SS_{ty} = 18{,}626.5 \qquad \Sigma\, y = 20{,}533$$

$$\hat{\beta}_1(10) = \frac{SS_{ty}}{SS_{tt}} = \frac{18{,}626.5}{82.5} = 225.8$$

$$\hat{\beta}_0(10) = \bar{y} - \hat{\beta}_1(10)\frac{n+1}{2} = \frac{20{,}533}{10} - 225.8 \times \frac{10+1}{2} = 811.4$$

$$\hat{T}_{10}(10) = \hat{\beta}_0(10) + 10\hat{\beta}_1(10) = 811.4 + 10 \times 225.8 = 3{,}069$$

$$\hat{y}_{11}(10) = \hat{T}_{10}(10) + \hat{\beta}_1(10) = 3{,}069 + 225.8 = 3{,}295$$

***At* t = *11:***

$$y_{11} = 3{,}467 \qquad e_{11} = 3{,}467 - 3{,}295 = 172$$

$$\hat{T}_{11}(11) = 3{,}295 + .19 \times 172 = 3{,}328$$

$$\hat{\beta}_1(11) = 225.8 + .01 \times 172 = 227.5$$

$$\hat{y}_{12}(11) = 3{,}328 + 227.5 = 3{,}556$$

***At* t = *12:***

$$y_{12} = 3{,}528 \qquad e_{12} = -28$$

$$\hat{T}_{12}(12) = 3{,}556 + .19 \times -28 = 3{,}551$$

$$\hat{\beta}_1(12) = 227.5 + .01 \times -28 = 227.2$$

$$\hat{y}_{13}(12) = 3{,}551 + 227.2 = 3{,}778$$

And so on, giving $\hat{y}_{14}(13) = 3{,}938$, $\hat{y}_{15}(14) = 4{,}086$, $\hat{y}_{16}(15) = 4{,}238$, $\hat{y}_{17}(16) = 4{,}333$, $\hat{y}_{18}(17) = 4{,}481$. For these seven forecasts, $\bar{e}$ is $-262.1$, the MAPE is 8.6%, and Thiel's $U$ is .27. ●

A question still remains as to what smoothing constant $\alpha$ should be used with Brown's method for smoothing the Metro Video data. The choice could be based on the analyst's judgment of how much tracking is needed. If the trend is linear and we anticipate no marked change in the incremental growth, then smaller values of $\alpha$ would be appropriate. If changes in the incremental growth are possible or likely, higher values of $\alpha$ will do a better job of tracking these changes. Even then, however, too large a value will tend to produce the unstable forecasts that arise from tracking the random error too closely. We could also select the value of $\alpha$ that would have optimized, say, the RMSE over the history (or a recent portion of the history) of the series.

For example, the RMSEs computed for forecast errors from $t = 6$ to $t = 17$ using initialization method 2 are graphed in Figure 3.18. (Only $e_6$ through $e_{17}$ were used because the actual smoothing did not start until $t = 5$, when $\hat{y}_6(5)$ was computed.) Notice that for $\alpha$ values in the vicinity of the optimal value, the RSME changes relatively little. A smoothing constant of about .35 would be a reasonable choice ($\bar{e} = -58.3$, MAPE $= 4.97\%$, $U = .78$). Poor initialization could have raised this value a bit, so a slightly smaller value, say about .25 to .30, may be preferred for ongoing forecasting.

If part of the history is used to generate the initial trend and slope es-

**Figure 3.18** RMSE for Different Values of α for Metro Video Data

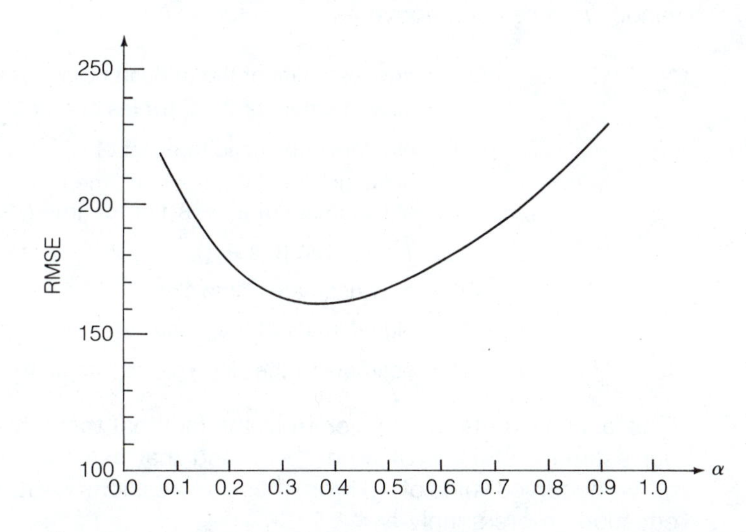


timates, the accuracy measures should be based on portions of the data that were not used in the initialization. If initial values are obtained using least squares methods on $y_1$ through $y_{10}$, for example, $\alpha$ should be chosen based on $e_{11}$ through $e_{17}$. When this is done, examination of the RMSEs shows $\alpha_{opt}$ to be about .43. Using either initialization method (2 or 4) leads to a rather large smoothing constant, mostly because the incremental growth seems to be decreasing in the latter part of the series, due, perhaps, to competition or saturation of the video rental market.

### 3.9.6 Linear Exponential Smoothing (LES), Holt's Method

Another method of expanding simple exponential smoothing to linear trend models is called **linear exponential smoothing (LES)** or **Holt's two-parameter method** (Holt, 1957). Holt's technique also uses double smoothing, but it smoothes the trend and slope directly and allows different smoothing constants for each. One of the shortcomings of Brown's scheme is that, in order to track changes in the slope even moderately, the trend must be tracked much more quickly, causing the estimated trend values to be very sensitive to noise. A two-parameter model gives more flexibility in selecting the rates at which the two series (trend and slope) are tracked.

The *smoothing* (*updating*) *equations* are

$$\hat{T}_t(t) = (1 - \alpha) \cdot \hat{y}_t(t - 1) + \alpha y_t \tag{3.17}$$

$$\hat{\beta}_1(t) = (1 - \gamma) \cdot \hat{\beta}_1(t - 1) + \gamma[\hat{T}_t(t) - \hat{T}_{t-1}(t - 1)] \tag{3.18}$$

where

$\alpha$ = smoothing constant for the trend

$\gamma$ = smoothing constant for the slope

It will help in understanding the updating if, at time $t$, we think of the various expressions above as

$$\begin{aligned}
\hat{T}_t(t) &= \text{new estimate of the trend at time } t \text{ (as of time } t) \\
&= \text{new estimate of } \beta_0 + \beta_1 t \text{ (as of time } t) \\
\hat{y}_t(t-1) &= \text{old one-step-ahead forecast of } y_t \\
&= \text{old estimate of the trend at time } t \\
&= \text{old estimate of } \beta_1 + \beta_1 t \text{ (as of time } t-1) \\
&= \hat{T}_t(t-1) + \hat{\beta}_1(t-1) \\
\hat{\beta}_1(t) &= \text{new estimate of the slope} \\
\hat{\beta}_1(t-1) &= \text{old estimate of the slope} \\
\hat{T}_t(t) - \hat{T}_t(t-1) &= \text{estimate of the slope based primarily on } y_t
\end{aligned}$$

(The last interpretation derives from the fact that the only difference between our estimate of $T_t$ as of time $t - 1$ and that as of time $t$ is caused by the newly observed value of $y_t$.) From these expressions we may state the updating equations more simply as

$$\text{(new trend)} = (1 - \alpha) \cdot \text{(old forecast)} + \alpha \cdot \text{(new observation)} \tag{3.19}$$

$$\text{(new slope)} = (1 - \gamma) \cdot \text{(old slope)} + \gamma \cdot \text{(new change in trend)} \tag{3.20}$$

There are also *error correction forms* for Equations 3.17 and 3.18. By rearranging Equation 3.17 we get

$$\hat{T}_t(t) = (1 - \alpha) \cdot \hat{y}_t(t-1) + \alpha y_t = \hat{y}_t(t-1) + \alpha[y_t - \hat{y}_t(t-1)]$$

or

$$\hat{T}_t(t) = \hat{y}_t(t-1) + \alpha e_t \tag{3.21}$$

where

$$\begin{aligned}
e_t &= y_t - \hat{y}_t(t-1) \\
&= \text{the one-step-ahead forecast error}
\end{aligned}$$

which means

$$\text{(new trend)} = \text{old forecast} + \alpha \cdot \text{(current forecast error)}$$

And from Equation 3.18 we get

$$\hat{\beta}_1(t) = (1 - \gamma) \cdot \hat{\beta}_1(t-1) + \gamma[\hat{T}_t(t) - \hat{T}_{t-1}(t-1)]$$

or

$$\hat{\beta}_1(t) = \hat{\beta}_1(t-1) + \gamma(\underbrace{[\hat{T}_t(t) - \hat{T}_{t-1}(t-1)]}_{Q_1} - \underbrace{\hat{\beta}_1(t-1)}_{Q_2}) \tag{3.22}$$

If we note that

$Q_1$ represents the estimate of $\beta_1$ based on the current observation

$Q_2$ represents the estimate of $\beta_1$ based on old observations ($y_{t-1}, y_{t-2}, \ldots$)

then $Q_1 - Q_2$ may be regarded as the current estimate of the error in $\hat{\beta}_1(t-1)$. So

(new slope) = (old slope) + $\gamma \cdot$ (new estimated error in slope)

These error correction forms are a bit cumbersome, but if we notice that $\hat{y}_t(t-1) = \hat{T}_{t-1}(t-1) + \hat{\beta}_1(t-1)$ in Equation 3.15, then we get

$$\hat{T}_t(t) = \hat{T}_{t-1}(t-1) + \hat{\beta}_1(t-1) + \alpha e_t$$

or

$$Q_1 = \hat{T}_t(t) - \hat{T}_{t-1}(t-1) = \hat{\beta}_1(t-1) + \alpha e_t$$

Substituting this $Q_1$ into Equation 3.22 we finally have a more convenient (and interesting) error correction form for updating the slope estimate, i.e.,

$$\hat{\beta}_1(t) = \hat{\beta}_1(t-1) + \alpha \cdot \gamma \cdot e_t \tag{3.23}$$

So

(new slope) = (old slope) + $\alpha \cdot \gamma \cdot$ (new forecast error)

**Linear Exponential Smoothing (LES)** Holt's Method

**Given:**

The current observation of the time series $y_t$ and

$\hat{T}_{t-1}(t-1)$ = estimated linear trend at time $t-1$ as of time $t-1$

$\hat{\beta}_1(t-1)$ = estimated slope as of time $t-1$

So

$\hat{y}_t(t-1)$ = one-step-ahead forecast for time $t$ as of time $t-1$

$= \hat{T}_{t-1}(t-1) + \hat{\beta}_1(t-1)$

**Updating Equations:**

**Smoothing Form:**

$$\hat{T}_t(t) = (1-\alpha) \cdot \hat{y}_t(t-1) + \alpha y_t$$

$$\hat{\beta}_1(t) = (1-\gamma) \cdot \hat{\beta}_1(t-1) + \gamma[\hat{T}_t(t) - \hat{T}_{t-1}(t-1)]$$

**Error Correction Form:**

$$\hat{T}_t(t) = \hat{y}_t(t-1) + \alpha e_t$$

$$\hat{\beta}_1(t) = \hat{\beta}_1(t-1) + \alpha \cdot \gamma \cdot e_t$$

where

$\alpha$ = smoothing constant for the trend

$\gamma$ = smoothing constant for the slope

$e_t = y_t - \hat{y}_t(t-1)$

= one-step-ahead forecast error

**Forecasts:**

(At time $t$ for $p$ periods ahead)

$$\hat{y}_{t+p} = \hat{T}_t(t) + p \cdot \hat{\beta}_1(t)$$

It is informative for understanding the interaction between the two smoothing constants ($\alpha$ and $\gamma$) and its effect on the stream of trend and slope estimates, that the trend is modified, in the same way as with SES, by a portion $\alpha \cdot 100\%$ of the forecast error, while the slope is modified by $\alpha \cdot \gamma \cdot 100\%$ of the forecast error. The immediate implications of this are:

- The slope is always smoothed more heavily than the trend (unless $\gamma = 1$).
- The amount of tracking for the slope can be set and the tracking of the trend can then be adjusted by varying $\alpha$ while holding $\alpha \cdot \gamma$ constant.

Finally, if one is using Brown's method (DES) with smoothing constant $\alpha_{DES}$, *exactly* the same forecasts could be obtained by using Holt's method (LES) with $\alpha_{LES} = 1 - (1 - \alpha_{DES})^2 = \alpha_{DES} \cdot (2 - \alpha_{DES})$ and $\gamma_{LES} = \alpha_{DES}/(2 - \alpha_{DES})$.

Both initialization and choice of smoothing constants may be accomplished in manners similar to those used for DES, one of the main differences being that *two* smoothing parameters must be specified. It simplifies matters when choosing $\alpha$ and $\gamma$ subjectively that their roles in the LES procedure are more directly comparable to the role of the SES smoothing constant than is $\alpha$ in Brown's method.

**EXAMPLE 3.18** ***Holt's Method, Metro Video Society***

Using Metro Video as an example once again, suppose we use initialization method 1 and $\alpha = \gamma = 10\%$.

*At t = 1:*

$y_1 = 999$

$\hat{T}_1(1) = 999$

$\hat{\beta}_1(1) = 0$

$\hat{y}_2(1) = \hat{T}_1(1) + \hat{\beta}_1(1) = 999 + 0 = 999$

*At t = 2:*

$y_2 = 1{,}123 \qquad e_2 = 124$

$$\hat{T}_2(2) = (1 - \alpha) \cdot \hat{y}_2(1) + \alpha \cdot y_2 = .9 \times 999 + .1 \times 1{,}123 = 1{,}011$$

$\hat{T}_2(2) - \hat{T}_1(1) = 1{,}061 - 1{,}011 = 12$

$$\hat{\beta}_1(2) = (1 - \gamma) \cdot \hat{\beta}_1(1) + \gamma[\hat{T}_2(2) - \hat{T}_1(1)] = .9 \times 0 + .1 \times 12 = 1.2$$

$$\hat{y}_3(2) = \hat{T}_2(2) + \hat{\beta}_1(2) = 1{,}011 + 1.2 = 1{,}012$$

*At t = 3:*

$y_3 = 1{,}503 \qquad e_3 = 491$

$$\hat{T}_3(3) = (1 - \alpha) \cdot \hat{y}_3(2) + \alpha \cdot y_3 = .9 \times 1{,}012 + .1 \times 1{,}503 = 1{,}061$$

$\hat{T}_3(3) - \hat{T}_2(2) = 1{,}061 - 1{,}011 = 50$

$$\hat{\beta}_1(3) = (1 - \gamma) \cdot \hat{\beta}_1(2) + \gamma[\hat{T}_3(3) - \hat{T}_2(2)] = .9 \times 1.2 + .1 \times 50 = 6.1$$

$$\hat{y}_4(3) = \hat{T}_3(3) + \hat{\beta}_1(3) = 1{,}061 + 6.1 = 1{,}067$$

***At* t = *4:*** $y_4 = 1{,}762 \qquad e_4 = 695$

$$\hat{T}_4(4) = .9 \times 1.067 + .1 \times 1{,}762 = 1{,}137$$

$$\hat{T}_4(4) - \hat{T}_3(3) = 1{,}137 - 1{,}061 = 76$$

$$\hat{\beta}_1(4) = .9 \times 6.1 + .1 \times 76 = 13.1$$

$$\hat{y}_5(4) = 1{,}137 + 13.1 = 1{,}150$$

And so forth.

Now that we have gone through the updating computations a few times, it will help us understand the updating process if we look at each quantity computed as the process proceeds. To do this suppose we are at $t = 5$ above. Then

$$\begin{aligned}
y_5 &= \text{estimate of } (\beta_0 + 5\beta_1) \text{ based only on } y_5 \\
&= 2{,}126 \\
\hat{T}_4(4) &= \text{estimate of } (\beta_1 + 4\beta_1) \text{ based on } y_1, \ldots, y_4 \\
&= 1{,}137 \\
\hat{y}_5(4) &= \text{estimate of } (\beta_0 + 5\beta_1) \text{ based on } y_1, \ldots, y_4 \\
&= 1{,}150 \\
\hat{T}_5(5) &= \text{estimate of } (\beta_0 + 5\beta_1) \text{ based on } y_1, \ldots, y_5 \\
&= \text{average of estimate of } (\beta_0 + 5\beta_1) \text{ based on } y_1, \ldots, y_4 \text{ and} \\
&\quad\ \text{estimate of } (\beta_0 + 5\beta_1) \text{ based only on } y_5 \\
&= .9 \cdot \hat{y}_5(4) + .1 \cdot y_5 = .9 \times 1{,}150 + .1 \times 2{,}126 = 1{,}248 \\
\hat{\beta}_1(4) &= \text{estimate of } \beta_1 \text{ based on } y_1, \ldots, y_4 \\
&= 13.0 \\
\hat{T}_5(5) - \hat{T}_4(4) &= \text{estimate of } (\beta_0 + 5\beta_1) \text{ based on } y_1, \ldots, y_5 - \text{estimate of} \\
&\quad\ (\beta_0 + 4\beta_1) \text{ based on } y_1, \ldots, y_4 \\
&= \text{estimate of } \beta_1 \text{ based primarily on } y_5 \\
&= 1{,}248 - 1{,}137 = 111 \\
\hat{\beta}_1(5) &= \text{estimate of } \beta_1 \text{ based on } y_1, \ldots, y_5 \\
&= \text{average of estimate of } \beta_1 \text{ based on } y_1, \ldots, y_4 \text{ and estimate} \\
&\quad\ \text{of } \beta_1 \text{ based primarily on } y_5 \text{ (weights } \gamma \text{ and } 1 - \gamma) \\
&= .9 \times 13.1 + .1 \times 111 = 22.9 \\
\hat{y}_6(5) &= \text{estimate of } (\beta_0 + 6\beta_1) \text{ based on } y_1, \ldots, y_5 \\
&= \text{estimate of } (\beta_0 + 5\beta_1) \text{ based on } y_1, \ldots, y_5 + \text{estimate of } \beta_1 \\
&\quad\ \text{based on } y_1, \ldots, y_5 \\
&= 1{,}247 + 22.8 = 1{,}270
\end{aligned}$$

which is our next one-step-ahead forecast. If, instead, we were using the error correction forms, Equations 3.21 and 3.23, at, say, time $t = 4$, then:

***At* t = *4:*** $y_4 = 1{,}762 \qquad e_4 = 695 \qquad$ (as before)

$$\hat{T}_4(4) = \hat{y}_4(3) + \alpha \cdot e_4 = 1{,}067 + .1 \times 695 = 1{,}137$$

$$\hat{\beta}_1(4) = \hat{\beta}_1(3) + \alpha \cdot \gamma \cdot e_4 = 6.1 + .01 \times 695 = 13.1$$

$$\hat{y}_5(4) = \hat{T}_4(4) + \hat{\beta}_1(4) = 1{,}137 + 13.1 = 1{,}150$$

which is the same forecast we obtained using the smoothing version of the updating equations.

Table 3.19 Forecast for Metro Video Data Using Holt's Two-Parameter Smoothing, $\alpha = \gamma = .1$

| $t$ | $y_t$ | $\hat{y}_t$ | $\hat{T}_t(t)$ | $\hat{\beta}_1(t)$ | $\hat{y}_{t+1}(t)$ | $e_t$ |
|---|---|---|---|---|---|---|
| 1 | 999 | | 999 | 0.0 | 999 | |
| 2 | 1,123 | 999 | 1,011 | 1.2 | 1,012 | 124 |
| 3 | 1,503 | 1,012 | 1,061 | 6.1 | 1,067 | 491 |
| 4 | 1,762 | 1,067 | 1,137 | 13.1 | 1,150 | 695 |
| 5 | 2,126 | 1,150 | 1,248 | 22.9 | 1,271 | 976 |
| 6 | 2,315 | 1,271 | 1,375 | 33.3 | 1,408 | 1,044 |
| 7 | 2,239 | 1,408 | 1,491 | 41.6 | 1,533 | 831 |
| 8 | 2,655 | 1,533 | 1,645 | 52.8 | 1,698 | 1,122 |
| 9 | 2,787 | 1,698 | 1,807 | 63.7 | 1,871 | 1,089 |
| 10 | 3,024 | 1,871 | 1,986 | 75.2 | 2,061 | 1,153 |
| 11 | 3,467 | 2,061 | 2,202 | 89.3 | 2,291 | 1,406 |
| 12 | 3,528 | 2,291 | 2,415 | 101.7 | 2,517 | 1,237 |
| 13 | 3,441 | 2,517 | 2,609 | 110.9 | 2,720 | 924 |
| 14 | 3,558 | 2,720 | 2,804 | 119.3 | 2,923 | 838 |
| 15 | 3,746 | 2,923 | 3,005 | 127.5 | 3,133 | 823 |
| 16 | 3,628 | 3,133 | 3,183 | 132.6 | 3,316 | 495 |
| 17 | 4,021 | 3,316 | 3,387 | 139.7 | 3,527 | 705 |
| 18 | ? | 3,527 | | | | |

The complete set of forecasts that would have resulted are shown in Table 3.19. Notice that all the forecast errors are positive again. The RMSE is 899, and Theil's $U$ is 3.58. The initial estimate $\hat{\beta}_1(1) = 0$ is at fault (again). Using larger values of $\alpha$ and $\gamma$ for the first few forecasts would improve the situation. For example, using $\alpha = \gamma = .3$ for the first four forecasts drops the RMSE to 363 and Theil's $U$ to 1.45. Using initialization method 2 yields even better results, with RMSE = 295 and Theil's $U = 1.18$. •

## Choosing the Values of $\alpha$ and $\gamma$

As with Brown's double smoothing, Holt's $\alpha$ and $\gamma$ may be chosen subjectively or by minimizing some performance measure such as the MAPE or RMSE. The problem here is more complex, however, since two values are involved. We could insist that $\alpha = \gamma$ if desired, thus providing equal amounts of smoothing for the slope and trend. Or we could search for jointly optimal values by first setting $\alpha$ and then finding the best $\gamma$ for that $\alpha$. Choosing several different values for $\alpha$ and finding their corresponding optimal values for $\gamma$, we could then select the pair that gives the best results.

To help judge the results of an optimality search it is good to remember that small optimal values tend to indicate that the trend model is linear and relatively stable and thus might be better handled with a single forecast technique. Large values, on the other hand, may indicate a lack of linear model fit.

Finally, initial values may have an inordinate influence in optimality searches of this type. Inaccuracies in these values will produce larger optimal values than are actually needed. It is advisable to calculate the accuracy measures for a "test set" that does not include the initial and early forecasts, or to use smoothing constants a bit smaller than the search seems to indicate. In addition, since there are two smoothing constants, a large value for one may compensate for a small value of the other.

**EXAMPLE 3.19** ***Choice of $\alpha$ and $\gamma$, Metro Video***

To select values for the smoothing constants in the case of Metro Video, initialization method 2 was used, and the "test set" consisted of observations 5 through 17. Theil's uncertainty coefficients $U$ were computed for $\alpha = .1(.1).9$ and $\gamma = .1(.1).9$, with the results shown in Table 3.20. The best results are obtained when $\alpha = .7$ and $\gamma = .1$, although for any $.5 \leq \alpha \leq .9$ and $.1 \leq \gamma \leq .3$ the results are nearly as good. If we wished $\alpha$ to be the same as $\gamma$, then the best choice would be $\alpha = \gamma = .5$. ●

## 3.9.7 Moving Average Percent Change (MPC)

Constant-growth-rate (simple exponential) models may also be updated. A very simple and intuitive way to do this is to estimate the growth rate at time $t$ based on the average of the growth rates of the most recent $k$ periods, i.e., a moving average percent change of length $k$ (Gross and Peterson, 1983). If we let

$$\begin{aligned} \text{PC}_t &= \text{percent change at time } t \\ &= \frac{y_t - y_{t-1}}{y_{t-1}} = \frac{\Delta y_t}{y_{t-1}} \times 100\% \end{aligned}$$

then

$$\begin{aligned} \text{MPC}_t &= \text{moving percent change at time } t \\ &= \frac{\sum_1^k \text{PC}_{t-i+1}}{k} \end{aligned}$$

**Table 3.20 Values of Theil's $U$ for $\alpha = .1(.1).9$ and $\gamma = .1(.1).9$ Using Holt's Method for the Metro Video Data**

| | $\gamma$ | | | | | | | | |
|---|---|---|---|---|---|---|---|---|---|
| $\alpha$ | .1 | .2 | .3 | .4 | .5 | .6 | .7 | .8 | .9 |
| .1 | 1.203 | 1.272 | 1.343 | 1.394 | 1.438 | 1.453 | 1.463 | 1.454 | 1.444 |
| .2 | 1.054 | 1.105 | 1.124 | 1.115 | 1.087 | 1.049 | 1.011 | 0.977 | 0.949 |
| .3 | 0.923 | 0.937 | 0.923 | 0.901 | 0.879 | 0.860 | 0.848 | 0.845 | 0.846 |
| .4 | 0.832 | 0.833 | 0.823 | 0.814 | 0.810 | 0.816 | 0.826 | 0.842 | 0.861 |
| .5 | 0.776 | 0.780 | 0.782 | 0.788 | 0.800 | 0.815 | 0.837 | 0.858 | 0.881 |
| .6 | 0.748 | 0.758 | 0.768 | 0.781 | 0.803 | 0.823 | 0.844 | 0.865 | 0.885 |
| .7 | 0.736 | 0.752 | 0.770 | 0.790 | 0.812 | 0.834 | 0.856 | 0.880 | 0.899 |
| .8 | 0.738 | 0.759 | 0.782 | 0.807 | 0.832 | 0.858 | 0.886 | 0.913 | 0.942 |
| .9 | 0.749 | 0.774 | 0.802 | 0.832 | 0.864 | 0.895 | 0.930 | 0.967 | 1.008 |

To obtain one-step-ahead forecasts, we can multiply $y_t$ by $(1 + MPC_t/100)$, or for $p$-step-ahead forecasts

$$\hat{y}_{t+p}(t) = y_t \cdot \left(1 + \frac{MPC_t}{100}\right)^p \qquad p \geq 1$$

### Moving Average Percent Change (MPC) of Length $k$

**Given:** Most recent $k$ observations $y_t, y_{t-1}, \ldots, y_{t-k+1}$

**Model:** $y_t = \beta_0 \cdot (\beta_1)^t \cdot \epsilon_t$

**Estimates:**

$$PC_t = \text{percent change from time } t-1 \text{ to time } t = \frac{\Delta y_t}{y_{t-1}} \times 100\%$$

$$MTPC_t = \text{moving total percent change at time } t = \sum_1^k PC_{t-i+1}$$

$$MPC_t = \text{moving average percent change at time } t = \frac{MTPC_t}{k}$$

$$\hat{T}_t(t) = y_t \qquad \hat{\beta}_1(t) = \frac{1 + MPC_t}{100}$$

**Updating Equations:**

$$MTPC_t = MTPC_{t-1} + (PC_t - PC_{t-k})$$

$$MPC_t = \frac{MTPC_t}{k}$$

**Forecasts:** (At time $t$ for $p$ steps ahead)

$$\hat{y}_{t+p}(t) = \hat{T}_t(t) \cdot \hat{\beta}_1(t)^p = y_t \cdot \left(1 + \frac{MPC_t}{100}\right)^p$$

To illustrate the method, suppose we are forecasting Metro Video's revenues starting in September 1985 ($t = 13$) with an MPC of length 4 (let MTPC be the moving total percent change).

*At* t = *13:*

| $t$ | $y_t$ | $\Delta y_t$ | $PC_t$ | $MTPC_t$ | $MPC_t$ |
|---|---|---|---|---|---|
| 9 | 2,787 | | | | |
| 10 | 3,024 | 237 | 8.50% | | |
| 11 | 3,467 | 443 | 14.65 | | |
| 12 | 3,528 | 61 | 1.76 | | |
| 13 | 3,444 | −81 | −2.47 | 22.44% | 5.61% |

$$\hat{y}_{14}(13) = y_{13} \times \left(1 + \frac{MPC_{13}}{100}\right) = 3{,}441 \times 1.0561 = 3{,}634$$

*At* t = *14:*

$$y_{14} = 3{,}558 \qquad \Delta y_{14} = 3{,}558 - 3{,}441 = 117$$

$$PC_{14} = \frac{117}{3{,}441} \times 100 = 3.40\%$$

$$MTPC_{14} = 22.44 + 3.40 - 8.50 = 17.34\%$$

$$MPC_{14} = \frac{17.34}{4} = 4.33\%$$

$$\hat{y}_{15}(14) = y_{14} \cdot \left(1 + \frac{MPC_{14}}{100}\right) = 3{,}558 \times 1.0433 = 3{,}712$$

*At* t = *15:* $y_{15} = 3{,}746 \qquad \Delta y_{15} = 188$

$$PC_{15} = \frac{188}{3{,}558} \times 100 = 5.28\%$$

$$MTPC_{15} = 17.34 + 5.28 - 14.65 = 7.97\%$$

$$MPC_{15} = \frac{7.97}{4} = 1.99\%$$

$$\hat{y}_{16}(15) = 3{,}746 \times 1.0199 = 3{,}821$$

*At* t = *16:* $y_{16} = 3{,}628 \qquad PC_{16} = -3.15\% \qquad MTPC_{16} = 3.06\%$

$MPC_{16} = 0.77\%$

$\hat{y}_{17}(16) = 3{,}656$

*At* t = *17:* $y_{17} = 4{,}021 \qquad PC_{17} = 10.83\% \qquad MTPC_{17} = 16.36\%$

$MPC_{17} = 4.09\%$

$\hat{y}_{18}(17) = 4{,}185$

The main advantages of the moving percent change method are that it is easy to understand, the calculations are simple, and it permits forecasts of a trended series without the inherent complexity of DMA's, DES, or LES. The method may thus be more comprehensible to those who must use the forecasts in decision making. In addition, if the rate of growth is not too large (less than 20% or so per period) and if the series is not too noisy, the MPC should yield better results than linear-based updating methods, which tend to lag a constant-growth-rate (simple exponential) model. The greatest problem with the method is that the forecasts, having been obtained by multiplying $y_t$ by the MPC, have in them the irregular movement that occurred at time $t$. This, when added to the irregular movement that occurs at time $t + 1$, leads to larger average squared forecast errors. If one tries smoothing the $y_t$ values and multiplying the smoothed value (rather than $y_t$) by the MPC, the level of complexity increases to the point where procedures with a sounder theoretical base are preferable. A final note of interest is that the MPC procedure discussed here is equivalent to correcting for trend by taking link relatives of the $y_t$ values and forecasting the link relative series using the simple-moving-average techniques discussed in Chapter 2.

### 3.9.8 Smoothing the Logarithms

The more common way of handling a simple exponential trend is to apply linear methods to the series of logarithms, $y'_t = \log y_t$, and take antilogs of the resulting forecasts. The method is straightforward, and a similar conversion for the single forecast case was discussed earlier in this chapter (Section 3.8.5).

As an illustration of the method, suppose we were starting to forecast the Procter & Gamble data in 1980. Let $t = 1$ in 1980, $\alpha = .3$, $\gamma = .3$, and (using the prime notation—$y'$, $T'$, etc.—for logarithms) $\hat{T}_1'(1) = y_1'$, $\hat{\beta}_1'(1) = 0$.

*At* t = *1 (1980):*

$$y_1 = 640 \qquad y_1' = \log 640 = 6.461$$
$$\hat{T}_1'(1) = 6.461$$
$$\hat{\beta}_1'(1) = 0$$
$$\hat{y}_2'(1) = \hat{T}_1'(1) + \hat{\beta}_1'(1) = 6.461 + 0 = 6.461$$
$$\hat{y}_2(1) = e^{6.461} = 640$$

*At* t = *2 (1981):*

$$y_2 = 668 \qquad y_2' = \log 668 = 6.504$$
$$\hat{T}_2'(2) = (1 - \alpha) \cdot \hat{y}_2'(1) + \alpha \cdot y_2' = .7 \times 6.461 + .3 \times 6.504 = 6.474$$
$$\hat{\beta}_1'(2) = (1 - \gamma) \cdot \hat{\beta}_1'(1) + \gamma \cdot [\hat{T}_2'(2) - \hat{T}_1'(1)]$$
$$= .7 \times 0 + .3 \times (6.474 - 6.461) = .004$$
$$\hat{y}_3'(2) = \hat{T}_2'(2) + \hat{\beta}_1'(2) = 6.474 + .004 = 6.478$$
$$\hat{y}_3(2) = e^{6.478} = 651$$

*At* t = *3 (1982):*

$$y_3 = 777 \qquad y_3' = 6.655$$
$$\hat{T}_3'(3) = .7 \times 6.478 + .3 \times 6.655 = 6.531$$
$$\hat{\beta}_1'(3) = .7 \times .004 + .3 \times (6.531 - 6.474) = .020$$
$$\hat{y}_4'(3) = 6.531 + .020 = 6.551$$
$$\hat{y}_4(3) = e^{6.551} = 700$$

*At* t = *4 (1983):*

$$y_4 = 866 \qquad y_4' = 6.763$$
$$\hat{T}_4'(4) = .7 \times 6.551 + .3 \times 6.763 = 6.615$$
$$\hat{\beta}_1'(4) = .7 \times .020 + .3 \times (6.615 - 6.531) = .039$$
$$\hat{y}_5'(4) = 6.615 + .039 = 6.654$$
$$\hat{y}_5(4) = e^{6.654} = 776$$

which is our forecast for 1984.

## 3.9.9 Other Updating Schemes

Moving average and exponential-smoothing schemes for more complex trend models are possible. In practice such schemes can lead to highly fluctuating forecasts and should be used only with care and experience. Computationally the schemes are quite complex as well, losing much of their intuitive motivation. For example, Brown's triple smoothing for quadratic trends involves the following updating equations:

$$A_t = (1 - \alpha) \cdot A_{t-1} + \alpha \cdot y_t$$
$$A_t' = (1 - \alpha) \cdot A_{t-1}' + \alpha \cdot A_t$$
$$A_t'' = (1 - \alpha) \cdot A_{t-1}'' + \alpha \cdot A_t'$$

This is simple enough. But the estimates are

$$\hat{T}_t(t) = 3A_t - 3A_t' + A_t''$$

$$\hat{\beta}_1(t) = \frac{\alpha}{2(1-\alpha)^2}[(6-5\alpha)A_t - (10-8\alpha)A_t' + (4-3\alpha)A_t'']$$

$$\hat{\beta}_2'(t) = \frac{\alpha^2}{(1-\alpha)^2}(A_t - 2A_t' + A_t'')$$

And, finally,

$$\hat{y}_{t+p}(t) = \hat{T}_t(t) + p \cdot \hat{\beta}_1(t) + \frac{1}{2}p^2 \cdot \hat{\beta}_2(t)$$

The procedures we have presented here should provide reasonable results for the majority of practical applications. It is possible to fashion a method of smoothing the ratio of successive link relatives to obtain updated forecasts of modified exponential trends, and exponential or moving average smoothing schemes exist for cubic and higher-order polynomial trends. However, treatment of these and other more complex methods is beyond the scope of our discussion.

## 3.10 Diagnostics

### 3.10.1 Introduction

The general remarks about model appraisal in Chapter 2 also apply to trend models. The most important additional consideration is that the model describe the trend in the series as accurately as possible. Seasonal or cyclical movement that shows up in the residuals may be handled using the methods in Chapter 5 or 6. However, the methods used there will not help account for trend that has not been modeled. Another word of warning: Extrapolation for trended models is generally more dangerous than for untrended ones, since long-term forecasts are sensitive to the functional form of the model as well as to any errors in the several parameter estimates.

### 3.10.2 Subjective Knowledge of the Series

Again, under subjective scrutiny the model should be consistent with your knowledge of the series. Along these lines, the model should be more closely examined if any of the following hold true.

- Extrapolated forecasts seem unrealistic.
- Estimates of parameter values conflict with your understanding of the process generating the series (growth rates too large, optimal moving average lengths too short, smoothing constants too large or too small, etc.).
- Successive estimates of $\beta_1$ in linear updating show a prolonged trend.
- One or two observations seem to have undue influence on the parameter estimates.
- Fitting the model to large subsets of the history produces very different results.

**Figure 3.19** Effect of Partial Cycle on Trend Fit

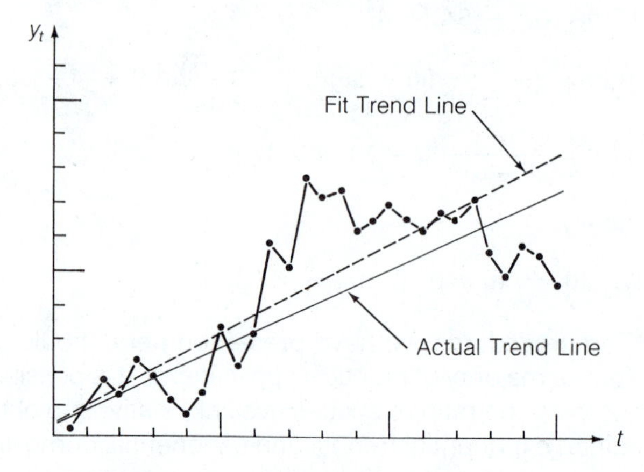


Attention must also be paid to the effect that other components in the series could have on the model fit. If the history contains only part of a cycle, for example, the trend model will be biased in the direction of the incomplete cycle, as illustrated in Figure 3.19. Years with incomplete monthly or quarterly data can also have this effect if seasonal movement is present in the series.

### 3.10.3 Empirical Knowledge of the Series

The residuals from the model should be computed and the steps suggested in Section 2.11 followed. Specifically,

- Construct the residual graph.
- Construct the $y$-versus-$\hat{y}$ graph.
- Compute the sample autocorrelation function.
- Compute Thiel's $U$ and decomposition for the updating models.

The residuals $e_t$ should form a random no-trend series; any departures from this behavior are grounds for inquiry about possible model improvements. Here are some areas of specific concern for trend modeling:

- In the single forecast case, fitting a linear model to a curvilinear series will tend to cause cyclical behavior in the residuals, as illustrated in Figure 3.20. This, in turn, will cause at least the shorter lag autocorrelations to be nonzero, and the $y$-versus-$\hat{y}$ graph will look similar to the residual graph tilted at a 45° angle. If an updating scheme is used, the forecasts will lag the series, producing nonzero autocorrelations and large values of $U_M$ and $U_R$. Large values of $\alpha$ (or $\alpha$ and $\gamma$) will tend to result from optimal selection procedures.
- *Seasonal movement,* if present and unmodeled, will remain in the residuals, causing the lag 12 (yearly) or lag 4 (quarterly) autocorrelations and their multiples to be positive. Updating schemes will lag the seasonal movement as well, with approximately the same results. Moving averages of length 12 (or 4) will tend to remove this effect. If no seasonal movement is present, however, the Yule-Slutsky effect may cause apparent (but induced) autocorrelation in the residuals from a smoothing scheme.

**Figure 3.20 Residuals from a Linear Model Fit to a Curvilinear Trend**

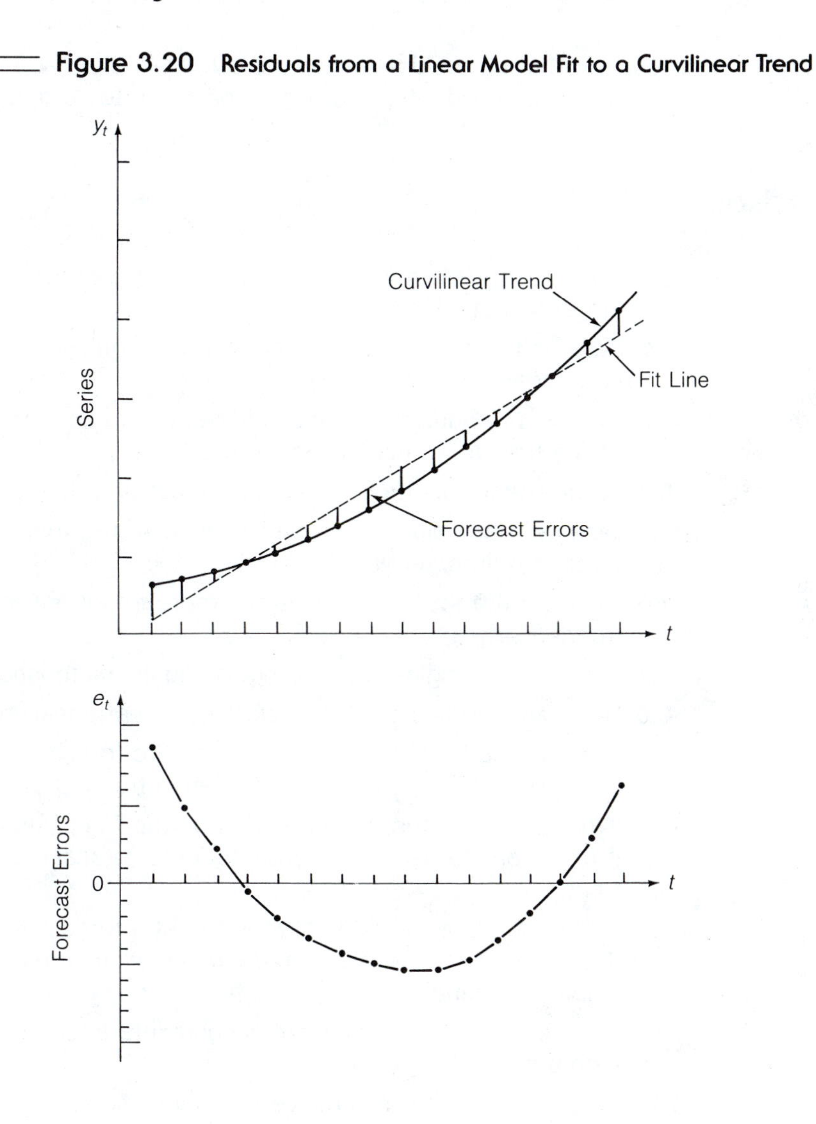


- *Cyclical movement,* if present and unmodeled, will remain in the residuals. But again, the Yule-Slutsky effect may induce such cyclical behavior.
- A *change in* the *slope* of a linear model (while the trend remains linear) will also cause apparent cycles in the residuals. Examining the graph of $y_t$ versus $t$ should help decide if this has occurred. If so, one can discard that portion of the history where the old value of the slope was active or use the dummy variable technique to be described in Chapter 4.

This list is not intended to be complete, but it gives an idea of the kind of residual analysis that should be done. The interaction between the various components (seasonal, cyclical, and irregular) in the series is quite complex. Some considerable experience (and no small artistic touch) may be required to accurately analyze it. The central idea, though, is that if some nonrandomness is apparent in the residual series, an effort should made to understand why it is there. If desired, steps may then be taken to correct the model.

## 3.11 Exercises

**3.1** What is the main difference between a trended time series and an untrended one?

**3.2** What is the main characteristic that distinguishes trend from the other components of times series models?

**3.3** How is a trended time series different from an untrended one in terms of the first differences of the series?

**3.4** What effect does trend have on the acf of a time series?

**3.5** How will the value of the test statistic in the runs test be affected by trend in a time series?

**3.6** Why are the scale and units of $t$ more important for trended than untrended series?

**3.7** What is the single most distinguishing characteristic of linear trend?

**3.8** How can one distinguish empirically between linear and curvilinear trend?

**3.9** What would be the general effect on the forecasts and forecast errors if one were to fit a linear trend to an untrended series?

**3.10** What would be the general effect on the forecasts and forecast errors if one were to use a no-trend model to forecast a series with a linear trend?

**3.11** What would be the general effect on the forecasts and forecast errors if one were to use a linear trend model to forecast a series with a curvilinear trend?

**3.12** What behavior of the first and second differences characterizes a series with a quadratic trend?

**3.13** What is the most distinctive characteristic of a series with a simple exponential trend?

**3.14** Describe the behavior of the link relatives of a series with a simple exponential trend.

**3.15** What specific characteristic would lead one to prefer a modified exponential over a simple exponential trend.

**3.16** From a mathematical point of view, how are the modified exponential, Gompertz, and logistic trend models related?

**3.17** What is the primary difference between a modified exponential and a Gompertz or logistic trend curve?

**3.18** What difficulty arises when one tries to use simple moving averages to forecast a trended series?

**3.19** What is meant by a $3 \times 5$ moving average?

**3.20** In what ways are Brown's and Holt's smoothing methods similar? How do they differ?

**3.21** For what sort of trend should the moving percent change method be used? What are its main advantages and disadvantages?

**3.22** Why would one not generally want to use a DMA, DES, or LES for a no-trend series (just in case a trend might develop)?

**3.23** The following is a trend equation for per capita weekly gasoline consumption in a particular area:

$T_t = 15.7 + .07t$ Origin: 1980
Units ($t$): Yearly
Units ($y$): Gallons

**a.** Describe the model in words.

**b.** Calculate the trend value for:
  **(i)** 1988
  **(ii)** January 1980
  **(iii)** March 1984
  **(iv)** I, 1984
  **(v)** 5/1/86

**3.24** For the following estimated trend equations, perform the indicated shifts of origin and scale:

**a.** $\hat{T}_t = 200 + 180t$ Origin: 1985
Units ($t$): Yearly

Change the origin to 1990, then change the units to monthly.

**b.** $\hat{T}_t = 43.6 + 4.65t$ Origin: January 1988
Units ($t$): Monthly

Change the origin to 1989 and the units to yearly.

**c.** $\hat{T}_t = 1{,}248 + 26.4t$ Origin: I, 1987
Units ($t$): Quarterly

Change the origin to III, 1991 and the units to monthly, then change the origin to January 1991.

**d.** $\hat{T}_t = 10.6(1.22)^t$ Origin: 1980s
Units ($t$): Decades

Change the origin to 1988 and the units to yearly.

**e.** $\hat{T}_t = 17.6 + 3.4t + .064t^2$ Origin: June 1989
Units ($t$): Monthly

Change the origin to 1990 and the units to yearly.

**f.** $\hat{T}_t = 1{,}200 - 950(.8)^t$ Origin: 1985
Units ($t$): Yearly

Change the origin to I, 1985 and the units to semiannual.

**3.25** A computer software manufacturer has estimated the following trend model for the monthly unit demand for its FORECASTER software package:

$$\hat{T}_t = 75 + 20t \qquad \text{Origin: February 1989} \quad \text{Units } (t)\text{: Monthly}$$

**a.** Give a verbal description of this trend model.
**b.** Use the equation as it now stands, by combining quarterly aggregates, to determine the aggregate trend for the year 1990.
**c.** Change to quarterly $t$ units with a I, 1990 origin and change the aggregate units to quarterly demand.
**d.** Use the equation obtained in (c) to give the trend for 1990.

**3.26** Economists employed by an electronics firm have fit the following trend model for quarterly sales ($1,000,000) of its compact disc players:

$$\hat{T}_t = 10(1.028)^t \qquad \text{Origin: I, 1989} \quad \text{Units } (t)\text{: Quarterly} \quad \text{Units } (y)\text{: \$1,000,000}$$

**a.** Give a verbal description of this trend model.
**b.** Use the equation as it now stands, by combining quarterly aggregates, to determine the aggregate trend value for the year 1990.
**c.** Change to yearly units with a 1989 origin and change to yearly aggregate units.
**d.** Use the equation obtained in (c) to give the trend value for 1990.

**3.27** What are the relative advantages and disadvantages of the following methods of fitting a linear trend model?

I Free-hand, or "eyeball"
II Selected points
III Semiaverages
IV Least squares

**3.28** A small independent computer retailer has experienced the following gross sales ($1,000) over the past 11 months:

| Month | Gross Sales | Month | Gross Sales |
|---|---|---|---|
| Feb | 19.2 | Jul | 83.3 |
| Mar | 40.6 | Aug | 122.9 |
| Apr | 41.7 | Sep | 66.9 |
| May | 63.2 | Oct | 200.5 |
| Jun | 37.6 | Nov | 177.3 |
| | | Dec | 114.0 |

**a.** Graph $y_t$ versus $t$ to begin to familiarize yourself with the series.
**b.** Graph $\Delta y_t$ versus $t$. Do you think a linear model might be appropriate? Why/why not?
**c.** Does Daniels' test indicate the presence of trend ($\alpha = 5\%$)?
**d.** Does the test for $\rho = 0$ indicate the presence of trend?

**e.** Fit a linear trend using the free-hand method, and calculate $\bar{e}$, the RMSE, and the MAPE.
**f.** Fit a linear trend using the selected points method (first and last points), and calculate $\bar{e}$, the RMSE, and the MAPE.
**g.** Fit a linear trend using the semiaverages method, discarding the oldest point, and calculate $\bar{e}$, the RMSE, and the MAPE.
**h.** Fit a linear trend using the least squares method, and calculate $\bar{e}$, the RMSE, and the MAPE.

**3.29** The quarterly revenues of a small medical supply company for the last several years (27 quarters) were plotted and seen to have a roughly linear trend component. The information below was computed for $y_1, y_2, \ldots, y_{27}$ using one of the company's spreadsheet computer programs.

$$y_t = \text{quarterly gross revenues (\$10,000)} \qquad t = 1, 2, \ldots, 27$$

$$\Sigma y = 7{,}146 \qquad \Sigma ty = 113{,}615 \qquad SS_{yy} = 120{,}640$$

**a.** Using the method of least squares, fit a linear trend model to the revenue data.
**b.** Test for significance of the trend component assuming independent, normally distributed error terms.
**c.** Forecast revenues for $t = 28$ and $t = 29$ using the model fit in part (a).
**d.** Construct 95% prediction intervals for $y_{28}$ and $y_{29}$.
**e.** Construct 90% prediction intervals for $y_{28}$ and $y_{29}$.

**3.30** A member of a citywide chain of low-cost retail optical stores has experienced the following demand for its eyeglasses over the past four and a half years:

$y_t$ = number of pairs of eyeglasses sold

| Year | Qtr | $t$ | $y_t$ | Year | Qtr | $t$ | $y_t$ |
|---|---|---|---|---|---|---|---|
| 1984 | III | 1 | 316 | | IV | 10 | 427 |
| | IV | 2 | 358 | 1987 | I | 11 | 448 |
| 1985 | I | 3 | 404 | | II | 12 | 533 |
| | II | 4 | 415 | | III | 13 | 541 |
| | III | 5 | 346 | | IV | 14 | 511 |
| | IV | 6 | 372 | 1988 | I | 15 | 548 |
| 1986 | I | 7 | 422 | | II | 16 | 528 |
| | II | 8 | 437 | | III | 17 | 550 |
| | III | 9 | 428 | | IV | 18 | 574 |

**a.** Fit a linear trend model using the method of semiaverages, and use this model to forecast the four quarters of 1989.
**b.** Fit a linear trend model using the method of least squares, and use this model to forecast the four quarters of 1989.
**c.** How do the forecasts in part (a) compare with those in part (b)?
**d.** Assuming independent, normally distributed errors, construct a 95% prediction interval for the first quarter of 1989.

**e.** Repeat part (d) for the first quarter of 1990. Compare the width of the two intervals, and explain any difference.

**3.31** A firm that manages and maintains several apartment house complexes has experienced the following maintenance/operating expenses for these complexes over the last five years:

$y_t$ = quarterly maintenance/operating expense ($1,000)

| Quarter | $y_t$ | Quarter | $y_t$ | Quarter | $y_t$ |
|---|---|---|---|---|---|
| 1 | 27 | 8 | 63 | 15 | 128 |
| 2 | 39 | 9 | 73 | 16 | 147 |
| 3 | 41 | 10 | 92 | 17 | 155 |
| 4 | 50 | 11 | 94 | 18 | 158 |
| 5 | 44 | 12 | 102 | 19 | 164 |
| 6 | 47 | 13 | 123 | 20 | 187 |
| 7 | 58 | 14 | 119 | | |

**a.** By plotting $y_t$, $\Delta y_t$, and $\Delta^2 y_t$, determine whether or not a curvilinear model is needed.

**b.** Fit a quadratic trend model to the series using the requested methods:

**(i)** Selected points, using observations 1, 10, and 20.

**(ii)** Semiaverages, omitting the two oldest observations.

**(iii)** Least squares.

**c.** Forecast $y_{24}$ using the models fit in part (b).

**d.** Examining the link relatives $y_t/y_{t-1}$, does it appear that a simple exponential growth model might serve? Explain.

**e.** Fit a simple exponential growth model using the requested methods:

**(i)** Free-hand.

**(ii)** Selected points, first and last points.

**(iii)** Semiaverages.

**(iv)** Least squares.

**f.** Forecast $y_{24}$ using the models fit in part (e).

**g.** Of the two trend models, quadratic and simple exponential, is one clearly superior?

**3.32** For the following time series:

| $t$ | 1 | 2 | 3 | 4 | 5 | 6 | 7 | 8 | 9 | 10 |
|---|---|---|---|---|---|---|---|---|---|---|
| $y_t$ | 65 | 80 | 80 | 96 | 105 | 112 | 133 | 135 | 120 | 167 |

**a.** Based on the $y_t$-versus-$t$ graph, would you consider using a linear trend model?

**b.** Plot the link relatives $y_t/y_{t-1}$, and compute the average link relative.

**c.** Fit a simple exponential growth model using:
  **(i)** The method of selected points, first and last points.
  **(ii)** The method of semiaverages.
  **(iii)** The method of least squares.

**d.** Based on the results of part (c), what would you say is the rate of growth of this series?

**3.33** The past 12 quarters of a large business office's copier paper usage are shown in the table below:

$y_t$ = copier paper used (reams)

| Week | $y_t$ | Week | $y_t$ |
|---|---|---|---|
| 1 | 247 | 7 | 355 |
| 2 | 248 | 8 | 368 |
| 3 | 297 | 9 | 364 |
| 4 | 276 | 10 | 387 |
| 5 | 266 | 11 | 406 |
| 6 | 367 | 12 | 386 |

**a.** Forecast copier usage for weeks 13 and 14 using SES with a 10% smoothing constant and $\hat{y}_{\text{init}} = y_1$. Compute the MSE for weeks 6–12.

**b.** Forecast copier usage for weeks 13 and 14 using a $2 \times 2$ DMA and a $3 \times 3$ DMA. Calculate the MSE for times 6–12 for each. Which value of $k$ do you prefer?

**c.** Do you prefer DMAs to SES? Why or why not?

**d.** Forecast copier usage for weeks 13 and 14 using Brown's Smoothing (DES), a 10% smoothing constant, and the following initializations.
  **(i)** Assume you had started forecasting at time $t = 2$ with only $y_1$ available for initialization. Compute the MSE for weeks 6–12.
  **(ii)** Use method 1 for a history of $n$ observations, i.e., $\hat{T}_1(1) = y_1$, $\hat{\beta}_1(1) = 0$ ($A_1 = A_1' = y_1$), and $\alpha = .5$ for the first three forecasts. Compute the MSE for weeks 6–12.
  **(iii)** Use method 2 for a history of $n$ observations. Compute the MSE for weeks 6–12.
  **(iv)** Use semiaverages estimates of the slope and intercept at $t = 4$ to initiate smoothing at time $t = 5$. Compute the MSE for weeks 6–12.
  **(v)** Use the method in part (i) above to initiate backcasting, and use the backcast estimates of $\hat{T}_1(1)$ and $\hat{\beta}_1(1)$ to initiate forecasting. Compute the MSE (of the forecasts) for weeks 6–12.

**e.** Forecast copier usage for weeks 13 and 14 using Holt's Smoothing (LES) and a 10% smoothing constant for both trend and slope. Initialize by using $\hat{T}_1(1) = y_1$ and $\hat{\beta}_1(1) = 0$, and 50% smoothing constants for the forecasts of $y_2$ through $y_4$. Compute the MSE for weeks 6–12.

**3.34** For the computer sales series in Exercise 3.28, use a $4 \times 4$ DMA to forecast January and February sales.

**3.35** Use Brown's Smoothing (DES) with a 20% smoothing constant to forecast January and February sales for the series in Exercise 3.28. (Use simple initialization, i.e., $\hat{T}_1(1) = y_1$ and $\hat{\beta}_1(1) = 0$.)

**3.36** Use Holt's Smoothing (LES) with $\alpha = .2$ and $\gamma = .2$ and simple initialization to forecast January and February sales for the series in Exercise 3.28.

**3.37** Use a moving percent change of length 4 to forecast the series in Exercise 3.32. Do you think $k = 4$ is about right? Or should a different value of $k$ be used?

**3.38** A minor league baseball team has recorded the following total season paid attendance at its ball games over the past twelve years:

| Year | Attendance (1000's) | Year | Attendance (1000's) |
|---|---|---|---|
| 1977 | 20.7 | 1983 | 504.5 |
| 1978 | 55.4 | 1984 | 651.1 |
| 1979 | 114.4 | 1985 | 769.1 |
| 1980 | 206.0 | 1986 | 781.7 |
| 1981 | 334.3 | 1987 | 949.9 |
| 1982 | 410.4 | 1988 | 948.9 |

**a.** Plot $y_t$ versus $t$.
**b.** Fit a Gompertz trend model to the series, and use it to forecast 1989.
**c.** Fit a Logistic trend model to the series, and use it to forecast 1989.

**3.39** A firm's after-tax profits ($100,000) for the last 15 years are shown in the table below:

| Year | Profits | Year | Profits | Year | Profits |
|---|---|---|---|---|---|
| 1 | 18.6 | 6 | 59.6 | 11 | 61.2 |
| 2 | 30.1 | 7 | 57.4 | 12 | 71.8 |
| 3 | 32.9 | 8 | 58.9 | 13 | 71.5 |
| 4 | 41.9 | 9 | 59.5 | 14 | 77.4 |
| 5 | 55.1 | 10 | 69.6 | 15 | 68.4 |

**a.** Suppose a modified exponential trend model is to be fit to this series. By inspecting the $y_t$-versus-$t$ graph of the series, what could you say about the values of $\beta_0$, $\beta_1$, and $\beta_2$ to expect?
**b.** Fit a modified exponential trend model to the series, and use this model to forecast sales for year 16.
**c.** What difficulties in terms of forecasts and forecast errors would you anticipate if a linear trend model were fit to the data? How about a simple exponential trend?
**d.** Use an MPC of length 5 to forecast the last 5 years of the series and year 16.

## References

Bell, W. R., and Hillmer, S. C. (1984). "Issues Involved with the Seasonal Adjustment of Economic Time Series," with comments. *J. Bus. Econ. Stat.* 2(4): 291–320, 349.

Brown, R. G. (1963). *Smoothing, Forecasting, and Prediction of Discrete Time Series.* Englewood Cliffs, N.J.: Prentice-Hall.

Gardner, E. S., and McKenzie, E. (1985). "Forecasting Trends in Time Series." *Management Sci.* 31(10): 1237–1246.

Gilchrist, W. (1976). *Statistical Forecasting.* New York: Wiley.

Gross, C. W., and Peterson, R. T. (1983). *Business Forecasting,* 2nd ed. Boston: Houghton Mifflin.

Holt, C. C. (1957). "Forecasting Seasonal and Trends by Exponentially Weighted Moving Averages." Office of Naval Research, Research Memorandum No. 52.

Holt, C. C., Modigliani, F., Muth, J. F., and Simon, H. A. (1960). *Planning Production Inventories and Work Force.* Englewood Cliffs, N.J.: Prentice-Hall.

Kendall, M. G. (1973). *Time Series.* London: Griffin.

Meade, N. (1984). "The Use of Growth Curves in Forecasting Market Development—A Review and Appraisal." *J. Forecasting* 3: 429-451.

Mendenhall, W., Reinmuth, J. E., Beaver, R., and Duhan, D. (1986). *Statistics for Management and Economics,* 5th ed. Boston: Duxbury.

Roberts, S. (1982). "A General Class of Holt-Winters Type Forecasting Models." *Management Sci.* 28: 67–82.

Thomopoulos, N. T. (1980). *Applied Forecasting Methods.* Englewood Cliffs, N.J.: Prentice-Hall.

# Using Regression in Forecasting

**Objective:** *To study the statistical methodology of regression analysis and its applications to time series.*

## 4.1 Introduction

**Regression,** or **"least squares" analysis,** is a statistical method for estimating the functional relationship between a response variable and one or more independent, predictor variables. Because it applies in all kinds of statistical modeling, regression occupies a major role in many areas of research, including time series analysis. Whenever regression analysis is used, there are two important things to remember:

- The specified mathematical form of the relationship must be based on judgment and experience.
- The relationship is *stochastic;* that is, an error term is included in the model.

For example, in modeling sales $y_t$ as a function of advertising expenditure $x_t$, we might write

$$y_t = f(x_t) + \epsilon_t$$

where $f$ is some function, such as $\beta_0 + \beta_1 x_t$, that we think expresses the relationship between these two variables. The error term $\epsilon_t$ is included to account for the effects of variables we forgot to consider, to allow for errors in the measurement of the variables, and to simply allow for some unexpected randomness in human buying behavior.

In this chapter we will:

- Explain the basic notation and assumptions of regression analysis.

- Present formulas for carrying out *simple linear regression analysis* (regression involving only one predictor variable).
- Illustrate the use of computer packages in performing *multiple regression analysis* (regression involving more than one predictor variable).
- Show special applications of regression techniques to time-series/forecasting situations.
- Explain the use of diagnostic tests designed to check the adequacy of estimated regression models.

## 4.2 General Regression Models

### 4.2.1 Model and Notation

Although the emphasis in this text is on time series data, regression analysis has much broader applications and does not require the use of only time series variables. In general regression models the response variable $y$ and the independent variables $x_1, x_2, \ldots, x_k$ may or may not be time series. When time series variables *are* involved, we will use the notation introduced in Section 1.4., i.e., $y_t, x_{1t}, x_{2t}, \ldots, x_{kt}$, when referring to the values of the variables at a particular time $t$. In this chapter, however, use of the subscript $t$ will be minimized, for two reasons. First, when the example or application involves only time series variables, the dependence on time will be clear enough from context without continually using the subscript $t$. Second, by suppressing $t$ when possible, the presentation of the formulas for regression calculations will be simplified.

In regression analysis, the mathematical form of the relationship between a response variable $y$ and the independent variables $x_1, x_2, \ldots, x_k$, is assumed to be

$$y = \beta_0 + \beta_1 x_1 + \beta_2 x_2 + \cdots + \beta_k x_k + \epsilon \qquad (4.1)$$

where $\beta_0, \beta_1, \beta_2, \ldots, \beta_k$, are fixed constants called *regression parameters* and $\epsilon$ is the error term. Equation 4.1 essentially says that the behavior of $y$ may be modeled as a "deterministic" part plus a "stochastic," or random, part. The deterministic part, $\beta_0 + \beta_1 x_1 + \beta_2 x_2 + \cdots + \beta_k x_k$, explains or predicts the values of $y$. That part of the behavior of $y$ that is not explained by the deterministic element is relegated to the stochastic term $\epsilon_t$.

Equation 4.1 is called the **general linear model,** and it is the purpose of **linear regression analysis** to generate estimates of the parameters $\beta_0, \beta_1, \beta_2, \ldots, \beta_k$ from the data on the variables $y, x_1, x_2, \ldots, x_k$.

The power and versatility of linear regression analysis as a modeling technique, though not immediately apparent from Equation 4.1, becomes more evident in light of the following.

- **independent variables** The $x$ variables can be just about anything, even functions of other $x$ variables. For example, $x$, $x^2$, and log $(x)$ can all appear in the same model. Thus, the following are all examples of linear regression models:

$$y = \beta_0 + \beta_1 x + \beta_2 x^2 + \epsilon$$

$$y = \beta_0 + \beta_1 x_1 + \beta_2 x_2 + \beta_3 x_1 x_2 + \epsilon$$

$$y = \beta_0 + \beta_1 x_1 + \beta_2 \sqrt{x_1} + \beta_3 x_2 + \beta_4 \log(x_2) + \epsilon$$

- **linear** The word *linear* in linear regression analysis refers to the *parameters,* not the $x$ variables. Thus, the parameters $\beta_0, \beta_1, \beta_2, \ldots, \beta_k$ are only allowed to appear with an exponent of one. There is no such restriction on the $x$ variables (as can be seen in the models above).
- **intrinsically linear models** Many relationships between variables are not of the form in Equation 4.1 but may still be *transformed* into a linear regression model. For example, the model

$$y = \beta_0 \cdot \beta_1^x \cdot \epsilon \tag{4.2}$$

may be transformed, by taking logarithms, into

$$\log y = \log \beta_0 + x \cdot \log \beta_1 + \log \epsilon \tag{4.3}$$

- **qualitative data** Qualitative information such as "season of the year" or "occurrence of a strike" can be incorporated into a linear regression model by means of "indicator" or "dummy" variables (see Section 4.7).

### 4.2.2 Types of Independent Variables

As mentioned above, there is a lot of freedom in choosing the $x$ variables for a regression model. When using time series variables, there are even more choices:

- **functions of other $x$ variables** Functions such as $x^2$ and $\log x$ can be used. This was mentioned above, but is included here for completeness.
- **lagged values of the $x$ variables** $x_{t-1}$, $x_{t-2}$, $x_{t-3}$, and so forth may be used as predictor variables. Models using only $x$ and/or lagged values of $x$ are called *distributed* lag models.
- **lagged values of $y$** $y$ can be regressed on $y_{t-1}$, $y_{t-2}$, etc. Models using only lagged values of $y$ as predictor variables are called *autoregressive* models.
- **mixed lag models** Lagged values of both $x$ and $y$ are used in the model. Estimation for such *mixed* models is part of econometrics and is not treated in this text.
- **time $t$** The time variable $t$ can be used as one of the $x$ variables. For example, $y = \beta_0 + \beta_1 t$ is often used to model linear trends in data.
- **dummy variables** Through the use of indicator variables, qualitative data (e.g., strikes, seasons) can be included in the model.

A review of the discussion of causal models and time series models in Chapter 1 would be instructive at this point. From the list above you can see that regression techniques apply to both types of models.

To guide decisions about which variables to select, refer to the general cost/benefit considerations in Chapter 1. In a later section, we will present some diagnostic tests for evaluating your choices.

## 4.3 Examples

### 4.3.1 Introduction

In the examples that follow, the mechanics of estimating regression coefficients (*model parameters*) and statistically summarizing the results of a regression analysis are not shown. Those procedures will be discussed later. Our purpose now is only to illustrate the use of regression, to wit:

- The search for useful explanatory (predictor) variables.
- Decisions about the form of the model.
- Interpreting the resulting model.
- Using the model to predict or forecast.

### 4.3.2 Example 4.1 *Practical Medicine* Magazine

*Practical Medicine* is a relatively new monthly magazine devoted to the dissemination of medical information to an increasingly health-conscious society. A typical issue contains articles on generic drugs, choosing a physician, new medical technologies, answers to reader questions, and health maintenance tips.

#### Predictor Variables

The magazine has been in operation for two years, and the marketing staff is now interested in fine-tuning the marketing plan by trying to relate sales data to local-area demographics. A brainstorming session was held in an effort to identify key variables thought to influence local sales. Sales to institutions such as hospitals and health clubs were easy to predict, but there was much less certainty about individual subscribers.

One variable in particular, "subscriber's age," was singled out for an exploratory study because it was thought that, except for a few young aspiring doctors, the typical individual subscriber tended to be older. The response variable chosen for the study was

$y$ = subscription rate per 1000 individuals in a region

It was decided that age would be measured by

$x$ = percent of population over 50 years of age in a region

Data on $x$ could easily be obtained from U.S. Census publications. Measurements of $x$ and $y$ for several sales regions are shown in Table 4.1.

**Table 4.1** ***Practical Medicine*** **Magazine**

| Sales Region | Subscription Rate (per 1000 people) | Percent of Population Over 50 | Sales Region | Subscription Rate (per 1000 people) | Percent of Population Over 50 |
|---|---|---|---|---|---|
| 1 | 12 | 5.0% | 7 | 22 | 20.1 |
| 2 | 11 | 2.1 | 8 | 10 | 4.4 |
| 3 | 12 | 5.4 | 9 | 12 | 2.2 |
| 4 | 14 | 10.2 | 10 | 28 | 24.9 |
| 5 | 18 | 14.8 | 11 | 40 | 30.2 |
| 6 | 13 | 7.5 | 12 | 36 | 28.0 |

#### Form of the Model

The next step in the study was to decide on the form of the regression model. Two simple models were chosen:

$$y = \beta_0 + \beta_1 x + \epsilon \tag{4.4}$$

and

$$y = \beta_0 + \beta_1 x + \beta_2 x^2 + \epsilon \tag{4.5}$$

Equation 4.4 was chosen to see just how well a simple straight line model would do, and Equation 4.5 was included because some suspected that sales increased much faster when the percent of the population over 50 reached higher levels. It was decided to compare the two models.

## Model Interpretation

A computer package capable of regression calculations was used to estimate the model parameters (using the data in Table 4.1). The resulting estimates for the two models were:

$$\hat{y} = 6.601 + 0.961x \tag{4.6}$$

and

$$\hat{y} = 11.691 - 0.187x + 0.037x^2 \tag{4.7}$$

Even though both models contain the predictor variable $x$, the estimated coefficient of $x$ is not the same in the two models ($\hat{\beta}_1 = 0.961$ in one and $-0.187$ in the other). This often happens when the same predictor variable appears in multiple models. In fact, only when a predictor variable is completely *uncorrelated* with all other predictors in the model(s) will the estimated coefficients turn out to be the same from model to model (Neter, Wasserman, and Kutner, 1983, p. 274).

Statistical analysis of Equations 4.6 and 4.7 revealed that both models fit the data well. However, a graph of residuals versus $x$ showed that the straight line model was not sufficient, and the final decision was to use only Equation 4.7 for forecasting. (This graph appears later, in Figure 4.17, when these data are analyzed in more detail.) The model in Equation 4.7 and a plot of the data are shown in Figure 4.1.

**Figure 4.1** Data and Estimated Regression Model, *Practical Medicine* Magazine

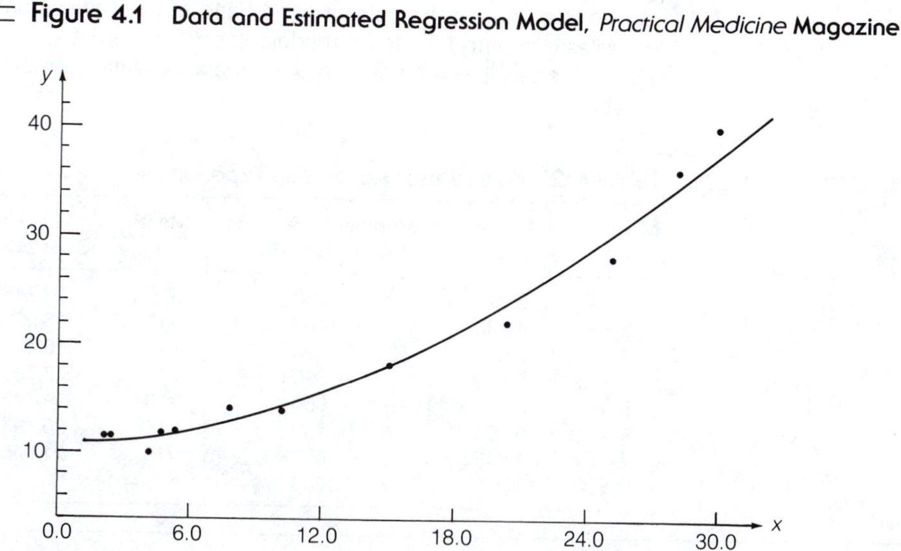

### Prediction/Forecasting

Forecasting with regression equations requires only that values of the predictor variables be plugged into the estimated equation(s). For example, suppose that a given sales region is predicted to have about 20 percent of its inhabitants over age 50 next month. Substituting $x = 20$ into Equation 4.7 yields a predicted number of subscriptions of $\hat{y} = 22.67$ (that is, between 22 and 23 subscriptions per 1000 people).

Notice that, in order to get $\hat{y}$, you must first obtain forecasts of the $x$ variable(s). The ease and accuracy with which the $x$ variables may be predicted will have a direct bearing on whether or not they are included in a regression model.

At this point, the *Practical Medicine* staff has a number of options. They could consider additional explanatory variables, examine other curvilinear models, or simply stop now and use Equation 4.7.

## 4.3.3 Example 4.2 Sales Versus Advertising Expenditures

Most companies eventually inquire about the degree to which money spent on advertising affects sales. Of course, there is no absolute relationship between sales dollars and advertising dollars, since the same advertising budget spent two different ways will result in different sales. However, for a fixed set of marketing channels, modeling the relationship between

$y_t$ = monthly sales

and

$x_t$ = monthly expenditure on advertising

can provide insights about the current marketing plan

### Predictor Variables

Money spent on advertising in one month usually has its effect on sales in a later month, since it takes time for the ads to reach the target audience. Conversely, the effect of an ad doesn't last forever, so its influence will probably extend only a few periods ahead. Suppose a company believes that its sales are dependent on advertising expenditures during the current period and the last two months. This means that $y_t$ = "sales in month $t$" is to be modeled as a function of $x_t$ = "advertising expense in month $t$" as well as a function of $x_{t-1}$ and $x_{t-2}$. Data from the last 15 months appear in Table 4.2.

**Table 4.2 Sales versus Advertising Expenditures**

| Month $t$ | Monthly Sales ($\$10^3$) | Monthly Advertising Expenditure ($\$10^3$) | Month $t$ | Monthly Sales ($\$10^3$) | Monthly Advertising Expenditure ($\$10^3$) |
|---|---|---|---|---|---|
| 1 | 2,945 | 280 | 9 | 13,745 | 1,050 |
| 2 | 4,295 | 400 | 10 | 15,095 | 1,200 |
| 3 | 5,645 | 450 | 11 | 16,445 | 1,250 |
| 4 | 6,995 | 590 | 12 | 17,795 | 1,350 |
| 5 | 8,345 | 650 | 13 | 19,145 | 1,460 |
| 6 | 9,695 | 750 | 14 | 20,495 | 1,500 |
| 7 | 11,045 | 890 | 15 | 21,845 | 1,650 |
| 8 | 12,395 | 1,000 | | | |

### Form of the Model

A distributed lag model (see Section 4.2) is the appropriate choice in this example:

$$y_t = \beta_0 + \beta_1 x_t + \beta_2 x_{t-1} + \beta_3 x_{t-2} + \epsilon_t \tag{4.8}$$

A constant term, $\beta_0$, has been included in Equation 4.8 but need not always appear in a model. Later sections discuss the effect of including or excluding the constant term.

### Model Interpretation

Using a statistical computer package and the data in Table 4.2, the estimated model is

$$\hat{y}_t = 522.1 + 3.681x_t + 4.966x_{t-1} + 5.200x_{t-2} \tag{4.9}$$

Further statistical analysis indicates that $x_t$ may not add much predictive power to a model containing $x_{t-1}$ and $x_{t-2}$, so another regression model, one that uses only $x_{t-1}$ and $x_{t-2}$, is estimated:

$$\hat{y}_t = 1{,}161.6 + 5.873x_{t-1} + 7.945x_{t-2} \tag{4.10}$$

Testing shows that each variable above should remain in the model, so Equation 4.10 will be used for forecasting instead of Equation 4.9.

### Forecasting

Equation 4.10 is especially easy to use for one-step-ahead forecasts, since it has no $x_t$ term. This means it is not necessary to generate an estimate of $x_t$ in order to forecast $y_t$. The forecast of $y_{16}$, for example, is simply

$$\hat{y}_{16} = 1{,}161.6 + 5.873 \times 1{,}650 + 7.945 \times 1{,}500 = 22{,}770$$

Similarly, during month 16 the value of $x_{16}$ will become available, and the forecast of $y_{17}$ may be made. ●

# 4.4 The Decision to Use Regression Models

## 4.4.1 Predictor Variables and Model

Regression analysis, in any of its many forms, may be applied whenever you want to use any other variables, that is, in addition to *y* itself, to predict future values of *y*. Regression methods are also used to test theories concerning how or why a response variable depends on various predictor variables.

The successful use of regression analysis depends, more heavily than with other models, on the person doing the modeling. That person must:

- Select relevant predictor variables.
- Decide on the form of the model.

With most other time series techniques, these decisions are not so completely in the forecaster's hands. Such decisions will have a great effect on the conclusions reached when testing the adequacy of the regression model.

For example, suppose we fit a particular model to some data and find that the tests of hypothesis concerning the model's effectiveness are not statistically significant. What conclusion can be drawn? In regression one cannot simply conclude "there is no relationship between $y$ and the $x_i$'s"; the conclusion could be any of the following:

- $y$ *isn't related* significantly to the $x_i$'s.
- $y$ may actually be related to the $x_i$'s, but the *wrong model* has been used.
- $y$ may be related to the predictor variables, but those *variables* were *quantified the wrong way.* (For instance, in Example 4.1, $x$ = number of people over 50 in a region is not as good a quantification of subscriber's age as $x$ = percent of people over 50 in a region.)

On the other hand, when the regression results are statistically significant, concluding that the modeler has done well is not warranted until a thorough diagnostic check has been performed. Hypothesis tests in regression analysis are based on assumptions that if not met can cause spurious "significant" results.

## 4.4.2 Regression Assumptions

The **least squares procedure** for estimating the regression parameters $\beta_0$, $\beta_1$, $\beta_2$, . . . , $\beta_k$ requires no probabilistic assumptions whatsoever concerning the variables involved. The least squares estimates are simply the quantities $\hat{\beta}_0$, $\hat{\beta}_1$, . . . , $\hat{\beta}_k$ that minimize the expression

$$\sum_t (y - \hat{\beta}_0 - \hat{\beta}_1 x_1 - \cdots - \hat{\beta}_k x_k)^2 \qquad (4.11)$$

Calculus and methods of matrix algebra are used to find the estimates. Figure 4.2 shows the estimated regression line for a one-variable model and the resulting estimated error terms $\hat{\epsilon} = e = y - \hat{y}$ (the *residuals*), for a particular set of data.

**Figure 4.2** An Estimated Regression and Its Residuals

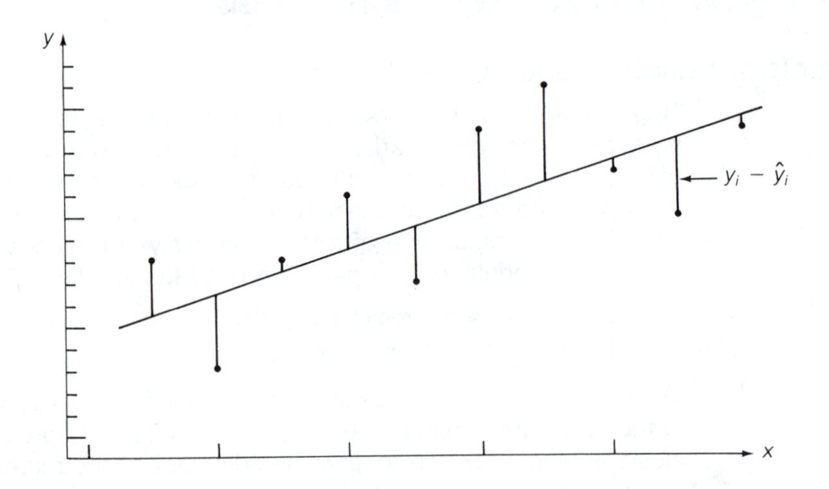

In order to perform hypothesis tests and find prediction intervals for regression parameters, probabilistic assumptions *are* required. In this setting, $y$ and $\epsilon$ are assumed to be random variables, while the predictor variables are not. The probabilistic assumptions are as follows.

**ASSUMPTION 1** *Zero Mean*

For any fixed combination of levels of $x_1, x_2, \ldots, x_k$, the random variable $\epsilon$ has an expected value of zero. This is equivalent to saying that the expected value of $y$, for fixed levels of $x_i$'s, is $\beta_0 + \beta_1 x_1 + \cdots + \beta_k x_k$.

**ASSUMPTION 2** *Normality*

For any combination of levels of the $x$ variables, $\epsilon$ follows a normal distribution.

**ASSUMPTION 3** *Homoscedasticity*

The variance of $\epsilon$ is always the same, regardless of what the levels of the $x$ variables may be. This means that the variance $\sigma_\epsilon^2$ is a constant that does not depend on the levels of the predictor variables.

**ASSUMPTION 4** *Independence*

The error terms $\epsilon_i$ and $\epsilon_j$, associated with two different settings of the predictor variables, are statistically independent.

In time series analysis, assumptions 3 and 4 are very likely to be violated. For example, consider Figure 4.3, which shows a graph of data with $y$ = personal consumption expenditure and $x$ = disposable income. Assumption 3 is violated, in these data, since the variability in $y$ when $x = x_a$ is much smaller than when $x = x_b$. Hypothesis tests and prediction intervals based on regressing $y$ on $x$ would be invalid and misleading. A common remedy in this situation is first to transform the data to stabilize the variance and then to proceed with a regression analysis.

**Figure 4.3** **Unequal Variances, a Violation of Assumption 3**

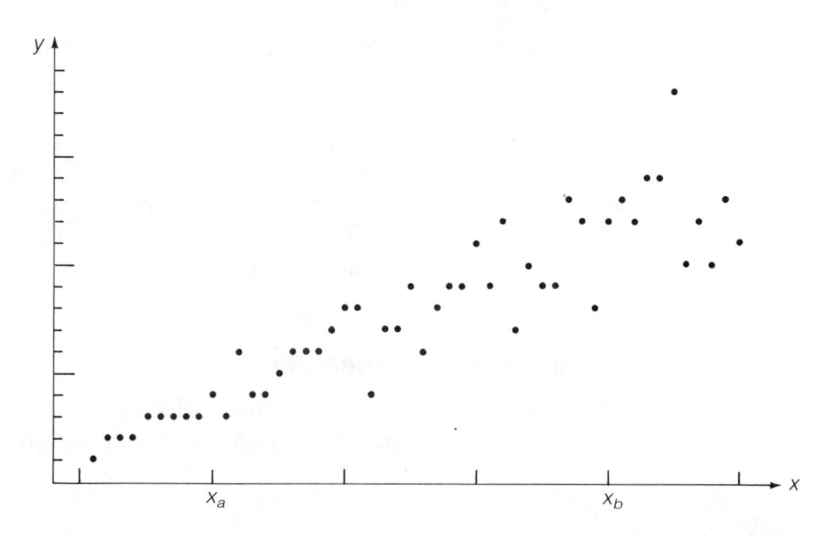

A situation in which assumption 4 is violated is illustrated in Figure 4.4, where the error terms are not independent because of high positive autocorrelation. This can lead to regression models that fit "too well," hypothesis tests with inflated significance levels, and prediction intervals that are too narrow.

**Figure 4.4 Autocorrelated Error Terms, a Violation of Assumption 4**

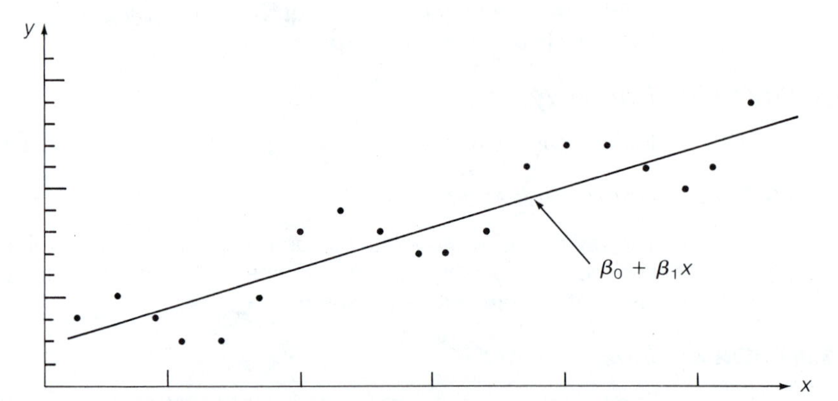


The diagnostic tests (Section 4.8) used in regression analysis are designed to check the regression assumptions above. If a diagnostic test indicates a serious departure from one or more of the assumptions, then doubt is cast upon the statistical tests and estimates, and some type of remodeling effort is necessary.

## 4.4.3 Data Requirements

There are some requirements on the type and amount of data needed to perform regression analyses.

### Number of Data Points

To obtain estimates of the regression parameters, the number of data points, $n$, must *exceed* the number of predictor variables $k$. Estimates cannot be found if $n$ is less than $k$.

The requirement that $n$ exceed $k$ may affect the types of models that can be used. For example, if you have 10 predictor variables and only four observations on these variables, it will be impossible to use all 10 variables in the regression model. This situation can easily occur when one wants to include in a model terms such as $x_1^2$, $x_1x_2$, etc. as well as the original variables $x_1, x_2, \ldots, x_k$.

### Forecasting Predictor Variables

If many of the predictor variables themselves must be forecast prior to using a regression equation (such as "percent of population over 50" in Example

4.1), then the forecasts of those variables should be easy to obtain. Modeling $y$ as a function of variables that are themselves difficult to estimate will only decrease the accuracy of the $y$ forecasts.

## Diagnostic Tests

To estimate regression parameters it is only required that $n$ equal or exceed $k$. To do hypothesis tests, calculate prediction intervals, or perform any diagnostic procedures, you need to have $n$ exceed $k$ by a reasonable amount. The larger $n - k$ is, the more accurate the estimates will be and the less subject to error the conclusions drawn from the data will be. The reason is that, besides estimating $\beta_0, \ldots, \beta_k$, we need to estimate $\sigma_\epsilon^2$ in assumption 3, since all interval estimates and hypothesis tests in regression analysis are based on $\sigma_\epsilon^2$.

To illustrate, consider Figure 4.5. In Figure 4.5(a), the simple one-variable model

$$y = \beta_0 + \beta_1 x_1 + \epsilon$$

has been fit using just two data points. No testing is possible, since we have no way to estimate the variability around the regression line. Figure 4.5(b), on the other hand, shows the situation after many more data points have been added to the sample. The additional data improve the estimates of $\beta_0$ and $\beta_1$ and allow estimation of the variation $\sigma_\epsilon^2$ (the formula for estimating $\sigma_\epsilon^2$ is given in Section 4.5.3).

**Figure 4.5** Estimating the Error Variance

**(a) Number of Data Points Does Not Exceed Number of Parameters, No Estimate of $\sigma_\epsilon$ Is Possible**

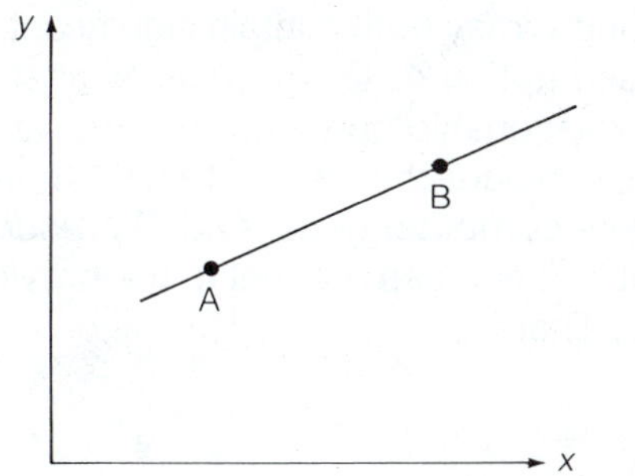


**(b) Number of Data Points Exceeds Number of Parameters, Estimate of $\sigma_\epsilon$ Is Possible**

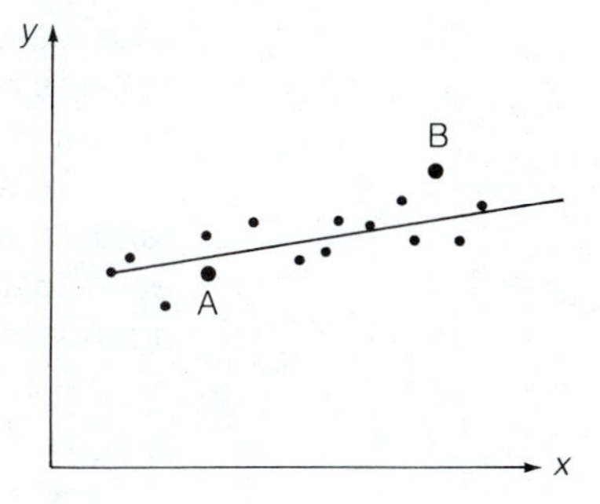


## 4.4.4 Precautions

The following are some common model-building traps to avoid when using regression analysis.

## Overtesting

Many algorithms have been devised for sifting through large lists of variables and selecting the "best" set of predictor variables. Most prominent among these are *stepwise, forward,* and *backward regression.* These techniques are discussed in most texts on regression (see, for example, Draper and Smith, 1981).

Whatever approach is taken, keep in mind that each time a model is tested (using a significance level $\alpha$) the total number of hypothesis tests performed increases. One expects to find $\alpha \times 100\%$ *false* indications of significance in the long run. In other words, any set of variables tested enough times will likely yield "significant" results even when there is no real relationship between $y$ and the $x$ variables (Dunnett, 1964).

The model-testing process can be thought of as panning for gold: Panning once or twice in a stream and finding some specks of gold is a good indication of the presence of gold. Panning every day for a week in the same spot before finally finding some gold specks is not convincing evidence of a real find.

## Large Number of Variables

Adding more and more variables to a model can lead to the problem of "overfitting." If $n$ periods of data are available and variables are added to the model long enough, the situation in Figure 4.5(a) will eventually occur; the model fits *too* well.

The goal of a good regression analysis is to find a simple model with a reasonable number of predictor variables that make sense. You should feel comfortable using the model and be able to explain why each variable used contributes to the model's predictive power.

## Regressing One Time Series on Another

Some caution is needed when regressing one time series, $y_t$, on another, $x_t$. One reason is that $y_t$ and $x_t$ may both contain significant autocorrelation, which violates regression assumption 4. The regression analysis could then lead to false indications of a relationship between the two variables.

To demonstrate, consider the data in Table 4.3. In the table, $y$ and $x$ are random walks that were computer generated. By *random walk* we mean that each successive point in the series is equal to the previous period's value plus a random "shock," $\epsilon_t$. That is,

$$y_t = y_{t-1} + \epsilon_t \tag{4.12}$$

where the $\epsilon_t$'s are independent and come from a normal distribution. Equivalently, a *random walk* is a time series whose first differences form a random sample from a normal distribution.

In Table 4.3, $x$ does not have any real relationship to $y$. In a regression of $y$ on $x$, however, hypothesis tests indicate that $x$ *is* a significant variable to use in predicting $y$. This occurs because random walks usually contain significant positive autocorrelation. Transforming the data before regression by differencing is one way to lessen the problem of spurious significant results (Granger and Newbold, 1986, p. 207).

Similar problems can occur when regressing one trended series on another. In fact, if $y$ and $x$ are any series exhibiting linear trend, a regression of $y$ on $x$ will usually indicate a relationship between the series when none exists. The reason is that the size of the correlation coefficient $r_{x,y}$ between the two series is governed by the correlations each separate series has with time. That is, if series $y$ and $x$ have linear trend, $r_{x,t}$ and $r_{y,t}$ will be large. In that case (Draper and Smith, 1981, p. 266), $r_{x,y}$ will be approximately equal to the product $r_{x,t} \cdot r_{y,t}$. "Detrending" data prior to an analysis is the usual approach to avoiding these spurious results.

**Table 4.3 Random Walks**

| Observation | $y$ | $x$ | Observation | $y$ | $x$ |
|---|---|---|---|---|---|
| 1 | 49.7 | 50.2 | 11 | 48.6 | 50.0 |
| 2 | 50.5 | 50.3 | 12 | 50.0 | 51.0 |
| 3 | 52.5 | 50.9 | 13 | 49.8 | 48.8 |
| 4 | 51.7 | 51.0 | 14 | 49.1 | 47.4 |
| 5 | 51.7 | 51.1 | 15 | 49.9 | 48.8 |
| 6 | 51.6 | 50.8 | 16 | 49.2 | 47.9 |
| 7 | 48.5 | 48.6 | 17 | 50.0 | 46.8 |
| 8 | 50.0 | 49.3 | 18 | 49.9 | 45.9 |
| 9 | 47.4 | 48.5 | 19 | 47.8 | 46.4 |
| 10 | 47.8 | 48.5 | 20 | 48.6 | 48.3 |

## 4.5 Simple Linear Regression

### 4.5.1 Introduction

Regression models are *simple* when only *one* predictor variable is involved. They are *linear* when they are linear in the *parameters,* as described in Section 4.2.1. Models such as the following would all qualify as simple linear models:

$$y = \beta_0 + \beta_1 x + \epsilon$$

$$y = \beta_0 + \beta_1 x^2 + \epsilon$$

$$y = \beta_0 + \beta_1 \log(x) + \epsilon$$

$$y = \beta_0 \beta_1^x \epsilon \quad \text{(\textit{intrinsically linear}, see Section 4.2.1)}$$

The development that follows will be presented in terms of the standard model

$$y = \beta_0 + \beta_1 x + \epsilon \tag{4.13}$$

since every one-variable linear regression model, with the right definition of the $x$ variable, can be put into this form.

## 4.5.2 Parameter Estimates

The estimates $\hat{\beta}_0$ and $\hat{\beta}_1$, of $\beta_0$ and $\beta_1$ above, arise from the least squares criterion (Equation 4.11). When there is only one predictor variable, the formulas for the estimates $\hat{\beta}_0$ and $\hat{\beta}_1$ have a particularly simple form, i.e.,

$$\hat{\beta}_1 = \frac{SS_{xy}}{SS_{xx}} \quad \text{and} \quad \hat{\beta}_0 = \bar{y} - \hat{\beta}_1\bar{x}$$

**Simple Linear Regression** Parameter Estimates

**Given:** A sample of $n$ pairs observations on $x$ and $y$, $(x_1, y_1), (x_2, y_2), \ldots, (x_n, y_n)$

**Estimates:**

$$\hat{\beta}_1 = \frac{SS_{xy}}{SS_{xx}} \qquad \hat{\beta}_0 = \bar{y} - \hat{\beta}_1\bar{x}$$

where

$$SS_{xx} = \sum x^2 - \frac{(\sum x)^2}{n} \qquad \bar{x} = \frac{\sum x}{n}$$

$$SS_{yy} = \sum y^2 - \frac{(\sum y)^2}{n} \qquad \bar{y} = \frac{\sum y}{n}$$

$$SS_{xy} = \sum xy - \frac{(\sum x)\cdot(\sum y)}{n}$$

**Estimated Regression Model:** $\hat{y} = \hat{\beta}_0 + \hat{\beta}_1 x$

**Forecasts (at time *t* for *p* periods ahead):** $\hat{y}_{t+p} = \hat{\beta}_0 + \hat{\beta}_1 x_{t+p}$

***Note:*** $SS_{yy}$, used in estimating $\sigma^2_\epsilon$, should be calculated now.

---

Notice that if $t$ is the predictor variable, the notation becomes $SS_{tt}$ and $SS_{ty}$, which is exactly the same notation used in Chapter 3 for linear trends.

Performing regression calculations by hand can be a little tedious and is subject to error, so the work is usually done by computer. But if hand calculations are necessary, the following guidelines will help.

- Put the data into *column* form. This facilitates calculation of the five sums needed for the estimates, i.e., $\sum x$, $\sum y$, $\sum xy$, $\sum x^2$, and $\sum y^2$.
- Don't round off the results of intermediate calculations severely. Regression estimates are sensitive to round-off error.
- Always calculate $SS_{yy}$. It isn't needed for the parameter estimates, but it will be needed for the hypothesis tests and prediction intervals.

Plotting the estimated regression line on a graph with the data is a good idea. The plot of $\hat{y} = \hat{\beta}_0 + \hat{\beta}_1 x$ versus $x$ is usually easy to obtain because the point $(\bar{x}, \bar{y})$ *must* be on the *estimated* regression line (the equation for $\hat{\beta}_0$, when rearranged, yields $\bar{y} = \hat{\beta}_0 + \hat{\beta}_1\bar{x}$). In plotting the line, it usually works simply to plot two points: $(\bar{x}, \bar{y})$ and the intercept, $(0, \hat{\beta}_0)$ The straight line through these points will be the estimated regression line.

As an illustration of the estimation procedure, we will find $\hat{\beta}_0$ and $\hat{\beta}_1$ for the model in Equation 4.4 of the *Practical Medicine* magazine example.

### EXAMPLE 4.3 Practical Medicine *Magazine, Parameter Estimates*

In Table 4.4, the five sums needed for the parameter estimates have been computed as follows:

$$SS_{yy} = \Sigma y^2 - \frac{(\Sigma y)^2}{n} = 5506.0 - \frac{(228)^2}{12} = 1174.0$$

$$SS_{xy} = \Sigma xy - \frac{(\Sigma x) \cdot (\Sigma y)}{n} = 4080.4 - \frac{(154.8)(228)}{12} = 1139.2$$

$$SS_{xx} = \Sigma x^2 - \frac{(\Sigma x)^2}{n} = 3182.2 - \frac{(154.8)^2}{12} = 1185.28$$

So

$$\hat{\beta}_1 = \frac{SS_{xy}}{SS_{xx}} = \frac{1139.2}{1185.28} = .96112$$

and

$$\hat{\beta}_0 = \bar{y} - \hat{\beta}_1\bar{x} = \frac{228}{12} - .96112 \times \frac{154.8}{12} = 6.601.$$

The estimated regression line (Equation 4.6) is then

$$\hat{y} = 6.601 + 0.961x$$

**Table 4.4** ***Practical Medicine*** **Magazine, Parameter Estimates**

| $y$ | $x$ | $xy$ | $y^2$ | $x^2$ |
|---|---|---|---|---|
| 12 | 5.0 | 60.0 | 144 | 25.00 |
| 11 | 2.1 | 23.1 | 121 | 4.41 |
| 12 | 5.4 | 64.8 | 144 | 29.16 |
| 14 | 10.2 | 142.8 | 196 | 104.04 |
| 18 | 14.8 | 266.4 | 324 | 219.04 |
| 13 | 7.5 | 97.5 | 169 | 56.25 |
| 22 | 20.1 | 442.2 | 484 | 404.01 |
| 10 | 4.4 | 44.0 | 100 | 19.36 |
| 12 | 2.2 | 26.4 | 144 | 4.84 |
| 28 | 24.9 | 697.2 | 784 | 620.01 |
| 40 | 30.2 | 1208.0 | 1600 | 912.04 |
| 36 | 28.0 | 1008.0 | 1296 | 784.00 |
| 228 | 154.8 | 4080.4 | 5506.0 | 3182.2 |
| $\Sigma y$ | $\Sigma x$ | $\Sigma xy$ | $\Sigma y^2$ | $\Sigma x^2$ |

### 4.5.3 Estimating the Variance of the Error Terms, $\sigma_\epsilon^2$

In regression analysis we make the assumption **(homoscedasticity)** that the variance of the error term $\epsilon$ is a constant that is independent of the levels of the $x$ variable. Estimates of this variance, $\sigma_\epsilon^2$, are required for all hypothesis tests and prediction intervals concerning the regression parameters.

The estimated regression parameters are those that minimize the expression in Equation 4.11. The minimum value of the expression is referred to as the **sum of squares error** (or error sum of squares) and is denoted **SSE.** Since SSE can also be written as $\sum_t e_t^2$ (the sum of squares of residuals), which is an aggregate measure of deviation about the regression model, we see that information about the size of $\sigma_\epsilon^2$ is contained in SSE. The estimate $s_\epsilon^2$ of $\sigma_\epsilon^2$ is given by the equation

$$s_\epsilon^2 = \frac{\text{SSE}}{n-2} \tag{4.14}$$

The calculation of $s_\epsilon^2$ normally does *not* proceed by using the calculated regression parameters in the expression

$$\text{SSE} = \sum_t (y - \hat{\beta}_0 - \hat{\beta}_1 x)^2$$

**Simple Linear Regression** Calculation of $s_\epsilon$

**Given:** $\text{SS}_{xy}$, $\text{SS}_{yy}$, and $\hat{\beta}_1$ (from previous calculations)

**Estimate:**

$$s_\epsilon = \sqrt{\frac{\text{SSE}}{n-2}}$$

where

$$\text{SSE} = \text{SST} - \text{SSR} \qquad \text{SST} = \text{SS}_{yy} \qquad \text{SSR} = \hat{\beta}_1 \cdot \text{SS}_{xy}$$

Instead, a more computationally efficient approach is to use

$$\text{SSE} = \text{SST} - \text{SSR} \tag{4.15}$$

where **SST** stands for the **sum of squares total** and **SSR** is the **sum of squares regression.** The components SST and SSR are easily calculated by using some of the intermediate results in the estimation of $\hat{\beta}_0$ and $\hat{\beta}_1$. That is,

$$\text{SST} = \text{SS}_{yy} \tag{4.16}$$

and

$$\text{SSR} = \hat{\beta}_1 \cdot \text{SS}_{xy} \tag{4.17}$$

The decomposition

$$\text{SST} = \text{SSR} + \text{SSE} \tag{4.18}$$

that results from Equation 4.15 indicates that the variation in the response variable, measured by SST, can be broken into two parts:

- SSR, the variation due to the relationship of $y$ and $x$
- SSE, the variation due to random fluctuations

SSR is often called the part of the total variation in $y$ "explained" by the model, while SSE is the "unexplained" part. A more detailed examination of Equation 4.18 will be given later in this section.

**EXAMPLE 4.4** *Estimating* $\sigma_\epsilon$, **Practical Medicine** *Magazine*

Using the results obtained in Example 4.3, we find

$$\text{SST} = \text{SS}_{yy} = 1{,}174.0$$

$$\text{SSR} = \hat{\beta}_1 \cdot \text{SS}_{xy} = .96112 \cdot 1{,}139.2 = 1{,}094.91$$

$$\text{and} \quad \text{SSE} = \text{SST} - \text{SSR} = 1{,}174.0 - 1{,}094.91 = 79.09$$

The estimate of $\sigma_\epsilon$ is then

$$s_\epsilon = \sqrt{\frac{\text{SSE}}{n-2}} = \sqrt{\frac{79.09}{12-2}} = 2.812$$

•

As a rough approximation, 95% prediction limits can be put around any predicted value $\hat{y}$ by using the formula $\hat{y} \pm 2s_\epsilon$. These limits should be used only as quick approximations. A more exact prediction interval formula will be introduced later.

### 4.5.4 Hypothesis Tests Concerning the Slope, $\beta_1$

In one-variable regression, the hypothesis of chief concern is $H_0$: $\beta_1 = 0$. For $x$ to be useful in predicting $y$, we must be able to reject the hypothesis that $\beta_1 = 0$. The reason is that $\beta_1$ measures the rate of change of $y$ with respect to $x$; so if $\beta_1$ is zero, changes in $x$ will have no effect on $y$.

---

**Test Procedure:** *Testing* $\beta_1$ *in Simple Linear Regression*

$H_0$: $y$ does not depend on $x$.

$H_a$: $y$ does depend on $x$.

or $H_0$: $\beta_1 = 0$

$H_a$: $\beta_1 \neq 0$

**Test Statistic:**

$$t = \frac{\hat{\beta}_1 \sqrt{\text{SS}_{xx}}}{s_\epsilon}$$

**Decision Rule:** (Level $\alpha$)

Reject $H_0$ if $|t| > t_{\alpha/2}$. **(Appendix GG)**

$t_{\alpha/2}$ = upper $\alpha/2 \cdot 100\%$ point of Student's $t$-distribution with $n - 2$ degrees of freedom

**Conclusion:** If $H_0$ is rejected, we conclude with $(1 - \alpha) \cdot 100\%$ confidence that the slope $\beta_1$ is not zero and that $x$ contributes to the prediction of $y$.

If $H_0$ is not rejected, we cannot conclude that $x$ is useful in predicting $y$.

---

Hypothesis tests on $\beta_1$ are carried out by using the statistic

$$t = \frac{\hat{\beta}_1 \sqrt{SS_{xx}}}{s_\epsilon}$$

which follows a $t$-distribution with $n - 2$ degrees of freedom.

There are two alternative, but equivalent, ways to test the hypothesis that $\beta_1$ is zero. One method involves simply calculating the sample correlation coefficient $r$ between $x$ and $y$ and then testing whether or not the population correlation coefficient $\rho$ is zero. The correlation test proceeds exactly like the special case (correlation between $t$ and $y_t$) that was covered in Section 3.4.6. The test statistic is

$$t_r = \frac{r\sqrt{n-2}}{\sqrt{1-r^2}}$$

which has a $t$-distribution with $n - 2$ degrees of freedom when $\rho$ is zero and $y$ is normally distributed.

The second method, also equivalent to the $t$-test on $\beta_1$, uses the $F$-distribution. The test statistic in this case is

$$F = \frac{SSR}{SSE/(n-2)}$$

which, when $\beta_1$ is zero, has an $F$-distribution with 1 degree of freedom in the numerator and $n - 2$ degrees of freedom in the denominator.

---

**Test Procedure:** ***Correlation Between y and x***

$H_0$: $\rho = 0$ (equivalently, $H_0$: $\beta_1 = 0$)

$H_a$: $\rho \neq 0$ (equivalently, $H_a$: $\beta_1 \neq 0$)

**Test Statistic:** $$t_r = \frac{r\sqrt{n-2}}{\sqrt{1-r^2}}$$

where $r$ = correlation coefficient between $x$ and $y$.

**Decision Rule:** Reject $H_0$ if $|t| > t_{\alpha/2}$. **(Appendix GG)**

$t_{\alpha/2}$ = upper $\alpha/2 \cdot 100\%$ point of Student's $t$-distribution with $n - 2$ degrees of freedom

**Conclusion:** If $H_0$ is rejected, we conclude with $(1 - \alpha) \times 100\%$ confidence that $\rho$ is nonzero (equivalently, that $\beta_1$ is nonzero) and that $x$ contributes to the prediction of $y$.

If $H_0$ is not rejected, we cannot conclude that there is a relationship between $x$ and $y$.

---

**Test Procedure:** *F-Test in Simple Linear Regression*

$H_0$: $\beta_1 = 0$

$H_a$: $\beta_1 \neq 0$

**Test Statistic:** $F = \dfrac{\text{SSR}}{\text{SSE}/(n-2)}$

where SSE and SSR are calculated from Equations 4.15 and 4.17, respectively.

**Decision Rule:** (Level $\alpha$)

Reject $H_0$ if $F > F_\alpha$. **(Appendix HH)**

($F_\alpha$ = upper $\alpha \times 100\%$ point of the $F$-distribution with $(1, n-2)$ degrees of freedom)

**Conclusion:** If $H_0$ is rejected, we conclude with $(1 - \alpha) \times 100\%$ confidence that $\beta_1$ is nonzero and that $x$ contributes to the prediction of $y$.

If $H_0$ is not rejected, we have no evidence of a relationship between $x$ and $y$.

---

**EXAMPLE 4.5** *Testing* $H_0$: $\beta_1 = 0$, Practical Medicine *Magazine*

To demonstrate the application and equivalence of the $\beta_1$, correlation, and $F$-tests, we will run each on the model in Equation 4.4. The necessary intermediate calculations have already been done in Examples 4.3 and 4.4

$H_0$: $\beta_1 = 0$ $\quad$ ($H_0$: $\rho = 0$)

$H_a$: $\beta_1 \neq 0$ $\quad$ ($H_a$: $\rho \neq 0$)

**Test Statistics:**

$$t = \frac{\hat{\beta}_1\sqrt{SS_{xx}}}{s_\epsilon} = \frac{.96112\sqrt{1185.28}}{2.812} = 11.77$$

$$t_r = \frac{r\sqrt{n-2}}{\sqrt{1-r^2}} = \frac{.9657\sqrt{12-2}}{\sqrt{1-(.9657)^2}} = 11.77$$

$$F = \frac{\text{SSR}}{\text{SSE}/(n-2)} = \frac{1094.91}{79.09/(12-2)} = 138.4$$

**Decision Rules:** ($\alpha = 5\%$)

Reject $H_0$ if $|t| > t_{.025} = 2.228$ (df = 12 − 2 = 10). **(Appendix GG)**

Reject $H_0$ if $|t_r| > t_{.025} = 2.228$ (df = 12 − 2 = 10). **(Appendix GG)**

Reject $H_0$ if $F > F_{.05} = 4.96$ (df = [1, 10]). **(Appendix HH)**

**Conclusions:** Since $|t| = |t_r| > 2.228$ and $F > 4.96$, we reject $H_0$ in all three cases and conclude with 95% confidence that $\beta_1$ is nonzero and that $x$ contributes to the prediction of $y$. ●

***Note:*** It can be shown that $t$ always equals $t_r$. Furthermore, $F_\alpha = t^2_{\alpha/2}$ when $F$ has [1, $m$] degrees of freedom and $t$ has $m$ degrees of freedom.

## 4.5.5 Confidence Intervals for the Slope, $\beta_1$

If $x$ is found to be useful in predicting $y$, often an interval estimate for the *rate* at which $y$ changes in response to $x$ is also wanted. A $(1 - \alpha) \times 100\%$ confidence interval for this rate, i.e., $\beta_1$, can be found from the formula:

$$\hat{\beta}_1 \pm t_{\alpha/2} \cdot \frac{s_\epsilon}{\sqrt{SS_{xx}}}$$

**EXAMPLE 4.6** *95% Confidence Interval for* $\beta_1$*,* **Practical Medicine** *Magazine*

Using the results from Examples 4.3 and 4.4 along with $t_{.025} = 2.228$ (df = 10), the interval estimate is

$$.96112 \pm 2.228 \frac{2.812}{\sqrt{1{,}185.28}} = .96112 \pm .1820 = (.779, 1.143)$$

The data indicate (with 95% confidence) that the true regression slope is somewhere between .779 and 1.143. ●

**Simple Linear Regression** Confidence Interval for $\beta_1$

**Given:** $\hat{\beta}_1$, $s_\epsilon$, and $\sqrt{SS_{xx}}$ from regression calculations

**$(1 - \alpha) \times 100\%$ Confidence Interval for $\beta_1$:**

$$\hat{\beta}_1 \pm t_{\alpha/2} \cdot \frac{s_\epsilon}{\sqrt{SS_{xx}}}$$

where $t_{\alpha/2}$ is the $\alpha/2 \times 100\%$ point of the $t$-distribution with $n - 2$ degrees of freedom. **(Appendix GG)**

## 4.5.6 The Constant Term, $\beta_0$

All the regression models presented have included a constant term. There has been no mention of how to fit regression models without constant terms. Nor has there been any discussion of hypothesis tests and interval estimates for $\beta_0$.

Procedures do exist both for testing $\beta_0$ and for fitting models without it, but they will not be discussed here (see Weisberg, 1980, sec. 2.4). Instead, it is recommended that a regression study be started with the constant term included in the model. When $\beta_0$ is not used, the regression line will be forced through the point (0, 0). This may have a detrimental effect on $\hat{\beta}_1$, which has to compensate for the lack of the $\beta_0$ term. It is often necessary to keep $\beta_0$ in the model even in situations where it is known in advance that $y$ should be zero when $x$ is zero (e.g., if $y$ = gallons of fuel used and $x$ = distance traveled). In many of these cases, data are not gathered for $x$ values near zero, and the nature of the relationship between $x$ and $y$ may be very different away from the origin.

## 4.5.7 Prediction Intervals for $y$ at $x = x_0$

After a regression model has been selected and fit to the data, that model may be used to forecast the $y$ variable. The predicted value of $y$ will depend on what level $x_0$ of the $x$ variable is used. The point estimate for the predicted value of $y$ when $x = x_0$ is found by putting $x_0$ into the estimated regression equation

$$\hat{y}_0 = \hat{\beta}_0 + \hat{\beta}_1 x_0$$

The interval estimate, however, will depend on the variability of $y$ around the regression line (estimated by $s_\epsilon$) and upon how far $x_0$ is from the center of the data, $\bar{x}$. The $(1 - \alpha) \times 100\%$ prediction interval for $y_0$ is given by

$$\hat{y}_0 \pm t_{\alpha/2} s_\epsilon \sqrt{1 + \frac{1}{n} + \frac{(x_0 - x)^2}{SS_{xx}}} \quad (4.19)$$

where the associated $t$-distribution has $n - 2$ degrees of freedom.

**Simple Linear Regression** Prediction Interval for $y_0$ When $x = x_0$

**Given:** $\hat{\beta}_0$, $\hat{\beta}_1$, $s_\epsilon$, $\bar{x}$, and $SS_{xx}$ from regression calculations

**Point Estimate:** $\hat{y}_0 = \hat{\beta}_0 + \hat{\beta}_1 x_0$

**Prediction Interval, $(1 - \alpha) \times 100\%$ Confidence:**

$$\hat{y}_0 \pm t_{\alpha/2} s_\epsilon \sqrt{1 + \frac{1}{n} + \frac{(x_0 - \bar{x})^2}{SS_{xx}}}$$

where $t_{\alpha/2}$ = upper $(1 - \alpha) \cdot 100\%$ point of the $t$-distribution with $n - 2$ degrees of freedom.

**(Appendix GG).**

**EXAMPLE 4.7** *Prediction Intervals, Q-Burger Franchise*

Q-Burger, Inc., is a large chain of fast-food restaurants specializing in hamburgers, fries, and shakes designed to taste and look like those from 1950s-era hamburger stands. Each restaurant promotes a '50s image, complete with posters and furniture of that period. As a service to investors interested in obtaining a Q-Burger franchise, the company provides estimates of the average number of customers per week that would frequent prospective restaurants. These estimates help investors decide whether or not a franchise in a particular area will be financially attractive.

To predict $y$ = average number of customers per week, the company uses $x$ = number of residences within a two-mile radius of the restaurant. Analysts on Q-Burger's corporate staff chose this predictor variable because they believed most fast-food customers come from the immediate vicinity of a restaurant and that higher population densities would generate higher weekly numbers of customers.

Data from 14 *Q-Burger* locations are given in Table 4.5. The company calculates 95% prediction intervals from these data. The following necessary preliminary calculations were done with a regression analysis computer package:

$$\hat{y} = 732.3 + 1.8811x$$
$$s_\epsilon = 463.2$$
$$\bar{x} = 2927.9$$
$$SS_{xx} = 20{,}576{,}114.9$$

An investor interested in, say, a location having approximately 4000 residences within a two-mile radius would then find the estimated average weekly number of customers to be

$$\hat{y}_0 = 732.3 + 1.8811 \times 4{,}000 = 8{,}257.$$

with a 95% prediction interval given by ($t_{.025} = 2.179$, df $= 12$)

$$8{,}257 \pm (2.179)(463.2)\sqrt{1 + \frac{1}{14} + \frac{(4{,}000 - 2{,}927.9)^2}{20{,}576{,}114.9}}$$
$$= 8{,}257 \pm 107.2 = (7{,}185;\ 9{,}329)$$

The investor has 95% confidence, then, that the weekly number of customers would be somewhere between 7,185 and 9,329.

**Table 4.5 Q-Burger, Inc.**

| Location | Residences Within a Two-Mile Radius $x$ | Customers per Week $y$ |
|---|---|---|
| 1 | 974 | 2,550 |
| 2 | 1,302 | 3,180 |
| 3 | 1,586 | 3,460 |
| 4 | 1,891 | 4,090 |
| 5 | 2,229 | 4,680 |
| 6 | 2,454 | 5,640 |
| 7 | 2,733 | 6,720 |
| 8 | 3,019 | 7,100 |
| 9 | 3,383 | 6,550 |
| 10 | 3,625 | 7,210 |
| 11 | 4,004 | 8,260 |
| 12 | 4,214 | 8,280 |
| 13 | 4,629 | 8,990 |
| 14 | 4,948 | 10,650 |

Since prediction interval calculations can be done for *any* value of $x$, it is informative to plot a number of intervals on the same graph. Drawing a line through all the lower and upper endpoints of the intervals forms a $(1 - \alpha) \times 100\%$ prediction band around the estimated line. For example, using the Q-Burger data, a 95% prediction band is shown in Figure 4.6. Notice that the width of the intervals increases the further $x_0$ is from $\bar{x}$ (cf. Example 3.12). ●

Figure 4.6 Ninety-Five Percent Prediction Band, Q-Burger Data

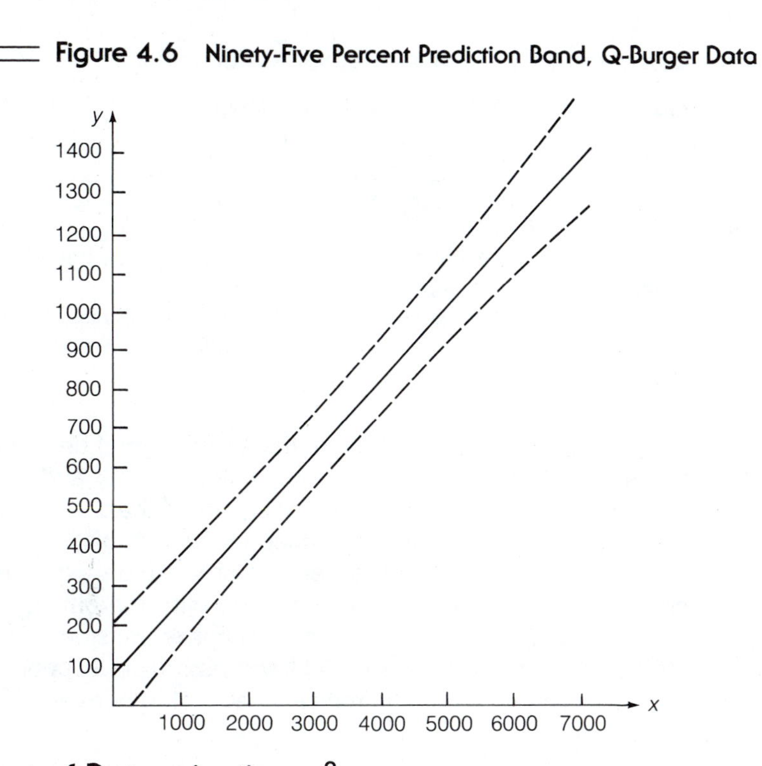


## 4.5.8 Coefficient of Determination, $r^2$

The decomposition of the total variation SST into "explained" and "unexplained" parts was discussed earlier, where it was written as (Equation 4.18)

$$\text{SST} = \text{SSR} + \text{SSE}$$

Dividing this equation through by SST we get

$$1 = \frac{\text{SSR}}{\text{SST}} + \frac{\text{SSE}}{\text{SST}} \tag{4.20}$$

which allows us to express explained and unexplained variation as proportions of the total variation. In particular, the proportion of the total variation in $y$ explained by the model is called the **coefficient of determination** and is denoted $r^2$, where

$$r^2 = \frac{\text{SSR}}{\text{SST}}$$

From Equation 4.20 it is apparent that $r^2$ is always between 0 and 1.

In regression analysis, the coefficient of determination serves as a quick summary measure of how well a model fits the data. Values of $r^2$ close to 1 indicate a close agreement between model and data, while values close to 0 indicate a poor fit. One of the goals of regression analysis is to seek models and variables that result in large values of $r^2$ while simultaneously passing all the diagnostic tests for a "good" model.

**EXAMPLE 4.8** *Calculating* $r^2$, Practical Medicine *Magazine*

From the results in Example 4.4, we know that SSR = 1,094.91 and SST = $SS_{yy}$ = 1,174.0. So

$$r^2 = \frac{SSR}{SST} = \frac{1{,}094.91}{1{,}174.0} = .9326$$

About 93% of the variation in subscription rates ($y$) has been "explained" by changes in the percentage of population over 50 ($x$).

The value of $r^2$ always depends on the model used. So if the quadratic model in Equation 4.7 has been used instead, the value of $r^2$ would have been different (even though the same $x$ variable is used in both models).

●

In simple linear regression, the coefficient of determination is related to the sample correlation coefficient, $r$, between $x$ and $y$. The coefficient of determination is always equal to the square of the correlation coefficient, which explains why the notation $r^2$ is used in the first place.

Since $r^2$ measures the percent of variation explained by a model, the size of $r^2$ should have something to do with the outcome of the test of $H_0$: $\beta_1 = 0$ versus $H_a$: $\beta_1 \neq 0$ in simple linear regression. That is, large values of $r^2$ should support $H_a$, while small values should support $H_0$. This is, in fact, the case. There is a mathematical relationship between $r^2$ and the value of $F$ used to test $H_0$, namely,

$$F = \frac{r^2}{(1 - r^2)/(n - 2)} \tag{4.21}$$

Thus, in Example 4.8, where $r^2 = .9326$ for the *Practical Medicine* Magazine data, we would have

$$F = \frac{.9326}{(1 - .9326)/(12 - 2)} = 138.4$$

which agrees with the calculated $F$-value in Example 4.5.

The relationship between $F$ and $r^2$ is used mainly in the interpretation of *partial* summaries of regression studies. For example, if you read a report that says a simple linear model was fit to 40 data points and gives the coefficient of determination as .75, you can use Equation 4.21 to find that the value of $F$ in the study was 114. Similarly, given only the $F$-value, you can find $r^2$.

## 4.6 Multiple Linear Regression

### 4.6.1 Introduction

When more than one predictor variable is used, the analysis of the general linear model

$$y = \beta_0 + \beta_1 x_1 + \beta_2 x_2 + \cdots + \beta_k x_k + \epsilon$$

is referred to as **multiple linear regression** analysis. Conceptually, there is no great difference between this and fitting one-variable models (simple linear regression). The parameter estimates still arise from the least squares criterion, and the hypothesis tests and interval estimates still depend on estimating $\sigma_{\epsilon}^2$. Computationally, however, there is a major difference between simple and multiple regression: Multiple regression requires matrix algebra calculations, which are extremely lengthy if done by hand and so are best left to a computer.

Our discussion of multiple regression will center around the use of computer packages for performing the needed calculations. We will use printouts from some of the major statistical packages (e.g., SAS, BMDP, Minitab),, but nearly any statistical software would suffice, since regression analysis is common to almost all such programs.

### 4.6.2 ANOVA Table

One of the most useful parts of the computer output from a regression analysis is the ANOVA table (the acronym comes from the term ANalysis Of VAriance). The ANOVA table summarizes information concerning the decomposition of the total variation in *y* into "explained" and "unexplained" parts: SST = SSR + SSE. A standard form of the ANOVA table is shown in Figure 4.7.

Each of the three sums of squares has its associated degrees of freedom. SST always has $n - 1$ degrees of freedom, regardess of how many predictor

**Figure 4.7** **Multiple Linear Regression** ANOVA Table

ANALYSIS OF VARIANCE

| SOURCE OF VARIATION | DF | SUM OF SQUARES | MEAN SQUARE | F-VALUE |
|---|---|---|---|---|
| REGRESSION | K | SSR | MSR | F |
| ERROR | N - (K + 1) | SSE | MSE | |
| TOTAL VARIATION | N - 1 | SST | | |

where

K = number of predictor variables in the model

N = number of data points

DF = degrees of freedom

MSR = mean square regression = SSR/K

MSE = mean square error = SSE/(N − (K + 1))

F = calculated F-value = MSR/MSE

variables are in the model. SSR has $k$ degrees of freedom, where $k$ is the number of predictor variables used. And SSE has the remaining degrees of freedom, $n - (k + 1)$. The degrees of freedom follow the same type of decomposition as the sums of squares:

| | | | | | |
|---|---|---|---|---|---|
| sums of squares | SST | = | SSR | + | SSE |
| degrees of freedom | $n - 1$ | = | $k$ | + | $n - (k + 1)$ |

This decomposition is shown in the DF column of Figure 4.7.

When sums of squares are divided by their corresponding degrees of freedom, the resulting quantities are called **mean squares.** Thus,

$$\text{mean square for regression} = \text{MSR} = \frac{\text{SSR}}{k}$$

$$\text{mean square error} = \text{MSE} = \frac{\text{SSE}}{n - (k + 1)}$$

The MSE is always an unbiased estimate of the variation $\sigma_\epsilon^2$ around the regression model and is used in all hypothesis tests and interval estimates.

***Note:*** We will use the notation $s_\epsilon^2$ and MSE interchangably throughout this chapter. Remember to distinguish these from the purely descriptive MSE's of earlier chapters. Notice also, when $k = 1$, MSE $=$ SSE/$(n - 2)$, which is the estimate $s_\epsilon^2$ introduced in Equation 4.14.

In the ANOVA table, the F-VALUE column contains the calculated $F$ statistic (MSR/MSE) for testing the hypothesis

$$H_0\text{:}\ \beta_1 = \beta_2 = \cdots = \beta_k = 0$$

which is the extension of $H_0$: $\beta_1 = 0$ in simple regression. In multiple regression, the $F$-test is often called the *global* or *overall F-test* because it tests the utility of all the predictor variables at once. It is discussed in detail in the next section (4.6.3).

**EXAMPLE 4.9** *ANOVA Table,* **Practical Medicine** *Magazine*

Figure 4.8 contains the ANOVA table from a simple regression using the BMDP computer package on the *Practical Medicine* data. Compare the numbers in the table to the results obtained in Examples 4.3 through 4.5 (there are some slight differences due to round-off error). Notice that BMDP assumes the user knows about the decomposition of SST and of its degrees of freedom, so the printout doesn't include the SST value or the total degrees of freedom, $n - 1$.

**Figure 4.8 Multiple Linear Regression** ANOVA Table from BMDP Printout

```
ANALYSIS OF VARIANCE

                    SUM OF SQUARES     DF    MEAN SQUARE     F-VALUE       P(TAIL)
      REGRESSION      1094.9484         1     1094.9484      138.511        .0000
      RESIDUAL          79.0516        10        7.9052
```

### 4.6.3 Hypothesis Tests

In regression, simple or multiple, we need to determine statistically whether or not the $x$ variables are useful in predicting $y$. For simple regression this means testing hypotheses about the coefficient of a single $x$ variable. For multiple regression there is a wider range of tests available, since it is possible to test hypotheses about any collection of the predictor variables. We will restrict our attention to two of these tests: a test of the usefulness of the *entire* collection of $x$ variables, and a test concerning the usefulness of any *single* variable in the model.

Testing the usefulness of the entire collection of predictor variables is accomplished by using the **overall** or **global *F*-test.** The null hypothesis tested is that *none* of the $x$ variables contributes to predicting $y$; that is, $H_0$: $\beta_1 = \beta_2 = \cdots = \beta_k = 0$. If $H_0$ is rejected, then at least one variable contributes to the prediction of $y$. Thus, the alternative hypothesis is written as $H_a$: at least one $\beta_i$ is nonzero. The statistic used in this test is the extension of the $F$-statistic used in simple regression, i.e.,

---

**Test Procedure:** *Overall* F*-test in Multiple Regression*

$H_0$: none of the variables contribute to predicting $y$.

$H_a$: at least one of the $x$ variables is useful.

Or

$H_0$: $\beta_1 = \beta_2 = \cdots = \beta_k = 0$

$H_a$: at least one $\beta_i$ is nonzero.

**Test Statistic:**

$$F = \frac{\text{MSR}}{\text{MSE}} = \frac{\text{SSR}/k}{\text{SSE}/[n - (k + 1)]}$$

**Decision Rule:** (Level $\alpha$)

Reject $H_0$ if $F > F_\alpha$

where $F_\alpha$ = upper $\alpha \cdot 100\%$ point of the $F$-distribution with $[k, n - (k + 1)]$ degrees of freedom." **(Appendix HH)**

**Conclusion:** If $H_0$ is rejected, we conclude with $(1 - \alpha) \times 100\%$ confidence that at least one of the $\beta_i$'s is nonzero and that, collectively, the $x$ variables contribute to the prediction of $y$.

If $H_0$ is not rejected, we do not have sufficient evidence to conclude that any of the $x$ variables are useful for predicting $y$.

---

**Figure 4.9** Multiple Linear Regression SAS Printout, *Practical Medicine* Magazine

REGRESSION OF Y ON X AND X-SQUARED
GENERAL LINEAR MODEL PROCEDURE
DEPENDENT VARIABLE: Y

| SOURCE | DF | SUM OF SQUARES | MEAN SQUARE | F VALUE | PR GT F | R-SQUARE | C. V. |
|---|---|---|---|---|---|---|---|
| MODEL | 2 | 1164.57290845 | 582.28645422 | 555.91 | 0.0001 | 0.991970 | 5.3366 |
| ERROR | 9 | 9.42709155 | 1.04745462 | | ROOT MSE | | Y MEAN |
| CORRECTED TOTAL | 11 | 1174.00000000 | | | 1.02345230 | | 19.000000000 |

| SOURCE | DF | TYPE I SS | F VALUE | PR GT F | DF | TYPE III SS | F VALUE | PR GT F |
|---|---|---|---|---|---|---|---|---|
| X | 1 | 1094.94839864 | 1045.34 | 0.0001 | 0.2255 | 1.77342498 | 1.690 | 0.2255 |
| XSQ | 1 | 69.62450981 | 66.47 | 0.0001 | 1 | 69.62450981 | 66.47 | 0.0001 |

| PARAMETER | ESTIMATE | T FOR H0: PARAMETER=0 | PR GT ABS(T) | STD ERROR OF ESTIMATE |
|---|---|---|---|---|
| INTERCEPT | 11.69134869 | 14.80 | 0.0001 | 0.79003914 |
| X | -0.18733261 | -1.30 | 0.2255 | 0.14397085 |
| XSQ | 0.03667412 | 8.15 | 0.0001 | 0.00449828 |

$$F = \frac{\text{MSR}}{\text{MSE}} = \frac{\text{SSR}/k}{\text{SSE}/[n - (k + 1)]}$$

which, when $H_0$ is true, follows the $F$-distribution with $[k, n - (k + 1)]$ degrees of freedom.

**EXAMPLE 4.10** *Overall* **F-*test,*** **Practical Medicine *Magazine***

Figure 4.9 contains part of the SAS output from fitting the second-order model $y = \beta_0 + \beta_1 x + \beta_2 x^2$ (Equations 4.5 and 4.7) to the *Practical Medicine* data.

$H_0$: $\beta_1 = \beta_2 = 0$

$H_a$: $\beta_1 \neq 0$ and/or $\beta_2 \neq 0$

**Test Statistic:** $F = 555.91$ (from printout, Figure 4.9)

**Decision Rule:** ($\alpha = 5\%$)

Reject $H_0$ if $F > F_{.05} = 4.26$ (df = [2, 9])

**Conclusion:** $F > 4.26$, so we reject $H_0$ and conclude with 95% confidence that the second-order model is useful for predicting the subscription rate.

In Example 4.10, the $F$-test indicated that a model containing $x$ and $x^2$ is useful for predicting $y$. However, the rejection of $H_0$ in this example does *not* imply that *both* variables are needed in the model. The $F$-test only implies that, as a group, $x$ and $x^2$ contribute to the prediction.

One might think at first that an easy way to check each variable separately is to regress $y$ on that variable and see if the $F$- or $t$-test for simple regression is significant. Indeed, for the *Practical Medicine* data, both the simple regressions of $y$ on $x$ and $y$ on $x^2$ yield significant $F$- (or $t$-) values. But this doesn't mean that both $x$ and $x^2$ are needed *together* in a single model. What is needed at this point is a way of testing the contribution a variable makes over and above the contribution made by the other variables in the model.

To test each variable individually, a *partial t-test* is used. It tests whether the variable adds significantly to a model already containing the remaining predictor variables. Specifically, the test is concerned with the null hypothesis that $\beta_i = 0$, where $x_i$ is the variable of interest. The test statistic is

$$t = \frac{\hat{\beta}_i}{s_{\hat{\beta}_i}}$$

where $s_{\hat{\beta}_i}$ is the estimated standard deviation of the statistic $\hat{\beta}_i$. Normally, the $t$-statistic is part of the computer printout, so no calculation is necessary. The degrees of freedom associated with the partial $t$-test is always the same as the MSE degrees of freedom, i.e., $n - (k + 1)$ (available in the ANOVA table).

**Test Procedure:** *Partial t-test in Multiple Regression*

$H_0$: $x_i$ does not contribute additional predictive power to a model already containing the other $x$ variables.

$H_a$: $x_i$ does contribute additional predictive power to the model.

or

$H_0$: $\beta_i = 0$
$H_a$: $\beta_i \neq 0$

**Test Statistic:** $t = \dfrac{\hat{\beta}_i}{s_{\hat{\beta}_i}}$ (normally obtained from computer printout)

**Decision Rule:** Reject $H_0$ if $|t| > t_{\alpha/2}$

where $t_{\alpha/2}$ = upper $(\alpha/2) \cdot 100\%$ point of the $t$-distribution with $n - (k + 1)$ degrees of freedom. **(Appendix GG)**

**Conclusion:** If $H_0$ is rejected, we conclude with $(1 - \alpha) \times 100\%$ confidence that the variable $x_i$ is useful in a model already containing the other $x$ variables.

If $H_0$ is not rejected, we cannot conclude that $x_i$ adds any predictive power beyond that already provided by the other variables in the model.

---

**EXAMPLE 4.11** *Partial t-tests,* Practical Medicine *Magazine*

We can now resolve the issue of whether to include both the $x$ and $x^2$ terms in a single model by using a partial $t$-test on each variable. The necessary inputs for these tests can be found in the printout of Figure 4.9: there are 9 degrees of freedom for the $t$-test (i.e., MSE df = 9) and the calculated $t$-statistics are in the T FOR HO column of the SAS printout.

$H_0$: $\beta_i = 0$
$H_a$: $\beta_i \neq 0$

**Test Statistics:**

$t = -1.30$ (for $\beta_1$)

$t = 8.15$ (for $\beta_2$)

**Decision Rule:** $(\alpha = 5\%)$

Reject $H_0$ if $|t| > t_{.025} = 2.262$ (df = 9) **(Appendix GG)**

**Conclusions:**

$\beta_1$ $|t| < 2.262$, so we do *not* have evidence, at the 5% level, that $x$ contributes significantly to a model already containing $x^2$.

$\beta_2$ $|t| > 2.262$, and we conclude that $x^2$ does contribute additional predictive power to a model already containing $x$. ●

We can now summarize all of the conclusions about the *Practical Medicine* example. First, the simple regression of $y$ on $x$ (Example 4.5) showed that $x$ was a reasonable predictor of $y$ ($r^2 = .9326$, from Example 4.8). The overall *F*-test (Example 4.10) indicated that the model with both $x$ and $x^2$ was also useful (which we already knew, since Example 4.5 showed that $x$ alone was useful). Finally, Example 4.11 produced the somewhat surprising result that $x$ was *not* needed in a model if $x^2$ was already present. All these tests point to the same conclusion: $x$ is a reasonably good predictor, but $x^2$ is even stronger. The appropriate model should contain an $x^2$ term, and there is even justification for eliminating the $x$ term (Example 4.11) altogether and just using the model $y = \beta_0 + \beta_1 x^2$. (In Section 4.8.3 we will see that the diagnostic tests also point strongly to using the $x^2$ term.)

## 4.6.4 Prediction Intervals for $y$

Forecasts of the $y$ variable depend on the particular levels of the predictor variables (as in simple regression). The point estimate for the predicted value of $y$ when $x_1 = c_1, x_2 = c_2, \ldots, x_k = c_k$ is

$$\hat{y} = \beta_0 + \beta_1 c_1 + \beta_2 c_2 + \cdots + \beta_k c_k$$

where $c_1, c_2, \ldots, c_k$ are the specified levels of the $x$ variables.

To obtain prediction intervals, the computer program used must have this option (not all packages do). Some programs automatically calculate prediction intervals for each combination of the $x$ variables in the data set, as in the SAS printout in Figure 4.10 (in case the user doesn't want to specify the particular $x$ levels). Others allow one to specify the $x$ levels.

**Figure 4.10 Prediction Intervals from SAS Printout Second-order model, *Practical Medicine* Magazine**

| X | OBSERVED VALUE | PREDICTED VALUE | LOWER 95% CL INDIVIDUAL | UPPER 95% CL INDIVIDUAL |
|---|---|---|---|---|
| 5 | 12.00000000 | 11.67153853 | 9.20281055 | 14.14026652 |
| 2.1 | 11.00000000 | 11.45968306 | 8.82235277 | 14.09701335 |
| 5.4 | 12.00000000 | 11.74916981 | 9.28715783 | 14.21118179 |
| 10.2 | 14.00000000 | 13.59613106 | 11.05697560 | 16.13528651 |
| 14.8 | 18.00000000 | 16.95192435 | 14.31016449 | 19.59368420 |
| 7.5 | 13.00000000 | 12.34927312 | 9.87883412 | 14.81971212 |
| 20.1 | 22.00000000 | 22.74267268 | 20.13138789 | 25.35395747 |
| 4.4 | 10.00000000 | 11.57709609 | 9.09180386 | 14.06238831 |
| 2.2 | 12.00000000 | 11.45671967 | 8.82932281 | 14.08411653 |
| 24.9 | 28.00000000 | 29.76508512 | 27.21484587 | 32.31532437 |
| 30.2 | 40.00000000 | 39.48216427 | 36.60047058 | 42.36385797 |
| 28 | 36.00000000 | 35.19854225 | 32.54208451 | 37.85499999 |
| ↓ $x$ | ↓ $y$ | ↓ $\hat{y}$ | | |

For example, when $x = 5$, $\hat{y} = 11.69 - .1873 \times 5 + .03667 \times 5^2 = 11.67$ (see Figure 4.9 for the regression equation) and a 95% prediction interval for $y$ when $x = 5$ is given by (9.203, 14.140).

## 4.6.5 Multiple Coefficient of Determination, $R^2$

The proportion of variation explained by a multiple regression model is the same, in concept, as in simple regression, but the notation changes from $r^2$ to $R^2$:

$$R^2 = \text{coefficient of determination} = \frac{\text{explained variation}}{\text{total variation}} = \frac{\text{SSR}}{\text{SST}}$$

With more than one variable in a model, the coefficient of determination now becomes the square of the **multiple correlation coefficient *R*,** which measures the strength of the relationship between $y$ and *all* the $x$ variables ($R$ also turns out to be equal to the sample correlation coefficient between $y$ and $\hat{y}$, that is, between the *observed* and the *predicted* $y$ values in the regression model).

Some caution should be exercised when interpreting $R^2$ because it is a mathematical fact that $R^2$ must increase with every new variable added to the model. It is possible to make $R^2$ artificially large by just adding more and more variables to a model (cf. Section 4.4.4). Inexperienced users of regression software sometimes fall into the trap of only trying to find a model that maximizes $R^2$. To avoid this mistake, remember that the goals of a "good" regression analysis are:

- a simple model with a reasonable number of predictor variables that make sense
- a statistically significant overall $F$-value
- statistically significant partial $t$-values for all variables in the model
- a high $R^2$ value
- residuals that pass all the diagnostic tests

Analogous to simple regression, there is an exact relationship between the overall $F$-test and $R^2$, i.e.,

$$F = \frac{R^2/k}{(1 - R^2)/[n - (k + 1)]}$$

If, for example, the printout in Figure 4.9 were to be summarized in a report by simply stating that "a quadratic model $y = \beta_0 + \beta_1 x + \beta_2 x^2$ was fit to 12 data points and the coefficient of determination was .99197," then anyone reading the report could calculate $F$:

$$F = \frac{.99197/2}{(1 - .99197)/[12 - (2 + 1)]} = 555.9$$

## 4.6.6 Adjusted Coefficient of Determination, $R_a^2$

As mentioned above, there is a problem with using $R^2$ to judge the predictive value of a newly added variable, since, after adding the $x$ variable, one question always faces the researcher: Was the resulting increase in $R^2$ due to the mathematical fact that $R^2$ *had to* increase, or did $R^2$ increase because $x$ was indeed a good predictor variable?

In an attempt to get around this problem, researchers have developed a modified $R^2$ measure designed to increase only when the newly added variable contributes something to the prediction of $y$. This new measure is called the **adjusted $R^2$** and is denoted $R_a^2$. It is defined by modifying the equation $R^2 = 1 - \text{SSE/SST}$ to

$$R_a^2 = 1 - \frac{n-1}{n-(k+1)} \cdot \frac{\text{SSE}}{\text{SST}}$$

or, equivalently,

$$R_a^2 = 1 - \frac{\text{MSE}}{s_y^2}$$

where $s_y^2$ is the sample variance of the $y$ data. Most regression programs routinely calculate both $R^2$ and $R_a^2$.

From the first of these two equations for $R_a^2$ it is not hard to see that $R_a^2$ never exceeds $R^2$. Also, consider the effect of adding a new variable, $x$. For $x$ to increase $R_a^2$, it must decrease the MSE. Such $x$ variables are desirable, of course, because small values of MSE (which measures the spread of the data about the regression model) are desirable. Conversely, if adding an $x$ variable does not decrease the MSE, then $R_a^2$ will decrease.

Since $R_a^2$ overcomes some of the drawbacks associated with $R^2$, it is natural now to ask whether the goal of a good regression analysis might simply be to search for models/variables that maximize $R_a^2$. Indeed, many analysts do exactly that, retaining only those $x$ variables that increase $R_a^2$ when added to the model. This is not a bad practice, since, from the second equation for $R_a^2$, we note that the goal of maximizing $R_a^2$ is equivalent to minimizing MSE. There is, however, one minor problem: Maximizing $R_a^2$ may not lead to models in which all the $x$ variables have significant partial $t$-values. For example, it is known (Haitovsky, 1969; Edwards, 1969) that $R_a^2$ must increase whenever an $x$ variable with $|t| < 1$ is removed from the model. Thus, to maximize $R_a^2$, $x$ variables with $|t| > 1$ should be kept in the model even though these $t$-values may not be statistically significant.

The adjusted $R^2$ is not meant to be a replacement for $R^2$. For one thing, $R_a^2$ does not have the intuitive interpretation as a percent of explained variation that $R^2$ does. Instead, $R_a^2$ should be used as designed, to aid in the selection of good predictor variables. After using $R_a^2$ to arrive at a suitable model, the researcher can have more faith that the $R^2$ associated with that model is more realistic than an $R^2$ made artificially high by adding superfluous $x$ variables.

## 4.7 Time Series Applications

### 4.7.1 Introduction

The regression procedures presented so far can be applied to any set of $y$ and $x$ variables, whether or not they are time series. There are other regression techniques, however, that have been adapted specifically for time series problems. In this section we discuss some of those methods.

## 4.7.2 Indicator Variables

As mentioned in Section 4.2.1, *qualitative* information such as season of the year, geographical region, or occurrence of a specific event can be incorporated into a regression model. To do this, indicator, or dummy, variables are used. **Indicator variables** convert qualitative information into quantitative information by means of a coding scheme. The most common coding scheme simply uses a 1 to indicate the occurrence of an event of interest and a 0 to indicate its nonoccurrence. When modeling monthly sales, for example, the effect of a competitor's store opening could be studied by using the following dummy variable:

$$x = \begin{cases} 1 & \text{for months after the competing store opens} \\ 0 & \text{for months up to the opening of the new store} \end{cases}$$

**EXAMPLE 4.12** *Indicator Variables, Metro Video Data*

Many of the examples in Chapter 3 concerned the Metro Video Society data (cf. Example 3.16 and Table 3.14). Suppose it is now known that a competing video rental store in the vicinity of Metro Video opened for business in June 1985. Since Metro Video's sales may be affected by the presence of the competing store, instead of modeling sales as a simple function of time ($t$) alone, an indicator variable ($x$) can be used to separate months after June 1985 from June 1985 or prior months.

The form of the regression equation to use depends on whether or not we want to allow for possible changes in the slope of the trend. If there is no difference in the rates of sales growth before and after June 1985, then the following model would suffice:

$$y = \beta_0 + \beta_1 t + \beta_2 x + \epsilon \qquad \text{Origin: August 1984} \quad \text{Units } (t)\text{: Monthly}$$

where

$$x = \begin{cases} 1 & \text{for } t > 10 \\ 0 & \text{for } t \leq 10 \end{cases}$$

Otherwise, we could use the following model to fit one line ($\beta_0 + \beta_1 t$) over the first portion, and another line ($\beta_0^* + \beta_1^* t$) over the second portion:

$$y = (\beta_0 + \beta_1 t)(1 - x) + (\beta_0^* + \beta_1^* t)x + \epsilon$$

By gathering terms and noting that the two lines must intersect at $t = 10$, the model reduces to (Draper and Smith, 1981)

$$y = \beta_0 + \beta_1 t + \beta_2 (t - 10)x + \epsilon$$

where $\beta_2$ is the difference between the two slopes ($\beta_1 - \beta_1^*$).

For illustration we will fit both models, even though the second model seems more appropriate here since it could easily be the case that Metro Video's rate of growth (slope) is affected by the competition. Fitting the two models to the data of Table 4.6, we find that

$$\hat{y} = 1{,}018.3 + 188.2t - 25.8x$$

and

$$\hat{y} = 764.70 + 238.55t - 127.70(t - 10)x$$

**Table 4.6** Using Indicator Variables with Metro Video Society Data

| Month | Revenues $y_t$ | Time $t$ | Indicator $x$ | $(t - 10)x$ |
|---|---|---|---|---|
| Sep 84 | 999 | 1 | 0 | 0 |
| Oct | 1,123 | 2 | 0 | 0 |
| Nov | 1,503 | 3 | 0 | 0 |
| Dec | 1,762 | 4 | 0 | 0 |
| Jan 85 | 2,126 | 5 | 0 | 0 |
| Feb | 2,315 | 6 | 0 | 0 |
| Mar | 2,239 | 7 | 0 | 0 |
| Apr | 2,655 | 8 | 0 | 0 |
| May | 2,787 | 9 | 0 | 0 |
| Jun | 3,024 | 10 | 0 | 0 |
| Jul | 3,467 | 11 | 1 | 1 |
| Aug | 3,528 | 12 | 1 | 2 |
| Sep | 3,441 | 13 | 1 | 3 |
| Oct | 3,558 | 14 | 1 | 4 |
| Nov | 3,746 | 15 | 1 | 5 |
| Dec | 3,628 | 16 | 1 | 6 |
| Jan 86 | 4,021 | 17 | 1 | 7 |

In the first model the partial $t$-value for $x$ is not significant, indicating a poor model choice, while in the second the variables $t$ and $(t - 10)x$ both have significant ($\alpha$ = 1%) partial $t$-values, indicating that there *has* been a shift in the rate of growth after June 1985.

Substituting the values of $x$ into the fitted equations we get the following.

**FIRST MODEL** ***No Change in Slope Allowed***

*June 1985 or prior* ($x = 0$):

$$\hat{y} = 1{,}018.3 + 188.2t - 25.8 \times (0) = 1{,}018.3 + 188.2t$$

*After June 1985* ($x = 1$):

$$\hat{y} = 1{,}018.3 + 188.2t - 25.8 \times (1) = 992.5 + 188.2t$$

**SECOND MODEL** *Change in Slope Allowed*

*June 1985 or prior* ($x = 0$):
$$\hat{y} = 764.70 + 238.55t - 127.70(t - 10) \times 0 = 764.70 + 238.55t$$

*After June 1985* ($x = 1$):
$$\hat{y} = 764.70 + 238.55t - 127.70(t - 10) \times 1 = 2041.7 + 110.85t$$

Both models and the Metro Video data are plotted in Figure 4.11. From the graphs it is clear that the second model is again the better choice and is the one to be used for forecasting. Forecasts for February 1985 and March 1985 would then be

*Feb 85:* $\hat{y}_{18}(17) = 2041.7 + 110.85(18) = 4037.0$

*Mar 85:* $\hat{y}_{19}(17) = 2041.7 + 110.85(19) = 4147.9$

**Figure 4.11** **Regression Model Using Indicator Variables with Metro Video Society Data**

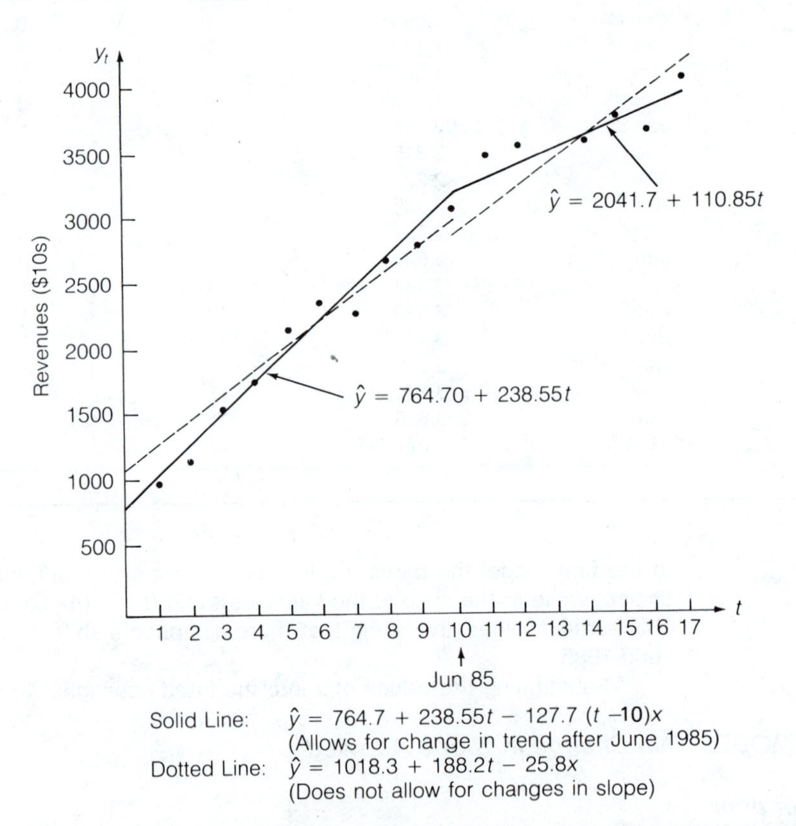


A natural question that arises during this analysis is: Why not just fit a simple trend line to the post-June 1985 data, since it is the trend over that region that is extrapolated to give forecasts? Of course, this can be done. But the dummy variable approach is usually preferred because it gives a better estimate of the variation $\sigma_\epsilon$ in the series by using *all* the available data. ●

Another common use of indicator variables is in modeling seasonal variation. Because there are usually more than two seasons (e.g., 12 months, four quarters), more than one indicator variable must be used. For example, with monthly data it is necessary to include dummy variables to account for all 12 months. This can be done by using

$$x_1 = \begin{cases} 1 & \text{Jan} \\ 0 & \text{otherwise} \end{cases}$$

$$x_2 = \begin{cases} 1 & \text{Feb} \\ 0 & \text{otherwise} \end{cases}$$

$$\vdots$$

$$x_{11} = \begin{cases} 1 & \text{Nov} \\ 0 & \text{otherwise} \end{cases}$$

Notice that no $x_{12}$ variable is used, since data for December are "indicated" when $x_1 = x_2 = \cdots = x_{11} = 0$. This illustrates the following general rule when using dummy variables in regression.

Use $c - 1$ dummy variables to separate $c$ different categories.

In fact, if you attempt to use the same number of dummy variables as categories, the least squares method will be overconstrained and will not yield any parameter estimates at all.

### Using Qualitative Variables in Regression

**Given:** A qualitative variable that can fall into any of $c$ categories

**Indicator (Dummy) Variables:** $x_1, x_2, \ldots, x_{c-1}$

where each

$$x_i = \begin{cases} 1 & \text{if the variable falls in the } i\text{th category} \\ 0 & \text{otherwise} \end{cases}$$

One way to model monthly data, then, is to use an equation like

$$y_t = \underbrace{\beta_0 + \beta_1 t}_{\substack{\text{trend} \\ \text{component}}} + \underbrace{M_1 x_1 + M_2 x_2 + \cdots + M_{11} x_{11}}_{\substack{\text{seasonal} \\ \text{component}}} + \underbrace{\epsilon_t}_{\substack{\text{random} \\ \text{component}}}$$

For simplicity, $M_1, M_2, \ldots, M_{11}$ are used to denote the parameters for the monthly seasonal component and $\beta$'s for the trend parameters. More details of the application of seasonal indicator variables in regression models are discussed in Chapter 5.

### 4.7.3 Time as a Predictor Variable

When the response variable is a time series, the time variable $t$ can be used in a regression model even in the absence of other explanatory variables (cf. Section 4.2). Modeling $y_t$ as a function of $t$ alone is done in order to identify the nature of the trend in the series. This was done extensively (although not always in a regression context) in Chapter 3. Familiar trend models that may be fit using regression include:

$$y_t = \beta_0 + \beta_1 t \qquad \text{(linear trend)}$$

$$y_t = \beta_0 + \beta_1 t + \beta_2 t^2 \qquad \text{(quadratic trend)}$$

$$y_t = \beta_0 \cdot \beta_1^t \qquad \text{(exponential trend)}$$

For example, in Chapter 3 the least squares method was used to fit all three of the above trend models.

In addition to describing the nature of the trend, a trend model also allows one way of detrending a series. A *detrended* series, as the name implies, is just the original series with the trend removed. For example, if we fit a linear trend $\hat{y} = \hat{\beta}_0 + \hat{\beta}_1 t$ to the series $y_t$, then the detrended series would be

$$\text{detrended } y_t = y_t - (\hat{\beta}_0 + \hat{\beta}_1 t)$$

which is another way of saying that the detrended series consists of the residuals $(y_t - \hat{y}_t)$ from the regression of $y_t$ on some function of $t$.

As mentioned at the end of Section 4.4.3, detrending is a convenient way to handle the problem of spurious correlation between trended series (Granger and Newbold, 1986, sec. 6.4). Recall that it is misleading to simply regress one trended series $(y_t)$ on another $(x_t)$, since the correlation $r_{x,y}$ will usually be significantly different from zero due to the fact that $r_{x,y} \simeq r_{x,t} \cdot r_{y,t}$ even when the $x$ and $y$ series are completely unrelated.

**EXAMPLE 4.13** ***Spurious Correlation, Personal Consumption Expenditure Versus Public Highway Construction***

Table 4.7 contains data on personal consumption expenditure $(y_t)$ and public highway expenditures $(x_t)$ in the United States over a 12-month period. Since personal consumption expenditure is a measure of spending after taxes have been taken out, it should not have any real relationship to public highway expenditures, which are largely funded by tax monies. However, if we naively regress $y_t$ on $x_t$, we find just the opposite. The sample correlation coefficient between the series turns out to be $r_{x,y} = .897$, while the $t$-statistic for testing $H_0$: $\beta_1 = 0$ is $t = 6.43$, easily significant at the 1% level.

The problem is that both $y_t$ and $x_t$ are strongly trended series (see Figure 4.12). In fact, simple regressions of $y_t$ and $x_t$ on linear functions of $t$ yield

$$r_{y,t} = .988 \qquad \text{and} \qquad r_{x,t} = .879$$

Thus, even before regressing the two series we could have predicted that $r_{x,y}$ would be approximately

$$r_{x,y} \approx r_{x,t} \cdot r_{y,t} = .879 \times .988 = .869$$

Regressing trended series, then, can easily lead to false indications of a relationship between causally unrelated series.

To properly analyze the data in Table 4.7, some sort of detrending will be necessary

**Table 4.7** U.S. Personal Consumption and Public Highway Construction Expenditures

| Month | Personal Consumption ($ billions) | Highway Construction ($ billions) | Month | Personal Consumption ($ billions) | Highway Construction ($ billions) |
|---|---|---|---|---|---|
| Sep 84 | 2383.7 | 16.5 | Mar | 2451.1 | 19.2 |
| Oct | 2378.1 | 16.2 | Apr | 2483.0 | 19.9 |
| Nov | 2395.9 | 16.9 | May | 2495.6 | 22.3 |
| Dec | 2415.5 | 16.9 | Jun | 2500.4 | 21.1 |
| Jan 85 | 2432.6 | 17.8 | Jul | 2506.9 | 19.7 |
| Feb | 2455.8 | 18.4 | Aug | 2536.1 | 20.3 |

*Source: Survey of Current Business*

**Figure 4.12** Graphs of (a) Personal Consumption Expenditure and (b) Public Highway Construction

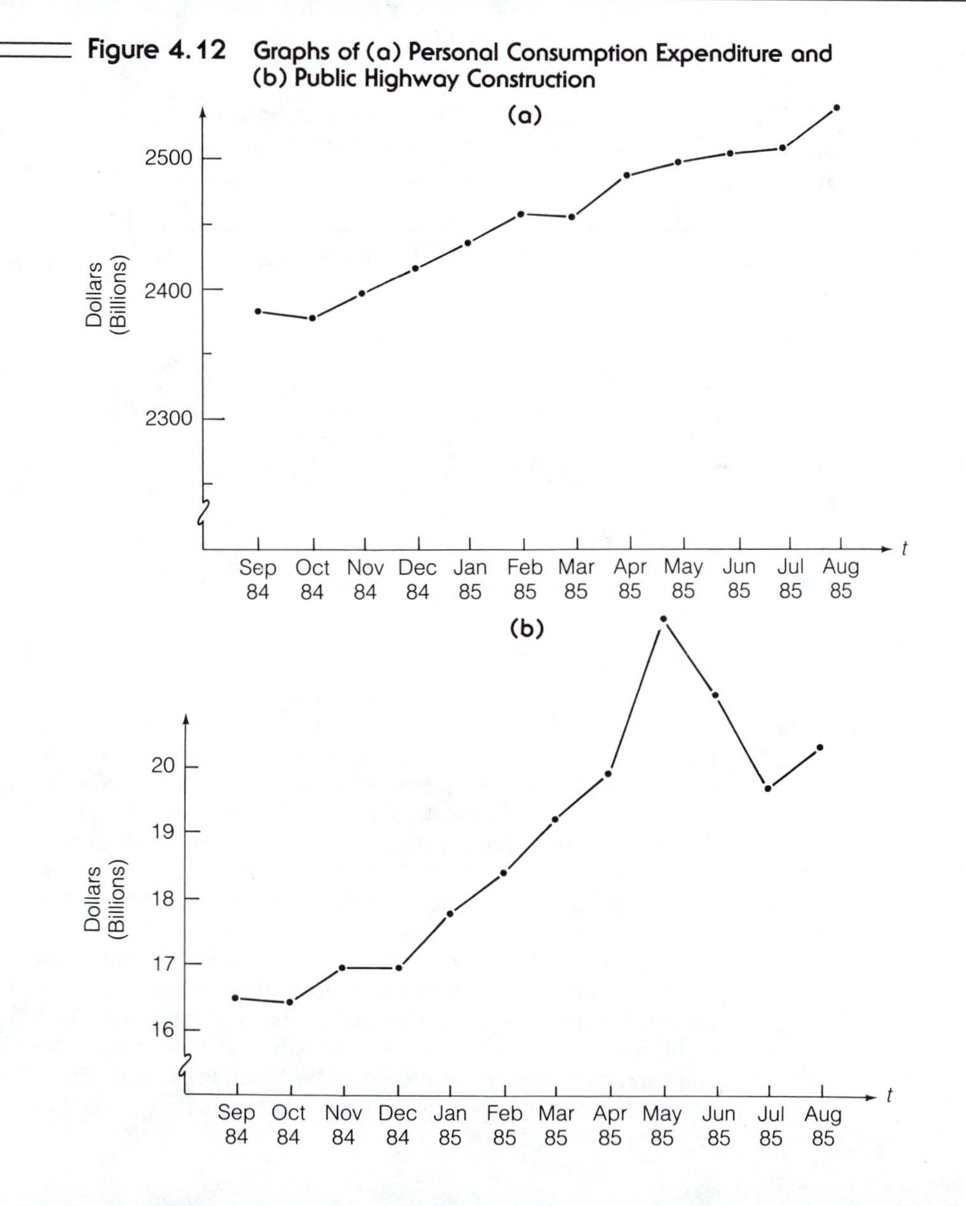

prior to regression. This could be accomplished by either differencing or fitting a trend model. We will focus on the trend-fitting approach here since it is more closely connected with regression using time as a predictor variable, but the reader should remember in the following discussion that series detrended by differencing or some other method could have been used as well.

When detrending, the first question is what to detrend. Do we regress $y_t$ on a detrended $x_t$, or do we detrend both series before regressing? In Table 4.8 the various regression equations needed to answer this question are collected. In Cases 1 and 2, linear trends were fit and the detrended series (detrended $y_t$ and detrended $x_t$) were formed. The subsequent regression of the original $y_t$ on the detrended $x_t$ in Case 3 results in a partial *t*-value of 0.19, which shows that, after detrending, $x_t$ is not useful as a predictor of $y_t$. This corresponds to our intuitive knowledge of these two series: Highway construction should have little to do with personal (after-tax) spending; indeed, it was only the common factor of trend that caused the misleading correlation in Example 4.13.

**Table 4.8 Detrending Personal Consumption and Public Highway Construction Expenditures**

| Case | Equation |
|---|---|
| 1 | $\hat{y} = 2359.4 + 14.4t \rightarrow$ detrended $y_t = y_t - \hat{y}_t$ <br> (20.7) |
| 2 | $\hat{x} = 15.7 + .478t \rightarrow$ detrended $x_t = x_t - \hat{x}_t$ <br> (5.82) |
| 3 | $\hat{y} = 2452.9 + 3.38 \cdot$ (detrended $x_t$) <br> (0.19) |
| 4 | $\hat{y} = 2359.4 + 14.4t + 3.38 \cdot$ (detrended $x_t$) <br> (21.5) (1.31) |
| 5 | detrended $\hat{y}_t = 0.00 + 3.38 \cdot$ (detrended $x_t$) <br> (1.31) |
| 6 | $\hat{y} = 2306.5 + 12.8t + 3.38x$ <br> (9.10) (1.31) |

*Note:* Partial *t*-statistics are in parentheses under their corresponding parameter estimates.

For $x_t$ to be a useful predictor of $y_t$, we have to ascertain whether $x_t$ contributes any predictive power over and above the trend already exhibited by $y_t$. Cases 4, 5, and 6 represent equivalent approaches to handling the trend in $y_t$ and the additional predictive power (if any) of $x_t$. In fact, all three equations lead to the same model [by substituting $y_t - (2359.4 + 14.4t)$ and $x - (15.7 + .478t)$ for detrended $y_t$ and detrended $x_t$, Cases 4 and 5 both reduce to Case 6]. Thus, whether we regress $y_t$ on $t$ and a detrended $x_t$ (Case 4), detrend both series before regressing (Case 5), or simply perform a multiple regression of $y_t$ on $t$ and $x_t$ (Case 6), we will correctly model the relationship between the two series.

Since the three approaches are mathematically equivalent, we recommend the simplest: Perform a multiple regression of $y_t$ on $t$ and $x_t$, which avoids any need to actually calculate the detrended series. The only drawback is that the trend component in the resulting equation will represent some combination of the trends in both series and cannot be simply interpreted as the trend in $y_t$ alone. Case 4, on the other hand, *will* yield a trend component that is the same as for $y_t$, but at the price of first having to detrend the $x_t$ series.

In general, when more than one $x$ variable is used and the series involved are trended, then an extension of the above procedure would be to regress $y_t$ on $t$ and the various $x$'s. ●

**Regressing Trended Series**

Step 1 Identify the functional form $f(t)$ of the trend component in $y_t$

Step 2 Perform a multiple regression of $y$ on $f(t)$ and $x_1, x_2, \ldots, x_k$

## 4.7.4 Autoregressive Models

Besides trend modeling, another method of fitting that is based only on the information in the $y_t$ series itself uses lagged values of $y_t$ in regression models (cf. Section 4.2.2). Such models have the form

$$y_t = \beta_0 + \beta_1 y_{t-1} + \beta_2 y_{t-2} + \cdots + \beta_p y_{t-p} + \epsilon_t \tag{4.22}$$

and are called **autoregressive models,** where $p$, the number of lagged terms, is the *order* of the model (see also Section 7.2.3). Autoregressive models are used when the current level of the series is thought to depend on the recent history of the series. For example, series that measure production and spending are often modeled as autoregressive processes since the amount produced or spent in one period may well affect the amounts produced or spent in successive periods. Alternatively, autoregressive models are appropriate when it is believed that the effects of the random component $\epsilon_t$ may be felt for a few periods beyond the current one (notice that because $y_t$ becomes $y_{t-1}$ next time, etc., $\epsilon_t$ is included in $y_t, y_{t+1}, \ldots, y_{t+p}$). This might be true when modeling inventory levels, where a large increase in inventories one month can affect the inventory level in the next month or two.

Three problems immediately arise when considering the use of autoregressive models:

- If $y_t$ has a trend, so will $y_{t-1}, y_{t-2}, \ldots$, leading to the problems associated with regressing trended series (see the previous discussion of time as a predictor variable).
- Because the independent variables are actually previous values of the dependent variable, there are problems with regression assumption 4.
- The order of the model, $p$, has to be selected.

If $y_t$ is trended, then a regression on $y_{t-1}, y_{t-2}, \ldots, y_{t-p}$ can easily give misleading results. Before using an autoregressive model, then, you must first make sure that $y_t$ is *stationary.* If it isn't, then either differencing or another detrending method will have to be used, and the autoregressive model fit to the resulting stationary series.

Next, the problems with the regression assumptions can cause the least squares estimates $\hat{\beta}_1, \hat{\beta}_2, \ldots, \hat{\beta}_p$ to be *biased downwards.* That is, the fitted regression parameters will, on the average, *underestimate* the actual parameters. However, if the number of data points (or the number of predictor variables) is large, this problem becomes negligible, so if possible, try to use larger data sets when working with autoregressive models. As an illustration, later in this section we will examine the bias in $\beta_1$ for the first-order autoregressive model $y_t = \beta_0 + \beta_1 y_{t-1} + \epsilon_t$.

## 4.7.5 Partial Autocorrelation Function (pacf)

The appropriate number of lags to use (i.e., the order $p$ of the model) can be determined by analyzing the partial autocorrelation coefficients of the series. The *partial autocorrelation coefficient of lag k,* denoted $\phi_{kk}$, is a measure of the correlation between $y_t$ and $y_{t-k}$ *after* adjusting for the presence of all the $y_t$'s of shorter lag, i.e., $y_{t-1}, y_{t-2}, \ldots, y_{t-k+1}$. This adjustment is done to see if there is any additional correlation between $y_t$ and $y_{t-k}$ above and beyond that induced by the correlation $y_t$ has with $y_{t-1}, y_{t-2}, \ldots, y_{t-k+1}$. When $\phi_{kk}$'s are graphed for lag 1, lag 2, . . . , the result is the **partial autocorrelation function (pacf)** of the series. The pacf (such as in Figure 4.13) is identical in format to the autocorrelation function introduced in Chapter 2.

**Figure 4.13** Example of a Typical Partial Autocorrelation Function

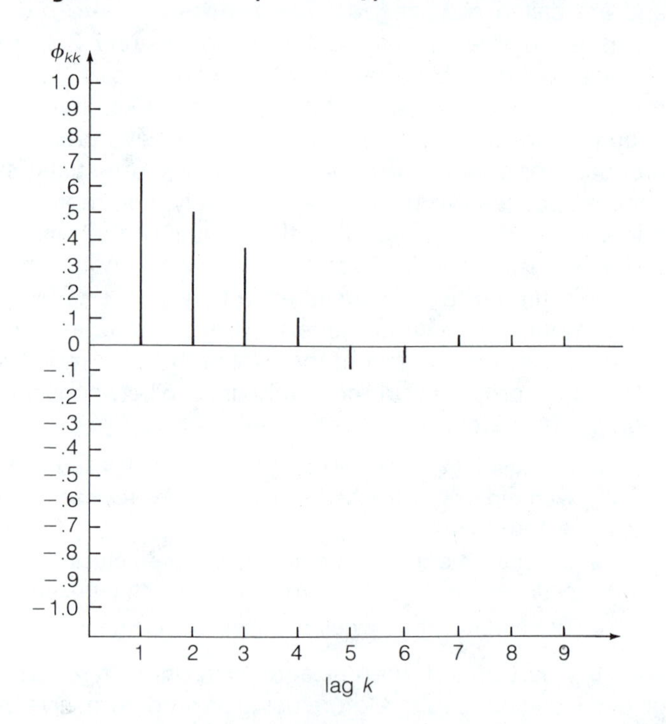


There are two methods of estimating the partial autocorrelations in a series. One is to simply perform a regression of $y_t$ on $y_{t-1}$ through $y_{t-k}$ and use the resulting coefficient of the $y_{t-k}$ term as the estimate of $\phi_{kk}$ (which works only for stationary series):

$$\hat{y}_t = \hat{\beta}_0 + \hat{\beta}_1 y_{t-1} + \cdots + \underset{\displaystyle \hat{\phi}_{kk}}{\underset{\downarrow}{\hat{\beta}_{t-k} y_{t-k}}}$$

This method is in keeping with the definition of $\hat{\phi}_{kk}$, since multiple regression coefficients are always "adjusted" for the presence of the other variables in the model (see Section 4.8). The other approach to estimating the pacf is to use an algorithm that recursively generates the partial autocorrelation coefficients from a knowledge of the autocorrelation coefficients $r_k$. Because of its recursive nature, most computer programs use this second method to calculate the pacf. Both methods are outlined in Table 4.9

**Table 4.9 Calculating the Partial Autocorrelation Coefficients**

**Method 1:**

- Regress $y_t$ on $y_{t-1}, y_{t-2}, \ldots, y_{t-k}$.
- $\hat{\phi}_{kk}$ is the coefficient of $y_{t-k}$ in the fitted equation. (*Note:* this method requires performing a different regression for each $\hat{\phi}_{kk}$.)

**Method 2:***

- Initial values: Let $N_1(j) = D_1(j) = r_j$, $j = 0, 1, 2, \ldots$ ($r_j$ = autocorrelation coefficient of lag $j$, $r_0 = 1$).
- Let $\hat{\phi}_{11} = N_1(1)/D_1(0)$; i.e., let $\hat{\phi}_{11} = r_1$.
- Recursively, for k = 2, 3, . . . let

$$N_k(j) = N_{k-1}(j+1) - D_{k-1}(j) \cdot \hat{\phi}_{k-1,k-1}$$

$$D_k(j) = D_{k-1}(j) - N_{k-1}(j+1) \cdot \hat{\phi}_{k-1,k-1}$$

- Calculate $\hat{\phi}_{kk} = N_k(1)/D_k(0)$.

* *Reference:* Weiss, G. (1983). "Computing Partial Autocorrelations by Recursion." *Amer. Statistician* 37(4): 324.

Testing whether a particular $\hat{\phi}_{kk}$ is significantly different from zero is done with the same rule of thumb procedure used in testing the autocorrelation coefficients. That is, if the true $\phi_{kk}$ is equal to zero, then $\hat{\phi}_{kk}$ will approximately follow a normal distribution with a standard deviation of $1/\sqrt{n}$ ($n$ is the length

**Test Procedure** *Testing the Partial Autocorrelation Coefficient*

$$H_0: \quad \phi_{kk} = 0$$

$$H_a: \quad \phi_{kk} \neq 0$$

**Test Statistic:** $\hat{\phi}_{kk}$ = estimated partial autocorrelation coefficient of lag $k$

**Decision Rule:** ($\alpha = 5\%$, $n$ large)

Reject $H_0$ if $|\hat{\phi}_{kk}| > 2/\sqrt{n}$.

**Conclusion:** If $H_0$ is rejected, we conclude with approximately 95% confidence that the partial autocorrelation of lag $k$ is nonzero.

If $H_0$ is not rejected, we have support for the lag $k$ partial autocorrelation being zero.

of the series), and we will conclude that $\phi_{kk}$ is nonzero whenever $|\hat{\phi}_{kk}|$ exceeds $2/\sqrt{n}$.

Determining whether an autoregressive model is appropriate and then determining its order can be accomplished quickly with the aid of the pacf. An autoregressive model of order $p$ may be used when the pacf is significant for lags 1 through $p$ and then cuts off sharply (i.e., is not significant) after lag $p$. Moreover, the acf of an autoregressive series should drop off exponentially to zero because the series must be stationary.

**Forecasting with Autoregressive Models**

**Given:** A sample of $n$ observations $y_1, y_2, \ldots, y_n$ from a stationary series. (If $y_t$ is not stationary, difference or detrend and forecast the transformed series.)

**Model:** $y_t = \beta_0 + \beta_1 y_{t-1} + \beta_2 y_{t-2} + \cdots + \beta_p y_{t-p}$
where the number $p$ of autoregressive terms is determined from the partial autocorrelation function; the pacf should cut off sharply after lag $p$.

**Estimates:** Peform a multiple regression of $y_t$ on $y_{t-1}, \ldots, y_{t-p}$ to obtain estimates of $\beta_0, \beta_1, \ldots, \beta_p$.

**Forecasts:** (At time $t$ for $p$ periods ahead)
Substitute the values of $y_{t-1}, \ldots, y_{t-p}$ into the regression equation $\hat{y}_t = \hat{\beta}_0 + \hat{\beta}_1 y_{t-1} + \cdots + \hat{\beta}_p y_{t-p}$; whenever a value of $y_{t-k}$ is itself a future value, use the forecast of $y_{t-k}$ in the equation.

**EXAMPLE 4.14** ***Autoregressive Model, CTV Subscription Service***

The monthly number of subscribers in one of the west coast sales regions of the CTV cable television service has been monitored for the last 25 months. These data appear in Table 4.10 and are plotted in Figure 4.14. The series drifts above and below a mean of about 50.6, and a glance at the acf (Figure 4.15) confirms that the series is indeed stationary. An autoregressive model is a logical candidate here, since the number of subscriptions in one month can be thought of as coming from two sources: (1) subscribers from the previous month who continue their subscriptions into the current

**Table 4.10 Monthly Numbers of Subscriptions, CTV Cable Television Service**

| Month $t$ | Subscriptions (1,000's) $y_t$ | Month $t$ | Subscriptions (1,000's) $y_t$ |
|---|---|---|---|
| 1 | 50.8 | 14 | 51.8 |
| 2 | 50.3 | 15 | 53.6 |
| 3 | 50.2 | 16 | 53.1 |
| 4 | 48.7 | 17 | 51.6 |
| 5 | 48.5 | 18 | 50.8 |
| 6 | 48.1 | 19 | 50.6 |
| 7 | 50.1 | 20 | 49.7 |
| 8 | 48.7 | 21 | 49.7 |
| 9 | 49.2 | 22 | 50.3 |
| 10 | 51.1 | 23 | 49.9 |
| 11 | 50.8 | 24 | 51.8 |
| 12 | 52.8 | 25 | 51.0 |
| 13 | 53.0 | | |

**Figure 4.14** Monthly Number of Subscriptions, CTV Cable Television Service

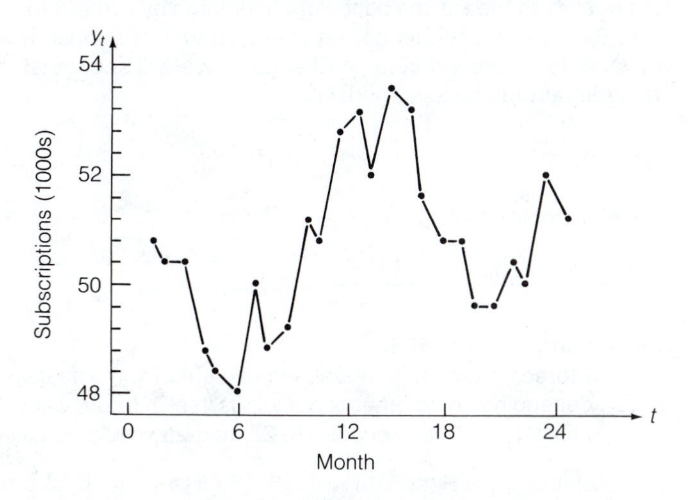


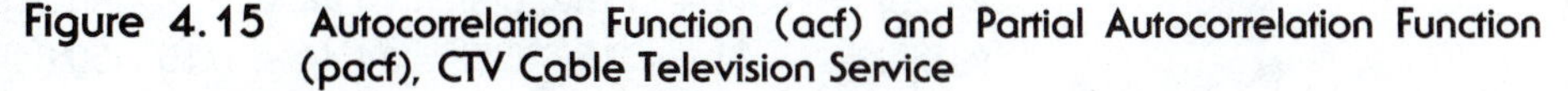

**Figure 4.15** Autocorrelation Function (acf) and Partial Autocorrelation Function (pacf), CTV Cable Television Service

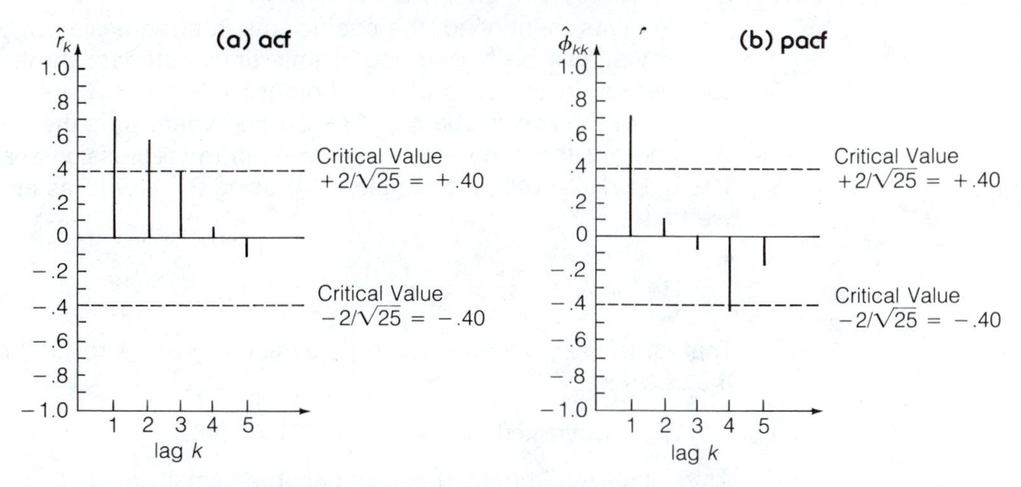


month, and (2) a random number of new subscribers in the current month. This scenario could be modeled as

| $y_t$ | = | $\beta_1 y_{t-1}$ | + | $\epsilon_t$ |
|---|---|---|---|---|
| subscribers in month $t$ | | old subscribers from month $t - 1$ ($\beta_1 < 1$) | | random number of new subscribers in month $t$ |

From the graphs of the acf and pacf in Figure 4.15 we see that both conditions for a first-order autoregressive model are satisfied. The acf decays exponentially, and the pacf cuts off sharply after lag 1 (the pacf does have a spike at lag 4 that barely crosses the rule-of-thumb line, but since $\hat{\phi}_{44}$ is on the borderline we will ignore it for the sake of parsimony).

Performing a regression of $y_t$ on $y_{t-1}$ gives the estimated first-order model as

$$\hat{y}_t = 14.44 + .715y_{t-1}$$

Notice that $\hat{\beta}_1 = .715$, which agrees closely with the result $\hat{\phi}_{11} = .713$ from Figure 4.15 (for which the acf and pacf were generated by computer). In addition, the constant term, $\hat{\beta}_0 = 14.44$, should not be confused with the mean level of the series, which we know to be around 50.6. In Chapter 7, we will show that the expected value of a first-order autoregressive model is

$$E(y_t) = \frac{\beta_0}{1 - \beta_1}$$

In this example, then,

$$E(y_t) \approx \frac{\hat{\beta}_0}{1 - \hat{\beta}_1} = \frac{14.44}{1 - .715} = 50.67$$

which is very close to 50.6.

To forecast with this model, we use either the *actual* value of $y_{t-1}$ or *forecasts* of $y_{t-1}$, depending on whether or not $y_{t-1}$ is itself a future value. For example, forecasts made at time $t = 25$ for periods 26, 27, and 28 would be calculated as follows:

$$\begin{aligned}
\hat{y}_{26}(25) &= 14.44 + .715y_{25} &&= 14.44 + .715 \cdot (51.0) = 50.91 \\
\hat{y}_{27}(25) &= 14.44 + .715\hat{y}_{26}(25) &&= 14.44 + .715 \cdot (50.91) = 50.84 \\
\hat{y}_{28}(25) &= 14.44 + .715\hat{y}_{27}(25) &&= 14.44 + .715 \cdot (50.84) = 50.79
\end{aligned}$$

●

As was mentioned, the coefficients in a regression of $y_t$ on $y_{t-1}, y_{t-2}, \ldots, y_{t-p}$ can be biased and slightly *underestimate* the actual regression parameters. In the case of a first-order model $y_t = \beta_0 + \beta_1 y_{t-1} + \epsilon_t$, the amount of the bias is about $-(1 + 3\beta_1)/n$, where $\beta_1$ is the true coefficient of $y_{t-1}$ and $n$ is the number of terms used in the regression analysis (Kennedy, 1983, pp. 102–103). In Example 4.14, using $\hat{\beta}_1 = .715$ as an estimate of $\beta_1$, we find

$$\text{bias in } \hat{\beta}_1 \approx \frac{-(1 + 3 \times (.715))}{24} = -.131$$

That is, $\hat{\beta}_1$ may underestimate $\beta_1$ by about .131. A revised estimate would then be

$$\hat{\beta}_1 + (\text{estimated bias}) = .715 + .131 = .846$$

As $n$ increases, though, the bias becomes small, in which case we generally accept the regression estimates without revising them.

## 4.8 Diagnostics

### 4.8.1 Introduction

As always, assessment of a forecasting model's adequacy is based on two things:

- **subjective knowledge** the analyst's understanding of the mechanism generating the series and the factors influencing the series
- **empirical knowledge** an evaluation of the model's performance over part of the history of the series

At a general level, the diagnostic practices outlined in Section 2.11 can be used in the evaluation of regression models. However, more specialized procedures are needed to check the various regression assumptions and to analyze the functional form of the models we use. Besides these new procedures, many of the diagnostic tests and graphs from previous chapters are also extended or altered to handle regression diagnostics.

## 4.8.2 Subjective Knowledge of the Series

An important tool for the subjective evaluation of regression models lies in the way in which the regression coefficients can be interpreted as partial rates of change. That is, $\hat{\beta}_i$ represents the estimated amount $y$ changes for each unit increase in $x_i$, if all other independent variables can be held constant. As subjective criteria, then, both the *sign* and *magnitude* of each $\hat{\beta}_i$ should agree with the analyst's knowledge of the series.

To illustrate, consider the model used for predicting sales volume from advertising expenditure in Example 4.2 (Equation 4.10):

$$\hat{y}_t = 1{,}161.6 + 5.783x_{t-1} + 7.945x_{t-2}$$

Using the partial rate-of-change interpretation, the analyst must decide if the following statement makes sense: Each $1,000 (i.e., one-unit) increase in last month's ad expense, two-month old expenditures being constant, should result in about a $5,783 (i.e., 5.783-unit) increase in sales. The fact that the coefficients of $x_{t-1}$ and $x_{t-2}$ are positive corresponds to our intuitive notion that increases in advertising should cause increases in sales. A negative coefficient here would need some explaining and could be reason to question the model's adequacy. The size of the above coefficients also seems believable, since advertising in adjacent months should have slightly different impacts on sales but should otherwise be of the *same order of magnitude* (a result like $\hat{\beta}_1 = .57$ and $\hat{\beta}_2 = 79.4$ would have immediately been suspicious).

Interpretation using partial rates of change is best done only when all the predictor variables are distinct. When different functions of the *same* variable are used more than once in a model, the practice is less helpful. An example of this occurred when the second-order model

$$\hat{y} = 11.691 - 0.187x + 0.037x^2$$

was fit to the *Practical Medicine* data. In this model it is impossible to let $x$ vary while holding all else ($x^2$) constant, so the rate-of-change interpretation doesn't apply, and a negative $\hat{\beta}_1$ coefficient should not necessarily be of concern.

## 4.8.3 Empirical Knowledge of the Series, Residual Analysis

### Objectives of Diagnostic Tests

Regression diagnostics are designed to test both for deviations from regression assumptions and for possible misspecification of the form of the model. These tests search primarily for:

- *nonrandomness* in the residuals
- *heteroscedasticity*—changes in the variances of the residuals
- *nonnormality* of the residuals
- *multicollinearity*—highly correlated predictor variables
- *misspecification*—wrong functional form of the model, and/or important predictor variables missing in the model
- *outliers*—observations inconsistent with the majority of the data

## Graphical Methods

Graphical diagnostic methods include the plots discussed in Section 2.11.4 (residuals versus time; acf of the residuals), graphs involving the predictor variables (residuals versus each $x_i$), and graphs using the predicted series (residuals versus $\hat{y}_t$). Plots of residuals versus the actual series are *not* used though, because $e_t$ and $y_t$ are always correlated (in fact, $r_{e,y} = \sqrt{1 - R^2}$; Draper and Smith, 1981, sec. 3.3).

Interpreting these graphs is done by referring to a set of standard patterns, each typical of a different problem or circumstance (see Figure 4.16). Essentially, a good regression model should result in residuals whose graphs against $t$, $\hat{y}_t$, and each $x_i$, look like random, no-trend series (Figure 4.16(a)). Patterns in these graphs that are similar to those in parts (b), (c), and (d) indicate problems with the model. Table 4.11 summarizes the conclusions we can draw when any of the standard plots in Figure 4.16 are encountered.

**Figure 4.16 Standard Shapes of Graphs of Residuals from Regression Analyses**

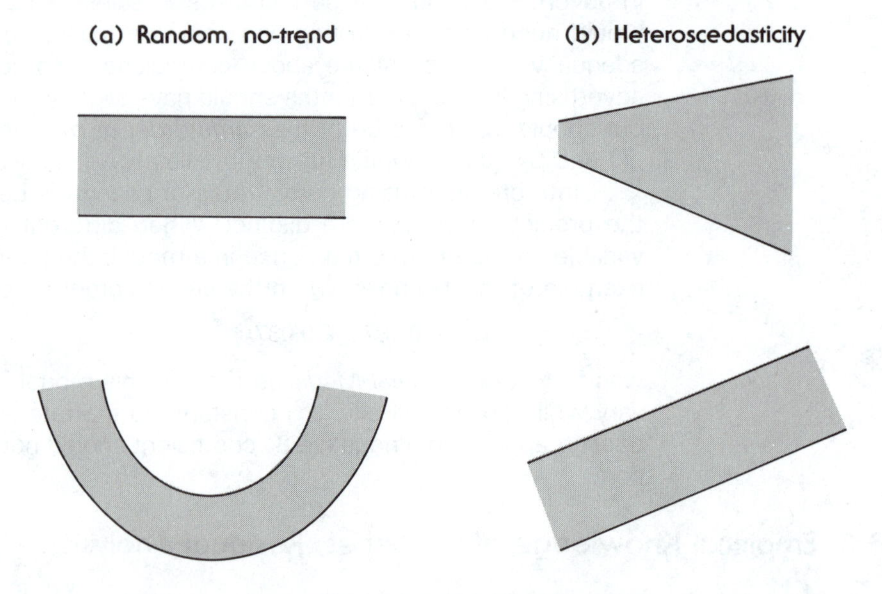


As an example of the application of these graphs, recall the first- and second-order models developed for the *Practical Medicine* data. The first-order model of sales ($y$) regressed on percent of population over 50 ($x$) was

$$\hat{y} = 6.601 + 0.961x$$

Residuals from this model plotted against $t$, $x$, and $\hat{y}$ are given in Figure 4.17. Notice that the graph of $e_t$ versus $t$ has a slight dip, which by itself is not strong evidence of a problem. The graphs of $e_t$ versus $x$ and $\hat{y}$, on the other hand, provide very convincing indications of the need for higher-order terms (Table 4.11(c), (2) and (3)). Thus, even though both the $t$- and $F$-statistics are significant (even at $\alpha = 1\%$), the first-order model is inappropriate, because the residual analysis contains strong evidence of misspecification in the form of a lack of higher-order terms in the model. After fitting the second-order model

$$\hat{y} = 11.691 - 0.187x + 0.037x^2$$

all three residual graphs resemble Figure 4.16(a), and we can conclude that simply adding $x^2$ to the model seems to have accounted for the missing higher-order terms.

**Table 4.11 Interpretation of Standard Residual Plots**

| Variables Plotted | Resulting Pattern: | Implications |
|---|---|---|
| **(a)** (1) $e_t$ vs $t$<br>(2) $e_t$ vs $x_{it}$<br>(3) $e_t$ vs $\hat{y}_t$ |  | (a) *No Problems Indicated*<br>Goal is to have all residual plots look like random no-trend series. |
| **(b)** (1) $e_t$ vs $t$<br>(2) $e_t$ vs $x_{it}$<br>(3) $e_t$ vs $\hat{y}_t$ | 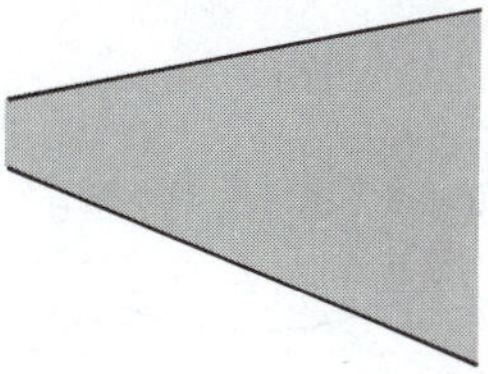 | (b) *Heteroscedasticity*<br>In each case, some transformation of $y$ and/or $x$ data is needed prior to regression. |
| **(c)** (1) $e_t$ vs $t$<br>(2) $e_t$ vs $x_{it}$<br>(3) $e_t$ vs $\hat{y}_t$ | 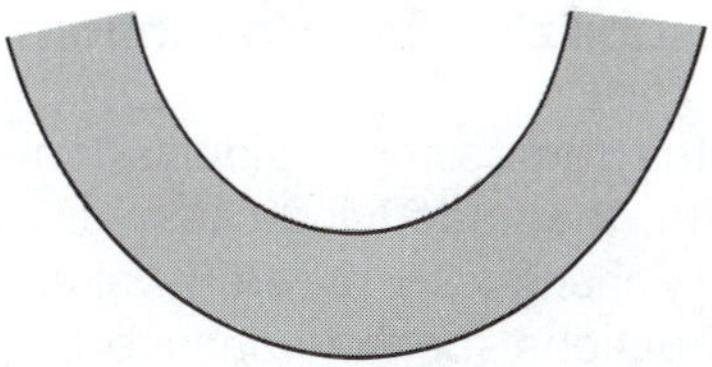 | (c) *Misspecification*<br>(1) Linear and higher-order terms in *time* are needed.<br>(2) Linear and higher-order terms in $x_i$ are needed.<br>(3) Higher-order terms are needed, or $y$ data need transforming prior to regression. |
| **(d)** (1) $e_t$ vs $t$<br>(2) $e_t$ vs $x_{it}$<br>(3) $e_t$ vs $\hat{y}_t$ | 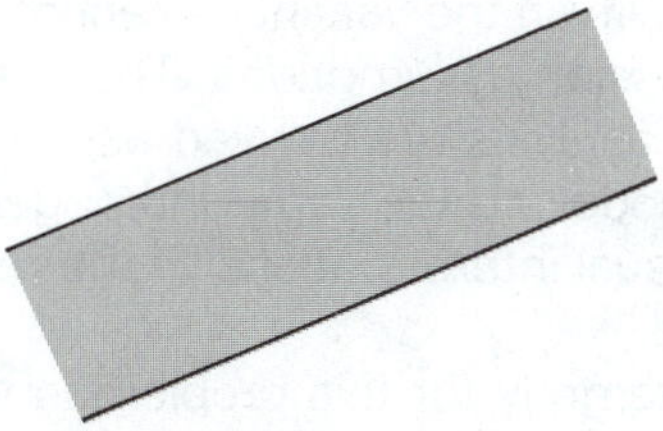 | (d) *Misspecification*<br>(1) Linear term in *time* is needed.<br>(2) Linear term in $x_i$ is needed.<br>(3) If a model without $\beta_0$ term was used, this plot indicates that $\beta_0$ is needed. |

**Figure 4.17** Regression Residual Plots for *Practical Medicine,* First-Order Model

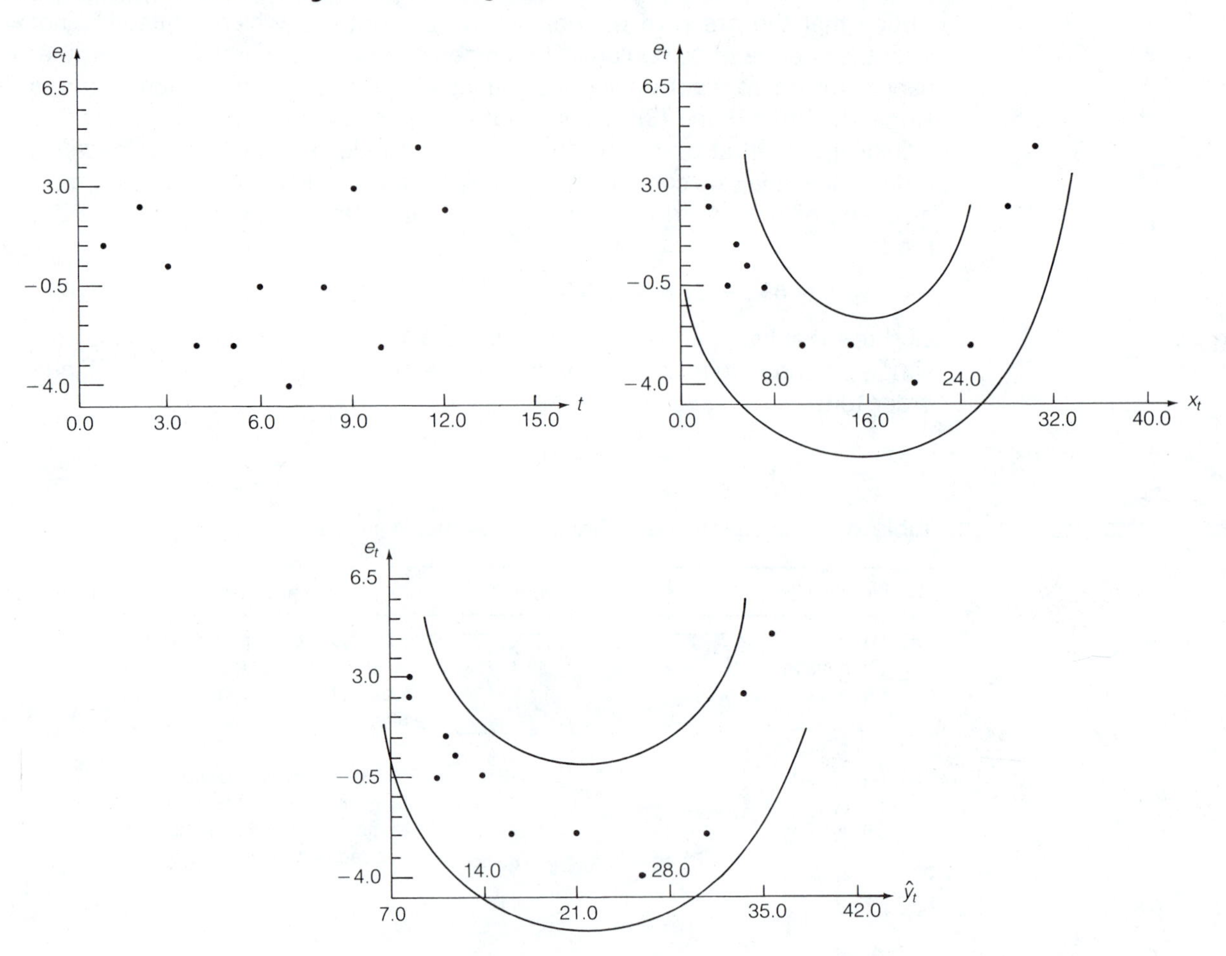


## Variance Stabilizing Transformations

When plotted regression residuals look like the plots in Figure 4.16(b) and Table 4.11(b), the underlying cause is usually heteroscedasticity; that is, the variability in $y$ changes for different values of the $x$ variable(s). It is most common in these situations for the variance of $y$ to grow as the average level of $y$ increases.

Instability in the variances cannot be ignored, since it violates one of the basic regression assumptions. The statistical tests and interval estimates in regression analysis are derived assuming that the variances around the regression model are the same (homogeneous). When that assumption fails to hold, the usual interpretations of the resulting statistics and estimates are no longer valid.

One remedy for this problem is to apply a *variance-stabilizing transformation* to the $y$ variable and then refit the regression model. If the right transformation is used, the residuals from the new model should not show signs of heteroscedasticity. The transformation is then said to have "stabilized" the variances. For a majority of cases met in practice, one of three transformations is used to achieve this result: $\sqrt{y}$, log ($y$), and $1/y$. There are many other types of transformations available, most notably the Box-Cox power transformations $\frac{y^\lambda - 1}{\lambda}$ (Box and Cox, 1964), but they are not discussed here. The table nearby lists these three transformations and the situations to which each is best suited. Since error variances that change with different levels of $x$ must also change with different levels of $\hat{y}$, the table expresses the variance of the residuals in terms of $\hat{y}_i$, the regression estimate of $y$ when $x = x_i$. In essence, $\sqrt{y}$ is used when the error variances don't change too rapidly, log ($y$) is used to handle more rapid changes, and $1/y$ is reserved for situations in which the variances grow explosively with changing $\hat{y}_i$ values. For series based on economic data, the log ($y$) transformation often turns out to be the best choice.

**Variance-Stabilizing Transformations**

| Pattern in the Residuals | Description | Transformation |
|---|---|---|
| $\sigma_{e_i} = k\sqrt{\hat{y}_i}$ | Variability grows at a rate proportional to the square root of $y_i$. | $\sqrt{y}$ |
| $\sigma_{e_i} = k\hat{y}_i$ | Variability grows in direct proportion to the size of the $y$ variable. | log ($y$) |
| $\sigma_{e_i} = k(\hat{y}_i)^2$ | Variability grows at a rate proportional to the square of $y_i$. | $\frac{1}{y}$ |

## Normality

Regression analysis assumes that the residuals follow a normal distribution with a mean of zero. In addition, when the regression model includes a constant term $\beta_0$, the least squares estimates yield residuals that sum to zero, that is, $\Sigma\, e_t = 0$. Together these facts imply that, at least for fairly large samples, a histogram of the residuals should resemble a normal distribution centered at zero. In most practical situations, it suffices to create a histogram of the $e_t$'s and examine it for any obvious departures from normality, such as being highly skewed or not being unimodal.

In another version of the histogram method, which facilitates checking for nonnormality, all the residuals are divided by the standard error $s_\epsilon$ to form **standardized residuals:**

$$\frac{e_t}{s_\epsilon} = \frac{y_t - \hat{y}_t}{s_\epsilon}$$

If the residuals follow a normal distribution, then the standardized residuals follow the standard normal distribution (mean = 0, standard deviation = 1), and the familiar standard normal distribution percentiles serve as a reference:

- About 99.7% of the standardized residuals should fall between +3 and −3.
- About 95% should be between +2 and −2.
- About 68% should be between +1 and −1.

With the histogram method, as with most tests of normality, the larger the data set, the more confidence we can have in the resulting diagnostics. For this reason, and because with small data sets it is hard to decide about the normality hypothesis, we will confine ourselves to using histograms only when searching for serious departures from normality.

### 4.8.4 Durbin-Watson Test

When error terms in regression are not independent, violating assumption 4, it is most often due to the presence of autocorrelation. The main difficulty caused by autocorrelation is that $s_\epsilon$ tends to underestimate $\sigma_\epsilon$, leading to unduly large values of $t$, $F$, and $R^2$ statistics. These inflated values can in turn yield overly optimistic conclusions about how well the model fits the data (cf. Section 4.4).

One method of checking for autocorrelated residuals is the **Durbin-Watson test** (Durbin and Watson, 1951). It is very closely related to the test based on von Neumann's ratio (the mean square successive difference test) discussed in Section 2.6.3. The Durbin-Watson test is used only on regression residuals, though, while the von Neumann test can apply to any time series other than regression residuals:

| Durbin-Watson Statistic $d$ | von Neumann Ratio $M$ |
|---|---|
| *Regression Residuals* | *Any Other Series* |
| $d = \dfrac{\sum_{2}^{n} (e_t - e_{t-1})^2}{\sum_{1}^{n} e_t^2}$ | $M = \dfrac{\sum_{2}^{n} (y_t - y_{t-1})^2}{\sum_{1}^{n} (y_t - \bar{y})^2}$ |

**Test Procedure:** *Durbin-Watson Test*

$H_0$: Residuals are independent.
$H_a$: Residuals are positively autocorrelated.

**Test Statistic:** $d = \dfrac{SS_{\Delta e}}{SS_{ee}}$

where $SS_{\Delta e} = \sum_{2}^{n} = (e_t - e_{t-1})^2 \qquad SS_{ee} = \sum_{1}^{n} e_t^2$

(*Note:* $SS_{ee}$ = SSE from ANOVA table.)

**Decision Rule:** (Level = $\alpha$, sample size = $n$, number of predictors = $k$)

Reject $H_0$ if $d < d_L$ (Appendix II)

Do not reject $H$ if $d > d_U$.

Otherwise, if $d_L < d < d_U$, the test is inconclusive.

**Conclusion:** If $H_0$ is rejected with $d < d_L$, we conclude there is positive autocorrelation in the residuals.

If $H_0$ is not rejected with $d > d_U$, we have some support for concluding the error terms are independent.

The test is inconclusive if $d_L < d < d_U$. Some other method of testing for autocorrelation should be used.

---

***Note:*** When testing for negative autocorrelation, $4 - d$ is substituted for $d$ in the above procedure. Also, a two-sided test may be performed by running both one-sided tests, which doubles $\alpha$.

If applied to the same series of regression residuals, the two statistics are identical, i.e., $d = M$, which can be seen by substituting $e_t$ for $y_t$ in $M$ and using the fact that $\bar{e} = 0$ for regression models with a $\beta_0$ term. The distinction between the two statistics lies in the tables used for hypothesis testing. The Durbin-Watson tables allow for $k$ predictor variables, while the von Neumann tables do not. For this reason, of the two tests, only the Durbin-Watson test is used for analyzing regression residuals.

The Durbin-Watson statistic *d always lies between 0 and 4.* Its interpretation is similar to that of the von Neumann ratio. Values close to 0 indicate positive autocorrelation, values close to 4 indicate negative autocorrelation, and values close to 2 indicate independent error terms.

Hypothesis tests involving $d$ are slightly different from most hypothesis tests in that there is no clear boundary between the rejection and nonrejection regions. Instead, there is a zone between these regions where a decision cannot be reached (see Figure 4.18).

**Figure 4.18 Indecision Zone for the Durbin-Watson Test**

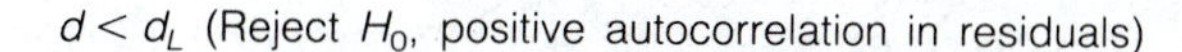

**Note:** $d_L$ and $d_U$ (the lower and upper boundaries of the indecision zone) depend on sample size $n$, significance level $\alpha$, and number of predictor variables $k$. Values for $d_L$ and $d_U$ are found in the Durbin-Watson tables, Appendix II. For negative autocorrelation, $4 - d$ is used in place of $d$.

**EXAMPLE 4.15** ***Durbin-Watson Test, Random Walk Data***

The random walks in Table 4.3 were used to illustrate some of the dangers of regressing one time series on another. The residuals from a regression of $y$ on $x$ for these data are recorded in Table 4.12. Here we use the Durbin-Watson test to check for positive autocorrelation in the $e_t$ series.

**Table 4.12 Residuals from Regression of $y$ on $x$ (Data from Table 4.3)**

| Observation $t$ | Observed Value $y_t$ | Predicted Value $\hat{y}_t$ | Residual $e_t$ | Observation $t$ | Observed Value $y_t$ | Predicted Value $\hat{y}_t$ | Residual $e_t$ |
|---|---|---|---|---|---|---|---|
| 1 | 49.7 | 50.330 | −.630 | 11 | 48.6 | 50.226 | −1.626 |
| 2 | 50.5 | 50.383 | .117 | 12 | 50.0 | 50.750 | −.750 |
| 3 | 52.5 | 50.697 | 1.803 | 13 | 49.8 | 49.597 | .203 |
| 4 | 51.7 | 50.750 | .950 | 14 | 49.1 | 48.864 | .236 |
| 5 | 51.7 | 50.802 | .898 | 15 | 49.9 | 49.597 | .303 |
| 6 | 51.6 | 50.645 | .955 | 16 | 49.2 | 49.126 | .074 |
| 7 | 48.5 | 49.492 | −.992 | 17 | 50.0 | 48.549 | 1.451 |
| 8 | 50.0 | 49.859 | .141 | 18 | 49.9 | 48.078 | 1.822 |
| 9 | 47.4 | 49.440 | −2.040 | 19 | 47.8 | 48.340 | −.540 |
| 10 | 47.8 | 49.440 | −1.640 | 20 | 48.6 | 49.335 | −.735 |

$H_0$: Residuals are independent.

$H_a$: Residuals are positively autocorrelated.

**Test Statistic:**

$$d = \frac{SS_{\Delta e}}{SS_{ee}} = \frac{23.509}{23.794} = .988$$

$$SS_{\Delta e} = \sum_{2}^{20} (e_t - e_{t-1})^2 = [.117 - (-.630)]^2 + (1.803 - .117)^2 + \cdots + [-.735 - (-.540)]^2 = 23.509$$

$$SS_{ee} = \sum_{1}^{20} e_t^2 = (-.630)^2 + (.117)^2 + \cdots + (-.735)^2 = 23.794$$

**Decision Rule:** ($\alpha = 5\%$, $n = 20$, $k = 1$)

Reject $H_0$ if $d < d_L = 1.20$. (Appendix II)

Do not reject if $d > d_U = 1.41$.

No conclusion if $1.20 < d < 1.41$.

**Conclusion:** Since $d = .988 < 1.20$, we reject $H_0$ and conclude that the residuals are positively autocorrelated. ●

As a final note, we point out that the Durbin-Watson test is actually a test for first-order (i.e., lag 1) autocorrelation. Because of this, just as with the von Neumann ratio, there is an approximate relationship between the lag 1 autocorrelation coefficient $r_1$ and $d$:

$$r_1 \simeq 1 - d/2$$

Since almost all computer programs for regression analysis also calculate the Durbin-Watson statistic, another way of *quickly* testing for autocorrelation is to use the rule-of-thumb test of Chapter 2 with $1 - d/2$ as an estimate of $r_1$. In the last example, then, we would have estimated $r_1 = 1 - .988/2 = .51$. Comparing this to the rule-of-thumb value of $\pm 2/\sqrt{20} = .45$, we would have concluded that positive autocorrelation was present. The virtue of this quick procedure is that no tables are needed to perform the test.

## 4.8.5 Remedies for Autocorrelation

If either the Durbin-Watson test or the acf of the residuals indicates autocorrelation, the next step is to discover the reason for this problem and take corrective measures. Common causes of autocorrelation are:

- The wrong trend model was used (e.g., a linear trend fit to curvilinear data will produce positively autocorrelated residuals).
- An important variable was omitted from the model.
- The series were not detrended prior to regression.

Discovering that the wrong trend model was used is usually just a matter of looking at the graph of $e_t$ versus $t$. The pattern in this graph will also help determine a more appropriate trend function to use.

Knowing which predictor variables to include is, of course, the key to building a good regression model. If autocorrelation exists and if all other precautions (detrending prior to regression, correctly modeling trends, etc.) have been taken, then the only recourse is to continue the search for good predictor variables. Before abandoning the current set of variables, however, the possibility of reexpressing them as rates, percentages, etc. should be examined.

### Generalized Differencing

Finally, a common procedure for removing autocorrelation is to make sure that the data are detrended before regression (either by differencing or fitting a trend model). **Generalized differencing** is one of the most popular detrending methods. To discuss this technique, we will need to temporarily break with our usual subscripting convention and use $x_{i,t}$, rather than $x_{it}$, to denote the value of the series $x_i$ at time $t$.

The rationale for generalized differencing is this: Lag 1 autocorrelation in the residuals means that there is a nonzero correlation $\rho$ between the errors $\epsilon_t$ and the lag 1 errors $\epsilon_{t-1}$, and we can write

$$\epsilon_t = \rho\epsilon_{t-1} + u_t$$

where the $u_t$'s are independent, with a mean of zero. Taking the original model

$$y_t = \beta_0 + \beta_1 x_{1,t} + \cdots + \beta_k x_{k,t} + \epsilon_t$$

and lagging it one period, we get

$$y_{t-1} = \beta_0 + \beta_1 x_{1,t-1} + \cdots + \beta_k x_{k,t-1} + \epsilon_{t-1}$$

Multiplying the lagged equation by $\rho$ and subtracting the result from the original model yields

$$y_t - \rho y_{t-1} = \beta_0(1 - \rho) + \beta_1(x_{1,t} - \rho x_{1,t-1}) + \cdots + \beta_k(x_{k,t} - \rho x_{k,t-1}) + (\epsilon_t - \rho\epsilon_{t-1})$$

This equation represents a regression of $y_t^* = y_t - \rho y_{t-1}$ on the variables

$$x_{1,t}^* = x_{1,t} - \rho x_{1,t-1}, \quad x_{2,t}^* = x_{2,t} - \rho x_{2,t-1}, \ldots, \quad x_{k,t}^* = x_{k,t} - \rho x_{k,t-1}$$

where the error terms $u_t = \epsilon_t - \rho\epsilon_{t-1}$ are independent, i.e., no longer autocorrelated. Variables of the form $y_t^* = y_t - \rho y_{t-1}$ and $x_{1,t}^* = x_{i,t} - \rho x_{i,t-1}$ are called *generalized differences.* In practice, when performing this regression, $\rho$ is usually replaced by an estimate, $r_1$.

**Generalized Differencing** To Eliminate First-Order Autocorrelation in Regression Residuals

**Given:** A sample of $n$ observations on the variables $y, x_1, x_2, \ldots, x_k$

**Generalized Differences:** Let

$$y_t^* = y_t - r_1 y_{t-1}$$
$$x_{1,t}^* = x_{1,t} - r_1 x_{1,t-1}$$
$$\vdots$$
$$x_{k,t}^* = x_{k,t} - r_1 x_{k,t-1}$$

where $r_1$ is an estimate of the first-order autocorrelation coefficient of the residuals from regressing $y$ on $x$'s. ($r_1$ is obtained from the acf of the residuals or by using the approximation $r_1 \approx 1 - d/2$, where $d$ is the Durbin-Watson statistic.)

**Estimates:** Regress $y^*$ on $x_1^*, x_2^*, \ldots, x_k^*$ to obtain the fitted equation

$$\hat{y}^* = \hat{\beta}_0 + \hat{\beta}_1 x_1^* + \hat{\beta}_2 x_2^* + \cdots + \hat{\beta}_k x_k^*$$

**Forecasts:** Substitute $y_t - r_1 y_{t-1}$ for $y^*$ and $x_{i,t} - r_1 x_{i,t-1}$ for $x_i^*$ (for $i = 1$ to $k$) to obtain the forecasting equation.

## EXAMPLE 4.16 *Generalized Differencing*

Suppose, to explain the behavior of a response variable $y$, an analyst has considered the use of a certain predictor variable $x$. After gathering some data on $x$ and $y$ (Table 4.13), a regression of $y$ on $x$ was performed and the following results obtained:

$$\hat{y} = 6.26 + .866x$$

$t = 10.53 \quad F = 110.9 \quad R^2 = .798$

Durbin-Watson statistic $d = .96$

Table 4.13 Data on $y$ and $x$ for Example 4.16

| $t$ | $y_t$ | $x_t$ | $t$ | $y_t$ | $x_t$ |
|---|---|---|---|---|---|
| 1 | 12.06 | 10.1 | 16 | 41.38 | 39.1 |
| 2 | 17.59 | 15.5 | 17 | 41.41 | 43.5 |
| 3 | 23.28 | 16.3 | 18 | 53.35 | 44.3 |
| 4 | 23.59 | 19.9 | 19 | 51.05 | 48.2 |
| 5 | 20.99 | 18.3 | 20 | 47.00 | 54.0 |
| 6 | 15.98 | 21.3 | 21 | 50.26 | 51.6 |
| 7 | 20.85 | 25.2 | 22 | 56.54 | 56.3 |
| 8 | 32.91 | 26.2 | 23 | 58.85 | 53.8 |
| 9 | 35.59 | 29.1 | 24 | 49.18 | 54.9 |
| 10 | 33.42 | 24.8 | 25 | 37.79 | 63.0 |
| 11 | 29.44 | 34.2 | 26 | 46.67 | 65.8 |
| 12 | 35.36 | 33.0 | 27 | 57.33 | 60.7 |
| 13 | 47.31 | 41.2 | 28 | 67.16 | 66.3 |
| 14 | 46.01 | 39.5 | 29 | 76.44 | 65.6 |
| 15 | 47.73 | 40.9 | 30 | 77.02 | 67.5 |

While the $t$- and $F$-statistics were significant ($\alpha = 1\%$) and $R^2$ was fairly large, the Durbin-Watson statistic signalled the presence of positive autocorrelation in the residuals ($d_L = 1.35$ for $k = 1$, $n = 30$, $\alpha = .05$). In addition, the acf of the residuals gave $r_1 = .475$, which also pointed to autocorrelation, since .475 exceeds the rule-of-thumb value of $2/\sqrt{30} = .365$.

To remove the autocorrelation, generalized differencing was applied, with

$$y_t^* = y_t - .475y_{t-1} \qquad \text{and} \qquad x_t^* = x_t - .475x_{t-1}$$

The results of regressing $y^*$ on $x^*$ were:

$$\hat{y}^* = 5.20 + .796x^*$$

$$t = 5.61 \qquad F = 31.5 \qquad R^2 = .539$$

Durbin-Watson statistic $d = 1.42$

Since the Durbin-Watson statistic now fell in the indecision zone ($d_L = 1.34$, $d_U = 1.48$ for $n = 29$, $k = 1$, $\alpha = 5\%$), there was no longer conclusive evidence of autocorrelated residuals. Checking the acf, $r_1$ was .271, which was also not significant (rule-of-thumb value $= 2/\sqrt{29} = .371$).

The forecasting model was, then,

$$\hat{y}_t - .475y_{t-1} = 5.20 + .796(x_t - .475x_{t-1})$$

or

$$\hat{y}_t = 5.20 + .475y_{t-1} + .796x_t - .378x_{t-1}$$

and a forecast for time period 31, if $x_{31}$ were 66.9, would be

$$\hat{y}_{31} = 5.20 + .475 \times 77.02 + .796 \times 66.9 - .378 \times 67.5 = 69.5$$

### 4.8.6 Theil's Decomposition

Whenever a least squares regression model is used to generate the predicted series, Theil's decomposition of the MSE (Section 1.5.8) takes on a special form. The values of $U_M$ (bias), $U_R$ (regression), and $U_D$ (disturbance) are 0, 0, and 1, respectively, provided a constant term is included in the model. Essentially, regression models avoid the systematic biases that $U_M$ and $U_R$ are designed to detect.

The $U_M$ component is zero because the average regression residual $\bar{e}$ is zero (for models with a $\beta_0$ term), so

$$0 = \bar{e} = \frac{1}{n}\Sigma\, e_t = \frac{1}{n}\Sigma\,(y_t - \hat{y}_t)$$
$$= \frac{1}{n}\Sigma\, y_t - \frac{1}{n}\Sigma\, \hat{y}_t = \bar{y} - \bar{\hat{y}}$$

and

$$U_M = \frac{(\bar{\hat{y}} - \bar{y})^2}{\text{MSE}} = \frac{0^2}{\text{MSE}} = 0$$

To show that $U_R = 0$, we write

$$R^2 = \frac{\text{SSR}}{\text{SST}} = \frac{\Sigma\,(\hat{y}_t - \bar{y})^2}{\Sigma\,(y_t - \bar{y})^2}$$

Then, because $\bar{y} = \bar{\hat{y}}$, as shown above,

$$R^2 = \frac{\Sigma\,(\hat{y}_t - \bar{\hat{y}})^2}{\Sigma\,(y_t - \bar{y})^2} = \frac{(1/n)\,\Sigma\,(\hat{y}_t - \bar{\hat{y}})^2}{(1/n)\,\Sigma\,(y_t - \bar{y})^2} = \frac{(s'_{\hat{y}})^2}{(s'_y)^2}$$

Finally, $R^2$ is always the square of the correlation $r$ between $y_t$ and $\hat{y}_t$, so

$$r = \frac{s'_{\hat{y}}}{s'_y}$$

and

$$U_R = \frac{(s'_{\hat{y}} - rs'_y)^2}{\text{MSE}} = \frac{(s'_{\hat{y}} - (s'_{\hat{y}}/s'_y)s'_y)^2}{\text{MSE}} = \frac{(s'_{\hat{y}} - s'_{\hat{y}})^2}{\text{MSE}} = 0$$

It now follows that $U_D = 1$, since $U_M + U_B + U_D = 1$.

Given these results for Theil's decomposition, it is natural to wonder if regression models are also optimal for Theil's uncertainty coefficient $U$. That is, will a regression model *always* outperform a no-change model and lead to $U < 1$? In general, the answer is no. Series that exhibit large cyclical swings, for example, would be better handled by the no-change model than by a simple regression model.

There is one case, though, in which regression will always do at least as well as the no-change model, and that is when an autoregressive model is used. Autoregressive models minimize the sum of squared forecast errors for the class of models

$$\hat{y}_t = \hat{\beta}_0 - \hat{\beta}_1 y_{t-1} - \cdots - \hat{\beta}_p y_{t-p}$$

of which the no-change model $\hat{y}_t = 0 + 1 \cdot y_{t-1}$ is a member. Therefore, $U$, being the ratio of the model sum of squares to the no-change sum of squares, cannot exceed 1.0.

## 4.8.7 Multicollinearity

In regression studies it can easily happen that two or more of the predictor variables in a model are highly correlated with one another. Highly correlated $x$ variables can arise (Kennedy, 1983, sec 9.1) in models that use any one of the following:

- too many lagged values of the same variable
- too many powers (i.e., $x^2$, $x^3$, etc.) of the same variable
- variables having a strong trend in common
- data collected over too small a range of values
- two predictor variables that are, in fact, linearly related

The main problem, under these circumstances, is that the least squares estimates may be imprecise. Knowing that the parameter estimates may be poor, we should then have less faith in the resulting forecasting equation. Correlated predictor variables and the estimation problems they cause are collectively referred to as *multicollinearity* (Judge et al., 1980).

To see why the parameter estimates may be bad, remember that, in estimating a particular $\beta_i$, the least squares procedure uses only the information in the data over and above that provided by the other model variables. Thus, if the other predictors account for much of the same variation in $y$ that $x_i$ accounts for, there is little information left for accurately estimating $x_i$'s contribution. The result is the same as for any estimate based on few data: The variance of the estimate will be high, causing interval estimates to be unacceptably wide and hypothesis tests to have little power.

In practice, the problems caused by multicollinearity include:

- Important predictors may be dropped from the model because their partial $t$-values are not significant.
- Because $\hat{\beta}_i$ is imprecise, interpreting it as a partial rate of change will be a much less reliable technique.
- Sharp changes in the parameter estimates may occur when a predictor is dropped from the model.
- Interval estimates may be too large to be of practical value.

Unfortunately, there are no proven statistical tests for detecting multicollinearity, but there are a few widely used rules of thumb. For example, we might conclude that multicollinearity is present if any one of the following is true:

- The simple correlation coefficient $r$ between two predictors *exceeds* the multiple correlation coefficient $R$ in the regression analysis.
- A variable you know to be important has an insignificant partial $t$-value.
- The $R^2$- and $F$-values are high, but most of the partial $t$-values are not significant.

As with any rule of thumb, the above rules have exceptions, but in any event, their use will most likely lead to an increased understanding of the forecasting model.

If it is decided that multicollinearity is present, there are two main remedies:

- Gather more data, and/or use "ridge regression" (see Judge et al, 1980).
- Drop one or more variables from the model.

If possible, gathering more data is preferred, since doing so will either increase the precision in the estimates or, failing that, pinpoint more clearly the cause of the multicollinearity. The second alternative, dropping variables, requires care, since when multicollinearity *is* present, it is not entirely clear exactly which variable should be dropped. Dropping the wrong variable can cause model misspecification, which biases estimates of the remaining parameters.

### 4.8.8 Outliers

An *outlier* is an observation in the data that differs noticeably from the other observations. Outliers, often called "extreme" values, can be detrimental to a statistical analysis. Because it is difficult to give exact criteria for deciding when a value is too big, too small, or, in general, too extreme, much research and even entire books have been devoted to the study of outliers. Such an undertaking is not warranted here, but a few important facts about the effects of outliers on regression models should be mentioned.

In regression, outliers are considered not so much extreme values as *influential* ones (Belsley, Kuh, and Welsch, 1980, ch. 2). That is, an outlier in regression could very well lie near the middle of the data set but still exert an unusually large influence on the regression line. This situation is shown in Figure 4.19. When the outlier (point A) is removed from the data set, the new regression line changes markedly, and it is in this sense that point A is said to be an influential observation. In addition, regression outliers tend to inflate $s_\epsilon$, which causes a reduction in the precision of the other estimates.

**Figure 4.19 Outliers in Regression Analysis**

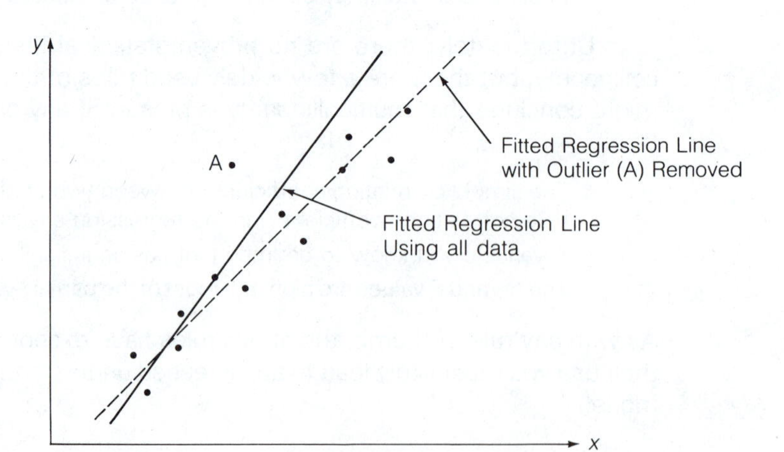

Outliers that result from data-recording errors or bad measurement techniques should be deleted from the data. The problem with deleting suspected outliers, however, is that you must be sure that the observation is really a bad one. Otherwise, if it was measured and recorded correctly, the reading may represent one of the most *important* pieces of data, perhaps pointing to some special, as yet undiscovered, feature of the relationship between the variables. A general rule for handling suspected outliers is:

Remove an observation from the data only if you have good reason to believe it is faulty (e.g., an obvious recording error), not solely because it differs from the rest of the data.

On intuitive grounds, an observation ought to be considered suspicious if any one of the following is true:

- A regression analysis using data with the observation removed results in a substantial change in the fitted equation.
- The standard error $s_\epsilon$ drops noticeably when the observation is removed.
- The standardized residual $e_i/s_\epsilon$ corresponding to the suspected outlier is large (say, 3 or more).

## 4.9 Exercises

**4.1** Why is $y_t = \beta_0 + \beta_1 x_t + \beta_2 x_t^2$ considered to be a *linear* regression model?

**4.2** Suppose that for a given set of data a particular explanatory variable $x_1$ is used in more than one model. Would you expect the coefficient of $x_1$ (i.e., $\beta_1$) to be the same from model to model?

**4.3** Suppose $y_t$ and $x_t$ are series having strong linear trends, with $r_{y,t} = .91$ and $r_{x,t} = .85$. If the $y_t$ series were regressed on $x_t$, what would you expect $r_{y,x}$ to be approximately?

**4.4** For the following data:

| $t$ | $y_t$ | $x_t$ | $t$ | $y_t$ | $x_t$ |
|---|---|---|---|---|---|
| 1 | 15 | 25 | 9 | 10 | 16 |
| 2 | 15 | 24 | 10 | 12 | 17 |
| 3 | 19 | 27 | 11 | 9 | 14 |
| 4 | 4 | 12 | 12 | 16 | 20 |
| 5 | 25 | 32 | 13 | 8 | 12 |
| 6 | 14 | 21 | 14 | 16 | 19 |
| 7 | 16 | 22 | 15 | 12 | 14 |
| 8 | 16 | 22 | | | |

a. Find the estimated regression model $\hat{y}_t = \hat{\beta}_0 + \hat{\beta}_1 x_t$.

b. Present the ANOVA table for this model.

c. Perform the *F*-test and *t*-test to see if $x_t$ is a significant predictor of $y_t$.

**4.5** In Problem 4.4, what is the relationship between the calculated *F*-statistic and the calculated *t*-statistic?

**4.6** A simple linear regression involving 15 observations is performed. Part of the resulting ANOVA table is given on page 310.

| SOURCE OF VARIATION | DF | SS | MS | F |
|---|---|---|---|---|
| REGRESSION | _____ | _____ | 850 | _____ |
| RESIDUALS | _____ | _____ | _____ | |
| TOTAL | _____ | 920 | | |

**a.** Fill out the remaining parts of the ANOVA table.
**b.** Find the calculated *t*-statistic for testing $H_0$: $\beta_1 = 0$.
**c.** Find the coefficient of determination.

**4.7** The coefficient of determination in a regression of *y* on *x* was .89 and it was noticed that *y* tended to decrease as *x* increased. What was the correlation coefficient between *x* and *y*?

**4.8** When trying to predict a response variable *y*, suppose that you consider 20 different independent variables and want to obtain the computer printouts for all possible regression models involving these variables (i.e., using each of the different subsets of the *x* variables). If, on the average, a printout for a model takes 15 seconds, what is the total time involved in printing out these results? (Disregard any time between printouts.)

**4.9** In Exercise 4.8, suppose you are leafing through the printouts for just the *one-variable* models and you are using $\alpha = .05$ to test each such model for significance. Suppose further that, in actuality, none of the variables is individually significant (i.e., that $\beta_1 = 0$ in each one-variable model). What is the probability that you will, nonetheless, find at least one printout that is significant, thus rejecting $H_0$: $\beta_1 = 0$ for that variable?

**4.10** An autoregressive model $y_t = 0.9y_{t-1} + 0.5y_{t-2}$ has been fit to $y_t$ = the number of monthly subscriptions to a local-area business periodical. The number of subscriptions in January and February were 6,000 and 6,185, respectively. What are the forecasts for the subscription levels in March, April, and May?

**4.11** One and a half year's worth of monthly data are used to fit a regression model for predicting a company's sales. A total of 16 independent variables (which incorporate general economic conditions, estimates of the competition, and internal company figures) are used. The resulting $R^2$ turns out to be .92. Using a 5% significance level, test whether the independent variables as a group contribute significantly to the prediction of total sales.

**4.12** The estimated regression line $\hat{y} = 5.98 + 0.10x$ was fit to the following data:

| *y* | *x* | *y* | *x* |
|---|---|---|---|
| 6.1 | 1 | 7.0 | 9 |
| 6.0 | 2 | 6.7 | 10 |
| 6.1 | 3 | 6.8 | 11 |
| 6.3 | 4 | 7.0 | 12 |
| 6.8 | 5 | 7.2 | 13 |
| 6.8 | 6 | 7.4 | 14 |
| 7.0 | 7 | 7.5 | 15 |
| 7.1 | 8 | | |

At the 5% significance level, test for positive autocorrelation in the residuals from this regression line.

**4.13** Regress $y$ on $x$ for each of the following four sets of data (adapted from *The Visual Display of Quantitative Information,* by E. R. Tufte, Graphics Press, 1983, pp. 13–14):

| Data Set 1 | | Data Set 2 | | Data Set 3 | | Data Set 4 | |
|---|---|---|---|---|---|---|---|
| $x$ | $y$ | $x$ | $y$ | $x$ | $y$ | $x$ | $y$ |
| 10.0 | 8.04 | 10.0 | 9.14 | 10.0 | 7.46 | 8.0 | 6.58 |
| 8.0 | 6.95 | 8.0 | 8.14 | 8.0 | 6.77 | 8.0 | 5.76 |
| 13.0 | 7.58 | 13.0 | 8.74 | 13.0 | 12.74 | 8.0 | 7.71 |
| 9.0 | 8.81 | 9.0 | 8.77 | 9.0 | 7.11 | 8.0 | 8.84 |
| 11.0 | 8.33 | 11.0 | 9.26 | 11.0 | 7.81 | 8.0 | 8.47 |
| 14.0 | 9.96 | 14.0 | 8.10 | 14.0 | 8.84 | 8.0 | 7.04 |
| 6.0 | 7.24 | 6.0 | 6.13 | 6.0 | 6.08 | 8.0 | 5.25 |
| 4.0 | 4.26 | 4.0 | 3.10 | 4.0 | 5.39 | 19.0 | 12.50 |
| 12.0 | 10.84 | 12.0 | 9.13 | 12.0 | 8.15 | 8.0 | 5.56 |
| 7.0 | 4.82 | 7.0 | 7.26 | 7.0 | 6.42 | 8.0 | 7.91 |
| 5.0 | 5.68 | 5.0 | 4.74 | 5.0 | 5.73 | 8.0 | 6.89 |

**a.** Construct the ANOVA table for each data set, then compute $r^2$, $s_\epsilon$, and the $t$-statistic.

**b.** Plot a scatter diagram for each data set. From these plots and the results in part (a), what can you conclude?

**4.14** For the following data:

| $t$: | 1 | 2 | 3 | 4 | 5 | 6 |
|---|---|---|---|---|---|---|
| $y_t$: | 13 | 40 | 20 | 35 | 30 | 45 |
| $x_t$: | 10 | 20 | 30 | 40 | 50 | 60 |

**a.** Calculate the Durbin-Watson statistic for the regression of $y$ on $x$.

**b.** Use the result of part (a) to estimate the autocorrelation coefficient of lag 1 ($r_1$).

**c.** Use the estimate from part (b) and generalized differencing to find a more suitable regression model for these data.

**4.15** As one adds more variables to a regression model, $R^2$ cannot decrease; it usually increases. Is there a corresponding statement that can be made concerning the overall $F$-statistic?

**4.16** In a simple linear regression, show that changing the units of measurement of the $x$ variable has no effect on SSR and, hence, no effect on the resulting ANOVA table. (*Hint:* changing the measurement units amounts to multiplying each value by some constant $c$.)

**4.17** Suppose one tries to estimate the trend in a series $y_t$ by simply regressing $y_t$ on $t$ (i.e., by fitting the model $y_t = \beta_0 + \beta_1 t$). Furthermore, suppose the ultimate goal is to estimate the incremental growth $\beta_1$ as accurately as possible. When requesting data to carry out this project, would it be best to get data at equally spaced time periods over the time span of interest or to request that more observations be taken near the extremes of the time span and fewer toward the middle? Justify your answer by using appropriate regression quantities, i.e., SSR, SSE, $SS_{tt}$, $SS_{yy}$, $s_\epsilon$, etc.

**4.18** For a time series of length 16, the following autoregressive models were fit:

$$\hat{y}_t = 73.5 - 0.603y_{t-1}$$
$$\hat{y}_t = 68.6 - 0.552y_{t-1} + 0.107y_{t-2}$$
$$\hat{y}_t = 49.1 - 0.446y_{t-1} + 0.262y_{t-2} + 0.158y_{t-3}$$
$$\hat{y}_t = 65.4 - 0.363y_{t-1} + 0.280y_{t-2} - 0.032y_{t-3} - 0.278y_{t-4}$$

From this information, evaluate the first four partial autocorrelation coefficients and draw this part of the pacf.

**4.19** Test for first-order positive autocorrelation in the residuals when $y_t$ is regressed on $x_t$ for the following data:

| $t$ | $x_t$ | $y_t$ | $t$ | $x_t$ | $y_t$ |
|---|---|---|---|---|---|
| 1 | 15 | 6 | 14 | 13 | 17 |
| 2 | 21 | 13 | 15 | 13 | 18 |
| 3 | 21 | 14 | 16 | 26 | 32 |
| 4 | 16 | 10 | 17 | 17 | 24 |
| 5 | 18 | 13 | 18 | 21 | 29 |
| 6 | 15 | 11 | 19 | 21 | 30 |
| 7 | 23 | 20 | 20 | 22 | 32 |
| 8 | 12 | 10 | 21 | 26 | 37 |
| 9 | 23 | 22 | 22 | 14 | 26 |
| 10 | 21 | 21 | 23 | 21 | 34 |
| 11 | 22 | 23 | 24 | 24 | 38 |
| 12 | 30 | 32 | 25 | 13 | 28 |
| 13 | 23 | 26 | | | |

## References


Belsley, D. A., Kuh, E., and Welsch, R. E. (1980). *Regression Diagnostics.* New York: Wiley.

Box, G. E. P., and Cox, D. R. (1964). "An Analysis of Transformations." *J. Royal Stat. Soc. B* 26: 211–243.

Draper, N., and Smith, H. (1981). *Applied Regression Analysis,* 2nd ed. New York: Wiley.

Dunnett, C. W. (1964). "New Tables for Multiple Comparisons with a Control." *Biometrics* 20: 482–491.

Durbin, J., and Watson, G. S. (1950). "Testing for Serial Correlation in Least-Squares Regression: I." *Biometrika* 37: 409–428.

Durbin, J., and Watson, G. S. (1951). "Testing for Serial Correlation in Least-Squares Regression: II." *Biometrika* 38: 159–178.

Durbin, J., and Watson, G. S. (1971). "Testing for Serial Correlation in Least-Squares Regression: III." *Biometrika* 58: 1–19.

Edwards, J. B. (1969). "The Relation Between the *F*-test and $\bar{R}^2$." *Amer. Statistician* 23(4): 28.

Granger, C. W. J., and Newbold, P. (1986). *Forecasting Economic Time Series,* 2nd ed. New York: Academic Press.

Haitovsky, Y. (1969). "A Note on the Maximization of $\bar{R}^2$." *Amer. Statistician* 23(1): 20–21.

Judge, G. G. et al, (1980). *The Theory and Practice of Econometrics.* New York: Wiley.

Kennedy, P. (1983). *A Guide to Econometrics.* Cambridge, Mass.: MIT Press.

Neter, J., Wasserman, W., and Kutner, M. H. (1983). *Applied Linear Regression Models.* Homewood, Ill.: Richard D. Irwin.

Weisberg, S. (1980). *Applied Linear Regression.* New York: Wiley.

# Forecasting Seasonal Series

**Objective:** *To examine procedures for forecasting time series whose average levels are affected by climatic and other influences arising at regular intervals during the calendar year.*

## 5.1 Introduction

Time series observed at shorter than yearly intervals often display a regular pattern of fluctuations that repeats from year to year. This *periodic* pattern is usually called **seasonal movement, seasonality,** or simply **seasonal.** Seasonality is especially prominent in economic series, which are often affected by annual changes in the weather and by social convention. In addition, one often wants to know whether a change in a series is attributable simply to normal seasonal fluctuation or perhaps to more important, nonseasonal factors. It is for this reason that, for example, *seasonally adjusted* unemployment rates are used to reflect "true" changes in economic health. It is not at all unusual for seasonal movement to constitute a large portion of a series' variability, often effectively masking trend or cyclical movement. When seasonality is present, failing to acknowledge it or trying to use a nonseasonal model (and thus allocating a large portion of the series' movement to an error term) is likely to result in highly variable parameter estimates and, ultimately, lead to poor forecasts.

When forecasting subannual observations, then, any seasonal influences should be recognized and models adapted to take them into account. Building a seasonal model usually involves selecting either a multiplicative or additive structure and then estimating a set of seasonal indexes from the history of the series. These indexes may then be used to incorporate seasonality in forecasts (to *seasonalize*) or to remove such effects from the observed values (to *deseasonalize,* or *seasonally adjust,* the data).

In this chapter, we will turn our attention to

- detecting seasonal movement in a series, both graphically and statistically
- the use of multiplicative and additive models to describe seasonal effects
- both single and updating procedures for estimating seasonal parameters for no-trend series
- both single and updating procedures for estimating seasonal parameters for trended series
- special aspects of diagnostics when applied to seasonal models

## 5.2 The Nature of Seasonal Movement

### 5.2.1 Introduction

Earth's movement around the sun each year is the principal cause of a type of nonstationarity characterized by

- a *regular,* predictable *pattern* that
- *repeats yearly* in such a way that
- *net movement* away from trend over a yearly period is zero.

---

**DEFINITION:** A time series $y_t$, observed $L$ times per year at times $t = 1, 2, \ldots$ is said to have constant **seasonality** (additive/multiplicative) if the average value of $y_t$ changes over time such that

*Additive:*

$$E(y_t) = f(\beta_0, \beta_1, \ldots; t) + S_t \qquad t = 1, 2, \ldots \tag{5.1}$$

*Multiplicative:*

$$E(y_t) = f(\beta_0, \beta_1, \ldots; t) \cdot S_t \qquad t = 1, 2, \ldots \tag{5.2}$$

where

$$S_t = S_{t+L} = S_{t+2L} = \cdots$$

and

*Additive:*

$$\sum_1^L S_t = 0$$

*Multiplicative:*

$$\sum_1^L S_t = L$$

and $f(\beta_0, \beta_1, \ldots; t)$ is a function describing the trend. Each observation period is called a **season,** and $L$, the **length of the seasonality,** is the number of seasons in a year. The $S_t$, whether, additive or multiplicative, are called **seasonal indexes.**

---

This type of movement is called *seasonality* and generally derives from annual weather patterns and the social behavior resulting from adherence to an annual calendar. Similar types of regular patterns may be found within weekly periods (e.g., television programming), monthly periods (e.g., payroll), and even longer than yearly periods (e.g., presidential elections), but the term usually implies annual repetition. Assuming this latter context and, in addition, that the seasonal effect does not change from year to year, we adopt the following definition.

### 5.2.2 Characteristics of Seasonal Series

There are several possibilities for general seasonal models, depending on which of the forms (Equation 5.1 or Equation 5.2) are to be used. If we let

$y_t$ = the actual value of the series at time $t$

$L$ = the length (number of seasons) of the seasonality ($L = 12$ for monthly, 6 for bimonthly, 4 for quarterly, and 2 for semiannual seasons)

$S_t$ = the seasonal effect at time $t$
$= S_{t+L} = S_{t+2L} = \cdots$ (*constant seasonality*)

$T_t = f(\beta_0, \beta_1, \ldots ; t)$ = the trend at time $t$

$\epsilon_t$ = the irregular fluctuation away from the trend and seasonal effects at time $t$

then the two most widely used models are those given in the following table.

**General Seasonal Models**

| | | | |
|---|---|---|---|
| **Additive:** | $y_t = T_t + S_t + \epsilon_t$ | $t = 1, 2, \ldots ;$ | $\underset{\text{any year}}{\Sigma S_t = 0}$ |
| **Multiplicative:** | $y_t = T_t \cdot S_t \cdot \epsilon_t$ | $t = 1, 2, \ldots ;$ | $\underset{\text{any year}}{\Sigma S_t = L}$ |

A seasonal series may be either trended or untrended. During any two time periods that are apart by multiples of $L$, the series experiences identical upward or downward influences on its average due solely to the fact the periods are during the same season of the year. Again, the $\epsilon_t$'s are usually assumed to form a random stationary series (with mean 0 for the additive model and mean 1 for the multiplicative model). Because $S_t$ is the same for all times that are $L$ periods apart, the seasonal effects are adequately described by a set of $L$ seasonal indexes, $S_1, S_2, \ldots, S_L$.

As an example of how the model works, let us suppose that the quarterly sales of camping equipment for a certain department store are described by an additive seasonal model with an additive linear trend component so that

$y_t$ = quarterly sales of camping equipment

$T_t = 4.8 + .4t$ Origin: IV, 1982
Units ($t$): Quarterly
Units ($y$): \$100,000

**Table 5.1 Department Store Camping Equipment Sales Series with Trend, Seasonal, and Irregular Components**

| Year | Qrtr | $t$ | $T_t$ | $S_t$ | $\epsilon_t$ | $T_t + S_t$ | $y_t = T_t + S_t + \epsilon_t$ |
|---|---|---|---|---|---|---|---|
| 1983 | I | 1 | 5.2 | −2.6 | .54 | 2.6 | 3.14 |
| | II | 2 | 5.6 | 1.6 | .34 | 7.2 | 7.54 |
| | III | 3 | 6.0 | .8 | −.83 | 6.8 | 5.97 |
| | IV | 4 | 6.4 | .2 | −.06 | 6.6 | 6.54 |
| 1984 | I | 5 | 6.8 | −2.6 | −1.12 | 4.2 | 3.08 |
| | II | 6 | 7.2 | 1.6 | .25 | 8.8 | 9.05 |
| | III | 7 | 7.6 | .8 | .20 | 8.4 | 8.60 |
| | IV | 8 | 8.0 | .2 | −.98 | 8.2 | 7.22 |
| 1985 | I | 9 | 8.4 | −2.6 | .93 | 5.8 | 6.73 |
| | II | 10 | 8.8 | 1.6 | −.28 | 10.4 | 10.12 |
| | III | 11 | 9.2 | .8 | −.16 | 10.0 | 9.84 |
| | IV | 12 | 9.6 | .2 | −.76 | 9.8 | 9.04 |
| 1986 | I | 13 | 10.0 | −2.6 | .27 | 7.4 | 7.67 |
| | II | 14 | 10.4 | 1.6 | −.10 | 12.0 | 11.90 |
| | III | 15 | 10.8 | .8 | −.90 | 11.6 | 10.70 |
| | IV | 16 | 11.2 | .2 | .38 | 11.4 | 11.78 |

$T_t = 4.8 + 4t$ Origin: IV, 1982; Units ($t$): Quarterly; Units ($y$): \$100,000

$S_1 = -2.60$ $S_2 = +1.60$ $S_3 = +.80$ $S_4 = +.20$

and

$$S_1 = -2.60 \qquad S_2 = +1.60 \qquad S_3 = +.80 \qquad S_4 = +.20$$

Notice that, since observations are made quarterly ($L = 4$), there are only four seasonal indexes, $S_1(= S_5 = S_9 = \cdots)$, $S_2(= S_6 = S_{10} = \cdots)$, $S_3(= S_7 = S_{11} = \cdots)$, and $S_4(= S_8 = S_{12} = \cdots)$. Note, also, that the units of additive seasonal indexes are the same as those of $y$. In this case, $y$ is in \$100,000 units; so $S_1 = -2.60$ means that camping equipment sales experience an average downward movement of \$260,000 in the first quarter, $S_2 = +1.60$ means an average upward movement of \$160,000 in the second quarter, and so forth. Finally, the sum of the seasonals is

$$\sum_{1}^{4} S_t = -2.60 + 1.60 + .80 + .20 = 0.0$$

making the *yearly net* average seasonal deviation from trend equal to zero. In Table 5.1 and Figure 5.1, the separate trend, seasonal, and hypothetical irregular components are given for I, 1983 through IV, 1986 along with the series that results from combining these components. In I, 1984 ($t = 5$), for example,

Figure 5.1 Quarterly Sales of Camping Equipment

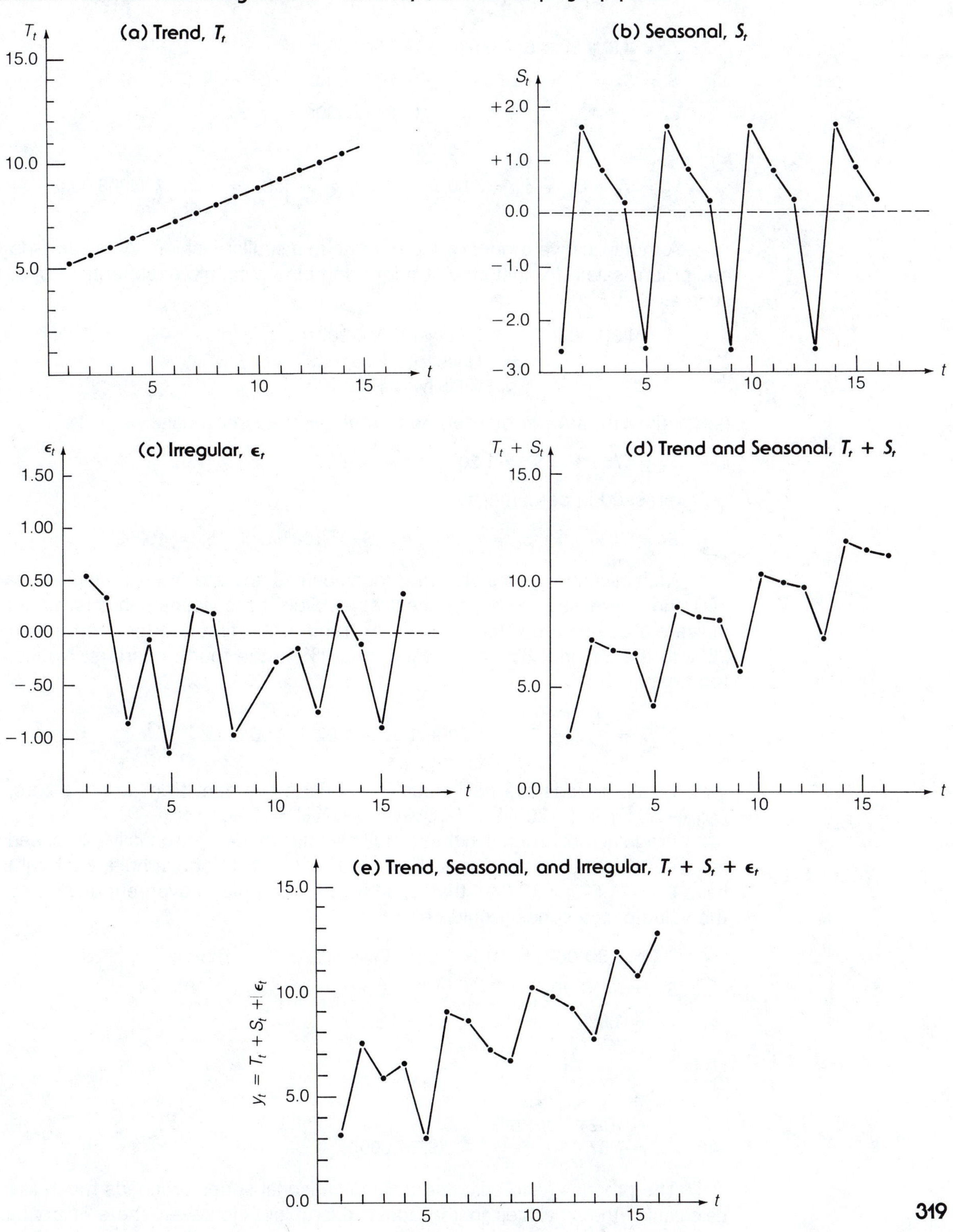

$$T_5 = 4.8 + .4 \times 5 = 6.80 \qquad (+\$680{,}000)$$

$$S_t = S_5 = S_1 = -2.60 \qquad (-\$260{,}000)$$

$$\epsilon_5 = -1.12 \qquad (-\$112{,}000)$$

and

$$y_5 = T_5 + S_5 + \epsilon_5 = 6.80 + (-2.60) + (-1.12) = 3.08 \qquad (\$308{,}000)$$

A multiplicative model would operate in a similar fashion. With camping equipment sales, for instance, it might combine a simple exponential trend, such as

$$T_t = 4.8(1.06)^t$$

Origin: IV, 1982
Units ($t$): Quarterly
Units ($y$): $100,000

(a 6% growth rate per quarter), with multiplicative seasonals,

$$S_1 = .70 \qquad S_2 = 1.20 \qquad S_3 = 1.08 \qquad S_4 = 1.02$$

or, expressed in percentages,

$$S_1 = 70\% \qquad S_2 = 120\% \qquad S_3 = 108\% \qquad S_4 = 102\%$$

Multiplicative seasonals are dimensionless and are often multiplied by 100 and expressed in percentages, as above. The seasonals in this model indicate a downward effect of 30% of trend in the first quarter and upward 20% in the second, 8% in the third, and 2% in the fourth quarters. Notice, too, that

$$\sum_1^L S_t = \sum_1^4 S_t = .70 + 1.20 + 1.08 + 1.02 = 4.00 = L$$

which again means the net annual seasonal movement from trend is zero, i.e., $(-30\%) + (+20\%) + (+8\%) + (+2\%) = 0\%$.

Irregular movements in these multiplicative models are usually expressed as dimensionless indexes as well. In I, 1986 ($t = 13$), for example, we might have $\epsilon_{13} = 1.07$ (or 107%), that is, an upward irregular movement of 7%. So the value of the series would be

$$T_{13} = 4.8(1.06)^{13} = 10.24 \qquad (\$1{,}024{,}000)$$

$$S_{13} = S_1 = .70 \qquad (-30\%)$$

$$\epsilon_{13} = 1.07 \qquad (+7\%)$$

and

$$\begin{aligned} y_{13} &= T_{13} \cdot S_{13} \cdot \epsilon_{13} \\ &= 10.24(.70)(1.07) \\ &= 7.67 \qquad (\$767{,}000) \end{aligned}$$

The process of forecasting a trend-seasonal series proceeds much like calculating the $y_t$ values in the above examples. However, there is a vital difference: The form of the model must be specified and the model parameters (trend parameters and seasonals) must be available before the calculations can be done. It was our tacit assumption in the above examples that kind Providence had supplied this elusive knowledge. In practice the information must be gleaned from the history of the series and from the analyst's understanding of the mechanisms underlying the series.

Models, then, may be either additive or multiplicative. They may include a trend component or not. Single or updating procedures may be desired. The appropriate methodology for each of these choices will be discussed as the chapter progresses.

### 5.2.3 Reasons for Seasonality

In many ways seasonality is easier to handle than other time series components, because it has a clear and proximate cause: Earth's annual orbit. Trend components may have any one of a myriad of functional forms, cycles can be any length and repeat at irregular intervals, but if there is seasonal, it will have a regular, repetitive, 12-month pattern. The presence of this pattern can often be anticipated, and the forecaster knows precisely the pattern for which to search.

For the most part, seasonal effects are due to either one or a combination of the following two influences:

- **weather and temperature related factors** items, sales and services related to coping with different types of weather (hot or cold) and with agricultural growing seasons: air conditioning, heating, adjusting to snow, rain, or heat, heating fuels, equipment and services, natural gas and electricity consumption, summer/winter sports and hobby activities, clothing, vehicle types and accessories, food consumption patterns, certain types of illnesses (even birth patterns to some degree), and many, many others.
- **calendar related factors** collective social behavior that develops because humanity lives and works by an annual calendar. Included would be holidays and religious events (such as Christmas, Thanksgiving, Independence Day, Passover, three-day weekends, and so on), school calendars and attendance patterns, tax filing schedules, television programming practices, and so forth.

On occasion the effect may be subtle (as with birth patterns). More often it is apparent and the job of forecasting consists mainly of estimating the nature and magnitude of the effect rather than questioning its existence (Bell and Hilmer, 1984; Burman, 1979).

## 5.3 Examples

Because seasonal effects follow such well-defined patterns, it is often easy to anticipate and recognize them. Some series are very dependent on seasonal effects and thus are dominated by seasonal variability. In others, the influence may be more moderate, so the effects themselves are hidden or distorted by noise in the series.

### 5.3.1 Example 5.1—Mid-Town Heating and Air Conditioning, Inc.

The service industries associated with heating (and cooling) equipment are especially sensitive to weather conditions. Mid-Town Heating and Air Conditioning Service is such a company. It operates a fleet of on-call service vans in a large city, where it installs new heating systems and repairs and maintains existing ones. Quarterly data on gross receipts for its furnace installation and repair operations for the past several years are given in Table 5.2. The graph of the series, shown in Figure 5.2, shows a substantial

**Table 5.2 Receipts for Mid-Town Heating and Air Conditioning, Inc.**
$y_t$ = quarterly gross receipts ($1,000)

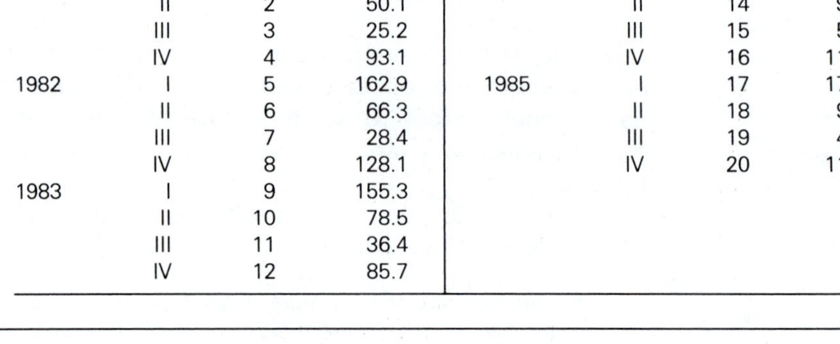

| Year | Qrtr | $t$ | $y_t$ | Year | Qrtr | $t$ | $y_t$ |
|---|---|---|---|---|---|---|---|
| 1981 | I | 1 | 135.9 | 1984 | I | 13 | 175.7 |
| | II | 2 | 50.1 | | II | 14 | 90.7 |
| | III | 3 | 25.2 | | III | 15 | 50.4 |
| | IV | 4 | 93.1 | | IV | 16 | 114.1 |
| 1982 | I | 5 | 162.9 | 1985 | I | 17 | 178.4 |
| | II | 6 | 66.3 | | II | 18 | 93.0 |
| | III | 7 | 28.4 | | III | 19 | 43.3 |
| | IV | 8 | 128.1 | | IV | 20 | 116.6 |
| 1983 | I | 9 | 155.3 | | | | |
| | II | 10 | 78.5 | | | | |
| | III | 11 | 36.4 | | | | |
| | IV | 12 | 85.7 | | | | |

**Figure 5.2 Gross Receipts, Mid-Town Heating**

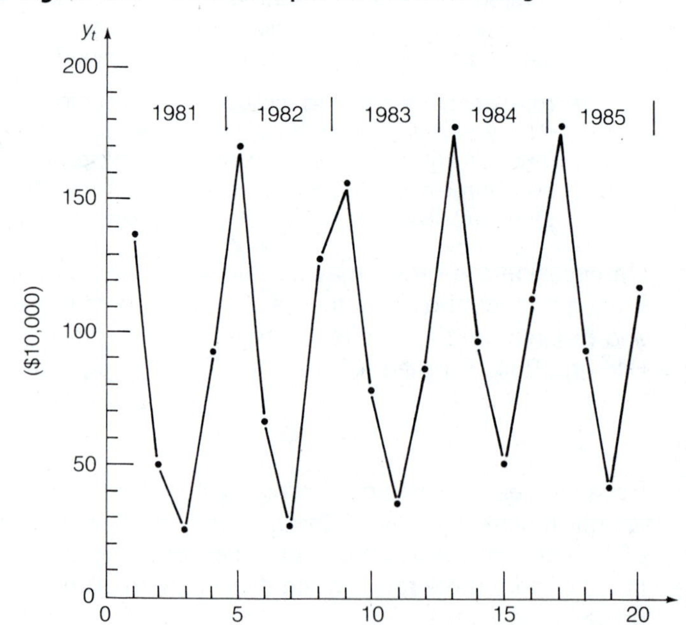

amount of variability, swinging all the way from \$250–\$300 thousand in some quarters to \$1.5–\$2.0 million in others. However, these high and low values occur during the same quarters each year, high in the fall/winter quarters (IV, I) and low in the spring/summer ones (II, III). This is the regular seasonal pattern that would be anticipated in such a business. There is so much seasonal variability, in fact, that possible growth patterns are effectively hidden from casual view.

## 5.3.2 Example 5.2—Retail Trade Earnings (U.S. Return Rates)

Another area that exhibits strong seasonal patterns, due mainly to calendar influences, is the retail trade industry. A major contributor here is the Christmas buying season, which begins to be felt around the beginning of the fourth quarter, builds strongly, and finally wanes during the early part of the following first quarter, the latter being generally the slowest quarter of the year. The annualized return rates on stockholder's equity for retail trade corporations in the United States for several years are given in Table 5.3 and graphed in Figure 5.3 (Source: *Quarterly Financial Report for Manufacturing, Mining and Trade*). Again, the regular seasonal pattern in the series is easily recognized. Pronounced peaks during the fourth quarter and troughs during the first quarter of each year are readily apparent.

**Figure 5.3** **Annual Return Rates (after Taxes) on Stockholder's Equity, U.S. Retail Mining, Manufacturing, and Trade Corporations, 1979–1984**

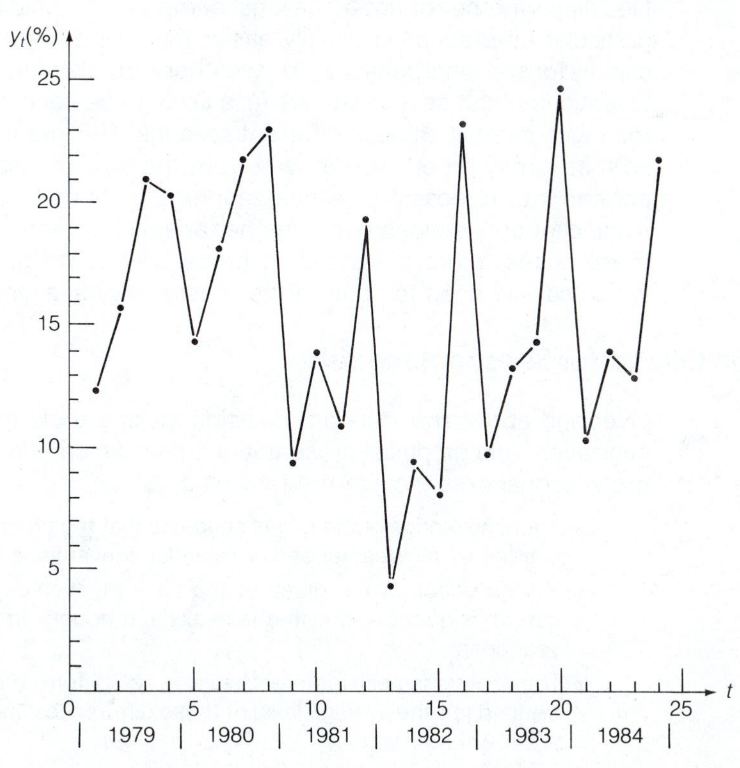

**Table 5.3 Annual Return Rates (after Taxes) on Stockholder's Equity for U.S. Retail Mining, Manufacturing, and Trade Corporations, 1979–1984**
$y_t$ = return rate (%)

| Year | Qrtr | $t$ | $y_t$ | Year | Qrtr | $t$ | $y_t$ |
|---|---|---|---|---|---|---|---|
| 1979 | I | 1 | 12.24 | 1982 | I | 13 | 4.40 |
| | II | 2 | 15.70 | | II | 14 | 9.39 |
| | III | 3 | 20.97 | | III | 15 | 8.00 |
| | IV | 4 | 20.32 | | IV | 16 | 23.07 |
| 1980 | I | 5 | 14.31 | 1983 | I | 17 | 9.64 |
| | II | 6 | 18.00 | | II | 18 | 13.24 |
| | III | 7 | 21.66 | | III | 19 | 14.14 |
| | IV | 8 | 22.88 | | IV | 20 | 24.61 |
| 1981 | I | 9 | 9.31 | 1984 | I | 21 | 10.25 |
| | II | 10 | 13.26 | | II | 22 | 13.93 |
| | III | 11 | 10.96 | | III | 23 | 12.68 |
| | IV | 12 | 19.11 | | IV | 24 | 21.72 |

## 5.4 The Decision to Use Seasonal Models

### 5.4.1 Introduction

Deciding whether or not a seasonal component would be appropriate for any particular time series is usually easier than for other components, since the causes for seasonal behavior, i.e., weather and calendar, are so straightforward. Questions about seasonality are less likely to be concerned with its presence than with its magnitude in different seasons. The intent is to isolate the effect so that it may be either removed from the data (deseasonalization) or incorporated into forecasts (seasonalization). Situations do arise, however, when significant connections with weather or social custom are open to debate. In these cases, there are standard procedures, both graphical and statistical, which can be used to confirm the presence of seasonal movements.

### 5.4.2 Specific Criteria for Seasonal Models

Over and above the standard decision-making tools (familiarization, data organization, and graphical presentation), specific criteria that may be applied in the seasonal case include the following.

- Your knowledge of the series suggests that the process generating the series is affected by weather and/or calendar conventions.
- Via inspection of the graph of the series ($y_t$ versus $t$), a repetitive up/down pattern is discernible, with the peaks and troughs in the movement spaced a year apart.
- The $L$ periods apart differences, $y_t - y_{t-L}$, form a horizontal series (for untrended $y_t$'s the average level of these differences should be zero; for trended $y_t$'s it will be nonzero).

- The $L$ periods apart link relatives, $y_t/y_{t-L}$, form a horizontal series (if the $y_t$'s are untrended, the average value will be 100%; for trended $y_t$'s it will be different from 100%).

One way to check graphically for seasonality is to compare the graph of the whole series to graphs of individual seasons. If the variability of the individual season graphs is noticeably smaller than for the whole series, seasonality is present. If the variability is about the same, then there is no seasonality. Notice in the strongly seasonal profit rate data from Example 5.2, for example, that there is much greater variability in the graph of the whole series, Figure 5.4(a), than in the graph of only the fourth quarters, Figure 5.4(b). In essence, the statistical tests that follow are based on quantitative measures of this type of difference in variability.

**Figure 5.4** Annual Return Rates on Stockholder's Equity

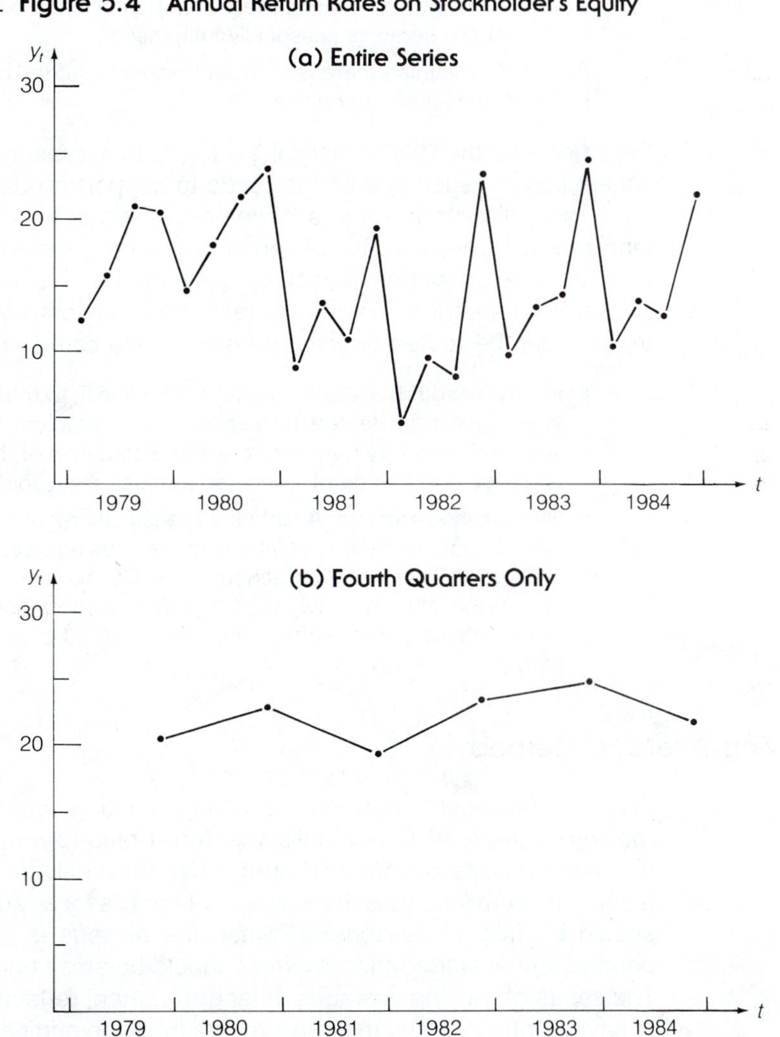

### 5.4.3 Statistical Tests for Seasonality

Generically, hypothesis tests for seasonality take the form

$H_0$: The series is stationary/random (has no seasonality).

$H_a$: The series has seasonality.

or, more formally, in a seasonal context,

$$H_0\colon\ S_1 = S_2 = S_3 = S_4 = 0 \quad \text{(additive)}$$
$$= 1 \quad \text{(multiplicative)}$$

$$H_a\colon\ S_i \neq 0 \quad \text{for some seasons} \quad \text{(additive)}$$
$$\neq 1 \quad \text{for some seasons} \quad \text{(multiplicative)}$$

and the decision may be either:

Do not reject $H_0$, concluding there is not sufficient evidence, at the $\alpha \times 100\%$ level, that the series is seasonally influenced.

Reject $H_0$, meaning there is sufficient evidence, at the $\alpha \times 100\%$ level, that the series is seasonally influenced.

Rejecting $H_0$, then, is statistical evidence that seasonality is present in the series. Lack of such a rejection tends to support models without seasonals.

The null hypothesis in seasonal tests is that the series consists only of random error. The presence of nonrandom components may cause these tests to reject $H_0$ even when seasonality is absent. For this reason, a trended series should be detrended *before* it is tested for seasonality. If the series has no trend, it can be tested directly, otherwise one can do the following.

- **test the residuals** A trend model can first be fit to the data. Then the residuals, $y_t - \hat{T}_t$, can be tested for seasonality. A problem here is that unmodeled seasonal variability may interfere with estimation of the trend parameters and tests for significance of trend can produce misleading results.
- **filter out the seasonal** A combined seasonal/irregular component may be separated from, or *filtered out* of, the series by using an appropriate-length moving average or other averaging scheme. This filtered component can then be tested for seasonality. The advantage here is that there are no assumptions about the functional form of the trend, nor do trend parameters need to be estimated.

### 5.4.4 Moving Average Method

Probably the most common approach to isolating seasonal movement is the *moving average,* or (in multiplicative form) *ratio-to-moving average,* method mentioned above. Because the cumulative seasonal effect for any yearly period is zero, the sum (and thus the average) of the $y_t$'s for any consecutive $L$ seasons should be free of seasonality. Since the average is taken over several ($L$) periods, the average random effect should be small enough to be neglected. The series of moving averages of length $L$, then, reflects only trend and cycle (if any). Subtracting the moving average from the original series in the additive case (or dividing by it in the multiplicative case) gives a series of *specific seasonals* (see table) that reflect only seasonal and random movements. [It is division by the moving average in the multiplicative case from which the term "ratio" (of the original series) "-to-" (the) "moving average" is derived.]

**Moving Average Filter**

| | Additive Model: | Multiplicative Model: |
|---|---|---|
| **Original Series:** | $y = T + S + C + I$ | $y = T \cdot S \cdot C \cdot I$ |
| **Moving Average:** | $\text{MA} = T + C$ | $MA = T \cdot C$ |
| **Specific Seasonal:** | $y - \text{MA} = S + I$ | $\dfrac{y}{\text{MA}} = S \cdot I$ |

A minor difficulty arises in the actual calculation of the specific seasonals because of the centering of the moving average. For example, with quarterly seasons the first term of the moving average series would be

$$\frac{y_1 + y_2 + y_3 + y_4}{4}$$
$$\downarrow$$
$$\text{center} = 2.5$$

which is centered at $t = 2.5$ and therefore is representative of the trend-cycle at $t = 2.5$; but to be accurate, we need a trend-cycle value for $t = 2$. To get this value, the two trend-cycle values centered at $t = 2.5$ and $t = 3.5$ are averaged. Thus

$$\underset{\text{center} = 2.5}{\frac{y_1 + y_2 + y_3 + y_4}{4}} + \underset{\text{center} = 3.5}{\frac{y_2 + y_3 + y_4 + y_5}{4}}$$
$$\downarrow$$
$$\text{center} = 2.0$$

which has been variously described as computing a "moving average of length 2 of the length-$L$ moving averages," or a "centered moving average of length $L$," or a "$2 \times L$ moving average." In point of fact this is a "weighted moving average of length $L + 1$," with weight 1 given to each end point and weight 2 given to the points in between, i.e.,

$$\frac{(y_1 + y_2 + y_3 + y_4)/4 + (y_2 + y_3 + y_4 + y_5)/4}{2} = \frac{(y_1 + 2y_2 + 2y_3 + 2y_4 + y_5)}{8}$$

This "centering" needs to be done only if $L$ is even, which, unfortunately, is the case for semiannual, quarterly, and monthy data (the kind usually seen in practice).

Another difficulty with the method is that the series of specific seasonals is about a year shorter than the series being filtered ($L$ observations shorter if $L$ is even, $L - 1$ shorter if $L$ is odd).

The actual calculation of the *centered moving average* (*CMA*) is easy but tedious. It can be performed in the three steps shown in the following table.

**Calculation of Centered Moving Average**

| | |
|---|---|
| **Step 1:** | Calculate $L$-period moving totals of the original data. |
| **Step 2:** | Calculate 2-period moving totals of the moving totals resulting from Step 1. |
| **Step 3:** | Divide the result of Step 2 by $2L$, giving the required centered moving average. |

The Mid-Town Heating data of Example 5.1 has so much variability that any trend in the series is probably hidden. As a matter of fact, the parametric test for $\rho = 0$ using Pearson's $r$ (see Section 3.4.6) gives $t_r = +.54$, which is not significant at any reasonable level. However, because of the strongly seasonal nature of the heating business, this result should be suspect. Although statistical justification for adding a seasonal component is not really needed for these data, calculating the CMA series illustrates the moving average filtering process. It also reveals information about the trend-cycle element of the series. The calculations are shown in Table 5.4. Since the data are quarterly, $L = 4$ and $2L = 8$. The value of the first moving total is

$$y_1 + y_2 + y_3 + y_4 = 135.9 + 50.1 + 25.2 + 93.1 = 304.3$$

The next value is conveniently found by subtracting the oldest observation and adding the new one. So

$$\begin{aligned}(y_1 + y_2 + y_3 + y_4) - y_1 + y_5 &= y_2 + y_3 + y_4 + y_5 \\ &= 304.3 - 135.9 + 162.9 \\ &= 331.3\end{aligned}$$

This procedure continues until all the moving totals are obtained.

Notice in Table 5.4 that the moving totals are written in between time periods to emphasize that they are off center. The *centered moving total* (*CMT*), directly across from the time period in which it is centered, is calculated by adding two consecutive moving totals together. Thus:

*For $t = 3$:* $\text{CMT} = 304.5 + 331.3 = 635.6$

*For $t = 4$:* $\text{CMT} = 331.3 + 347.5 = 678.8$

and so forth. To get the CMA, the CMT needs only to be divided by $2L = 8$. Remember that this CMA describes the trend-cycle in the original series (with, perhaps, a little leftover random error). When the CMA's are graphed (Figure 5.5), the trend, which had been masked by seasonal variability in the original series, can be clearly seen.

**Table 5.4** Computation of Centered Moving Average for Mid-Town Data

| Year | Qrtr | $t$ | $y_t$ | Moving Total | Centered Moving Total | | Centered Moving Average |
|---|---|---|---|---|---|---|---|
| 1981 | I | 1 | 135.9 | | | | |
| | II | 2 | 50.1 | 304.3 | | | |
| | III | 3 | 25.2 | 331.3 | 635.6 | ÷ 8 = | 79.45 |
| | IV | 4 | 93.1 | 347.5 | 678.8 | ÷ 8 = | 84.85 |
| 1982 | I | 5 | 162.9 | 350.7 | 698.2 | ÷ 8 = | 87.28 |
| | II | 6 | 66.3 | 385.7 | 736.4 | | 92.05 |
| | III | 7 | 28.4 | 378.1 | 763.8 | | 95.47 |
| | IV | 8 | 128.1 | 390.3 | 768.4 | | 96.05 |
| 1983 | I | 9 | 155.3 | 398.3 | 788.6 | | 98.58 |
| | II | 10 | 78.5 | 355.9 | 754.2 | | 94.28 |
| | III | 11 | 36.4 | 376.3 | 732.2 | | 91.53 |
| | IV | 12 | 85.7 | 388.5 | 764.8 | | 95.60 |
| 1984 | I | 13 | 175.7 | 402.5 | 791.0 | | 98.88 |
| | II | 14 | 90.7 | 430.9 | 833.4 | | 104.18 |
| | III | 15 | 50.4 | 433.6 | 864.5 | | 108.06 |
| | IV | 16 | 114.1 | 435.9 | 869.5 | | 108.69 |
| 1985 | I | 17 | 178.4 | 428.8 | 864.7 | | 108.09 |
| | II | 18 | 93.0 | 431.3 | 860.1 | | 107.51 |
| | III | 19 | 43.3 | | | | |
| | IV | 20 | 116.6 | | | | |

**Figure 5.5** Trend-Cycle Component (CMA) of Mid-Town Data

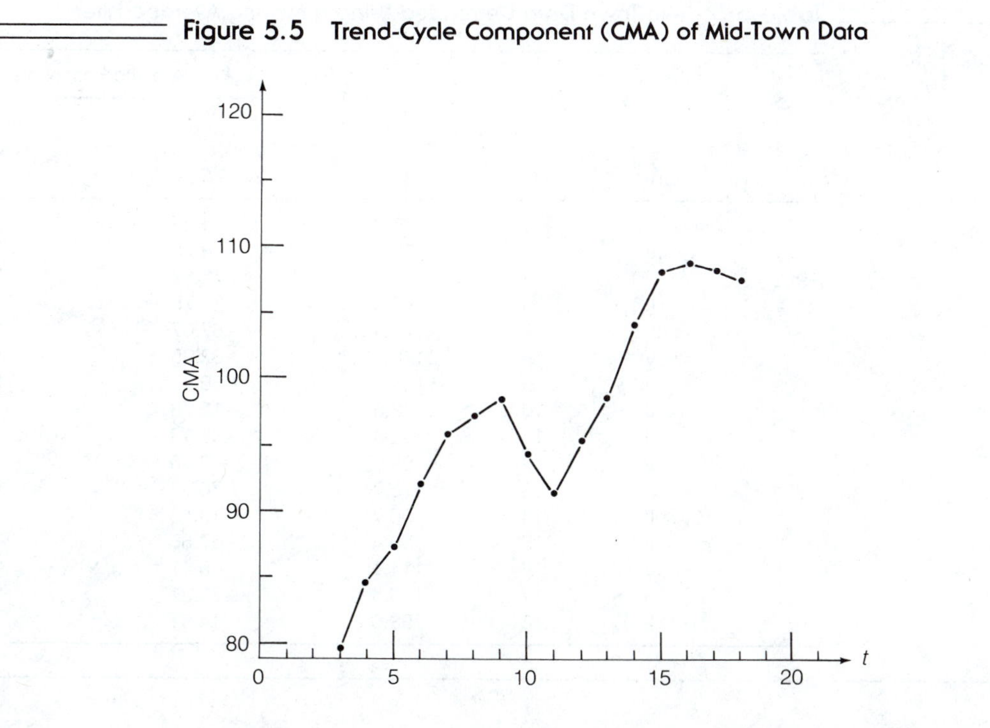

Another thing to notice about the CMA's is that they start at $t = 3$ and end at $t = 18$. A half year is lost at the beginning and another at the end of the series. As a consequence, considerably more than one year's worth of data are needed for CMA's to be used effectively. In fact, any serious investigation of the seasonal behavior of a series will require several years of data.

The procedure in the final stage of filtering depends upon whether an additive or multiplicative model is being considered.

*Additive Case:* (specific seasonal) = (original value) − (trend + cycle)

$$\hat{S}_t + \hat{I}_t = y_t - (\hat{T}_t + \hat{C}_t)$$

So

*For III, 81:* $\hat{S}_3 + \hat{I}_3 = y_3 - (\hat{T}_3 + \hat{C}_3) = 25.2 - 79.45 = -54.25$ (−$54,250)

*For IV, 81:* $\hat{S}_4 + \hat{I}_4 = y_4 - (\hat{T}_4 + \hat{C}_4) = 93.1 - 84.85 = +8.25$ (+$8,250)

and so forth.

*Multiplicative Case:* (specific seasonal) = (original value)/(trend · cycle)

$$\hat{S}_t \cdot \hat{I}_t = \frac{y_t}{\hat{T}_t \cdot \hat{C}_t}$$

So

*For III, 81:* $\hat{S}_3 \cdot \hat{I}_3 = \dfrac{y_3}{\hat{T}_3 \cdot \hat{C}_3} = \dfrac{25.2}{79.45} = .317$ (31.7%, or −68.3%)

**Table 5.5 Mid-Town Data Detrended Using a Moving Average Filter**

| | | | | Specific Seasonals | |
|---|---|---|---|---|---|
| Year | Qrtr | $t$ | $y_t$ | Additive Model $y_t' = \hat{S}_t + \hat{I}_t$ | Multiplicative Model $y_t' = \hat{S}_t \cdot \hat{I}_t$ |
| 1981 | III | 3 | 25.2 | −54.25 | .317 |
| | IV | 4 | 93.1 | +8.25 | 1.097 |
| 1982 | I | 5 | 162.9 | +75.62 | 1.867 |
| | II | 6 | 66.3 | −25.75 | .720 |
| | III | 7 | 28.4 | −67.07 | .297 |
| | IV | 8 | 128.1 | +32.05 | 1.334 |
| 1983 | I | 9 | 155.3 | +56.72 | 1.575 |
| | II | 10 | 78.5 | −15.78 | .833 |
| | III | 11 | 36.4 | −55.13 | .398 |
| | IV | 12 | 85.7 | −9.90 | 8.96 |
| 1984 | I | 13 | 175.7 | +76.82 | 1.777 |
| | II | 14 | 90.7 | −13.48 | .871 |
| | III | 15 | 50.4 | −57.66 | .466 |
| | IV | 16 | 114.1 | +5.41 | 1.050 |
| 1985 | I | 17 | 178.4 | +70.31 | 1.651 |
| | II | 18 | 93.0 | −14.51 | .865 |

*For IV, 81:* $$\hat{S}_4 \cdot \hat{I}_4 = \frac{y_4}{\hat{T}_4 \cdot \hat{C}_4} = \frac{93.1}{84.85} = 1.097 \qquad (109.7\%, \text{ or } +9.7\%)$$

and so forth.

Both sets of specific seasonals are given in Table 5.5. They show the same pattern of seasonality, where the additive ones are negative (positive), the multiplicative ones are less than 1.0 (greater than 1.0). The difference between them is due to differing judgment about the nature (additive or multiplicative) of the underlying seasonality. A hypothesis test may now be performed with the specific seasonals ($y_t''$s) to decide whether they contain seasonal effect or only random movement.

### 5.4.5 A Nonparametric Test, Kruskal-Wallis

While some of the nonparametric tests mentioned in Section 2.5 can detect seasonality (specifically, the turning points and runs tests), there is a more sensitive procedure that applies more directly to the seasonal case. This procedure is an application of a test called the **Kruskal-Wallis one-way analysis of variance,** or more simply, the Kruskal-Wallis test (Kruskal and Wallis, 1952; Marascuilo and McSweeney, 1977, ch. 12). It does not require that the error terms be normally distributed. The rationale for the test is that if the specific seasonals are purely random with no seasonality, their distribution should be the same in all $L$ seasons. Then, if ranks are assigned to the $y_t''$s (in the same manner as when computing Spearman's rho), a given rank should be as likely to fall in one season as in another. There should not be a preponderance of, say, low ranks in any one season, so the average rank in all seasons should be about the same, i.e., within sampling variability of one another. If we let

$$\begin{aligned} R_i &= \text{the sum of the ranks of the } y_t'' \text{s in the } i\text{th season} \\ n_i &= \text{the number of specific seasonals in the } i\text{th season} \\ n &= \text{the total number of specific seasonals} \\ &= n_1 + n_2 + \cdots + n_L \end{aligned}$$

then the statistic used to test the randomness hypothesis is the weighted sum of squared differences between the average ranks within seasons and the overall average rank.

Suppose we let

$$\begin{aligned} \bar{R}_i &= \text{the average rank in the } i\text{th season} \\ &= \frac{R_i}{n_i} \\ \bar{R} &= \text{the overall average rank} \\ &= \frac{\Sigma R_i}{n} = \frac{n(n+1)/2}{n} = \frac{n+1}{2} \end{aligned}$$

then the statistic $H$ is

$$H = n_1(\bar{R}_1 - \bar{R})^2 + n_2(\bar{R}_2 - \bar{R})^2 + \cdots + n_L(\bar{R}_L - \bar{R})^2$$
$$= \Sigma\, n_i(\bar{R}_i - \bar{R})^2$$

which, with some algebra, can be shown to be

$$H = \frac{12}{n(n+1)}\left[\Sigma \frac{R_i^2}{n_i}\right] - 3(n+1)$$

If the specific seasonals are random, $H$ will likely be fairly small; so if $H$ is too large we would reject randomness in favor of seasonality. When the null hypothesis is true and the $n_i$'s are even moderately large, $H$ is known to follow approximately a chi-square distribution with $L - 1$ degrees of freedom, so the critical values for the test may be found in Appendix FF.

---

**Test Procedure:** *Kruskal-Wallis Test*

$H_0$: $S_1 = S_2 = \cdots = S_L = 0$
$H_a$: $S_i \neq 0$ for some seasons

**Test Statistic:**

$$H = \frac{12}{n(n+1)}\left[\Sigma \frac{R_i^2}{n_i}\right] - 3(n+1)$$

$n_i$ = number of observations in $i$th season
$n$ = total number of specific seasonals
$= \Sigma\, n_i$
$y'_t$ = specific seasonal for time $t$
$R_i = \Sigma$ Rank$(y'_t)$
$i$th season

**Decision Rule:** (Level $\alpha$)

Reject $H_0$ if $H > \chi^2_\alpha(L - 1)$

where $\chi^2_\alpha(L - 1)$ is the upper $\alpha \times 100\%$ point of the chi-square distribution with $L - 1$ degrees of freedom. **(Appendix FF)**

**Conclusion:** If $H_0$ is rejected, we conclude with $(1 - \alpha) \cdot 100\%$ confidence that the series has seasonality.

If $H_0$ is not rejected, we have some support for models with no seasonal component.

---

**EXAMPLE 5.3** *Kruskal-Wallis Test, Mid-Town Heating*

In order to test for seasonality in the Mid-Town Heating series, we must first rank the series of specific seasonals, $y'_t$. For illustrative purposes, this has been done for both the additive and multiplicative cases in Table 5.5 and the results presented in Table

## Table 5.6 Ranks of Specific Seasonals, Mid-Town Heating

| | | Additive | | Multiplicative | |
|---|---|---|---|---|---|
| $t$ | Season | $y_t'$ | Rank($y_t'$) | $y_t'$ | Rank($y_t'$) |
| 3 | 3 | −54.25 | 4 | .317 | 2 |
| 4 | 4 | 8.25 | 11 | 1.097 | 11 |
| 5 | 1 | 75.62 | 15 | 1.867 | 16 |
| 6 | 2 | −25.75 | 5 | .720 | 5 |
| 7 | 3 | −67.07 | 1 | .297 | 1 |
| 8 | 4 | 32.05 | 12 | 1.334 | 12 |
| 9 | 1 | 56.72 | 13 | 1.575 | 13 |
| 10 | 2 | −15.78 | 6 | .833 | 6 |
| 11 | 3 | −55.13 | 3 | .398 | 3 |
| 12 | 4 | −9.9 | 9 | .896 | 9 |
| 13 | 1 | 76.82 | 16 | 1.777 | 15 |
| 14 | 2 | −13.48 | 8 | .871 | 8 |
| 15 | 3 | −57.66 | 2 | .466 | 4 |
| 16 | 4 | 5.41 | 10 | 1.050 | 10 |
| 17 | 1 | 70.31 | 14 | 1.651 | 14 |
| 18 | 2 | −14.51 | 7 | .865 | 7 |

5.6. (If tied observations, i.e., observations of the same numerical value, should occur, each tied observation is assigned the average of the tied ranks.) Notice that for simplicity we refer to quarters I, II, III, and IV as seasons 1, 2, 3, and 4. After ranking, the Kruskal-Wallis statistic can be calculated by separating the ranks for each season and summing. In the first season (quarter I), $t = 5, 9, 13, 17$, the ranks for the additive model are 15, 13, 16, and 14, and $R_1 = 15 + 13 + 16 + 14 = 58$. Continuing for the other seasons, and the multiplicative model, we get:

## Computation of Kruskal-Wallis Statistic

| | Season | Ranks | | | | $\Sigma R_i$ |
|---|---|---|---|---|---|---|
| **Additive Case:** | 1 | 15 | 13 | 16 | 14 | $15 + 13 + 16 + 14 = 58$ |
| | 2 | 5 | 6 | 8 | 7 | $5 + 6 + 8 + 7 = 26$ |
| | 3 | 4 | 1 | 3 | 2 | $4 + 1 + 3 + 2 = 10$ |
| | 4 | 11 | 12 | 9 | 10 | $11 + 12 + 9 + 10 = 42$ |
| **Multiplicative Case:** | 1 | 16 | 13 | 15 | 14 | $16 + 13 + 15 + 14 = 58$ |
| | 2 | 5 | 6 | 8 | 7 | $5 + 6 + 8 + 7 = 26$ |
| | 3 | 2 | 1 | 3 | 4 | $2 + 1 + 3 + 4 = 10$ |
| | 4 | 11 | 12 | 9 | 10 | $11 + 12 + 9 + 10 = 42$ |

Notice that, although there is some switching around of ranks within seasons, the rank sums happen to be the same in both cases, which means that the test statistic and conclusions will be the same for both models. The test goes as follows:

$$H_0\!: \quad S_1 = S_2 = S_3 = S_4 = 0 \quad \text{(additive)}$$
$$= 1 \quad \text{(multiplicative)}$$

$$H_a\!: \quad S_i \neq 0 \quad \text{for some seasons} \quad \text{(additive)}$$
$$\neq 1 \quad \text{for some seasons} \quad \text{(multiplicative)}$$

**Test Statistic:**

$$n_1 = n_2 = n_3 = n_4 = 4 \qquad \text{(All seasons have 4 observations.)}$$
$$N = 4 + 4 + 4 + 4 = 16 \qquad \text{(total number of observations)}$$
$$L - 1 = 4 - 1 = 3 \qquad \text{(number of seasons less 1)}$$
$$H = \frac{12}{16(16 + 1)}\left(\frac{58^2}{4} + \frac{26^2}{4} + \frac{10^2}{4} + \frac{42^2}{4}\right) - 3(16 + 1) = 14.12$$

**Decision Rule:** ($\alpha = 5\%, L - 1 = 3$)

Reject $H_0$ if $H > \chi^2_{.05}(3) = 7.81$ **(Appendix FF)**

**Conclusion:** Since $H = 14.12 > 7.81$, we reject $H_0$ and conclude that there is seasonality in the Mid-Town series. ●

That similar decisions are reached for both additive and multiplicative models in the above example is not unusual. The detection of the *presence* of seasonal effect is not especially sensitive to which model is used. It is in estimating the *nature* of the effect that the model plays a more important role. However, a different conclusion for the two models, should it occur, will give an indication of which model is more appropriate.

### 5.4.6 The "Rule of Thumb" for Seasonality

A very straightforward parametric test for seasonality is to compute the lag $L$ sample autocorrelation coefficient, $r_L$, and test it using the "rule of thumb" procedure (Section 2.6.5). The test procedure, adapted especially for the seasonality case, is shown below. A one-tailed version is used, because seasonality, consisting of like movements $L$ periods apart, induces positive lag $L$ autocorrelation (not negative).

Trend is not as likely to give false indications of seasonal movement when $r_L$ is used, but the series should be detrended first anyway.

The lag 4 sample autocorrelation coefficient for the two sets of specific seasonals from the Mid-Town series are $r_4 = .700$ for the additive set and $r_4 = .690$ for the multiplicative set (Table 5.7). The "rule of thumb" test for seasonality in the additive case (the multiplicative case is similar) would be as follows:

$$H_0\!: \quad \rho_4 = 0$$
$$H_a\!: \quad \rho_4 > 0$$

**Test Statistic:** $r_4 = .700$

**Decision Rule:** ($\alpha \simeq 5\%, n = 16$)

Reject $H_0$ if $r_4 > 1.645/\sqrt{16} = .500$.

**Test Procedure:** *"Rule of Thumb" Test for Seasonality*

$H_0$: $\rho_L = 0$
$H_a$: $\rho_L > 0$

**Test Statistic:** $r_L$ = sample autocorrelation coefficient of lag $L$

**Decision Rule:** (approximate level $\alpha$)

Reject $H_0$ if $r_L > z_\alpha/\sqrt{n}$ **(Appendix BB)**

where

$z_\alpha$ = the upper $\alpha \cdot 100\%$ point of the standard normal distribution

$n$ = the total number of specific seasonals

**Conclusion:** If $H_0$ is rejected, we conclude, with $(1 - \alpha) \cdot 100\%$ confidence that there is positive lag $L$ autocorrelation (seasonality) in the series.

If $H_0$ is not rejected, we have some support for models with no seasonal component.

**Table 5.7 Sample Autocorrelation Coefficients of Specific Seasonals for Mid-Town Data**

| Additive | | Multiplicative | |
|---|---|---|---|
| $k$ | $r_k$ | $k$ | $r_k$ |
| 1 | −.052 | 1 | −.050 |
| 2 | −.837 | 2 | −.837 |
| 3 | .028 | 3 | .035 |
| 4 | .700 | 4 | .690 |
| 5 | .001 | 5 | −.001 |
| 6 | −.548 | 6 | −.547 |

**Conclusion:** Since .700 > .500, we reject $H_0$ and conclude (just as before with the Kruskal-Wallis test) that seasonal effects are present in the series.

***Note:*** The significantly high negative autocorrelation at lag 2 is a coincidental consequence of the shape of the seasonality in the data.

## 5.4.7 A Least Squares Test for Seasonality

A more complex, but powerful, parametric test for seasonality uses least squares regression analysis and seasonal indicator variables (Section 4.7; Green and Doll, 1974). In effect, the method fits a least squares linear trend to either $y_t$ (additive case) or log $y_t$ (multiplicative case) to detrend (if necessary), then fits seasonal indicator variables to the residuals and tests the model fit. The following regression models are used.

### Seasonal Regression Models

**Additive Case:**

$$y_t = \underbrace{\beta_0 + \beta_1 t}_{T_t} + \underbrace{S_1x_1 + S_2x_2 + \cdots + S_Lx_L}_{S_t} + \underbrace{\epsilon_t}_{I_t} \qquad (5.3)$$

**Multiplicative Case:**

$$y_t' = \log y_t = \beta_0' + \beta_1't + S_1'x_1 + S_2'x_2 + \cdots + S_L'x_L + \epsilon_t' \qquad (5.4)$$

So

$$y_t = \underbrace{\beta_0(\beta_1)^t}_{T_t} \cdot \underbrace{S_1^{x_1} \cdot S_2^{x_2} \cdot \cdots \cdot S_L^{x_L}}_{S_t} \cdot \underbrace{\epsilon_t}_{I_t} \qquad (5.5)$$

where

$\beta_0 = e^{\beta_0'}$ $\quad\beta_1 = e^{\beta_1'}$ $\quad\epsilon_t = e^{\epsilon_t'}$ $\quad S_i = e^{S_i'}$ $\quad i = 1, 2, \ldots, L$

We require that

$$\Sigma S_i = 0 \quad \text{(additive)}$$
$$\Sigma S_i' = 0 \quad \text{(multiplicative)} \qquad (5.6)$$

and define $x_1, x_2, \ldots, x_L$ as seasonal indicator variables so that

$$x_i = \begin{cases} 1 \text{ in the } i\text{th season} \\ 0 \text{ otherwise} \end{cases} \qquad i = 1, 2, \ldots, L$$

The usual regression programs (and formulas) do not work for these models because the set of seasonal indicators is redundant (Section 4.7); that is, once $L - 1$ of the indicators are known, the $L$th is determined. The zero-sum constraints, Equation 5.6, have to be used in order to solve for $S_1, S_2, \ldots, S_L$ or $S_1', S_2', \ldots, S_L'$. There are a number of ways to get around the problem, the easiest of which is to use only $L - 1$ of the seasonal dummies. The $L$th seasonal is then absorbed into the other coefficients and regular regression can be done. There is a relatively simple method to recover $S_L$ and convert the other estimates so the conventional seasonal structure is maintained. It will be discussed later. For now, though, the value of $S_L$ is not really needed for testing, nor for that matter are any of the other parameter estimates.

To understand the testing procedure, recall from Chapter 4 that the regression technique decomposes the total variability (SST) in a series into a part that is "explained" by the model (SSR) and a part which remains "unexplained" (SSE). The main idea when testing for seasonality is to see if adding a seasonal component to the model significantly increases the explained variability (Kleinbaum, Kupper, and Muller, 1988, secs. 9.3–9.5). The procedure is best understood if undertaken in steps, as follows.

### Step 1

Fit a regression model with (1) a linear trend and (2) $L - 1$ seasonal dummy variables; and determine

SSR(trend + seasonal) = variability explained by the full trend/seasonal model

SSE(trend + seasonal) = variability left unexplained by the full trend/seasonal model

### Step 2

Using a $t$-test with $n - L - 1$ degrees of freedom, determine if the linear trend coefficient ($\beta_1$ or $\beta_1'$ in the seasonal models) is significant, as follows:

---

**Test Procedure:** *Test for Significance of Trend*

$H_0$: $\beta_1 = 0$ (or $\beta_1' = 0$)

$H_a$: $\beta_1 \neq 0$ (or $\beta_1' \neq 0$)

**Test Statistic:** $t = \dfrac{\hat{\beta}_1}{s_{\hat{\beta}_1}} \quad \left(\text{or } t = \dfrac{\hat{\beta}_1'}{s_{\hat{\beta}_1'}}\right)$

**Decision Rule:** Reject $H_0$ if $|t| > t_{\alpha/2}$ (df $= n - L - 1$).
Otherwise do not reject $H_0$.

**Conclusion:** Reject $H_0 \rightarrow$ trend is significant ($\beta_1$ or $\beta_1' \neq 0$).
Do not reject $H_0 \rightarrow$ trend is not significant.

---

### Step 3

If the trend is not significant, refit the model without a trend component ($\beta_1 t$ or $\beta_1' t$), that is, with $L - 1$ seasonals only, yielding

SSR(seasonal only) = variability explained by the seasonal-only model

SSE(seasonal only) = variability left unexplained by the seasonal-only model

Then use an $F$-test (Section 4.6) to test overall model fit. Otherwise go to Step 4. (To use a no-trend/no-seasonal model, retest the trend component without the seasonal element in the model to make sure it is still not significant.)

**Test Procedure:** ***Trend Not Significant***

$H_0$: $S_1 = S_2 = \cdots = S_{L-1} = 0$ (or $S_i'$'s $= 0$)
$H_a$: $S_i \neq 0$ for some seasons (or $S_i' \neq 0$)

**Test Statistic:**

$$F = \frac{\text{MSR(seasonal)}}{\text{MSE(seasonal)}}$$

where

$$\text{MSR(seasonal only)} = \frac{\text{SSR(seasonal only)}}{L - 1}$$

$$\text{MSE(seasonal only)} = \frac{\text{SSE(seasonal only)}}{n - L}.$$

**Decision Rule:** Reject $H_0$ if $F > F_\alpha$ with df $= L - 1, n - L$.
Otherwise do not reject $H_o$. **(Appendix HH)**

**Conclusion:** Reject $H_0 \rightarrow$ conclude with $(1 - \alpha) \times 100\%$ confidence that seasonality is present.
Do not reject $H_0 \rightarrow$ some support for models with no seasonal.

### Step 4

Fit a regression model with only linear trend ($\beta_0 + \beta_1 t$ or $\beta_0' + \beta_1' t$), no seasonal dummies, to determine

SSE(trend) = variability explained by a model with only linear trend

and calculate the amount of improvement obtained by adding a seasonal component as

SSR(seasonal) = amount of improvement provided by the seasonal component
= SSR(trend + seasonal) − SSR(trend)

### Step 5

Test the significance of the seasonal by calculating

$$F = \frac{\text{MSR(seasonal)}}{\text{MSE(trend + seasonal)}}$$

where

$$\text{MSR(seasonal)} = \frac{\text{SSR(seasonal)}}{L - 1}$$

$$\text{MSE(trend + seasonal)} = \frac{\text{SSE(trend + seasonal)}}{n - L - 1}$$

using critical values of the $F$-distribution with df $= (L - 1, n - L - 1)$.

**Test Procedure:** ***Trend Is Significant***

$H_0$: $S_1 = S_2 = \cdots = S_{L-1} = 0$ (or $S_i''$s $= 0$)
$H_a$: $S_i \neq 0$ for some season (or $S_i' \neq 0$)

**Test Statistic:**

$$F = \frac{\text{MSR(seasonal)}}{\text{MSE(trend + seasonal)}}$$

**Decision Rule:** Reject $H_0$ if $F > F_\alpha$ with df $= L - 1, n - L - 1$.
Otherwise do not reject $H_0$. **(Appendix HH)**

**Conclusion:** Reject $H_0 \rightarrow$ conclude with $(1 - \alpha) \cdot 100\%$ confidence that seasonality is present.
Do not reject $H_0 \rightarrow$ some support for models with no seasonal.

**EXAMPLE 5.4** ***Least Squares Test (Additive Model), Mid-Town Heating Data***

The values of the variables $t$ and $y_t$ as well as those of the indicators $x_1$, $x_2$, $x_3$ to be entered into the computer regression program to perform the least squares procedure for the Mid-Town series, assuming an additive model, are given in Table 5.8.

## Step 1

After the data are entered and the program does the fitting, a printout similar to that in Figure 5.6 should result. From this printout,

SSR(trend + seasonal) = 44,052.7

SSE(trend + seasonal) = 1,886.5

**Table 5.8** **Computer Data for Least Squares Procedure, Mid-Town Heating, Additive Model**

| $t$ | $y_t$ | $x_1$ | $x_2$ | $x_3$ | $t$ | $y_t$ | $x_1$ | $x_2$ | $x_3$ |
|---|---|---|---|---|---|---|---|---|---|
| 1 | 135.9 | 1 | 0 | 0 | 11 | 36.4 | 0 | 0 | 1 |
| 2 | 50.1 | 0 | 1 | 0 | 12 | 85.7 | 0 | 0 | 0 |
| 3 | 25.2 | 0 | 0 | 1 | 13 | 175.7 | 1 | 0 | 0 |
| 4 | 93.1 | 0 | 0 | 0 | 14 | 90.7 | 0 | 1 | 0 |
| 5 | 162.9 | 1 | 0 | 0 | 15 | 50.4 | 0 | 0 | 1 |
| 6 | 66.3 | 0 | 1 | 0 | 16 | 114.1 | 0 | 0 | 0 |
| 7 | 28.4 | 0 | 0 | 1 | 17 | 178.4 | 1 | 0 | 0 |
| 8 | 128.1 | 0 | 0 | 0 | 18 | 93.0 | 0 | 1 | 0 |
| 9 | 155.3 | 1 | 0 | 0 | 19 | 43.3 | 0 | 0 | 1 |
| 10 | 78.5 | 0 | 1 | 0 | 20 | 116.6 | 0 | 0 | 0 |

**Figure 5.6** Computer Printout for Mid-Town Heating Data, Additive, Full Trend/ Seasonal Model

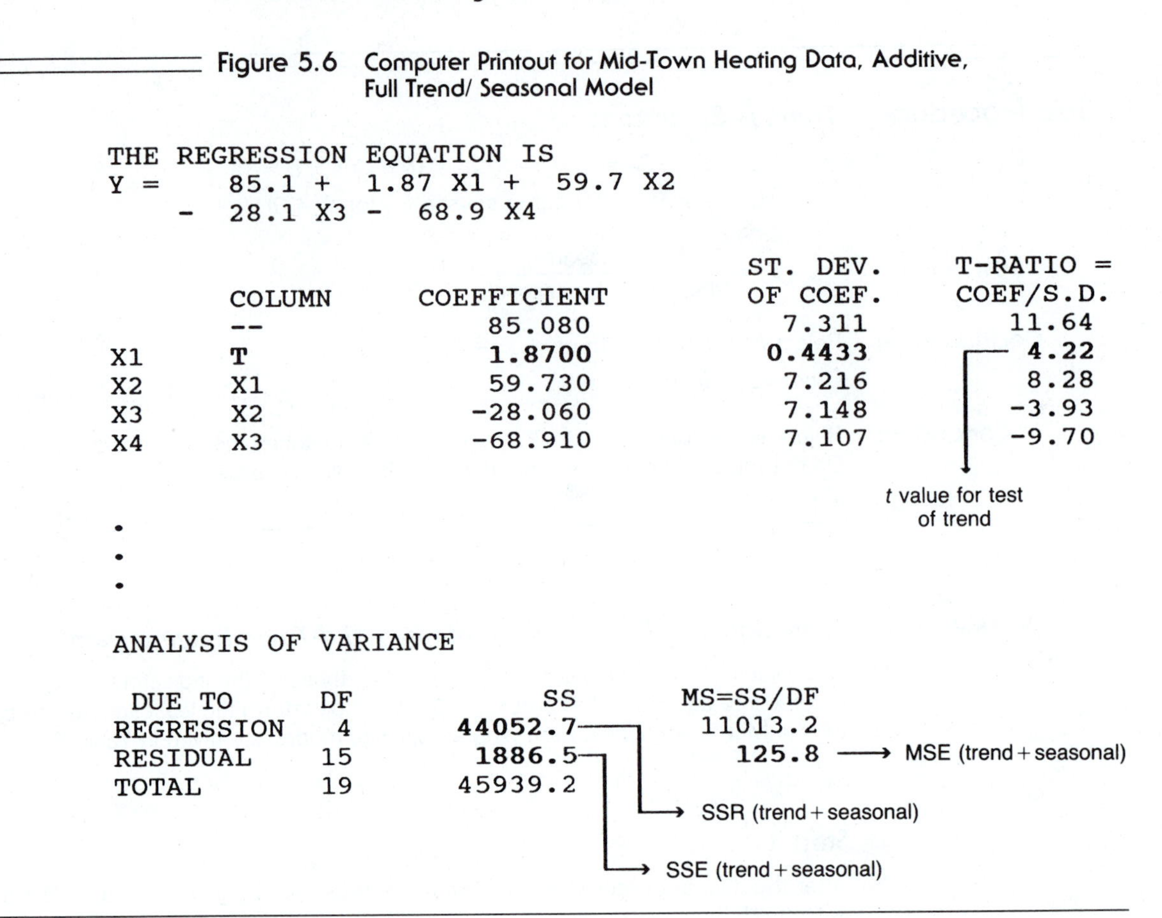

```
THE REGRESSION EQUATION IS
Y =     85.1 +   1.87 X1 +   59.7 X2
      - 28.1 X3 -   68.9 X4

                                          ST. DEV.     T-RATIO =
          COLUMN       COEFFICIENT        OF COEF.     COEF/S.D.
          --              85.080           7.311        11.64
X1        T               1.8700           0.4433        4.22
X2        X1              59.730           7.216         8.28
X3        X2             -28.060           7.148        -3.93
X4        X3             -68.910           7.107        -9.70

•
•
•

ANALYSIS OF VARIANCE

 DUE TO        DF            SS          MS=SS/DF
REGRESSION      4         44052.7        11013.2
RESIDUAL       15          1886.5          125.8
TOTAL          19         45939.2
```



### Step 2

To test the trend for significance:

$H_0$: $\beta_1 = 0$

$H_a$: $\beta_1 \neq 0$

**Test Statistic:** $t = 4.22$ (from printout)

**Decision Rule:** ($\alpha = 5\%$, df $= n - L - 1 = 20 - 4 - 1 = 15$)

Reject $H_0$ if $|t| > t_{.025} = 2.131$. **(Appendix GG)**

**Conclusion:** Since $4.22 > 2.131$, we reject $H_0$ and conclude that the Mid-Town series has trend.

Since the trend component is significant, we now proceed to Step 4 of the least squares procedure.

### Step 4

The first two columns of data in Table 5.8 are now used to fit a simple linear trend to the Mid-Town series. The printout in Figure 5.7 results. The only thing needed from

Figure 5.7 Computer Printout for Mid-Town Heating Additive, Trend-Only Model

```
THE REGRESSION EQUATION IS
Y =     84.5 +  1.04 X1

                                          ST. DEV.     T-RATIO =
         COLUMN      COEFFICIENT          OF COEF.     COEF/S.D.
         --               84.46             23.28          3.63
X1       T                1.043             1.944          0.54

ANALYSIS OF VARIANCE

 DUE TO       DF             SS       MS=SS/DF
REGRESSION     1            723            723
RESIDUAL      18          45216           2512
TOTAL         19          45939
```

SSR (trend)

this printout is SSR(trend) = 723, which is the amount of variability that could be explained by trend only. The amount of improvement attributable to adding a seasonal component is then:

$$\text{SSR(seasonal)} = \text{SSR(trend + seasonal)} - \text{SSR(trend)} = 44{,}052.7 - 723 = 43{,}329.7$$

$$\text{MSR(seasonal)} = \frac{\text{SSR(seasonal)}}{L - 1} = \frac{43{,}329.7}{3} = 14{,}443.2$$

MSE(trend + seasonal) may be taken from the printout in Figure 5.6 or computed:

$$\text{MSE(trend + seasonal)} = \frac{\text{SSE(trend + seasonal)}}{n - L - 1} = \frac{1886.5}{20 - 4 - 1} = 125.8$$

And, finally, the $F$-statistic is

$$F = \frac{\text{MSR(seasonal)}}{\text{MSE(trend + seasonal)}} = \frac{14{,}443.2}{125.8} = 112.40$$

## Step 5

The $F$-test for seasonality may now be completed as follows:

$H_0$: $S_1 = S_2 = S_3 = 0$

$H_a$: $S_i \neq 0$ for some seasons

**Test Statistic:** $F = 112.40$

**Decision Rule:** ($\alpha = .05$, df $= L - 1 = 3$, $n - L - 1 = 15$)

Reject $H_0$ if $F > F_{.05} = 3.29$. (Appendix GG)

**Conclusion:** Since $112.4 > 3.29$, we reject $H_0$ and conclude that there is seasonality in the Mid-Town series.

This conclusion is consistent with the results of the other tests already conducted on the data. ●

**EXAMPLE 5.5** *Least Squares Test (Multiplicative Model), U.S. Return Rate Data*

Let us now apply the least squares testing procedure to the U.S. Return Rate data, assuming a multiplicative model. The progression and details for the multiplicative case are very similar to those for the additive one. Keep in mind that the values of the parameter estimates are immaterial at this stage, only the explained and unexplained variability are needed. To test the multiplicative model we first take the logarithms of the $y_t$'s: Thus, for $t = 1$, $y'_1 = \log y_1 = \log 12.24 = 2.505$; for $t = 2$, $y'_2 = \log y_2 = \log 15.70 = 2.754$; and so forth. The logarithms of the series and the other values to be input to the regression are given in Table 5.9, and a portion of the computer output is shown in Figure 5.8. We first test the significance of the trend component ($\beta'_1$):

$$H_0\colon\ \beta'_1 = 0$$
$$H_a\colon\ \beta'_1 \neq 0$$

**Test Statistic:** $t = -1.65$ (from printout)

**Decision Rule:** ($\alpha = 5\%$, df $= n - L - 1 = 24 - 4 - 1 = 19$)

Reject $H_0$ if $|t| > t_{.025} = 2.093$.

**Conclusion:** Since $|-1.65| < 2.093$, we do not reject $H_0$, which supports not including a trend component in the model.

Since the trend component is not significant, the data are refit to a model without trend; only $x_1$, $x_2$, and $x_3$ are included. The computer printout for this fit is given in

**Table 5.9 Computer Data for Least Squares Procedure, U.S. Return Rates, Multiplicative Model**

| $t$ | $y'_t$ | $x_1$ | $x_2$ | $x_3$ | $t$ | $y'_t$ | $x_1$ | $x_2$ | $x_3$ |
|---|---|---|---|---|---|---|---|---|---|
| 1 | 2.505 | 1 | 0 | 0 | 13 | 1.482 | 1 | 0 | 0 |
| 2 | 2.754 | 0 | 1 | 0 | 14 | 2.240 | 0 | 1 | 0 |
| 3 | 3.043 | 0 | 0 | 1 | 15 | 2.079 | 0 | 0 | 1 |
| 4 | 3.012 | 0 | 0 | 0 | 16 | 3.139 | 0 | 0 | 0 |
| 5 | 2.661 | 1 | 0 | 0 | 17 | 2.266 | 1 | 0 | 0 |
| 6 | 2.890 | 0 | 1 | 0 | 18 | 2.583 | 0 | 1 | 0 |
| 7 | 3.075 | 0 | 0 | 1 | 19 | 2.649 | 0 | 0 | 1 |
| 8 | 3.130 | 0 | 0 | 0 | 20 | 3.203 | 0 | 0 | 0 |
| 9 | 2.231 | 1 | 0 | 0 | 21 | 2.327 | 1 | 0 | 0 |
| 10 | 2.585 | 0 | 1 | 0 | 22 | 2.634 | 0 | 1 | 0 |
| 11 | 2.394 | 0 | 0 | 1 | 23 | 2.540 | 0 | 0 | 1 |
| 12 | 2.950 | 0 | 0 | 0 | 24 | 3.078 | 0 | 0 | 0 |

**Figure 5.8** Computer Printout for U.S. Return Rate Data, Multiplicative, Full Trend/Seasonal Model

```
THE REGRESSION EQUATION IS
Y =      3.29 - .0144 X1 -    .883 X2
     -  .500 X3 -  .470 X4

                                         ST. DEV.      T-RATIO =
       COLUMN       COEFFICIENT          OF COEF.      COEF/S.D.
       --              3.2868             0.1704        19.29
X1     T             -0.014393           0.008711      -1.65
X2     X1             -0.8832             0.1703        -5.18
X3     X2             -0.4998             0.1692        -2.95
X4     X3             -0.4697             0.1685        -2.79
```

*t* value for testing trend

Figure 5.9. The *F*-statistic can be computed directly from this output (the boldface is ours, not the computer's), where SSR(seasonal only), SSE(seasonal only), MSR(seasonal only), and MSE(seasonal only) are already computed:

$$F = \frac{\text{MSR(seasonal only)}}{\text{MSE(seasonal only)}} = 7.684$$

The last step, then, is to conduct the *F*-test for significance of seasonality, as below:

$H_0$: $S_1' = S_2' = S_3' = 0$

$H_a$: $S_i' \neq 0$ for some seasons

**Figure 5.9** Computer Printout for U.S. Return Rate Data, Multiplicative, Seasonal-Only Model

```
THE REGRESSION EQUATION IS
Y =      3.09 -   .840 X1 -   .471 X2
     -   .455 X3
              :
ANALYSIS OF VARIANCE

 DUE TO        DF         SS            MS=SS/DF
REGRESSION      3       2.12872          0.70957
RESIDUAL       20       1.84686          0.09234
TOTAL          23       3.97557
```

SSR (seasonal only) = 2.12872 SSE (seasonal only) = 1.84686
MSR (seasonal only) = .70957 MSE (seasonal only) = .09234

**Test Statistic:** $F = 7.684$

**Decision Rule:** ($\alpha = 5\%$, df $= L - 1 = 3$, $n - L = 20$)

Reject $H_0$ if $F > F_{.05} = 3.10$.

**Conclusion:** Since $7.648 > 3.10$, we reject $H_0$ and conclude that there is seasonality in the U.S. Return Rate series.

The overall conclusion from least squares testing, then, is that, if a multiplicative model of the form in Equation 5.4 is to be used for forecasting U.S. Return Rates, it should include a seasonal component but probably does not need a trend component. ●

## 5.4.8 Summary

In deciding whether a seasonal element should be added to a time series model, it is important to understand how the series is affected by repetitive annual phenomena. The pattern is often obvious, needing no formal statistical inquiry other than graphical examination of the data. In some more subtle instances, a hypothesis test may be necessary. The series should then be detrended before being tested.Three tests were discussed.

### Kruskal-Wallis Test

This test does not require that the error terms be normally distributed and is relatively easy to complete. It lacks some power compared to the other two tests and requires several years' worth of observations, but may be preferred in time series situations where normality is often in doubt. If this test concludes that seasonality is present, it is quite likely that the results of the other two tests will agree.

### "Rule of Thumb" for Seasonality

This test is an approximate one, normally distributed errors are assumed, and about four years' worth of data are needed, but the test is by far the simpler, both in concept and execution, of the two parametric procedures discussed. It should prove adequate in many practical situations when a desire for simplicity dominates.

### Least Squares Test for Seasonality

Normal errors are required, but the test can be run (theoretically) with a minimum of $L + 2$ observations. Provided the assumptions of regression analysis are realistic, this is the most sensitive of the three alternatives, but it is by far the most complex. At least a rudimentary knowledge of regression principles is necessary to its proper use and interpretation. Its complexity is mitigated by the fact that calculations can usually be computer aided.

## 5.5 Single Forecast Models

### 5.5.1 Introduction

The body of methodology used to handle seasonality in a time series has been developed with several objectives in mind:

- **filtering** to *isolate* seasonal changes from other components of variability so that purely seasonal characteristics may be studied
- **modeling** to *describe and explain* the nature of seasonal influences
- **seasonal adjustment** to *adjust observed series* for seasonality (deseasonalize) so that the nonseasonal features are more clearly revealed for study and analysis
- **seasonalization** to *incorporate* seasonality into forecasts

Filtering has already been encountered in connection with testing for seasonality. General seasonal models were also examined in that context. We now consider the problems involved with estimating the seasonal parameters (seasonal indexes) $S_1, S_2, \ldots, S_L$ of these models and the use of these estimates in seasonal adjustment and forecasting.

Starting as usual with the single forecast case, the simplest of circumstances is treated first, i.e., a horizontal series with random error terms. The additive and multiplicative models are given in the following table.

| | Horizontal Seasonal Models |
|---|---|
| **Additive:** | $y_t = \beta_0 + S_t + \epsilon_t$ |
| **Multiplicative:** | $y_t = \beta_0 \cdot S_t \cdot \epsilon_t$ |

The $S_t$'s are additive (or multiplicative) seasonal indexes that sum annually to zero (or $L$), with $y_t$, $\beta_0$, and $\epsilon_t$ having their usual meanings.

The strategy when fitting seasonal models is to first use an appropriate technique to estimate the seasonal indexes, then use them to seasonally adjust the original series. After seasonal adjustment, the model's trend element, which is now unobscured by unmodeled seasonality, is estimated. In broad terms the procedure is as follows.

- Describe the seasonality.
- Remove it from the series.
- Fit a trend to the seasonality-free series.

### 5.5.2 The Method of Simple Averages, An Intuitive Procedure

Suppose there are $L$ seasons and a history of $n$ consecutive $y_t$'s. Let $n_i$ = the number of observations in the $i$th season. To simplify the discussion let us also suppose that there are $m$ complete years of data, with observation $y_1$

falling in the first season, $y_2$ in the second, and so on through $y_L$ in the $L$th, $y_{L+1}$ in the first, $y_{L+2}$ in the second, and so forth. We can derive estimates of the seasonal indexes for an additive model by noting that the average of the $y_t$'s in the first season, call it $\bar{y}_1$, would be

$$\bar{y}_1 = \frac{\sum_{\text{1st season}} y_t}{n_1} = \frac{y_1 + y_{L+1} + y_{2L+1} + \cdots}{n_1}$$

$$= \frac{\beta_0 + S_1 + \epsilon_1 + \beta_0 + S_1 + \epsilon_{L+1} + \beta_0 + S_1 + \epsilon_{2L+1} + \cdots}{n_1}$$

$$= \frac{n_1\beta_0 + n_1 S_1 + (\epsilon_1 + \epsilon_{L+1} + \epsilon_{2L+1} + \cdots)}{n_1}$$

$$= \beta_0 + S_1 + \bar{\epsilon}_1 \tag{5.7}$$

where $\bar{\epsilon}_1$ is the average of the error terms in the first season.

Likewise,

$$\bar{y}_2 = \beta_0 + S_2 + \bar{\epsilon}_2$$

$$\bar{y}_3 = \beta_0 + S_3 + \bar{\epsilon}_3$$

$$\vdots$$

$$\bar{y}_L = \beta_0 + S_L + \bar{\epsilon}_L \tag{5.8}$$

By adding all the Equations 5.7 and 5.8 together and averaging, we get

$$\bar{y} = \text{overall average}$$

$$= \frac{\sum y_t}{n}$$

$$= \frac{\bar{y}_1 + \bar{y}_2 + \cdots + \bar{y}_L}{L}$$

$$= \beta_0 + \left(\sum_1^L S_i\right) + \bar{\epsilon} \tag{5.9}$$

where $\bar{\epsilon}$ is the overall average error.

Now, since $\sum S_i$ is 0 in the additive case, and $\bar{\epsilon}$, being the average of several additive error terms, should be near zero, then $\bar{y}$ should be a good estimate of $\beta_0$, i.e., $\hat{\beta}_0 = \bar{y}$. Finally, if we suppose that $\bar{\epsilon}_i$'s, again averages of additive error terms, are also near zero, then $\bar{y}_i - \bar{y}$ should be an estimate of $S_i$, i.e., $\hat{S}_i = \bar{y}_i - \bar{y}$. The only complication that arises is when the $n_i$ are not all equal. In that case $\bar{y}$ will have to be computed using Equation 5.9, that is by averaging the $\bar{y}_i$'s rather than by simply averaging all the observations.

The derivation is similar for the multiplicative model, that is,

$$\bar{y}_1 = \frac{\beta_0 \cdot S_1 \cdot \epsilon_1 + \beta_0 \cdot S_1 \cdot \epsilon_{L+1} + \beta_0 \cdot S_1 \cdot \epsilon_{2L+1} + \cdots}{n_1}$$

$$= \frac{\beta_0 \cdot S_1 \cdot (\epsilon_1 + \epsilon_{L+1} + \epsilon_{2L+1} + \cdots)}{n_1}$$

$$= \beta_0 \cdot S_1 \cdot \bar{\epsilon}_1$$
$$\bar{y}_2 = \beta_0 \cdot S_2 \cdot \bar{\epsilon}_2$$
$$\vdots$$
$$\bar{y}_L = \beta_0 \cdot S_L \cdot \bar{\epsilon}_L \quad \text{(5.10)}$$

and

$$\bar{y} = \frac{\beta_0 \cdot (S_1 \cdot \bar{\epsilon}_1 + S_2 \cdot \bar{\epsilon}_2 + \cdots + S_L \cdot \bar{\epsilon}_L)}{L}$$

But since the $\epsilon_i$'s are random multiplicative errors, they should average about 1.0 in any season. Thus

$$\bar{y} \simeq \frac{\beta_0 \cdot (S_1 + S_2 + \cdots + S_L)}{L}$$

The $S_i$'s are multiplicative indexes, so they sum to $L$; therefore, $\bar{y}$ should be a reasonable estimate of $\beta_0$. From Equations 5.10, since the $\bar{\epsilon}_i$'s should be approximately 1.0 and $\hat{\beta}_0 = \bar{y}$,

$$\bar{y}_1 \simeq \beta_0 \cdot S_1 \rightarrow \hat{S}_1 = \frac{\bar{y}_1}{\bar{y}}$$
$$\bar{y}_2 \simeq \beta_0 \cdot S_2 \rightarrow \hat{S}_2 = \frac{\bar{y}_2}{\bar{y}}$$
$$\vdots$$
$$\bar{y}_L \simeq \beta_0 \cdot S_L \rightarrow \hat{S}_L = \frac{\bar{y}_L}{\bar{y}}$$

Forecasting with these models can be thought of as simply forecasting with the horizontal model from Chapter 2, $\hat{y}_{t+p} = \hat{\beta}_0$, and then seasonalizing the forecast by either adding or multiplying by the appropriate seasonal index; that is,

$$\hat{y}_{t+p} = \hat{\beta}_0 + \hat{S}_t \qquad \text{or} \qquad \hat{y}_{t+p} = \hat{\beta}_0 \cdot \hat{S}_t$$

which means the forecast for all periods in the $i$th season would be

$$\hat{y}_{t+p} = \bar{y} + (\bar{y}_i - \bar{y}) = \bar{y}_i \qquad \text{or} \qquad \hat{y}_{t+p} = \bar{y}\left(\frac{\bar{y}_i}{\bar{y}}\right) = \bar{y}_i$$

Thus, the forecast would be the same for both models. As a matter of fact, the forecasting process using simple averages to estimate the seasonal indexes in a horizontal model amounts to simply forecasting each of the $L$ individual seasons with their historical season averages (see Thomopoulos, 1980, ch. 9).

### Single Forecasts Using Simple Averages, Horizontal Seasonal Model

**Given:** A sample of $n$ observations $\{y_t;\ t = 1, 2, \ldots, n\}$ from $L$ seasons, $n_i$ observations in the $i$th season

**Model:**

$$y_t = \beta_0 + S_t + \epsilon_t \quad \text{(additive)}$$

or

$$y_t = \beta_0 \cdot S_t \cdot \epsilon_t \quad \text{(multiplicative)}$$

**Estimates:**

$$\bar{y} = \frac{(\bar{y}_1 + \cdots + \bar{y}_L)}{L} \qquad \bar{y}_i = \frac{\sum\limits_{i\text{th season}} y_t}{n_i}$$

$$\hat{\beta}_0 = \bar{y}$$

$$\hat{S}_i = \bar{y}_i - \bar{y} \quad \text{(additive)}$$

$$\hat{S}_i = \frac{\bar{y}_i}{\bar{y}} \quad \text{(multiplicative)}$$

**Forecasts:** (At time $t$ for $p$ periods ahead, in the $i$th season)

$$\hat{y}_{t+p} = \hat{\beta}_0 + \hat{S}_i \quad \text{(additive)}$$

$$\hat{y}_{t+p} = \hat{\beta}_0 \cdot \hat{S}_i \quad \text{(multiplicative)}$$

$$= \bar{y}_i \quad \text{(either additive or multiplicative)}$$

## EXAMPLE 5.6 *Simple Averages, U.S. Return Rates*

Previous examination of the U.S. Return Rate series (Example 5.5) indicated that it is untrended. Suppose we wish to forecast and/or seasonally adjust this series using a simple horizontal model and the method of simple averages. The preliminary calculations, which are quite straightforward, are shown in Table 5.10. From the results in that table the estimated seasonal indexes can be determined:

$$n_1 = n_2 = n_3 = n_4 = 6 \qquad L = 4$$

$$\bar{y}_1 = \frac{60.15}{6} = 10.03 \qquad \bar{y}_3 = \frac{88.41}{6} = 14.74$$

$$\bar{y}_2 = \frac{83.52}{6} = 13.92 \qquad \bar{y}_4 = \frac{131.71}{6} = 21.95$$

$$\bar{y} = \frac{10.03 + 13.92 + 14.74 + 21.95}{4} = 15.16$$

$$\hat{\beta}_0 = \bar{y} = 15.16$$

*Additive Seasonal Indexes:*

$$\hat{S}_1 = \bar{y}_1 - \bar{y} = 10.03 - 15.16 = -5.13$$

$$\hat{S}_2 = \bar{y}_2 - \bar{y} = 13.92 - 15.16 = -1.24$$

$$\hat{S}_3 = \bar{y}_3 - \bar{y} = 14.74 - 15.16 = -.42$$

$$\hat{S}_4 = \bar{y}_4 - \bar{y} = 21.95 - 15.16 = +6.79$$

**Table 5.10** Computations for Simple Averages Estimates of Seasonal Indexes, U.S. Return Rate Series

| | Season | | | |
|---|---|---|---|---|
| | 1 | 2 | 3 | 4 |
| 1979 | 12.24 | 15.70 | 20.97 | 20.32 |
| 1980 | 14.31 | 18.00 | 21.66 | 22.88 |
| 1981 | 9.31 | 13.26 | 10.96 | 19.11 |
| 1982 | 4.40 | 9.39 | 8.00 | 23.07 |
| 1983 | 9.64 | 13.24 | 14.14 | 24.61 |
| 1984 | 10.25 | 13.93 | 12.68 | 21.72 |
| Sum | 60.15 | 83.52 | 88.41 | 131.71 |

*Multiplicative Seasonal Indexes:*

$$\hat{S}_1 = \frac{\bar{y}_1}{\bar{y}} = \frac{10.03}{15.16} = .662\ (-33.8\%)$$

$$\hat{S}_2 = \frac{\bar{y}_2}{\bar{y}} = \frac{13.92}{15.16} = .918\ (-8.2\%)$$

$$\hat{S}_3 = \frac{\bar{y}_3}{\bar{y}} = \frac{14.74}{15.16} = .972\ (-2.8\%)$$

$$\hat{S}_4 = \frac{\bar{y}_4}{\bar{y}} = \frac{21.95}{15.16} = 1.448\ (+44.8\%)$$

The forecasts for each season (quarter) (for either model, of course) are:

*Forecasts:*

Season 1: $\bar{y}_1 = 10.03$

Season 2: $\bar{y}_2 = 13.92$

Season 3: $\bar{y}_3 = 14.74$

Season 4: $\bar{y}_4 = 21.95$

If we wanted to seasonally adjust (deseasonalize) the data for, say, 1984, we would either subtract the additive seasonal index or divide by the multiplicative one, as follows. ●

**Seasonally Adjusted Data for 1984**

| Additive Model: | Multiplicative Model: |
|---|---|
| **I, 84:** $y_{21} - \hat{S}_1 = 10.25 - (-5.13) = 15.38$ | **I, 84:** $\frac{y_{21}}{\hat{S}_1} = \frac{10.25}{.662} = 15.48$ |
| **II, 84:** $y_{22} - \hat{S}_2 = 13.93 - (-1.24) = 15.17$ | **II, 84:** $\frac{y_{22}}{\hat{S}_2} = \frac{13.93}{.918} = 15.17$ |
| **III, 84:** $y_{23} - \hat{S}_3 = 12.68 - (-.42) = 13.10$ | **III, 84:** $\frac{y_{23}}{\hat{S}_3} = \frac{12.68}{.972} = 13.04$ |
| **IV, 84:** $y_{24} - \hat{S}_4 = 21.72 - (+6.79) = 14.96$ | **IV, 84:** $\frac{y_{24}}{\hat{S}_4} = \frac{21.72}{1.448} = 15.00$ |

### 5.5.3 The Method of Simple Averages with Trend

Provided several full years of data are available (suppose we have $m$ full years of data, so $n = mL$), the simple averages method may be easily modified to handle additive models with linear trend. By taking logarithms before fitting, this modified method may also be used to fit multiplicative models with simple exponential growth. These two models are given in the table below. The primed parameters in the additive logarithmic model are restored to the unprimed ones to give the multiplicative model after fitting is complete.

**Trended Seasonal Models (Simple Averages)**

**Additive—Linear Trend:**

$y_t = \beta_0 + \beta_1 t + S_t + \epsilon_t$

**Multiplicative—Simple Exponential Trend:**

$y_t' = \log y_t = \beta_0' + \beta_1' t + S_t' + \epsilon_t'$

or

$y_t = \beta_0(\beta_1)^t \cdot S_t \cdot \epsilon_t$

#### Linear Trend

Trend causes difficulty in the simple averages method because the series grows (or declines) in successive seasons quite apart from any seasonality.

**Single Forecasts Using Simple Averages, Additive Model with Linear Trend**

**Given:** A sample of $n$ observations $y_t$; $t = 1, 2, \cdots, n$, from $L$ seasons in $m$ full years ($n = mL$)

**Model:** $y_t = \beta_0 + \beta_1 t + S_t + \epsilon_t$ where $\Sigma S_t = 0$

**Estimates:**

$$\bar{y}_i = \frac{\underset{i\text{th season}}{\Sigma y_t}}{m} \qquad i = 1, 2, \cdots, L$$

$$\bar{y}_i(\text{adj}) = \bar{y}_i - \frac{(i-1)\hat{\beta}_1^*}{L} \qquad i = 1, 2, \cdots, L$$

where $\hat{\beta}_1^*$ is the slope of a line fit to the yearly averages of the $y_t$'s.

$$\bar{y}(\text{adj}) = \frac{\bar{y}_1(\text{adj}) + \bar{y}_2(\text{adj}) + \cdots + \bar{y}_L(\text{adj})}{L}$$

$$\hat{S}_i = \bar{y}_i(\text{adj}) - \bar{y}(\text{adj}) \qquad i = 1, 2, \cdots, L$$

**Forecasting Model:**

$\hat{y}_t = \hat{\beta}_0 + \hat{\beta}_1 t + \hat{S}_t$

$\hat{S}_t = \hat{S}_i$ for $t$ in the $i$th season

where $\hat{\beta}_0$ and $\hat{\beta}_1$ are, respectively, the slope and intercept of a line fit to the $(y_t - \hat{S}_t)$'s, i.e., the seasonally adjusted $y_t$'s.

This can be corrected by adjusting the seasonal averages for the amount of trend change to be expected from one season to the next throughout the year. For example, if the series were growing at the linear rate of 100 units per year and if we average consecutive full years, Quarter II's average could be expected to be 25 units (one-fourth of the year's growth) larger than Quarter I's, Quarter III's 50 units larger, and Quarter IV's 75 units larger. Subtracting 25 units from $\bar{y}_2$, 50 units from $\bar{y}_3$, and 75 from $\bar{y}_4$ would then eliminate the differences due to trend, and we could proceed pretty much as before to estimate the seasonal indexes.

To make the adjustment, an estimate of the slope of the trend line is needed. This is obtained by fitting a line to the *yearly averages* of the data, with time scale $t = 1$ in the first year, $t = 2$ in the second, etc. Seasonal effects are eliminated in these averages and therefore do not interfere with the trend estimates.

Suppose $\hat{\beta}_1^*$ is the slope estimate obtained above. Then the seasonal averages are adjusted by subtracting $\hat{\beta}_1^*/L$ from $\bar{y}_2$, $2\hat{\beta}_1^*/L$ from $\bar{y}_3$, on through $(L - 1)\hat{\beta}_1^*/L$ from $\bar{y}_L$. The seasonal indexes are then computed using these adjusted averages exactly as they were for the horizontal model; that is,

$$\bar{y}_i(\text{adj}) = \bar{y}_i - \frac{(i-1)\hat{\beta}_1^*}{L} \qquad i = 1, 2, \cdots, L$$

$$\bar{y}(\text{adj}) = \frac{\Sigma\, \bar{y}_i(\text{adj})}{L}$$

$$\hat{S}_i = \bar{y}_i(\text{adj}) - \bar{y}(\text{adj})$$

To fit the linear trend part of the model, the above seasonal indexes are used to seasonally adjust the original series and a line is fit to the deseasonalized data.

**EXAMPLE 5.7** *Simple Averages with Linear Trend, Mid-Town Heating*

To apply the simple averages technique to the Mid-Town series ($n = 20$, $m = 4$, and $L = 4$), which is trended, the yearly averages $y_t^*$ are computed and the slope of the least squares line is found as shown in Table 5.11. From this table, $\hat{\beta}_1^* = 7.470$ (which, incidently, has a significant $t$-value). So $\hat{\beta}_1^*/4 = 1.87$ The rest of the simple averages calculations can now be done (Table 5.12). Then the seasonal indexes are calculated:

$$\hat{S}_1 = \bar{y}_1(\text{adj}) - \bar{y}(\text{adj}) = 161.64 - 92.60 = +69.04$$

$$\hat{S}_2 = \bar{y}_2(\text{adj}) - \bar{y}(\text{adj}) = 73.85 - 92.60 = -18.75$$

$$\hat{S}_3 = \bar{y}_3(\text{adj}) - \bar{y}(\text{adj}) = 33.00 - 92.60 = -59.60$$

$$\hat{S}_4 = \bar{y}_4(\text{adj}) - \bar{y}(\text{adj}) = 101.91 - 92.60 = +9.31$$

To complete the model estimation process, the seasonal indexes are used to deseasonalize the original series, then a least squares linear trend is fit to the deseasonalized data. The data and a computer printout of the fit are shown in Table 5.13.

**Table 5.11 Linear Fit to Mid-Town Heating, Yearly Averages**
$y_t^*$ = year's quarterly average ($1,000)

| Year | $t$ | $y_t^*$ |
|---|---|---|
| 1981 | 1 | 76.1 |
| 1982 | 2 | 96.4 |
| 1983 | 3 | 89.0 |
| 1984 | 4 | 107.7 |
| 1985 | 5 | 107.8 |

```
THE REGRESSION EQUATION IS
Y =      73.0 +   7.47 X1
                                           ST. DEV.     T-RATIO =
        COLUMN      COEFFICIENT            OF COEF.     COEF/S.D.
        --               72.990              7.695           9.49

X1      TIME              7.470 ──→ β̂₁*      2.320           3.22

THE ST. DEV. OF Y ABOUT REGRESSION LINE IS
S = 7.337
WITH (   5- 2) =   3 DEGREES OF FREEDOM

R-SQUARED = 77.6 PERCENT
R-SQUARED = 70.1 PERCENT, ADJUSTED FOR D.F.

ANALYSIS OF VARIANCE

 DUE TO      DF          SS          MS=SS/DF
REGRESSION    1       558.01          558.01
RESIDUAL      3       161.49           53.83
TOTAL         4       719.50
```

***Note:*** $y_1^* = (135.9 + 50.1 + 25.2 + 93.1)/4 = 76.1$

**Table 5.12 Simple Averages Calculations, Linear Trend, Additive Model**

| | Season | | | | | |
|---|---|---|---|---|---|---|
| Year | 1 | 2 | 3 | 4 | Yearly Total | Yearly Average |
| 1981 | 135.9 | 50.1 | 25.2 | 93.1 | 304.3 | 76.1 |
| 1982 | 162.9 | 66.3 | 28.4 | 128.1 | 385.7 | 96.4 |
| 1983 | 155.3 | 78.5 | 36.4 | 85.7 | 355.9 | 89.0 |
| 1984 | 175.7 | 90.7 | 50.4 | 114.1 | 430.9 | 107.7 |
| 1985 | 178.4 | 93.0 | 43.3 | 116.6 | 431.3 | 107.8 |
| Sum | 808.2 | 378.6 | 183.7 | 537.6 | | |
| $\bar{y}_i$ | 161.64 | 75.72 | 36.74 | 107.52 | $\hat{\beta}_1^* = 7.47$ | |
| Trend | 0.00 | 1.87 | 3.74 | 5.61 | $\hat{\beta}_1^*/4 = 1.87$ | |
| $\bar{y}_i$(adj) | 161.64 | 73.85 | 33.00 | 101.91 | | |

$$\bar{y}(\text{adj}) = \frac{161.64 + 73.85 + 33.00 + 101.91}{4} = \frac{370.4}{4} = 92.60$$

**Table 5.13 Least Squares Trend Fit to Deseasonalized Series (Simple Averages), Mid-Town Heating** $y_t - \hat{S}_t$ = seasonally adjusted receipts ($1,000)

| | Quarter: | | | |
|---|---|---|---|---|
| | 1 | 2 | 3 | 4 |
| 1981 | 66.86 | 68.85 | 84.80 | 83.79 |
| 1982 | 93.86 | 85.05 | 88.00 | 118.79 |
| 1983 | 86.26 | 97.25 | 96.00 | 76.39 |
| 1984 | 106.66 | 109.45 | 110.00 | 104.79 |
| 1985 | 109.36 | 111.75 | 102.90 | 107.29 |

```
THE REGRESSION EQUATION IS
Y =     75.8 +   1.87 X1
                                                ST. DEV.     T-RATIO =
        COLUMN        COEFFICIENT               OF COEF.     COEF/S.D.

        --                 75.770 ⟶ β̂0           4.756          15,93

X1      TIME               1.8700 ⟶ β̂1          0.3970           4.71

THE ST. DEV. OF Y ABOUT REGRESSION LINE IS
S = 10.24
WITH (  20- 2) =   18 DEGREES OF FREEDOM

R-SQUARED = 55.2 PERCENT

ANALYSIS OF VARIANCE

 DUE TO       DF          SS        MS=SS/DF
REGRESSION     1       2325.4        2325.4
RESIDUAL      18       1886.5         104.8
TOTAL         19       4212.0
```

The complete model can now be assembled by combining the trend from the last fit with the additive seasonal indexes obtained before, giving:

*Final Model:*

$$\hat{y}_t = \hat{\beta}_0 + \hat{\beta}_1 t + \hat{S}_t$$
$$= 75.77 + 1.870t + \hat{S}_t$$

Origin: IV, 80
Units ($t$): Quarters
Units ($y$): $1,000

where

$\hat{S}_1 = +69.04 \qquad \hat{S}_2 = -18.75 \qquad \hat{S}_3 = -59.60 \qquad \hat{S}_4 = -9.31$ •

## Simple Exponential Trend

While there is no simple way to correct the seasonal averages for the effects of simple exponential trend, the problem can be avoided by changing from $y_t$ to $y_t' = \log y_t$, using the simple averages/linear-trend method on the $y_t'$'s to get an *additive logarithmic* model, and then restoring the original multiplicative model by exponentiating. In summary, the steps (assuming $n$ observations from $L$ seasons in $m$ full years) are as follows.

*Step 1:* Change to logarithms: $y'_t = \log y_t$.

*Step 2:* Fit a linear model to the yearly-average logarithms, giving slope $\hat{\beta}_1^*$.

*Step 3:* Perform the simple averages calculations using $\hat{\beta}_1^*/L$ to adjust seasonal-average logarithms, then compute logarithmic additive seasonal indexes as

$$\hat{S}'_i = \bar{y}'_i(\text{adj}) - \bar{y}'(\text{adj})$$

*Step 4:* Seasonally adjust the logarithms using the indexes obtained in Step 3 by computing $y'_t - \hat{S}'_t$, then fit a linear trend to the seasonally adjusted logarithms, obtaining $\hat{T}'_t = \hat{\beta}'_0 + \hat{\beta}'_1 t$ and the additive logarithmic forecasting model

$$\hat{y}'_t = \hat{T}'_t + \hat{S}'_t = \hat{\beta}'_0 + \hat{\beta}'_1 t + \hat{S}'_t \tag{5.11}$$

Once the additive logarithmic model has been fit, it is possible to forecast $y'_t$ from Equation 5.11 and then convert to a forecast of $y_t$ by exponentiating, i.e.,

$$\hat{y}_t = e^{\hat{y}'_t}$$

But, it is more conventional to restore the original multiplicative model expressed in terms of the original (unprimed) parameters.

*Step 5:* Restore the original unprimed parameters using

$$\hat{\beta}_0 = e^{\hat{\beta}'_0} \qquad \hat{\beta}_1 = e^{\hat{\beta}'_1} \qquad \hat{S}_i = e^{\hat{S}'_i} \qquad i = 1, 2, \cdots, L$$

Because the $\hat{S}'_i$'s sum to zero, the product of the $S_i$'s will be 1. However, the $S_i$'s will not satisfy the conventional constraint that multiplicative seasonals sum to $L$. To convert to conventional indexes we need only multiply each by an appropriate constant and absorb the effect into $\beta_0$. This will have no real effect on the model; it is only a reparameterization. The constant to use is $\bar{S} = (\Sigma\ \hat{S}_i)/L$, and the conversion is

$$\hat{S}_1 \rightarrow \frac{\hat{S}_1}{\bar{S}}$$

$$\hat{S}_2 \rightarrow \frac{\hat{S}_2}{\bar{S}}$$

$$\vdots$$

$$\hat{S}_L \rightarrow \frac{\hat{S}_L}{\bar{S}}$$

$$\hat{\beta}_0 \rightarrow \hat{\beta}_0 \cdot \bar{S}$$

with $\hat{\beta}_1$ remaining unchanged.

## Single Forecasts Using Simple Averages, Multiplicative Model with Simple Exponential Trend

**Given:** A sample of $n$ observations $y_1, y_2, \ldots, y_n$ from $L$ seasons in $m$ full years ($n = mL$).

**Model:** $y_t = \beta_0 \cdot (\beta_1)^t \cdot S_t \cdot \epsilon_t$ where $\Sigma S_t = L$

**Estimates:** Change $y_t$ to $y'_t = \log y_t$ and fit an additive model with linear trend to the $y'_t$'s, giving $\hat{\beta}'_0$, $\hat{\beta}'_1$, and $\hat{S}'_i$; $i = 1, 2, \ldots, L$ as parameter estimates for the additive logarithmic model ($\hat{\beta}'_0$ and $\hat{\beta}'_1$ being the result of fitting a linear trend to the seasonally adjusted logarithms). Then compute the estimates of the unprimed parameters as

$$\hat{\beta}_0 = e^{\hat{\beta}'_0} \qquad \hat{\beta}_1 = e^{\hat{\beta}'_1} \qquad \hat{S}_i = e^{\hat{S}'_i} \qquad i = 1, 2, \cdots, L$$

and the conversion factor $\bar{S} = (\Sigma \hat{S}_i)/L$ to reparameterize the model so it has conventional seasonal indexes ($\Sigma \hat{S}_i = L$):

$$\hat{S}_i \rightarrow \frac{\hat{S}_i}{\bar{S}} \qquad i = 1, 2, \cdots, L$$

$$\hat{\beta}_0 \rightarrow \hat{\beta}_0 \cdot \bar{S}$$

$$\hat{\beta}_1 \rightarrow \hat{\beta}_1$$

**Forecasting Model:**

$$\hat{y}_t = \hat{\beta}_0 \cdot (\hat{\beta}_1)^t \cdot \hat{S}_t$$

$$\hat{S}_t = \hat{S}_i \qquad \text{for } t \text{ in the } i\text{th season}$$

where $\hat{\beta}_0$, $\hat{\beta}_1$, and $\hat{S}_i$ are the converted parameter estimates.

### EXAMPLE 5.8 *Simple Averages with Simple Exponential Trend, Mid-Town Heating*

To illustrate the process, suppose we return to the Mid-Town Heating data once more and fit a multiplicative model with simple exponential growth. The various calculations required are summarized in Tables 5.14 and 5.15. We will refer to these tables as we go along.

*Step 1:* Change to the logarithms (Table 5.14):

$$y'_1 = \log(135.9) = 4.912$$

$$y'_2 = \log(50.1) = 3.914 \qquad \text{and so on.}$$

*Step 2:* Fit a linear trend (Table 5.14) to the yearly-average logarithms (4.417, 4.371, 4.363, 4.583, and 4.561), giving $\hat{\beta}^*_1 = .1040$.

*Step 3:* Perform the simple averages calculations (Table 5.14) using $\hat{\beta}^*_1/4 = .026$ to adjust the seasonal-average logarithms. Then compute the logarithmic additive seasonal indexes as:

$$\hat{S}'_1 = \bar{y}'_1(\text{adj}) - \bar{y}'(\text{adj}) = 5.081 - 4.366 = +.715$$

$$\hat{S}'_2 = \bar{y}'_2(\text{adj}) - \bar{y}'(\text{adj}) = 4.276 - 4.366 = -.090$$

$$\hat{S}'_3 = \bar{y}'_3(\text{adj}) - \bar{y}'(\text{adj}) = 3.519 - 4.366 = -.847$$

$$\hat{S}'_4 = \bar{y}'_4(\text{adj}) - \bar{y}'(\text{adj}) = 4.589 - 4.366 = +.223$$

**Table 5.14 Simple Averages Calculations, Simple Exponential Trend, Multiplicative Model, Mid-Town Heating**
$y'_t = \log y_t$ = logarithm of quarterly receipts ($1,000)

| | Season | | | | Yearly Total | Yearly Average |
|---|---|---|---|---|---|---|
| | 1 | 2 | 3 | 4 | | |
| 1981 | 4.912 | 3.914 | 3.227 | 4.534 | 16.587 | 4.147 |
| 1982 | 5.093 | 4.194 | 3.346 | 4.853 | 17.486 | 4.371 |
| 1983 | 5.045 | 4.363 | 3.595 | 4.451 | 17.454 | 4.363 |
| 1984 | 5.169 | 4.508 | 3.920 | 4.737 | 18.334 | 4.583 |
| 1985 | 5.184 | 4.533 | 3.768 | 4.759 | 18.244 | 4.561 |
| Sum | 25.403 | 21.512 | 17.856 | 23.334 | | |
| $\bar{y}'_i$ | 5.081 | 4.302 | 3.571 | 4.667 | $\hat{\beta}_1^* = .104$ | |
| Trend | .000 | .026 | .052 | .078 | $\hat{\beta}_1^*/4 = .026$ | |
| $\bar{y}'_i$(adj) | 5.081 | 4.276 | 3.519 | 4.589 | | |

$$\bar{y}'(\text{adj}) = \frac{5.081 + 4.276 + 3.519 + 4.589}{4} = \frac{17.465}{4} = 4.366$$

```
THE REGRESSION EQUATION IS
Y =     4.09 +  .104 X1
                                               ST. DEV.     T-RATIO =
       COLUMN       COEFFICIENT               OF COEF.     COEF/S.D.
       --              4.09300                0.07975         51.33

X1     T               0.10400 ——→ β̂1*        0.02404          4.33

THE ST. DEV. OF Y ABOUT REGRESSION LINE IS
S = 0.07604
WITH (   5- 2) =    3 DEGREES OF FREEDOM

R-SQUARED = 86.2 PERCENT

ANALYSIS OF VARIANCE

 DUE TO        DF          SS          MS=SS/DF
REGRESSION      1     0.108160         0.108160
RESIDUAL        3     0.017344         0.005781
TOTAL           4     0.125504
```

*Step 4:* Seasonally adjust the logarithms (Table 5.15) using the seasonals determined in Step 3:

$$y'_1 - \hat{S}'_1 = 4.912 - (+.715) = 4.197$$
$$y'_2 - \hat{S}'_2 = 3.914 - (-.090) = 4.004$$
$$y'_3 - \hat{S}'_3 = 3.227 - (-.847) = 4.074$$
$$y'_4 - \hat{S}'_4 = 4.534 - (+.223) = 4.311$$
$$y'_5 - \hat{S}'_5 = y'_5 - \hat{S}'_1 = 5.093 - (+.715) = 4.378$$

**Table 5.15** Least Squares Trend Fit to Deseasonalized Logarithm Series, Mid-Town Heating $y'_t - \hat{S}'_t$ = deseasonalized logs of receipts ($1,000)

| | Quarter | | | |
|---|---|---|---|---|
| | 1 | 2 | 3 | 4 |
| 1981 | 4.197 | 4.004 | 4.074 | 4.311 |
| 1982 | 4.378 | 4.284 | 4.193 | 4.630 |
| 1983 | 4.330 | 4.453 | 4.442 | 4.228 |
| 1984 | 4.454 | 4.598 | 4.767 | 4.514 |
| 1985 | 4.469 | 4.623 | 4.615 | 4.536 |

```
THE REGRESSION EQUATION IS
Y =     4.13 + .0260 X1
                                                 ST. DEV.     T-RATIO =
         COLUMN      COEFFICIENT                 OF COEF.     COEF/S.D.
         --               4.13186  ⟶ β̂'0         0.06189         66.76

X1       TIME            0.026014  ⟶ β̂'1        0.005166          5.04

THE ST. DEV. OF Y ABOUT REGRESSION LINE IS
S = 0.1332
WITH (  20- 2) =  18 DEGREES OF FREEDOM

R-SQUARED = 58.5 PERCENT

ANALYSIS OF VARIANCE

 DUE TO       DF          SS          MS=SS/DF
REGRESSION     1      0.45001          0.45001
RESIDUAL      18      0.31948          0.01775
TOTAL         19      0.76948
```

and so forth. Then fit a least squares linear trend (Table 5.15) to the deseasonalized logarithms, giving

$$\hat{T}'_t = \hat{\beta}'_0 + \hat{\beta}'_1 t = 4.132 + .0260t$$

and for the complete additive logarithmic model:

$$\hat{y}'_t = 4.132 + .0260t + \hat{S}'_t$$

Origin: IV, 80
Units ($t$): Quarters
Units ($y$): log($1,000)

$\hat{S}'_1 = +.715 \qquad \hat{S}'_2 = -.090 \qquad \hat{S}'_3 = -.847 \qquad \hat{S}'_4 = +.223$

*Step 5:* Restore the multiplicative model. First change to unprimed parameters:

$$\hat{\beta}_0 = e^{\hat{\beta}'_0} = e^{4.132} = 62.30$$

$$\hat{\beta}_1 = e^{\hat{\beta}'_1} = e^{.0260} = 1.026$$

$$\hat{S}_1 = e^{\hat{S}'_1} = e^{+.715} = 2.044$$

$$\hat{S}_2 = e^{\hat{S}'_2} = e^{-.090} = .914$$

$$\hat{S}_3 = e^{\hat{S}'_3} = e^{-.847} = .429$$

$$\hat{S}_4 = e^{\hat{S}'_4} = e^{+.223} = 1.249$$

Notice that these multiplicative seasonal indices are quite different from those given in Example 5.7 (1.746, .798, .356, and 1.101, respectively). The reason is that they have been constrained so that $\Pi\hat{S}_i = 1.0$, not so that $\Sigma\,\hat{S}_i = 4.0$. To get an equivalent model with *conventional* seasonals we compute

$$\bar{S} = \frac{\Sigma\,S_i}{L} = \frac{2.044 + .914 + .429 + 1.249}{4} = \frac{4.636}{4} = 1.159$$

and convert:

$$S_1 \rightarrow \frac{S_1}{\bar{S}} = \frac{2.044}{1.159} = 1.764$$

$$S_2 \rightarrow \frac{S_2}{\bar{S}} = \frac{0.914}{1.159} = .789$$

$$S_3 \rightarrow \frac{S_3}{\bar{S}} = \frac{0.429}{1.159} = .370$$

$$S_4 \rightarrow \frac{S_4}{\bar{S}} = \frac{1.249}{1.159} = 1.078$$

$$\hat{\beta}_0 \rightarrow \hat{\beta}_0 \cdot \bar{S} = 62.30 \times 1.159 = 72.19$$

$$\hat{\beta}_1 \rightarrow \hat{\beta}_1 = 1.026$$

to get the final multiplicative model.

*Final Multiplicative Model:*

$$\hat{y}_t = 72.19(1.026)^t \cdot \hat{S}_t$$

Origin: IV, 84
Units ($t$): Quarters
Units ($y$): $1,000

where

$$\hat{S}_1 = 1.764 \qquad \hat{S}_2 = .789 \qquad \hat{S}_3 = .370 \qquad \hat{S}_4 = 1.078$$

•

### 5.5.4 Moving Average and Ratio-to-Moving Average Methods

The moving average filter used in Section 5.4.4 to detrend a series before testing provides the basis for the most popular of the seasonal modeling techniques. Refinements (some rather elaborate) of the basic method described here are used in many highly regarded computer software packages, including the Census X-11 and SABL (Seasonal Adjustment, Bell Laboratories) programs.

A very attractive feature of the moving average filter is that it does not require knowledge of the functional form of the trend in the series, or even of whether the series has trend or not. It can also be used in the presence of cycles. The moving average (or centered moving average) is itself an estimate of the trend-cycle value at the center of the period over which it is computed and has even been said to define the "trend" or "empirical trend" of the series.

The essential rationale for the procedure is that the specific seasonals, $\hat{S}_t + \hat{\epsilon}_t$ (additive) or $\hat{S}_t \cdot \hat{\epsilon}_t$ (multiplicative), when averaged within a season will be accurate measures of the seasonal movement in that season. These averages may then be used to adjust the original series. The adjusted series, having much less variability, will then give a clearer picture of the nonseasonal behavior of the original series, leading to a better choice of functional form for the trend and allowing trend parameters to be estimated without seasonal interference.

Because the filtering procedure has already been discussed in Section 5.4.4, only the latter stages of the modeling process need be discussed here. We will therefore assume that the original series of $n$ observations has been filtered and that the $n - L$ (or, if $L$ should happen to be odd, $n - L + 1$) specific seasonals are available. There are two steps which remain:

*Step 1:* Compute the season averages (in very much the same way as for the simple averages method), i.e., letting $n_i$ be the number of specific seasonals in the $i$th season. Then

$$\hat{S}_i = \bar{y}_i = \frac{\sum\limits_{i\text{th season}} (\hat{S}_t + \hat{I}_t)}{n_i} \qquad \text{(additive)}$$

or

$$\hat{S}_i = \bar{y}_i = \frac{\sum\limits_{i\text{th season}} (\hat{S}_t \cdot \hat{I}_t)}{n_i} \qquad \text{(multiplicative)}$$

Then compute a conversion factor and convert the indexes so they sum to 0.0 (additive) or $L$ (multiplicative); i.e., compute $\bar{S} = (\Sigma\, \hat{S}_i)/L$ and convert:

$$\hat{S}_i \to \hat{S}_i - \bar{S} \qquad i = 1, 2, \cdots, L \qquad \text{(additive)}$$

$$\hat{S}_i \to \frac{\hat{S}_i}{\bar{S}} \qquad i = 1, 2, \cdots, L \qquad \text{(multiplicative)}$$

***Note:*** There are no $\beta$ parameters to be converted here because there is no assumed trend function.

*Step 2:* Deseasonalize the original series with the adjusted seasonal indexes $y_t - \hat{S}_t$ (additive) or $y_t/\hat{S}_t$ (multiplicative), and fit the desired trend model, $f(t; \beta_0, \beta_1, \ldots)$, to the seasonally adjusted series, giving $\hat{T}_t = f(t; \hat{\beta}_0, \hat{\beta}_1, \ldots)$ and the forecasting model

$$\hat{y}_t = \hat{T}_t + \hat{S}_t = f(t; \hat{\beta}_0, \hat{\beta}_1, \cdots) + \hat{S}_t \qquad \text{(additive)}$$

or

$$\hat{y}_t = \hat{T}_t \cdot \hat{S}_t = f(t; \hat{\beta}_0, \hat{\beta}_1, \cdots) \cdot \hat{S}_t \qquad \text{(multiplicative)}$$

**EXAMPLE 5.9** *Moving Average Method, Mid-Town Heating*

Suppose we wish to fit a seasonal model to the Mid-Town Heating series using a moving average filter. The specific seasonals (both additive and multiplicative) have already been obtained (Table 5.5). An additive model is considered first. To estimate seasonal indexes, we arrange the additive-specific seasonals, $S_t + \hat{\epsilon}_t$, by season and compute season averages (unadjusted indexes) as shown in Table 5.16. The averages are then adjusted so they sum to zero:

**Table 5.16 Calculation of Seasonal Averages (Additive) for Moving Average Filter, Mid-Town Heating** $\hat{S}_t + \hat{\epsilon}_t$ = additive specific seasonals ($1,000)

| | $\hat{S}_t + \hat{\epsilon}_t$ for Quarter: | | | |
|---|---|---|---|---|
| | 1 | 2 | 3 | 4 |
| 1981 | — | — | −54.25 | +8.25 |
| 1982 | +75.62 | −25.75 | −67.07 | +32.05 |
| 1983 | +56.72 | −15.78 | −55.13 | −9.90 |
| 1984 | +76.82 | −13.48 | −57.66 | +5.41 |
| 1985 | +70.31 | −14.51 | — | — |
| $\Sigma (\hat{S}_t + \hat{\epsilon}_t)$ | +279.47 | −69.52 | −234.11 | +35.81 |
| $\hat{S}_t$ | +69.87 | −17.38 | −58.53 | +8.95 |

*Season Averages* ($\hat{S}_i$): $\hat{S}_1 = +69.87 \quad \hat{S}_2 = -17.38 \quad \hat{S}_3 = -58.53 \quad \hat{S}_4 = +8.95$

*Adjustment of* $\hat{S}_i$:

$$\bar{S} = \frac{69.87 - 17.38 - 58.93 + 8.95}{4} = \frac{2.91}{4} = +.73$$

$$\hat{S}_1 \rightarrow \hat{S}_1 - \bar{S} = +69.87 - .73 = +69.14$$

$$\hat{S}_2 \rightarrow \hat{S}_2 - \bar{S} = -17.38 - .73 = -18.11$$

$$\hat{S}_3 \rightarrow \hat{S}_3 - \bar{S} = -58.53 - .73 = -59.26$$

$$\hat{S}_4 \rightarrow \hat{S}_4 - \bar{S} = +8.95 - .73 = +8.22$$

We can now seasonally adjust the original series with these indexes so that a trend model can be fit. The data and computer printout for such a trend fit are given in Table 5.17. (The least squares method and linear trend are chosen, although any trend fitting method and any appropriate functional form could be used.) From this fit, the estimated trend is

$$\hat{T}_t = 75.62 + 1.88t$$

which, combined with the seasonal indexes, produces the forecasting model.

*Forecasting Model:* $\hat{y}_t = \hat{T}_t + \hat{S}_t = 75.62 + 1.88t + \hat{S}_t$

Origin: IV, 80
Units ($t$): Quarters
Units ($y$): $1,000

$\hat{S}_1 = +69.14$ $\hat{S}_2 = -18.11$ $\hat{S}_3 = -59.26$ $\hat{S}_4 = +8.22$

**Table 5.17 Least Squares Linear Trend Fit to Deseasonalized Series (Centered Moving Average, Additive), Mid-Town Heating**

$y_t - \hat{S}_t$ = seasonally adjusted receipts ($1,000)

| | Quarter: | | | |
|---|---|---|---|---|
| | 1 | 2 | 3 | 4 |
| 1981 | 66.76 | 68.21 | 84.46 | 84.88 |
| 1982 | 93.76 | 84.41 | 87.66 | 119.88 |
| 1983 | 86.16 | 96.61 | 95.66 | 77.48 |
| 1984 | 106.56 | 108.81 | 109.66 | 105.88 |
| 1985 | 109.26 | 111.11 | 102.56 | 108.38 |

```
THE REGRESSION EQUATION IS
Y =      75.6 +   1.88 X1
                                              ST. DEV.     T-RATIO =
       COLUMN      COEFFICIENT                OF COEF.     COEF/S.D.
       --               75.620 ──→ β̂0           4.766         15.87

X1     TIME             1.8845 ──→ β̂1          0.3979          4.74

THE ST. DEV. OF Y ABOUT REGRESSION LINE IS
S = 10.26
WITH (   20- 2)  =   18 DEGREES OF FREEDOM

R-SQUARED = 55.5 PERCENT

ANALYSIS OF VARIANCE

 DUE TO        DF          SS         MS=SS/DF
REGRESSION      1       2361.8          2325.4
RESIDUAL       18       1895.0           105.3
TOTAL          19       4256.8
```

For a multiplicative model the procedure is essentially the same except that multiplicative-specific seasonals, $\hat{S}_t \cdot \hat{\epsilon}_t$, are used. Table 5.18 shows the computation of the season averages, which, in this particular case, do not need adjustment because they already sum to $L = 4$.

**Table 5.18 Calculation of Seasonal Averages (Multiplicative) for Moving Average Filter, Mid-Town Heating** $\hat{S}_t \cdot \hat{\epsilon}_t$ = multiplicative specific seasonals

| | Quarter | | | |
|---|---|---|---|---|
| | 1 | 2 | 3 | 4 |
| 1981 | — | — | .317 | 1.097 |
| 1982 | 1.867 | .720 | .297 | 1.334 |
| 1983 | 1.575 | .833 | .398 | .896 |
| 1984 | 1.777 | .871 | .466 | 1.050 |
| 1985 | 1.651 | .865 | — | — |
| $\Sigma(\hat{S}_t \cdot \hat{\epsilon}_t)$ | 6.870 | 3.289 | 1.478 | 4.377 |
| $\hat{S}_i$ | 1.716 | .822 | .369 | 1.093 |

*Multiplicative Seasonal Indexes:*

$\hat{S}_1 = 1.716\ (171.6\%)$ $\hat{S}_2 = .822\ (82.2\%)$

$\hat{S}_3 = .369\ (36.9\%)$ $\hat{S}_4 = 1.093\ (109.3\%)$

Because of the multiplicative model structure, a simple exponential trend is selected, although linear or other functional forms could be used as well. Least squares is used to fit a linear trend (Table 5.19) to the logarithms of the deseasonalized series, $y'_t = \log(y_t/\hat{S}_t)$, and converted to simple exponential form by taking antilogs:

$$\hat{T}'_t = \hat{\beta}'_0 + \hat{\beta}'_1 t = 4.276 + .0257t$$
$$\hat{\beta}_0 = e^{\hat{\beta}'_0} = e^{4.276} = 71.95$$
$$\hat{\beta}_1 = e^{\hat{\beta}'_1} = e^{.0257} = 1.026$$
$$\hat{T}_t = \hat{\beta}_0(\hat{\beta}_1)^t = 71.95(1.026)^t$$

The forecasting model then becomes the following.

*Forecasting Model:*

$$\hat{y}_t = \hat{T}_t \cdot \hat{S}_t = 71.95(1.026)^t \cdot \hat{S}_t$$

Origin: IV, 80
Units ($t$): Quarters
Units ($y$): $1,000

$\hat{S}_1 = 1.716$ $\hat{S}_2 = .822$ $\hat{S}_3 = .369$ $\hat{S}_4 = 1.093$ •

## Resistant Averaging

Since season averages are frequently taken over only a few years' worth of data (only a few observations in each season) and since the arithmetic mean is sensitive to extreme values, an unusually large irregular movement in any one season can cause estimates of all of the seasonal indexes to be in error. To avoid this problem, other methods of averaging the irregular movements have been devised, methods that are "resistant" to the effects of atypical values. Two popular methods that are easy to implement follow.

**Table 5.19** **Least Squares Exponential Trend Fit to Deseasonalized Series (Centered Moving Average, Multiplicative), Mid-Town Heating**

$(y_t/\hat{S}_t)$ = seasonally adjusted receipts ($1,000)

$y'_t = \log(y_t/\hat{S}_t)$

| | Quarter | | | |
|---|---|---|---|---|
| | 1 | 2 | 3 | 4 |
| 1981 | (79.20) 4.372 | (60.95) 4.110 | (68.29) 4.074 | (85.18) 4.445 |
| 1982 | (94.93) 4.553 | (80.66) 4.390 | (76.96) 4.343 | (117.20) 4.764 |
| 1983 | (90.50) 4.505 | (95.50) 4.559 | (98.64) 4.591 | (78.41) 4.362 |
| 1984 | (102.39) 4.629 | (110.34) 4.704 | (136.59) 4.917 | (104.39) 4.648 |
| 1985 | (103.96) 4.644 | (113.14) 4.729 | (117.34) 4.765 | (106.68) 4.670 |

```
THE REGRESSION EQUATION IS
Y =     4.28 + .0257 X1
                                                ST. DEV.      T-RATIO =
         COLUMN        COEFFICIENT              OF COEF.      COEF/S.D.
         --                4.27618 ⟶ β̂0'        0.06306          67.82

X1       TIME            0.025716 ⟶ β̂1'       0.005264           4.89

THE ST. DEV. OF Y ABOUT REGRESSION LINE IS
S = 0.1357
WITH (   20- 2) =   18 DEGREES OF FREEDOM

R-SQUARED = 57.0 PERCENT

ANALYSIS OF VARIANCE

 DUE TO         DF             SS        MS=SS/DF
REGRESSION       1        0.43977         0.43977
RESIDUAL        18        0.33167         0.01843
TOTAL           19        0.77143
```

- Use the *median* of the specific seasonals rather than the mean.
- Trim the specific seasonals in each season by removing the largest and smallest before computing the mean. The resulting average is variously called the *trimmed mean,* the *medial average,* or, more simply, the *modified mean.*

If the modified mean is used when fitting the multiplicative model in Example 5.9, the computations in Table 5.18 change to those in Table 5.20. The estimated seasonals, which are the averages of the observations that are not struck out, no longer sum quite to 4 and so require some minor adjustment.

**Table 5.20 Calculation of Seasonal Modified Means for Moving Average Filter, Mid-Town Heating** $\hat{S}_t \cdot \hat{\epsilon}_t$ = multiplicative specific seasonals

| | $\hat{S}_t \cdot \hat{\epsilon}_t$ for Quarter | | | |
|---|---|---|---|---|
| | 1 | 2 | 3 | 4 |
| 1981 | — | — | .317 | 1.097 |
| 1982 | ~~1.867~~ | ~~.720~~ | ~~.297~~ | ~~1.334~~ |
| 1983 | ~~1.575~~ | .833 | .398 | ~~.896~~ |
| 1984 | 1.777 | ~~.871~~ | ~~.466~~ | 1.050 |
| 1985 | 1.651 | .865 | — | — |
| Modified mean | 1.714 | .849 | .358 | 1.074 |

*Unadjusted Seasonal Indexes:*

$$\hat{S}_1 = 1.714 \qquad \hat{S}_2 = .849 \qquad \hat{S}_3 = .358 \qquad \hat{S}_4 = 1.074$$

$$\bar{S} = \frac{1.714 + .849 + .358 + 1.074}{4} = \frac{3.995}{4} = .99875$$

*Adjusted Seasonal Indexes:*

$$\hat{S}_1 \rightarrow \frac{\hat{S}_1}{\bar{S}} = \frac{1.714}{.99875} = 1.716$$

$$\hat{S}_2 \rightarrow \frac{\hat{S}_2}{\bar{S}} = \frac{.849}{.99875} = .850$$

$$\hat{S}_3 \rightarrow \frac{\hat{S}_3}{\bar{S}} = \frac{.358}{.99875} = .358$$

$$\hat{S}_4 \rightarrow \frac{\hat{S}_4}{\bar{S}} = \frac{1.074}{.99875} = 1.075$$

Comparing these with the indexes computed from the usual averages (1.716, .822, .369, 1.093), the differences are relatively minor, as no particular season seems to have one clearly aberrant observation.

The median could also be used in deriving these indexes. But in this example there are four specific seasonals for each season. Therefore, striking out the largest and smallest and computing the average is the same as averaging the two middle observations, making the median and modified mean equivalent.

### 5.5.5 Least Squares Seasonal Modeling

Regression analysis can be used not only to test for seasonality, but also to fit trended or untrended seasonal models to a series. Although complex in its mathematics, the method is relatively easy to implement if a regression program is available (Menezes, 1971; Wilst, 1977).

As mentioned in Section 5.4.7, a slight difficulty is encountered because of the $L$ seasonal indicator variables Equations 5.4 and 5.5, but using $L - 1$

### Single Forecasts Using a Moving Average Filter, Multiplicative and Additive Models

**Given:** A sample of $n$ observations $y_t$; $t = 1, 2, \ldots, n$ from $L$ seasons

**Model:** $y_t = T_t + S_t + \epsilon_t$ where $\Sigma S_i = 0$
or
$y_t = T_t \cdot S_t \cdot \epsilon_t$ where $\Sigma S_i = L$

**Estimates:** Specific seasonals, $\hat{S}_t + \hat{\epsilon}_t$ or $\hat{S}_t \cdot \hat{\epsilon}_t$, are obtained from a moving average filter, then grouped by season. The indexes are then estimated by computing one of the following.

**Season Mean:** $\hat{S}_i$ = the mean of the specific seasons in the $i$th season

**Season Median:** $\hat{S}_i$ = the median of the specific seasons in the $i$th season

**Season Modified Mean:** $\hat{S}_i$ = the mean of the specific seasonals in the $i$th season, excluding the largest and smallest

The indexes are adjusted so they sum to 0 or $L$ with

$$\bar{S} = \frac{\Sigma \hat{S}_i}{L} \qquad \hat{S}_i \rightarrow \frac{\hat{S}_i}{\bar{S}} \quad \text{or} \quad \hat{S}_i \rightarrow \hat{S}_i - \bar{S} \qquad i = 1, 2, \cdots, L$$

and used to seasonally adjust the original series.
Finally, a trend model is fit to the deseasonalized series using standard methods to give

$$\hat{T}_t = f(\hat{\beta}_0, \hat{\beta}_1, \cdots; t)$$

**Forecasting Model:** $\hat{y}_t = \hat{T}_t + \hat{S}_t$ or $\hat{y}_t = \hat{T}_t \cdot \hat{S}_t$

where $\hat{S}_t = \hat{S}_i$ for $t$ in the $i$th season.

---

indicators circumvents the problem and conventional seasonals can be recovered quite easily.

Suppose the deterministic component of a model with $L - 1$ dummies is

$$\hat{\beta}_0^* + \hat{\beta}_1^* t + \hat{S}_1^* x_1 + \hat{S}_2^* x_2 + \cdots + \hat{S}_{L-1}^* x_{L-1} \tag{5.12}$$

and with $L$ dummies

$$\hat{\beta}_0 + \hat{\beta}_1 t + \hat{S}_1 x_1 + \hat{S}_2 x_2 + \cdots + \hat{S}_L x_L \tag{5.13}$$

Regression can be used to produce $\hat{\beta}_0^*, \hat{\beta}_1^*, \hat{S}_1^*, \hat{S}_2^*, \ldots, \hat{S}_{L-1}^*$. To convert these to $\hat{\beta}_0, \hat{\beta}_1, \hat{S}_1, \hat{S}_2, \ldots, \hat{S}_L$ we note that if $\hat{S}_L^* = 0$, Equation 5.12 can be written as

$$\hat{\beta}_0^* + \hat{\beta}_1^* t + \hat{S}_1^* x_1 + \hat{S}_2^* x_2 + \cdots + \hat{S}_{L-1}^* x_{L-1} x_{L-1} + S_L^* x_L \tag{5.14}$$

without changing the model. The set of indexes $S_1^*, S_2^*, \ldots, S_L^*$ in Equation 5.14 is merely a set of $L$ unconventional seasonals (they do not sum to zero). To normalize them we let $\bar{S}^* = (\Sigma \hat{S}_1^*)/L$, and

$$\hat{S}_i = \hat{S}_i^* - \bar{S}^* \qquad i = 1, 2, \cdots, L \tag{5.15}$$

$$\hat{\beta}_0 = \hat{\beta}_0^* + \bar{S}^* \tag{5.16}$$

$$\hat{\beta}_1 = \hat{\beta}_1^* \tag{5.17}$$

Notice that the values of the two models, Equations 5.13 and 5.14, are the same in all seasons, that is, in season $i$, Equation 5.14 is $\hat{\beta}_0^* + \hat{\beta}_1^* t + \hat{S}_i^*$ and Equation 5.13 is $\hat{\beta}_0 + \hat{\beta}_1 t + \hat{S}_i$. But from Equations 5.15–5.17,

$$\hat{\beta}_0 + \hat{\beta}_1 t + \hat{S}_i = (\hat{\beta}_1^* + \bar{S}^*) + \hat{\beta}_1^* t + (\hat{S}_i^* - \bar{S}^*) = \hat{\beta}_1^* + \hat{\beta}_1^* t + \hat{S}_i^*$$

So no change has been made in the model. In terms of the $L - 1$ starred seasonals derived from the regression, then:

$$\begin{aligned} \hat{S}_i &= \hat{S}_i^* - \bar{S}^* \qquad i = 1, 2, \cdots, L - 1 \\ \hat{S}_L &= -\bar{S}^* \\ \hat{\beta}_0 &= \hat{\beta}_0^* + \bar{S}^* \\ \hat{\beta}_1 &= \hat{\beta}_1^* \end{aligned} \tag{5.18}$$

To find the least squares seasonal model, then, we regress $y_t$ on $t, x_1, x_2, \ldots, x_{L-1}$ to give $\hat{\beta}_0^*, \hat{\beta}_1^*, \hat{S}_1^*, \hat{S}_2^*, \ldots, \hat{S}_{L-1}^*$ and use the relationships Equation 5.18 to convert to conventional parameters, $\hat{\beta}_0, \hat{\beta}_1, \hat{S}_1, \hat{S}_2, \ldots, \hat{S}_L$. For the multiplicative model, $\log y_t$ is used instead of $y_t$, and $\hat{\beta}_0', \hat{\beta}_1', \hat{S}_1', \ldots, \hat{S}_L'$ result. The unprimed parameters may be recovered in the same way as when the simple averages method is used on the logarithms, i.e.,

$$\text{(unprimed estimate)} = e^{\text{(primed estimate)}}$$

and $\beta_0, \hat{S}_1, \hat{S}_2, \ldots, \hat{S}_L$ are adjusted so that $\Sigma\, \hat{S}_i = L$.

### Single Forecasts Using Least Squares Regression, Multiplicative and Additive Models

**Given:** A sample of $n$ observations $y_t$; $t = 1, 2, \ldots, n$ from $L$ seasons

**Model:**

$$y_t = \beta_0 + \beta_1 t + S_1 x_1 + \cdots + S_L x_L \qquad \text{where } \Sigma\, S_i = 0$$

or

$$\begin{aligned} y_t' &= \log y_t \\ &= \beta_0' + \beta_1' t + S_1' x_1 + \cdots + S_L' x_L + \epsilon_t' \qquad \text{where } \Sigma\, S_i' = 0 \\ y_t &= \beta_0 (\beta_1)^t \cdot (S_1)^{x_1} \cdots (S_L)^{x_L} \cdot \epsilon_t \qquad \text{where } \Sigma\, S_i = L \end{aligned}$$

$$x_i = \begin{cases} 1 & \text{for } t \text{ in the } i\text{th season} \\ 0 & \text{otherwise} \end{cases} \qquad i = 1, 2, \cdots, L$$

**Estimates:** $y_t$ or $y_t'$ is regressed on $t, x_1, x_2, \ldots, x_{L-1}$, giving estimates $\hat{\beta}_0^*, \hat{\beta}_1^*, \hat{S}_1^*, \ldots, \hat{S}_{L-1}^*$, which are converted to conventional estimates so that $\Sigma\, \hat{S}_i = 0$ or $\Sigma\, \hat{S}_i' = 0$ using:

$\bar{S}^* = (\Sigma\, \hat{S}_i^*)L$:

$$\begin{aligned} \hat{S}_i &= \hat{S}_i^* - \bar{S}^* &\quad \text{or} \quad \hat{S}_i' &= \hat{S}_i^* - \bar{S}^* \qquad i = 1, 2, \cdots, L - 1 \\ \hat{S}_L &= -\bar{S}^* &\quad \text{or} \quad \hat{S}_L' &= -\bar{S}^* \\ \hat{\beta}_0 &= \hat{\beta}_0^* + \bar{S}^* &\quad \text{or} \quad \hat{\beta}_0' &= \hat{\beta}_0^* + \bar{S}^* \\ \hat{\beta}_1 &= \hat{\beta}_1^* &\quad \text{or} \quad \hat{\beta}_1' &= \hat{\beta}_1^* \end{aligned}$$

In the multiplicative case the primed parameters can be changed to unprimed ones by taking antilogs, then normalizing the resulting seasonals using

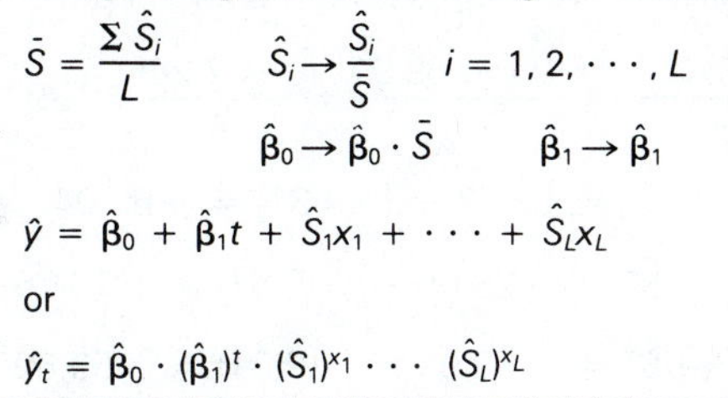

$$\bar{S} = \frac{\Sigma \hat{S}_i}{L} \qquad \hat{S}_i \rightarrow \frac{\hat{S}_i}{\bar{S}} \qquad i = 1, 2, \cdots, L$$

$$\hat{\beta}_0 \rightarrow \hat{\beta}_0 \cdot \bar{S} \qquad \hat{\beta}_1 \rightarrow \hat{\beta}_1$$

**Forecasting Model:**

$$\hat{y} = \hat{\beta}_0 + \hat{\beta}_1 t + \hat{S}_1 x_1 + \cdots + \hat{S}_L x_L$$

or

$$\hat{y}_t = \hat{\beta}_0 \cdot (\hat{\beta}_1)^t \cdot (\hat{S}_1)^{x_1} \cdots (\hat{S}_L)^{x_L}$$

---

**EXAMPLE 5.10** *Least Squares Models, Mid-Town Heating*

The data and computer printout for a least squares model fit to the Mid-Town Heating series ($L = 4$, $n = 20$) were given in Table 5.8 and Figure 5.6. For easy reference, we reproduce the printout here, in Figure 5.10. The values of the starred estimates may be read from the printout as follows:

$$\hat{\beta}_0^* = 85.08 \qquad \hat{\beta}_1^* = 1.870$$

$$\hat{S}_1^* = 59.73 \qquad \hat{S}_2^* = -28.06 \qquad \hat{S}_3^* = -68.91$$

**Figure 5.10** Computer Printout for Mid-Town Heating, Least Squares Additive Model

```
THE REGRESSION EQUATION IS
Y =     85.1 +  1.87 X1 +  59.7 X2
   -    28.1 X3 -   68.9 X4
```

| | COLUMN | COEFFICIENT | | ST. DEV. OF COEF. | T-RATIO = COEF/S.D. |
|---|---|---|---|---|---|
| | -- | 85.080 | $\longrightarrow \hat{\beta}_0^*$ | 7.311 | 11.64 |
| X1 | T | 1.8700 | $\longrightarrow \hat{\beta}_1^*$ | 0.4433 | 4.22 |
| X2 | X1 | 59.730 | $\longrightarrow \hat{S}_1^*$ | 7.216 | 8.28 |
| X3 | X2 | -28.060 | $\longrightarrow \hat{S}_2^*$ | 7.148 | -3.93 |
| X4 | X3 | -68.910 | $\longrightarrow \hat{S}_3^*$ | 7.107 | -9.70 |

•
•
•

ANALYSIS OF VARIANCE

| DUE TO | DF | SS | MS=SS/DF |
|---|---|---|---|
| REGRESSION | 4 | 44052.7 | 11013.2 |
| RESIDUAL | 15 | 1886.5 | 125.8 |
| TOTAL | 19 | 45939.2 | |

From previous testing we know that both the linear trend and seasonal portions of the model are significant. To express the model in conventional form:

$$\bar{S}^* = \frac{59.73 - 28.06 - 68.91}{4} = \frac{-37.24}{4} = -9.31$$

$$\hat{S}_1 = \hat{S}_1^* - \bar{S}^* = 59.73 - (-9.31) = +69.04$$

$$\hat{S}_2 = \hat{S}_2^* - \bar{S}^* = -28.06 - (-9.31) = -18.75$$

$$\hat{S}_3 = \hat{S}_3^* - \bar{S}^* = -68.91 - (-9.31) = -59.60$$

$$\hat{S}_4 = -\bar{S}^* = -(-9.31) = +9.31$$

$$\hat{\beta}_0 = \hat{\beta}_0^* + \bar{S}^* = 85.08 + (-9.31) = 75.77$$

$$\hat{\beta}_1 = \hat{\beta}_1^* = 1.870$$

So the forecasting model is as follows.

***Forecasting Model:***

$$\hat{y}_t = 75.77 + 1.87t + 69.04x_1 - 18.75x_2 - 59.60x_3 + 9.31x_4$$

Origin: IV, 80
Units ($t$): Quarters
Units ($y$): $1,000

A multiplicative model with simple exponential trend could be quickly fit by changing from $y_t$ to $y'_t = \log y_t$. The computer output in Figure 5.11 results from regressing $y'_t$ on $t$ and $x_1$ through $x_{L-1}$. The parameter estimates are

$$\hat{\beta}_0^* = 4.355 \qquad \hat{\beta}_1^* = .0260$$

$$\hat{S}_1^* = +.492 \qquad \hat{S}_2^* = -.312 \qquad \hat{S}_3^* = -1.070$$

from which the primed (logarithmic) parameters can be calculated:

$$\bar{S}^* = \frac{.492 - .312 - 1.070}{4} = \frac{-.890}{4} = -.223$$

$$\hat{S}'_1 = \hat{S}_1^* - \bar{S}^* = +.492 - (-.223) = +.715$$

$$\hat{S}'_2 = \hat{S}_2^* - \bar{S}^* = -.312 - (-.223) = -.089$$

$$\hat{S}_3 = \hat{S}_3^* - \bar{S}^* = -1.070 - (-.223) = -.847$$

$$\hat{S}'_4 = -\bar{S}^* = -(-.223) = +.223$$

$$\hat{\beta}'_0 = \hat{\beta}_0^* + \bar{S}^* = 4.355 + (-.223) = 4.132$$

$$\hat{\beta}'_1 = \hat{\beta}_1^* = .0260$$

These estimates may be used to forecast the logarithms of Mid-Town's receipts. To convert out of the logarithmic form we first change to unprimed parameters:

$$\hat{\beta}_0 = e^{\hat{\beta}'_0} = e^{4.132} = 62.30 \qquad \hat{\beta}_1 = e^{\hat{\beta}'_1} = e^{.0260} = 1.026$$

$$\hat{S}_1 = e^{\hat{S}'_1} = e^{+.715} = 2.044$$

$$\hat{S}_2 = e^{\hat{S}'_2} = e^{-.089} = .915$$

Figure 5.11 Computer Printout for Mid-Town Heating, Least Squares Multiplicative Model

```
THE REGRESSION EQUATION IS
Y =     4.35 + .0260 X1 +  .492 X2
    -  .312 X3 -  1.07 X4

                                          ST. DEV.    T-RATIO =
       COLUMN      COEFFICIENT            OF COEF.    COEF/S.D.
       --              4.35465 ⟶ β̂0*      0.09514        45.77

X1     T              0.026013 ⟶ β1*     0.005769         4.51

X2     S1              0.49184 ⟶ Ŝ1*      0.09391         5.24

X3     S2             -0.31237 ⟶ Ŝ2*      0.09302        -3.36

X4     S3             -1.06959 ⟶ Ŝ3*      0.09348       -11.57
•
•
•

ANALYSIS OF VARIANCE

 DUE TO        DF          SS       MS=SS/DF
REGRESSION      4     6.58667        1.64667
RESIDUAL       15     0.31947        0.02130
TOTAL          19     6.90615
```

$$\hat{S}_3 = e^{\hat{S}'_3} = e^{-.847} = .429$$

$$\hat{S}_4 = e^{\hat{S}'_4} = e^{+.223} = 1.249$$

Remember that although the logarithmic indexes $\hat{S}'_i$ sum to zero, the final indexes $\hat{S}_i$ should be adjusted to sum to $L = 4$:

$$\bar{S} = \frac{2.044 + .915 + .429 + 1.249}{4} = \frac{4.637}{4} = 1.159$$

$$S_1 \rightarrow \frac{S_1}{\bar{S}} = \frac{2.044}{1.159} = 1.764$$

$$S_2 \rightarrow \frac{S_2}{\bar{S}} = \frac{.915}{1.159} = .789$$

$$S_3 \rightarrow \frac{S_3}{\bar{S}} = \frac{.429}{1.159} = .370$$

$$S_4 \rightarrow \frac{S_4}{\bar{S}} = \frac{1.249}{1.159} = 1.078$$

$$\hat{\beta}_0 \rightarrow \hat{\beta}_0 \cdot \bar{S} = 62.30 \cdot 1.159 = 72.20$$

$$\hat{\beta}_1 \rightarrow \hat{\beta}_1 = 1.026$$

The forecasting model is as follows.

*Forecasting Model:* $\hat{y}_t = 72.20(1.026)^t \cdot (1.764)^{x_1} \cdot (.789)^{x_2} \cdot (.370)^{x_3} \cdot (1.078)^{x_4}$

Origin: IV, 80
Units ($t$): Quarters
Units ($y$): $1,000 ●

**EXAMPLE 5.11** *Least Squares Model, No Trend, U.S. Return Rates*

In testing for seasonality in the U.S. Return Rate series, the trend component was found not to be significant. Least squares no-trend models for this data would be

$$\hat{y}_t = \hat{\beta}_0 + \hat{S}_1 x_1 + \hat{S}_2 x_2 + \hat{S}_3 x_3 + \hat{S}_4 x_4 \qquad \text{(additive)}$$

$$\hat{y}_t = \hat{\beta}_0 \cdot (\hat{S}_1)^{x_1} \cdot (\hat{S}_2)^{x_2} \cdot (\hat{S}_3)^{x_3} \cdot (\hat{S}_4)^{x_4} \qquad \text{(multiplicative)}$$

Taking an additive model to begin with, we regress $y_t$ on $x_1$, $x_2$, $x_3$ as shown in Figure 5.12, which gives us the following least squares parameters:

$$\hat{\beta}_0^* = 21.95 \qquad \hat{S}_1^* = -11.93 \qquad \hat{S}_2^* = -8.03 \qquad \hat{S}_3^* = -7.22$$

which are converted to a conventional model:

**Figure 5.12** Computer Printout for U.S. Return Rates, Least Squares Additive Model

```
THE REGRESSION EQUATION IS
Y =      22.0 -  11.9 X1 -   8.03 X2
       - 7.22 X3
```

| | COLUMN | COEFFICIENT | | ST. DEV. OF COEF. | T-RATIO = COEF/S.D. |
|---|---|---|---|---|---|
| | -- | 21.952 | ⟶ $\hat{\beta}_0^*$ | 1.494 | 14.69 |
| X1 | S1 | -11.927 | ⟶ $\hat{S}_1^*$ | 2.133 | -5.64 |
| X2 | S2 | -8.032 | ⟶ $\hat{S}_2^*$ | 2.133 | -3.80 |
| X3 | S3 | -7.217 | ⟶ $\hat{S}_3^*$ | 2.133 | -3.42 |

•
•
•

ANALYSIS OF VARIANCE

| DUE TO | DF | SS | MS=SS/DF |
|---|---|---|---|
| REGRESSION | 3 | 445.28 | 148.43 |
| RESIDUAL | 20 | 267.84 | 13.39 |
| TOTAL | 23 | 713.12 | |

$$\bar{S}^* = \frac{-11.93 - 8.03 - 7.22}{4} = \frac{-27.18}{4} = -6.80$$

$$\hat{S}_1 = \hat{S}_1^* - \bar{S}^* = -11.93 - (-6.80) = -5.13$$

$$\hat{S}_2 = \hat{S}_2^* - \bar{S}^* = -8.03 - (-6.80) = -1.23$$

$$\hat{S}_3 = \hat{S}_3^* - \bar{S}^* = -7.22 - (-6.80) = -.42$$

$$\hat{S}_4 = -\bar{S}^* = -(-6.80) = +6.80$$

$$\hat{\beta}_0 = \hat{\beta}_0^* + \bar{S}^* = 21.95 + (-6.80) = 15.15$$

The forecasting model is as follows.

***Forecasting Model:*** $\hat{y}_t = 15.15 - 5.13x_1 - 1.23x_2 - .42x_3 + 6.80x_4$

Origin: IV, 78
Units ($t$): Quarters
Units ($y$): (%)

If this model is used to forecast 1985, the forecasts are

***For I, 85 (t = 25):*** $\hat{y}_{25} = 15.15 - 5.13 \times 1 - 1.23 \times 0 - .42 \times 0 + 6.80 \times 0 = 15.15 - 5.13$
$= 10.02$

***For II, 85 (t = 26):*** $\hat{y}_{26} = 15.15 - 5.13 \times 0 - 1.23 \times 1 - .42 \times 0 + 6.80 \times 0 = 15.15 - 1.23$
$= 13.92$

***For III, 85 (t = 27):*** $\hat{y}_{27} = 15.15 - 5.13 \times 0 - 1.23 \times 0 - .42 \times 1 + 6.80 \times 0 = 15.15 - .42$
$= 14.73$

***For IV, 85 (t = 28):*** $\hat{y}_{28} = 15.15 - 5.13 \times 0 - 1.23 \times 0 - .42 \times 0 + 6.80 \times 1 = 15.15 + 6.80$
$= 21.95$

Figure 5.13 shows a multiplicative model fit with the following parameter estimates:

$\hat{\beta}_0^* = 3.085$ $\quad \hat{S}_1^* = -.840$ $\quad \hat{S}_2^* = -.471$ $\quad \hat{S}_3^* = -.455$

which gives $\bar{S}^* = -.442$ and

$\hat{\beta}_0' = 2.643$ $\quad \hat{S}_1' = -.398$ $\quad \hat{S}_2' = -.029$ $\quad \hat{S}_3' = -.013$
$\hat{S}_4' = +.442$

Then, taking antilogs,

$\hat{\beta}_0 = 14.06$ $\quad \hat{S}_1 = .672$ $\quad \hat{S}_2 = .971$ $\quad \hat{S}_3 = .987$
$\hat{S}_4 = 1.556$

and, finally, with $\bar{S} = 1.0465$, adjusting so $\Sigma\, \hat{S}_i = 4$:

$\hat{\beta}_0 = 14.71$ $\quad \hat{S}_1 = .642$ $\quad \hat{S}_2 = .928$ $\quad \hat{S}_3 = .943$
$\hat{S}_4 = 1.487$

The forecasting model, then, is as follows.

***Forecasting Model:*** $\hat{y}_t = 14.71(.642)^{x_1} \cdot (.928)^{x_2} \cdot (.943)^{x_3} \cdot (1.487)^{x_4}$

Origin: IV, 78
Units ($t$): Quarters
Units ($y$): (%)

Figure 5.13 Computer Printout for U.S. Return Rates, Least Squares Multiplicative Model

```
THE REGRESSION EQUATION IS
Y =     3.09 +   .840 X1 +   .471 X2
    -   .455 X3

                                               ST. DEV.     T-RATIO =
        COLUMN      COEFFICIENT                OF COEF.     COEF/S.D.
        --               3.0853 ---> β̂0*        0.1241        24.87
X1      S1              -0.8400 ---> Ŝ1*        0.1754        -4.79
X3      S2              -0.4710 ---> Ŝ2*        0.1754        -2.68
X4      S3              -0.4553 ---> Ŝ3*        0.1754        -2.60
•
•
•
ANALYSIS OF VARIANCE

 DUE TO       DF          SS      MS=SS/DF
REGRESSION     3     2.12872       0.70957
RESIDUAL      20     1.84686       0.09234
TOTAL         23     3.97557
```

This model would give the following forecasts for 1985.

*For I, 85 (t = 25):* $\hat{y}_{25} = 14.71(.642)^1 \cdot (.928)^0 \cdot (.943)^0 \cdot (1.487)^0 = 14.71 \times .642 = 9.44$

*For II, 85 (t = 26):* $\hat{y}_{26} = 14.71(.642)^0 \cdot (.928)^1 \cdot (.943)^0 \cdot (1.487)^0 = 14.71 \times .928 = 13.65$

*For III, 85 (t = 27):* $\hat{y}_{27} = 14.71(.642)^0 \cdot (.928)^0 \cdot (.943)^1 \cdot (1.487)^0 = 14.71 \times .943 = 13.87$

*For IV, 85 (t = 28):* $\hat{y}_{28} = 14.71(.642)^0 \cdot (.928)^0 \cdot (.943)^0 \cdot (1.487)^1 = 14.71 \times 1.487 = 21.87$

Some insight into the least squares modeling process can be gained if we remark here that fitting a no-trend additive model amounts to forecasting each season with its historical arithmetic mean, while a no-trend multiplicative model forecasts with the geometric mean (the $n_i$th root of the product or, equivalently, the antilog of the average logarithm of the season's historical observations). In this example, then, 10.02 is the average of the observed first quarter return rates, and 9.44 is the geometric mean of those rates. ●

Although linear and simple exponential trends are the most common (and convenient) in least squares seasonal modeling, quadratic (polynomial) and other models could be used as long as they lend themselves to standard regression analysis, that is, if they have a constant term $\beta_0$ and are linear in, or can be transformed so they are linear in, the unknown parameters.

If a nonlinear (in the parameters) trend model such as the modified exponential is proposed, it is better to use the moving average filter to get seasonals and deseasonalize, then use the methods of Chapter 3.

### 5.5.6 Overview

We have discussed three different methods for incorporating a seasonal element into forecasting models. All of these methods can be (and have been) refined to reflect the peculiarities of specific modeling environments, but the concepts involved are basic and, if understood, should provide solid intuitive grounds for more elaborate efforts.

*Simple averaging* requires knowledge of the functional form of the trend component, something that may be difficult to obtain if seasonal effects are strong. But the method is very heuristic and the calculations are not complex. A trend model other than those mentioned (linear, simple exponential) can be used if it can be fit without interference from the seasonal variability (e.g., to yearly averages). The model can be evaluated for each value of $t$ and the data detrended observation by observation. After that, no-trend simple averages will yield indexes, and the data can be deseasonalized and the trend model refit. (If a multiplicative model is being used, incidentally, the detrending should be done by taking the ratio to trend, $y_t/\hat{T}_t$, rather than subtracting, $y_t - \hat{T}_t$.) Used in this way, however, the method loses much of its simplicity (and its major appeal).

The outstanding advantage of the *moving average filter* is that it requires no explicit trend model; i.e., a well-fit trend is not needed to get good estimates of the seasonal indexes. Although hand calculation of the averages is tedious, it is not challenging. Moreover the calculation is usually done by computer. The filter can also be made resistant to occasional aberrant behavior in the series. If desired, any trend model can be fit to the adjusted series using only methods of Chapter 3 and without the fit-deseasonalize-refit interaction of simple averaging. The method is somewhat demanding of data. Since a whole year's worth of observations is lost in the averaging process, a minimum of two years' data is required (and several more for any real accuracy).

*Least squares* has the almost mystical appeal of its "optimality" with respect to historical squared forecast errors. Regression analysis is also familiar to, even revered by, many. One can get by with a bare minimum of $L + 1$ successive observations if there is no trend ($L + 2$, if there is). However, most analysts recognize that, in reality, about four or more years' worth of data is needed. It is also difficult to deal with aberrant data (see Chapter 4). Finally, some have claimed that the statistical testing procedures are assumption-ridden, the estimates obtained are not really intuitive to many, and that a lack of understanding caused by the underlying mathematical complexity can lead to blunders, gross misinterpretations, or a reluctance, on the part of those who matter, to implement the analytical results.

## 5.6 Updating Procedures

### 5.6.1 Introduction

Problems with seasonality are common when the updating schemes discussed in Chapters 2 and 3 are used for subannual data. The schemes tend to treat unmodeled seasonality as though it reflected underlying parameter changes, constantly revising the estimates in the direction indicated by the most recent seasonal effects, and causing the forecasts to bounce around—chasing, but always lagging behind, the seasonal pattern. If the seasonality is recognized and provision made for it, the forecasts will likely improve markedly.

When the seasonal pattern is *constant* and the necessary computations are not a barrier, one corrective approach would be to use a single forecast method (preferably the moving average filter because of the dynamism implied when updating is being used) to estimate a set of seasonal indexes, use these to seasonally adjust the incoming observations, and then forecast the seasonally adjusted series. Using these same indexes, seasonality may be reincorporated into the forecasts after the updating scheme has finished its work.

### 5.6.2 Exponential Smoothing with Seasonal Models

Another approach is to make the seasonal element part of the updating scheme itself so that for each new observation estimates of the level, slope (if trended), and seasonal indexes are all updated. This method would be preferred if the seasonal pattern is *changing* or when it is impractical to do the analysis necessary to implement and maintain the single estimation scheme mentioned above.

Exponential smoothing is generally used to do the updating. The idea is pretty much the same as for simple smoothing (only a level involved) or double smoothing (both a level and a slope). The main difference is that there is now a third stream of estimates to be maintained, the seasonal indexes. The basic concept is identical. That is,

(updated estimate) = (constant) × (estimate based on new data)
+ (1 − constant) × (old estimate)

or, in error correction form,

(updated estimate) = (old estimate) + (constant) × (forecast error)

Although there are more estimates involved for seasonal models, each new estimate is, as in previous chapters, simply a weighted average of new and old information, with the smoothing constant being the weight (out of a total of 1.0) given to the new. The updating also remains equivalent to adjusting old estimates in the direction that would have reduced the current forecast error.

## Notation

Previous notation will be kept; i.e.,

$\hat{T}_t(t)$ = estimate, as of time $t$, of (unseasonalized) trend component (or level) of the series at time $t$

$\hat{\beta}_1(t)$ = estimate, as of time $t$, of the slope of a linear trend model

And, additionally,

$\hat{S}_i(t)$ = estimate, as of time $t$, of the seasonal index for the $i$th season

$\hat{S}_t(t)$ = estimate, as of time $t$, of the seasonal index for time $t$

Note here that $\hat{S}_t(\text{any time}) = \hat{S}_i(\text{any time})$ for $t$ in the $i$th season, consistent with previously adopted notation.

## Models and Updating Equations

Although four different models are used (horizontal or trended, additive or multiplicative) the updating equations are similar in form. We let

$\alpha$ = smoothing constant for the level

$\gamma$ = smoothing constant for the slope

$\delta$ = smoothing constant for the seasonals

Then assuming that $y_t$ has just become available, the general equations (smoothing form) for the three sets of estimates are as follows.

*For the level, $T_t$:*

$$\hat{T}_t(t) = \alpha \cdot (\text{seasonally adjusted } y_t) + (1 - \alpha) \cdot \hat{T}_t(t-1)$$

where

$\hat{T}_t(t)$ = updated estimate of the current unseasonalized level of the series (based on $y_1, y_2, \cdots, y_t$)

$\hat{T}_t(t-1)$ = old estimate of the current unseasonalized level of the series (based on $y_1, y_2, \cdots, y_{t-1}$)

$$(\text{seasonally adjusted } y_t) = y_t - \hat{S}_t(t-1) \qquad \text{(additive models)}$$

$$= \frac{y_t}{\hat{S}_t(t-1)} \qquad \text{(multiplicative models)}$$

This seasonally adjusted $y_t$ is an estimate of the current unseasonalized level of the series based primarily on $y_t$. The form of the old estimate $T_t$ depends on whether a horizontal model is being used:

$$\hat{T}_t(t-1) = \hat{T}_{t-1}(t-1) \qquad \text{(horizontal model)}$$

$$= \hat{T}_{t-1}(t-1) + \hat{\beta}_1(t-1) \qquad \text{(trended model)}$$

*For the slope (if needed):*

$$\hat{\beta}_1(t) = \gamma[\hat{T}_t(t) - \hat{T}_{t-1}(t-1)] + (1-\gamma) \cdot \hat{\beta}_1(t-1)$$

This last is exactly as it was for Linear Exponential Smoothing in Chapter 3. Remember that $\hat{T}_t(t) - \hat{T}_{t-1}(t-1)$ is an estimate of $\beta_1$ based primarily on $y_t$.

***For the seasonals, $S_1, S_2, \cdots, S_L$:***

$$\hat{S}_i(t) = \hat{S}_i(t-1) \qquad \text{if } t \text{ is } \textit{not} \text{ in the } i\text{th season}$$
$$= \gamma \cdot (\text{detrended } y_t) + (1-\gamma)\cdot \hat{S}_i(t-1) \qquad \text{if } t \textit{ is} \text{ in the } i\text{th season}$$

where detrended $y_t$ is $y_t - \hat{T}_t(t)$ or $y_t/\hat{T}_t(t)$, depending on whether the model is additive or multiplicative, respectively, and is an estimate of the seasonal index at time $t$ based primarily on $y_t$. One can also think of detrended $y_t$ as the specific seasonal at time $t$. Notice that the seasonal index for the $i$th season can change only when an observation in the $i$th season is obtained.

These smoothing forms are of interest mostly because they show that the updated estimates are really weighted averages of new and old information. The error correction forms, which are much more convenient for computation, are also similar. Letting $e_t = y_t - \hat{y}_t(t-1)$ be the one-step-ahead forecast error, then gives the following.

***Level:***
$$\hat{T}_t(t) = \hat{T}_t(t-1) + \alpha \cdot (\text{seasonally adjusted } e_t)$$

***Slope:***
$$\hat{\beta}_1(t) = \hat{\beta}_1(t-1) + \alpha \cdot \gamma \cdot (\text{seasonally adjusted } e_t)$$

***Seasonal:***
$$\hat{S}_i(t) = \hat{S}_i(t-1) + \delta \cdot (1-\alpha) \cdot (\text{seasonally adjusted } e_t) \qquad \text{for } t \text{ in the } i\text{th season}$$
$$= \hat{S}_i(t-1) \qquad \text{otherwise } (\textit{as before})$$

where seasonally adjusted $e_t$ is $e_t$ or $e_t/\hat{S}_t(t)$, depending on whether the model is additive or multiplicative, respectively. In the additive case, $e_t$ is naturally free of seasonal; i.e.,

$$e_t = y_t - \hat{y}_t(t-1) = y_t - [\hat{T}_t(t-1) + \hat{S}_t(t-1)]$$
$$= \underbrace{[y_t - \hat{S}_t(t-1)]}_{\text{deseasonalized}} - \underbrace{\hat{T}_t(t-1)}_{\text{no seasonal}}$$

But in the multiplicative case, $e_t$ must be divided by the seasonal index to deseasonalize it:

$$e_t = y_t - \hat{y}_t(t-1) = y_t - \hat{T}_t(t-1) \cdot \hat{S}_t(t-1)$$
$$\frac{e_t}{\hat{S}_t(t-1)} = \underbrace{\frac{y_t}{\hat{S}_t(t-1)}}_{\text{deseasonalized}} - \underbrace{\hat{T}_t(t-1)}_{\text{no seasonal}}$$

### Initialization

Starting values for the process may be found in various ways, depending on both how much of the series' history is available and how much analytical effort is to be expended. If no data are available, initial values, $\hat{T}_0(0)$, $\hat{S}_1(0)$, . . . , $\hat{S}_L(0)$ (and $\hat{\beta}_1(0)$ if trended), may be available from a previously operating forecasting system (even a subjective one) or an outside source. Since seasonal patterns for general categories of products, services, and so forth are often alike, usable sets of seasonal indexes for similar series may be available from government or other industry publications.

Suppose that a history $y_1, y_2, \ldots, y_n$ is available. Since it is probably best to use a nonseasonal model until at least a year of history is available, we will assume that $n \geq L$. A very simple way to start is to use

$$\hat{T}_1(1) = y_1$$

$$\hat{S}_1(1) = \cdots = \hat{S}_L(1) \quad = 0.0 \quad \text{(additive)}$$

$$= 1.0 \quad \text{(multiplicative)}$$

$$\hat{\beta}_1(1) = 0 \quad \text{(if trended)}$$

and then smooth through the series from $t = 2, \ldots, n$. This method will not generally begin to give serviceable forecasts until at least one revision of each of the estimates has occurred (including all the seasonal indexes), and it is a good idea to use large values of the smoothing constants for the first year or so to overcome bias caused by the poor initial values. As a rule, it is far better to obtain at least a couple of years' data and start the process more intelligently.

If between one and two years' worth of data is available, a modification of the simple averages method may be used. Simple averages cannot be used directly if a trend is involved, because there is only one yearly average for trend fitting. Suppose we let

$$\overline{y1} = \text{average of the first } L \text{ observations}$$

$$= \sum_{1}^{L} \frac{y_t}{L}$$

$$\overline{y2} = \text{average of the last } L \text{ observations}$$

$$= n - \sum_{L+1}^{n} \frac{y_t}{L}$$

Then, for horizontal models use

$$\hat{S}_i(L) = y_i - \overline{y1} \qquad i = 1, 2, \cdots, L \qquad \text{(additive)} \tag{5.19}$$

$$= y_i / \overline{y1} \qquad i = 1, 2, \cdots, L \qquad \text{(multiplicative)} \tag{5.20}$$

$$\hat{T}_L(L) = \overline{y1} \qquad \text{(both)}$$

Then smooth from $t = L + 1$ to $n$ (which will give $\hat{T}_n(n)$, $\hat{\beta}_1(n)$, and $\hat{S}_1(n), \ldots, \hat{S}_L(n)$ to use in forecasting $y_{n+1}$.

For a linear trend model, initial level and slope estimates are

$$\hat{\beta}_1(L) = \frac{\overline{y1} - \overline{y2}}{n - L}$$

$$\hat{T}_L(L) = \overline{y1} + \frac{L - 1}{2} \cdot \hat{\beta}_1(L)$$

Since simplicity is the overriding concern, Equations 5.19 and 5.20 are usually used for initial seasonal values, although they are only rough approximations. For more accuracy, the simple averages method should be used to determine initial values with

$$\hat{\beta}_1^* = L \cdot \hat{\beta}_1(L)$$

If a more complex initialization is desired, least squares can be used, provided that $n \geq L + 1$ (horizontal model) or $n \geq L + 2$ (linear trend).

For two or more years' worth of data, one of the single forecast methods (simple averages, moving averages, or least squares) or the backcasting technique described in earlier chapters may be used to generate the initial estimates. Where feasible, at least two years' data should be reserved for use as a *test set* to determine desirable values of $\alpha$, $\gamma$, and $\delta$. If it is felt that the structure of the series is quite dynamic, backcasting is preferred or the single forecast model can be fit to the first two years and the estimates smoothed through the remaining history.

## Renormalization

Although the initial estimates of the seasonal indexes are adjusted to sum to zero or to $L$, after revision this normalization is lost. The indexes may be renormalized, if desired, by subtracting or dividing by $\bar{S} = [\Sigma\, \hat{S}_i(t)]/L$, the same as in single forecast models. To prevent a discontinuity in the forecast stream—that is, to avoid causing the forecasts to jump—because of what amounts to merely a reparameterization of the model, care should be taken to revise the estimated level (and slope in multiplicative models) to reflect the changed indexes and retain model equivalence. Suppose renormalization is done at time $t$. Then

$$\bar{S} = \sum_{1}^{L} \frac{\hat{S}_i(t)}{L}$$

| **Additive** | **Multiplicative** |
|---|---|
| $\hat{S}_i(t) \rightarrow \hat{S}_i(t) - \bar{S}$ | $\hat{S}_i(t) \rightarrow \dfrac{\hat{S}_i(t)}{\bar{S}}$ |
| $\hat{T}_t(t) \rightarrow \hat{T}_t(t) + \bar{S}$ | $\hat{T}_t(t) \rightarrow \hat{T}_t(t) \cdot \bar{S}$ |
| $\hat{\beta}_1(t) \rightarrow \hat{\beta}_1(t)$ | $\hat{\beta}_1(t) \rightarrow \hat{\beta}_1(t) \cdot \bar{S}$ |

**(5.21)**

To see why the revision of the level (and slope) is necessary, note that in the additive case the one-step-ahead forecast is

$$\hat{y}_{t+1}(t) = \hat{T}_t(t) + \hat{\beta}_1(t) + \hat{S}_{t+1}(t)$$

When we reparameterize, $\hat{S}_{t+1}(t) \rightarrow \hat{S}_{t+1}(t) - \bar{S}$, that is, we subtract $\bar{S}$. So to keep the forecast the same, we must add $\bar{S}$ as well; i.e.,

$$\hat{y}_{t+1}(t) = \underbrace{[\hat{T}_t(t) + \bar{S}]}_{\text{new level}} + \underbrace{\hat{\beta}_1(t)}_{\text{same slope}} + \underbrace{[\hat{S}_{t+1}(t) - \bar{S}]}_{\text{new seasonal}}$$

For the multiplicative model,

$$\hat{y}_{t+1}(t) = [\hat{T}_t(t) + \hat{\beta}_1(t)] \cdot \hat{S}_{t+1}(t)$$

$\hat{S}_{t+1}(t) \rightarrow \hat{S}_{t+1}(t)/\bar{S}$, and

$$\hat{y}_{t+1}(t) = [\hat{T}_t(t) + \hat{\beta}_1(t)] \cdot \bar{S} \cdot \frac{\hat{S}_{t+1}(t)}{\bar{S}}$$

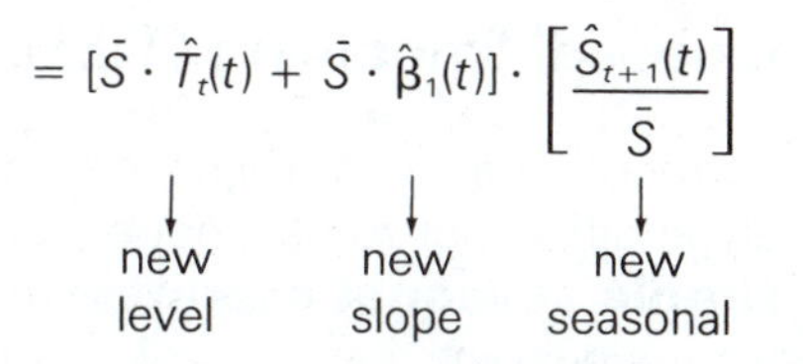


Renormalization of the indexes is not really necessary from a forecasting point of view, but if the indexes are to be used in other contexts, such as seasonal adjustment or standard description of seasonal effects, they will need to be converted to their conventional form (McKenzie, 1986).

## Choice of $\alpha$, $\gamma$, and $\delta$

Remarks in previous chapters about the choice of smoothing constants are also valid for seasonal smoothing schemes. It is popular in practice to (almost arbitrarily) select values for $\alpha$ and $\gamma$ between .1 and .3 while setting $\delta$, the seasonal smoothing constant, to about .4. Some recent studies indicate that this practice is a poor one (Gardner, 1985). One could search among a matrix of values, say $\alpha, \gamma, \delta = .01(.01).99$ and select the set that gives acceptable values of one (or several) of the goodness measures (MSE, MAD, U, etc.). To prevent interference from poor starting values (which tend to demand tracking power that may not be needed in the long run), a test set should be used. As an alternative, the goodness criterion could be calculated only for forecasts generated after the influence of the starting values has diminished.

The size of the complete search (about $1{,}000^3 = 1$ billion different sets of $\alpha, \gamma, \delta$) taxes even the speed of modern computers, and the matter is worsened if there are (as is often the case) many series to be forecast. It would appear, however, that at least some abbreviated search for decent values of $\alpha, \gamma, \delta$ is worth the effort. We might try $\alpha, \gamma, \delta = .1, .2, .3$, which would require only 27 evaluations, or .1(.1).9, which would need only 729.

Another factor that may help in the choice is to note the differing effective amounts of smoothing for the three different streams of estimates by examining the error correction forms of the updating equations.

| | Smoothing Constants for Errors |
|---|---|
| **Level:** | $\alpha$ |
| **Slope:** | $\alpha \cdot \gamma$ |
| **Seasonal:** | $\delta \cdot (1 - \alpha)$ |

For example, if $\alpha = \gamma = .1$ and $\delta = .4$, the effective smoothing constants for the different estimates in terms of error correction are

Level: .1 Slope: .01 Seasonal: .36

We will now discuss each of the four seasonal updating models in more detail and give examples of their use (see also Pegels, 1969).

### 5.6.3 Simple Seasonal Exponential Smoothing (SSES), Additive Model

Of the four procedures to be discussed, the one with the simplest structure assumes an additive horizontal model. We will refer to this procedure as (additive) **simple seasonal exponential smoothing (SSES).** The level of the series is constant; i.e., $T_t = \beta_0$ for all $t$, so $\hat{T}_t(t) = \hat{\beta}_0(t)$ for all values of $t$.

In its updating form, the model is designed for a series that has seasonality but whose expected level is otherwise constant (or only slowly changing), with, perhaps, autocorrelation in the error structure (drift). Updating is relatively simple, since only two equations are needed. In addition, $p$-step-ahead forecasts are constant except for seasonality.

**Simple Seasonal Exponential Smoothing (SSES), Additive Model**

**Given:** The current observation of the time series $y_t$, and

$\hat{T}_{t-1}(t-1)$ = estimated unseasonalized level as of time $t-1$

$\hat{S}_i(t-1)$ = estimated index for season $i$ as of time $t-1$ $\quad i = 1, \ldots, L$

**Model:**

$$y_t = \beta_0 + S_t + \epsilon_t$$

So

$$\hat{T}_{t+p}(t) = \hat{T}_t(t) \qquad \text{for } p \geq 1$$

**Updating Equations:**

Smoothing Form:

$$\hat{T}_t(t) = \alpha \cdot [y_t - \hat{S}_t(t-1)] + (1-\alpha) \cdot \hat{T}_t(t-1)$$

$$\hat{S}_i(t) = \delta \cdot [y_t - \hat{T}_t(t)] + (1-\delta) \cdot \hat{S}_i(t-1) \qquad \text{if } t \text{ in season } i$$

$$= \hat{S}_i(t-1) \qquad \text{if } t \text{ not in season } i$$

Error Correction Form:

$$\hat{T}_t(t) = \hat{T}_t(t-1) + \alpha \cdot e_t$$

$$\hat{S}_i(t) = \hat{S}_i(t-1) + \delta \cdot (1-\alpha) \cdot e_t \qquad \text{if } t \text{ in season } i$$

$$= \hat{S}_i(t-1) \qquad \text{if } t \text{ not in season } i$$

where

$\hat{S}_t(\text{any time}) = \hat{S}_i(\text{any time}) \qquad$ for $t$ in season $i$

$e_t$ = one-step-ahead forecast error

$= y_t - \hat{y}_t(t-1)$

$\alpha$ = smoothing constant for the level

$\delta$ = smoothing constant for the seasonal

**Forecasts:** (At time $t$ for $p$ periods ahead)

$$\hat{y}_{t+p}(t) = \hat{T}_t(t) + \hat{S}_{t+p}(t)$$

where

$$\hat{S}_{t+p}(t) = \hat{S}_i(t) \qquad \text{for } t+p \text{ in season } i$$

**EXAMPLE 5.12** *Additive SSES, Stilling School District*

As part of its budgeting efforts, a small school district, Stilling Unified Schools, wishes to forecast its water usage charges for next year, updating each forecast as charges become known so as to have planning flexibility during the course of the year.

The usage consists of student demands during the school year and care/maintenance of the school grounds year round. The local water district bills bimonthly, and data for the past 15 months' billing periods are available for analysis (Table 5.21). The data are distinctly seasonal; but no trend is apparent, so none will be included at this time. The forecasting model will be used for a year or so, after which time its performance will be reviewed. The seasonal indexes will also be used to seasonally adjust the data so the nonseasonal behavior of the series may be studied, especially with an eye toward spotting periods of excessive usage, which will likely appear as large residual movement.

**Table 5.21** **Water Usage Charges, Stilling Unified School District**
$y_t$ = bimonthly charges ($)

| | Jan/Feb | Mar/Apr | May/Jun | Jul/Aug | Sep/Oct | Nov/Dec |
|---|---|---|---|---|---|---|
| 1984 | 193.6 | 154.2 | 207.9 | 343.6 | 290.7 | 178.0 |
| 1985 | 139.7 | 166.3 | 213.4 | 286.1 | 314.4 | 183.6 |
| 1986 | 166.6 | 160.8 | 181.2 | | | |

Simple averages are used with the first year of data to obtain initial estimates, which are then smoothed through the remaining eight observations to provide the forecast for July/August 1986. To compute the initial estimates, we average the 1984 data and calculate the seasonal differences from the average:

$$\bar{y}_{1984} = \frac{y_1 + y_2 + y_3 + y_4 + y_5 + y_6}{6}$$

$$= \frac{193.6 + 154.2 + 207.9 + 343.6 + 290.7 + 179.0}{6} = \frac{1368.0}{6} = 228.0$$

Then, letting January/February be season 1, March/April be season 2, and so forth,

$$\hat{S}_1(6) = 193.6 - 228.0 = -34.4$$

$$\hat{S}_2(6) = 154.2 - 228.0 = -73.8$$

$$\hat{S}_3(6) = 207.9 - 228.0 = -20.1$$

$$\hat{S}_4(6) = 343.6 - 228.0 = 115.6$$

$$\hat{S}_5(6) = 290.7 - 228.0 = 62.7$$

$$\hat{S}_6(6) = 178.0 - 228.0 = -50.0$$

and the initial estimate of the level $\beta_0$ is

$$\hat{T}_6(6) = \hat{\beta}_0(6) = \bar{y}_{1984} = 228.0$$

Since the model is horizontal,

$$\hat{T}_7(6) = \hat{T}_6(6) = 228.0$$

So the forecast for January/February 1985 (season 1 of 1985) is

$$\hat{y}_7(6) = \hat{T}_7(6) + \hat{S}_7(6) = \hat{T}_7 + \hat{S}_1(6)$$
$$= 228.0 + (-34.4) = 193.6 \qquad \text{(Jan/Feb 1985)}$$

This really means that, to begin with, we forecast that season 1 of 1985 will be just like season 1 of 1984 (a consequence of the simple averages method being used for only one year's data). Selecting $\alpha = .1$ and $\delta = .3$ so that $\delta \cdot (1 - \alpha) = .27$, we now smooth through the rest of the series using the error correction forms of the updating equations:

***For Jan/Feb 85 (t = 7, i = 1):***

$$y_7 = 139.7$$
$$e_7 = y_7 - \hat{y}_7(6) = 139.7 - 193.6 = -53.9$$
$$\hat{T}_7(7) = \hat{T}_7(6) + \alpha \cdot e_7 = 228.0 + .1(-53.9) = 222.6$$
$$\hat{T}_8(7) = \hat{T}_7(7) = 222.6$$
$$\hat{S}_1(7) = \hat{S}_1(6) + \delta \cdot (1 - \alpha) \cdot e_7 = -34.4 + .27(-53.9) = -49.0$$
$$\hat{S}_2(7) = \hat{S}_2(6) \quad \hat{S}_3(7) = \hat{S}_3(6) \quad \cdots \quad \hat{S}_6(7) = \hat{S}_6(6) \text{ (since } t = 7 \text{ is in season 1)}$$
$$\hat{y}_8(7) = \hat{T}_8(7) + \hat{S}_8(7) = \hat{T}_8(7) + \hat{S}_2(7)$$
$$= 222.6 + (-73.8) = 148.8 \text{ (Mar/Apr 1985) (since } t = 8 \text{ is in season 2)}$$

Note that $\hat{S}_1(7)$ will not be used until $t = 12$ (November/December 1985), when we again forecast a season 1 (January/February 1986) and will remain unchanged until $t = 13$ (January/February 1986), when we again update for season 1. In addition, no seasonal estimates change at this time except $S_1$, a fact that will henceforth be understood without being stated.

***For Mar/Apr 85 (t = 8, i = 2):***

$$y_8 = 166.3$$
$$e_8 = 166.3 - 148.8 = 17.5$$
$$\hat{T}_8(8) = \hat{T}_9(8) = \hat{T}_8(7) + \alpha \cdot e_8 = 222.6 + .1 \times 17.5 = 224.4$$
$$\hat{S}_2(8) = \hat{S}_2(7) + \delta \cdot (1 - \alpha) \cdot e_8 = -73.8 + .27 \times 17.5 = -69.1$$
$$\hat{y}_9(8) = \hat{T}_9(8) + \hat{S}_9(8) = \hat{T}_9(8) + \hat{S}_3(8) = 224.4 + (-20.1)$$
$$= 204.3 \text{ (May/Jun 1985)}$$

***For May/Jun 85 (t = 9, i = 3):***

$$y_9 = 213.4$$
$$e_9 = 213.4 - 204.3 = 9.1$$
$$\hat{T}_9(9) = \hat{T}_{10}(9) = 224.4 + .1 \times 9.1 = 225.3$$

$$\hat{S}_3(9) = \hat{S}_3(8) + \delta \cdot (1 - \alpha) \cdot e_9 = -20.1 + .27 \times 9.1 = -17.6$$

$$\hat{y}_{10}(9) = \hat{T}_{10}(9) + \hat{S}_{10}(9) = \hat{T}_{10}(9) + \hat{S}_4(9) = 225.3 + 115.6$$
$$= 340.9 \text{ (Jul/Aug 1985)}$$

and so on,

*For May/Jun 86 ($t = 15$, $i = 3$):*

$$y_{15} = 181.2$$

$$e_{15} = 181.2 - 206.4 = -25.2$$

$$\hat{T}_{15}(15) = \hat{T}_{16}(15) = \hat{T}_{15}(14) + \alpha \cdot e_{15} = 224.0 + .1(-25.2) = 221.5$$

$$\hat{S}_3(15) = -17.6 + .27(-25.2) = -24.4$$

$$\hat{y}_{16}(15) = \hat{T}_{16}(15) + \hat{S}_{16}(15) = \hat{T}_{16}(15) + \hat{S}_4(15) = 221.5 + 100.8$$
$$= 322.3 \text{ (Jul/Aug 1986)}$$

***Note:*** $\hat{S}_4(15) = \hat{S}_4(14) = \cdots = \hat{S}_4(11) = \hat{S}_4(10)$, since the index for season 4 was last updated at $t = 10$ (Jul/Aug 1985).

The complete set of updates for the series is given in Table 5.22. The first actual forecast is the last one in the table $\hat{y}_{16}(15)$ for July/August 1986, since $y_1, y_2, \ldots, y_{15}$ had already been observed. The value of $\hat{y}_{16}(15)$ depends on what values of $\alpha$ and $\delta$ are used. In order to make a reasonable, not arbitrary, choice, the entire process was repeated for $\alpha, \delta = .1(.1).9$ and the RMSE computed for each different set. The optimal value among the sets was for $\alpha = .1$ and $\delta = .4$, which gave an RMSE $= 30.09$, $\hat{T}_{15}(15) = 221.6$, and $\hat{y}_{16}(15) = 317.5$. The final set of seasonal indices was

$$\hat{S}_1(15) = -55.1 \qquad \hat{S}_2(15) = -65.8 \qquad \hat{S}_3(15) = -26.2$$

$$\hat{S}_4(15) = +95.9 \qquad \hat{S}_5(15) = +74.2 \qquad \hat{S}_6(15) = -46.2$$

Since we wish to seasonally adjust the data, these indexes must be normalized. To do this we compute

$$\bar{S} = \frac{-55.1 - 65.8 - 26.2 + 95.9 + 74.2 - 46.2}{6} = \frac{-23.2}{6} = -3.9$$

$$\hat{S}_1(15) \rightarrow \hat{S}_1(15) - \bar{S} = -55.1 - (-3.9) = -51.2$$

$$\hat{S}_2(15) \rightarrow \hat{S}_2(15) - \bar{S} = -65.8 - (-3.9) = -61.9$$

$$\hat{S}_3(15) \rightarrow \hat{S}_3(15) - \bar{S} = -26.2 - (-3.9) = -22.3$$

$$\hat{S}_4(15) \rightarrow \hat{S}_4(15) - \bar{S} = +95.9 - (-3.9) = +99.8$$

$$\hat{S}_5(15) \rightarrow \hat{S}_5(15) - \bar{S} = +74.2 - (-3.9) = +78.1$$

$$\hat{S}_6(15) \rightarrow \hat{S}_6(15) - \bar{S} = -46.2 - (-3.9) = -42.3$$

If it is desired to use these new indexes in future forecasting (otherwise the old ones will have to be used), we compute

**Table 5.22 Smoothing Calculations, Stilling Unified School District, Additive Horizontal Model ($\alpha$ = .1, $\delta$ = .3)**

| | Season | $t$ | $y_t$ | $\hat{y}_t(t-1)$ | $e_t$ | $\hat{T}_t(t-1)$ | $\hat{T}_t(t)$ | Season $i$ | $\hat{S}_i(t)$ | $\hat{y}_{t+1}(t)$ |
|---|---|---|---|---|---|---|---|---|---|---|
| 1984 | Jan/Feb | 1 | 193.6 | | | | | 1 | −34.4 | |
| | Mar/Apr | 2 | 154.2 | | | | | 2 | −73.8 | |
| | May/Jun | 3 | 207.9 | | | | | 3 | −20.1 | |
| | Jul/Aug | 4 | 343.6 | | | | | 4 | +115.6 | |
| | Sep/Oct | 5 | 290.7 | | | | | 5 | +62.7 | |
| | Nov/Dec | 6 | 178.0 | | | | 228.0 | 6 | −50.0 | 193.6 |
| 1985 | Jan/Feb | 7 | 139.7 | 193.6 | −53.9 | 228.0 | 222.6 | 1 | −49.0 | 148.8 |
| | Mar/Apr | 8 | 166.3 | 148.8 | 17.5 | 222.6 | 224.4 | 2 | −69.1 | 204.3 |
| | May/Jun | 9 | 213.4 | 204.3 | 9.1 | 224.4 | 225.3 | 3 | −17.6 | 340.9 |
| | Jul/Aug | 10 | 286.1 | 340.9 | −54.8 | 225.3 | 219.8 | 4 | +100.8 | 282.5 |
| | Sep/Oct | 11 | 314.4 | 282.5 | 31.9 | 219.8 | 223.0 | 5 | +71.3 | 173.0 |
| | Nov/Dec | 12 | 183.6 | 173.0 | 10.6 | 223.0 | 224.1 | 6 | −47.1 | 175.1 |
| 1986 | Jan/Feb | 13 | 166.6 | 175.1 | −8.5 | 224.1 | 223.3 | 1 | −51.3 | 154.2 |
| | Mar/Apr | 14 | 160.8 | 154.2 | 6.6 | 223.3 | 224.0 | 2 | −67.3 | 206.4 |
| | May/Jun | 15 | 181.2 | 206.4 | −25.2 | 224.0 | 221.5 | 3 | −24.4 | 322.3 |
| | Jul/Aug | 16 | ? | 322.3 | | 221.5 | | 4 | | |
| | Sep/Oct | 17 | ? | 292.8 = $\hat{y}_{17}(15)$ | | | | 5 | | |
| | Nov/Dec | 18 | ? | 174.4 = $\hat{y}_{18}(15)$ | | | | 6 | | |
| 1987 | Jan/Feb | 19 | ? | 170.2 = $\hat{y}_{19}(15)$ | | | | 1 | | |
| | Mar/Apr | 20 | ? | 154.2 = $\hat{y}_{20}(15)$ | | | | 2 | | |
| | May/Jun | 21 | ? | 197.1 = $\hat{y}_{21}(15)$ | | | | 3 | | |

$$\hat{T}_{15}(15) \rightarrow \hat{T}_{15}(15) + \bar{S} = 221.6 + (-3.9) = 217.7$$

At this point, forecasts for, say, the next three seasons would be

*For Jul/Aug 86 (t = 16, i = 4):* $\hat{y}_{16}(15) = 217.7 + 99.8 = 317.5$

*For Sep/Oct 86 (t = 17, i = 5):* $\hat{y}_{17}(15) = 217.7 + 78.1 = 295.8$

*For Nov/Dec 86 (t = 18, i = 6):* $\hat{y}_{18}(15) = 217.7 - 42.3 = 175.4$

To seasonally adjust, say, the last three observations:

*For Jan/Feb 86 (t = 13, i = 1):* $y_{13} - \hat{S}_1(15) = 166.6 - (-51.2) = 217.8$

*For Mar/Apr 86 (t = 14, i = 2):* $y_{14} - \hat{S}_2(15) = 160.8 - (-61.9) = 222.7$

*For May/Jun 86 (t = 15, i = 3):* $y_{15} - \hat{S}_3(15) = 181.2 - (-22.3) = 203.5$ •

### 5.6.4 Simple Seasonal Exponential Smoothing, Multiplicative Models

A multiplicative seasonal structure should be used if it is felt that the magnitude of the seasonal movement is proportional to the level of the series. Under these circumstances, if a step change in the level of the series ($\beta_0$) occurs, multiplicative indexes will do a better job of tracking the seasonality than will additive ones. The model usually assumed is $y_t = \beta_0 \cdot S_t + \epsilon_t$.

The main differences in the updating procedures from those for an additive model are that $y_t$ is detrended or deseasonalized by dividing by the trend or seasonal estimate rather than subtracting it. Also, $e_t$ must be deseasonalized before it is used.

***Note:*** For a horizontal series "detrending" means that an estimate of the level is removed.

### EXAMPLE 5.13 *Multiplicative SSES, U.S. Return Rates*

To illustrate the model, let us return to the U.S. Return Rate series (Table 5.3), where $y_t$ = the rate of return on stockholders' equity (%). Suppose that the forecasting model is being developed during the fourth quarter of 1983. It will be used to forecast $y_t$ for the four quarters of 1984. In quarter IV, 1983, the twenty values $y_1, y_2, \ldots, y_{20}$ are available for fitting. To initialize the model we will backcast the series starting at time 20 and use the estimates at the end of the backcasting as initial values for forecasting through the series. The values of $\alpha$ and $\delta$ will be chosen as the ones, among all the pairs of values .1(.1).9, that have the smallest RMSE for the forecasting part of the initialization process. Because the subscripts and time indexing in backcasts can be confusing, the calculations will be discussed later. For now we need only know that the final backcast values are

$$\hat{T}_1(1) = 18.59$$

$$\hat{S}_1(1) = .702 \qquad \hat{S}_2(1) = .864 \qquad \hat{S}_3(1) = .945 \qquad \hat{S}_4(1) = 1.226$$

These are the initial values for the level and seasonal indexes to be used in the forecasting part of the procedure, and they give the initial forecast for quarter II, 1979 ($t = 2$) as

$$\hat{y}_2(1) = \hat{T}_2(1) \cdot \hat{S}_2(1) = \hat{T}_1(1) \cdot \hat{S}_2(1) = 18.59 \times .864 = 16.06$$

The search for optimal $\alpha$ and $\delta$ gave $\alpha = .3$ and $\delta = .6$ as having the smallest RMSE (3.67%), so they are used to demonstrate the updating procedure.

***For II, 1979 ($t = 2, i = 2$):***

$$y_2 = 15.70 \qquad e_2 = 15.07 - 16.06 = -.36$$

$$\hat{T}_2(2) = \hat{T}_2(1) + \frac{\alpha \cdot e_2}{\hat{S}_2(1)}$$

$$= 18.59 + \frac{.3(-.36)}{.864} = 18.47 = \hat{T}_3(2)$$

$$\hat{S}_2(2) = \hat{S}_2(1) + \frac{\delta \cdot (1 - \alpha) \cdot e_2}{\hat{T}_2(2)}$$

$$= .864 + \frac{.6 \times .7(-.36)}{18.47} = .856$$

$$\hat{y}_3(2) = \hat{T}_3(2) \cdot \hat{S}_3(2) = 18.47 \times .945 = 17.45$$

## Simple Seasonal Exponential Smoothing (SSES), Multiplicative Model

**Given:** The current observation of the time series $y_t$ and

$\hat{T}_{t-1}(t-1)$ = estimated unseasonalized level as of time $t-1$

$\hat{S}_i(t-1)$ = estimated index for season $i$ as of time $t-1$ $\quad i = 1, \ldots, L$

**Model:** $y_t = \beta_0 \cdot S_t + \epsilon_t$

So

$\hat{T}_{t+p}(t) = \hat{T}_t(t) \quad$ for $p \geq 1$

**Updating Equations:**

Smoothing Form:

$$\hat{T}_t(t) = \frac{\alpha \cdot y_t}{\hat{S}_t(t-1)} + (1-\alpha) \cdot \hat{T}_t(t-1)$$

$$\hat{S}_i(t) = \frac{\delta \cdot y_t}{\hat{T}_t(t)} + (1-\delta) \cdot \hat{S}_i(t-1) \quad \text{if } t \text{ in season } i$$

$$= \hat{S}_i(t-1) \quad \text{if } t \text{ not in season } i$$

Error Correction Form:

$$\hat{T}_t(t) = \hat{T}_t(t-1) + \frac{\alpha \cdot e_t}{\hat{S}_t(t-1)}$$

$$\hat{S}_i(t) = \hat{S}_i(t-1) + \delta \cdot (1-\alpha) \cdot \frac{e_t}{\hat{S}_i(t-1)} \quad \text{if } t \text{ in season } i$$

$$= \hat{S}_i(t-1) \quad \text{if } t \text{ not in season } i$$

where

$\hat{S}_t(\text{any time}) = \hat{S}_i(\text{any time}) \quad$ for $t$ in season $i$

$e_t$ = one-step-ahead forecast error

$= y_t - \hat{y}_t(t-1)$

$\alpha$ = smoothing constant for the level

$\delta$ = smoothing constant for the seasonal

**Forecasts:** (At time $t$ for $p$ periods ahead)

$$\hat{y}_{t+p}(t) = \hat{T}_t(t) \cdot \hat{S}_{t+p}(t)$$

where

$\hat{S}_{t+p}(t) = \hat{S}_i(t) \quad$ for $t + p$ in season $i$

***For III, 1979 ($t = 3$, $i = 3$):***

$$y_3 = 20.97 \qquad e_3 = 20.97 - 17.45 = 3.52$$

$$\hat{T}_3(3) = \hat{T}_3(2) + \frac{\alpha \cdot e_3}{\hat{S}_3(2)}$$

$$= 18.47 + \frac{.3 \times 3.52}{.945} = 19.59 = \hat{T}_4(3)$$

$$\hat{S}_3(3) = \hat{S}_3(2) + \frac{\delta \cdot (1-\alpha) \cdot e_3}{\hat{T}_3(3)}$$

$$= .945 + \frac{.6 \times .7 \times 3.52}{19.59} = 1.020$$

$$\hat{y}_4(3) = \hat{T}_4(3) \cdot \hat{S}_4(3) = 19.59 \times 1.226 = 24.02$$

*For IV, 1979 (t = 4, i = 4):*

$$y_4 = 20.32 \qquad e_4 = 20.32 - 24.02 = -3.70$$

$$\hat{T}_4(4) = 19.59 + \frac{.3(-3.70)}{1.226} = 18.68 = \hat{T}_5(4)$$

$$\hat{S}_4(4) = 1.226 + \frac{.42(-3.70)}{18.68} = 1.143$$

$$\hat{y}_5(4) = \hat{T}_5(4) \cdot \hat{S}_5(4) = \hat{T}_5(4) \cdot \hat{S}_1(4) = 18.68 \times .702 = 13.11$$

*For I, 80 (t = 5, i = 1):*

$$y_5 = 14.31 \qquad e_5 = 14.31 - 13.11 = 1.20$$

$$\hat{T}_5(5) = 18.68 + \frac{.3 \times 1.20}{.702} = 19.19 = \hat{T}_6(5)$$

$$\hat{S}_1(5) = .702 + \frac{.42 \times 1.20}{19.19} = .728$$

$$\hat{y}_6(5) = 19.19 \times .856 = 16.43$$

*For II, 80 (t = 6, i = 2):*

$$y_6 = 18.00 \qquad e_6 = 18.00 - 16.43 = 1.57$$

$$\hat{T}_6(6) = 19.19 + \frac{.3 \times 1.57}{.856} = 19.74 = \hat{T}_7(6)$$

$$\hat{S}_2(6) = .856 + \frac{.42 \times 1.57}{19.74} = .889$$

$$\hat{y}_7(6) = 19.74 \times 1.020 = 20.13$$

and so on. The complete set of updates is given in Table 5.23.

The forecasts for the four quarters of 1984 as of quarter IV, 1983, are

$$\hat{y}_{21}(20) = 17.03 \times 0.533 = 9.08$$
$$\hat{y}_{22}(20) = 17.03 \times 0.792 = 13.48$$
$$\hat{y}_{23}(20) = 17.03 \times 0.797 = 13.57$$
$$\hat{y}_{24}(20) = 17.03 \times 1.455 = 24.78$$

The last three of these will change when the quarter I, 1984, return rate becomes available.

Since the return rates for 1984 are actually known, we can see how the forecasting model would have done with data that were not actually used to help fit the model. The one-step-ahead forecasts and forecast errors for 1984 would be as in the following table.

| Forecast | Actual | Error | Relative Error |
|---|---|---|---|
| $\hat{y}_{21}(20) = 9.08\%$ | $y_{21} = 10.25\%$ | $e_{21} = +1.17\%$ | (+11.4%) |
| $\hat{y}_{22}(21) = 14.01\%$ | $y_{22} = 13.93\%$ | $e_{22} = -0.08\%$ | (−0.6%) |
| $\hat{y}_{23}(22) = 14.08\%$ | $y_{23} = 12.68\%$ | $e_{23} = -1.40\%$ | (−11.0%) |
| $\hat{y}_{24}(23) = 24.92\%$ | $y_{24} = 21.72\%$ | $e_{24} = -3.20\%$ | (−14.7%) |

**Table 5.23 Smoothing Calculations, U.S. Return Rates, Multiplicative Horizontal Model ($\alpha = .3$, $\delta = .6$)**

| | Qrtr | $t$ | $y_t$ | $\hat{y}_t(t-1)$ | $e_t$ | $\hat{T}_t(t)$ | Season $i$ | $\hat{S}_i(t)$ | $\hat{y}_{t+1}(t)$ |
|---|---|---|---|---|---|---|---|---|---|
| 1979 | I | 1 | 12.24 | | | 18.59 | 1 | | 16.06 |
| | II | 2 | 15.70 | 16.06 | −.36 | 18.47 | 2 | .856 | 17.45 |
| | III | 3 | 20.97 | 17.45 | 3.52 | 19.59 | 3 | 1.020 | 24.02 |
| | IV | 4 | 20.32 | 24.02 | −3.70 | 18.68 | 4 | 1.143 | 13.11 |
| 1980 | I | 5 | 14.31 | 13.11 | 1.20 | 19.19 | 1 | .728 | 16.43 |
| | II | 6 | 18.00 | 16.43 | 1.57 | 19.74 | 2 | .889 | 20.13 |
| | III | 7 | 21.66 | 20.13 | 1.53 | 20.19 | 3 | 1.052 | 23.08 |
| | IV | 8 | 22.88 | 23.08 | −.20 | 20.14 | 4 | 1.139 | 14.66 |
| 1981 | I | 9 | 9.31 | 14.66 | −5.35 | 17.94 | 1 | .603 | 15.95 |
| | II | 10 | 13.26 | 15.95 | −2.69 | 17.03 | 2 | .823 | 17.92 |
| | III | 11 | 10.96 | 17.92 | −6.96 | 15.05 | 3 | .858 | 17.14 |
| | IV | 12 | 19.11 | 17.14 | 1.97 | 15.57 | 4 | 1.192 | 9.39 |
| 1982 | I | 13 | 4.40 | 9.39 | −4.99 | 13.09 | 1 | .443 | 10.77 |
| | II | 14 | 9.39 | 10.77 | −1.38 | 12.59 | 2 | .777 | 10.80 |
| | III | 15 | 8.00 | 10.80 | −2.80 | 11.61 | 3 | .757 | 13.84 |
| | IV | 16 | 23.07 | 13.84 | 9.23 | 13.93 | 4 | 1.470 | 6.17 |
| 1983 | I | 17 | 9.64 | 6.17 | 3.47 | 16.28 | 1 | .533 | 12.65 |
| | II | 18 | 13.24 | 12.65 | .59 | 16.51 | 2 | .792 | 12.50 |
| | III | 19 | 14.14 | 12.50 | 1.64 | 17.16 | 3 | .797 | 25.23 |
| | IV | 20 | 24.61 | 25.23 | −.62 | 17.03 | 4 | 1.455 | 9.08 |
| 1984 | I | 21 | ? | 9.08 | ? | ? | 1 | ? | ? |

Initial values from backcasting: $\hat{T}_1(1) = 18.59$, $\hat{S}_1(1) = .702$, $\hat{S}_2(1) = .864$, $\hat{S}_3(1) = .945$, $\hat{S}_4(1) = 1.226$. So $\hat{y}_2(1) = 18.59 \times .864 = 16.06$, $\hat{y}_3(2) = 18.47 \times .945 = 17.45$.

The seasonal indexes at time 20 do not sum to 4, nor do the initial seasonal estimates. Either set could easily be made conventional by computing $\bar{S}$ and using the conversion formulas (Equation 5.21). For the quarter IV, 1983, indexes, for example,

$$\bar{S} = \frac{.533 + .792 + .797 + 1.455}{4} = \frac{3.577}{4} = .894$$

$$\hat{S}_1(20) \rightarrow \frac{.533}{.894} = .596$$

$$\hat{S}_2(20) \rightarrow \frac{.792}{.894} = .886$$

$$\hat{S}_3(20) \rightarrow \frac{.797}{.894} = .891$$

$$\hat{S}_4(20) \rightarrow \frac{1.455}{.894} = 1.628$$

$$\hat{T}_{20}(20) \rightarrow 17.03 \times .894 = 15.22$$

•

Having spent some time with the forecasting formulas, you should now more easily understand the backcasting. The backcasting error correction (backward updating) formulas are

$$\hat{T}_t(t) = \hat{T}_t(t+1) + \frac{\alpha \cdot e_t}{\hat{S}_i(t+1)}$$

$$\hat{S}_i(t) = \hat{S}_i(t+1) + \frac{\delta \cdot (1-\alpha) \cdot e_t}{\hat{T}_t(t)} \qquad \text{for } t \text{ in the } i\text{th season}$$

To do the backcasting, smoothing constants and initial backcast values are needed. For the return rate series, $\alpha = .3$, $\delta = .3$, and the simple initial values were used:

$$\hat{T}_{20}(20) = \hat{y}_{19}(20) = y_{20} = 24.61$$

$$\hat{S}_1(20) = \hat{S}_2(20) = \hat{S}_3(20) = \hat{S}_4(20) = 1.00$$

Backcasting proceeded as follows.

*At* $t = 19$, $i = 3$:

$$y_{19} = 14.14 \qquad e_{19} = 14.14 - 24.61 = -10.47$$

$$\hat{T}_{19}(19) = \hat{T}_{19}(20) + \frac{\alpha \cdot e_{19}}{\hat{S}_3(20)} = 24.61 + \frac{.3(-10.47)}{1.000} = 21.47 = \hat{T}_{18}(19)$$

$$\hat{S}_3(19) = \hat{S}_3(20) + \frac{\delta \cdot (1-\alpha) \cdot e_{19}}{\hat{T}_{19}(19)} = 1.00 + \frac{.21(-10.47)}{21.47} = .898$$

$$\hat{y}_{18}(19) = \hat{T}_{18}(19) \cdot \hat{S}_{18}(19) = \hat{T}_{18}(19) \cdot \hat{S}_2(19) = 21.47 \times 1.000 = 21.47$$

*At* $t = 18$, $i = 2$:

$$y_{18} = 13.24 \qquad e_{18} = 13.24 - 21.47 = -8.23$$

$$\hat{T}_{18}(18) = 21.47 + \frac{.3(-8.23)}{1.000} = 19.00 = \hat{T}_{17}(18)$$

$$\hat{S}_2(18) = \hat{S}_2(19) + \frac{\delta \cdot (1-\alpha) \cdot e_{18}}{\hat{T}_{18}(18)} = 1.000 + \frac{.21(-8.23)}{19.00} = .909$$

$$\hat{y}_{17}(18) = \hat{T}_{17}(18) \cdot \hat{S}_{17}(18) = \hat{T}_{17}(18) \cdot \hat{S}_1(18) = 19.00 \times 1.000 = 19.00$$

*At* $t = 17$, $i = 1$:

$$y_{17} = 9.64 \qquad e_{17} = 9.64 - 19.00 = -9.36$$

$$\hat{T}_{17}(17) = 19.00 + \frac{.3(-9.36)}{1.000} = 16.19 = \hat{T}_{16}(17)$$

$$\hat{S}_1(17) = \hat{S}_1(18) + \frac{\delta \cdot (1-\alpha) \cdot e_{17}}{\hat{T}_{17}(17)} = 1.000 + \frac{.21(-9.36)}{16.19} = .879$$

$$\hat{y}_{16}(17) = \hat{T}_{16}(17) \cdot \hat{S}_4(17) = 16.19 \times 1.000 = 16.19$$

*At* $t = 16$, $i = 4$:

$$y_{16} = 23.07 \qquad e_{16} = 23.07 - 16.19 = 6.88$$

$$\hat{T}_{16}(16) = 16.19 + \frac{.3 \times 6.88}{1.000} = 18.25 = \hat{T}_{15}(16)$$

$$\hat{S}_4(16) = \hat{S}_4(17) + \frac{\delta \cdot (1-\alpha) \cdot e_{16}}{\hat{T}_{16}(16)} = 1.000 + \frac{.21 \times 6.88}{18.25} = 1.079$$

$$\hat{y}_{15}(16) = \hat{T}_{15}(16) \cdot \hat{S}_3(16) = 18.25 \times .898 = 16.39$$

*At t = 15, i = 3:* $y_{15} = 8.00 \qquad e_{15} = 8.00 - 16.39 = -8.39$

$$\hat{T}_{15}(15) = 18.25 + \frac{.3(-8.39)}{.898} = 15.45 = \hat{T}_{14}(15)$$

$$\hat{S}_3(15) = .898 + \frac{.21(-8.39)}{15.45} = .784$$

$$\hat{y}_{14}(15) = 15.45 \times .909 = 14.04$$

*At t = 14, i = 2:* $y_{14} = 9.39 \qquad e_{14} = 9.39 - 14.04 = -4.65$

$$\hat{T}_{14}(14) = 15.45 + \frac{.3(-4.65)}{.909} = 13.92 = \hat{T}_{13}(14)$$

$$\hat{S}_2(14) = .909 + \frac{.21(-4.65)}{13.92} = .839$$

$$\hat{y}_{13}(14) = 13.92 \times .879 = 12.24$$

and so on. The complete set of up(back)dates is shown in Table 5.24.

The last entry in Table 5.24 is for $t = 1$, where $\hat{T}_1(1) = 18.59$, $\hat{S}_1(1) = .702$, $\hat{S}_2(1) = .864$, $\hat{S}_3(1) = .945$, and $\hat{S}_4(1) = 1.226$ are obtained. These

**Table 5.24** Backcasting Calculations, U.S. Return Rates, Multiplicative Horizontal Model ($\alpha = .3$, $\delta = .3$)

| | Qrtr | $t$ | $y_t$ | $\hat{y}_t(t+1)$ | $e_t$ | $\hat{T}_t(t)$ | Season $i$ | $\hat{S}_i(t)$ | $\hat{y}_{t+1}(t)$ |
|---|---|---|---|---|---|---|---|---|---|
| 1983 | IV | 20 | 24.61 | | | 24.61 | 4 | | 24.61 |
| | III | 19 | 14.14 | 24.61 | −10.47 | 21.47 | 3 | .898 | 21.47 |
| | II | 18 | 13.24 | 21.47 | −8.23 | 19.00 | 2 | .909 | 19.00 |
| | I | 17 | 9.64 | 19.00 | −9.36 | 16.19 | 1 | .879 | 16.19 |
| 1982 | IV | 16 | 23.07 | 16.19 | 6.88 | 18.25 | 4 | 1.079 | 16.39 |
| | III | 15 | 8.00 | 16.39 | −8.39 | 15.45 | 3 | .784 | 14.04 |
| | II | 14 | 9.39 | 14.04 | −4.65 | 13.92 | 2 | .839 | 12.24 |
| | I | 13 | 4.40 | 12.24 | −7.84 | 11.24 | 1 | .739 | 12.13 |
| 1981 | IV | 12 | 19.11 | 12.13 | 6.98 | 13.18 | 4 | 1.190 | 10.33 |
| | III | 11 | 10.96 | 10.33 | .63 | 13.42 | 3 | .794 | 11.26 |
| | II | 10 | 13.26 | 11.26 | 2.00 | 14.14 | 2 | .869 | 10.36 |
| | I | 9 | 9.31 | 10.36 | −1.05 | 13.71 | 1 | .717 | 16.31 |
| 1980 | IV | 8 | 22.88 | 16.31 | 6.57 | 15.37 | 4 | 1.280 | 12.20 |
| | III | 7 | 21.66 | 12.20 | 9.46 | 18.94 | 3 | .899 | 16.46 |
| | II | 6 | 18.00 | 16.46 | 1.54 | 19.47 | 2 | .886 | 13.96 |
| | I | 5 | 14.31 | 13.96 | .35 | 19.62 | 1 | .721 | 25.11 |
| 1979 | IV | 4 | 20.32 | 25.11 | −4.79 | 18.50 | 4 | 1.226 | 16.63 |
| | III | 3 | 20.97 | 16.63 | 4.34 | 19.95 | 3 | .945 | 17.68 |
| | II | 2 | 15.70 | 17.68 | −1.98 | 19.28 | 2 | .864 | 13.90 |
| | I | 1 | 12.24 | 13.90 | −1.66 | 18.59 | 1 | .702 | — |

Initial values: $\hat{T}_{20}(20) = 24.61$, $\hat{S}_1(20) = \cdots = \hat{S}_4(20) = 1.000$.

become the starting values for the forecasting part of the initialization procedure. The moderately high values (.3, .3) for $\alpha$ and $\delta$ and the fact that the estimates have been smoothed through 19 observations should make these initial forecasting values relatively free of the effect of the poor initial backcasting values ($\hat{T}_{20}(20) = y_{20}$, $\hat{S}_1(20) = \hat{S}_2(20) = \cdots = \hat{S}_4(20) = 1.00$). Consequently, good values for $\alpha$ and $\delta$ should be chosen in the forward part of the process, reducing any fear that inappropriately large values might result.

Should the time indexing above become too confusing, backcasting may be accomplished quite easily by writing the series in reverse order and using the regular updating formulas on the reversed series. When this is done, however, remember that the values of the subscripts of the seasonal indexes will be in *reverse* order and need to be corrected before starting the forward steps.

### 5.6.5 Additive Holt-Winters Models

Seasonal smoothing combined with Holt's linear-trend updating is usually called the Holt-Winters model or Holt-Winters smoothing (HWS). With an additive seasonal structure the model is

$$y_t = (\beta_0 + \beta_1 t) + S_t + \epsilon_t$$

and

$$\hat{T}_t(t) = \text{estimate of } \beta_0 + \beta_1 t \text{ as of time } t$$

So

$$\begin{aligned}\hat{T}_{t+p}(t) &= \text{estimate of } \beta_0 + \beta_1 (t + p) \text{ as of time } t \\ &= \text{estimate of } (\beta_0 + \beta_1 t) + p \cdot \beta_1 \text{ as of time } t \\ &= \hat{T}_t(t) + p \cdot \hat{\beta}_1(t)\end{aligned}$$

**Holt-Winters Exponential Smoothing (HWS), Additive Model**

**Given:** The current observation of the time series $y_t$ and

$\hat{T}_{t-1}(t-1)$ = estimated unseasonalized level as of time $t - 1$

$\hat{\beta}_1(t-1)$ = estimated slope of the trend line as of time $t - 1$

$\hat{S}_i(t-1)$ = estimated index for season $i$ as of time $t - 1$ $\quad i = 1, \ldots, L$

**Model:** $y_t = (\beta_0 + \beta_1 t) + S_t + \epsilon_t$

So

$\hat{T}_{t+p}(t) = \hat{T}_t(t) + p \cdot \hat{\beta}_1(t) \quad \text{for } p \geq 1$

(*Table continues*)

**Holt-Winters Exponential Smoothing (HWS), Additive Model** *(Continued)*

**Updating Equations:**

Smoothing Form:

$$\hat{T}_t(t) = \alpha[y_t - \hat{S}_t(t-1)] + (1-\alpha)\cdot \hat{T}_t(t-1)$$

$$\hat{\beta}_1(t) = \gamma[\hat{T}_t(t) - \hat{T}_t(t-1)] + (1-\gamma)\cdot \hat{\beta}_1(t-1)$$

$$\hat{S}_i(t) = \delta[y_t - \hat{T}_t(t)] + (1-\delta)\cdot \hat{S}_i(t-1) \qquad \text{if } t \text{ in season } i$$

$$= \hat{S}_i(t-1) \qquad \text{if } t \text{ not in season } i$$

Error Correction Form:

$$\hat{T}_t(t) = \hat{T}_t(t-1) + \alpha\cdot e_t$$

$$\hat{\beta}_1(t) = \hat{\beta}_1(t-1) + \alpha\cdot\gamma\cdot e_t$$

$$\hat{S}_i(t) = \hat{S}_i(t-1) + \delta\cdot(1-\alpha)\cdot e_t \qquad \text{if } t \text{ in season } i$$

$$= \hat{S}_i(t-1) \qquad \text{if } t \text{ is not in season } i$$

where

$\hat{S}_t(\text{any time}) = \hat{S}_i(\text{any time})$ for $t$ in season $i$

$e_t$ = one-step-ahead forecast error = $y_t - \hat{y}_t(t-1)$

$\alpha$ = smoothing constant for the level

$\gamma$ = smoothing constant for the slope

$\delta$ = smoothing constant for the seasonal

**Forecasts:**

(At time $t$ for $p$ periods ahead)

$$\hat{y}_{t+p}(t) = \hat{T}_{t+p}(t) + \hat{S}_{t+p}(t) = \hat{T}_t(t) + p\cdot\hat{\beta}_1(t) + \hat{S}_{t+p}(t)$$

where

$\hat{S}_{t+p}(t) = \hat{S}_i(t)$ for $t + p$ in season $i$

The mechanics of the updating are much like those for the additive horizontal model, except that there is a stream of slope estimates and the one-step-ahead trend is calculated by adding the slope to the estimate of the current trend (Holt, 1957; Winters, 1960).

**EXAMPLE 5.14** *Additive HWS, Quantitec Corp.*

The net quarterly income history for Quantitec Corp., one of many newly formed subsidiaries of a multinational corporation, are shown in Table 5.25. The parent company wishes to forecast the net incomes of its several subsidiaries using smoothing techniques. It is expected that the income series for each company will have both seasonality and trend and that the series are likely to be quite changeable as the newly formed companies grow. The forecasting will be relatively short term, so an additive seasonal

**Table 5.25** **Net Income for Quantitec Corp.** $y_t$ = quarterly net income ($100,000)

| Year | Quarter | $y_t$ | Year | Quarter | $y_t$ |
|---|---|---|---|---|---|
| 1985 | I | 216.1 | 1986 | I | 232.8 |
| | II | 222.2 | | II | 228.4 |
| | III | 224.5 | | | |
| | IV | 205.2 | | | |

structure is deemed adequate. There are very few data for these companies, so a single forecast method will be used to get initial values. Smoothing constants $\alpha = .2$, $\gamma = .2$, and $\delta = .4$ will then be used to update the estimates each quarter as new data become available.

A standard regression program is used to fit additive seasonals (using dummy variables $x_1$, $x_2$, $x_3$ for quarters I, II, III) and a linear trend to the Quantitec data, with the following results:

$$\hat{y}_t = 193.8 + 2.9t + 22.1x_1 + 20.1x_2 + 22.2x_3$$

Origin: IV, 84
Units ($t$): Quarters
Units ($y$): $100,000

which may be normalized (with a fourth dummy variable, $x_4$, for quarter IV) to give

$$\hat{y}_t = 211.9 + 2.9t + 6.0x_1 + 4.0x_2 + 6.1x_3 - 16.1x_4$$

Origin: IV, 84
Units ($t$): Quarters
Units ($y$): $100,000

(Although not actually necessary for forecasting, this normalization makes the indexes more easily interpretable.) The initial estimates (at $t = 6$) are

$$\hat{\beta}_1(6) = 2.9 \qquad \hat{\beta}_0(6) = 211.9$$

$$\hat{T}_6(6) = \hat{\beta}_0(6) + \hat{\beta}_1(6) \cdot 6$$

$$= 211.9 + 2.9 \times 6 = 229.3$$

$$\hat{S}_1(6) = +6.0 \qquad \hat{S}_2(6) = +4.0 \qquad \hat{S}_3(6) = +6.1 \qquad \hat{S}_4(6) = -16.1$$

$$\hat{T}_7(6) = \hat{T}_6(6) + \hat{\beta}_1(6) = 229.3 + 2.9 = 232.2$$

$$\hat{y}_7(6) = \hat{T}_7(6) + \hat{S}_7(6) = \hat{T}_7(6) + \hat{S}_3(6) = 232.2 + 6.1 = 238.3$$

Suppose now that Quantitec's next four quarters of net income are $y_7 = 243.6$, $y_8 = 206.4$, $y_9 = 247.1$, and $y_{10} = 245.6$. The updates and forecasts would be ($\alpha = .2$, $\alpha \cdot \gamma = .2 \times .2 = .04$, $\delta(1 - \alpha) = .4 \times (1 - .2) = .32$) as follows.

*For III, 86 ($t = 7$, $i = 3$):*

$$y_7 = 243.6 \qquad e_7 = 243.6 - 238.3 = 5.3$$

$$\hat{T}_7(7) = \hat{T}_7(6) + \alpha \cdot e_7 = 232.2 + .2 \times 5.3 = 233.3$$

$$\hat{\beta}_1(7) = \hat{\beta}_1(6) + \alpha \cdot \gamma \cdot e_7 = 2.9 + .04 \times 5.3 = 3.1$$

$$\hat{S}_3(7) = \hat{S}_3(6) + \delta(1 - \alpha) \cdot e_7 = 6.1 + .32 \times 5.3 = 7.8$$

$$\hat{T}_8(7) = \hat{T}_7(7) + \hat{\beta}_1(7) = 233.3 + 3.1 = 236.4$$

$$\hat{y}_8(7) = \hat{T}_8(7) + \hat{S}_8(7) = \hat{T}_8(7) + \hat{S}_4(7) = 236.4 - 16.1 = 220.3$$

*For II, 86*
*(t = 8, i = 4):*

$$y_8 = 206.4 \qquad e_8 = 206.4 - 220.3 = -13.9$$
$$\hat{T}_8(8) = \hat{T}_8(7) + \alpha \cdot e_8 = 236.4 + .2(-13.9) = 233.6$$
$$\hat{\beta}_1(8) = \hat{\beta}_1(7) + \alpha \cdot \gamma \cdot e_8 = 3.1 + .04(-13.9) = 2.5$$
$$\hat{S}_4(8) = \hat{S}_4(7) + \delta(1 - \alpha) \cdot e_8 = -16.1 + .32(-13.9) = -20.5$$
$$\hat{T}_9(8) = \hat{T}_8(8) + \hat{\beta}_1(8) = 233.6 + 2.555 = 236.1$$
$$\hat{y}_9(8) = \hat{T}_9(8) + \hat{S}_9(8) = \hat{T}_9(8) + \hat{S}_1(8) = 236.1 + 6.0 = 242.1$$

*For I, 87*
*(t = 9, i = 1):*

$$y_9 = 247.1 \qquad e_9 = 247.1 - 242.1 = 5.0$$
$$\hat{T}_9(9) = 236.1 + .2 \times 5.0 = 237.1$$
$$\hat{\beta}_1(9) = 2.5 + .04 \times 5.0 = 2.7$$
$$\hat{S}_1(9) = 6.0 + .32 \times 5.0 = 7.6$$
$$\hat{T}_{10}(9) = 237.1 + 2.7 = 239.8$$
$$\hat{y}_{10}(9) = 239.8 + 4.0 = 243.8$$

*For II, 87*
*(t = 10, i = 2):*

$$y_{10} = 245.6 \qquad e_{10} = 1.8$$
$$\hat{T}_{10}(10) = 239.8 + .2 \times 1.8 = 240.2$$
$$\hat{\beta}_1(10) = 2.7 + .04 \times 1.8 = 2.8$$
$$\hat{S}_2(10) = 4.0 + .32 \times 1.8 = 4.6$$
$$\hat{T}_{11}(10) = 240.2 + 2.8 = 243.0$$
$$\hat{y}_{11}(10) = 243.0 + 7.8 = 250.8$$

At this time, forecasts for the next four quarters would be

$$\hat{y}_{11}(10) = 243.0 + 7.8 = 250.8$$
$$\hat{y}_{12}(10) = (239.8 + 2.8) - 20.5 = 245.8 - 20.5 = 225.3$$
$$\hat{y}_{13}(10) = (245.8 + 2.8) + 7.6 = 248.6 + 7.6 = 256.2$$
$$\hat{y}_{14}(10) = (248.6 + 2.8) + 4.6 = 251.4 + 4.6 = 256.0$$

If we wished to normalize the seasonals, we would get

$$\bar{S} = \frac{7.6 + 4.6 + 7.8 - 20.5}{4} = -.1$$

$$\hat{S}_1(10) \rightarrow 7.7$$
$$\hat{S}_2(10) \rightarrow 4.6$$
$$\hat{S}_3(10) \rightarrow 7.9$$
$$\hat{S}_4(10) \rightarrow -20.4$$
$$\hat{T}_{10}(10) \rightarrow 240.1$$
$$\hat{\beta}_1(10) \rightarrow 2.8$$

which can be used in future forecasting. •

### 5.6.6 Multiplicative Holt-Winters Models

The only change from the additive updating procedures for a multiplicative Holt-Winters model is in the way the seasonal adjustment and detrending is done, by division rather than subtraction. The updating procedures are otherwise identical.

**Holt-Winters Exponential Smoothing (HWS), Multiplicative Model**

**Given:** The current observation of the time series $y_t$ and

$\hat{T}_{t-1}(t-1)$ = estimated unseasonalized level as of time $t-1$

$\hat{\beta}_1(t-1)$ = estimated slope of the trend line as of time $t-1$

$\hat{S}_i(t-1)$ = estimated index for season $i$ as of time $t-1$ $\quad i = 1, \ldots, L$

**Model:** $y_t = (\beta_0 + \beta_1 t) \cdot S_t + \epsilon_t$

So

$$\hat{T}_{t+p}(t) = \hat{T}_t(t) + p \cdot \hat{\beta}_1(t) \qquad \text{for } p \geq 1$$

**Updating Equations:**

Smoothing Form:

$$\hat{T}_t(t) = \frac{\alpha \cdot y_t}{\hat{S}_t(t-1)} + (1-\alpha) \cdot \hat{T}_t(t-1)$$

$$\hat{\beta}_1(t) = \gamma[\hat{T}_t(t) - \hat{T}_t(t-1)] + (1-\gamma) \cdot \hat{\beta}_1(t-1)$$

$$\hat{S}_i(t) = \frac{\delta \cdot y_t}{\hat{T}_t(t)} + (1-\delta) \cdot \hat{S}_i(t-1) \qquad \text{if } t \text{ in season } i$$

$$= \hat{S}_i(t-1) \qquad \text{if } t \text{ not in season } i$$

Error Correction Form:

$$\hat{T}_t(t) = \hat{T}_t(t-1) + \frac{\alpha \cdot e_t}{\hat{S}_t(t-1)}$$

$$\hat{\beta}_1(t) = \hat{\beta}_1(t-1) + \frac{\alpha \cdot \gamma \cdot e_t}{\hat{S}_t(t-1)}$$

$$\hat{S}_i(t) = \hat{S}_i(t-1) + \frac{\delta(1-\alpha) \cdot e_t}{\hat{S}_t(t-1)} \qquad \text{if } t \text{ in season } i$$

$$= \hat{S}_i(t-1) \qquad \text{if } t \text{ is not in season } i$$

where

$\hat{S}_t(\text{any time}) = \hat{S}_i(\text{any time}) \qquad$ for $t$ in season $i$

$e_t$ = one-step-ahead forecast error = $y_t - \hat{y}_t(t-1)$

$\alpha$ = smoothing constant for the level

$\gamma$ = smoothing constant for the slope

$\delta$ = smoothing constant for the seasonal

**Forecasts:** (At time $t$ for $p$ periods ahead)

$$\hat{y}_{t+p}(t) = \hat{T}_{t+p}(t) \cdot \hat{S}_{t+p}(t) = [\hat{T}_t(t) + p \cdot \hat{\beta}_1(t)] \cdot \hat{S}_{t+p}(t)$$

where

$$\hat{S}_{t+p}(t) = \hat{S}_i(t) \qquad \text{for } t + p \text{ in season } i$$

**EXAMPLE 5.15** *Multiplicative HWS, Mid-Town Heating*

Let us suppose that a multiplicative HWS model is to be used to forecast the Mid-Town Heating series, beginning in 1986. A moving average filter is used on the 1981–1983 data to get initial values of the seasonal indexes. The 1981–1983 data are then seasonally adjusted with these seasonals, and the semiaverages method from Chapter 3 is used to fit a linear trend model to the adjusted series. Using the last two years' history as a test set, values of $\alpha$, $\gamma$, and $\delta$ are chosen from .1(.1).4 based on the RMSE of the forecast errors over the two years of ex post facto forecasts for 1984 and 1985. The smallest RMSE occurs for $(\alpha, \gamma, \delta) = (.1, .1, .4)$. The forecasts using these values are then the forecasts for 1986.

- **Calculate the initial values of the seasonal indexes.** The first three years' worth of data from the Mid-Town series gives two complete years of specific seasonals. (The calculations were already discussed in Section 5.4.4 and are shown in Tables 5.4 and 5.5.) Using these eight specific seasonals we get the values shown in the following table.

| | Quarter | | | |
|---|---|---|---|---|
| | I | II | III | IV |
| 1981 | | | .317 | 1.097 |
| 1982 | 1.867 | .720 | .297 | 1.334 |
| 1983 | 1.575 | .833 | | |
| $\Sigma$ | 3.442 | 1.553 | .614 | 2.431 |
| $\hat{S}_i$ | 1.721 | .776 | .307 | 1.216 |

The indexes are adjusted to sum to 4, giving

$$\hat{S}_1(12) = 1.712 \qquad \hat{S}_2(12) = .772 \qquad \hat{S}_3(12) = .307 \qquad \hat{S}_4(12) = 1.216$$

- **Fit the linear trend using semiaverages.** Using the indexes above, the 1981–1983 observations are deseasonalized to give the values in the following table.

| | Quarter | | | |
|---|---|---|---|---|
| | I | II | III | IV |
| 1981 | 79.4 | 64.9 | 82.6 | 76.9 |
| 1982 | 95.2 | 85.9 | 93.1 | 105.9 |
| 1983 | 90.7 | 101.7 | 119.3 | 70.8 |

The semiaverages calculations are:

$$n = 12$$

$$\overline{y1} = \frac{79.4 + 64.9 + 82.6 + 76.9 + 95.2 + 85.9}{6} = \frac{484.9}{6} = 80.82$$

$$\overline{y2} = \frac{93.1 + 105.9 + 90.7 + 101.7 + 119.3 + 70.8}{6} = \frac{581.5}{6} = 96.92$$

$$\hat{\beta}_1 = \frac{2(\overline{y2} - \overline{y1})}{n} = \frac{2(96.92 - 80.82)}{12} = 2.7$$

$$\hat{\beta}_0 = \overline{y1} - \frac{n+2}{4}\hat{\beta}_1 = 80.8 - \frac{14}{4} \times 2.7 = 71.4$$

- **Determine initial values for the slope, level, and forecast.** The starting values for forecasting ex post facto through 1984 and 1985 can now be found:

$$\hat{T}_{12}(12) = \hat{\beta}_0 + 12\hat{\beta}_1 = 71.4 + 12 \times 2.7 = 103.8$$
$$\hat{\beta}_1(12) = 2.7$$
$$\hat{T}_{13}(12) = \hat{T}_{12}(12) + \hat{\beta}_1(12) = 103.8 + 2.7 = 106.5$$

and the first forecast (for quarter I, 1984) is

$$\hat{y}_{13}(12) = \hat{T}_{13}(12) \cdot \hat{S}_{13}(12) = \hat{T}_{13}(12) \cdot \hat{S}_1(12) = 106.5 \times 1.712 = 182.3$$

- **Smooth the initial estimates through the two years' worth of data to obtain the first true forecast (for quarter I, 1986).** Using the values $\alpha = .1$, $\gamma = .1$, and $\delta = .4$ (so that $\alpha \cdot \gamma = .01$ and $\delta(1 - \alpha) = .36$), we get the following.

*For I, 84 ($t = 13$, $i = 1$):*

$$y_{13} = 175.7 \qquad e_{13} = 175.7 - 182.3 = -6.6$$

$$\hat{T}_{13}(13) = \hat{T}_{13}(12) + \frac{\alpha \cdot e_{13}}{\hat{S}_1(12)} = 106.5 + \frac{.1 \times (-6.6)}{1.712} = 106.1$$

$$\hat{\beta}_1(13) = \hat{\beta}_1(12) + \frac{\alpha \cdot \gamma \cdot e_{13}}{\hat{S}_1(12)} = 2.7 + \frac{.01 \times (-6.6)}{1.712} = 2.7$$

$$\hat{S}_1(13) = \hat{S}_1(12) + \frac{\delta(1-\alpha) \cdot e_{13}}{\hat{T}_{13}(13)} = 1.712 + \frac{.36(-6.6)}{106.1} = 1.690$$

$$\hat{T}_{14}(13) = \hat{T}_{13}(13) + \hat{\beta}_1(13) = 106.1 + 2.7 = 108.8$$

$$\hat{y}_{14}(13) = \hat{T}_{14}(13) \cdot \hat{S}_{14}(13) = \hat{T}_{14}(13) \cdot \hat{S}_2(13) = 108.8 \times .772 = 84.0$$

*For II, 84 ($t = 14$, $i = 2$):*

$$y_{14} = 90.7 \qquad e_{14} = 90.7 - 84.0 = 6.7$$

$$\hat{T}_{14}(14) = \hat{T}_{14}(13) + \frac{\alpha \cdot e_{14}}{\hat{S}_2(13)} = 108.8 + \frac{.1 \times 6.7}{.772} = 109.7$$

$$\hat{\beta}_1(14) = \hat{\beta}_1(13) + \frac{\alpha \cdot \gamma \cdot e_{14}}{\hat{S}_2(13)} = 2.7 + \frac{.01 \times 6.7}{.772} = 2.8$$

$$\hat{S}_2(14) = \hat{S}_2(13) + \frac{\delta(1-\alpha) \cdot e_{14}}{\hat{T}_{14}(14)} = .772 + \frac{.36 \times 6.7}{109.7} = .794$$

$$\hat{T}_{15}(14) = \hat{T}_{14}(14) + \hat{\beta}_1(14) = 109.7 + 2.8 = 112.5$$

$$\hat{y}_{15}(14) = \hat{T}_{15}(14) \cdot \hat{S}_{15}(14) = \hat{T}_{15}(14) \cdot \hat{S}_3(14) = 112.5 \times .307 = 34.5$$

***For III, 84 (t = 15, i = 3):***

$$y_{15} = 50.4 \qquad e_{15} = 15.9$$

$$\hat{T}_{15}(15) = 112.5 + \frac{.1 \times 15.9}{.307} = 117.7$$

$$\hat{\beta}_1(15) = 2.8 + \frac{.01 \times 15.9}{.307} = 3.3$$

$$\hat{S}_3(15) = .307 + \frac{.36 \times 15.9}{117.7} = .356$$

$$\hat{T}_{16}(15) = 117.7 + 3.3 = 121.0$$

$$\hat{y}_{16}(15) = 121.0 \times 1.210 = 146.4$$

***For IV, 84 (t = 16, i = 4):***

$$y_{16} = 114.1 \qquad e_{16} = -32.3$$

$$\hat{T}_{16}(16) = 121.0 + \frac{.1(-32.3)}{1.210} = 118.3$$

$$\hat{\beta}_1(16) = 3.3 + \frac{.01(-32.3)}{1.210} = 3.0$$

$$\hat{S}_4(16) = 1.210 + \frac{.36(-32.3)}{118.3} = 1.112$$

$$\hat{T}_{17}(16) = 118.3 + 3.0 = 121.3$$

$$\hat{y}_{17}(16) = 121.3 \times 1.690 = 205.0$$

and so forth. The complete set of updates is shown in Table 5.26.

**Table 5.26 Multiplicative Holt-Winters Smoothing for Mid-Town Heating Data**

| | Qrtr | $y_t$ | $\hat{y}_t(t-1)$ | $e_t$ | $\hat{T}_t(t)$ | $\hat{\beta}_1(t)$ | $\hat{T}_{t+1}(t)$ | $i$ | $\hat{S}_i(t)$ |
|---|---|---|---|---|---|---|---|---|---|
| 1984 | I | 175.7 | 182.3 | −6.6 | 106.1 | 2.7 | 108.8 | 1 | 1.690 |
| | II | 90.7 | 84.0 | 6.7 | 109.7 | 2.8 | 112.5 | 2 | .794 |
| | III | 50.4 | 34.5 | 15.9 | 117.7 | 3.3 | 121.0 | 3 | .356 |
| | IV | 114.1 | 146.4 | −32.3 | 118.3 | 3.0 | 121.3 | 4 | 1.112 |
| 1985 | I | 178.4 | 205.0 | −26.6 | 119.7 | 2.8 | 122.5 | 1 | 1.610 |
| | II | 93.0 | 97.3 | −4.3 | 122.0 | 2.7 | 124.7 | 2 | .781 |
| | III | 43.3 | 44.4 | −1.1 | 124.4 | 2.7 | 127.1 | 3 | .353 |
| | IV | 116.6 | 141.3 | −24.7 | 124.9 | 2.5 | 127.4 | 4 | 1.041 |
| 1986 | I | ? | 201.1 | | | | | | |

The 1-, 2-, 3-, and 4-step ahead forecasts for the four quarters of 1986 are

I, 86: $\hat{y}_{21}(20) = (124.9 + 2.5) \times 1.610 = 201.1$

II, 86: $\hat{y}_{22}(20) = (124.9 + 5.0) \times 0.781 = 101.5$

III, 86: $\hat{y}_{23}(20) = (124.9 + 7.5) \times 0.353 = 46.7$

IV, 86: $\hat{y}_{24}(20) = (124.9 + 10.0) \times 1.041 = 140.4$ •

### 5.6.7 Other Models

The four models we have discussed are the most widely known. Others have been developed and are used to varying degrees. These include an analog to Brown's double exponential smoothing of Section 3.9.5, which can be achieved for given values of $\alpha_{DES}$, $\delta_{DES}$ by choosing the HWS parameters as $\alpha_{HWS} = 1 - (1 - \alpha_{DES})^2 = \alpha_{DES} \cdot (2 - \alpha_{DES})$, $\gamma_{HWS} = \alpha_{DES}/(2 - \alpha_{DES})$, and $\delta_{HWS} = \delta_{DES}$.

For a simple exponential trend with multiplicative structure ($y_t = \beta_0 \cdot \beta_1^t \cdot S_t + \epsilon_t$ or $y_t = \beta_0 \cdot \beta_1^t \cdot S_t \cdot \epsilon_t$), one can smooth the logarithms of the series or use the following formulas:

$$\hat{T}_t(t) = \hat{T}_{t-1}(t-1) \cdot \hat{\beta}_1(t-1) + \frac{\alpha \cdot e_t}{\hat{S}_t(t-1)}.$$

$$\hat{\beta}_1(t) = \hat{\beta}_1(t-1) + \frac{\alpha \cdot \gamma \cdot e_t}{\hat{T}_{t-1}(t-1) \cdot \hat{S}_t(t-1)}$$

$$\hat{S}_i(t) = \hat{S}_i(t-1) + \frac{\delta \cdot (1-\alpha) \cdot e_t}{\hat{T}_t(t)}$$

$$\hat{y}_{t+p} = \hat{T}_t(t) \cdot [\hat{\beta}_1(t)]^p \cdot \hat{S}_t(t)$$

There are also so-called "damped" trend models (developed to avoid a tendency to overshoot for larger values of $p$, which has been noted in Holt-Winters models) and, finally, mathematically complex general exponential smoothing, which uses trigonometric functions to model the seasonal structure (see also Pegels, 1969; Gardner, 1985; Holt et al., 1960).

## 5.7 Diagnostics

### 5.7.1 General Issues

With seasonality, other than the usual concerns about model fit, special attention should be given to the present and future accuracy of the estimates of the seasonal indexes. The index values for different seasons should reflect movements that are roughly comparable to your subjective understanding of how the series should respond to annual phenomena. An index value that conflicts with this understanding in terms of magnitude or, especially, direction warrants more detailed examination. Remember, too, that since most indexes are forced to sum to a given value (0 or $L$), an error in the value for one season will cause errors in the values of the others and may lead, as well, to poor estimates of the level of the series.

#### Parameter Evolution

Estimates based on fits to historical data can be rendered obsolete by slowly or precipitously changing consumer behavior and other societal patterns, or by advancing technology. In such cases, single forecast models can be periodically refit to recent data, or updating schemes can be used. It is also possible, by plotting each season's specific seasonals, to subjectively or empirically estimate future indexes. For example, one could fit a linear trend by eye, say, for each season, and forecast future indexes based on these trend fits (making sure to normalize afterwards).

#### Cycles

By becoming confounded with seasonal movement, cycles, especially incomplete ones, can also result in poor index estimates. The moving average filter is the least sensitive to this problem and is probably the method of choice for series with a cyclical component, although the use of updating may help.

#### Outliers

Aberrant observations (outliers) are also a problem when dealing with seasonality because there are often only a few observations for each season. Resistant averaging or use of the median is suggested under these circumstances. Outliers with assignable causes should be discarded before fitting a single forecast model. With updating schemes, no estimate revision should be done in the period when the outlier occurs.

### 5.7.2 Empirical Analysis

From an empirical point of view, remember that the residuals ($e_t$'s) from a well-fit model should be random. For seasonality this means the following.

- There should be no seasonal pattern in the historical forecast errors. Any seasonal ups and downs apparent in plots of the original series should have disappeared from plots of the $e_t$'s or of the seasonally adjusted series.
- The lag $L$ sample autocorrelation coefficient of the $e_t$'s should not be significant. In fact, $r_L$ will likely be quite small if the model has been properly fit.
- Plots of a given season's specific seasonals should show a random no-trend series. If a multiplicative (additive) model is fit when an additive (multiplicative) one would be more appropriate, the specific seasonals ($y_t - T_t$ or $y_t/T_t$) will tend to show a trend. To some degree one can tell which model is appropriate by whether the trend disappears when switching from one model to the other. A trend can also result from changing seasonal patterns, as mentioned earlier.

***Caution:*** When plotting specific seasonals from an updating scheme, remember that successive sets of updated indexes are comparable only if both sets have been normalized.

## 5.8 Exercises

**5.1** What is the distinguishing characteristic of seasonal movement?

**5.2** What is the root cause of all seasonal movement?

**5.3** What is the difference between the nature of seasonal effects as described by an additive model and those described by a multiplicative model?

**5.4** How do the units of measurement for additive and multiplicative seasonal indexes differ?

**5.5** What specifically does *seasonally adjusted* mean? Why is seasonal adjustment done?

**5.6** Specifically, how are additive and multiplicative seasonal indexes used in computing forecasts?

**5.7** Suppose the value of a centered moving average for monthly data is being computed as of the December 1988 observation, which has just become available.

a. Which month's observations influence this CMA value?
b. What is the weight, out of 100%, that each of these months gets?
c. At what point in time is this value of the CMA centered?

**5.8** What main advantage does a centered moving average have over fitting a trend model when detrending a series to test for seasonal?

**5.9** On what rationale is the use of a moving average filter based?

**5.10** A men's clothing store has found that its quarterly sales ($1,000) can be adequately described by the following multiplicative seasonal model:

$$\hat{T}_t = 128.2(1.02)^t \qquad \text{Origin: IV, 1987} \quad \text{Units } (t)\text{: Quarterly}$$

$$\hat{S}_1 = .81 \qquad \hat{S}_2 = 1.09 \qquad \hat{S}_3 = .91 \qquad \hat{S}_4 = 1.19$$

a. Forecast sales for 1990.
b. Seasonally adjust the store's 1989 sales figures: qtr I: 112.3; qtr II: 162.1; qtr III: 135.4; qtr IV: 171.6.

**5.11** Quarterly orders for a certain part from a regional auto parts supply house are described by the following additive seasonal model:

$$\hat{T}_t = 40 + 3t \qquad \text{Origin: II, 1988} \quad \text{Units } (t)\text{: Quarterly}$$

$$\hat{S}_1 = -8 \qquad \hat{S}_2 = +3 \qquad \hat{S}_3 = +18 \qquad \hat{S}_4 = -13$$

a. Forecast sales for 1990.
b. Seasonally adjust the store's 1989 orders figures: qtr I: 49; qtr II: 74; qtr III: 88; qtr IV: 55.

**5.12** A quarterly time series is detrended using a moving average filter, then the specific seasonals are ranked, giving the following results:

| | Rank | | | |
|---|---|---|---|---|
| | Qtr I | Qtr II | Qtr III | Qtr IV |
| 1984 | 6 | 12 | 16 | 2 |
| 1985 | 8 | 10 | 19 | 9 |
| 1986 | 4 | 17 | 14 | 3 |
| 1987 | 1 | 15 | 18 | 5 |
| 1988 | 13 | 11 | 20 | 7 |

Do these data constitute statistical evidence, at the 5% level, that there is seasonality in the series?

**5.13** The autocorrelation coefficient of the specific seasonals ranked in Exercise 5.12 is:

| $j$ | 1 | 2 | 3 | 4 | 5 | 6 | 7 | 8 |
|---|---|---|---|---|---|---|---|---|
| $r_j$ | −.096 | −.358 | −.108 | .486 | −.005 | −.571 | −.013 | .332 |

Is there statistical evidence here, at the 5% level, that seasonality is present?

**5.14** The original series in Exercise 5.12 (not the specific seasonals) is entered into a regression program, and a linear trend model with seasonal dummy variables is fit, with the following results:

For linear trend: $t = +1.44$

SSR(trend) = 256.0 SSE(trend) = 6,489.9

SSR(trend + seasonal) = 5,018.5 SSE(trend + seasonal) = 1,727.4

SSR(seasonal only) = 4,779.8 SSE(seasonal only) = 1,966.0

Is there statistical evidence here, at the 5% level, that seasonality is present in the series?

**5.15** Given the following time series:

| | Qtr I | Qtr II | Qtr III | Qtr IV |
|---|---|---|---|---|
| 1983 | 242 | 279 | 263 | 233 |
| 1984 | 241 | 253 | 254 | 249 |
| 1985 | 240 | 277 | 273 | 229 |
| 1986 | 232 | 276 | 271 | 236 |
| 1987 | 235 | 251 | 255 | 230 |
| 1988 | 223 | 278 | 243 | 248 |

**a.** Asuming there is no trend, test for seasonality using the Kruskal-Wallis test ($\alpha = .05$).

**b.** Assuming there is no trend, test for seasonality using the "rule-of-thumb."

**c.** Test for seasonality using least squares regression with dummy variables ($\alpha = .05$).

**5.16** For the data in Exercise 5.15: Assuming no trend, find seasonal indexes, deseasonalize the 1988 data, and forecast 1989 using (a) simple averages with additive seasonals, and (b) simple averages with multiplicative seasonals.

**5.17** For the data in Exercise 5.15: Use the ratio-to-moving average method with seasonal means to determine seasonal indexes, deseasonalize the 1988 data, and forecast the series for 1989, assuming no-trend, using $\hat{\beta}_0 = \bar{y}$, and (a) an additive model, and (b) a multiplicative model.

**5.18** For the data in Exercise 5.15: Use the least squares method to determine seasonal indexes (normalized) and to forecast 1989 using (a) an additive model, then (b) a multiplicative model.

**5.19** The following data represent quarterly electricity usage (thousands of kwh) over several years at a certain manufacturing plant. The data are clearly seasonal.

**Electricity Usage**

| | Quarter | | | |
|---|---|---|---|---|
| | I | II | III | IV |
| 1984 | 611 | 1042 | 460 | 1304 |
| 1985 | 578 | 976 | 489 | 1358 |
| 1986 | 585 | 932 | 473 | 1293 |
| 1987 | 633 | 1026 | 489 | 1390 |
| 1988 | 609 | 950 | 455 | 1342 |

**a.** Use a moving average to filter the seasonality out of the data. Do the data have trend? Justify your answer.
**b.** Use the ratio-to-moving average method and seasonal means to determine additive seasonal indexes.
**c.** Use the ratio-to-moving average method and seasonal means to determine multiplicative seasonal indexes.
**d.** Repeat part (c) using seasonal medians.
**e.** Repeat part (d) using seasonal modified means.

**5.20** The following data represent the quarterly number of persons (10,000's) passing through the metal detection security gates at a large municipal airport.

**Number of Persons Through Security Gates**

| | Quarter | | | |
|---|---|---|---|---|
| | I | II | III | IV |
| 1984 | 318 | 380 | 358 | 423 |
| 1985 | 379 | 394 | 412 | 439 |
| 1986 | 413 | 458 | 492 | 493 |
| 1987 | 461 | 468 | 529 | 575 |
| 1988 | 441 | 548 | 561 | 620 |

**a.** Fit an additive seasonal model with linear trend using the method of simple averages, and forecast the four quarters of 1989.

**b.** Fit a multiplicative seasonal model with simple exponential trend using the method of simple averages, and forecast the four quarters of 1989.

**5.21** For the data in Exercise 5.20: Fit a multiplicative model with linear trend using the ratio-to-moving average method with seasonal means and a least squares trend fit to the deseasonalized data. Use this model to forecast the four quarters of 1989.

**5.22** For the data in Exercise 5.20: Fit an additive model with linear trend using the least squares method. Use this model to forecast the four quarters of 1989.

**5.23** For the data in Exercise 5.20: Fit a multiplicative model with simple exponential trend using the least squares method. Express the model with normalized indexes in both logarithmic and nonlogarithmic form. Use the model to forecast the four quarters of 1989.

**5.24** Which of the two models in Exercises 5.22 and 5.23 do you prefer? Why?

**5.25** The table below lists the quarterly number of diskettes shipped (1,000's) by a mail order firm over a four-year period. Fit a multiplicative model to these data using the ratio-to-moving average method for seasonal indexes and a least squares simple exponential trend. Express the model both in logarithmic and nonlogarithmic form with normalized seasonals. Use the model to forecast the four quarters of 1989.

**Number of Diskettes Shipped**

| | Quarter | | | |
|---|---|---|---|---|
| | I | II | III | IV |
| 1985 | 86.1 | 74.4 | 52.4 | 88.2 |
| 1986 | 91.7 | 82.5 | 59.3 | 97.8 |
| 1987 | 101.7 | 81.8 | 61.3 | 112.6 |
| 1988 | 113.8 | 96.4 | 63.2 | 114.4 |

**5.26** Using the data in Exercise 5.25: Fit the most appropriate model using least squares techniques, and forecast the four quarters of 1990.

**5.27** The following table shows the net after-taxes monthly profits ($1,000's) for a manufacturer of small recreational camping trailers. Using least squares techniques, find the most suitable model and use it to forecast sales for the summer quarter (III) of 1990.

**Net Profits after Taxes**

| | 1982 | 1983 | 1984 | 1985 | 1986 | 1987 |
|---|---|---|---|---|---|---|
| Jan | 620 | 771 | 782 | 894 | 999 | 1153 |
| Feb | 720 | 831 | 965 | 1128 | 1243 | 1267 |
| Mar | 809 | 901 | 968 | 1106 | 1303 | 1535 |
| Apr | 885 | 960 | 1156 | 1212 | 1360 | 1489 |
| May | 957 | 1150 | 1257 | 1318 | 1545 | 1697 |
| Jun | 1108 | 1298 | 1347 | 1520 | 1822 | 1984 |
| Jul | 1036 | 1269 | 1286 | 1426 | 1651 | 1777 |
| Aug | 1147 | 1227 | 1355 | 1502 | 1811 | 1725 |
| Sep | 853 | 955 | 1065 | 1251 | 1317 | 1439 |
| Oct | 838 | 968 | 1050 | 1168 | 1233 | 1412 |
| Nov | 842 | 882 | 1007 | 1049 | 1151 | 1337 |
| Dec | 651 | 769 | 872 | 1001 | 1152 | 1267 |

**5.28** For the data of Exercise 5.20: Use the first three years' data and simple averages to initialize an additive horizontal model, then use additive SSES with $\alpha = .1$, $\delta = .1$, and $\gamma = .4$ to forecast the four quarters of 1989.

**5.29** Repeat Exercise 5.28 using a multiplicative model.

**5.30** For the data of Exercise 5.25: Use the first two years' data and simple averages to initialize an additive linear trend model, then use additive Holt-Winters smoothing with $\alpha = .1$, $\delta = .1$, and $\gamma = .4$ to forecast the four quarters of 1989.

**5.31** Repeat Exercise 5.30 using multiplicative Holt-Winters smoothing.

## References

Bell, W. R., and Hilmer, S. C. (1984). "Issues Involved with the Seasonal Adjustment of Economic Time Series," with commentary. *J. Bus. and Econ. Stat.* 2(4): 291–342.

Burman, J. P. (1979). "Seasonal Adjustment—A Survey." *TIMS Studies in Management Sciences* 12: 45–57.

Gardner, E.S. (1985). "Exponential Smoothing: The State of the Art." *J. Forecasting* 4(1): 1–28.

Green, R. D., and Doll, J. P. (1974). "Dummy Variables and Seasonality." *Amer. Statistician* 28(2): 60–62.

Holt, C. C. (1957). "Forecasting Seasonal and Trends by Exponentially Weighted Moving Averages." Office of Naval Research, Research Memorandum No. 52.

Holt, C. C., Modigliani, F., Muth, J. F., and Simon, H. A. (1960). *Planning Production Inventories and Work Force.* Englewood Cliffs, N.J.: Prentice-Hall.

Kleinbaum, D. G., Kupper, L. C., and Muller, K. E. (1988). *Applied Regression Analysis and Other Multivariate Methods.* Boston: PWS-Kent Publishing Co.

Kruskal, W. H., and Wallis, W. A. (1952). "Use of Ranks in One-Criterion Variance Analysis." *J. Amer. Stat. Assoc.* 67: 401–412.

Marascuilo, L. A., and McSweeney, M. (1977). *Nonparametric and Distribution-Free Methods for the Social Sciences.* Monterey, Calif.: Brooks/Cole.

McKenzie, E. (1986). "Renormalization of Seasonals in Winter's Forecasting Systems: Is It Necessary?" *Operations Research* 34(1): 174–176.

Menezes, O. J. (1971). "The Dummy Variable Method for Jointly Estimating Seasonal Indexes and Regression on Non-Seasonal Variables." *Amer. Statistician* 25(4): 32–36.

Pegels, C. L. (1969). "Exponential Forecasting: Some New Variations." *Management Sci.* 12(5): 311–315.

Thomopoulos, N. T. (1980). *Applied Forecasting Methods.* Englewood Cliffs, N.J.: Prentice-Hall.

Wilst, A. R. (1977). "Estimating Models of Seasonal Market Response Using Dummy Variables." *J. Marketing Research* 14: 34–41.

Winters, P. R. (1960). "Forecasting Sales by Exponentially Weighted Moving Averages." *Management Sci.* 6: 324–342.

# Forecasting Cyclical Series

**Objective:** *To study procedures for forecasting time series when forces other than seasonal ones cause the level to swing above and below the trend.*

## 6.1 Introduction

Cyclical movements in a time series can be described as long swings away from trend that are due to factors other than seasonality. Cycles generally occur over a number of years; however, the methods we discuss may also be applied to shorter time intervals. Cyclical components are difficult to model because their patterns are not as stable as trend and seasonal ones are. The up-down oscillations of a cycle rarely repeat at fixed intervals of time, and the amplitude of the fluctuations can also vary.

Decomposition methods can be extended to include cycles. However, because of the irregular behavior of cycles, analyzing and forecasting the cyclical component of a series $y_t$ often requires finding a closely related series $x_t$ whose own cyclical swings tend to occur a few periods before the swings in $y_t$. Such *leading indicators* can be used not only to forecast the magnitude of the cycle, but also to identify or predict turning points.

The topics of concern in this chapter include:

- the nature and causes of cyclical movements
- graphical and statistical methods for detecting cycles
- leading, lagging, and coincident indicators
- the use of diffusion indexes and rates of change in analyzing the business cycle
- procedures for forecasting the cyclical component of a series

## 6.2 The Nature of Cyclical Movement

### 6.2.1 Introduction

Cyclical movement in a time series is characterized by oscillating swings that do not arise from seasonal influences. Among its prominent features are:

- oscillations away from trend that occur over time spans generally longer than a year, whose
- amplitudes and periods may not be fixed.

Unlike seasonality, cycles have many different causes. Thus, explanations of why cycles occur depend on the particular time series studied. For a series that has trend, seasonal, and cyclical movement, the following definition will be used.

---

**DEFINITION:** A time series $y_t$ is said to have **cyclical movement** if the average value of $y_t$ changes over time such that

$$E(y_t) = f(\beta_0, \beta_1, \ldots; t) + S_t + C_t \qquad t = 1, 2, \ldots \tag{6.1}$$

or

$$E(y_t) = f(\beta_0, \beta_1, \ldots; t) \cdot S_t \cdot C_t \qquad t = 1, 2, \ldots \tag{6.2}$$

where

$S_t$ = the seasonal effect at time $t$
$C_t$ = the cyclical component at time $t$

where $f(\beta_0, \beta_1, \ldots; t)$ is a function describing the trend.

---

### 6.2.2 Characteristics of Cyclical Series

To estimate the cyclical component, a model (either Equation 6.1 or Equation 6.2) needs to be specified, using the following notation:

$y_t$ = the actual value of the series at time $t$
$L$ = the length of the seasonality
$S_t$ = the seasonal effect at time $t$
$T_t$ = the trend at time $t$
= $f(\beta_0, \beta_1, \ldots; t)$
$C_t$ = the cyclical effect at time $t$
$\epsilon_t$ = the random movement in the series at time $t$ not accounted for by the effects of trend, seasonality, or cycle

The additive and multiplicative models are defined as in the following table.

**Cyclical Models**

| | | | |
|---|---|---|---|
| **Additive:** | $y_t = T_t + S_t + C_t + \epsilon_t$ | $t = 1, 2, \ldots$ | $\sum_{\text{any year}} S_t = 0$ |
| **Multiplicative:** | $y_t = T_t \cdot S_t \cdot C_t \cdot \epsilon_t$ | $t = 1, 2, \ldots$ | $\sum_{\text{any year}} S_t = L$ |

As with the seasonal component, the units of measure for $C_t$ depend upon whether the additive or multiplicative model is used. For the additive model, $C_t$ is in the same units as $y_t$ ($C_t = 0$ corresponds to no effect); for the multiplicative model, $C_t$ is a dimensionless index, or percent of trend with a base value of 100% ($C_t = 1$ corresponds to no effect).

Graphically, a series with all four components might appear as in Figure 6.1, which shows a quarterly series with a seasonally strong fourth quarter

**Figure 6.1** Time Series with Trend, Seasonal, Cyclical, and Irregular Components

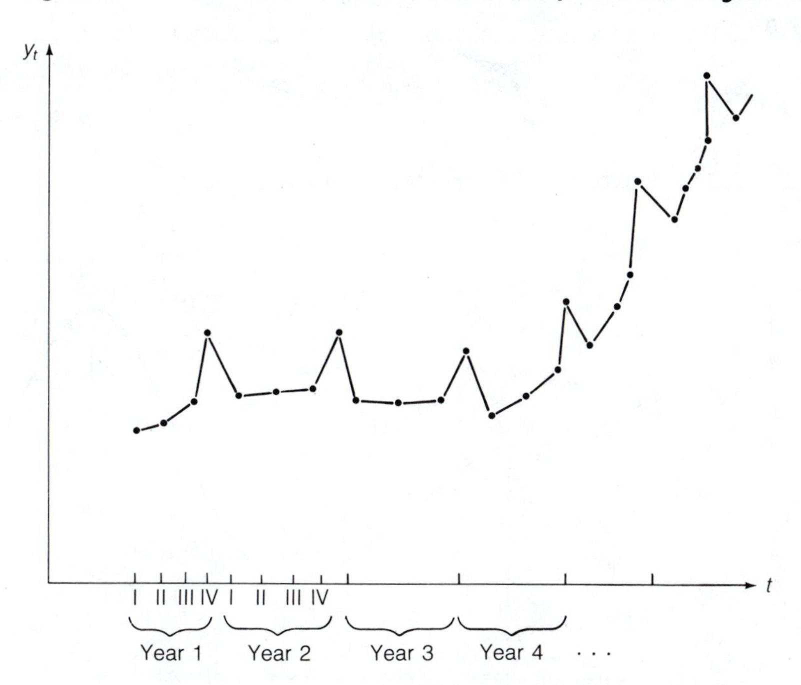


and cyclical swings above and below an apparent trend. Figure 6.2 shows the same series with the trend and cyclical components superimposed. Looking at the cyclical component alone, with trend and seasonality removed, as in Figure 6.3, we can define the various phases of a cycle as follows:

- **growth** the time period during which there is upward movement in the cycle.
- **peak** a turning point marking the end of a growth phase and the beginning of a decline.

**Figure 6.2 The Series of Figure 6.1 with Trend and Cyclical Components Identified**

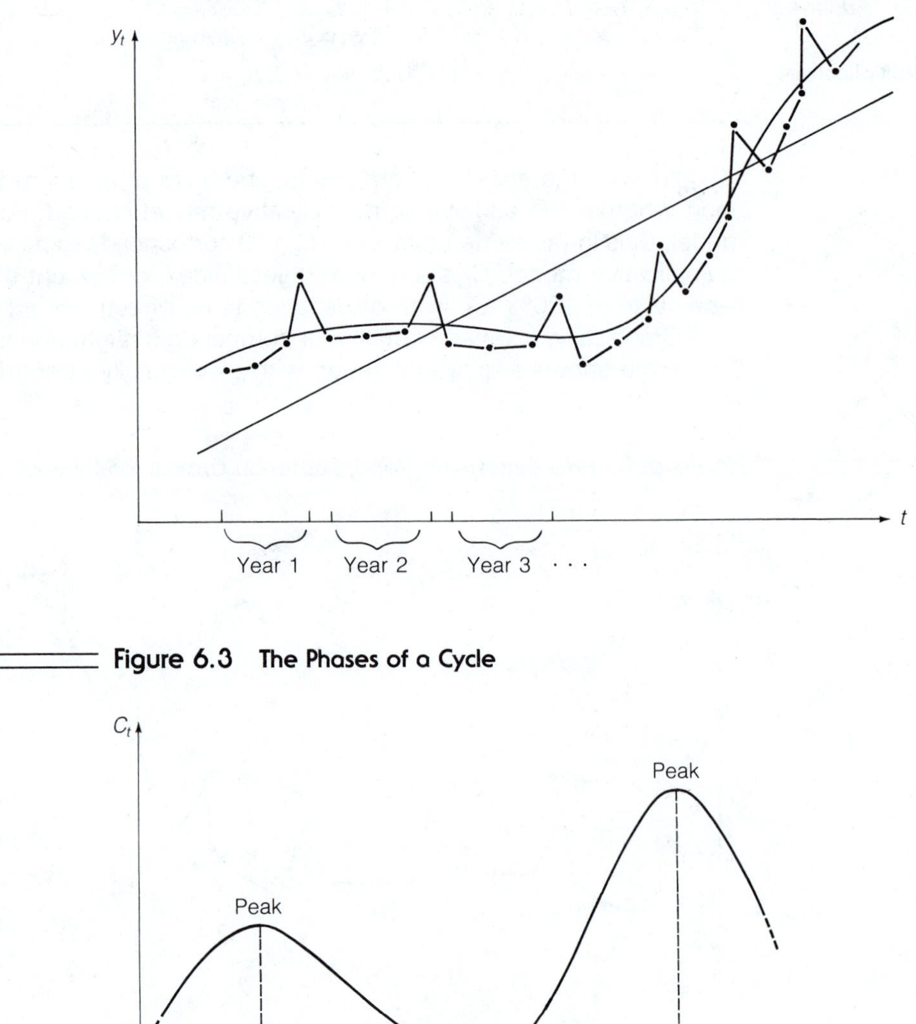


**Figure 6.3 The Phases of a Cycle**

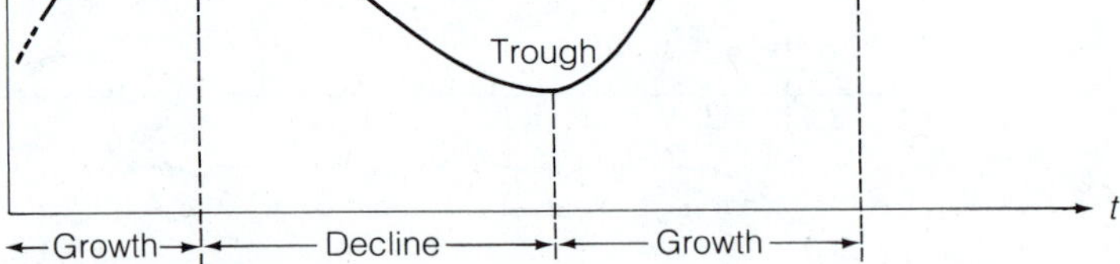


- **decline** the time period during which there is a downward swing in the cycle.
- **trough** a turning point marking the end of a decline and the beginning of a new growth phase.

The phases of a cycle may be given other names, as when discussing the *business cycle,* that is, periods of growth and decline affecting many key variables in the economy simultaneously (see Section 6.5). In that context, the growth phase is called an *expansion,* which is often subdivided into periods of *upswing* and then *slowdown* just prior to a peak. The decline phase is referred to as a *contraction,* which itself consists of periods of *recession* and eventual *leveling off* before a trough is encountered.

There is sometimes a question about how to distinguish trend from cycle. Whereas the distinction between cycle and seasonality is clear (cyclical swings generally do not occur within a year and, although recurrent, are not regular), distinguishing cycle from trend is not so easy (some believe it is impossible). Consider Figure 6.1 again. The series shown there could be the result of cyclical swings around a linear trend or could simply be exhibiting quadratic or exponential trend and no cycle. It generally takes several years' worth of data (i.e., a few complete cycles) before it becomes possible to distinguish cycle from trend. Over the short term, then, it is often difficult to say whether or not the trend is really just part of a long cycle. For this reason, some authors do not even attempt to separate the two components; instead, they refer to the *trend-cycle* of the series (Makridakis, Wheelwright, and McGee, 1983; Shiskin, 1967). For long cycles, the techniques of Chapters 2, 3, and 5 should be used, since for short-term forecasting it doesn't matter much whether the trend component is considered to be trend or trend-cycle.

### 6.2.3 Reasons for Cycles

The first attempts at estimating components of movement in a series began in the early 1800s. From that time to the present, a number of explanations have been offered for why cycles occur (Nerlove, Grether, and Carvalho, 1979), including everything from planetary movements and sunspots to agricultural yields, wars, population shifts, and psychological forces. Early studies, for example, suggested that movements in the planets and moon affected the weather, which in turn influenced crop yields that, in agriculturally based economies, could have a great effect on business. Below are listed a few of the possible causes of cycles that have been presented over the years.

- **psychological forces** Cyclical swings in series linked to popular tastes (e.g., fashions, music, food) have always occurred. A typical cycle starts with a few adherents to a particular fashion whose ranks eventually swell. The growing familiarity then breeds disinterest, causing a decline in popularity. Businesses tied to such societal shifts experience corresponding swings. Some attempts to study this behavior mathematically have included models initially designed for the study of epidemics, which follow similar patterns.
- **population** Populations naturally undergo successive booms and declines. Extreme events (e.g., wars, famines, epidemics, natural disasters) cause significant declines (during the event) and sharp increases (after the event) in birthrates, whose effects ripple through a series as those born in the boom or decline begin, themselves, to have children. Population-dependent series will, as a matter of course, reflect such cycles.
- **institutional causes** Changes in public policy or business practices can produce cycles. For example, a decision to increase a city's police force, when gradually

implemented, causes a reduction in crime. Reduced crime rates, though, eventually lead to arguments for reducing the police force, and the cycle starts anew.

- **replacement cycles** New products usually experience periods of rapid sales that eventually level off when market saturation approaches. Eventually, the many early customers find themselves owning old or broken products needing replacement, thus spurring a new round of growth.
- **education** Students often congregate in fields of study in which the prospects for future employment seem high (this is in addition to any effect of population swings). After perhaps four years, however, the demand in a particular vocation may be satisfied, leaving a glut of qualified applicants and causing new students to avoid that area of study.
- **predator/prey relationships** Certain animal prey are known to have distinct predators (e.g., the snowshoe rabbit and its predator, the lynx). In such situations, too few predators lead to increased breeding and proliferation of the prey; this increases the predator's food supply; too many predators eventually compete for and reduce the number of prey, reducing the food supply and leading to a diminished number of predators. Mathematical models have been developed for predicting the cycles in predator-prey relationships (Rubinow, 1975).
- **combined causes** It is possible to have more than one cause acting at a time. Education, for example, can be affected by cyclical changes in population, perceived "fashionable" fields of study, institutional policies (e.g., increased government spending in specific fields), and retraining later in life (a sort of replacement cycle).

## 6.3 Examples

### 6.3.1 Introduction

When incorporating a cyclical factor into forecasts, three situations commonly occur:

- **forecasting *yearly* series** Since seasonality does not play a part in yearly series, the task simplifies to deciding whether a cycle is present (in the *detrended* data) and, if so, estimating its period and amplitude.
- **forecasting *seasonal* series** In this case, the data must be *deseasonalized and detrended* prior to looking for the cycle.
- **using *leading indicators*** The concern here is to identify possible leading indicators and to estimate the extent to which they lead the series.

The following examples illustrate some approaches to these situations.

### 6.3.2 Example 6.1—U.S. Wheat Production

The annual production of wheat (in millions of bushels) for 1971–1985 is given in Table 6.1. Since seasonality is not involved in yearly data, forecasting this series amounts to first detrending, then identifying the cyclical factor, and finally combining the two factors (additively or multiplicatively) to create forecasts.

**Table 6.1 Annual U.S. Wheat Production, 1971–1985** $y_t$ = production ($10^6$ bushels)

| Year | $t$ | $y_t$ | Year | $t$ | $y_t$ |
|---|---|---|---|---|---|
| 1971 | 1 | 1618 | 1979 | 9 | 2134 |
| 1972 | 2 | 1545 | 1980 | 10 | 2374 |
| 1973 | 3 | 1705 | 1981 | 11 | 2799 |
| 1974 | 4 | 1796 | 1982 | 12 | 2765 |
| 1975 | 5 | 2122 | 1983 | 13 | 2420 |
| 1976 | 6 | 2142 | 1984 | 14 | 2596 |
| 1977 | 7 | 2036 | 1985 | 15 | 2425 |
| 1978 | 8 | 1798 | | | |

*Source: Survey of Current Business.*

**Figure 6.4 U.S. Wheat Production ($10^6$ bushels), 1971–1985**

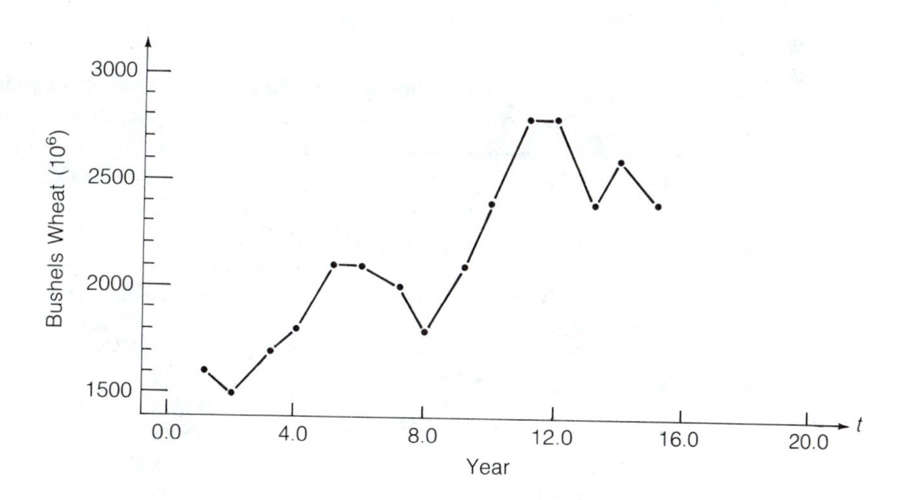


Over the period 1971–1985, the wheat production data plotted in Figure 6.4 appear to have a linear trend. This trend can be estimated by simply fitting the least squares model

$$T_t = \beta_0 + \beta_1 t$$

which for these data turns out to be

$$\hat{T}_t = 1523.1 + 78.57t \qquad \text{Origin: } 1970$$
$$\text{Units } (t)\text{: Yearly}$$

Next, detrending the data will depend on the form of the model used (additive or multiplicative). Suppose, for example, that a multiplicative model is used. Then, the series can be written

$$y_t - \hat{T}_t = \hat{C}_t + \hat{\epsilon}_t$$

and the detrended series is found by dividing each $y_t$ by its corresponding estimated trend component $\hat{T}_t$. The detrended series and its autocorrelation function are shown

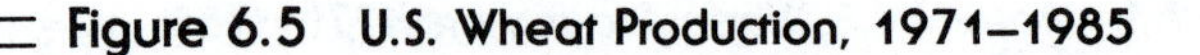
**Figure 6.5 U.S. Wheat Production, 1971–1985**

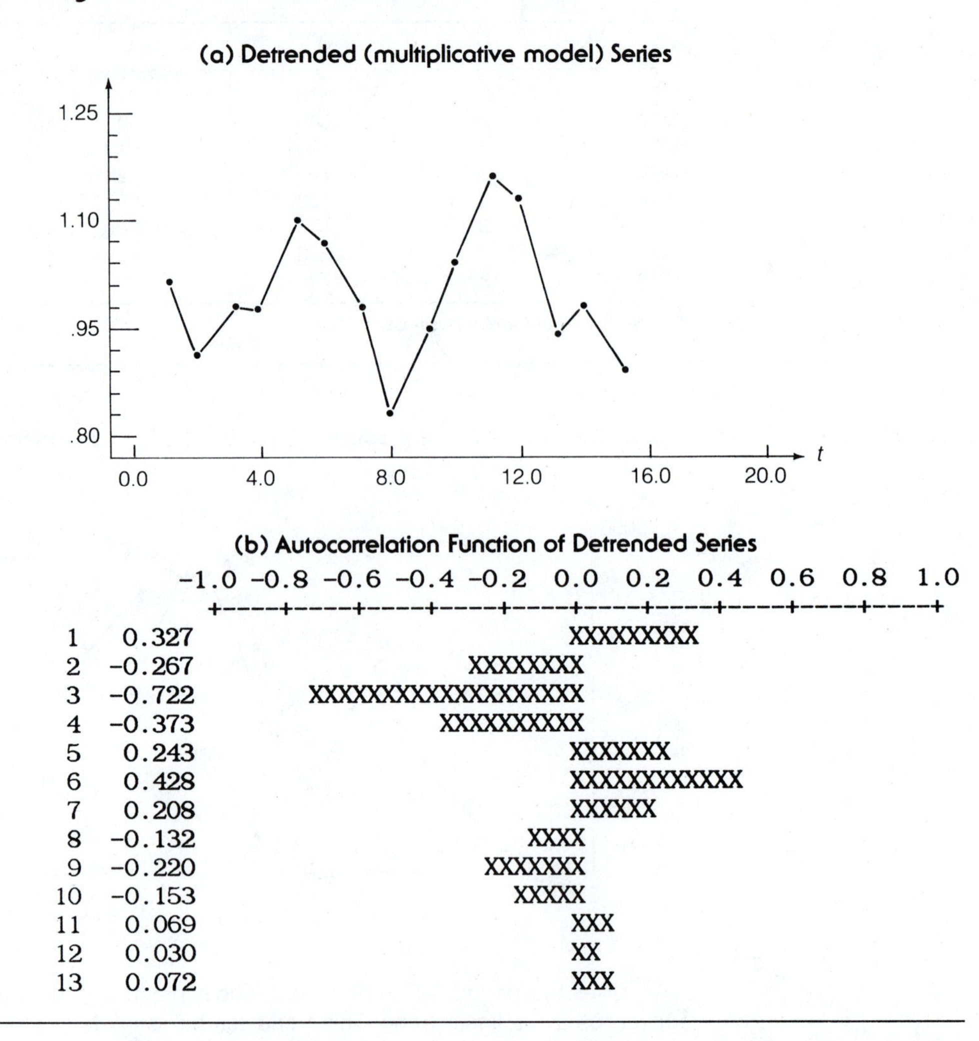


in Figure 6.5. That a cyclical pattern is discernible in Figure 6.5(a) is reinforced by an examination of the acf in Figure 6.5(b). The large lag-3 autocorrelation coefficient ($r_3 = -.721$) indicates that the detrended series tends to swing to opposite sides of its mean every three or so time periods (years).

To forecast annual wheat production for 1986 ($t = 16$), estimates of the trend, $T_{16}$, and cycle, $C_{16}$, are needed. From the fitted trend line, we see that

$$\hat{T}_{16} = 1523.1 + 78.57(16) = 1257.12$$

Estimating $C_{16}$ requires two things: getting the *direction* of the change right (i.e., will $C_{16}$ be higher or lower than $C_{15}$?) and estimating the *magnitude* of this change. Since the last three detrended values were below their overall mean, and since $r_3$ is negative, there is good reason to conclude that $C_{16}$ will be higher than $C_{15}$. As for the magnitude of the change, we note that the past changes in $C_t$ rarely exceeded .10 in either direction, so the increase in period 16 will probably also be in the .00–.10 range. To pick a specific value in this range, any special knowledge of the series should be used (proposed grain deals with other countries, impending farm legislation, and the like). In this example, suppose we estimate the change in the cyclical index as about .08 in magnitude. Then, since

$$\hat{C}_{15} = \frac{y_{15}}{\hat{T}_{15}} = \frac{2425}{1523.1 + 78.57(15)} = \frac{2425}{2701.65} = .898$$

the estimate of $C_{16}$ will be

$$\hat{C}_{16} \simeq \hat{C}_{15} + (\text{change}) = .898 + .08 = .978$$

Combining trend and cycle, the forecast for 1986 would then be

$$\hat{y}_{16} = \hat{T}_{16} \cdot \hat{C}_{16} \cdot \hat{\epsilon}_{16} = (1257.12)(.978)(1) = 1229.5$$

Notice that in obtaining this forecast we used the usual convention of forecasting the irregular term to be its nominal level (i.e., $\epsilon_{16} = 1$ in a multiplicative model). ●

### 6.3.3 Example 6.2—Seasonal Data

Table 6.2 contains four years of quarterly data from a series with a strong seasonal component. The seasonally strong fourth quarter is recognizable in the plotted data (Figure 6.6), but the presence of a cycle is not at all apparent. To check for a cycle, the trend and seasonality must first be screened out.

**Table 6.2** Quarterly Time Series for Example 6.2

| | Qtr | $t$ | $y_t$ | | Qtr | $t$ | $y_t$ |
|---|---|---|---|---|---|---|---|
| 1984 | I | 1 | 56 | 1986 | I | 9 | 131 |
| | II | 2 | 70 | | II | 10 | 150 |
| | III | 3 | 76 | | III | 11 | 152 |
| | IV | 4 | 97 | | IV | 12 | 184 |
| 1985 | I | 5 | 76 | 1987 | I | 13 | 138 |
| | II | 6 | 99 | | II | 14 | 171 |
| | III | 7 | 120 | | III | 15 | 200 |
| | IV | 8 | 164 | | IV | 16 | 265 |

**Figure 6.6** Plot of Data in Table 6.2

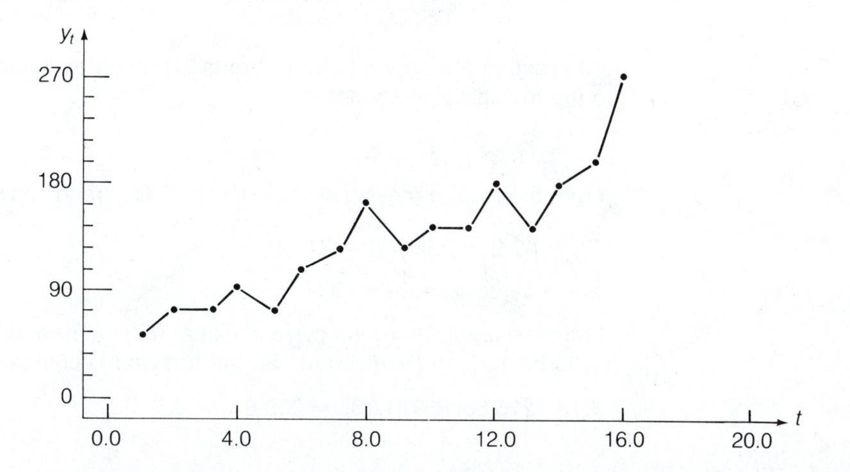

Using a multiplicative model, the ratio-to-moving average method (Chapter 5) applied to these data yields normalized seasonal indexes:

$$\hat{S}_1 = .857 \qquad \hat{S}_2 = .959 \qquad \hat{S}_3 = .993 \qquad \hat{S}_4 = 1.191$$

and an estimated trend line,

$$\hat{T}_t = 49.2 + 9.85t$$

Since the multiplicative model assumes $y_t = T_t \cdot S_t \cdot C_t \cdot \epsilon_t$, the detrended, deseasonalized series is

$$\frac{y_t}{T_t \cdot S_t} = C_t \cdot \epsilon_t$$

This series is plotted in Figure 6.7. The cyclical element, which isn't easily discerned in the plot of the $y_t$'s, now becomes obvious. The length of the cycle seems to be about eight quarters (two years).

**Figure 6.7 Data from Table 6.2 (Detrended and Deseasonalized)**

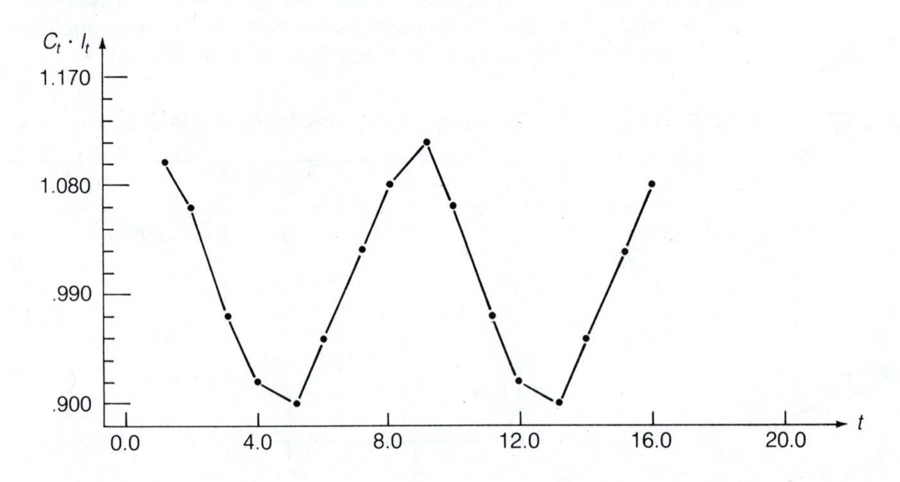


Forecasting the next period, $y_{17}$, entails putting estimates of $T_{17}$, $S_{17}$, $C_{17}$, and $\epsilon_{17}$ into the multiplicative model

$$\hat{y}_{17} = \hat{T}_{17} \cdot \hat{S}_{17} \cdot \hat{C}_{17} \cdot \hat{\epsilon}_{17}$$

From the ratio-to-moving average method, the trend and seasonal estimates are

$$\hat{T}_{17} = 49.2 + 9.85(17) = 216.65$$

$$\hat{S}_{17} = \hat{S}_{13} = \hat{S}_9 = \hat{S}_1 = .857$$

From the regularity of the cycle in Figure 6.7, we have little difficulty estimating $C_{17} \cdot \epsilon_{17}$ to be 1.13, or thereabouts. So the forecast becomes

$$\hat{y}_{17} = 216.65(.857)(1.13) = 209.8$$

●

### 6.3.4 Example 6.3—Leading Indicators, U.S. Corn and Hog Production

It has long been held that the yearly production of hogs in the United States is a cyclical series, the reason presumably being that since corn is the major component in hog feeds, fluctuations in the production (and hence the price) of corn will induce fluctuations in hog production (Valentine, 1987; Young, 1987). Essentially, large corn crops lead to lower corn prices, which in turn make hog production more profitable, leading to larger pig crops.

Suppose one had wanted to forecast hog production for 1986 from records of the annual production of corn and hogs shown in Table 6.3 for the period 1971–1985. One approach would be to simply proceed as with the wheat production data in Example 6.1, using the hog production data alone to estimate trend and cycle. A problem arises, however, in that the cyclical component is not as easily recognized as it was for wheat production.

**Table 6.3 Annual U.S. Corn and Hog Production, 1971–1985**
$y_t$ = hog production (1,000's)
$x_t$ = corn production ($10^6$ bu.)

| Year | $t$ | $y_t$ | $x_t$ |
|---|---|---|---|
| 1971 | 1 | 86,667 | 5,641.0 |
| 1972 | 2 | 78,759 | 5,573.0 |
| 1973 | 3 | 72,264 | 5,647.0 |
| 1974 | 4 | 77,071 | 4,663.6 |
| 1975 | 5 | 64,926 | 5,797.0 |
| 1976 | 6 | 70,454 | 6,266.4 |
| 1977 | 7 | 74,019 | 6,425.5 |
| 1978 | 8 | 74,139 | 7,086.7 |
| 1979 | 9 | 85,425 | 7,938.8 |
| 1980 | 10 | 91,882 | 6,644.8 |
| 1981 | 11 | 87,850 | 8,201.6 |
| 1982 | 12 | 79,328 | 8,235.1 |
| 1983 | 13 | 84,762 | 4,174.7 |
| 1984 | 14 | 82,478 | 7,656.2 |
| 1985 | 15 | 81,974 | 8,865.0 |

*Source: Survey of Current Business.*

Another approach would be to try to take advantage of the relationship between the hog and corn series. A close comparison of Figures 6.8(a) and (b) reveals that, with few exceptions, each increase (or decrease) in the corn series is followed the next year by an increase (decrease) in the hog series. Thus, the general belief that corn production "leads" hog production seems to be borne out in these data. (In Section 6.4.6 a method is given for investigating statistically whether or not, and by how many periods, one series leads another.) Since the most recent change in the corn series (from $t = 14$ to $t = 15$) has been an increase, it seems likely that the movement in the hog series a year later (i.e., from 1985 to 1986) will also be upward. Next, because the period-to-period changes in hog production appear to be on the order of 2,000–8,000 units, a reasonable forecast for 1986 hog production would be about 84,000–90,000 units (i.e., 1985 production plus an increase of 2,000–8,000 units). ●

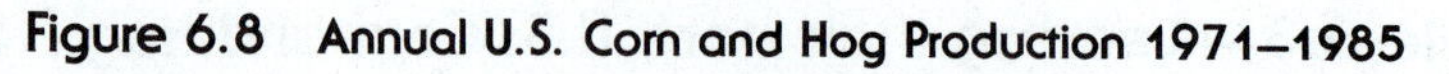

**Figure 6.8 Annual U.S. Corn and Hog Production 1971–1985**

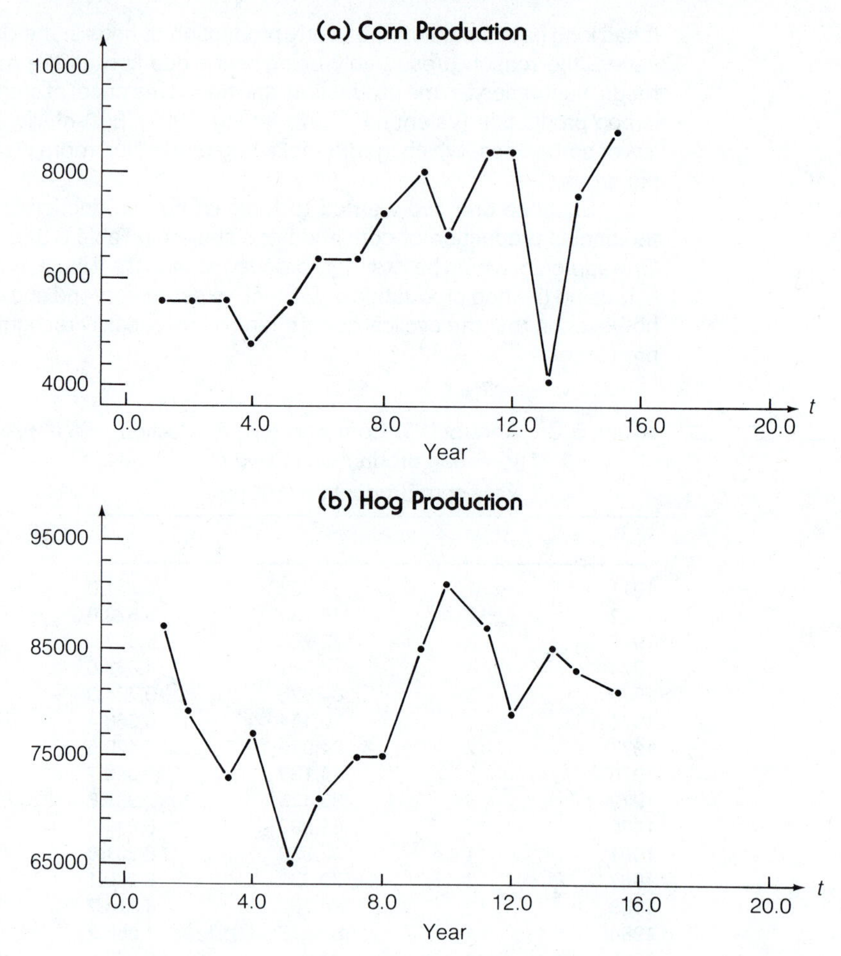


## 6.4 The Decision to Model Cycles

### 6.4.1 Introduction

Some forecasters decide, at the outset of their investigations, not to incorporate a cyclical component into their models because they believe one or both of the following:

- Due to their irregular nature, cyclical movements will be difficult, if not impossible, to effectively capture in a model.
- Since models that react to trends in a series will also react to long cycles, the forecasts resulting from simply modeling trend, or the *trend-cycle,* will probably be adequate for their purposes.

For some series this may in fact be a good approach. For others, though, there are good reasons to take the extra effort to separately consider a cyclical component:

- **the desire to use leading indicators** There is usually much more information at hand than merely the data on the particular series to be forecast. When another series, $x_t$, is known to affect or lead $y_t$, it is natural to want to use this information to anticipate $y_t$'s movements. Leading indicators provide a simple way (i.e., no equation relating $x_t$ to $y_t$ need be specified) to explain cyclical behavior in a series.
- **the prediction of turning points** Trend models generally describe only increasing or decreasing movements and so cannot predict turning points. The capability of forecasting turning points, of major importance in many cases, is one of the by-products of modeling cycles.
- **cyclical movements can dominate a series** In this case, modeling trend and seasonality is not enough. For example, large swings above and below trend constitute a large portion of the variability in yearly crop yields.

While it is hard to *prove* that cyclical forces are at work on a given series, there are some tests that will indicate when modeling more than just trend and seasonality is justified.

## 6.4.2 Testing for Cycles

The methods discussed in the pages that follow are designed to sense the possible presence of cycles in a series and to investigate the periodicity of these cycles. In cases where the recurrence interval is not regular and leading indicators are available, tests are described that indicate when, and by how many time periods, one series leads another.

In point of fact, none of the tests discussed in this section were developed specifically to see cycles; for the most part, they are tests for autocorrelation. The tests will detect cycles, however, since a cyclical component will cause the residuals (after removing trend and seasonality) to remain autocorrelated. On the other hand, there could be other reasons for autocorrelation in the residuals (e.g., if an inappropriate trend line is fit, or an autoregressive model should have been used), so caution is in order when interpreting the tests' results.

## 6.4.3 Graphical Methods

To see cycles in a series, graphically or analytically, the effects of trend and seasonality must first be removed. In turn, detrending and deseasonalizing depends upon which model, additive or multiplicative, is used. In either case, to test graphically for cycles, simply plot the detrended, deseasonalized series versus $t$. Nonrandom behavior in this plot may be interpreted as evidence that a cycle is present. Matters are simplified, of course, when using yearly data, since no seasonal component need be estimated. The general procedure is shown in the table.

**Graphical Procedure to Test for Cycles**

**Yearly Data:**

1. Specify a model:

Additive: $y_t = T_t + C_t + \hat{\epsilon}_t$

Multiplicative: $y_t = T_t \cdot C_t \cdot \hat{\epsilon}_t$

2. Estimate the trend component, $\hat{T}_t$.
3. Plot the residuals versus $t$:

Additive model: Plot $y_t - \hat{T}_t$ versus $t$.

Multiplicative model: Plot $y_t/\hat{T}_t$ versus $t$.

Nonrandom behavior in this plot may be evidence of a cycle.

**Nonyearly Data:**

1. Specify a model:

Additive: $y_t = T_t + S_t + C_t + \epsilon_t$

Multiplicative: $y_t = T_t \cdot C_t \cdot \epsilon_t$

2. Estimate the trend, $\hat{T}_t$, and seasonal, $\hat{S}_t$, components.
3. Plot the residuals versus $t$:

Additive model: Plot $y_t - \hat{T}_t - \hat{S}_t$ versus $t$.

Multiplicative model: Plot $y_t/\hat{T}_t \cdot \hat{S}_t$ versus $t$.

Nonrandom behavior in this plot may be evidence of a cycle.

**EXAMPLE 6.4** ***Residual Plot, Additive Model, U.S. Wheat Production***

The U.S. wheat production data of Example 6.1, being a yearly series, did not require estimating a seasonal component. For comparison with the multiplicative model of that example, let us now examine the residuals from an additive model. Using the trend line from Example 6.1, the additive residuals

$$y_t - \hat{T}_t = \hat{C}_t + \hat{\epsilon}_t$$

are plotted in Figure 6.9. From these plots it is apparent that, regardless of the choice of model, there are cyclical swings away from a linear trend. ●

Using the cyclical-irregular term in forecasting requires a knowledge of the relative importance of its two components, $C_t$ and $\epsilon_t$. If the cycle dominates the noise, then the cyclical-irregular term will be useful for forecasting. If $\epsilon_t$ dominates, then the cyclical-irregular term, being primarily noise, is of no value in forecasting.

One way of evaluating the relative strength of the cyclical component is by comparing the trend-cycle to the noise in the series. Since longer and longer moving averages will eventually smooth out the irregular term, one way to measure the strength of the trend-cycle is to find the minimum-length moving average for which the trend-cycle begins to dominate (accounts for more variability than) the irregular term. For monthly data, the length of this minimum-length moving average is called the **months for cyclical dom-**

**Figure 6.9** U.S. Wheat Production, 1971–1985, Detrended (Additive Model) Series

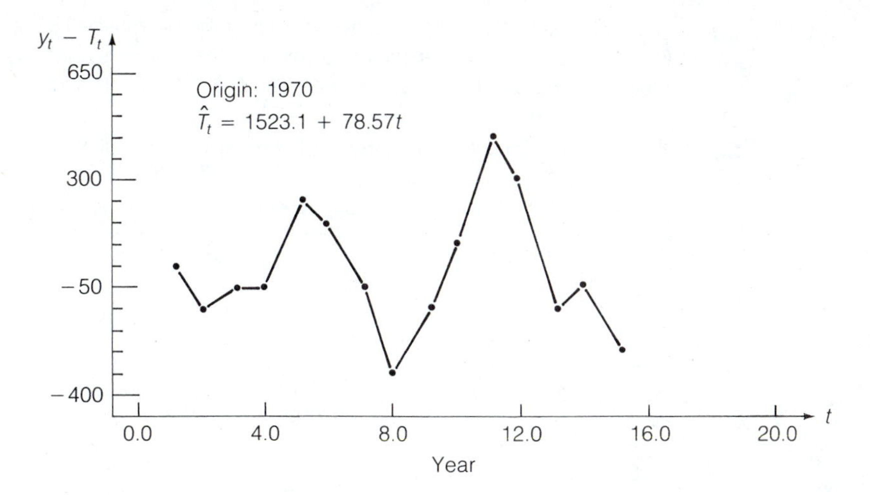


**inance (MCD);** for quarterly data it is called the **quarters for cyclical dominance (QCD).** The procedure for calculating the MCD or QCD is given in the table nearby.

**Calculating the Months for Cyclical Dominance (MCD) or Quarters for Cyclical Dominance (QCD)**

**Given:**

A sample of $n$ observations, $y_t$; $t = 1, 2, \ldots, n$
A set of normalized seasonal indexes, $S_1, S_2, \ldots, S_L$.
The set of centered moving averages, $MA_t$, of the $y_t$'s.

**Calculate Percentage Changes:**

Calculate the average absolute % changes in trend-cycle and irregular terms for 1, 2, 3, . . . period time spans.

Multiplicative Model:

$T_t \cdot C_t \simeq MA_t$ — Calculate % changes in $MA_t$

$$S_t \cdot \epsilon_t = \frac{T_t \cdot C_t \cdot S_t \cdot \epsilon_t}{T_t \cdot C_t} \simeq \frac{y_t}{MA_t}$$

$$\epsilon_t = \frac{S_t \cdot \epsilon_t}{S_t} \approx \frac{y_t}{S_t \cdot MA_t}$$ — Calculate % changes in $\epsilon_t$.

Additive Model:

$T_t + C_t \approx MA_t$ — Calculate % changes in $MA_t$.

$$S_t + \epsilon_t = T_t + C_t + S_t + \epsilon_t - (T_t + C_t) \simeq y_t - MA_t$$

$$\epsilon_t = S_t + \epsilon_t - S_t \approx y_t - MA_t - S_t$$ — Calculate % changes in $\epsilon_t$.

**MCD/QCD:**

The MCD or QCD is the shortest time span for which the ratio below is less than 1.

$$\frac{\text{average absolute \% change in irregular term}}{\text{average absolute \% change in trend-cycle}}$$

**Interpretation:**

MCD's (QCD's) of 1 indicate that the trend-cycle is strong; MCD's (QCD's) beyond 6 indicate very weak or nonexistent cyclical movement.

*Source:* Shiskin, J., *Handbook of Cyclical Indicators* (1984), U.S. Department of Commerce, Bureau of Economic Analysis.

**EXAMPLE 6.5** ***Calculating the QCD, Data from Example 6.2***

Assuming a multiplicative model and using the normalized seasonals given in Example 6.4, we calculate the irregular terms and the percent changes in the trend-cycle for spans of one and two months (see Table 6.4). From these calculations we find

**Table 6.4 Calculation of the QCD for Example 6.2**

| Period | $y_t$ | Centered $MA_t$ | $S_t \cdot \epsilon_t$ | $S_t$ | $\epsilon_t$ |
|---|---|---|---|---|---|
| 1 | 56 | — | — | — | — |
| 2 | 70 | — | — | — | — |
| 3 | 76 | 77.250 | .984 | .993 | .991 |
| 4 | 97 | 83.375 | 1.163 | 1.191 | .977 |
| 5 | 76 | 92.500 | .822 | .857 | .959 |
| 6 | 99 | 106.375 | .931 | .959 | .970 |
| 7 | 120 | 121.625 | .987 | .993 | .994 |
| 8 | 164 | 134.875 | 1.216 | 1.191 | 1.021 |
| 9 | 131 | 145.250 | .902 | .857 | 1.052 |
| 10 | 150 | 151.750 | .988 | .959 | 1.031 |
| 11 | 152 | 155.125 | .980 | .993 | .987 |
| 12 | 184 | 158.625 | 1.160 | 1.191 | .974 |
| 13 | 138 | 167.250 | .825 | .857 | .963 |
| 14 | 171 | 183.375 | .933 | .959 | .972 |
| 15 | 200 | — | — | — | — |
| 16 | 265 | — | — | — | — |

**Calculation of Absolute Percent (|%|) Change**

| $T_t \cdot C_t$ | 1-Quarter \|%\| Change | 2-Quarter \|%\| Change | $\epsilon_t$ | 1-Quarter \|%\| Change | 2-Quarter \|%\| Change |
|---|---|---|---|---|---|
| 77.250 | — | — | .991 | — | — |
| 83.375 | 7.93 | — | .977 | 1.40 | — |
| 92.500 | 10.95 | 19.74 | .959 | 1.86 | 3.23 |
| 106.375 | 15.00 | 27.59 | .970 | 1.22 | .65 |
| 121.625 | 14.34 | 31.49 | .994 | 2.38 | 3.64 |
| 134.875 | 10.89 | 26.79 | 1.021 | 2.75 | 5.20 |
| 145.250 | 7.69 | 19.43 | 1.052 | 3.08 | 5.92 |
| 151.750 | 4.48 | 12.51 | 1.031 | 2.06 | .96 |
| 155.125 | 2.22 | 6.80 | .987 | 4.27 | 6.24 |
| 158.625 | 2.26 | 4.53 | .974 | 1.30 | 5.51 |
| 167.250 | 5.44 | 7.82 | .963 | 1.15 | 2.43 |
| 183.375 | 9.64 | 15.60 | .972 | 1.00 | .16 |
| Average: | 8.26 | 17.23 | | 2.04 | 3.39 |

$$\frac{\text{Average } |\%| \text{ Change in } \epsilon}{\text{Average } |\%| \text{ Change in } T \cdot C}$$

$$\text{1-Quarter Spans: } \frac{2.04}{8.26} = .247$$

$$\text{2-Quarter Spans: } \frac{3.39}{17.23} = .197$$

$$\frac{\text{average \% change in irregular term}}{\text{average \% change in trend-cycle}} = .247 \qquad \text{(1-quarter spans)}$$

$$\frac{\text{average \% change in irregular term}}{\text{average \% change in trend-cycle}} = .197 \qquad \text{(2-quarter spans)}$$

Since the ratio of percent changes is less than one for 1-quarter spans, the process could have been stopped at that point and the QCD taken to be 1. The calculation for 2-quarter spans was included for purposes of illustration. The fact that the QCD is 1 indicates that the trend-cycle is very strong compared to the irregular component, which agrees with the conclusions reached in Example 6.2. ●

### 6.4.4 von Neumann's Rank Ratio, a Nonparametric Test

Testing for autocorrelation (and hence, cycles) can be accomplished by using a **nonparametric** version of **von Neumann's ratio** test (Section 2.6.3; Bartels, 1982). Simply put, this test is just von Neumann's ratio test applied to ranks of the data. Since von Neumann's ratio has been denoted as *M*, RM will be used when referring to the version using ranks. The interpretation of RM is similar to that of its parametric counterpart: Values of RM close to 0 indicate positive autocorrelation, while values close to 4 indicate negative autocorrelation.

---

**Test Procedure:** *von Neumann's Ratio Test, Ranked Data*

$H_0$: Residuals are independent.

$H_a$: Residuals are positively autocorrelated.

**Test Statistic:**

$$\text{RM} = \frac{\sum_{t=1}^{n-1} (R_{t+1} - R_t)^2}{\sum_{1}^{n} (R_t - \bar{R})^2} = \frac{\sum_{t=1}^{n-1} (R_{t+1} - R_t)^2}{n(n^2 - 1)/12}$$

where $R_t$ is the *rank* of $y_t$.

***Note:*** Since $\Sigma\,(R_t - \bar{R})^2 = \Sigma\, R_t^2 - (\Sigma R_t)^2/n$, the denominator of *RM* is unaffected by the ranking and is therefore always equal to $n(n^2 - 1)/12$.

**Decision Rule:** Reject $H_0$ if $\text{RM} < \text{RM}_\alpha$. **(Table 6.5)**

Otherwise, do not reject $H_0$.
[$\text{RM}_\alpha$ = lower $\alpha \times 100\%$ point in the distribution of von Neumann's rank ratio]

**Conclusion:** If $H_0$ is rejected, we conclude there is positive autocorrelation in the residuals that may be due to the presence of a cycle.

If $H_0$ is not rejected, we have some support for concluding that the residuals are independent and that no cyclical component need be modeled.

**Table 6.5 Critical Values for the von Neumann Rank Ratio Test**

| | α for One-Tailed Test | | α/2 for Two-Tailed Test | | |
|---|---|---|---|---|---|
| *n* | .005 | .01 | .025 | .05 | .10 |
| 10 | .62 | .72 | .89 | 1.04 | 1.23 |
| 11 | .67 | .77 | .93 | 1.08 | 1.26 |
| 12 | .71 | .81 | .96 | 1.11 | 1.29 |
| 13 | .74 | .84 | 1.00 | 1.14 | 1.32 |
| 14 | .78 | .87 | 1.03 | 1.17 | 1.34 |
| 15 | .81 | .90 | 1.05 | 1.19 | 1.36 |
| 16 | .84 | .93 | 1.08 | 1.21 | 1.38 |
| 17 | .87 | .96 | 1.10 | 1.24 | 1.40 |
| 18 | .89 | .98 | 1.13 | 1.26 | 1.41 |
| 19 | .92 | 1.01 | 1.15 | 1.27 | 1.43 |
| 20 | .94 | 1.03 | 1.17 | 1.29 | 1.44 |
| 21 | .96 | 1.05 | 1.18 | 1.31 | 1.45 |
| 22 | .98 | 1.07 | 1.20 | 1.32 | 1.46 |
| 23 | 1.00 | 1.09 | 1.22 | 1.33 | 1.48 |
| 24 | 1.02 | 1.10 | 1.23 | 1.35 | 1.49 |
| 25 | 1.04 | 1.12 | 1.25 | 1.36 | 1.50 |
| 26 | 1.05 | 1.13 | 1.26 | 1.37 | 1.51 |
| 27 | 1.07 | 1.15 | 1.27 | 1.38 | 1.51 |
| 28 | 1.08 | 1.16 | 1.28 | 1.39 | 1.52 |
| 29 | 1.10 | 1.18 | 1.30 | 1.40 | 1.53 |
| 30 | 1.11 | 1.19 | 1.31 | 1.41 | 1.54 |
| 32 | 1.13 | 1.21 | 1.33 | 1.43 | 1.55 |
| 34 | 1.16 | 1.23 | 1.35 | 1.45 | 1.57 |
| 36 | 1.18 | 1.25 | 1.36 | 1.46 | 1.58 |
| 38 | 1.20 | 1.27 | 1.38 | 1.48 | 1.59 |
| 40 | 1.22 | 1.29 | 1.39 | 1.49 | 1.60 |
| 42 | 1.24 | 1.30 | 1.41 | 1.50 | 1.61 |
| 44 | 1.25 | 1.32 | 1.42 | 1.51 | 1.62 |
| 46 | 1.27 | 1.33 | 1.43 | 1.52 | 1.63 |
| 48 | 1.28 | 1.35 | 1.45 | 1.53 | 1.63 |
| 50 | 1.29 | 1.36 | 1.46 | 1.54 | 1.64 |
| 55 | 1.33 | 1.39 | 1.48 | 1.56 | 1.66 |
| 60 | 1.35 | 1.41 | 1.50 | 1.58 | 1.67 |
| 65 | 1.38 | 1.43 | 1.52 | 1.60 | 1.68 |
| 70 | 1.40 | 1.45 | 1.54 | 1.61 | 1.70 |
| 75 | 1.42 | 1.47 | 1.55 | 1.62 | 1.71 |
| 80 | 1.44 | 1.49 | 1.57 | 1.64 | 1.71 |
| 85 | 1.45 | 1.50 | 1.58 | 1.65 | 1.72 |
| 90 | 1.47 | 1.52 | 1.59 | 1.66 | 1.73 |
| 95 | 1.48 | 1.53 | 1.60 | 1.66 | 1.74 |
| 100 | 1.49 | 1.54 | 1.61 | 1.67 | 1.74 |

*Source: J. Amer. Stat. Assoc.* 77:377 (1982), pp. 40–46. Reproduced by permission.

## EXAMPLE 6.6 *von Neumann's Rank Ratio Test, U.S. Wheat Production*

Using the additive model, we will analyze the detrended wheat production data. The detrended series and its ranks are shown in Table 6.6. Since $n$ is 15, the denominator of RM is

$$\sum_{1}^{15}(R_t - \bar{R})^2 = \frac{n(n^2-1)}{12} = \frac{15(15^2-1)}{12} = 280$$

Thus,

$$\text{RM} = \frac{\sum_{1}^{14}(R_{t+1} - R_t)^2}{\sum_{1}^{15}(R_t - \bar{R})^2} = \frac{404}{280} = 1.44$$

which is significant at a level close to 10% ($\text{RM}_{.10} = 1.36$ from Table 6.5). This is not overwhelming support for rejecting $H_0$, but coupled with the analysis from Example 6.1 it is reasonable to conclude that the residuals are positively autocorrelated and that cyclical behavior may be the cause.

**Table 6.6 Residuals and Ranked Residuals, Additive Model, Annual U.S. Wheat Production ($10^6$ bu.), 1971–1985**

| Period $t$ | Wheat Production $y_t$ | Detrended Series $y_t - T_t$ | Ranks $R_t$ | $R_{t+1} - R_t$ |
|---|---|---|---|---|
| 1 | 1618 | 16.308 | 10 | −7 |
| 2 | 1545 | −135.260 | 3 | 3 |
| 3 | 1705 | −53.827 | 6 | 1 |
| 4 | 1796 | −41.395 | 7 | 6 |
| 5 | 2122 | 206.037 | 13 | −1 |
| 6 | 2142 | 147.469 | 12 | −4 |
| 7 | 2036 | −37.099 | 8 | −7 |
| 8 | 1798 | −353.667 | 1 | 4 |
| 9 | 2134 | −96.235 | 5 | 6 |
| 10 | 2374 | 65.198 | 11 | 4 |
| 11 | 2799 | 411.630 | 15 | −1 |
| 12 | 2765 | 299.062 | 14 | −10 |
| 13 | 2420 | −124.506 | 4 | 5 |
| 14 | 2596 | −27.074 | 9 | −7 |
| 15 | 2425 | −276.642 | 2 | — |

**Calculations:** $\sum_{1}^{14}(R_{t+1} - R_t)^2 = 404 \qquad \sum_{1}^{15}(R_t - \bar{R})^2 = \frac{15(15^2-1)}{12} = 280$

$$\text{RM} = \frac{404}{280} = 1.44$$

●

### 6.4.5 Parametric Tests

The autocorrelation function and the parametric version of von Neumann's ratio test (both discussed in Section 2.6) are two familiar tools that will signal the presence of cycles. The acf has the added advantage, in cases where the cycle is fairly regular, of giving an indication of the period. For example, a cycle that repeats every three years (i.e., peak to peak or trough to trough) should lead to a significant lag-3 autocorrelation coefficient.

**EXAMPLE 6.7** ***acf Test for Cycles***

The quarterly data from Example 6.2 contained a very regular 8-quarter (2-year) cycle that was clearly evident in the plot of the detrended, deseasonalized series in Figure 6.7. The acf of these residuals shown in Figure 6.10 also gives a good indication of a cycle of length 8, since $r_4 = -.749$ and $r_8 = .499$. The lag-4 coefficient is significant at about the 5% level ($|r_4| > 2/\sqrt{n} = 2/\sqrt{16} = .50$). The value of $r_8$ borders on being significant, and in conjunction with $r_4$, supports the contention that an 8-year cycle is present.

**Figure 6.10** Autocorrelation Function of Detrended, Deseasonalized Series Using Data of Table 6.2, Multiplicative Model

(a) Residuals

| $t$ | $y_t/(T_t \cdot S_t)$ Residuals | $t$ | $y_t/(T_t \cdot S_t)$ Residuals |
|---|---|---|---|
| 1 | 1.106 | 9 | 1.109 |
| 2 | 1.059 | 10 | 1.059 |
| 3 | .971 | 11 | .972 |
| 4 | .919 | 12 | .923 |
| 5 | .901 | 13 | .909 |
| 6 | .953 | 14 | .953 |
| 7 | 1.023 | 15 | 1.023 |
| 8 | 1.076 | 16 | 1.076 |

(b) acf

```
                 -1.0  -.8   -.6   -.4   -.2    0    .2    .4    .6    .8   1.0
                   +----+----+----+----+----+----+----+----+----+----+
 1     .602                                     XXXXXXXXXXXXXXXX
 2    -.084                                   XXX
 3    -.621                    XXXXXXXXXXXXXXXXX
 4    -.749                 XXXXXXXXXXXXXXXXXXXX
 5    -.431                         XXXXXXXXXXXX
 6     .078                                     XXX
 7     .448                                     XXXXXXXXXXXX
 8     .499                                     XXXXXXXXXXXXX
 9     .251                                     XXXXXXX
10    -.081                                   XXX
11    -.268                              XXXXXXXX
12    -.243                               XXXXXXX
13    -.078                                   XXX
14     .081                                     XXX
```

**EXAMPLE 6.8** *von Neumann's Ratio Test, U.S. Wheat Production*

von Neumann's (parametric) ratio can also be computed from the data in Table 6.1. Following the test procedure in Section 2.6.3, and using the detrended data in Table 6.6, we find the following.

$H_0$: The series is random/stationary.

$H_a$: The series has autocorrelated errors.

**Test Statistic:**

$$M = \frac{SS_{\Delta e}}{SS_{ee}} = \frac{\sum_{2}^{n} (\Delta e)^2}{\sum_{1}^{n} (e_t - \bar{e})^2}$$

$$= \frac{704{,}897}{578{,}981} = 1.22$$

**Decision Rule:** ($\alpha = 10\%$, $n = 15$)

Reject $H_0$ if $M < M_{.90} = 1.370$. **(Appendix EE)**

**Conclusion:** Since $M = 1.22 < 1.370$, we conclude that there is significant ($\alpha = 10\%$) autocorrelation in the residuals, and that a cycle may be the cause. ●

## 6.4.6 Leading Indicators and the Cross-Correlation Function (ccf)

As already mentioned, the two main questions when dealing with leading indicators are:

1. Does a given series $x_t$ really lead $y_t$?
2. By how many periods does $x_t$ lead $y_t$?

There is a tool for quantitatively answering these questions (Box and Jenkins, 1970). It is called the cross-correlation function (ccf). We now turn to its calculation and interpretation.

The **cross-correlation function (ccf)** describes the extent to which two series $x_t$ and $y_t$ are correlated. It does this in exactly the same way the autocorrelation function describes a single series. That is, for each integer $k$ (positive or negative), the ccf estimates the correlation between $y_t$ and the shifted series $x_{t-k}$ (or, equivalently, between $y_{t+k}$ and $x_t$). The calculation of the ccf is also analogous to that of the acf; namely,

$r_{xy}(k)$ = sample cross-correlation coefficient of lag $k$

$$= \frac{\sum_{1}^{n-k} (x_t - \bar{x})(y_{t+k} - \bar{y})}{\sqrt{\sum_{1}^{n} (x_t - \bar{x})^2 \cdot \sum_{1}^{n} (y_t - \bar{y})^2}} \qquad k = \ldots -3, -2, -1, 0, 1, 2, 3, \ldots$$

We denote by $\rho_{xy}(k)$ the theoretical cross-correlation coefficient that $r_{xy}(k)$ is intended to estimate.

Continuing with the analogy to the acf, testing the coefficients in the ccf is also done by a rule of thumb. The test is based on the approximate standard deviation of $r_{xy}(k)$:

$$\sigma_{r_{xy}(k)} \simeq \frac{1}{\sqrt{n - |k|}}$$

Coefficients whose absolute values exceed 2 standard deviations are judged to be significant at approximately the 5% level. Remember, though, that the ccf (just like the acf) is easily interpreted only if both the $x_t$ and $y_t$ series are stationary. The test procedure can be summarized as follows.

---

**Test Procedure:** *Cross-Correlation Coefficient*

$H_0$: $\rho_{xy}(k) = 0$
$H_a$: $\rho_{xy}(k) \neq 0$

**Test Statistic:** $r_{xy}(k)$ = sample cross-correlation coefficient of lag $k$

**Decision Rule:** ($\alpha \simeq 5\%$, $n$ large)

Reject $H_0$ if $|r_{xy}(k)| > 2/\sqrt{n - |k|}$. Otherwise, do not reject $H_0$.

**Conclusion:** If $H_0$ is rejected, we conclude with approximately 95% confidence that there is cross-correlation at lag $k$.

If $H_0$ is not rejected, we have some support for concluding that there is no cross-correlation at lag $k$.

---

When testing indicates that $\rho_{xy}(k)$ is different from zero, one can say that $y_t$'s movements tend to be related to $x_t$'s movements $k$ periods ago ($k \geq 0$) or $k$ periods hence ($k \leq 0$). For example, suppose $r_{xy}(3) = .80$ for two series $x_t$ and $y_t$, each of length 20. Then the lag-3 cross-correlation is significant (since $.80 > 2/\sqrt{20 - 3} = .485$), and one could conclude that $x_t$ tends to lead $y_t$ by about three periods. That is, three periods after $x_t$ swings above (below) its mean, there is a tendency for $y_t$ to move above (below) its mean.

**EXAMPLE 6.9** *Cross-Correlation Function, U.S. Corn and Hog Production*

In Example 6.3, visual inspection of the graphs revealed that movements in corn production tend to be reflected one year later in hog production. Figure 6.11 shows the cross-correlation function of these data. The largest coefficient occurs at lag 1, with $r_{xy}(1) = .504$. This coefficient is close to being significant ($2/\sqrt{15 - 1} = .53$) and

**Figure 6.11** Cross-Correlation Function, U.S. Corn and Hog Production, 1971–1985

```
CCF - CORRELATES      CORN(T) AND      HOGS(T+K)
        -1.0  -.8  -.6  -.4  -.2   .0   .2   .4   .6   .8  1.0
          +----+----+----+----+----+----+----+----+----+----+
-6   -.200                    XXXXXX
-5    .099                         XXX
-4   -.176                     XXXXX
-3   -.352                XXXXXXXXXX
-2    .091                         XXX
-1    .366                         XXXXXXXXXX
 0    .252                         XXXXXXX
 1    .504                         XXXXXXXXXXXXXX
 2    .491                         XXXXXXXXXXXXX
 3    .346                         XXXXXXXXXX
 4    .276                         XXXXXXXX
 5    .014                         X
 6   -.150                     XXXXX
```

tends to support the contention that corn production leads hog production by about a year. The virtue of using the ccf, compared with the crude method used in Example 6.3, is that the ccf quickly detects possible lead/lag relationships, freeing the analyst from tedious point-by-point comparisons of the two series. ●

# 6.5 The Business Cycle

## 6.5.1 Introduction

Expansions and contractions in the United States economy have occurred many times over its some 200 years of history. In the context of the economy or business activity, contractions are called *recessions,* and describing the *business cycle* amounts to determining the dates and severity of the various U.S. recessions (Moore, 1967).

For these concepts to be useful, of course, they must be given precise meaning. Responsibility for this task falls on the National Bureau of Economic Research (NBER), which, in conjunction with the Bureau of Economic Analysis (BEA) of the Department of Commerce, makes available detailed data concerning the cyclical behavior of business activity and establishes dates at which the cycle peaks and troughs have occurred. For example, Table 6.7 shows a complete list of NBER business cycle reference dates from 1854 to the present. Underlying the determination of these dates is an operational definition of the business cycle, which is given shortly.

**Table 6.7 U.S. Business Cycle Expansions and Contractions**

| Business Cycle Reference Dates | | Duration, in Months | | | |
|---|---|---|---|---|---|
| | | Contraction | Expansion | Cycle | |
| Trough | Peak | (Trough from Previous Peak) | (Trough to Peak) | Trough from Previous Trough | Peak from Previous Peak |
| Dec 1854 | Jun 1857 | . . . . | 30 | . . . . | . . . . |
| Dec 1858 | Oct 1860 | 18 | 22 | 48 | 40 |
| Jun 1861 | Apr 1865 | 8 | 46 | 30 | 54 |
| Dec 1867 | Jun 1869 | 32 | 18 | 78 | 50 |
| Dec 1870 | Oct 1873 | 18 | 34 | 36 | 52 |
| Mar 1879 | Mar 1882 | 65 | 36 | 99 | 101 |
| May 1885 | Mar 1887 | 38 | 22 | 74 | 60 |
| Apr 1888 | Jul 1890 | 13 | 27 | 35 | 40 |
| May 1891 | Jan 1893 | 10 | 20 | 37 | 30 |
| Jun 1894 | Dec 1895 | 17 | 18 | 37 | 35 |
| Jun 1897 | Jun 1899 | 18 | 24 | 36 | 42 |
| Dec 1900 | Sep 1902 | 18 | 21 | 42 | 39 |
| Aug 1904 | May 1907 | 23 | 33 | 44 | 56 |
| Jun 1908 | Jan 1910 | 13 | 19 | 46 | 32 |
| Jan 1912 | Jan 1913 | 24 | 12 | 43 | 36 |
| Dec 1914 | Aug 1918 | 23 | 44 | 35 | 67 |
| Mar 1919 | Jan 1920 | 7 | 10 | 51 | 17 |
| Jul 1921 | May 1923 | 18 | 22 | 28 | 40 |
| Jul 1924 | Oct 1926 | 14 | 27 | 36 | 41 |
| Nov 1927 | Aug 1929 | 13 | 21 | 40 | 34 |
| Mar 1933 | May 1937 | 43 | 50 | 64 | 93 |
| Jun 1938 | Feb 1945 | 13 | 80 | 63 | 93 |
| Oct 1945 | Nov 1948 | 8 | 38 | 88 | 45 |
| Oct 1949 | Jul 1953 | 11 | 45 | 48 | 56 |
| May 1954 | Aug 1957 | 10 | 39 | 55 | 49 |
| Apr 1958 | Apr 1960 | 8 | 24 | 47 | 32 |
| Feb 1961 | Dec 1969 | 10 | 106 | 34 | 116 |
| Nov 1970 | Nov 1973 | 11 | 36 | 117 | 47 |
| Mar 1975 | Jan 1980 | 16 | 58 | 52 | 74 |
| Jul 1980 | Jul 1981 | 6 | 12 | 64 | 18 |
| Nov 1982 | . . . | | | | |

*Note:* Underlined figures are wartime expansions, postwar contractions, and the full cycles that include the wartime expansions.
*Source: Business Conditions Digest,* 1987.

## 6.5.2 National Bureau of Economic Research

The NBER was formed towards the beginning of this century and has been concerned with the study of the cyclical nature of business activity (Nerlove,

Grether, and Carvalho, 1979). One of the bureau's missions has been to develop measures capable of anticipating the start and signalling the end of recessionary periods. To do this, NBER uses the following working definition of the business cycle.

**DEFINITION:** **Business Cycle:** "Business cycles have been defined as sequences of expansion and contraction in various economic processes that show up as major fluctuations in aggregate economic activity—that is, in comprehensive measures of production, employment, income, and trade. While recurrent and pervasive, business cycles of historical experience have been definitely nonperiodic and have varied greatly in duration and intensity, reflecting changes in economic systems, conditions, policies, and outside disturbances." (*Source: Business Conditions Digest,* 1988.)

The key phrase in this definition is *aggregate economic activity,* which implies that *many,* not few, economic series should be involved and that their *collective,* not individual, movements should be used to ascertain peaks and troughs of business activity.

Data on many individual and aggregate series are a by-product of efforts to describe the business cycle. These data, along with other economic measures, are published monthly by the BEA in the *Business Conditions Digest* (*BCD*). Each issue of the *BCD* contains:

- data on a few hundred key economic series and *composite indexes* based upon these series
- official dates for the turning points of the business cycle
- graphs of leading, lagging, and coincident indicators of cyclical changes
- data on and graphs of *diffusion indexes* and *rates of change*

In the following pages each of these items is discussed along with how the NBER uses them to establish the turning points of the business cycle.

### 6.5.3 Composite Indexes

A simple way to monitor aggregate economic activity is by means of a weighted average, or **composite index,** of a group of series. If the component series are chosen carefully, the composite index based on those series can consolidate a lot of information into a single, easily monitored, measure that is more reliable than any of its individual components (Ratti, 1985; Shiskin, 1971; *Handbook of Cyclical Indicators,* 1984). The various composite indexes constructed by the NBER are classified by either *cyclical timing* (i.e., leading, coincident, or lagging) or *economic process* (e.g., employment, production, inventories, capital investment).

Three of the most important indexes are the *index of leading indicators,* the *index of roughly coincident indicators,* and the *index of lagging indicators.*

The components that comprise these indexes are shown in Table 6.8. According to the *Business Conditions Digest,* these indicators are evaluated on the basis of their performance with respect to "economic significance, statistical adequacy, consistency of timing at business cycle peaks and troughs, conformity to business cycle expansions and contractions, smoothness, and prompt availability (currency)." Being averages, the composite indexes are not only more reliable but also smoother than the noisy, individual components.

Actual construction of composite indexes is complicated. For example, each component series is *standardized* by dividing its month-to-month changes by the long-run series average to ensure that the more volatile series do not dominate the index. In addition, component series are adjusted for trend and seasonality and then given a weight, with the better-performing series getting

**Table 6.8 Components of the Indexes of Leading, Coincident, and Lagging Indicators**

| | Components | Series* |
|---|---|---|
| **Leading Index:** | Average weekly hours of production or nonsupervisory workers, manufacturing | 1 |
| | Average weekly initial claims for unemployment insurance, State programs | 5 |
| | Manufacturer's new orders in 1972 dollars, consumer goods and materials industries | 8 |
| | Vendor performance, percent of companies receiving slower deliveries | 32 |
| | Index of net business formation** | 12 |
| | Contracts and orders for plant and equipment in 1972 dollars | 20 |
| | Index of new private housing units authorized by local building permits | 29 |
| | Change in manufacturing and trade inventories on hand and on order in 1972 dollars, smoothed | 36 |
| | Change in sensitive materials prices, smoothed | 99 |
| | Index of stock prices, 500 common stocks | 19 |
| | Money supply M2 in 1972 dollars | 106 |
| | Change in business and consumer credit outstanding | 111 |
| **Coincident Index:** | Employees on nonagricultural payrolls | 41 |
| | Personal income less transfer payments in 1972 dollars | 51 |
| | Index of industrial production | 47 |
| | Manufacturing and trade sales in 1972 dollars | 57 |
| **Lagging Index:** | Average duration of unemployment in weeks | 91 |
| | Ratio, manufacturing and trade inventories to sales in 1972 dollars | 77 |
| | Index of labor cost per unit of output, manufacturing—actual data as a percent of trend | 62 |
| | Average prime rate charged by banks | 109 |
| | Commercial and industrial loans outstanding in 1972 dollars | 101 |
| | Ratio, consumer installment credit outstanding to personal income | 95 |

*Source:* From *Business Conditions Digest.*
* *BCD* series identification number.
** This series was eliminated from the leading index in March 1987.

the larger weights. After these adjustments, the composite (weighted) series is divided by its average value in a base year (1967 is currently used) and multiplying by 100. This last step causes the composite indicator to be indexed on the base year. Graphs of the indexes of leading, coincident, and lagging indicators are shown in Figure 6.12 (the vertical scale in each graph is logarithmic). Notice that each graph, being indexed on 1967, passes through the value 100 in that year.

Another important indicator is the *ratio of coincident to lagging indexes* (Moore, 1969). This indicator has generated much interest since first being studied in the late 1960s for the following reasons:

**Figure 6.12** **Composite Indexes of Leading, Roughly Coincident, and Lagging Indicators, 1950–1987**

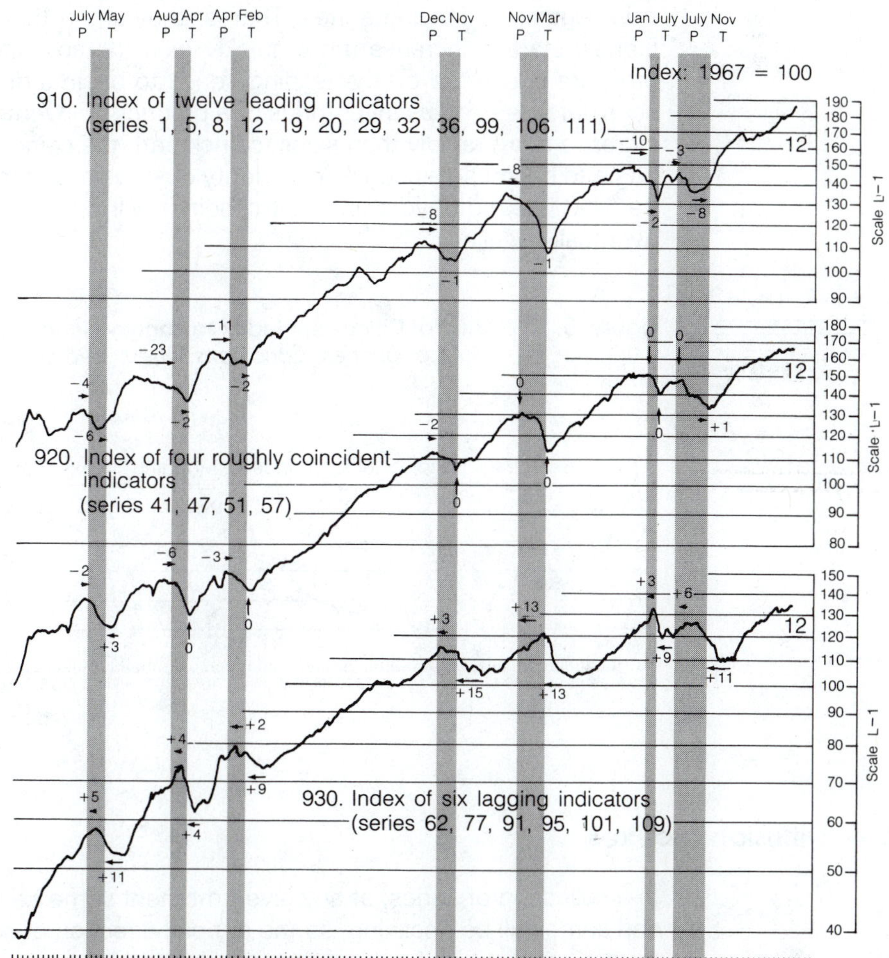

- It is a leading indicator (even though its components are only coincident and lagging series).
- It usually leads turns in business activity by longer intervals than does the index of leading indicators (i.e., it leads the leading indicators).

A graph of the coincident/lagging ratio is shown in Figure 6.13. Comparing this graph to that of the index of leading indicators (top graph in Figure 6.12), the two facts mentioned above become apparent. Since 1950 (and, in fact, throughout this century), the coincident/lagging ratio has led turns (peaks) in the U.S. business cycle by longer intervals than has the index of leading indicators.

The reasons why the coincident/lagging ratio should lead the economy are partially, but not fully, understood. On the surface, there is a natural tendency for any coincident/lagging ratio to peak before peaks in the coincident series. While the coincident series slows and flattens near its peak, the lagging series is still rising quickly, so the resulting *ratio* peaks before the coincident series. Beyond that, some researchers suggest that there is a causal connection *from* lagging *to* leading series. That is, they argue that as lagging indicators (series) start to increase more quickly than the coincident indicator (series), pressure is exerted on the leading series to begin a decrease. For example, consider sales versus inventories of a product: If inventories (lagging) start to increase more rapidly than sales (coincident), the coincident/lagging ratio will begin to decrease since there is plenty of stock and less fear that orders may be filled late, so buyers' interest (leading) will not be as keen and sales will eventually decline.

**Figure 6.13 Ratio of Coincident Index to Lagging Index, 1950–1987**
(*Source: Business Conditions Digest, 1987*)

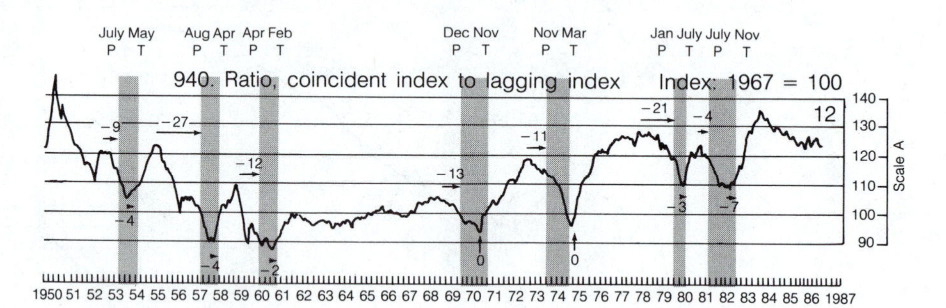


## 6.5.4 Diffusion Indexes

In a collection of series, at any given moment some series will be moving up and some will be declining, so the general direction of activity can be gauged by the proportion of series that are increasing. In a sense, this proportion measures how diffused the increase in activity is throughout the group (of series).

A **diffusion index** is the *percentage* of component series that are *rising* (Broida, 1955; *Handbook of Cyclical Indicators,* 1984). As a rule, if there are any series that have not changed in either direction, half are counted as rising and half as declining. Computationally, then, diffusion indexes are simple to construct, but they can tend to behave erratically if limited to very short time spans (e.g., monthly periods). To produce smooth series, some diffusion indexes are also calculated for six- and/or nine-month spans.

Diffusion indexes, along with the composite indicators and rates of change, play a part in the NBER's procedure for establishing dates of business cycle peaks and troughs. Their primary contribution is in showing how widely spread through the economy an increase or decline has been. Figure 6.14 shows graphs of the diffusion indexes of the leading, coincident, and lagging composite indicators. Notice that the diffusion index for the coincident series often hits its extremes (0% or 100%), since there are only four series comprising the coincident index.

**Figure 6.14** **Diffusion Indexes of Leading, Roughly Coincident, and Lagging Indicators, 1950–1987 (*Source: Business Conditions Digest,* 1987)**

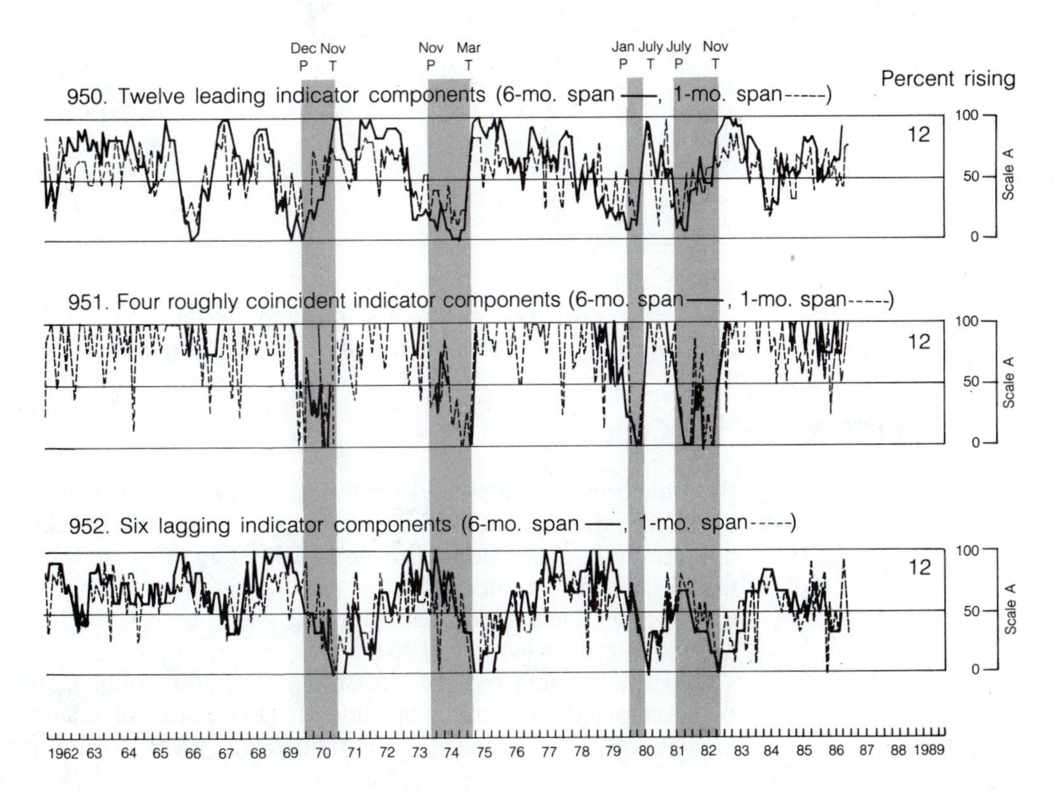

## 6.5.5 Rates of Change

While diffusion indexes gauge only the *direction* of business activity, *rates of change* measure both direction and *degree*. Rates of change address the question of how severe a contraction (or how intense an expansion) is. For example, a diffusion index may indicate that activity is slowing; but if the degree of slowing is very minor, it would not be evidence that the economy was slipping into a recession.

For a given series or composite index, the **rate of change** is simply the *percent change* in the series or index over a specified time span (*Handbook of Cyclical Indicators,* 1984). The rates of change monitored in the *Business Conditions Digest* are shown in Figure 6.15. Notice that they are reported as annualized rates. Looking at the index of leading indicators during 1986, for example, we get the history shown in the table.

**History of Leading Indicators for 1986**

| | Index | Change from Previous Month | % Change | Annualized % Change |
|---|---|---|---|---|
| Jan | 173.4 | — | — | — |
| Feb | 174.9 | 1.5 | .865% | 10.38% |
| Mar | 175.9 | 1.0 | .572 | 6.86 |
| Apr | 178.2 | 2.3 | 1.308 | 15.69 |
| May | 178.1 | −.1 | −.0561 | −6.73 |
| Jun | 177.7 | −.4 | −.225 | −2.70 |
| Jul | 179.3 | 1.6 | .900 | 10.81 |
| Aug | 179.1 | −.2 | −.112 | −1.34 |
| Sep | 179.4 | .3 | .168 | 2.01 |
| Oct | 180.6 | 1.2 | .669 | 8.03 |
| Nov | 182.2 | 1.6 | .886 | 10.63 |
| Dec | 186.1 | 3.9 | 2.141 | 25.69 |

For a particular series, the diffusion index and the rate of change will usually, though not always, move in the same direction at the same time.

## 6.5.6 NBER Reference Cycles

Reference dates (Table 6.7) for the peaks and troughs of the business cycle are determined by the NBER. Since contractions are usually of much shorter duration than expansions, *Business Conditions Digest* shows these peak-to-trough periods as darkly shaded regions on each of its charts (e.g., Figures 6.12–6.15). Each series can then be easily compared to these underlying *reference cycles* (Moore, 1967).

To determine the dates of the peaks and troughs, the NBER looks at composite indexes, diffusion indexes, and rates of change, together with knowledge of any special circumstances and governmental actions. Their two-

**Figure 6.15 Rates of Change of the Leading, Roughly Coincident, and Lagging Indicators, 1950–1987** (*Source: Business Conditions Digest,* 1987)

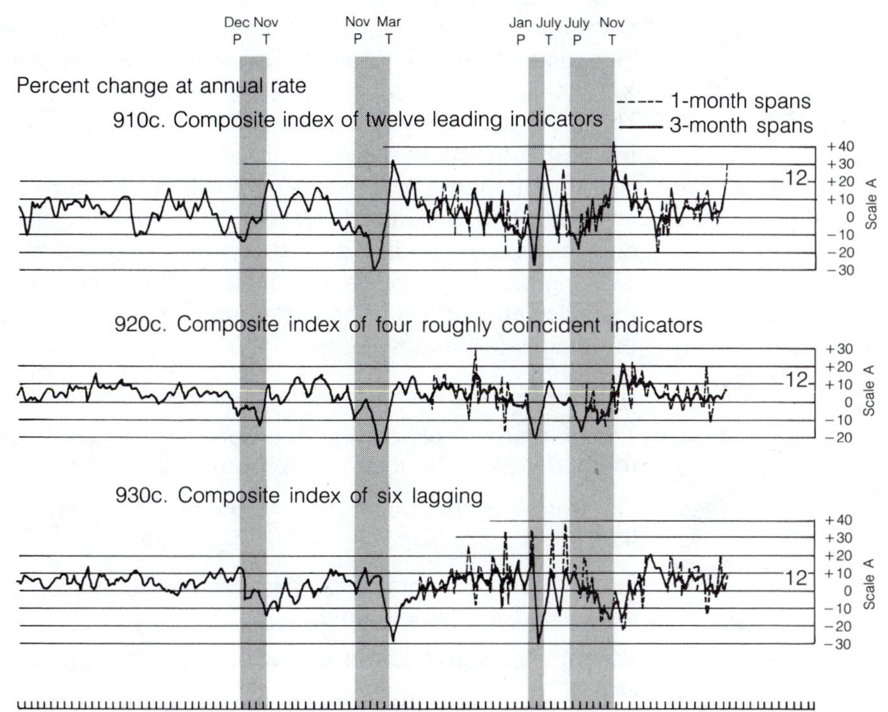


stage procedure begins by comparing current activity to past contractions (and expansions) in order to decide if a contraction (expansion) is in progress. Second, if necessary, they then decide on the most likely *month* that the contraction (expansion) began.

By calculating rates of change, the severity of a contraction can be compared to that of past cycles. If there is close agreement, a decision can quickly be made that a recession is beginning. Diffusion indexes add further evidence of the general direction of activity. Once a decision is made that a contraction is, indeed, occurring, then the date at which the peak occurred must be established. The leading, coincident, and lagging indicators narrow this choice, since the date of a peak should occur *near* the peak of the coincident series, *before* the peak in the lagging series, and *after* the peak in the leading series. Diffusion indexes then help pin down the date more precisely. For example, near a peak, a diffusion index should be approaching the 50% mark from above and then proceed to dip below 50% after the peak.

Sometimes the comparisons to past cycles are not completely clear, and the NBER may deliberate for a few months about the existence of a contraction before reaching a conclusion. In light of new data, they may also go back and make slight revisions to previously determined (published) dates.

### 6.5.7 X-11

In order to more clearly identify cyclical movements, most series in *Business Conditions Digest* are deseasonalized prior to the calculation of the various indexes and rates of change. The method NBER currently uses is called the *X-11 Variant of the Census II Seasonal Adjustment Program,* or, simply, X-11. This program is conventionally used by many agencies and businesses besides the NBER to create deseasonalized data.

Fundamentally, X-11 is just a refined version of the ratio-to-moving average method described in Section 5.5.4. The program allows for adjustments due to holidays, trading days (i.e., not all months have the same number of working days), and outliers and uses more specialized weighted averages in the calculation of its trend-cycle estimates. The following chronology shows the stages in the development of X-11.

*1922:* Ratio-to-moving average method was introduced by NBER.

*1954:* The first large-scale computer program based on the ratio-to-moving average method was introduced by the Bureau of the Census.

*1955:* The Bureau of the Census revised the procedure and named it Census Method II.

*1960:* Experimental variants of Census II began to appear, starting with version X-3 (X stands for "experimental").

*1961:* Variants X-9 and X-10 (for use with erratic series) were introduced.

*1965:* X-11 replaced all other variants of Census II.

We will not present the details of running an X-11 analysis, but refer the reader to the primary reference on the subject: "The X-11 Variant of the Census Method II Seasonal Adjustment Program," Technical Paper No. 15, 1967 revision, available from the U.S. Government Printing Office, Washington, D.C. Some major statistical packages, notably SAS, also offer X-11 as a supplemental package.

## 6.6 Forecasting the Cycle

### 6.6.1 Introduction

Forecasting the cyclical component in a series is one of the more difficult tasks a forecaster faces. When there is no other information available apart from the $y_t$ series itself, one can only hope that the cycle is regular enough to allow extrapolating a period or two into the future. Lacking such regularity, the safest bet when forecasting the next period is simply to assume no change in the cycle.

The acf and MCD (or QCD) will give indications of whether a cycle is regular enough to warrant extrapolation. For example, if the acf of the detrended, deseasonalized series looks random and if the MCD (QCD) is large, then the cycle is either weak or nonexistent. It doesn't make much sense, then, to spend time trying to accurately forecast the next value of $C_t$. On the other hand, if the MCD (QCD) is small (especially if it is 1), then there is evidence that the cycle is strong compared to the noise, so forecasting $C_t$ is worthwhile.

When the cyclical pattern is strong, the following forecasting procedure can be used: Forecast the direction (if possible) of the next movement in $C_t$ and then add or subtract, depending on that direction, the average amount of the period-to-period changes in the past cyclical-irregular terms. This procedure is summarized in the following table.

**Forecasting the Cycle**

1. If it is unclear what the direction of change will be, then simply use the current cyclical-irregular term as the forecast; otherwise, forecast the direction (+ or −) of the next movement in the cycle.
2. Calculate the average absolute change in the past cyclical-irregular terms.
3. Using the current cyclical-irregular term, add/subtract the average cyclical change to get a forecast of the next cyclical-irregular term.

**EXAMPLE 6.10** *Forecasting the Cycle for Data from Example 6.2*

From Example 6.5, QCD = 1 for the quarterly data of Example 6.2. This is an indication of a strong cyclical component, so estimating the next cyclical-irregular term, $C_{17} \cdot \epsilon_{17}$, is worthwhile. Using the $C_t \cdot \epsilon_t$ terms plotted in Figure 6.7, we get about .052 as the average absolute value of the period-to-period changes. Since the direction of the next move in the cycle appears to be *upward* (recall Examples 6.2 and 6.7),

$$\hat{C}_{17} \cdot \hat{\epsilon}_{17} \approx C_{16} \cdot \epsilon_{16} + \text{(average change)} = 1.076 + .052 = 1.13$$

(This is how the estimate of $\hat{C}_{17} \cdot \hat{\epsilon}_{17}$ was obtained for Example 6.2.) •

### 6.6.2 Leading Indicators

At the very least, a leading indicator can help anticipate the *direction* of the next move in the cycle. It is possible, though, to use the indicator to estimate the *magnitude* of the next movement as well. Two ways of doing this are given in what follows.

Suppose that a series $x_t$ has been found to lead $y_t$ by $k$ periods. This means that movements in $y_t$ are, to some extent, affected by movements in $x_{t-k}$. That is, what $x_t$ did $k$ periods ago should help explain what $y_t$ is doing now. Assuming that the two series have already been detrended and deseasonalized, one can then use regression analysis to find an equation for predicting $y_t$ from $x_{t-k}$:

$$\hat{y}_t = \hat{\beta}_0 + \hat{\beta}_1 x_{t-k}$$

Forecasting $y_{t+1}$ is then a matter of substituting $x_{t+1-k}$ into this equation, which gives

$$\hat{y}_{t+1} = \hat{\beta}_0 + \hat{\beta}_1 x_{t+1-k}$$

Notice that since $x_t$ leads $y_t$, $x_{t+1-k}$ will be known at time $t$.

**EXAMPLE 6.11** ***Leading Indicator Forecast via Regression, U.S. Corn and Hog Production***

In Example 6.9, the cross-correlation function indicated that corn production, $x_t$, tended to lead hog production, $y_t$, by about one year. Since these two series require no deseasonalization or detrending, $y_t$ can immediately be regressed on $x_{t-1}$. Using the data from Example 6.3, we find that this regression equation turns out to be

$$\hat{y}_t = 57{,}518 + 3.34x_{t-1}$$

So the forecast for the next year's hog production would be

$$\hat{y}_{16} = 57{,}518 + 3.34x_{15} = 57{,}518 + 3.34(8865) = 87{,}092$$

This forecast is in line with the rough estimates (from 84,000 to 90,000) found in Example 6.3. ●

There is also a simple heuristic method for forecasting with a leading indicator that doesn't require regression or any other statistical technique. The procedure consists of estimating the historical period-to-period percent changes in each series ($y_t$ and $x_t$) and using their relative magnitudes to forecast future percent changes in $y_t$. The procedure is described in the table. The rationale for this method is simple: The percent change in $y_t$ from period $t$ to period $t+1$ is *assumed* to be in approximately the same ratio to $\%\Delta x_{t-k}$ as the historical ratio of percent changes in the two series. That is,

$$\frac{(y_{t+1} - y_t)/y_t}{\%\Delta x_{t-k}} \simeq \frac{\text{average \% change in } y_t}{\text{average \% change in } x_t}$$

which, when solved for $y_{t+1}$, gives the forecasting equation in the table.

**Forecasting the Cycle Using Percentage Changes in $y_t$ and $x_t$**

**Given:** Deseasonalized, detrended series $y_t$ and $x_t$, where $x_t$ leads $y_t$ by $k$ periods

**Calculate % Changes:** Calculate the average absolute period-to-period percent changes in both series.

**Forecasts:** Find $\%\Delta x_{t-k}$, the signed percent change in $x_t$ from period $t-1-k$ to period $t-k$. The forecasting equation is then

$$\hat{y}_{t+1} = \left(1 + \frac{\text{average \% change in } y_t}{\text{average \% change in } x_t} \times \%\Delta x_{t-k}\right) \times y_t$$

**EXAMPLE 6.12** ***Leading Indicator Forecast, Heuristic Method, U.S. Corn and Hog Production***

Table 6.9 shows the period-to-period percent changes in the corn and hog production data of Example 6.3. This table shows that the absolute changes in these series have

**Table 6.9 Year-to-Year Percent Change, U.S. Corn and Hog Production, 1971–1985**

| $t$ | Hogs | % Change | \|% Change\| | Corn | % Change | \|% Change\| |
|---|---|---|---|---|---|---|
| 1 | 86667 | — | — | 5641.0 | — | — |
| 2 | 78759 | −9.1% | 9.1% | 5573.0 | −1.2% | 1.2% |
| 3 | 72264 | −8.2 | 8.2 | 5647.0 | 1.3 | 1.3 |
| 4 | 77071 | 6.7 | 6.7 | 4663.6 | −17.4 | 17.4 |
| 5 | 64926 | −15.8 | 15.8 | 5797.0 | 24.3 | 24.3 |
| 6 | 70454 | 8.5 | 8.5 | 6266.4 | 8.1 | 8.1 |
| 7 | 74019 | 5.1 | 5.1 | 6425.5 | 2.5 | 2.5 |
| 8 | 74139 | 0.2 | 0.2 | 7086.7 | 10.3 | 10.3 |
| 9 | 85425 | 15.2 | 15.2 | 7938.8 | 12.0 | 12.0 |
| 10 | 91882 | 7.6 | 7.6 | 6644.8 | −16.3 | 16.3 |
| 11 | 87850 | −4.4 | 4.4 | 8201.6 | 23.4 | 23.4 |
| 12 | 79328 | −9.7 | 9.7 | 8235.1 | 0.4 | 0.4 |
| 13 | 84762 | 6.9 | 6.9 | 4174.7 | −49.3 | 49.3 |
| 14 | 82478 | −2.7 | 2.7 | 7656.2 | 83.4 | 83.4 |
| 15 | 81974 | −0.6 | 0.6 | 8865.0 | 15.8 | 15.8 |
| | | Average: | 7.2% | | | 19.0% |

been running at around 19% and 7.2%, respectively. Corn production leads by one year, so $k$ is 1 in the forecasting equation. For period $t = 16$, $t - k = 16 - 1 = 15$, and

$$\%\Delta x_{t-k} = \%\Delta x_{15} = \frac{x_{15} - x_{14}}{x_{14}} = \frac{8865 - 7656.2}{7656.2} = .16, \text{ or } 16\%$$

and the forecast for $t = 16$ becomes

$$\hat{y}_{16} = \left(1 + \frac{7.2}{19} \times .16\right) \times 81{,}974 = 86{,}944$$

Note that this result closely agrees with the regression approach of Example 6.11. •

## 6.7 Diagnostics

To a large extent, modeling cycles involves searching for the *forces* (e.g., population shifts, wars) that affect a series and the *mechanism* through which these forces exert their influence. For example, corn production seems to be the major force behind cycles in hog production (Example 6.3), where the mechanism involves the price of corn and the fact that it is the primary component of hog feed.

If an attempt to explain the forces governing a cycle is weak or incorrect, then diagnostic checking should bring these problems to light. For models with a cyclical component, diagnostic tests include the following.

- **direction of the cycle** Does the model correctly predict the next move, up or down, in the series? Getting the direction right is the most important requirement of any cyclical model.

- **turning points accuracy** Do the predicted turning points closely match the actual turning points in the series?
- **strength of a leading indicator** When a leading indicator is used, the quality of the model will be directly related to the quality of the leading indicator.
- **magnitude of the residuals** The MSE (or MAD or RMSE) of the residuals from a cyclical model should not exceed that of a model without a cyclical component. In other words, modeling the cycle should not make the forecasts worse.
- **randomness of the residuals** As always, patterns in the residuals are signs of a problem with the model.

## 6.8 Exercises

**6.1** How is the cyclical component measured in additive and multiplicative models?

**6.2** Are the cyclical indexes constrained to add to 0 or *L*, as seasonal indexes are?

**6.3** In the ratio-to-moving average method applied to cyclical series, what component or components does the moving average attempt to estimate?

**6.4** Explain why researchers use aggregate economic activity, instead of a single time series, when describing the business cycle.

**6.5** According to Table 6.8, stock prices generally lead economic activity in the United States while the prime rate charged by banks tends to lag. What reasons can you give to explain this?

**6.6** How could you use the months for cyclical dominance (MCD) to decide whether it is worthwhile to model the cyclical element in a series?

**6.7** Can the cross-correlation function be used with trended series, or, as with the autocorrelation function, must the series first be detrended?

**6.8** Suppose you calculate the cross-correlation function of a time series with itself. What is the interpretation of the resulting coefficients?

**6.9** Does the *Composite Index of Leading Indicators* change more rapidly during a recession or during the following recovery?

**6.10** Suppose a diffusion index consisting of 25 individual time series is created to monitor a particular segment of the economy.

**a.** What are the possible values that the index may assume?

**b.** Suppose, for a period of six months, the individual series performed as follows:

| Month | Number of Series Rising | Month | Number of Series Rising |
|---|---|---|---|
| 1 | 10 | 4 | 5 |
| 2 | 15 | 5 | 12 |
| 3 | 10 | 6 | 25 |

Find the values of the diffusion index during these months.

c. Calculate the rates of change in the index over the six-month period in part (b).

**6.11** Sales and inventory levels for a certain company are as follows:

| Month | Sales (units) | Inventory (units) | Month | Sales (units) | Inventory (units) |
|---|---|---|---|---|---|
| 1 | 100 | 83 | 9 | 85 | 103 |
| 2 | 122 | 84 | 10 | 79 | 88 |
| 3 | 136 | 101 | 11 | 77 | 83 |
| 4 | 150 | 123 | 12 | 75 | 78 |
| 5 | 141 | 134 | 13 | 81 | 77 |
| 6 | 130 | 152 | 14 | 90 | 76 |
| 7 | 105 | 143 | 15 | 102 | 81 |
| 8 | 90 | 129 | 16 | 110 | 83 |

a. Plot sales versus time and inventory versus time on the same graph. By how many periods does inventory tend to lag sales?

b. Find the cross-correlation function of sales and inventory. Does the ccf support your conclusion in part (a)?

c. Calculate the ratio of sales to inventory for the 16 months.

d. Find the cross-correlation function of sales and the ratio found in part (d). By how many periods does the ratio tend to lead sales?

e. Perform a regression of sales on lagged values of the ratio in part (d). Use the resulting equation to predict sales for month 17.

## References


Bartels, R. (1982). "The Rank Version of von Neumann's Ratio Test for Randomness." *J. Amer. Stat. Assoc.* 77(377): 40–46.

Box, G. E. P., and Jenkins, G. M. (1970). *Time Series Analysis, Forecasting and Control.* San Francisco, Calif.: Holden-Day.

Broida, A. L. (1955). "Diffusion Indexes." *The American Statistician* 9: 7–16.

*Handbook of Cyclical Indicators* (1984). U.S. Dept. of Commerce, Bureau of Economic Analysis.

Makridakis, S., Wheelwright, S. C., and McGee, V. E. (1983). *Forecasting: Methods and Applications,* 2nd ed. New York: Wiley.

Malabre, A. L. (1987). "Business Cycle May Overwhelm Policies." *The Wall Street Journal,* Nov. 6, p. 9.

McLaughlin, R. L. (1982). "A Model of an Average Recession and Recovery." *J. Forecasting* 1(1): 55–65.

Moore, G. H. (1969a). "Generating Leading Indicators from Lagging Indicators." *Western Economics Journal* 7(2): 137–144.

Moore, G. H. (1969b). "Forecasting Short-Term Economic Change." *J. Amer. Stat. Assoc.* 64(325): 1–22.

Moore, G. H. (1967). "What Is a Recession?" *The American Statistician* 21: 16–20.

Nerlove, M., Grether, D. M., and Carvalho, J. L. (1979). *Analysis of Economic Time Series,* New York: Academic Press.

Ratti, R. A. (1985). "A Descriptive Analysis of Economic Indicators." *Federal Reserve Board of St. Louis, Review* 67(1): 14–24.

Rubinow, S. I. (1975). *Introduction to Mathematical Biology,* New York: Wiley, ch. 1, sec. 8.

Shiskin, J. (1971). "Modernizing Business Cycle Concepts." *The American Statistician* 25: 17–19.

Shiskin, J. (1967). "The X-11 Variant of the Census II Seasonal Adjustment Program." Technical paper no. 15, Bureau of the Census, U.S. Dept. of Commerce.

Valentine, L. M. (1987). *Business Cycles and Forecasting,* 7th ed. Cincinnati, Ohio: South-Western Publishing Co.

Young, T. (1987). "Not So Down on the Farm." *Fortune,* Sept. 14.

# CHAPTER 7 The Box-Jenkins Approach to Forecasting

**Objective:** *To study procedures and forecasting applications that use Box-Jenkins techniques.*

## 7.1 Introduction

The Box-Jenkins methodology consists, on the one hand, of a large class of time series models and, on the other, of a set of procedures for choosing models from this class to fit to data from a given series. In their original work, G. E. P. Box and G. M. Jenkins (1970) consolidated many commonly used time series techniques into a structured model-building process that emphasizes simple, parsimonious models.

The time series models used in Box-Jenkins forecasting are called **autoregressive–integrated–moving average** models, or **ARIMA** models for short. To encompass the diverse forecasting applications that arise in practice, this class of models has to be, and is, very large. We will see, for example, that exponential smoothing (Chapter 2), autoregressive models (Chapter 4), and random-walk models (Chapter 4) are all special forms of ARIMA models.

Box-Jenkins modeling relies heavily on the use of three familiar time series tools: differencing, the autocorrelation function (acf), and the partial autocorrelation function (pacf). Differencing is used to reduce nonstationary series to stationary ones. The acf and pacf are then used to identify an appropriate ARIMA model and the required number of parameters.

After the model is identified, parameter estimates are obtained: that is, the selected model is fit to the available data. The algorithm used is based on the least squares concept and usually requires several iterations before producing the desired estimates. It is necessary, therefore, to rely on computer programs to implement the Box-Jenkins procedure.

In this chapter the nature of ARIMA models will be discussed along with the step-by-step Box-Jenkins procedure. Specifically, we will examine:

- the class of ARIMA models and their associated parameters
- the role of differencing and other methods for achieving stationarity in a series
- how to use the acf and pacf to identify suitable ARIMA models for a given set of data
- the use of computer routines for model estimation and forecasting
- the role of diagnostic checking in model-selection

## 7.2 ARIMA Models

### 7.2.1 Introduction

Box-Jenkins models can only be applied to stationary series or series which have been made stationary by differencing. The models fall into one of the following three categories:

- purely **autoregressive (AR)** models
- purely **moving average (MA)** models
- mixed **autoregressive–moving average (ARMA)** models

If differencing is required to achieve stationarity, then the series will eventually have to be undifferenced or **integrated** before forecasting. In this case, an *I* is added to the names of the three models, giving **ARI** models, **IMA** models, and **ARIMA** models.

### 7.2.2 Notation

The different Box-Jenkins models are identified by the number of autoregressive parameters ($p$), the degree of differencing ($d$), and the number of moving average parameters ($q$). Any such model can be written using the uniform notation ARIMA($p,d,q$).

**ARIMA Model Notation**

**ARIMA($p,d,q$):** An autoregressive– (of order $p$) moving average (of order $q$) model applied to a series that has been differenced $d$ times

**Special Cases:**

ARIMA($p$,0,0) = an autoregressive model of order $p$; often abbreviated AR($p$)
ARIMA(0,0,$q$) = a moving average model of order $q$; often abbreviated MA($q$)
ARIMA($p$,0,$q$) = a mixed autoregressive–moving average model of order $p$, $q$; often abbreviated ARMA($p,q$)

### 7.2.3 Autoregressive Models

Autoregressive (AR) models, identical to those in Chapter 4, are based on the application of regression analysis to lagged values of the $y_t$ series. In the context of Box-Jenkins modeling, parameters of AR models are conventionally denoted by $\phi_i$'s instead of the $\beta_i$'s of regression analysis. Using this convention, an autoregressive model is expressed as follows:

$y_t$ = the actual value of the series at time $t$
$y_{t-i}$ = the actual value of the series at time $t - i$
$\phi_i$ = the autoregressive parameter for $y_{t-i}$
$\epsilon_t$ = the irregular fluctuation at time $t$, not correlated with past values of the $y_t$'s

**Autoregressive Model of Order $p$, AR($p$)**

$$y_t = \phi_0 + \phi_1 y_{t-1} + \phi_2 y_{t-2} + \phi_3 y_{t-3} + \cdots + \phi_p y_{t-p} + \epsilon_t$$

As mentioned in Chapter 4, since regressing a nonstationary series on itself produces misleading results, AR($p$) models should be fit only to stationary series.

### 7.2.4 Stationarity and the Yule-Walker Equations

Stationarity places restrictions on the parameters $\phi_1, \phi_2, \ldots, \phi_p$ (Abraham and Ledolter, 1983, p. 207). One of these restrictions is that the sum of the parameters must be smaller than 1: $\phi_1 + \phi_2 + \ldots + \phi_p < 1$. On intuitive grounds alone this condition is necessary, since models having only large parameter values will tend to be unstable and put increasing weights on past values of the series. To see the necessity of this condition, consider the following AR(1) series:

$$y_t = \phi_0 + \phi_1 y_{t-1} + \epsilon_t$$

When repeatedly substituted into itself, this series yields

$$\begin{aligned} y_t &= \phi_0 + \phi_1(\phi_0 + \phi_1 y_{t-2} + \epsilon_{t-1}) + \epsilon_t \\ &= \phi_0 + \phi_0\phi_1 + \phi_1^2 y_{t-2} + \underbrace{[\phi_1\epsilon_{t-1} + \epsilon_t]}_{\text{[error terms]}} \\ &= \phi_0 + \phi_0\phi_1 + \phi_1^2(\phi_0 + \phi_1 y_{t-3}) + [\text{error terms}] = \ldots \\ &= \phi_0\left(\sum_0^{k-1} \phi_1^i\right) + \phi_1^k y_{t-k} + [\text{error terms}] \end{aligned}$$

If $\phi_1$ (which is the sum of the parameters, since $p = 1$) were to exceed 1, the weight given to $y_{t-k}$ would increase with $k$, which runs counter to the intuitive feeling that the older values in a series be given lesser weight.

The theoretical autocorrelation coefficients $\rho_k$ of an AR($p$) series are related to the $\phi_i$'s as follows:

$$\rho_k = \phi_1\rho_{k-1} + \phi_2\rho_{k-2} + \cdots + \phi_p\rho_{k-p} \qquad k = 1, 2, \ldots \tag{7.1}$$

which, except for the $\phi_0$ term, is exactly the same as the model equation with $\rho_k$'s in place of $y_t$'s. This makes it easy to remember Equation 7.1. Since $\rho_{-k}$ and $\rho_k$ are really the same and $\rho_0$ is 1, the general form of the first $p$ equations (for $k = 1, 2, 3, \ldots, p$) in Equation 7.1 is

$$\begin{aligned}
\rho_k = {} & \phi_1\rho_{k-1} + \phi_2\rho_{k-2} + \cdots + \phi_{k-1}\rho_1 && \rightarrow \quad \rho_{k-1}, \ldots, \rho_3, \rho_2, \rho_1 \\
& + \phi_k\rho_0 && \rightarrow \quad \rho_0 \\
& + \phi_{k+1}\rho_1 + \phi_{k+2}\rho_2 + \cdots + \phi_p\rho_{p-k} && \rightarrow \quad \rho_1, \rho_2, \rho_3, \ldots, \rho_{p-k}
\end{aligned}$$

Using this general form, we can write the first $p$ equations as

$$\begin{array}{lllllll}
k = 1: & \rho_1 = \phi_1 & + \phi_2\rho_1 & + \phi_3\rho_2 + \cdots + & \phi_p\rho_{p-1} \\
k = 2: & \rho_2 = \phi_1\rho_1 & + \phi_2 & + \phi_3\rho_1 + \cdots + & \phi_p\rho_{p-2} \\
 & \vdots & & \vdots & \vdots \\
k = p: & \rho_p = \phi_1\rho_{p-1} & + \phi_2\rho_{p-2} & \quad + \cdots + & \phi_p
\end{array}$$

which are called the **Yule-Walker equations** (Yule, 1927; Walker, 1931). Since there are $p$ equations and $p$ unknown $\rho_k$'s, the Yule-Walker system can be solved for the first $p$ autocorrelation coefficients in terms of the $\phi_i$'s. The higher lag coefficients ($\rho_{p+1}, \rho_{p+2}, \ldots$) can then be found by putting these first $p$ coefficients back into Equation 7.1. The import of this is that the Yule-Walker equations together with Equation 7.1 guarantee that the acf of an AR($p$) series is completely determined by the $\phi_i$'s.

Together, the stationarity restrictions and the Yule-Walker equations ensure that the acf of any AR($p$) series drops off exponentially. This is one of the characteristics of AR series. Another is that the partial autocorrelation function (pacf) of an AR($p$) series cuts off sharply after lag $p$ (a fact discussed in Chapter 4). These two characteristics form a basic part of the model identification procedure discussed in Section 7.7.2.

The expected value $\mu$ of an AR($p$) series is also determined by the values of the $\phi_i$'s, as follows:

$$\begin{aligned}
\mu = E(y_t) &= E(\phi_0 + \phi_1 y_{t-1} + \cdots + \phi_p y_{t-p} + \epsilon_t) \\
&= \phi_0 + \phi_1 E(y_{t-1}) + \phi_2 E(y_{t-2}) + \cdots + \phi_p E(y_{t-p}) + E(\epsilon_t) \\
&= \phi_0 + \phi_1\mu + \phi_2\mu + \cdots + \phi_p\mu + 0
\end{aligned}$$

which can be easily solved for $\mu$ to give

$$\mu = \frac{\phi_0}{1 - (\phi_1 + \phi_2 + \cdots + \phi_p)} \tag{7.2}$$

Notice, in particular, that the average level of an AR($p$) series does not equal $\phi_0$, except in the special case when $\phi_0 = 0$.

### 7.2.5 Moving Average Models

Instead of depending on past levels of the series, $y_t$ may be most influenced by the recent "shocks" (i.e., random errors) to the series. That is, the current value of a series may be best explained by looking at the most recent $q$ error terms, as follows:

$y_t$ = the actual value of the series at time $t$

$\theta_i$ = the moving average parameter for $\epsilon_{t-i}$

$\epsilon_{t-i}$ = the error term at time $t - i$

$\epsilon_t$ = the error term at time $t$

**Note:** $\epsilon_t, \epsilon_{t-1}, \epsilon_{t-2}, \ldots$ are uncorrelated with one another.

**Moving Average Model of Order $q$, MA($q$)**

$$y_t = \theta_0 + \epsilon_t - \theta_1\epsilon_{t-1} - \theta_2\epsilon_{t-2} - \cdots - \theta_q\epsilon_{t-q}$$

Note that the average level $\mu$ of an MA($q$) series is equal to the constant term, $\theta_0$, in the model since $E(\epsilon_t) = 0$ for all values of $t$. Also, the negative signs on the parameters are simply a matter of convention.

Box and Jenkins use the term *moving average* differently from most other forecasting applications. Previously, moving average was used when referring to an average of the most recent *values* of a series. MA($q$) models, on the other hand, refer to weighted averages of the most recent *shocks.*

The acf of an MA($q$) series is more complex than for an AR($p$) series; i.e.,

$$\rho_k = \frac{-\theta_k + \theta_1\theta_{k+1} + \theta_2\theta_{k+2} + \cdots + \theta_{q-k}\theta_q}{1 + \theta_1^2 + \theta_2^2 + \cdots + \theta_q^2} \qquad k = 1, 2, \ldots, q$$
$$\rho_k = 0 \qquad k \geq q + 1 \tag{7.3}$$

There is no particular pattern to the first $q$ coefficients of the acf (they don't drop off exponentially, for example), but they do *cut off* sharply (to 0) after lag $q$. In contrast, the pacf of an MA($q$) series does *drop off exponentially* to 0. This characteristic behavior of the acf and pacf is used in the identification stage of modeling.

Unfortunately, without further restrictions the parameters of an MA($q$) model do not uniquely determine the acf. For example, using Equation 7.3, the MA(1) model

$$y_t = \theta_0 + \epsilon_t - \theta_1\epsilon_{t-1}$$

has the acf

$$\rho_1 = \frac{-\theta_1}{1 + \theta_1^2} \qquad \text{and} \qquad \rho_k = 0 \qquad \text{for } k \geq 2$$

and yet, using Equation 7.3 again, the model

$$y_t = \theta_0 + \epsilon_t - \frac{1}{\theta_1} \cdot \epsilon_{t-1}$$

gives rise to the same acf. To avoid this problem, a constraint called *invertibility* is imposed on MA models in Box-Jenkins analysis. We will not discuss all the details of the invertibility requirement, except to point out that it puts exactly the same restrictions on the parameters $\theta_1, \theta_2, \ldots, \theta_q$ that the stationarity requirement puts on the parameters of an AR process (Nelson, 1973, p. 46). In particular, we must have $\theta_1 + \theta_2 + \cdots + \theta_q < 1$.

The term *invertibility* was coined to highlight the fact that every invertible MA($q$) model can be written as an infinite-order AR model. To see this, consider the MA(1) model

$$y_t = \theta_0 + \epsilon_t - \theta_1\epsilon_{t-1} \tag{7.4}$$

which can be written as

$$\epsilon_t = y_t - \theta_0 + \theta_1\epsilon_{t-1}$$

Putting $t - 1$ in for $t$ above and substituting the result into Equation 7.4 gives

$$\begin{aligned} y_t &= \theta_0 + \epsilon_t - \theta_1(y_{t-1} - \theta_0 + \theta_1\epsilon_{t-2}) \\ &= [\theta_0 + \theta_0\theta_1] + \epsilon_t - \theta_1 y_{t-1} - \theta_1^2\epsilon_{t-2} \end{aligned}$$

Repeating this procedure with $t - 2, t - 3, \ldots$ we get

$$\begin{aligned} y_t &= [\theta_0 + \theta_0\theta_1 + \theta_0\theta_1^2 + \cdots] + \epsilon_t - \theta_1 y_{t-1} - \theta_1^2 y_{t-2} - \theta_1^3 y_{t-3} \cdots \\ &= \left[\frac{\theta_0}{1 - \theta_1}\right] - \theta_1 y_{t-1} - \theta_1^2 y_{t-2} - \theta_1^3 y_{t-3} - \cdots + \epsilon_t \end{aligned} \tag{7.5}$$

which is an AR series with an infinite number of autoregressive terms. In a sense, then, the MA(1) model has been inverted and written as an AR($\infty$) model. Notice also that the invertibility condition $\theta_1 < 1$ causes decreasing weights to be given to the past values of the series. To summarize, using invertible MA models guarantees that:

- There is a unique MA($q$) model for a given acf.
- An MA($q$) model can be inverted and written as an AR($\infty$) model.

A natural question to ask at this point is: If MA models are really AR models in disguise, why not use only AR models in the Box-Jenkins method? The answer is parsimony. While it may be possible to fit large-order AR models, it is much more efficient to fit corresponding MA models having only a few parameters.

## 7.2.6 ARMA and ARIMA Models

Mixing autoregressive and moving average terms in the same model is also allowed. The reason, again, is parsimony. A mixed (ARMA) model sometimes requires fewer parameters than a model with only AR or only MA terms. ARMA models are defined as follows:

$y_t$ = the actual value of the series at time $t$

$y_{t-i}$ = the value of the series at time $t - i$

$\epsilon_t$ = the error term at time $t$

$\epsilon_{t-i}$ = the error term at time $t - i$

$\phi_i$ = the autoregressive parameter for $y_{t-i}$

$\theta_i$ = the moving average parameter for $\epsilon_{t-1}$

**Autoregressive–Moving Average Model of Order $p$ and $q$, ARMA($p$,$q$)**

$$y_t = \phi_1 y_{t-1} + \phi_2 y_{t-2} + \cdots + \phi_p y_{t-p} + \theta_0 + \epsilon_t - \theta_1 \epsilon_{t-1} - \theta_2 \epsilon_{t-2} - \cdots - \theta_q \epsilon_{t-q}$$

***Note:*** $\epsilon_t, \epsilon_{t-1}, \epsilon_{t-2}, \ldots$ are uncorrelated with one another.

ARIMA model equations are the same as ARMA equations except $y_t$ is replaced by the differenced series $w_t$:

$$w_t = \Delta^d y_t = \text{the } y_t \text{ series differenced } d \text{ times}$$

**Autoregressive–Integrated–Moving Average Model of Order $p$, $d$, and $q$, ARIMA($p$,$d$,$q$)**

$$w_t = \phi_1 w_{t-1} + \phi_2 w_{t-2} + \cdots + \phi_p w_{t-p} + \epsilon_t + \theta_0 - \theta_1 \epsilon_{t-1} - \theta_2 \epsilon_{t-2} - \cdots - \theta_q \epsilon_{t-q}$$

***Note:*** $\epsilon_t, \epsilon_{t-1}, \epsilon_{t-2}, \ldots$ are uncorrelated with one another.

By convention, in mixed models the constant term is denoted by $\theta_0$ rather than $\phi_0$.

For the reasons given earlier, ARMA models are required to be both stationary and invertible. Because both AR and MA terms are involved, the acf and pacf have the behavior of both types of models. The net effect is that both the acf and the pacf drop off exponentially to zero for ARMA models. In Section 7.7.2 we discuss the shapes of the acf and the pacf in more detail and explain how these functions are used to select the orders $p$ and $q$ of the model.

## 7.3 Familiar ARIMA Models

### 7.3.1 Introduction

The class of ARIMA models derives its usefulness from the variety of forecasting situations in which these models can be applied. For example, trend models, exponential smoothing, random-walk models, and autoregression can all be shown to be special cases of ARIMA models.

### 7.3.2 Stochastic Versus Deterministic Trend

Much of this book has centered around *deterministic* trend models, that is, models in which the average level of the series grows or declines according to some specific function of time (e.g., $y_t = \beta_0 + \beta_1 t + \epsilon_t$, $y_t = \beta_0 \beta_1^t$). Many time series, however, don't exhibit the stability inherent in deterministic trends but instead have trends with variable rates of growth. Such series are said to have *stochastic* trend. Stochastic trend models can be thought of roughly as deterministic models in which the parameters randomly change over time.

Both deterministic and/or stochastic trend can be modeled using ARIMA models. Deterministic trend can be handled using ARIMA($p$,$d$,$q$) models with $d \geq 1$, $q \geq 1$, and $\theta_0 \neq 0$, where the level of differencing, $d$, is equal to the degree of a deterministic polynomial trend (Pankratz, 1983, p. 186). As an example, let us examine the ARIMA(0,1,1) model. This is a pure moving average model that can be written as

$$w_t = \theta_0 + \epsilon_t - \theta_1\epsilon_{t-1}$$

where $w_t$ represents the first differences ($d = 1$) of the original series $y_t$. Putting $y_t - y_{t-1}$ in for $w_t$, we have

$$y_t - y_{t-1} = \theta_0 + \epsilon_t - \theta_1\epsilon_{t-1} \tag{7.6}$$

Writing Equation 7.6 out for each of the next $j$ periods (breaking here from our usual convention of forecasting $p$ periods ahead, since in this chapter $p$ refers to the order of an autoregressive model) gives

$$\begin{aligned}
y_{t+1} - y_t &= \theta_0 + \epsilon_{t+1} - \theta_1\epsilon_t \\
y_{t+2} - y_{t+1} &= \theta_0 + \epsilon_{t+2} - \theta_1\epsilon_{t+1} \\
\vdots \qquad &\;\;\vdots \qquad\qquad \vdots \\
y_{t+j} - y_{t+j-1} &= \theta_0 + \epsilon_{t+j} - \theta_1\epsilon_{t+j-1}
\end{aligned}$$

Then summing, we have

$$y_{t+j} - y_t = j \cdot \theta_0 + [\text{sum of error terms}]$$

or,

$$y_{t+j} = y_t + j \cdot \theta_0 + [\text{sum of error terms}]$$

So

$$E(y_{t+j}) = E(y_t) + j \cdot \theta_0$$

Thus, the average level of the series grows with a deterministic linear trend.

The constant $\theta_0$ must be nonzero to model deterministic trend. For this reason $\theta_0$ is often called the *trend parameter* in an ARIMA model. Since the decision to use deterministic trend is up to the analyst, most commercial computer programs for Box-Jenkins analysis allow the user to choose whether or not to include a constant term in the model.

### 7.3.3 Exponential Smoothing

An ARIMA(0,1,1) model with no constant term generates the same forecasts as those in simple exponential smoothing. To see this, consider the general ARIMA(0,1,1) model with $\theta_0 = 0$ in Equation 7.6, where by putting $t + 1$ in for $t$ we get

$$y_{t+1} = y_t + \epsilon_{t+1} - \theta_1\epsilon_t$$

As of time $t$, $\epsilon_t$ is known and $\epsilon_{t+1}$ is not, so the forecast for $y_{t+1}$ would be (substituting 0 for $\epsilon_{t+1}$)

$$\begin{aligned}
\hat{y}_{t+1} &= y_t + 0 - \theta_1(y_t - \hat{y}_t) \\
&= (1 - \theta_1)y_t + \theta_1\hat{y}_t
\end{aligned} \tag{7.7}$$

Letting $\alpha = 1 - \theta_1$ makes Equation 7.7 identical to the smoothing form of the updating equation for a simple exponential smoothing scheme. Moreover, the Box-Jenkins estimation method, being based on least squares, gives an estimated smoothing constant $\hat{\alpha} = 1 - \hat{\theta}_1$ for which the mean square error is minimized (Muth, 1960).

### 7.3.4 Random Walks

In Chapter 4 we defined a *random walk* as a series whose first differences form a sample from a normal distribution; that is,

$$y_t - y_{t-1} = \epsilon_t$$

This is exactly in the form of an ARIMA(0,1,0) model with no constant term.

If a nonzero constant term, $\theta_0$, is used, the ARIMA(0,1,0) model is then called a *random walk with drift.* Using an argument similar to that used for deterministic trend, the expected value of the series $j$ steps ahead is just

$$E(y_{t+j}) = E(y_t) + j \cdot \theta_0$$

which indicates that the level of the series changes or drifts in one direction.

## 7.4 The Box-Jenkins Modeling Procedure

The stages in the Box-Jenkins procedure are: identification, estimation, diagnostic checking, and forecasting. We discuss each stage in detail in Section 7.7. At a more general level, the steps may be summarized as follows (in practice, each of these stages requires the use of a computer program).

- **identification** The series is differenced, if necessary, to make it stationary. Then the sample acf and pacf are calculated; the behavior of both the acf and pacf determine the number of autoregressive parameters ($p$) and/or moving average parameters ($q$).
- **estimation** Least squares estimates of the process parameters are generated.
- **diagnostic checking** The residuals from the estimated model should look like a random series; failing that, further analysis of the residuals leads to a respecification of the model.
- **forecasting** The fitted model, having first been "integrated" if necessary, is used to forecast the $y_t$'s.

## 7.5 Examples

### 7.5.1 Introduction

Having had a somewhat theoretical discussion in Section 7.2, let us now turn to an examination of data sets from some of the simpler ARIMA models. The goals of this section are to

- examine the relationships between model parameters and their associated acf's and pacf's
- demonstrate the use of the acf, the pacf, and differencing in model identification

- give examples comparing series having stochastic versus deterministic trend
- illustrate the mechanics of forecasting with fitted ARIMA models

In all the examples that follow, the average levels and important parameters of the series have been kept relatively constant in order to help the reader see the differences between the models and the effects of the parameters on acf's and pacf's.

### 7.5.2 Example 7.1—AR(1) Process: $y_t = \phi_0 + \phi_1 y_{t-1} + \epsilon_t \quad (\phi_1 > 0)$

The data in Table 7.1 are a computer-generated sample from the AR(1) process

$$y_t = 10 + .90y_{t-1} + \epsilon_t \tag{7.8}$$

Figure 7.1 shows a plot of these data. The average value of the series is 104.38, which agrees reasonably well with the expected level of $y_t$ in Equation 7.8, i.e.,

**Table 7.1 Sample from the AR(1) Process** $y_t = 10 + .90y_{t-1} + \epsilon_t$

| $t$ | $y_t$ | $t$ | $y_t$ | $t$ | $y_t$ | $t$ | $y_t$ | $t$ | $y_t$ | $t$ | $y_t$ |
|---|---|---|---|---|---|---|---|---|---|---|---|
| 1 | 99 | 14 | 97 | 27 | 109 | 40 | 124 | 53 | 100 | 66 | 104 |
| 2 | 100 | 15 | 103 | 28 | 105 | 41 | 120 | 54 | 99 | 67 | 102 |
| 3 | 95 | 16 | 105 | 29 | 114 | 42 | 120 | 55 | 96 | 68 | 103 |
| 4 | 88 | 17 | 101 | 30 | 115 | 43 | 130 | 56 | 96 | 69 | 100 |
| 5 | 83 | 18 | 94 | 31 | 111 | 44 | 126 | 57 | 94 | 70 | 90 |
| 6 | 95 | 19 | 102 | 32 | 115 | 45 | 126 | 58 | 93 | 71 | 90 |
| 7 | 100 | 20 | 105 | 33 | 122 | 46 | 128 | 59 | 85 | 72 | 90 |
| 8 | 108 | 21 | 110 | 34 | 124 | 47 | 117 | 60 | 84 | 73 | 90 |
| 9 | 109 | 22 | 109 | 35 | 129 | 48 | 113 | 61 | 91 | 74 | 92 |
| 10 | 109 | 23 | 105 | 36 | 122 | 49 | 99 | 62 | 91 | 75 | 97 |
| 11 | 108 | 24 | 105 | 37 | 115 | 50 | 94 | 63 | 88 | 76 | 94 |
| 12 | 105 | 25 | 106 | 38 | 119 | 51 | 102 | 64 | 92 | 77 | 97 |
| 13 | 101 | 26 | 111 | 39 | 126 | 52 | 99 | 65 | 102 | | |

**Figure 7.1 Plot of the Data in Table 7.1**

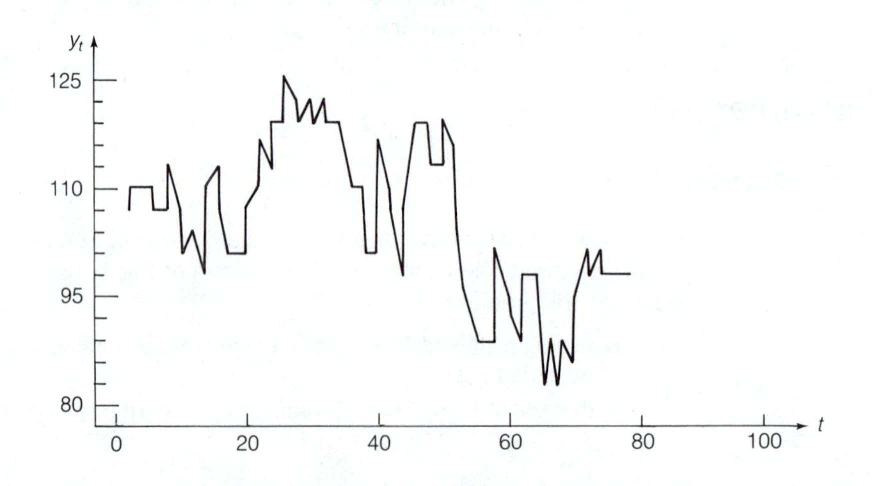

$$E(y_t) = \frac{\phi_0}{1 - \phi_1} = \frac{10}{1 - .90} = 100 \qquad \text{given by Equation 7.2.}$$

From Equation 7.1, the acf of an AR(1) series has the particularly simple form

$$\rho_k = \phi_1^k \qquad k = 1, 2, 3, \ldots \tag{7.9}$$

Thus, in this example the theoretical autocorrelation coefficients are $\rho_1 = .900$, $\rho_2 = (.90)^2 = .810$, $\rho_3 = (.90)^3 = .729$, $\rho_4 = (.90)^4 = .656$, $\rho_5 = (.90)^5 = .590$, and so on. Using a computer program, we have calculated the estimated coefficients shown in Figure 7.2(a): $r_1 = .904$, $r_2 = .782$, $r_3 = .694$, $r_4 = .609$, $r_5 = .528$, etc., which exhibit

**Figure 7.2** acf and pacf of the Data in Table 7.1

(a) acf

```
              -1.0  -.8  -.6  -.4  -.2   .0   .2   .4   .6   .8  1.0
                +----+----+----+----+----+----+----+----+----+----+
  1     .904                             XXXXXXXXXXXXXXXXXXXXXXXX
  2     .782                             XXXXXXXXXXXXXXXXXXXXX
  3     .694                             XXXXXXXXXXXXXXXXXX
  4     .609                             XXXXXXXXXXXXXXXX
  5     .528                             XXXXXXXXXXXXXX
  6     .459                             XXXXXXXXXXXX
  7     .415                             XXXXXXXXXXX
  8     .391                             XXXXXXXXXXX
  9     .373                             XXXXXXXXXX
 10     .346                             XXXXXXXXXX
 11     .286                             XXXXXXXX
 12     .230                             XXXXXXX
 13     .184                             XXXXXX
 14     .078                             XXX
 15    -.048                            XX
```

(b) pacf

```
              -1.0  -.8  -.6  -.4  -.2   .0   .2   .4   .6   .8  1.0
                +----+----+----+----+----+----+----+----+----+----+
  1     .904                             XXXXXXXXXXXXXXXXXXXXXXXX
  2    -.190                        XXXXXX
  3     .140                             XXXX
  4    -.095                           XXX
  5     .015                             X
  6    -.013                             X
  7     .102                             XXXX
  8     .048                             XX
  9     .025                             XX
 10    -.044                            XX
 11    -.192                        XXXXXX
 12     .044                             XX
 13    -.041                            XX
 14    -.361                    XXXXXXXXXX
 15    -.093                           XXX
```

the exponential decay typical of an AR(1) process. As a general rule, we are interested in the *way* in which the earlier (shorter-lag) $r_k$'s die out (exponentially or not). Coefficients at higher lags often are not statistically significant and are of less concern (this will be clarified later). The other identifying feature of an AR(1) process is that its pacf cuts off sharply after lag 1, as is indeed the case in Figure 7.2(b). ●

### 7.5.3 Example 7.2—AR(1) Process: $y_t = \phi_0 + \phi_1 y_{t-1} + \epsilon_t \quad (\phi_1 < 0)$

As in the previous example, the data (Table 7.2) are computer-generated from an AR(1) process. This sample was drawn from the process

$$y_t = 190 - .90y_{t-1} + e_t \tag{7.10}$$

For comparison with Example 7.1, the magnitude of $\phi_1$ was kept the same and the sign changed. In addition, $\phi_0 = 190$ was used so the resulting series would have the same expected level as that in Equation 7.8, i.e.,

$$E(y_t) = \frac{\phi_0}{1 - \phi_1} = \frac{190}{1 - (-.90)} = 100$$

**Table 7.2 Sample from the AR(1) Process** $y_t = 190 - .90y_{t-1} + \epsilon_t$

| $t$ | $y_t$ | $t$ | $y_t$ | $t$ | $y_t$ | $t$ | $y_t$ | $t$ | $y_t$ | $t$ | $y_t$ |
|---|---|---|---|---|---|---|---|---|---|---|---|
| 1 | 105 | 14 | 89 | 27 | 88 | 40 | 103 | 53 | 106 | 66 | 110 |
| 2 | 94 | 15 | 105 | 28 | 109 | 41 | 101 | 54 | 90 | 67 | 92 |
| 3 | 106 | 16 | 101 | 29 | 95 | 42 | 93 | 55 | 116 | 68 | 101 |
| 4 | 92 | 17 | 100 | 30 | 101 | 43 | 108 | 56 | 82 | 69 | 93 |
| 5 | 108 | 18 | 97 | 31 | 99 | 44 | 86 | 57 | 113 | 70 | 110 |
| 6 | 84 | 19 | 101 | 32 | 106 | 45 | 108 | 58 | 77 | 71 | 86 |
| 7 | 118 | 20 | 103 | 33 | 92 | 46 | 98 | 59 | 115 | 72 | 119 |
| 8 | 82 | 21 | 94 | 34 | 108 | 47 | 92 | 60 | 92 | 73 | 74 |
| 9 | 119 | 22 | 113 | 35 | 96 | 48 | 103 | 61 | 101 | 74 | 130 |
| 10 | 85 | 23 | 97 | 36 | 109 | 49 | 91 | 62 | 100 | 75 | 78 |
| 11 | 115 | 24 | 99 | 37 | 92 | 50 | 102 | 63 | 96 | 76 | 129 |
| 12 | 88 | 25 | 95 | 38 | 109 | 51 | 99 | 64 | 97 | 77 | 77 |
| 13 | 116 | 26 | 110 | 39 | 89 | 52 | 94 | 65 | 97 | | |

Since the theoretical acf must have the form given in Equation 7.9, we expect the first few coefficients to be $\rho_1 = -.90$, $\rho_2 = (-.90)^2 = .810$, $\rho_3 = (-.90)^3 = -.729$, $\rho_4 = (-.90)^4 = .656$, $\rho_5 = (-.90)^5 = -.590$, etc. These differ somewhat from the estimated coefficients $r_1 = -.881$, $r_2 = .743$, $r_3 = -.610$, $r_4 = .480$, $r_5 = -.353$ shown in Figure 7.3(a), but this much estimation error is not unusual for a series of length 77.

The pacf in Figure 7.3(b) cuts off after the first lag, which, along with the exponential decay seen in the acf, is fairly conclusive evidence of an AR(1) process. Furthermore, the plotted data (Figure 7.4) show the jagged features one expects in a series with negative first-order autocorrelation (see Section 2.2.2).

**Figure 7.3** acf and pacf of the Data in Table 7.2

(a) acf

```
             -1.0  -.8  -.6  -.4  -.2   .0   .2   .4   .6   .8  1.0
                +----+----+----+----+----+----+----+----+----+----+
 1   -.881         XXXXXXXXXXXXXXXXXXXXXXX
 2    .743                                XXXXXXXXXXXXXXXXXXXX
 3   -.610                XXXXXXXXXXXXXXXX
 4    .480                                XXXXXXXXXXXXX
 5   -.353                      XXXXXXXXXX
 6    .260                                XXXXXXXX
 7   -.160                           XXXXX
 8    .078                                XXX
 9   -.036                               XX
10    .048                                XX
```

(b) pacf

```
             -1.0  -.8  -.6  -.4  -.2   .0   .2   .4   .6   .8  1.0
                +----+----+----+----+----+----+----+----+----+----+
 1   -.881         XXXXXXXXXXXXXXXXXXXXXXX
 2   -.144                           XXXXX
 3    .051                                XX
 4   -.076                             XXX
 5    .072                                XXX
 6    .061                                XXX
 7    .134                                XXXX
 8    .011                                X
 9   -.093                             XXX
10    .167                                XXXXX
```

**Figure 7.4** Plot of the Data in Table 7.2

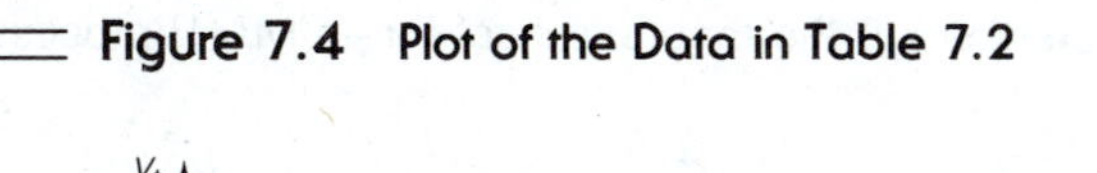

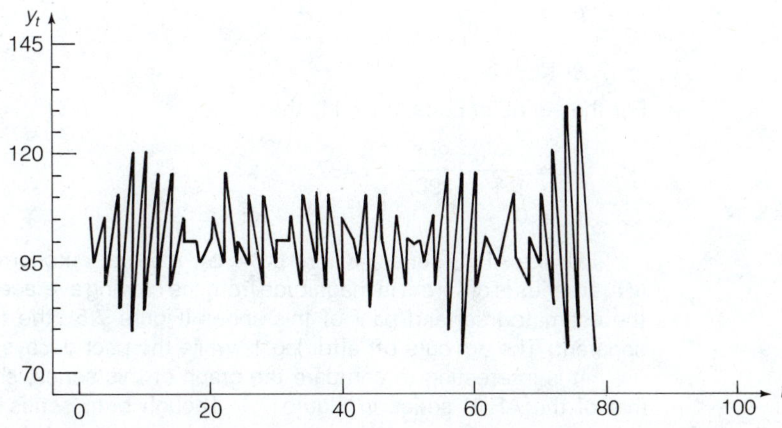

### 7.5.4 Example 7.3—MA(1) Process: $y_t = \theta_0 + \epsilon_t - \theta_1\epsilon_{t-1}$ $(\theta_1 < 0)$

Table 7.3 contains a series generated from the MA(1) process

$$y_t = 100 + \epsilon_t + .90\epsilon_{t-1} \tag{7.11}$$

Again, by appropriately choosing the constant term, the expected level of this series has been kept the same as in the previous series; i.e., $E(y_t) = \theta_0 = 100$. The magnitude of the process parameter was also kept the same as in the past, so we have $\theta_1 = -.90$ (due to the convention of using minus signs in the defining equations for MA models).

**Table 7.3** Sample from the MA(1) Process $y_t = 100 + \epsilon_t + .90\epsilon_{t-1}$

| $t$ | $y_t$ | $t$ | $y_t$ | $t$ | $y_t$ | $t$ | $y_t$ | $t$ | $y_t$ | $t$ | $y_t$ |
|---|---|---|---|---|---|---|---|---|---|---|---|
| 1 | 106 | 18 | 102 | 35 | 103 | 52 | 102 | 69 | 98 | 86 | 96 |
| 2 | 111 | 19 | 102 | 36 | 109 | 53 | 100 | 70 | 90 | 87 | 105 |
| 3 | 110 | 20 | 96 | 37 | 106 | 54 | 102 | 71 | 94 | 88 | 98 |
| 4 | 104 | 21 | 105 | 38 | 95 | 55 | 101 | 72 | 102 | 89 | 98 |
| 5 | 106 | 22 | 107 | 39 | 91 | 56 | 104 | 73 | 98 | 90 | 103 |
| 6 | 105 | 23 | 99 | 40 | 94 | 57 | 92 | 74 | 97 | 91 | 99 |
| 7 | 94 | 24 | 92 | 41 | 97 | 58 | 95 | 75 | 109 | 92 | 89 |
| 8 | 100 | 25 | 92 | 42 | 102 | 59 | 114 | 76 | 99 | 93 | 97 |
| 9 | 100 | 26 | 104 | 43 | 112 | 60 | 101 | 77 | 92 | 94 | 91 |
| 10 | 98 | 27 | 96 | 44 | 99 | 61 | 86 | 78 | 101 | 95 | 90 |
| 11 | 106 | 28 | 89 | 45 | 86 | 62 | 84 | 79 | 92 | 96 | 100 |
| 12 | 100 | 29 | 93 | 46 | 89 | 63 | 91 | 80 | 85 | 97 | 98 |
| 13 | 96 | 30 | 101 | 47 | 103 | 64 | 99 | 81 | 84 | 98 | 90 |
| 14 | 93 | 31 | 92 | 48 | 109 | 65 | 95 | 82 | 88 | | |
| 15 | 100 | 32 | 85 | 49 | 98 | 66 | 92 | 83 | 100 | | |
| 16 | 109 | 33 | 92 | 50 | 100 | 67 | 88 | 84 | 111 | | |
| 17 | 101 | 34 | 99 | 51 | 108 | 68 | 91 | 85 | 95 | | |

From Equation 7.3 the theoretical acf for an MA(1) process is given by

$$\rho_1 = \frac{-\theta_1}{1 + \theta_1^2} \qquad \rho_k = 0 \quad \text{for } k \geq 2 \tag{7.12}$$

For the series in Equation 7.11, then,

$$\rho_1 = \frac{-(-.90)}{1 + (-.90)^2} = .497$$

$$\rho_k = 0 \qquad \text{for } k \geq 2$$

Unlike AR(1) series, where $\rho_1 = \phi_1$, the first autocorrelation coefficient for an MA(1) series is different in magnitude from the moving average parameter $\theta_1$. Examining the estimated acf and pacf of this series (Figure 7.5), the typical MA(1) behavior is apparent: The acf cuts off after lag 1, while the pacf decays exponentially.

It is interesting to compare the graph of this series, shown in Figure 7.6, with that of the AR(1) series in Figure 7.1. Though both series have positive first-order

**Figure 7.5** acf and pacf of the Data in Table 7.3

(a) acf

```
           -1.0  -.8  -.6  -.4  -.2   .0   .2   .4   .6   .8  1.0
                 +----+----+----+----+----+----+----+----+----+----+
  1    .408                             XXXXXXXXXXX
  2   -.093                           XXX
  3    .035                             XX
  4    .072                             XXX
  5    .063                             XXX
  6    .070                             XXX
  7    .009                             X
  8   -.032                            XX
  9   -.060                            XX
 10   -.106                          XXXX
```

(b) pacf

```
           -1.0  -.8  -.6  -.4  -.2   .0   .2   .4   .6   .8  1.0
                 +----+----+----+----+----+----+----+----+----+----+
  1    .408                             XXXXXXXXXXX
  2   -.312                     XXXXXXXXX
  3    .282                             XXXXXXXX
  4   -.156                         XXXXX
  5    .196                             XXXXXX
  6   -.083                           XXX
  7    .047                             XX
  8   -.064                           XXX
  9   -.044                            XX
 10   -.093                           XXX
```

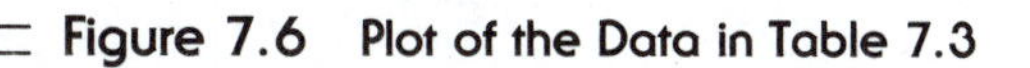

**Figure 7.6** Plot of the Data in Table 7.3

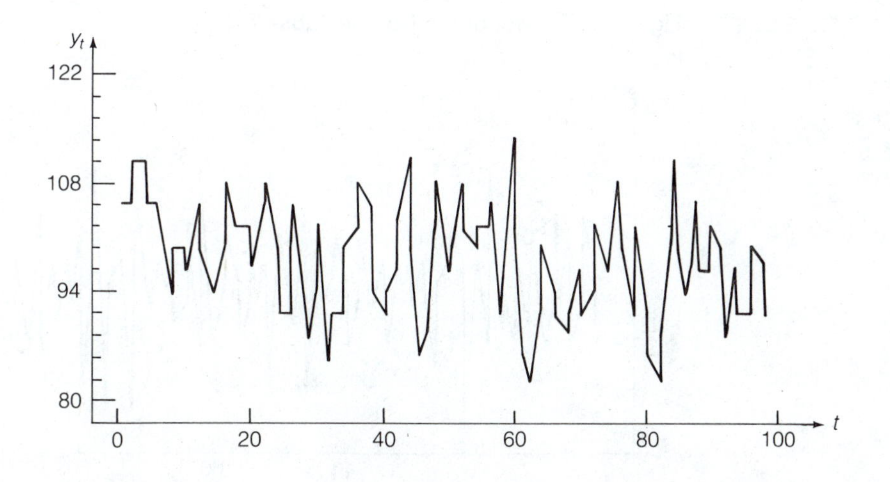

autocorrelation, the plot in Figure 7.1 has much longer swings above and below its mean level than does the plot in Figure 7.6 because the first set of data also has positive autocorrelation at *lags beyond* lag 1, while the second set does not. •

### 7.5.5 Example 7.4—MA(1) Process: $y_t = \theta_0 + \epsilon_t - \theta_1\epsilon_{t-1}$ $(\theta_1 > 0)$

The data in Table 7.4 are a sample from the MA(1) process

$$y_t = 100 + \epsilon_t - .90\epsilon_{t-1} \tag{7.13}$$

The magnitude of the parameter and the average level of the series have been kept the same as in the previous examples; i.e., $E(y_t) = \theta_0 = 100$ and $\theta_1 = .90$.

The jagged appearance of the plot in Figure 7.7 is characteristic of a series in

**Table 7.4** Sample from the MA(1) Process $y_t = 100 + \epsilon_t - .90\epsilon_{t-1}$

| $t$ | $y_t$ | $t$ | $y_t$ | $t$ | $y_t$ | $t$ | $y_t$ | $t$ | $y_t$ | $t$ | $y_t$ |
|---|---|---|---|---|---|---|---|---|---|---|---|
| 1 | 96 | 18 | 93 | 35 | 99 | 52 | 105 | 69 | 98 | 86 | 110 |
| 2 | 102 | 19 | 103 | 36 | 97 | 53 | 99 | 70 | 100 | 87 | 93 |
| 3 | 109 | 20 | 103 | 37 | 108 | 54 | 103 | 71 | 113 | 88 | 95 |
| 4 | 99 | 21 | 83 | 38 | 98 | 55 | 92 | 72 | 84 | 89 | 105 |
| 5 | 91 | 22 | 104 | 39 | 92 | 56 | 100 | 73 | 104 | 90 | 99 |
| 6 | 106 | 23 | 107 | 40 | 100 | 57 | 108 | 74 | 105 | 91 | 98 |
| 7 | 97 | 24 | 105 | 41 | 106 | 58 | 100 | 75 | 102 | 92 | 107 |
| 8 | 109 | 25 | 101 | 42 | 102 | 59 | 94 | 76 | 82 | 93 | 99 |
| 9 | 93 | 26 | 89 | 43 | 97 | 60 | 93 | 77 | 115 | 94 | 94 |
| 10 | 94 | 27 | 99 | 44 | 108 | 61 | 116 | 78 | 104 | 95 | 105 |
| 11 | 107 | 28 | 105 | 45 | 86 | 62 | 92 | 79 | 85 | 96 | 104 |
| 12 | 96 | 29 | 97 | 46 | 111 | 63 | 105 | 80 | 108 | 97 | 95 |
| 13 | 105 | 30 | 103 | 47 | 94 | 64 | 99 | 81 | 98 | 98 | 95 |
| 14 | 93 | 31 | 102 | 48 | 99 | 65 | 96 | 82 | 103 | | |
| 15 | 97 | 32 | 103 | 49 | 108 | 66 | 97 | 83 | 96 | | |
| 16 | 96 | 33 | 92 | 50 | 106 | 67 | 102 | 84 | 103 | | |
| 17 | 115 | 34 | 103 | 51 | 90 | 68 | 103 | 85 | 93 | | |

**Figure 7.7** Plot of the Data in Table 7.4

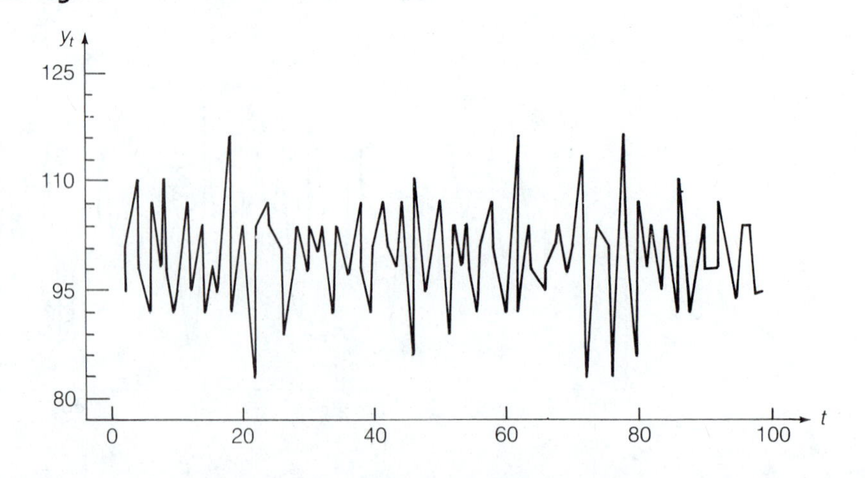

**Figure 7.8** acf and pacf of the Data in Table 7.4

(a) acf

```
         -1.0  -.8  -.6  -.4  -.2   .0   .2   .4   .6   .8  1.0
           +----+----+----+----+----+----+----+----+----+----+
  1  -.442               XXXXXXXXXXXX
  2  -.156                     XXXXX
  3   .087                         XXX
  4   .057                         XX
  5  -.123                      XXXX
  6   .103                         XXXX
  7   .070                         XXX
  8  -.143                     XXXXX
  9   .107                         XXXX
 10  -.104                      XXXX
```

(b) pacf

```
         -1.0  -.8  -.6  -.4  -.2   .0   .2   .4   .6   .8  1.0
           +----+----+----+----+----+----+----+----+----+----+
  1  -.442               XXXXXXXXXXXX
  2  -.437               XXXXXXXXXXXX
  3  -.316                  XXXXXXXXX
  4  -.196                     XXXXXX
  5  -.300                   XXXXXXXX
  6  -.191                     XXXXXX
  7  -.022                        XX
  8  -.084                       XXX
  9   .105                         XXXX
 10  -.045                        XX
```

which negative autocorrelation is the dominant feature. The acf and pacf plots in Figure 7.8 reinforce this conclusion. From these graphs it is easy to identify an MA(1) series, since the pacf dies out exponentially and the acf cuts off after lag 1. For further comparison, note that the estimated acf in Figure 7.8 is reasonably close to the expected acf given by Equation 7.12; i.e.,

$$\rho_1 = \frac{-\theta_1}{1+\theta_1^2} = \frac{-.90}{1+(.90)^2} = -.497$$

$$\rho_k = 0 \qquad \text{for } k \geq 2$$

$$r_1 = -.442$$

($r_k$ is small for $k \geq 2$.) •

### 7.5.6 Example 7.5—ARIMA(0,1,1) Process: $y_t - y_{t-1} = \theta_0 + \epsilon_t - \theta_1\epsilon_{t-1}$

In this example we look at two ARIMA(0,1,1) series: one with *stochastic* trend ($\theta_0 = 0$), the other with *deterministic* trend ($\theta_0 \neq 0$). (See Section 7.3.2.)

**Table 7.5 Sample from the ARIMA(0,1,1) Process** $y_t - y_{t-1} = \epsilon_t + .90\epsilon_{t-1}$

| $t$ | $y_t$ | $t$ | $y_t$ | $t$ | $y_t$ | $t$ | $y_t$ | $t$ | $y_t$ | $t$ | $y_t$ |
|---|---|---|---|---|---|---|---|---|---|---|---|
| 1 | 173 | 18 | 107 | 35 | 105 | 52 | 74 | 69 | 100 | 86 | 83 |
| 2 | 176 | 19 | 110 | 36 | 113 | 53 | 68 | 70 | 85 | 87 | 78 |
| 3 | 174 | 20 | 111 | 37 | 115 | 54 | 58 | 71 | 79 | 88 | 83 |
| 4 | 171 | 21 | 110 | 38 | 110 | 55 | 61 | 72 | 84 | 89 | 89 |
| 5 | 170 | 22 | 109 | 39 | 103 | 56 | 76 | 73 | 88 | 90 | 85 |
| 6 | 176 | 23 | 109 | 40 | 98 | 57 | 76 | 74 | 87 | 91 | 80 |
| 7 | 180 | 24 | 108 | 41 | 95 | 58 | 77 | 75 | 86 | 92 | 77 |
| 8 | 179 | 25 | 104 | 42 | 85 | 59 | 83 | 76 | 95 | 93 | 70 |
| 9 | 177 | 26 | 104 | 43 | 87 | 60 | 83 | 77 | 111 | 94 | 71 |
| 10 | 163 | 27 | 107 | 44 | 80 | 61 | 85 | 78 | 121 | 95 | 74 |
| 11 | 152 | 28 | 106 | 45 | 62 | 62 | 84 | 79 | 118 | 96 | 73 |
| 12 | 145 | 29 | 105 | 46 | 62 | 63 | 81 | 80 | 107 | 97 | 69 |
| 13 | 128 | 30 | 105 | 47 | 70 | 64 | 86 | 81 | 105 | 98 | 61 |
| 14 | 110 | 31 | 107 | 48 | 71 | 65 | 99 | 82 | 108 | 99 | 61 |
| 15 | 105 | 32 | 108 | 49 | 63 | 66 | 108 | 83 | 101 | | |
| 16 | 99 | 33 | 100 | 50 | 59 | 67 | 108 | 84 | 90 | | |
| 17 | 100 | 34 | 98 | 51 | 69 | 68 | 107 | 85 | 87 | | |

Table 7.5 contains data sampled from the process

$$y_t - y_{t-1} = \epsilon_t + .90\epsilon_{t-1} \tag{7.14}$$

which can also be written $y_t = y_{t-1} + \epsilon_t + .90\epsilon_{t-1}$

Since $\theta_0 = 0$, $\theta_1 = -.90$, and $d = 1$ in this model, the series should exhibit a stochastic or meandering trend, as indeed it does in Figure 7.9. The acf (Figure 7.10) shows strong positive autocorrelation or, perhaps, trend, and does not drop off rapidly to zero. Thus differencing is necessary to achieve stationarity. After taking the first differences of $y_t$, we see that the new acf and pacf (Figure 7.11) are typical of an MA(1) process.

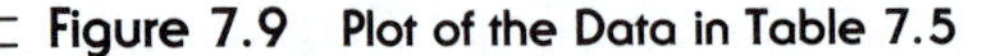

**Figure 7.9 Plot of the Data in Table 7.5**

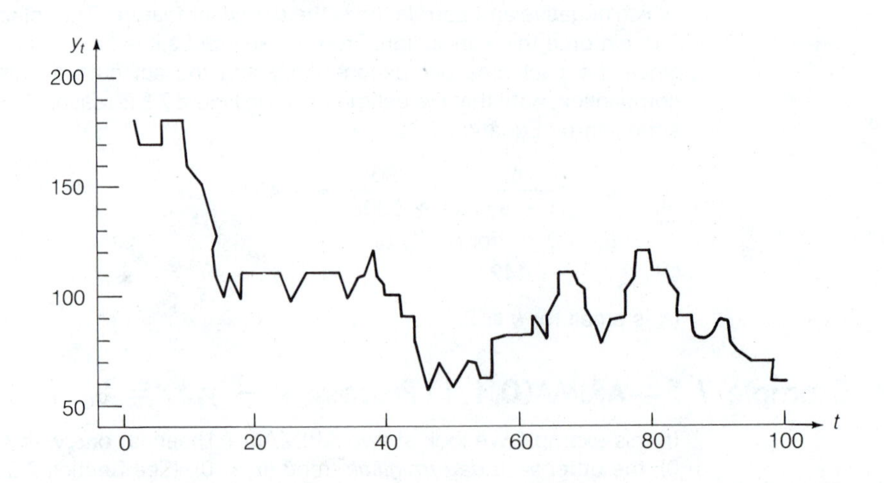

Figure 7.10 acf of the Data in Table 7.5

```
          -1.0  -.8  -.6  -.4  -.2   .0   .2   .4   .6   .8  1.0
             +----+----+----+----+----+----+----+----+----+----+
  1    .939                           XXXXXXXXXXXXXXXXXXXXXXXX
  2    .854                           XXXXXXXXXXXXXXXXXXXXXX
  3    .776                           XXXXXXXXXXXXXXXXXXXX
  4    .702                           XXXXXXXXXXXXXXXXXXX
  5    .624                           XXXXXXXXXXXXXXXXX
  6    .538                           XXXXXXXXXXXXXX
  7    .454                           XXXXXXXXXXXX
  8    .383                           XXXXXXXXXXX
  9    .318                           XXXXXXXXX
 10    .265                           XXXXXXXX
```

Figure 7.11 acf and pacf of the First Differences of the Data in Table 7.5

(a) acf

```
          -1.0  -.8  -.6  -.4  -.2   .0   .2   .4   .6   .8  1.0
             +----+----+----+----+----+----+----+----+----+----+
  1    .457                           XXXXXXXXXXXX
  2   -.085                         XXX
  3   -.021                          XX
  4    .080                           XXX
  5    .017                           X
  6   -.157                       XXXXX
  7   -.220                     XXXXXXX
  8   -.111                        XXXX
  9    .018                           X
 10    .023                           XX
```

(b) pacf

```
          -1.0  -.8  -.6  -.4  -.2   .0   .2   .4   .6   .8  1.0
             +----+----+----+----+----+----+----+----+----+----+
  1    .457                           XXXXXXXXXXXX
  2   -.371                  XXXXXXXXXX
  3    .296                           XXXXXXXX
  4   -.140                        XXXX
  5    .065                           XXX
  6   -.255                     XXXXXXX
  7    .010                           X
  8   -.083                         XXX
  9    .101                           XXXX
 10   -.071                         XXX
```

To show the effect of the trend parameter, consider the data in Table 7.6. These data differ from Table 7.5 only in that a trend parameter of $\theta_0 = -3$ was incorporated into the model (the same $\epsilon_t$'s were used). The model equation then became

$$y_t = y_{t-1} - 3 + \epsilon_t + .90\epsilon_{t-1} \tag{7.15}$$

The nonzero value for $\theta_0$ results in a pronounced linear trend (Figure 7.12). The first differences of this series have an acf and a pacf identical to those shown in Figure 7.11 because the same $\epsilon_t$'s were used to generate both series.

**Table 7.6 Data from Table 7.5 with a Trend Parameter of $\theta_0 = -3$ Included**

| $t$ | $y_t$ | $t$ | $y_t$ | $t$ | $y_t$ | $t$ | $y_t$ | $t$ | $y_t$ | $t$ | $y_t$ |
|---|---|---|---|---|---|---|---|---|---|---|---|
| 1 | 170 | 18 | 53 | 35 | 0 | 52 | −82 | 69 | −107 | 86 | −175 |
| 2 | 170 | 19 | 53 | 36 | 5 | 53 | −91 | 70 | −125 | 87 | −183 |
| 3 | 165 | 20 | 51 | 37 | 4 | 54 | −104 | 71 | −134 | 88 | −181 |
| 4 | 159 | 21 | 47 | 38 | −4 | 55 | −104 | 72 | −132 | 89 | −178 |
| 5 | 155 | 22 | 43 | 39 | −14 | 56 | −92 | 73 | −131 | 90 | −185 |
| 6 | 158 | 23 | 40 | 40 | −22 | 57 | −95 | 74 | −135 | 91 | −193 |
| 7 | 159 | 24 | 36 | 41 | −28 | 58 | −97 | 75 | −139 | 92 | −199 |
| 8 | 155 | 25 | 29 | 42 | −41 | 59 | −94 | 76 | −133 | 93 | −209 |
| 9 | 150 | 26 | 26 | 43 | −42 | 60 | −97 | 77 | −120 | 94 | −211 |
| 10 | 133 | 27 | 26 | 44 | −52 | 61 | −98 | 78 | −113 | 95 | −211 |
| 11 | 119 | 28 | 22 | 45 | −73 | 62 | −102 | 79 | −119 | 96 | −215 |
| 12 | 109 | 29 | 18 | 46 | −76 | 63 | −108 | 80 | −133 | 97 | −222 |
| 13 | 89 | 30 | 15 | 47 | −71 | 64 | −106 | 81 | −138 | 98 | −233 |
| 14 | 68 | 31 | 14 | 48 | −73 | 65 | −96 | 82 | −138 | 99 | −236 |
| 15 | 60 | 32 | 12 | 49 | −84 | 66 | −90 | 83 | −148 | | |
| 16 | 51 | 33 | 1 | 50 | −91 | 67 | −93 | 84 | −162 | | |
| 17 | 49 | 34 | −4 | 51 | −84 | 68 | −97 | 85 | −168 | | |

**Figure 7.12 Plot of the Data in Table 7.6**

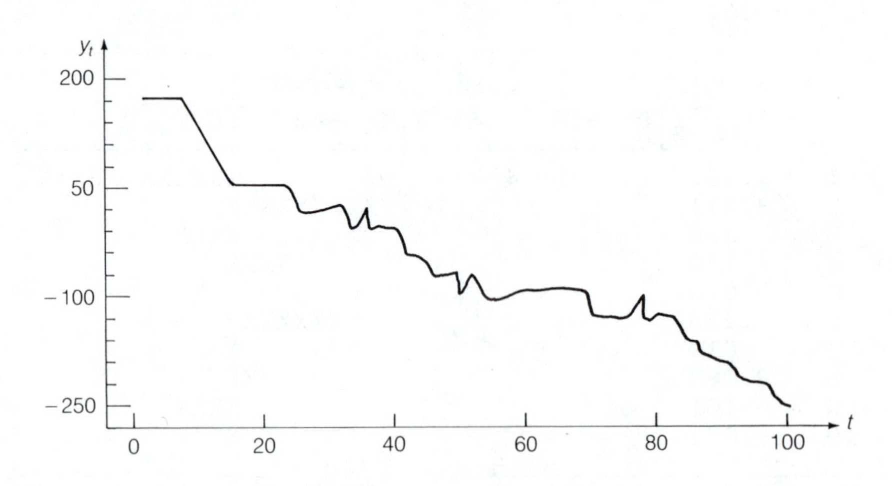

The usual tools for identifying Box-Jenkins models (differencing, acf, and pacf) would lead us to conclude that the series in Figures 7.9 and 7.12 were both of the same ARIMA(0,1,1) type. On the basis of these indicators alone, we cannot distinguish series with stochastic trends from those with deterministic trends. That determination must be based on other criteria. •

### 7.5.7 Example 7.6—Forecasting with ARIMA Models

A complete set of procedures for generating forecasts from fitted ARIMA models will be given later. For the moment, a few of the required techniques will be illustrated by generating forecasts for an ARIMA(0,1,1) model fit to the data of Table 7.5.

The first step is to take the general form of the model (which in this case has no trend parameter):

$$y_t - y_{t-1} = \epsilon_t - \theta_1\epsilon_{t-1}$$

and undifference it by writing

$$y_t = y_{t-1} + \epsilon_t - \theta_1\epsilon_{t-1} \qquad (7.16)$$

Then, $t + 1, t + 2, t + 3, \ldots$ are substituted for $t$ to yield

$$y_{t+1} = y_t + \epsilon_{t+1} - \theta_1\epsilon_t$$

$$y_{t+2} = y_{t+1} + \epsilon_{t+2} - \theta_1\epsilon_{t+1}$$

$$y_{t+3} = y_{t+2} + \epsilon_{t+3} - \theta_1\epsilon_{t+2}$$

and so on.

Next, a computer package is used to estimate $\theta_1$. Forecasts may be obtained from these equations by substituting *known* values in the right-hand sides whenever possible and using *expected* or *estimated/forecast* values otherwise. At time $t$, for example, $y_t$ would be known and $\epsilon_t$ could be estimated using the residual at time $t$ from the regression program, i.e., $\hat{\epsilon}_t = e_t = y_t - \hat{y}_t(t-1)$, while all future values ($y_{t+1}, y_{t+2}, \ldots$ and $\epsilon_{t+1}, \epsilon_{t+2}, \ldots$) would be as yet unknown. So forecast values would be substituted for the $y$'s and zeros for the $\epsilon$'s. Thus, the forecasting equations are

$$\hat{y}_{t+1}(t) = y_t + \hat{\theta}_1\epsilon_t$$

$$\hat{y}_{t+2}(t) = \hat{y}_{t+1}(t) + 0 + \hat{\theta}_1 \cdot 0 = \hat{y}_{t+1}(t)$$

$$\hat{y}_{t+3}(t) = \hat{y}_{t+2}(t) + 0 + \theta_1 \cdot 0 = \hat{y}_{t+2}(t) = \hat{y}_{t+1}(t)$$

$$\vdots \qquad\qquad \vdots$$

So, for this particular model the forecasts will be constant for all periods beyond time $t$.

Using the Minitab package with the data from Table 7.5, $\hat{\theta}_1$ is $-.9098$ and the residual at $t = 99$ is $e_{99} = 3.62$. The forecast for period 100 is, then,

$$\hat{y}_{100}(99) = y_{99} + .9098e_{99} = 61 + .9098(3.62) = 64.29$$

which also will be the forecast for periods 101, 102, and so on. •

## 7.6 The Decision to Use Box-Jenkins Methods

### 7.6.1 Introduction

When deciding whether the Box-Jenkins methodology is appropriate, the complexity of the technique plays a major role. As an overview of the Box-Jenkins method, we can say the following.

- It requires greater knowledge of special techniques/tools than other methods (e.g., acf, pacf, invertibility, nonlinear least squares, ARIMA models, Yule-Walker equations).
- In practice, computer programs are required for model identification and for the estimation of model parameters.
- It is very demanding of data (i.e., larger sample sizes are required).
- The additional complexity of the method does not necessarily guarantee better forecasts than those of simpler techniques (e.g., exponential smoothing).
- It works best when forecasting in the short term (i.e., only a few periods ahead).

Reading this list naturally evokes a question about whether the additional complexity and effort is really warranted, especially in light of recent studies indicating that ARIMA forecasting does not always outperform simpler forecasting methods. There is considerable support for Box-Jenkins procedures, though, because:

- ARIMA models can be shown to be *optimal,* since for a given model no other forecasting method can, on average, give forecasts with smaller MSE's (Pankratz, 1983, p. 258).
- Box-Jenkins analysis provides a *systematic* approach to model selection, utilizing all the information contained in the sample autocorrelation and partial autocorrelation functions.

However, the forecasts will be optimal only if the series is actually generated by an ARIMA model. In addition, the model must be correctly identified and its parameters accurately estimated. These are the critical issues that will determine whether the Box-Jenkins method is successful. In what follows we expand upon the requirements, strengths,and drawbacks of Box-Jenkins forecasting.

### 7.6.2 Data Requirements

Most forecasting methods in this book implicitly assume that data are collected at equally spaced time intervals (see Chapter 1), otherwise model parameters would have to change for each observation period. ARIMA models are no exception. Moreover, the Box-Jenkins identification stage involves the acf and the pacf, whose interpretation depends upon the use of equally spaced intervals.

As to the *minimum* amount of data needed, Box and Jenkins themselves originally recommended *at least 50 observations.* Many practitioners now recommend using more. Two important reasons for larger sample sizes are:

- With small data sets it is harder to see statistically which autocorrelation or partial autocorrelation coefficients are significant, making it difficult to identify the correct model.
- Larger data sets result in more accurate parameter estimates and, hence, more precise forecasts.

Because of the large amount of data needed for accurate identification and estimation, the Box-Jenkins method has become known as one of the large-sample procedures.

## 7.6.3 Lead-Time Considerations

As a result of seeking parsimony, many ARIMA models arising in practice have restrictions on their forecasting horizons. That is, the number of periods ahead, $j$, for which useful forecasts can be obtained is limited by the order of the ARIMA model used.

To demonstrate, suppose an MA(3) model is fit to a particular series, resulting in the estimated model

$$y_t = 100 + \epsilon_t - .4\epsilon_{t-1} - .5\epsilon_{t-2} - .3\epsilon_{t-3} \qquad \textbf{(7.17)}$$

Using the same procedure as in Example 7.6 (estimating *past* residuals and setting *future* residuals to zero), the forecasting equations would be

$$\hat{y}_{t+1}(t) = 100 + 0 - .4e_t - .5e_{t-1} - .3e_{t-2} = 100 - .4e_t - .5e_{t-1} - .3e_{t-2}$$

$$\hat{y}_{t+2}(t) = 100 + 0 - 0 - .5e_t - .3e_{t-1} = 100 - .5e_t - .3e_{t-1}$$

$$\hat{y}_{t+3}(t) = 100 + 0 - 0 - 0 - .3e_t = 100 - .3e_t$$

$$\hat{y}_{t+j}(t) = 100 + 0 - 0 - 0 - 0 = 100 \qquad \text{for } j \geq 4$$

This MA(3) model would then use the estimated residuals $e_t$, $e_{t-1}$, and $e_{t-2}$ from a computer program to produce forecasts for up to three periods ahead, while all forecasts more than three periods ahead would be constant. For this reason Box-Jenkins forecasts are best suited to short-term forecasting.

## 7.6.4 Updating Parameter Estimates

The estimation stage in Box-Jenkins analysis is similar to that in linear regression. Both methods employ the least squares criterion and neither is amenable to conveniently updating parameter estimates as new data become available. Thus, using the Box-Jenkins method necessitates periodically incorporating new data by repeating the identification and estimation stages.

There is a way of using new data to update existing forecasts. To use this method, though, you must assume that the model parameters don't need updating so that no reestimation is needed. Updating equations can then be written that allow for revised forecasts as new data become known. This practice (discussed in more detail in Section 7.7) represents a compromise between doing no updating of any kind and doing a complete model refit for each new observation.

### 7.6.5 Cost Factors

As a rule, Box-Jenkins forecasting is more costly than other techniques because of the added complexity of ARIMA modeling. When deciding whether to use the Box-Jenkins method, items whose costs should be considered include the following.

- **computer software** Computer programs are essential to the identification, estimation, and forecasting stages. Such programs must be purchased or rented, and some training in their operation is required.
- **a trained analyst** Someone trained in analyzing acf's, pacf's, ARIMA models, and the stages of the Box-Jenkins process is needed.
- **data base** Estimation of ARIMA-model parameters requires that reasonable amounts of data (a minimum of 50 observations per series) be stored. These data cannot be discarded, because of the need for periodic reestimation of the parameters.
- **scope of the forecasting project** Costs associated with the analyst's time and maintaining the data base increase in direct proportion to the number of variables being forecast.

### 7.6.6 Comparison with Other Methods

Box-Jenkins analysis is attractive to practitioners and academicians alike because in forecasting, a field where ad hoc procedures and techniques abound, Box and Jenkins have consolidated new and existing techniques into a single, systematic method. Furthermore, the class of ARIMA models is large enough to include many simpler forecasting models (e.g., exponential smoothing) as special cases. Box-Jenkins analysis, then, represents a unified, theoretically sound approach to forecasting any series.

On the other hand, recent research indicates that there is probably no single method best for all time series arising in practice (Makridakis et al., 1984). There are series, for example, where simple smoothing methods produce smaller average forecast errors than Box-Jenkins forecasts. The conclusion to be drawn is that the choice of a forecasting method, Box-Jenkins or otherwise, always depends on the characteristics of the series being forecast.

That ARIMA forecasts have been shown to be optimal (i.e., minimum MSE) has been used by many to justify the use of Box-Jenkins methods to the exclusion of simpler ones. However, let us reiterate that this optimality exists in practice only when three conditions hold:

1. The given series must be generated by a particular ARIMA model (remember, there are time series models not in this class).
2. The quality and amount of data available must be sufficient for accurate model identification.

3. The amount of data must be sufficient for precise estimation.

If any of these three requirements fails to hold, the resulting forecasts may not be optimal. This makes it easier to understand why simpler methods can sometimes outperform ARIMA forecasts and why such occurrences in no way contradict the theoretical soundness of Box-Jenkins analysis.

Finally, it should be kept in mind that many ARIMA models can be handled without specifically using Box-Jenkins methods. For example, the parameters of an AR($p$) model can be estimated using regression analysis (which is, perhaps, more widely understood). Similarly, exponential smoothing can always be used to forecast an ARIMA(0,1,1) series.

## 7.7 Box-Jenkins Models

### 7.7.1 Introduction

The four Box-Jenkins modeling phases—*identification, estimation, diagnostic checking,* and *forecasting*—are, in actuality, basic to all sound forecasting practice. Because of the complexity of Box-Jenkins modeling, however, it is better (and easier) to delineate the stages and undertake them one at a time.

### 7.7.2 Identification

There are two steps to the identification of an appropriate ARIMA model:

1. If necessary, the series is differenced to make it stationary.
2. The acf and the pacf of the resulting stationary series are analyzed to determine the orders of the autoregressive and moving average parts of the model.

#### Differencing

In the Box-Jenkins method the acf and the pacf are meaningful only when applied to stationary series. If the acf of the original data either cuts off or tails off rapidly to zero, that is good evidence (see Section 2.4.4) that the series is already stationary and we can proceed directly to the analysis of the acf and pacf. Otherwise, some detrending is necessary. Detrending is done by differencing the series repeatedly until the acf dies out rapidly. Since differencing increases the variation in the (differenced) series, the number of repetitions should be as small as possible. In practice, stationarity can usually be achieved within the first or second differencings of the original data (Box and Jenkins, 1970).

#### Testing the acf

The "rule-of-thumb" test discussed in Section 2.6.5 is adequate in many applications where we are looking for signals of *some* autocorrelation without necessarily being concerned as to the particular $\rho_k$'s that are nonzero. In Box-

Jenkins analysis, however, it is important to know exactly which coefficients are nonzero, since this will be instrumental in determining the order of the moving average part of the ARIMA model. Bartlett (1946) developed a test that is more appropriate than the "rule-of-thumb" when working with a moving average process. The test is based on the fact that, for a moving average process of order $q$, the standard deviation of the sample autocorrelation coefficient $r_k$ is approximately

$$\sigma_{r_k} \simeq \frac{1}{\sqrt{n}}\left(1 + 2 \cdot \sum_{1}^{q} r_i^2\right)^{1/2} \qquad \text{for } k > q \tag{7.18}$$

Notice that this is just the standard deviation for the "rule-of-thumb," with a correction that represents Bartlett's refinement. Because for large sample sizes ($n$) the $r_k$'s are also approximately normal, we can test $H_0$: $\rho_k = 0$ (at approximately the 5% level) by rejecting $H_0$ whenever $|r_k|$ exceeds $2\sigma_{r_k}$.

---

**Test Procedure:** ***Bartlett's Test for Autocorrelation***

$H_0$: $\rho_k = 0$

$H_a$: $\rho_k \neq 0$

**Test Statistic:** $r_k$ = sample autocorrelation coefficient of lag $k$

**Decision Rule:** ($\alpha \simeq 5\%$, $n$ large)

Reject $H_0$ if $|r_k| > \frac{2}{\sqrt{n}}\left(1 + 2 \cdot \sum_{1}^{k-1} r_i^2\right)^{1/2}$

**Conclusion:** If $H_0$ is rejected, we conclude with approximately 95% confidence that the autocorrelation at lag $k$ is nonzero.

If $H_0$ is not rejected, we have some support for the lag $k$ autocorrelation being zero.

---

With Bartlett's test, the critical values for rejecting $H_0$ increase with $k$, since the summation $\Sigma_1^{k-1} r_i^2$ must increase with $k$. Plotting the successive rejection regions on the acf leads to wider and wider rejection thresholds (many computer programs for Box-Jenkins analysis include these rejection lines on the acf plot). For example, suppose a series of 64 observations has the following sample autocorrelation coefficients:

| $k$ | 1 | 2 | 3 | 4 | 5 | 6 | 7 |
|---|---|---|---|---|---|---|---|
| $r_k$ | .91 | .79 | .15 | .35 | −.12 | .20 | −.09 |

Using Bartlett's test, the successive critical values (CV's) are

*k* = *1:* $$CV = \frac{2}{\sqrt{64}} = .25$$

*k* = *2:* $$CV = \frac{2}{\sqrt{64}}[1 + 2(r_1^2)]^{1/2} = .25[1 + 2(.91)^2]^{1/2} = .407$$

*k* = *3:* $$CV = \frac{2}{\sqrt{64}}[1 + 2(r_1^2 + r_2^2)]^{1/2}$$
$$= .25[1 + 2(.91)^2 + 2(.79)^2]^{1/2} = .494$$

*k* = *4:* $$CV = \frac{2}{\sqrt{64}}[1 + 2(r_1^2 + r_2^2 + r_3^2)]^{1/2}$$
$$= .25[1 + 2(.91)^2 + 2(.79)^2 + 2(.15)^2]^{1/2}$$
$$= .499$$

and so on.

The acf and the critical values for both Bartlett's test and the "rule-of-thumb" are shown in Figure 7.13. With the "rule-of-thumb," $r_4$ appears significant, while with the more precise critical values from Bartlett's test it does not. Ignoring $r_4$, then, we conclude that the acf in Figure 7.13 is characteristic of an MA(2) series.

**Figure 7.13** Bartlett's Test Versus the "Rule-of-Thumb" Test

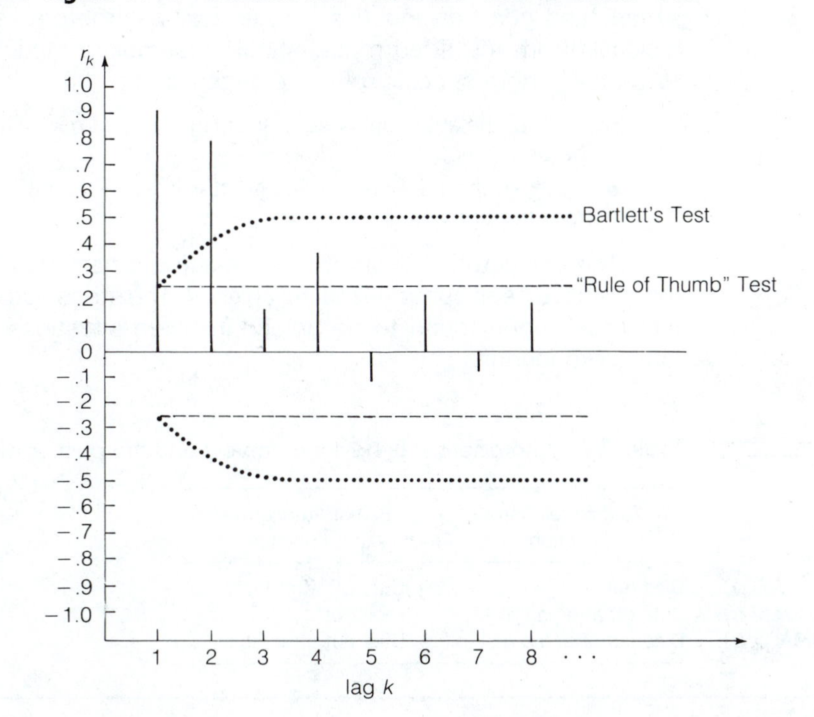

### Testing the pacf

The partial autocorrelation function is used to determine the order $p$ of an ARIMA model's autoregressive component. The notation for the partial autocorrelation, $\phi_{kk}$, and the procedure (simply the "rule-of-thumb") for testing $H_0$: $\phi_{kk} = 0$ are the same as in the original discussion of these concepts in Section 4.5.7. The double subscript in $\phi_{kk}$, which may have seemed superfluous before, now becomes a necessary device for distinguishing between these coefficients and the autoregressive parameters $\phi_1, \phi_2, \ldots, \phi_k$. We also note that no refinement like Bartlett's test for the acf has been developed for the pacf.

### Determining $p$ and $q$

The theoretical behavior of the acf's and pacf's of pure autoregressive or pure moving average models was described in Section 7.2. To repeat briefly, a stationary AR($p$) process gives rise to an acf that dies out exponentially to zero, while its pacf cuts off sharply after lag $p$. Stationary MA($q$) processes have pacf's that die out exponentially and acf's that cut off after lag $q$.

ARMA models have autoregressive *and* moving average characteristics, so both the acf and pacf drop off to zero. With such models it may also take a lag or two before the pattern of decay emerges. Trying to see cutoffs and decays is difficult, but the added complication of trying to determine at which lags these patterns start makes the identification of mixed models the most demanding part of Box-Jenkins analysis.

When identifying mixed models, the guiding principles should be *parsimony* and *good diagnostics* (i.e., as few parameters as possible, and the residuals from the fitted model should resemble a random no-trend series). Two other, more specific rules are also helpful:

- If $q \geq p$, the acf should decay after lag $q - p$; if $q < p$, it should drop off from the beginning.
- If $p \geq q$, the pacf should decay after lag $p - q$; otherwise it should decay from the beginning.

The characteristic patterns in the acf and pacf for AR, MA, and ARMA models have been summarized in Table 7.7. The steps outlined in the following table have been found to be helpful in those situations when $p$ and $q$ are difficult to identify.

**Table 7.7 Characteristic Behavior of the acf and the pacf of ARIMA Models**

| | Autocorrelation Function | Partial Autocorrelation Function |
|---|---|---|
| **AR($p$):** | Dies out | Cuts off after lag $p$ |
| **MA($q$):** | Cuts off after lag $q$ | Dies out |
| **ARIMA($p,q$):** | Dies out after lag $q - p$ | Dies out after lag $p - q$ |

**EXAMPLE 7.7** *Identifying a Mixed Model, Printer's, Inc.*

Printer's, Inc., is a medium-size retailer of computer printers and other small computer peripherals. Besides selling name brands, Printer's, Inc., makes and sells its own, lower-priced Lockstep printer. New models of the Lockstep are introduced every year to keep pace with the fast-changing printer market. Table 7.8 contains data on weekly shipments of Lockstep printers for the last year and a half.

**Table 7.8 Weekly Number of Lockstep Printers Shipped, January 1987–June 1988**

| Week $t$ | No. of Units $y_t$ | Week $t$ | No. of Units $y_t$ | Week $t$ | No. of Units $y_t$ | Week $t$ | No. of Units $y_t$ | Week $t$ | No. of Units $y_t$ | Week $t$ | No. of Units $y_t$ |
|---|---|---|---|---|---|---|---|---|---|---|---|
| 1 | 105 | 14 | 87 | 27 | 99 | 40 | 91 | 53 | 99 | 66 | 105 |
| 2 | 85 | 15 | 90 | 28 | 86 | 41 | 101 | 54 | 100 | 67 | 113 |
| 3 | 106 | 16 | 92 | 29 | 84 | 42 | 92 | 55 | 116 | 68 | 114 |
| 4 | 88 | 17 | 92 | 30 | 90 | 43 | 107 | 56 | 93 | 69 | 95 |
| 5 | 96 | 18 | 90 | 31 | 92 | 44 | 92 | 57 | 107 | 70 | 108 |
| 6 | 97 | 19 | 93 | 32 | 94 | 45 | 107 | 58 | 100 | 71 | 104 |
| 7 | 99 | 20 | 93 | 33 | 90 | 46 | 103 | 59 | 96 | 72 | 96 |
| 8 | 93 | 21 | 93 | 34 | 98 | 47 | 114 | 60 | 114 | 73 | 97 |
| 9 | 86 | 22 | 92 | 35 | 100 | 48 | 97 | 61 | 92 | 74 | 96 |
| 10 | 107 | 23 | 90 | 36 | 91 | 49 | 112 | 62 | 119 | 75 | 98 |
| 11 | 92 | 24 | 99 | 37 | 99 | 50 | 112 | 63 | 94 | 76 | 98 |
| 12 | 92 | 25 | 84 | 38 | 91 | 51 | 96 | 64 | 113 | 77 | 98 |
| 13 | 93 | 26 | 80 | 39 | 102 | 52 | 117 | 65 | 112 | 78 | 99 |

To identify a suitable ARIMA model for these data, we first examine the sample acf in Figure 7.14(a). The series appears to be stationary, since the acf dies out quickly, so no differencing will be necessary. Looking at the acf and pacf together, we see that the easily recognized features of pure AR or MA models do not seem to be present. In addition, $r_1$ (= .139) is small compared to $r_2$ (= .553), and $\hat{\phi}_{11}$ (= .139) is smaller than $\hat{\phi}_{22}$ (= .544). Ignoring $r_1$, we find that the acf seems to decay *after* lag 1, while the pacf seems to either cut off after lag 3 or, perhaps, decay in an oscillating fashion from the beginning.

This behavior seems to point to a mixed model. It also seems that $p$ and $q$ are different, otherwise $p - q$ would be zero and we would expect the acf and pacf to drop off from the start. Likely candidates, then, might be ARIMA(1,0,2) or ARIMA(2,0,1) models.

**Further Steps to Aid in Identifying $p$ and $q$**

**Step 1:** Start by fitting a low-order AR model to the data. This will often clear up the acf of the residuals and make it easier to determine the order of the MA component.

**Step 2:** Analyze the acf and pacf (Table 7.7) of the resulting residuals for additional AR or MA terms to add; let parsimony be the guide.

**Step 3:** Stop when the fitted model(s) have random residuals. (*Note:* More than one ARIMA model may adequately fit the data.)

**Figure 7.14** acf and pacf for Printer's, Inc., Data

(a) acf

```
            -1.0   -.8   -.6   -.4   -.2    .0    .2    .4    .6    .8   1.0
              +----+----+----+----+----+----+----+----+----+----+
 1    .139                                  XXXX
 2    .553                                  XXXXXXXXXXXXXXX
 3    .480                                  XXXXXXXXXXXXX
 4    .320                                  XXXXXXXXX
 5    .431                                  XXXXXXXXXXXX
 6    .263                                  XXXXXXXX
 7    .324                                  XXXXXXXXX
 8    .306                                  XXXXXXXXX
 9    .163                                  XXXXX
10    .356                                  XXXXXXXXXX
11    .185                                  XXXXXX
12    .253                                  XXXXXXX
13    .261                                  XXXXXXXX
14    .131                                  XXXX
15    .256                                  XXXXXXX
```

(b) pacf

```
            -1.0   -.8   -.6   -.4   -.2    .0    .2    .4    .6    .8   1.0
              +----+----+----+----+----+----+----+----+----+----+
 1    .139                                  XXXX
 2    .544                                  XXXXXXXXXXXXXXX
 3    .535                                  XXXXXXXXXXXXXX
 4    .148                                  XXXXX
 5   -.030                                 XX
 6   -.213                             XXXXXX
 7   -.128                               XXXX
 8    .096                                  XXX
 9   -.008                                  X
10    .160                                  XXXXX
11    .145                                  XXXXX
12    .072                                  XXX
13   -.002                                  X
14   -.229                            XXXXXXX
15   -.199                             XXXXXX
```

The ARIMA(1,0,2) is the more logical choice here, since then $q - p = 1$. It explains why the acf waits until after lag 1 before decaying. The aptness of this choice can be further seen by fitting an ARIMA(1,0,2) model and examining the residuals. Figure 7.15 shows the acf and pacf of these residuals. Both plots indicate a random no-trend series, so no further terms appear necessary. •

**Figure 7.15** acf and pacf of Residuals from an ARIMA(1,0,2) Model Fit to the Printer's, Inc., Data

(a) acf

```
              -1.0  -.8  -.6  -.4  -.2   .0   .2   .4   .6   .8  1.0
                 +----+----+----+----+----+----+----+----+----+----+
  1      .045                             XX
  2      .123                             XXXX
  3      .041                             XX
  4     -.157                         XXXXX
  5     -.039                           XX
  6     -.131                          XXXX
  7     -.028                           XX
  8     -.060                           XX
  9     -.203                        XXXXXX
 10      .065                             XXX
 11      .010                             X
 12      .137                             XXXX
 13      .173                             XXXXX
 14      .046                             XX
 15      .166                             XXXXX
```

(b) pacf

```
              -1.0  -.8  -.6  -.4  -.2   .0   .2   .4   .6   .8  1.0
                 +----+----+----+----+----+----+----+----+----+----+
  1      .045                             XX
  2      .121                             XXXX
  3      .032                             XX
  4     -.177                         XXXXX
  5     -.037                           XX
  6     -.090                          XXX
  7      .004                             X
  8     -.057                           XX
  9     -.211                        XXXXXX
 10      .060                             XX
 11      .055                             XX
 12      .119                             XXXX
 13      .092                             XXX
 14     -.006                             X
 15      .112                             XXXX
```

## 7.7.3 Estimation

After the identification stage, parameter estimates must be found. The least squares criterion is used to obtain these estimates. For AR($p$) models the parameter estimates could be found by regressing $y_t$ on $y_{t-1}, y_{t-2}, \ldots, y_{t-p}$ exactly as in Section 4.7.4, but with MA($q$) models or mixed models, the situation is more difficult. For example, we know from Equations 7.4 and 7.5 that the MA(1) model

$$y_t = \theta_0 + \epsilon_t - \theta_1 \epsilon_{t-1}$$

can be written as an infinite-order model

$$y_t = \frac{\theta_0}{1 - \theta_1} - \theta_1 y_{t-1} - \theta_1^2 y_{t-2} - \theta_1^3 y_{t-3} - \cdots + \epsilon_t$$

Because this latter expression has infinitely many terms, and since these terms contain powers of $\theta_1$, linear regression analysis won't work. Instead, parameter estimation must be based on *nonlinear* least squares algorithms. These algorithms usually consist of a combination of search routines and successive approximations, which makes computer programs a necessity.

## Starting Values

In order to calculate, and eventually minimize, the sum of squared errors, preliminary estimates of the parameters are needed, because algorithms involving successive approximations always require *starting values.* Most Box-Jenkins computer programs have a default feature that automatically generates these starting values if so desired. Some programs simply use a constant starting value (often taken to be 0.1) for each $\hat{\phi}_1, \ldots, \hat{\phi}_p$ and each $\hat{\theta}_1, \ldots, \hat{\theta}_q$. This practice seems to work most of the time; however, there are times when it fails (notably, when the successive approximation algorithms do not converge). The analyst must then become actively involved in specifying the initial values.

Reasonable starting values can be found by using the acf of the data. For AR($p$) processes, initial estimates are found by solving the (linear) Yule-Walker equations (Equation 7.1) for the $\phi_i$'s as functions of the $\rho_i$'s and then replacing each $\rho_i$ with its estimate, $r_i$, from the sample acf. With an AR(2) series, for example, the solutions to the Yule-Walker equations are

$$\phi_1 = \frac{\rho_1(1 - \rho_2)}{1 - \rho_1^2} \qquad \text{and} \qquad \phi_2 = \frac{\rho_2 - \rho_1^2}{1 - \rho_1^2}$$

So the starting values would be

$$\hat{\phi}_1 = \frac{r_1(1 - r_2)}{1 - r_1^2} \qquad \text{and} \qquad \hat{\phi}_2 = \frac{r_2 - r_1^2}{1 - r_1^2}$$

Moving average models present more of a problem, since their $\rho_k$'s are nonlinearly related to the $\theta_i$'s. For instance, using Equation 7.3, an MA(1) model yields

$$\rho_1 = \frac{-\theta_1}{1 + \theta_1^2}$$

So the quadratic equation

$$r_1 = \frac{-\hat{\theta}_1}{1 + \hat{\theta}_1^2}$$

must be solved for the initial estimate of $\theta_1$. (Of the two solutions to the quadratic, only the one satisfying the invertibility condition $|\theta_1| < 1$ would be used.) Higher-order MA models require solving a system of nonlinear equations. Finding exact solutions is not an easy task, but approximate ones that will probably suffice can be found by trial and error.

A general approach to setting parameter starting values follows.

1. Try the computer software's (default) starting values. These will probably be sufficient for most applications, since the Box-Jenkins algorithms often work well regardless of the initial values used.
2. If error diagnostics occur (e.g., a message such as CONVERGENCE CRITERION NOT MET or UNABLE TO REDUCE SUM OF SQUARES) then bypass the default feature and specify some approximate starting values as outlined above.

## Backcasting

Most Box-Jenkins algorithms employ backcasting in the parameter estimation stage. Backcasting, first encountered in Chapter 2, was originally introduced by Box and Jenkins (1970, pp. 199–200). It is based on the fact that data from any stationary series, when written in reverse order, have the same characteristics as in the original order. Working with the reversed series amounts to forecasting backwards in time, or *backcasting.*

As will be seen later, all ARIMA forecasts of stationary series eventually settle down and approach the estimated mean of the series. Applying this to the reversed series, we should find some time period $t = -M$ by which the backcasts have become fairly constant. Starting at $t = -M$ and forecasting, we can generate estimates of $y_t$ and $\epsilon_t$ for time periods $-M, -M + 1, \ldots, -3, -2, -1, 0$. These estimates are needed because all ARIMA models are represented as functions of *past* $y_t$'s and $\epsilon_t$'s, so forecasts for the first few periods require a knowledge of the series for a few periods *before* $t = 1$.

Taking as an example the ARIMA(0,1,1) model

$$y_t - y_{t-1} = \theta_0 + \epsilon_t - \theta_1\epsilon_{t-1}$$

and solving for $\epsilon_t$, we get

$$\epsilon_t = -y_t + y_{t-1} + \theta_0 - \theta_1\epsilon_{t-1}$$

It is apparent that calculation of the sum of squared errors $\Sigma\, \epsilon_t^2$ (which starts with $\epsilon_1 = -y_1 + y_0 + \theta_0 - \theta_1\epsilon_0$) requires estimates of $y_0$ and $\epsilon_0$. Backcasting provides these estimates.

Familiarity with the concept of backcasting is helpful because most Box-Jenkins computer programs make reference to backcasts in their printouts. When reporting the minimum sum of squared errors, programs often remind the user that the printed minimum is calculated only for the actual data and that the negative time periods $-M, -M + 1, \ldots, -3, -2, -1, 0$ (used for backcasting) were not included. Similarly, diagnostics such as BACKFORECASTS NOT DYING OUT RAPIDLY require some understanding on the part of the user so that proper corrective actions can be taken.

## Interpreting Computer Printouts

Software for Box-Jenkins analysis is commonly available for mainframes and microcomputers. Most major statistical packages (SAS, BMDP, SPSS, Minitab) support Box-Jenkins methods, some to the exclusion of any other forecasting method.

Like regression analysis, the format of Box-Jenkins printouts varies from program to program, but the content is similar. For this reason we will continue to consider Minitab as representative.

Figure 7.16 shows part of a printout from fitting an ARIMA(1,0,2) model to the Printer's, Inc., data. According to this printout, the estimated model is

$$y_t = .9011y_{t-1} + 9.7277 + \epsilon_t - 1.3995\epsilon_{t-1} + .9645\epsilon_{t-2}$$

**Figure 7.16** Minitab Printout of an ARIMA(1,0,2) Model Fit to the Printer's, Inc., Data

```
FINAL ESTIMATES OF PARAMETERS
NUMBER       TYPE       ESTIMATE        ST. DEV.   T-RATIO
   1       AR   1          .9011           .0524     17.18
   2       MA   1         1.3995           .0409     34.24
   3       MA   2         -.9645           .0428    -22.54
   4      CONSTANT        9.7277           .3613     26.92
          MEAN            98.323           3.652

RESIDUALS.       SS =         2372.44   (BACKFORECASTS EXCLUDED)
                 DF =     74   MS =        32.06
NO. OF OBS.       78
```

The printed standard deviations and *t-ratios* (i.e., the estimated parameter divided by the standard deviation of estimate) allow for quick determinations of whether a parameter is significantly different from zero or not and are used in the diagnostic checking phase. The estimated mean of the series (in this example, 98.323) should always agree, approximately, with that given by Equation 7.2; i.e.,

$$\hat{\mu} \simeq \frac{\hat{\theta}_0}{1 - \hat{\phi}_1} = \frac{9.7277}{1 - .9011} = 98.359$$

During its operation, the Minitab program uses the original series plus backcast periods in its attempt to find parameters that minimize the sum of squared errors. After these estimates are found, only the sum of squares for the original series is reported. (This accounts for the phrase BACKFORECASTS EXCLUDED in the printout.) Thus, in this example, 2372.44 is the program's estimate of the minimum value of the error sum of squares. The degrees of freedom associated with this sum of squares is found from the formula

$$\text{degrees of freedom} = n - (p + d + q) - 1$$

which explains the $78 - (1 + 0 + 2) - 1 = 74$ degrees of freedom in the printout. As usual, a mean square is calculated by dividing the sum of squares by its degrees of freedom

$$\text{MS} = \text{mean square} = \frac{\text{sum of squares}}{\text{degrees of freedom}} = \frac{\text{SS}}{\text{df}}$$

It plays the same role as the MSE in regression analysis (i.e., it is an estimate of the variation unexplained by the model). All interval estimates around ARIMA forecasts involve this mean square in their calculation.

## 7.7.4 Diagnostic Checking

Before forecasting with a fitted model, diagnostic tests are necessary. The tests serve two purposes: to validate the model, or, failing that, to point the way to a better model choice. Since it is often an important part of model respecification, we will discuss diagnostic checking here instead of at chapter's end.

For a good model, the guiding criteria in diagnostic checking are:

- The residuals should be approximately normal, with an acf and pacf indicative of a random no-trend series.
- All parameter estimates should have significant *t*-ratios.
- The model should be parsimonious.

### Analyzing the Residuals

The basic requirement that residuals behave like a random no-trend series is verified by examining the acf and pacf. There should be no significant autocorrelation or partial autocorrelation coefficients. This is illustrated in Figure 7.15 by the residuals from the ARIMA(1,0,2) model fit to the Printer's, Inc., data. None of the coefficients in either plot exceed the "rule-of-thumb" value $\pm 2/\sqrt{78} = \pm .226$. We note that the **Box-Ljung** (modified Box-Pierce) **test** (see Section 2.6.6) could also be used here to collectively test the autocorrelation (but *not* the partial autocorrelation coefficients) for significance (Ljung and Box, 1978).

It is often assumed that the residuals should be approximately normally distributed with a mean of zero. The standard deviation of the residuals is approximated by $\sqrt{\text{MS}}$, where MS is the aforementioned mean square from the computer printout. Histograms of the residuals are usually sufficient for checking for serious departures from normality, but any statistical test of normality may be applied.

When the residuals are not random, the acf and pacf will contain information about which alternate models to consider. If we treat the residual acf and pacf just as in the initial identification stage, the rules summarized in Table 7.7 provide guidance in picking other models. The estimation/diagnostic checking procedure is one of fitting, examining residuals, respecification, refitting, etc. until a suitable model is found. An example will illustrate the process.

**EXAMPLE 7.8** *Diagnostic Checking and Model Respecification, Printer's, Inc., Data*

The acf and pacf for the Printer's, Inc., data, given in Figure 7.14, contain patterns typical of a mixed model. Following the recommended identification procedure, we first fit a low-order [AR(1)] model to the data and examine the residuals. The resulting acf and pacf are given in Figure 7.17. Figure 7.17(a) shows little if any improvement

**Figure 7.17** **acf and pacf of the Residuals of an AR(1) Model Fit to the Printer's, Inc., Data**

**(a) acf**

```
          -1.0  -.8  -.6  -.4  -.2   .0   .2   .4   .6   .8  1.0
            +----+----+----+----+----+----+----+----+----+----+
  1   -.077                        XXX
  2    .486                          XXXXXXXXXXXXX
  3    .375                          XXXXXXXXXX
  4    .203                          XXXXXX
  5    .365                          XXXXXXXXXX
  6    .166                          XXXXX
  7    .256                          XXXXXXX
  8    .249                          XXXXXXX
  9    .075                          XXX
 10    .320                          XXXXXXXXX
 11    .106                          XXXX
 12    .200                          XXXXXX
 13    .217                          XXXXXX
 14    .063                          XXX
 15    .230                          XXXXXXX
```

**(b) pacf**

```
          -1.0  -.8  -.6  -.4  -.2   .0   .2   .4   .6   .8  1.0
            +----+----+----+----+----+----+----+----+----+----+
  1   -.077                        XXX
  2    .483                          XXXXXXXXXXXXX
  3    .566                          XXXXXXXXXXXXXXX
  4    .228                          XXXXXXX
  5    .032                          XX
  6   -.192                     XXXXXX
  7   -.168                      XXXXX
  8    .075                          XXX
  9   -.019                          X
 10    .138                          XXXX
 11    .154                          XXXXX
 12    .094                          XXX
 13    .044                          XX
 14   -.197                     XXXXXX
 15   -.215                     XXXXXX
```

over the original acf of Figure 7.14(a). There is little doubt now that a pure AR model would be inadequate. The decay in the residual acf *after* lag 1 implies (see Table 7.7) that MA terms are needed and that they probably differ in number from the AR terms. Adding the smallest possible number of MA terms consistent with these findings (parsimony), we are once again led to the ARIMA(1,0,2) model as a candidate. That this model adequately fits the data was shown in Example 7.7. ●

### Hypothesis Tests for the Parameters

The hypotheses $H_0$: $\phi_i = 0$; $i = 1, 2, \ldots, p$ and $H_0$: $\theta_i = 0$; $i = 1, 2, \ldots, q$ are tested with the *t*-ratios

$$t = \frac{\hat{\phi}_i}{\hat{\sigma}_{\hat{\phi}_i}} \qquad \text{and} \qquad t = \frac{\hat{\theta}_i}{\hat{\sigma}_{\hat{\theta}_i}}$$

which will be part of a Box-Jenkins computer printout (e.g, see Figure 7.16). Even though they are called *t*-ratios, the critical values from a normal table are routinely used for the tests, since the degrees of freedom in a Box-Jenkins analysis are usually large (ARIMA modeling requires large samples and fits few parameters). In fact, it is common practice to adopt an $\alpha$ of 5% and conclude that a parameter is nonzero if the magnitude of its *t*-ratio exceeds 2.

---

**Test Procedure:** *Testing Model Parameters*

$$H_0: \ \phi_i = 0 \qquad \qquad H_0: \ \theta_i = 0$$
$$\text{or}$$
$$H_a: \ \phi_i \neq 0 \qquad \qquad H_a: \ \theta_i \neq 0$$

**Test Statistic:** $t = \dfrac{\hat{\phi}_i}{\hat{\sigma}_{\hat{\phi}_i}}$ or $t = \dfrac{\hat{\theta}_i}{\hat{\sigma}_{\hat{\theta}_i}}$

**Decision Rule:** ($\alpha \simeq 5\%$)

Reject $H_0$ if $|t| > 2$.

**Conclusion:** If $H_0$ is rejected, we conclude with approximately 95% confidence that the parameter being tested ($\phi_i$ or $\theta_i$) is nonzero.

If $H_0$ is not rejected, we have some support for the parameter's being zero.

---

Although many models can be fit to a given data set, not all of the terms in each model may be necessary (i.e., statistically significant). Good modeling practice requires that we retain only models in which every term is significant. Terms whose *t*-ratios are small should be dropped and the model refit with the remaining terms.

## Overfitting

When an ARIMA($p,d,q$) model has been tentatively accepted, we can use **overfitting** to see if higher-order AR or MA terms would improve model performance, i.e., we can successively add and test AR *or* MA terms (but not both) to the model until the last term added is not significant. The technique is:

| | | | |
|---|---|---|---|
| | ⋮ | | ⋮ |
| | ↑ | | ↑ |
| *Overfit Models:* | $p + 2$ | | $q + 2$ |
| | ↑ | | ↑ |
| | $p + 1$ | | $q + 1$ |
| | ↑ | | ↑ |
| *Tentative Model:* | ARIMA($p,d,q$) | *or* | ARIMA($p,d,q$) |

AR and MA terms are not added simultaneously, because parameter redundancy (discussed shortly) may result.

***Note:*** Overfitting should not be used alone, but, rather, in conjunction with residual analysis. For example, simply fitting AR models of successively higher orders, eventually stopping, then successively adding MA terms does not work well with data from mixed models. Analysis of acf's and pacf's should guide initial choices of $p$ and $q$ *before* overfitting is used.

**EXAMPLE 7.9** *Overfitting, Printer's, Inc., Data*

In Example 7.8 residual analysis was the sole diagnostic method guiding the final selection of an ARIMA(1,0,2) model for the Printer's, Inc., data. Since the orders $p = 1$, $q = 2$ were the smallest that seemed consistent with the patterns in the acf and pacf, a question remains as to whether the fit might be improved by increasing $p$ and/or $q$. Increasing $p$ alone would lead to an ARIMA(2,0,2) model, and increasing $q$ alone suggests an ARIMA(1,0,3).

Figure 7.18 shows Minitab printouts from fitting both these models. In Figure 7.18(a), the additional AR term appears not to be needed, since its $t$-ratio is small ($t = -.39$). Likewise, the small $t$-ratio of $\hat{\theta}_3$ in Figure 7.18(b) means that further MA terms are unnecessary. As further evidence that neither model is viable, notice that in both printouts there has been negligible improvement in the sum of squares over that of the ARIMA(1,0,2) model.

To summarize: We have tried to go beyond the ARIMA(1,0,2) model by separately overfitting with AR and MA terms. These extra terms were not statistically significant, so once again the ARIMA(1,0,2) model proves to be a good choice. •

## Parameter Redundancy

Parsimony is recommended in Box-Jenkins analysis not only because simpler models are easier to fit and explain, but also because parsimonious models help avoid the problem of **parameter redundancy** (Pankratz, 1983, pp. 203–

Figure 7.18 ARIMA(2,0,2) and ARIMA(1,0,3) Models Fit to the Printer's, Inc., Data

(a) ARIMA(2,0,2)

FINAL ESTIMATES OF PARAMETERS

| NUMBER | TYPE | ESTIMATE | ST. DEV. | T-RATIO |
|---|---|---|---|---|
| 1 | AR 1 | .9407 | .1228 | 7.66 |
| 2 | **AR 2** | **−.0475** | **.1218** | **−.39** |
| 3 | MA 1 | 1.4020 | .0454 | 30.89 |
| 4 | MA 2 | −.9656 | .0478 | −20.20 |
| 5 | CONSTANT | 10.5182 | .3603 | 29.19 |
| | MEAN | 98.449 | 3.373 | |

```
RESIDUALS.     SS =       2367.29  (BACKFORECASTS EXCLUDED)
               DF =    73   MS =       32.43
NO. OF OBS.     78
```

(b) ARIMA(1,0,3)

FINAL ESTIMATES OF PARAMETERS

| NUMBER | TYPE | ESTIMATE | ST. DEV. | T-RATIO |
|---|---|---|---|---|
| 1 | AR 1 | .8950 | .0556 | 16.10 |
| 2 | MA 1 | 1.3740 | .0572 | 24.03 |
| 3 | MA 2 | −.9213 | .0879 | −10.49 |
| 4 | **MA 3** | **−.0309** | **.0637** | **−.48** |
| 5 | CONSTANT | 10.3174 | .3708 | 27.82 |
| | MEAN | 98.280 | 3.532 | |

```
RESIDUALS.     SS =       2367.02  (BACKFORECASTS EXCLUDED)
               DF =    73   MS =       32.42
NO. OF OBS.     78
```

206). Redundancy occurs when a higher-order model is used that, for all practical purposes, is indistinguishable from a model with fewer terms. While both models may be fit to the data, the parameter estimates in the higher-order model can be very unstable, changing dramatically even for different samples from the same process.

As an example of how parameter redundancy arises, consider the simple white-noise series

$$y_t = \epsilon_t \tag{7.19}$$

which, technically, is an ARIMA(0,0,0) series. It follows from Equation 7.19 that

$$y_t - y_{t-1} = \epsilon_t - \epsilon_{t-1}$$

So

$$y_t = y_{t-1} + \epsilon_t - \epsilon_{t-1} \tag{7.20}$$

which is an ARIMA(1,0,1)-model equation. The point is that we can easily fit an ARIMA(1,0,1) model to a series that, in actuality, has fewer AR and MA terms.

In practice, all candidate models should be checked for parameter redundancy. This is done by first fitting a model and then examining the coefficients to see if, perhaps, there is a nearly equivalent lower-order model (in the sense that Equation 7.20 came from Equation 7.19). When redundancy exists, the lower-order model should be used.

### 7.7.5 Forecasting

The discussion of forecasting that follows is intended to provide the details, both mechanical and conceptual, used in actually arriving at ARIMA forecasts. We hasten to point out that these manipulations are routinely carried out by most Box-Jenkins programs and that they are presented here to give the reader an understanding and appreciation of the principles underlying such programs' generation of these forecasts.

Some of the methods used in Box-Jenkins forecasting, as applied to an ARIMA(0,1,1) model, were illustrated in Example 7.6. In that example we followed a procedure that can be used with any ARIMA model. The steps are summarized in the Forecasting ARIMA Series table.

**Forecasting ARIMA Series**

| | |
|---|---|
| **Step 1:** | Write down the *estimated* model and solve for $y_t$ as a function of all other terms in the model. |
| **Step 2:** | Replace $t$ by $t + j$ in the equation resulting from Step 1. |
| **Step 3:** | To generate $j$-period-ahead forecasts $\hat{y}_{t+j}(t)$, use the following substitutions in the equation from Step 2: |
| **Past Values:** | Use the actual value, $y_{t-k}$, for $k = 0, 1, 2, \ldots$ |
| **Future Values:** | Replace $y_{t+k}$ by $\hat{y}_{t-k}(t)$, for $k = 1, 2, 3, \ldots$ |
| **Past Errors:** | Replace $\epsilon_{t-k}$ by $e_{t-k}$, for $k = 0, 1, 2, \ldots$ The $e_{t-k}$ are obtained from a computer program. |
| **Future Errors:** | Replace $\epsilon_{t+k}$ by 0, for $k = 1, 2, 3, \ldots$ |
| | Step 3 can be summarized simply: Use actual values of the series when possible; otherwise use forecasts. Use estimates (residuals) for *past* error terms, and 0 (their expected value) for *future* error terms. |

For stationary series (i.e., those requiring no initial differencing), it is a mathematical fact that the forecasts must converge to the *estimated mean* of the series (Pankratz, 1983, p. 244). That is, as of any time $t$, the $j$-period-ahead forecasts $\hat{y}_{t+j}(t)$ eventually approach a constant. This makes intuitive sense, since stationary ARIMA models only use information in the past few $y_t$'s and $\epsilon_t$'s, and the contribution of these terms is felt most in the first few forecasts.

**EXAMPLE 7.10** *Forecasting an ARIMA(1,1,1) Series*

Almost all the concepts and details that can arise in ARIMA forecasting can be illustrated with the ARIMA(1,1,1) model. We will proceed through the steps outlined above (for simplicity we refrain from using estimated parameters until the final step).

*Step 1:* The ARIMA(1,1,1) model, which has first-order differencing, can be written in the form

$$w_t = \phi_1 w_{t-1} + \theta_0 + \epsilon_t - \theta_1 \epsilon_{t-1}$$

where $w_t = \Delta y_t = y_t - y_{t-1}$. So in terms of the *original* series,

$$y_t - y_{t-1} = \phi_1(y_{t-1} - y_{t-2}) + \theta_0 + \epsilon_t - \theta_1 \epsilon_{t-1}$$

When solved for $y_t$, this becomes

$$y_t = \theta_0 + (1 + \phi_1)y_{t-1} - \phi_1 y_{t-2} + \epsilon_t - \theta_1 \epsilon_{t-1} \tag{7.21}$$

*Step 2:* We replace $t$ by $t + j$ in Equation 7.21:

$$y_{t+j} = \theta_0 + (1 + \phi_1)y_{t+j-1} - \phi_1 y_{t+j-2} + \epsilon_{t+j} - \theta_1 \epsilon_{t+j-1} \tag{7.22}$$

*Step 3:* The forecasts, as of time $t$, for $j = 1, 2, 3, \ldots$ periods ahead are as follows.

*For time t + 1:*

$$\hat{y}_{t+1}(t) = \hat{\theta}_0 + (1 + \hat{\phi}_1)y_t - \hat{\phi}_1 y_{t-1} - \hat{\theta}_1 e_t$$

$\epsilon_{t+1}$ is replaced by 0, $\epsilon_t$ is replaced by the computer program's estimate, $e_t$.

*For time t + 2:*

$$\hat{y}_{t+2}(t) = \hat{\theta}_0 + (1 + \hat{\phi}_1)\hat{y}_{t+1}(t) - \hat{\phi}_1 y_t$$

Both error terms are in the future and are replaced by 0; $y_{t+1}$ is also a future value and is replaced by its forecast value, $\hat{y}_{t+1}(t)$.

*For time t + 3:*

$$\hat{y}_{t+3}(t) = \hat{\theta}_0 + (1 + \hat{\phi}_1)\hat{y}_{t+2}(t) - \hat{\phi}_1 \hat{y}_{t+1}(t)$$

Every forecast from this point on depends on only the previous two forecasts. •

**EXAMPLE 7.11** *Forecasting, Printer's, Inc., Data*

In previous examples we have shown that an ARIMA(1,0,2) model adequately fits the Printer's, Inc., data. Following the forecasting procedure above, estimates of the number of Lockstep printers to be shipped in the next three weeks would be the following.

*Step 1:* The estimated model for the data (see Figure 7.16) is

$$y_t = .9011y_{t-1} + 9.7277 + \epsilon_t - 1.3995\epsilon_{t-1} + .9645\epsilon_{t-2}$$

Since $d = 0$, this equation is already in the form desired for forecasting.

*Step 2:* Replacing $t$ by $t + j$, we get

$$y_{t+j} = .9011y_{t+j-1} + 9.7277 + \epsilon_{t+j} - 1.3995\epsilon_{t+j-1} + .9645\epsilon_{t+j-2}$$

*Step 3:* Forecasts, as of time $t = 78$, for $j = 1, 2, 3$ periods ahead are, then, as follows.

*For t = 79:*

$$\hat{y}_{79}(78) = .9011y_{78} + 9.7277 + e_{79} - 1.3995e_{78} + .9645e_{77}$$
$$= .9011(99) + 9.7277 + 0 - 1.3995(1.3898) + .9645(3.0997) = 99.98$$

The estimated residuals $e_{78} = 1.3898$ and $e_{77} = 3.0997$ were obtained from the computer program used to find the estimated model.

*For t = 80:*

$$\hat{y}_{80}(78) = .9011\hat{y}_{79}(78) + 9.7277 + e_{80} - 1.3995e_{79} + .9645e_{78}$$
$$= .9011(99.98) + 9.7277 + 0 - 0 + .9645(1.3898) = 101.16$$

*For t = 81:*

$$\hat{y}_{81}(78) = .9011\hat{y}_{80}(78) + 9.7277 + e_{81} - 1.3995e_{80} + .9645e_{79}$$
$$= .9011(101.16) + 9.7277 + 0 - 0 + 0 = 100.88$$

The forecasts are not rounded during the calculations, since the equations can be sensitive to roundoff error. In summary:

$$\hat{y}_{79}(78) = 99.98 \simeq 100$$
$$\hat{y}_{80}(78) = 101.16 \simeq 101$$
$$\hat{y}_{81}(78) = 100.88 \simeq 101$$

•

## Updating Form

Updating ARIMA parameters can be performed only by completely refitting the model. When this is done, the new parameter estimates are often very close to the old ones, so a practical substitute (especially in the short term) for updating the parameters is to assume that they do not change and then derive updating equations using the current values of the series. To find the updated forecasts the model must first be written in random-shock form. A model is in *random-shock* form when it is written as a sum of error terms only; i.e.,

$$y_t = \mu + \epsilon_t + \psi_1\epsilon_{t-1} + \psi_2\epsilon_{t-2} + \psi_3\epsilon_{t-3} + \cdots \quad \textbf{(7.23)}$$

An MA($q$) model is already in random-shock form, but some work is necessary to convert AR($p$) and mixed models. A mechanical technique for doing this conversion is summarized in Table 7.9. For example, to convert an AR(1) process to random-shock form we would write

$$(1 + \psi_1 x + \psi_2 x^2 + \cdots)(1 - \phi_1 x) = 1 + (\psi_1 - \phi_1)x + (\psi_2 - \phi_1\psi_1)x^2 + (\psi_3 - \phi_1\psi_2)x^3 + \cdots$$

which (upon equating coefficients of $x$, $x^2$, $x^3$, . . . to 0) leads to

$$\psi_1 - \phi_1 = 0 \rightarrow \psi_1 = \phi_1$$
$$\psi_2 - \phi_1\psi_1 = 0 \rightarrow \psi_2 = \phi_1\psi_1 = \phi_1^2$$
$$\psi_3 - \phi_1\psi_2 = 0 \rightarrow \psi_3 = \phi_1\psi_2 = \phi_1^3$$

**Table 7.9 Procedure for Reducing ARIMA Models to Random-Shock Form**

| | |
|---|---|
| **Random-Shock Form:** | $y_t = \mu + \epsilon_t + \psi_1\epsilon_{t-1} + \psi_2\epsilon_{t-2} + \psi_3\epsilon_{t-3} + \cdots$ |
| MA($q$): | $y_t = \theta_0 + \epsilon_t - \theta_1\epsilon_{t-1} - \theta_2\epsilon_{t-2} - \cdots - \theta_q\epsilon_{t-q}$ <br> Any MA($q$) model is already in its random-shock form. |
| AR($p$): | In the expression <br> $(1 + \psi_1 x + \psi_2 x^2 + \psi_3 x^3 + \cdots)(1 - \phi_1 x - \phi_2 x^2 - \phi_3 x^3 - \cdots - \phi_p x^p)$ <br> set the coefficients of $x, x^2, x^3, \ldots$ equal to zero and solve the resulting equations for $\psi_1, \psi_2, \psi_3, \ldots$. Next let <br> $\mu = \dfrac{\phi_0}{1 - (\phi_1 + \phi_2 + \cdots + \phi_p)}$ |
| ARIMA($p,d,q$): | Equate the coefficients of $x, x^2, x^3, \ldots$ on both sides of <br> $(1 + \psi_1 x + \psi_2 x^2 + \psi_3 x^3 + \cdots)(1 - \phi_1 x - \phi_2 x^2 - \cdots - \phi_p x^p)(1 - x)^d = (1 - \theta_1 x - \theta_2 x^2 - \cdots - \theta_q x^q)$ <br> and solve for the $\psi_i$'s. |

and so on. So the random-shock form of an AR(1) model is

$$y_t = \frac{\phi_0}{1 - \phi_1} + \epsilon_t + \phi_1\epsilon_{t-1} + \phi_1^2\epsilon_{t-2} + \phi_1^3\epsilon_{t-3} + \cdots \qquad \textbf{(7.24)}$$

Updating equations are derived from Equation 7.24 as follows: Replacing $t$ by $t + j$, noting that $\mu = \phi_0/(1 - \phi_1)$, and using the above relationships between the $\psi$'s and $\phi$'s, we have

$$y_{t+j} = \mu + \epsilon_{t+j} + \psi_1\epsilon_{t+j-1} + \psi_2\epsilon_{t+j-2} + \cdots + \psi_{j-1}\epsilon_{t+1} + \psi_j\epsilon_t + \cdots$$

Then, as of time $t$, replacing future $\epsilon$'s by zeros, we see that

$$\hat{y}_{t+j}(t) = \mu + \psi_j e_t + \psi_{j+1}e_{t-1} + \cdots$$

and, as of time $t + 1$,

$$\hat{y}_{t+j}(t+1) = \mu + \psi_{j-1}e_{t+1} + \psi_j e_t + \cdots$$

So by subtraction,

$$\hat{y}_{t+j}(t+1) = \hat{y}_{t+j}(t) + \psi_{j-1}\cdot[y_{t+1} - \hat{y}_{t+1}(t)] \qquad \textbf{(7.25)}$$

Note that this is very similar to exponential smoothing; that is,

$$(\text{new forecast}) = (\text{old forecast}) + \psi_{j-1}\cdot(\text{current forecast error})$$

**Updating Equations for ARIMA Forecasting**

| | |
|---|---|
| **Step 1:** | Put the model in random-shock form (see Table 7.9): <br> $y_t = \mu + \epsilon_t + \psi_1\epsilon_{t-1} + \psi_2\epsilon_{t-2} + \cdots$ |
| **Step 2:** | As of time $t + 1$, with $y_{t+1}$ now available, to update the forecast of $y_{t+j}$, use <br> $\hat{y}_{t+j}(t+1) = \hat{y}_{t+j}(t) + \psi_{j-1}\cdot[y_{t+1} - \hat{y}_{t+1}(t)] \quad j = 2, 3, \ldots$ |

**EXAMPLE 7.12** *Updating Forecasts for an AR(1) Model*

The series in Table 7.1 was generated from an AR(1) process with $\phi_0 = 10$ and $\phi_1 = .90$. Using Minitab, the *estimated* model is

$$y_t = 9.1861 + .9109y_{t-1} + \epsilon_t \tag{7.26}$$

from which we obtain forecasts, at period 77:

$$\hat{y}_{78}(77) = 9.1861 + .9109y_{77} \qquad = 9.1861 + .9109 \times 97 \qquad = 97.54$$
$$\hat{y}_{79}(77) = 9.1861 + .9109\hat{y}_{78}(77) = 9.1861 + .9109 \times 97.54 = 98.03$$
$$\hat{y}_{80}(77) = 9.1861 + .9109\hat{y}_{79}(77) = 9.1861 + .9109 \times 98.03 = 98.48$$

and so on.

From Equations 7.24 and 7.26, the random-shock form of the model is

$$\begin{aligned} y_t &= \frac{9.1861}{1 - .9109} + \epsilon_t + .9109\epsilon_{t-1} + (.9109)^2\epsilon_{t-2} + \cdots \\ &= 103.098 + \epsilon_t + .9109\epsilon_{t-1} + .8297\epsilon_{t-2} + \cdots \end{aligned}$$

Suppose that at period 78 the actual value of the series turns out to be 99. Then the updated forecasts for periods 79, 80, . . . are

$$\begin{aligned} \hat{y}_{79}(78) &= \hat{y}_{79}(77) + \psi_1 \cdot [y_{78} - \hat{y}_{78}(77)] \\ &= 98.03 + .9109(99 - 97.54) = 99.37 \end{aligned}$$

***Note:*** $j = 2$ implies $\psi_{j-1} = \psi_1$ in Equation 7.25.

$$\begin{aligned} \hat{y}_{80}(78) &= \hat{y}_{80}(77) + \psi_2 \cdot [y_{78} - \hat{y}_{78}(77)] \\ &= 98.48 + .8297(99 - 97.54) = 99.70 \end{aligned}$$

***Note:*** $j = 3$ implies $\psi_{j-1} = \psi_2$ in Equation 7.25 and so on. •

## Prediction Intervals

Calculation of prediction intervals for ARIMA forecasts requires (1) that the model be put into random-shock form and (2) that the variability in the $\epsilon_t$ series be estimated. Table 7.9 is used to convert the model, while the estimated variance of the error terms is obtained from a Box-Jenkins computer printout (Figure 7.16).

From the random-shock form (Equation 7.23), we can put $t + j$ in place of $t$:

$$y_{t+j} = \mu + \epsilon_{t+j} + \psi_1\epsilon_{t+j-1} + \psi_2\epsilon_{t+j-2} + \cdots + \psi_j\epsilon_t + \psi_{j+1}\epsilon_{t-1} + \cdots$$

So (using zeros for future $\epsilon$'s) with $j$-step-ahead forecast is

$$\hat{y}_{t+j}(t) = \mu + \psi_j\epsilon_t + \psi_{j+1}\epsilon_{t-1} + \cdots$$

Subtracting these equations, we find that the *j-step-ahead forecast error* is

$$e_{t+j} = y_{t+j} - \hat{y}_{t+j}(t) = \epsilon_{t+j} + \psi_1\epsilon_{t+j-1} + \psi_2\epsilon_{t+j-2} + \cdots + \psi_{j-1}\epsilon_{t+1}$$

Since the error terms are assumed to be independent, the variance of the forecast errors is

$$\sigma^2_{e_{t+j}} = \text{variance of } (\epsilon_{t+j} + \psi_1\epsilon_{t+j-1} + \cdots + \psi_{j-1}\epsilon_{t+1})$$
$$= \sigma^2_\epsilon + \psi_1^2\sigma^2_\epsilon + \psi_2^2\sigma^2_\epsilon + \cdots + \psi_{j-1}^2\sigma^2_\epsilon$$
$$= \sigma^2_\epsilon(1 + \psi_1^2 + \psi_2^2 + \cdots + \psi_{j-1}^2) \quad \textbf{(7.27)}$$

Approximating $\sigma^2_\epsilon$ with the mean square (MS) from a computer printout, and using the estimated $\phi$'s and $\theta$'s to approximate the $\psi_i$'s, we find that the estimated standard deviation of the $j$-step-ahead forecast errors is

$$\hat{\sigma}_{e_{t+j}} = \sqrt{\text{MS}}(1 + \psi_1^2 + \psi_2^2 + \cdots + \psi_{j-1}^2)^{1/2}$$

For large samples the forecast errors will be approximately normal, so $(1 - \alpha) \times 100\%$ prediction limits can be found from

$$\hat{y}_{t+j}(t) \pm z_{\alpha/2}\sqrt{\text{MS}} \cdot (1 + \psi_1^2 + \psi_2^2 + \cdots + \psi_{j-1}^2)^{1/2} \quad \textbf{(7.28)}$$

**Prediction Intervals for ARIMA Forecasts**

**Step 1:** Put the model in random-shock form (see Table 7.9).

**Step 2:** Use a Box-Jenkins computer program to obtain parameter estimates (needed for estimating the $\psi_i$'s in the random-shock equation) and MS (mean square, estimate of error term variance).

**Step 3:** At time $t$, a $(1 - \alpha) \cdot 100\%$ prediction interval for $y_{t+j}$ is

$$\hat{y}_{t+j}(t) \pm z_{\alpha/2}\sqrt{\text{MS}} \cdot (1 + \psi_1^2 + \psi_2^2 + \cdots + \psi_{j-1}^2)^{1/2}$$

where $z_{\alpha/2}$ is the upper $(\alpha/2) \cdot 100\%$ point of the standard normal distribution (Appendix BB).

**EXAMPLE 7.13** *Prediction Intervals for Data of Table 7.1*

The random-shock form of an AR(1) model fit to the data in Table 7.1 was shown in Example 7.12 to be

$$y_t = 103.098 + \epsilon_t + .9109\epsilon_{t-1} + (.9109)^2\epsilon_{t-2} + \cdots$$

and the first few forecasts, at $t = 77$, were $\hat{y}_{78}(77) = 97.54$, $\hat{y}_{79}(77) = 98.03$, and $\hat{y}_{80}(77) = 98.48$.

Figure 7.19 contains part of the Minitab printout from the model fit. From the

**Figure 7.19** Minitab Printout, AR(1) Model Fit to the Data of Table 7.1

```
RELATIVE CHANGE IN EACH ESTIMATE LESS THAN    .0010


FINAL ESTIMATES OF PARAMETERS
NUMBER        TYPE       ESTIMATE        ST. DEV.   T-RATIO
     1       AR  1          .9109           .0485     18.76
     2     CONSTANT        9.1861           .5773     15.91
           MEAN           103.045           6.476

RESIDUALS.      SS =         1899.47    (BACKFORECASTS EXCLUDED)
                DF =      75    MS =          25.33
NO. OF OBS.       77
```

printout, MS = 25.33, so 95% prediction limits for the first few forecasts, at $t = 77$, are

$$\hat{y}_{78}(77) \pm z_{.025}\sqrt{\text{MS}} \cdot (1)^{1/2} = 97.54 \pm 1.96\sqrt{25.33}$$
$$= 97.54 \pm 9.86, \text{ or } (87.68, 107.40)$$
$$\hat{y}_{79}(77) \pm z_{.025}\sqrt{\text{MS}} \cdot (1 + \psi_1^2)^{1/2} = 98.03 \pm 1.96\sqrt{25.33} \cdot (1 + .9109^2)^{1/2}$$
$$= 98.03 \pm 13.34, \text{ or } (84.69, 111.37)$$
$$\hat{y}_{80}(77) \pm z_{.025}\sqrt{\text{MS}} \cdot (1 + \psi_1^2 + \psi_2^2)^{1/2}$$
$$= 98.48 \pm 1.96\sqrt{25.33} \cdot (1 + .9109^2 + .8297^2)^{1/2}$$
$$= 98.48 \pm 15.65, \text{ or } (82.83, 114.13)$$

In practice, most Box-Jenkins computer programs automatically calculate forecasts and prediction intervals. Some programs, though, don't calculate interval estimates, and of those that do, some only find 95% prediction intervals. In such cases you may find it necessary to calculate the intervals yourself.

For comparison with Example 7.13, we have included an extended Minitab printout (Figure 7.20) of the AR(1) model fit to the data of Table 7.1. The printout also shows forecasts and 95% prediction intervals for periods 78, 79, and 80.

**Figure 7.20** Minitab Printout, AR(1) Model Fit to the Data of Table 7.1 with Forecasts and Prediction Intervals

```
RELATIVE CHANGE IN EACH ESTIMATE LESS THAN    .0010


FINAL ESTIMATES OF PARAMETERS
NUMBER      TYPE      ESTIMATE      ST. DEV.   T-RATIO
     1     AR  1         .9109         .0485     18.76
     2    CONSTANT      9.1861         .5773     15.91
          MEAN        103.045          6.476

RESIDUALS.     SS =        1899.47   (BACKFORECASTS EXCLUDED)
               DF =    75   MS =         25.33
NO. OF OBS.      77

FORECASTS FROM PERIOD  77

                                     95 PERCENT LIMITS
PERIOD         FORECAST          LOWER           UPPER          ACTUAL
  78            97.539          87.673         107.405
  79            98.030          84.685         111.375
  80            98.477          82.822         114.132
```

## 7.8 Seasonal Box-Jenkins Models

### 7.8.1 Introduction

ARIMA models and the four phases of Box-Jenkins modeling can be extended to handle seasonal time series. The same tools are used (differencing, acf, pacf), but in slightly different ways than they are with nonseasonal data. As in Chapter 5, seasonality also raises the question of whether to use additive or multiplicative models.

### 7.8.2 Identification

Seasonal ARIMA models consist of two parts: terms that incorporate the season-to-season movements, and terms that model the nonseasonal, or within seasons, movements. Because of this, the identification stage requires two separate passes, one to determine the seasonal parameters, and one for the nonseasonal parameters.

The seasonal part of the model has its own autoregressive and moving average parameters with orders $P$ and $Q$, while the nonseasonal part has orders $p$ and $q$. In addition to the usual differencing, seasonal differencing, discussed below, is also necessary with seasonal data. The number of seasonal differences used, $D$, and the number of regular differences, $d$, are needed when specifying the model.

The determination of $P$, $D$, and $Q$ involves examining the acf and pacf just as with nonseasonal models, but with one difference: Only the seasonal or *near-seasonal* autocorrelation and partial autocorrelation coefficients are used to determine $P$, $D$, and $Q$. That is, if the length of the seasonality is $L$, then the acf and pacf are only examined at the seasonal lags $L$, $2L$, $3L$, . . . and at a couple of lags on either side of these lags (i.e., the near-seasonal lags). The identification then proceeds almost the same as for nonseasonal data. For example, if the acf (at the seasonal lags) dies out while the pacf cuts off after the first seasonal lag, then a seasonal autoregressive model is indicated, with $P = 1$ and $Q = 0$. The near-seasonal lags come into play when determining whether to use an additive or a multiplicative model. The specifics of the seasonal identification procedure are presented below.

#### Differencing

Seasonal series, like nonseasonal ones, must be stationary before the identification phase can proceed. This means that the acf must die off quickly not only for the nonseasonal lags, but also for the seasonal ones. It is best to view the acf as being composed of two acf's: a nonseasonal one, with the seasonal one overlaid. Thus, stationary seasonal series should have the early (nonseasonal) $r_k$'s damping out rapidly or cutting off the same as the seasonal ones.

If the acf indicates nonstationarity in either the seasonal or nonseasonal part of the series, then differencing will be required. Regular differencing is applied to induce stationarity in the nonseasonal part of the series, while seasonal differencing is used for the seasonal part. Seasonal and regular differencing are similar, except the seasonal differences are taken over a span of $L$ periods rather than one period:

*Regular Differencing:*

$$\Delta y_t = y_t - y_{t-1}$$

*Seasonal Differencing:*

$$\Delta_L y_t = y_t - y_{t-L}$$

Both types of differencing are necessary. For example, successive use of regular differencing will not have much effect on the seasonal part of the acf. Seasonal differencing, on the other hand, affects the acf primarily at the seasonal lags. Fortunately, the order in which seasonal and nonseasonal differencing are carried out is not important. That is, differencing a series regularly and then seasonally will yield the same result as when the seasonal differencing is done first. That this is so can be seen from the following.

$$\begin{aligned} \text{Regular First/Seasonal Second} \rightarrow \Delta_L(\Delta y_t) &= \Delta_L(y_t - y_{t-1}) \\ &= (y_t - y_{t-1}) - (y_{t-L} - y_{t-L-1}) \\ &= (y_t - y_{t-L}) - (y_{t-1} - y_{t-L-1}) \\ &= \Delta(\Delta_L y_t) \leftarrow \text{Seasonal First/Regular Second} \end{aligned}$$

As a rule, regular differencing is done first, since it tends to clear up the acf, making it easier to interpret at the seasonal lags. Just as in the nonseasonal case, in practice it is usually not necessary to seasonally difference a series more than twice (i.e., $D \leq 2$).

## Testing the acf and pacf

Testing the early, nonseasonal lags of the acf proceeds as before (Section 7.7). At the seasonal lags, though, there is a slight change. The standard deviation of the $r_k$'s at the seasonal lags is still computed using Bartlett's formula (Equation 7.18), but $t$-values that exceed 1.25 are now considered significant (Pankratz, 1983, p. 283). One reason for allowing smaller $t$-values is that, from Bartlett's formula, the early, nonseasonal $r_k$'s can tend to inflate $\sigma_{r_k}$ by the time $k$ reaches $L$. This sometimes masks the significance of $r_L$, $r_{2L}$, $r_{3L}$, . . . .

Turning to the pacf, for seasonal data the $\hat{\phi}_{kk}$'s at the seasonal lags are treated the same as those at the nonseasonal lags. That is, the "rule-of-thumb" test for the pacf (Section 7.7) applies to all lags of the pacf, whether seasonal or not.

## Additive and Multiplicative Models

Seasonal parameters are denoted $\Theta_i$ and $\Phi_i$, the uppercase versions of the nonseasonal ones. Thus, a purely seasonal (i.e., one with no nonseasonal terms) first-order moving average model would be written

$$y_t = \Theta_0 + \epsilon_t - \Theta_1 \epsilon_{t-L}$$

while a purely seasonal AR model of order 2 would be

$$y_t = \Phi_0 + \Phi_1 y_{t-L} + \Phi_2 y_{t-2L} + \epsilon_t$$

For seasonal models that also have nonseasonal terms, a choice exists for how the seasonal and nonseasonal terms are to be combined into a single model. For example, *additive* models simply add all the seasonal and nonseasonal terms together. An additive model with one seasonal MA term and one nonseasonal MA term is written

$$y_t = \Theta_0 + \epsilon_t - \theta_1 \epsilon_{t-1} - \Theta_1 \epsilon_{t-L}$$

*Multiplicative* models, on the other hand, are formed by applying the nonseasonal model to the terms from a purely seasonal model. Specifically, the nonseasonal MA terms are applied to the seasonal MA terms and the nonseasonal AR terms to the seasonal AR terms (Abraham and Ledolter, 1983, pp. 283–285). For example, a multiplicative model with one seasonal MA term and one nonseasonal MA term is formed as follows: Writing the seasonal MA part as

$$y_t = \Theta_0 + u_t - \Theta_1 u_{t-L}$$

then applying the nonseasonal MA term to the errors $u_t$, we have

$$u_t = \theta_0 + \epsilon_t - \theta_1 \epsilon_{t-1}$$

Substituting this expression for $u_t$ into the previous equation gives

$$y_t = \Theta_0 + (\theta_0 + \epsilon_t - \theta_1 \epsilon_{t-1}) - \Theta_1 \cdot (\theta_0 + \epsilon_{t-L} - \theta_1 \epsilon_{t-L-1})$$

After combining all the constants into one term, the multiplicative model becomes,

$$y_t = \Theta_0 + \epsilon_t - \theta_1 \epsilon_{t-1} - \Theta_1 \epsilon_{t-L} + \theta_1 \Theta_1 \epsilon_{t-L-1} \tag{7.29}$$

Multiplicative models derive their name from terms such as $\theta_1 \Theta_1 \epsilon_{t-L-1}$, in which the various model parameters are multiplied together.

The acf of a multiplicative model has a distinctive characteristic: Some of the near-seasonal autocorrelation coefficients may also be large. To see how this can happen, reconsider the model in Equation 7.29 and suppose that the seasonality is of length $L = 12$. The equation then becomes

$$y_t = \Theta_0 + \epsilon_t - \theta_1 \epsilon_{t-1} - \Theta_1 \epsilon_{t-12} + \theta_1 \Theta_1 \epsilon_{t-13} \tag{7.30}$$

Substituting $t - 11$ for $t$ gives

$$y_{t-11} = \Theta_0 + \epsilon_{t-11} - \theta_1 \epsilon_{t-12} - \Theta_1 \epsilon_{t-23} - \theta_1 \Theta_1 \epsilon_{t-24} \tag{7.31}$$

Notice that $\epsilon_{t-12}$ appears in both of Equations 7.30 and 7.31. This means that, to some extent, $y_t$ and $y_{t-11}$ will be correlated and so we can expect to see some autocorrelation at lag 11. Equation 7.30 for $y_t$ also contains an $\epsilon_{t-13}$ term, which occurs in $y_{t-13}$ (putting $t - 13$ in for $t$ in Equation 7.30), and this gives rise to some autocorrelation at lag 13.

As a general rule, each additional nonseasonal parameter in a model adds near-seasonal spikes to the acf. For example, a multiplicative model with two nonseasonal MA terms and one seasonal MA term would give rise to an acf similar to that shown in Figure 7.21, which shows two near-seasonal spikes surrounding each seasonal lag. With experience, one can learn to use such information to help in the identification of the nonseasonal part of the model.

Cataloging the possible patterns that can occur at the near-seasonal lags is beyond the scope of this chapter, and the reader is referred to other texts, such as Hoff, 1983, ch. 15. We will only use the presence of near-seasonal spikes in the acf as a signal to use a multiplicative, rather than additive, model. As a matter of fact, since multiplicative models tend to occur most often in practice, we will confine our attention to them in the rest of this chapter.

## Notation

After identifying the nonseasonal orders $(p,d,q)$ and the seasonal orders $(P,D,Q)$ for a given series, a multiplicative model is denoted as an **AR-**

**Figure 7.21** Typical acf for a Multiplicative Seasonal Model

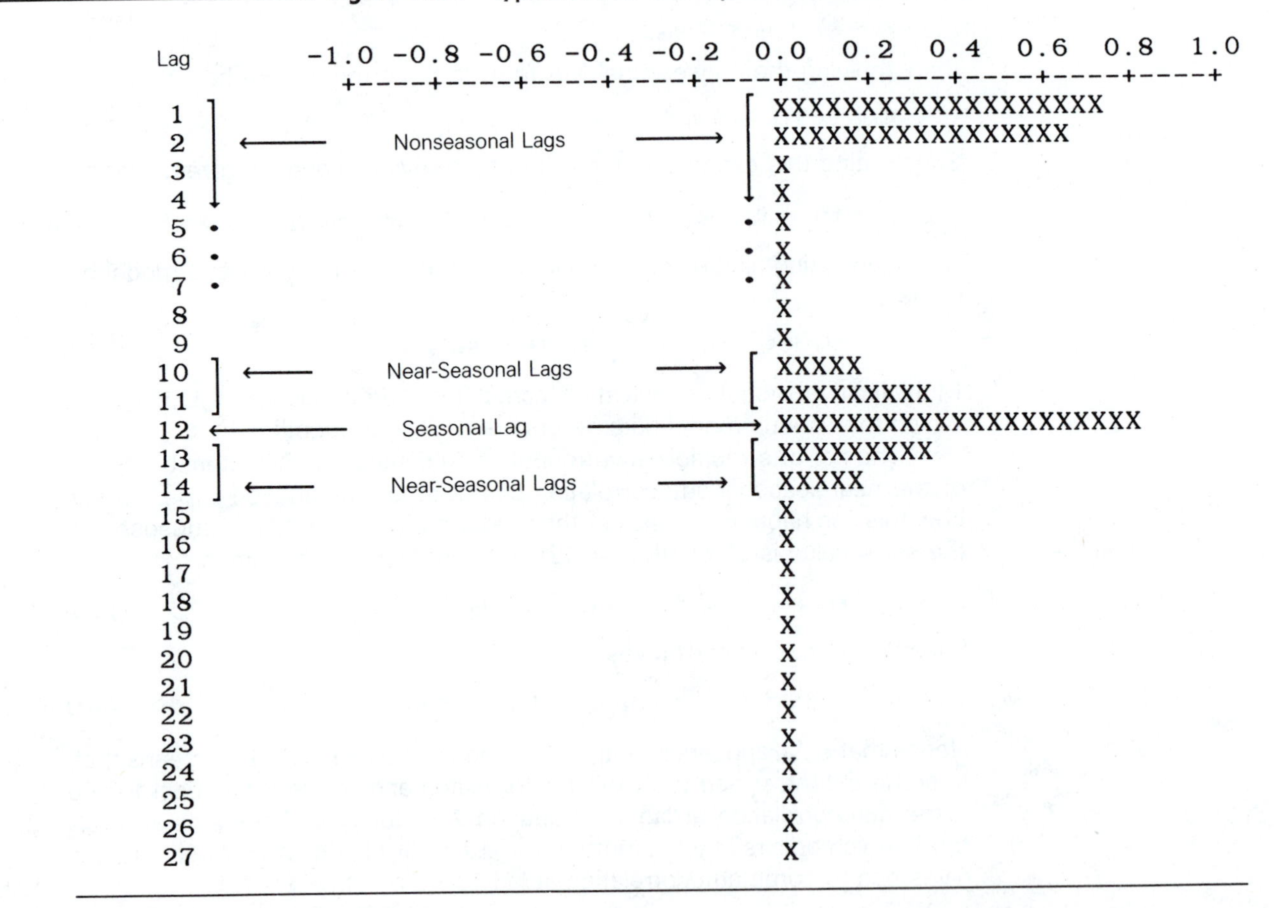

**IMA(*p,d,q*)(*P,D,Q*)$_L$** model. To illustrate: A monthly series with two regular AR terms and one seasonal MA term, which was differenced once seasonally and once nonseasonally, would be denoted ARIMA(2,1,0)(0,1,1)$_{12}$. Similarly, ARIMA(0,1,1)(0,1,2)$_4$ refers to a quarterly series that was differenced once seasonally and once nonseasonally and has one nonseasonal MA term and two seasonal AR terms.

### EXAMPLE 7.14 *Identifying a Seasonal Model, U.S. Retail Sales*

Retail sales normally exhibit very strong seasonal patterns. Table 7.10 lists the estimated monthly sales of all U.S. retail stores over the period 1980–1986. The graph of this series in Figure 7.22 shows the typical strong growth around November–December and sharp decline in January–February.

**Table 7.10 Estimated Monthly Sales ($$10^6$), All U.S. Retail Stores, 1980–1986**

| | 1980 | 1981 | 1982 | 1983 | 1984 | 1985 | 1986 |
|---|---|---|---|---|---|---|---|
| Jan | $69,449 | $ 77,361 | $ 76,508 | $ 81,342 | $ 93,089 | $ 98,817 | $105,642 |
| Feb | 69,575 | 73,727 | 75,557 | 78,884 | 93,686 | 95,585 | 99,661 |
| Mar | 74,942 | 83,921 | 86,129 | 93,760 | 104,294 | 110,167 | 114,236 |
| Apr | 74,209 | 85,210 | 87,502 | 93,970 | 104,344 | 113,107 | 115,710 |
| May | 78,215 | 86,899 | 90,347 | 97,840 | 111,312 | 120,337 | 125,421 |
| Jun | 76,442 | 87,309 | 88,426 | 100,611 | 111,980 | 114,962 | 120,351 |
| Jul | 78,937 | 88,248 | 90,600 | 99,563 | 106,553 | 115,490 | 120,736 |
| Aug | 80,780 | 89,046 | 89,130 | 100,228 | 110,650 | 121,122 | 124,059 |
| Sep | 76,650 | 85,522 | 87,755 | 97,970 | 103,932 | 114,171 | 124,645 |
| Oct | 82,997 | 88,779 | 90,877 | 100,665 | 109,229 | 116,144 | 123,055 |
| Nov | 82,835 | 87,331 | 93,878 | 103,865 | 113,276 | 118,556 | 120,789 |
| Dec | 99,588 | 106,069 | 112,348 | 125,666 | 131,790 | 139,467 | 151,493 |

**Figure 7.22 Minitab Graph of Estimated Monthly Sales ($$10^6$), All U.S. Retail Stores, 1980–1986**

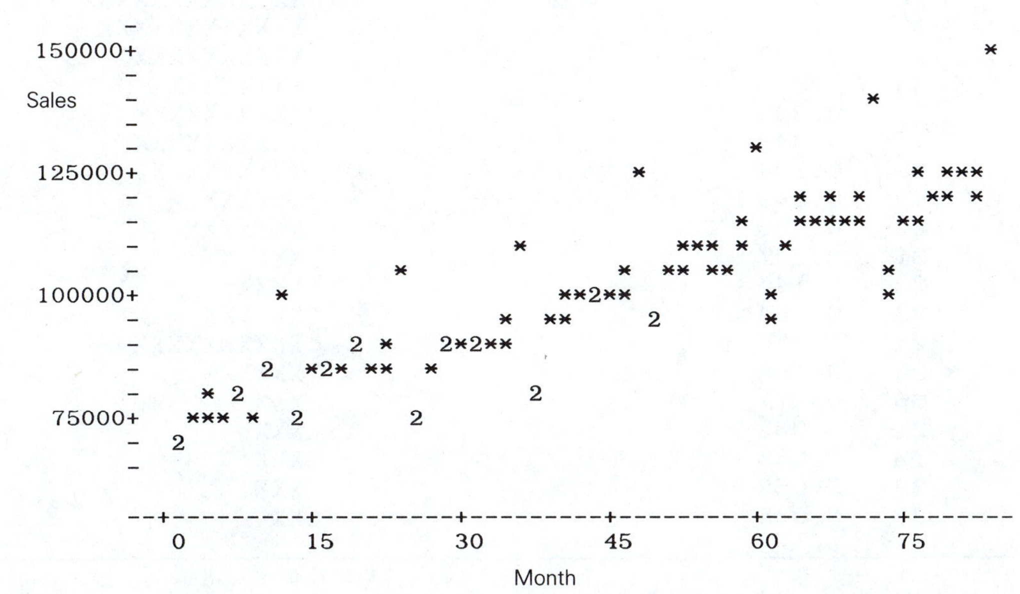

The first step in modeling this series is to examine its acf (Figure 7.23). The slow decline in the early, nonseasonal lags indicates the need for regular differencing. The seasonal spikes at lags 12 and 24, although noticeable, are somewhat masked by the effect of the nonseasonal part of the acf. We difference (regular) once; the acf of the new series appears in Figure 7.24. The nonseasonal part of the series now seems to be stationary, and the spikes at the seasonal lags are clear. The seasonal autocorrelation coefficients seem to be decaying slowly, which calls for differencing seasonally. Figures 7.25 and 7.26 show the acf and pacf of the series after seasonally differencing the (regularly) differenced series. From the acf in Figure 7.25, it seems that stationarity has now been achieved in both the seasonal and nonseasonal parts of the series, so we will use $d = 1$ and $D = 1$ in the rest of the identification process.

Concentrating on the nonseasonal lags, we see that the pacf appears to cut off sharply after lag 2, while the acf dies out rapidly. This behavior is characteristic of a

**Figure 7.23** acf of Data in Table 7.10, Undifferenced

```
                 -1.0 -0.8 -0.6 -0.4 -0.2  0.0  0.2  0.4  0.6  0.8  1.0
                   +----+----+----+----+----+----+----+----+----+----+
  1    0.721                                XXXXXXXXXXXXXXXXXXX
  2    0.632                                XXXXXXXXXXXXXXXXX
  3    0.634                                XXXXXXXXXXXXXXXXX
  4    0.631                                XXXXXXXXXXXXXXXXX
  5    0.627                                XXXXXXXXXXXXXXXXX
  6    0.581                                XXXXXXXXXXXXXXXX
  7    0.571                                XXXXXXXXXXXXXXX
  8    0.516                                XXXXXXXXXXXXXX
  9    0.470                                XXXXXXXXXXXXX
 10    0.405                                XXXXXXXXXXX
 11    0.456                                XXXXXXXXXXXX
 12    0.660                                XXXXXXXXXXXXXXXXX
 13    0.432                                XXXXXXXXXXXX
 14    0.343                                XXXXXXXXXX
 15    0.335                                XXXXXXXXX
 16    0.345                                XXXXXXXXXX
 17    0.343                                XXXXXXXXXX
 18    0.304                                XXXXXXXXX
 19    0.289                                XXXXXXXX
 20    0.246                                XXXXXXX
 21    0.207                                XXXXXX
 22    0.146                                XXXXX
 23    0.184                                XXXXXX
 24    0.355                                XXXXXXXXXX
 25    0.158                                XXXXX
 26    0.077                                XXX
 27    0.062                                XXX
 28    0.065                                XXX
 29    0.068                                XXX
 30    0.035                                XX
```

**Figure 7.24** acf of Data in Table 7.10, One Regular Differencing

```
          -1.0 -0.8 -0.6 -0.4 -0.2  0.0  0.2  0.4  0.6  0.8  1.0
            +----+----+----+----+----+----+----+----+----+----+
  1  -0.313                  XXXXXXXXX
  2  -0.220                    XXXXXXX
  3  -0.005                          X
  4   0.013                         X
  5   0.108                         XXXX
  6  -0.081                        XXX
  7   0.083                         XXX
  8   0.009                         X
  9   0.071                         XXX
 10  -0.215                     XXXXXX
 11  -0.352                 XXXXXXXXXX
 12   0.807                         XXXXXXXXXXXXXXXXXXXXX
 13  -0.249                    XXXXXXX
 14  -0.160                      XXXXX
 15  -0.033                         XX
 16  -0.007                          X
 17   0.115                         XXXX
 18  -0.057                         XX
 19   0.041                         XX
 20   0.010                         X
 21   0.072                         XXX
 22  -0.171                      XXXXX
 23  -0.290                   XXXXXXXX
 24   0.637                         XXXXXXXXXXXXXXXXX
 25  -0.195                     XXXXXX
 26  -0.117                       XXXX
 27  -0.018                          X
 28  -0.028                         XX
 29   0.099                         XXX
 30  -0.043                         XX
 31   0.012                         X
 32   0.035                         XX
 33   0.045                         XX
 34  -0.124                       XXXX
 35  -0.241                    XXXXXXX
 36   0.491                         XXXXXXXXXXXXX
 37  -0.140                       XXXX
 38  -0.075                        XXX
 39  -0.023                         XX
 40  -0.032                         XX
 41   0.103                         XXXX
 42  -0.050                         XX
 43   0.010                         X
 44   0.030                         XX
 45   0.043                         XX
 46  -0.077                        XXX
 47  -0.189                     XXXXXX
 48   0.330                         XXXXXXXXX
```

**Figure 7.25** acf of Data in Table 7.10, One Regular and One Seasonal Differencing

```
         -1.0 -0.8 -0.6 -0.4 -0.2  0.0  0.2  0.4  0.6  0.8  1.0
           +----+----+----+----+----+----+----+----+----+----+
 1  -0.375                 XXXXXXXXXX
 2  -0.137                       XXXX
 3   0.161                          XXXXX
 4  -0.036                         XX
 5   0.055                          XX
 6   0.064                          XXX
 7  -0.176                      XXXXX
 8   0.076                          XXX
 9   0.116                          XXXX
10  -0.247                    XXXXXXX
11   0.137                          XXXX
12  -0.081                        XXX
13  -0.026                         XX
14   0.071                          XXX
15  -0.052                         XX
16  -0.095                        XXX
17   0.159                          XXXXX
18  -0.111                       XXXX
19   0.013                          X
20  -0.008                          X
21   0.057                          XX
22  -0.111                       XXXX
23   0.071                          XXX
24  -0.073                        XXX
25   0.028                          XX
26   0.037                          XX
27  -0.131                       XXXX
28   0.078                          XXX
29   0.010                          X
30  -0.033                         XX
31  -0.050                         XX
32   0.122                          XXXX
33  -0.063                        XXX
34   0.016                          X
35  -0.006                          X
36  -0.031                         XX
37   0.099                          XXX
38  -0.017                          X
39  -0.061                        XXX
40   0.046                          XX
41  -0.023                         XX
42  -0.004                          X
43   0.055                          XX
44  -0.065                        XXX
45   0.030                          XX
46   0.070                          XXX
47  -0.103                       XXXX
48   0.044                          XX
```

**Figure 7.26** pacf of Data in Table 7.10, One Regular and One Seasonal Differencing

```
           -1.0 -0.8 -0.6 -0.4 -0.2  0.0  0.2  0.4  0.6  0.8  1.0
             +----+----+----+----+----+----+----+----+----+----+
 1  -0.375                   XXXXXXXXXX
 2  -0.324                    XXXXXXXXX
 3  -0.038                         XX
 4  -0.020                          X
 5   0.103                          XXXX
 6   0.165                          XXXXX
 7  -0.068                        XXX
 8  -0.032                         XX
 9   0.074                          XXX
10  -0.186                     XXXXXX
11  -0.021                         XX
12  -0.146                      XXXXX
13  -0.067                        XXX
14  -0.023                         XX
15   0.007                          X
16  -0.070                        XXX
17   0.058                          XX
18  -0.040                         XX
19   0.018                          X
20  -0.111                       XXXX
21   0.092                          XXX
22  -0.164                      XXXXX
23  -0.017                          X
24  -0.130                       XXXX
25  -0.030                         XX
26  -0.051                         XX
27  -0.088                        XXX
28  -0.054                         XX
29   0.004                          X
30  -0.053                         XX
31  -0.057                         XX
32   0.003                          X
33   0.045                          XX
34  -0.069                        XXX
35  -0.024                         XX
36  -0.104                       XXXX
37   0.019                          X
38  -0.043                         XX
39  -0.007                          X
40  -0.065                        XXX
41  -0.056                         XX
42  -0.091                        XXX
43   0.011                          X
44  -0.060                         XX
45   0.036                          XX
46   0.011                          X
47  -0.032                         XX
48  -0.009                          X
```

nonseasonal AR(2) model, so a reasonable choice for the nonseasonal part of the model would be $p = 2$, $d = 1$, and $q = 0$.

Examining the seasonal lags ($k = 12, 24, 36, \ldots$) and the near-seasonal lags, we see that neither the acf nor the pacf has any significant spikes, which means that no seasonal AR or MA terms appear to be necessary. Tentatively, then, we would choose $P = 0$, $D = 1$, and $Q = 0$ for the seasonal part of the model.

From this analysis, an ARIMA$(2,1,0)(0,1,0)_{12}$ model has been identified. It now remains to estimate the parameters of this model (with a computer program) and submit the fit to diagnostic testing. ●

### 7.8.3 Diagnostic Checking

Diagnostic checking is, again, similar for seasonal and nonseasonal models. The residuals from the fitted model are used to detect signs of inadequacy.

The Box-Ljung test (Sections 2.6.6 and 7.7.4) is conducted as before, but with two small changes. For an ARIMA$(p,d,q)(P,D,Q)_L$ model the statistic is calculated from

$$Q_m = n(n + 2) \cdot \sum_{1}^{m} \frac{r_k^2}{n - k}$$

where $n$ is the number of terms in the series *after* differencing and $m$ is the number of autocorrelation coefficients being tested. The number of degrees of freedom used in testing the chi-square statistic, $Q_m$, is

$$\text{degrees of freedom} = m - (p + q) - (P + Q)$$

For seasonal models, most computer programs calculate the Box-Ljung statistic at a few multiples of the seasonality $L$.

**EXAMPLE 7.15** *Diagnostic Checking, ARIMA$(2,1,0)(0,1,0)_{12}$ Model, U.S. Retail Sales*

Figure 7.27 shows part of a Minitab printout from fitting the ARIMA$(2,1,0)(0,1,0)_{12}$ model to the retail sales data of Table 7.10. The *t*-values for the parameters are significant (at $\alpha = .01$), and the Box-Ljung (modified Box-Pierce) test, which is calculated for lags 12, 24, 36, and 48, indicates that there is no significant autocorrelation in the residuals.

**Figure 7.27** Minitab Printout, ARIMA$(2,1,0)(0,1,0)_{12}$ Model for Data of Table 7.10

```
Estimates at each iteration
Iteration          SSE       Parameters
    0          778644224     0.100      0.100
    1          673737856    -0.050     -0.005
    2          594654208    -0.200     -0.110
    3          541531648    -0.350     -0.217
    4          514544768    -0.500     -0.326
    5          510975232    -0.571     -0.381
    6          510951808    -0.576     -0.386
    7          510951488    -0.576     -0.387
    8          510951488    -0.576     -0.387
Relative change in each estimate less than   0.0010
```

**Figure 7.27** Minitab Printout (*Continued*)

Final Estimates of Parameters

| Type | | Estimate | St. Dev. | t-ratio |
|---|---|---|---|---|
| AR | 1 | -0.5761 | 0.1188 | -4.85 |
| AR | 2 | -0.3870 | 0.1209 | -3.20 |

```
Differencing: 1 regular, 1 seasonal of order 12
No. of obs.:  Original series 84, after differencing 71
Residuals:    SS =  509203776  (backforecasts excluded)
              MS =    7379764  DF = 69
```

Continue?
Modified Box-Pierce chisquare statistic

| Lag | 12 | 24 | 36 | 48 |
|---|---|---|---|---|
| Chisquare | 7.5(DF=10) | 13.5(DF=22) | 18.0(DF=34) | 22.3(DF=46) |

Forecasts from period 84

| | | 95 Percent Limits | | |
|---|---|---|---|---|
| Period | Forecast | Lower | Upper | Actual |
| 85 | 113837 | 108511 | 119162 | |
| 86 | 106273 | 100488 | 112057 | |
| 87 | 123242 | 117134 | 129351 | |
| 88 | 123949 | 116997 | 130902 | |
| 89 | 133175 | 125734 | 140617 | |
| 90 | 128682 | 120827 | 136536 | |
| 91 | 128922 | 120567 | 137278 | |
| 92 | 132105 | 123321 | 140890 | |
| 93 | 132828 | 123649 | 142006 | |
| 94 | 131213 | 121633 | 140793 | |
| 95 | 128909 | 118951 | 138866 | |
| 96 | 159644 | 149328 | 169961 | |

Continuing with the analysis: A histogram (dot plot) of the residuals and a plot of the residuals versus *t* are shown in Figure 7.28. The histogram supports the contention that the residuals are approximately normal, and the plot of $e_t$ versus *t* doesn't show any signs of heteroscedasticity (changes in the variability of the error terms). In summary, the ARIMA(2,1,0)(0,1,0)$_{12}$ model appears to provide a reasonable fit to the data. ●

**Figure 7.28** Histogram of Residuals and Plot of Residuals vs Time, ARIMA(2,1,0)(0,1,0)$_{12}$ Model, Estimated Sales ($\$10^6$), All U.S. Retail Stores

(a) Dot Plot

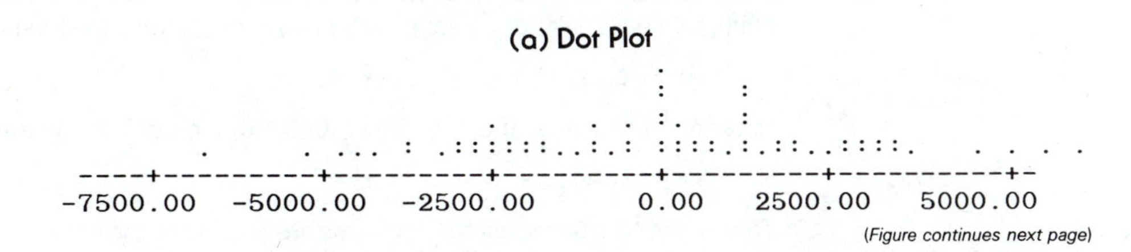


*(Figure continues next page)*

**Figure 7.28** Histogram of Residuals and Plot of Residuals vs Time (*Continued*)

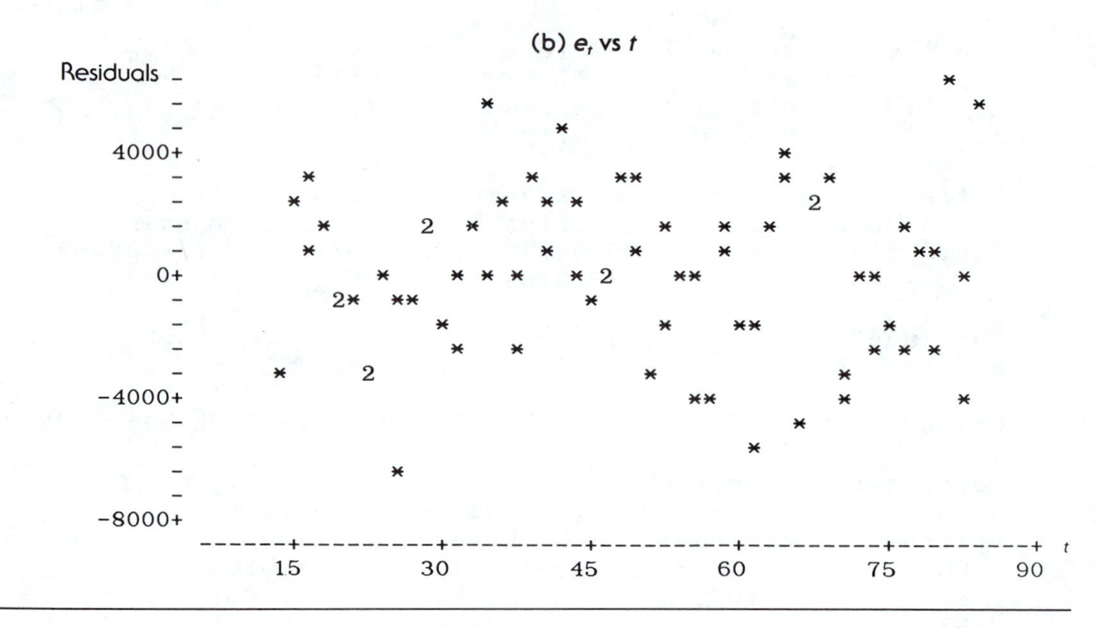


## 7.8.4 Forecasting

Once again, the procedure for forecasting with seasonal ARIMA models is similar to that for nonseasonal ones (see Section 7.7.5, "Forecasting ARIMA Series" table). In the seasonal case, the most difficult step is writing down the exact form of the model. To simplify things, it is usually easiest to write the model in terms of the *differenced* series. That is, if the original series $y_t$ was nonseasonally differenced $d$ times and seasonally differenced $D$ times, then it is best to write the model in terms of the differenced series

$$w_t = \Delta_L^D(\Delta^d y_t)$$

and to then substitute forecasts of the $w_t$'s into the resulting model to obtain forecasts of the $y_t$'s. The following example demonstrates the process.

**EXAMPLE 7.16** *Forecasting, ARIMA(2,1,0)(0,1,0)$_{12}$ Model, U.S. Retail Sales*

In Example 7.14 a multiplicative ARIMA(2,1,0)(0,1,0)$_{12}$ model was fit to the U.S. retail sales data. Since it was necessary to difference the original series once seasonally and once nonseasonally to achieve stationarity, the differenced series can be written

$$w_t = \Delta_L(\Delta y_t) = y_t - y_{t-1} - y_{t-L} + y_{t-L-1} \qquad (7.32)$$

In terms of the $w_t$'s, the ARIMA(2,1,0)(0,1,0)$_{12}$ model becomes simply

$$w_t = \phi_1 w_{t-1} + \phi_2 w_{t-2} + \epsilon_t$$

That is, the two nonseasonal AR terms are applied to the series $w_t$, and no nonseasonal

MA terms (since $q = 0$) are applied to the error terms (see discussion earlier in this section).

Putting in the estimated parameters from the computer printout (Figure 7.27) gives

$$w_t = -.5761w_{t-1} - .3870w_{t-2} \quad \textbf{(7.33)}$$

To forecast January 1987 sales, $y_{85}$, one only has to put $t = 85$ in Equation 7.33:

$$w_{85} = -.5761w_{84} - .3870w_{83}$$

and then convert this into an equation for forecasting $y_{85}$.

From Equation 7.32, with $L = 12$, we find

$$\begin{aligned} w_{85} &= y_{85} - y_{84} - y_{73} + y_{72} \\ &= y_{85} - 151{,}493 - 105{,}642 + 139{,}467 = y_{85} - 117{,}668 \end{aligned}$$

$$\begin{aligned} w_{84} &= y_{84} - y_{83} - y_{72} + y_{71} \\ &= 151{,}493 - 120{,}789 - 139{,}467 + 118{,}556 = 9{,}793 \end{aligned}$$

$$\begin{aligned} w_{83} &= y_{83} - y_{82} - y_{71} + y_{70} \\ &= 120{,}789 - 123{,}055 - 118{,}556 + 116{,}144 = -4{,}678 \end{aligned}$$

Substituting these values into Equation 7.33 gives

$$y_{85} - 117{,}668 = -.5761(9{,}793) - .3870(-4{,}678)$$

or,

$$\hat{y}_{85} = 117{,}668 - .5761(9{,}793) - .3870(-4{,}678) = 113{,}837$$

(Recall that the series is measured in units of $\$10^6$ when interpreting this forecast.) Most computer programs routinely do the above manipulations for multiplicative models and generate the forecasts, so in practice one doesn't have to go to the trouble of writing down the model equation. For example, in Figure 7.27 the Minitab printout gives the forecasts for each month of 1987 (i.e., $t = 85, 86, \ldots, 96$). For enhanced understanding of how these forecasts are generated, though, it is worthwhile to write down the forecasting equation. ●

# 7.9 Diagnostics

## 7.9.1 Characteristics of a Good Model

The diagnostics discussions in Sections 7.7.4 and 7.8.3 focused on the role of diagnostic checking in the identification/respecification stage of ARIMA modeling. The same tests also signal when further respecification is unnecessary. In summary, for a given series, a good model has the following characteristics.

- The model is parsimonious.
- The residuals resemble those from a random series.
- The *t*-ratios of all model parameters are statistically significant.
- Overfitting does not lead to additional model choices.

- The parameter estimates remain stable when the model is refit to additional data.
- No parameter redundancy appears to exist.

## 7.9.2 Problems in Practice

The four phases of the Box-Jenkins method do not always proceed smoothly. Below are listed some of the more commonly occurring implementation problems and some suggested remedies.

### Problem: Differencing Doesn't Produce a Stationary Series

Not all series can be made stationary by differencing. Series with nonstationary variances (variances that change over time) may behave badly when differenced. The usual remedy in such situations is to *transform* the data (e.g., taking logarithms, square roots) and then to difference the transformed series. Forecasts of the transformed series can later be untransformed.

### Problem: Identifying Appropriate Orders ($p$ and $q$) Is Difficult

This usually happens with mixed models because the behavior of their associated acf's and pacf's is complicated. With these models one must rely on the identification rules in Table 7.7 as a guide and try overfitting to find (possibly) better choices for $p$ and $q$. Keep in mind that more than one model may adequately fit the data.

### Problem: The Backcasts Don't Converge

Most Box-Jenkins computer programs print out diagnostic messages when the backcasts do not seem to be converging to a constant value. These programs assume convergence when successive backcasts eventually lie within some specified tolerance of one another. If this convergence criterion is not met within a specified number of steps (around 100), diagnostic messages are printed. There are two main causes of this problem. One is that the series is not stationary (only stationary series have backcasts that converge). You may have to do additional differencing or transforming to achieve stationarity. The other possibility, when stationarity isn't in question, is that a better set of starting values is needed (you then have to bypass the default values used by the program and set your own).

### Problem: The Sum of Squared Errors Does Not Reduce Quickly

After each iteration, most computer programs check the sum of squared errors to see if an improvement (a reduction) has been made. This search stops when the successive improvements become smaller than some prespecified amount (often .001) or when the number of steps exceeds some prespecified maximum. If the improvements don't become small enough before the search stops, a warning message such as UNABLE TO REDUCE SUM OF SQUARES

ANY FURTHER is printed. When this happens, changing the starting values often solves the problem.

### Problem: Whether to Use a Trend Parameter or Not

As shown in Example 7.5, the usual identification procedure (differencing/analyzing the acf and pacf) isn't sufficient to distinguish stochastic from deterministic trend. One way to make this distinction is simply to fit one model with a trend parameter and one without, then perform diagnostic tests on the two models. If a trend parameter is needed, its *t*-ratio should be significant and the sum of squared errors should be noticeably smaller than when no constant term is present (Pankratz, 1983, pp. 187–189).

### Problem: Seasonality Is Present in the Data

The Box-Jenkins procedures for handling seasonality explained in Section 7.8 are complex and require even more data than the minimum suggested in this chapter (50). Another way to approach seasonality is to seasonally adjust the series using Chapter 5 techniques and apply the nonseasonal methods of this chapter to the adjusted series.

## 7.10 Exercises

**7.1** In Box-Jenkins analysis, what does the term *moving average* mean? How does it differ from the moving average encountered in earlier chapters?

**7.2** In what sense are ARIMA models integrated?

**7.3** Is the MA(1) model $y_t = 11.2 + \epsilon_t + 1.03\epsilon_{t-1}$ invertible?

**7.4** Given an ARIMA model with $p = 1$, $d = 1$, $q = 0$ and parameter $\phi_1 = .30$, write the model in a form suitable for forecasting.

**7.5** Suppose the first few sample autocorrelation coefficients of a time series of length 100 are: $r_1 = -.87$, $r_2 = .41$, $r_3 = .15$, $r_4 = -.34$, $r_5 = .22$. Use Bartlett's formula to test the corresponding population autocorrelation coefficients for significance (use $\alpha \approx .05$).

**7.6** The stationarity conditions for an AR(2) model are:

$$\phi_1 + \phi_2 < 1,\ \phi_2 - \phi_1 < 1, \qquad \text{and} \qquad |\phi_2| < 1$$

Similarly, the invertibility requirements for an MA(2) model are:

$$\theta_1 + \theta_2 < 1,\ \theta_2 - \theta_1 < 1, \qquad \text{and} \qquad |\theta_2| < 1$$

On a graph of $\phi_2$ versus $\phi_1$ (or $\theta_2$ versus $\theta_1$), show the region where the stationarity (or invertibility) conditions are met.

**7.7** What ARIMA model(s) are most likely to be associated with the following acf and pacf? (Assume that a series of length 100 was used.)

```
                                   acf
        -1.0 -0.8 -0.6 -0.4 -0.2  0.0  0.2  0.4  0.6  0.8  1.0
         +----+----+----+----+----+----+----+----+----+----+
 1  -0.702              XXXXXXXXXXXXXXXXXX
 2   0.251                                XXXXXXX
 3  -0.006                                X
 4   0.012                                X
 5   0.101                                XXX
 6  -0.185                             XXXX
 7   0.082                                XXX
 8   0.007                                X
 9   0.076                                XXX
10  -0.213                           XXXXXX
```

```
                                   pacf
        -1.0 -0.8 -0.6 -0.4 -0.2  0.0  0.2  0.4  0.6  0.8  1.0
         +----+----+----+----+----+----+----+----+----+----+
 1  -0.702              XXXXXXXXXXXXXXXXXX
 2  -0.350                       XXXXXXXXX
 3  -0.105                             XXX
 4   0.013                                X
 5   0.103                                XXXX
 6   0.067                                XXX
 7   0.072                                XXX
 8   0.003                                X
 9   0.078                                XXX
10   0.115                                XXXX
```

---

**7.8** The model $y_t = 5.14 + 1.14y_{t-1} - .62y_{t-2}$ was fit to data from a stationary series. Estimate the average value (level) of this series.

**7.9** Suppose an ARIMA(0,1,1) model was fit to a stationary series and that the estimated moving average parameter $\hat{\theta}_1$ was .71. If simple exponential smoothing had been used instead, what value of the smoothing constant ($\alpha$) would minimize the sum of squared forecast errors?

**7.10** The following models were fit to different series, each of length 72. In each case, obtain forecasts for the next 6 periods.

**a.** ARIMA(1,0,0): $\hat{\phi}_1 = .4$, $y_{72} = 100$

**b.** ARIMA(2,0,0): $\hat{\phi}_1 = 1.3$, $\hat{\phi}_2 = -.7$, $y_{72} = 50$, $y_{71} = 45$

**c.** ARIMA(0,0,1): $\hat{\theta}_0 = 100$, $\hat{\theta}_1 = .8$, $e_{72} = 12.5$

**d.** ARIMA(0,0,2): $\hat{\theta}_0 = 10$, $\hat{\theta}_1 = 1.2$, $\hat{\theta}_2 = -.8$, $e_{72} = -5$, $e_{71} = 2$

**e.** ARIMA(0,1,1): $\hat{\theta}_0 = 20$, $\hat{\theta}_1 = .7$, $e_{72} = 10$

**f.** ARIMA(1,0,1): $\hat{\theta}_0 = 50$, $\hat{\theta}_1 = .8$, $\hat{\phi}_1 = .4$, $y_{72} = 90$, $e_{72} = -10$

**g.** ARIMA(2,1,0): $\hat{\phi}_1 = .9$, $\hat{\phi}_2 = -.4$, $y_{72} = 25$, $y_{71} = 30$

**7.11** Using a computer program, perform the indicated steps to fit a seasonal ARIMA model to the following table of data.

**Retail Sales, U.S. Apparel and Accessory Stores (in $\$10^6$), 1980–1986**

| Month | 1980 | 1981 | 1982 | 1983 | 1984 | 1985 | 1986 |
|---|---|---|---|---|---|---|---|
| 1 | 3061 | 3279 | 3302 | 3496 | 3765 | 4449 | 4694 |
| 2 | 2796 | 2911 | 3168 | 3203 | 3630 | 4260 | 4401 |
| 3 | 3351 | 3448 | 3729 | 4185 | 4413 | 5548 | 6128 |
| 4 | 3549 | 3972 | 4038 | 4327 | 4857 | 5780 | 5737 |
| 5 | 3608 | 3735 | 3934 | 4264 | 4846 | 5883 | 6313 |
| 6 | 3383 | 3632 | 3649 | 4178 | 4811 | 5465 | 5680 |
| 7 | 3343 | 3598 | 3812 | 4043 | 4296 | 5239 | 5533 |
| 8 | 4010 | 4126 | 4130 | 4555 | 4970 | 6372 | 6606 |
| 9 | 3664 | 3929 | 3919 | 4312 | 4789 | 5667 | 5992 |
| 10 | 4026 | 4234 | 4157 | 4617 | 4889 | 6062 | 6349 |
| 11 | 4288 | 4271 | 4495 | 5169 | 5680 | 7015 | 7067 |
| 12 | 6539 | 6659 | 6848 | 7910 | 8367 | 10158 | 10719 |

a. Plot $y_t$ versus $t$.
b. Does the plot in step (a) indicate that the variability is changing over time? If so, transform the data (e.g., using logarithms) and check the plot of the transformed series to see if the variability has been stabilized.
c. Examine the acf of the (transformed) series and determine what (if any) orders of regular and seasonal differencing are needed to achieve stationarity.
d. Examine the acf and pacf of the stationary series in step (c) and identify a tentative ARIMA model to fit.
e. Fit the model in step (d) and obtain the residuals (from the computer program).
f. Diagnostic checking:
   (i) Create a histogram of the residuals.
   (ii) Check the $t$-ratios of the estimated parameters.
   (iii) Plot the residuals versus $t$ (a visual check for autocorrelation and nonstability in the variance of the residuals).
g. Based on the results of step (f), revise the model, if necessary, and repeat the diagnostic checks.
h. Write the forecasting equation and obtain forecasts for each month of 1987.

**7.12** An ARIMA$(0,0,1)(0,1,1)_{12}$ model and an ARIMA$(1,0,0)(0,1,1)_{12}$ model were fit to the same set of data, and in both cases the diagnostic checks showed that the fits were good.
a. Write the forecasting equations for both models.
b. Use the equations in part (a) to explain why both models may have fit the same data well.
c. From part (b), show that $\theta_1$ will approximately equal $-\phi_1$ (when either of these parameters is small).

## References

Abraham, B., and Ledolter, J. (1983). *Statistical Methods for Forecasting.* New York: Wiley.

Bartlett, M. S. (1946). "On the Theoretical Specification of Sampling Properties of Autocorrelated Time Series." *J. Royal Stat. Soc.* B8:27–41.

Box, G. E. P., and Jenkins, G. M. (1970). *Time Series Analysis, Forecasting and Control.* San Francisco: Holden-Day.

Cryer, J. D. (1986). *Time Series Analysis.* Boston: Duxbury Press.

Dickey, D. A., and Pantula, S. G. (1987). "Determining the Order of Differencing in Autoregressive Processes," *J. Bus. Econ. Stat.* 5(4): 455–461.

Hoff, J. C. (1983). *A Practical Guide to Box-Jenkins Forecasting.* Belmont, Calif.: Wadsworth.

Ljung, G. M., and Box, G. E. P. (1978). "On a Measure of Lack of Fit in Time Series Models." *Biometrika* 65: 67–72.

Makridakis, S., et al. (1984). *The Forecasting Accuracy of Major Time Series Methods.* Chichester, UK: Wiley.

Muth, J. F. (1960). "Optimal Properties of Exponentially Weighted Forecasts of Time Series with Permanent and Transitory Components." *J. Amer. Stat. Assoc.* 55: 299–306.

Nelson, C. R. (1973). *Applied Time Series Analysis for Managerial Forecasting.* San Francisco: Holden-Day.

Pankratz, A. (1983). *Forecasting with Univariate Box-Jenkins Methods: Concepts and Cases.* New York: Wiley.

Walker, G. (1931). "On Periodicity in Series of Related Terms." *Proc. Royal Soc.* A131: 518.

Yule, G. U. (1927). "On a Method of Investigating Periodicities in Disturbed Series, with Special Reference to Wölfer's Sunspot Numbers." *Phil. Trans.* A226: 267.

CHAPTER 8

# Monitoring Forecasts

**Objective:** *To study methods for monitoring the accuracy of ongoing forecasting activity, allowing reaction and intervention when apparent faults appear.*

## 8.1 Introduction

Once a forecasting scheme is in place, generating forecasts that are being acted upon, there is still a need to make sure that things are going well. The final test of a forecasting scheme must be, after all, that it sustain production of acceptably accurate forecasts.

Over time, the forces driving the time series may evolve, making models obsolete and causing forecast accuracy to deteriorate. The level of a horizontal series may change, for instance, or a trend may develop. The parameter values or functional form of a trend may change. Seasonal patterns may be volatile as well. From time to time then, the model may need to be altered to reflect new conditions.

Updating can help in such situations, but even these schemes will tend to lag changes considerably, or require very large smoothing constants (which can be undesirable because they track the random errors so closely). It is better, as a rule, to build into the forecasting scheme itself some sort of accuracy monitor or *tracking signal,* and then take whatever corrective steps may be appropriate when the monitor signals a loss of accuracy. The process of recognizing problems and altering the method of generating forecasts to compensate for them is thought of as adapting to the changes, and the term *adaptive forecasting* is often used to describe such activity.

In the process of developing monitoring devices, we will discuss in more detail:

- what *adapting* means, both intuitively and mathematically, as applied to forecasting schemes
- practical considerations calling for the use of adaptive methods

- different types of quantities that may be used as monitoring devices (tracking signals), and, in terms of these monitors, what constitutes a distress or out-of-control signal
- how to react to a loss-of-control signal
- the consequences, both desirable and undesirable, of building adaptivity into forecasting schemes

# 8.2 Adaptivity and Forecast Monitors

## 8.2.1 Introduction

In a way, any scheme may be considered adaptive if it compares forecast and actual values as they become available and then uses the comparison to influence future forecasts. Such a process is said to incorporate *feedback* or a *feedback loop,* since a measure of performance of the system's output is returned as input (see Figure 8.1), thereby influencing future output. Updating schemes such as SES, Brown's DES, Holt's LES, and the Holt-Winters method, which employ exponential smoothing, are adaptive in this sense since the forecast errors are fed back (see Figure 8.2) to become part of future forecasts.

It is more common, however, to reserve use of the terms *adaptive* and *adaptive control* for schemes that use feedback to enable the adaptive fore-

**Figure 8.1 An Adaptive Forecasting Process**

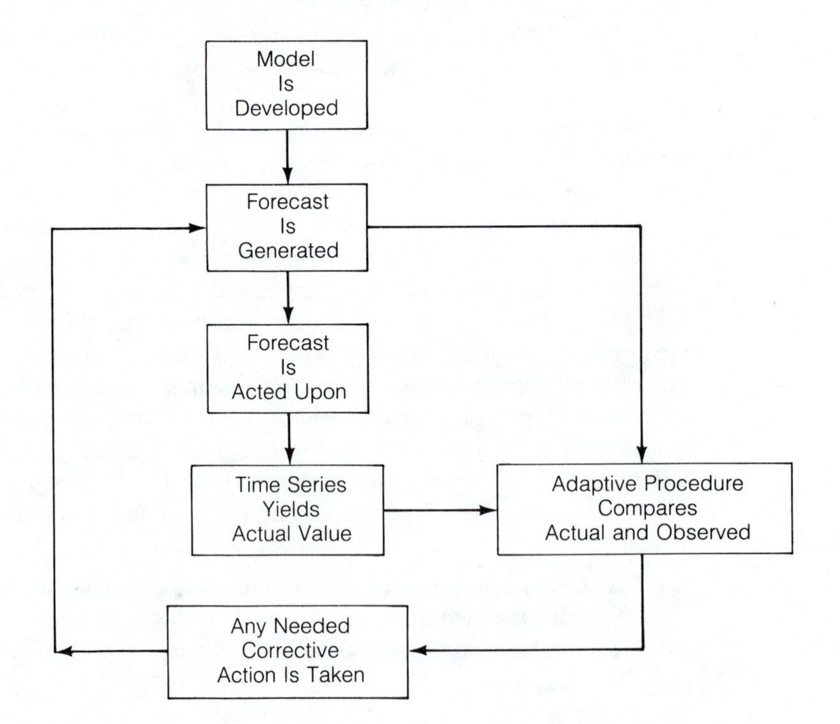

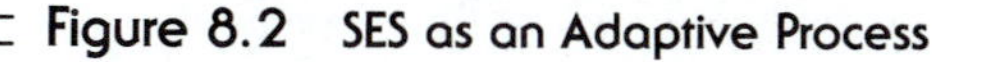

Figure 8.2 SES as an Adaptive Process

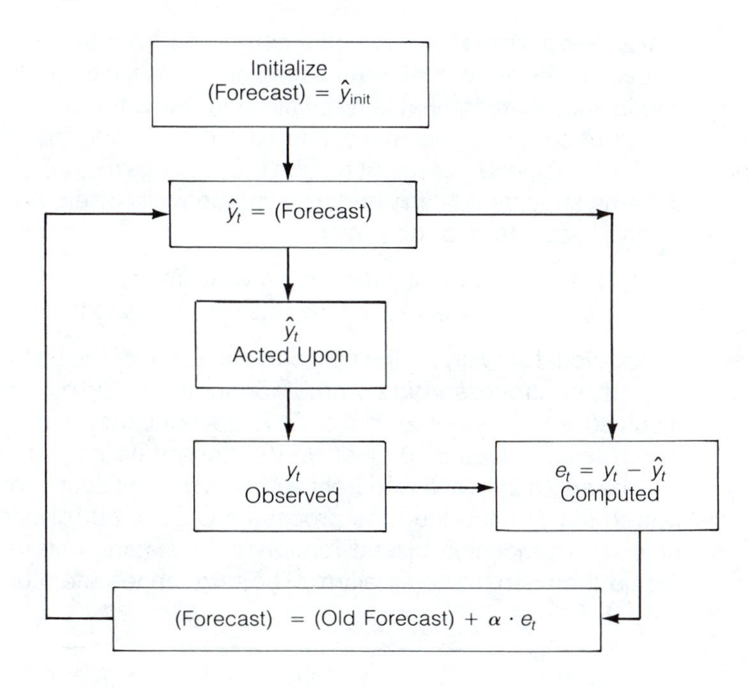


casting mechanisms to respond more rapidly than their nonadaptive counterparts to marked changes in the characteristics of a series. *Adaptivity* for single forecasts (though not often seen in that context) may be taken to imply that feedback determines whether it is necessary to refit a model or to reestimate the model parameters. The term is mostly reserved for use with updating schemes. Such schemes already use feedback to alter estimates of *model* parameters ($\hat{\beta}_0, \hat{\beta}_1, \ldots$). They may be considered adaptive schemes if, in addition, feedback is used to alter the *forecasting* parameters (the length of a moving average, the smoothing constants, etc.), thereby changing the responsiveness (tracking power) of the system (see Gilchrist, 1976).

## 8.2.2 Notation

In the discussion of the techniques involved in adaptive schemes it is assumed that a forecasting procedure is in place. For simplicity, the time scale is adjusted so that the first forecast is $\hat{y}_1$ (for $y_1$) and is made at time $t = 0$. As of general time $t$, then, the procedure will have produced forecasts $\hat{y}_1, \hat{y}_2, \ldots, \hat{y}_t$ for $y_1, y_2, \ldots, y_t$, resulting in forecast errors $e_1, e_2, \ldots, e_t$, and will be generating a forecast for $y_{t+1}$ or later values of the series.

All the measures of forecasting performance used to monitor the process are based on the sequence of $e_t$'s. There are two types of adaptive mechanisms: (1) tracking signals and (2) dynamic response rates.

### 8.2.3 Tracking Signals

A **tracking signal** is a quantity computed from the $e_t$'s that should take on values in one range if the forecasting process is in control (i.e., acting properly) and in another range if it is not. Falling in the latter range constitutes an *alarm* or *out-of-control* signal, indicating corrective action may be needed.

For instance, we might reason that if things are going well, the forecasting scheme should over- and underforecast equally often. A tracking signal based on this rationale might be, say,

$TS_t$ = the value of the tracking signal at time $t$
= the number of *over*forecasts out of the last 10 forecasts

which would actually be the number of $e_t$'s, out of the last 10, that are negative.

If the process is *in* control, $TS_t$ should be 5; but sampling variability is involved, so values of 2, 3, 4 or 6, 7, 8 would not be unduly alarming. On the other hand, values of 0, 1 or 9, 10, since their combined probability would only be about 2% for a well-behaved process, indicate something has probably gotten out of hand (i.e., the process has gone out of control) and, for some reason, is producing biased forecasts. Observing one of these latter values would then constitute an alarm. The two ranges mentioned above would be

$$\underbrace{0 \quad 1}_{\text{alarm}} \qquad \underbrace{2 \quad 3 \quad 4 \quad 5 \quad 6 \quad 7 \quad 8}_{\text{in control}} \qquad \underbrace{9 \quad 10}_{\text{alarm}}$$

The monitoring device could be stated more concisely as follows.

**Simple Forecast Monitor:**

1. At each time $t$, compute

   $TS_t$ = the number of overforecasts out of the last 10 forecasts

2. Then,

   $2 \leq TS_t \leq 8$ in control
   $TS_t < 2$ or $TS_t > 8$ alarm

Many other more sophisticated and more sensitive types of tracking signals are possible, of course, and they will be discussed as the chapter progresses.

An alarm generally calls for some sort of response. The responses can be classified into three broad categories:

- **intervention** The forecaster whose attention has been drawn by the alarm analyzes the current forecasting picture and decides which (or whether) actions might correct the fault.
- **semiautomatic** A predetermined course of action is undertaken at the discretion of the forecaster, who first judges the validity of the alarm and any apparent reasons for it.
- **automatic** A predetermined course of action is carried out without further investigation of the reason(s) for the alarm. The action should be selected so it tends to correct the bad forecasts regardless of the nature of the disturbance

but not so drastically as to badly degrade the forecasts if the alarm happens to be erroneous.

### 8.2.4 Dynamic Response Rates

When using updating schemes, tracking ability, often called **response rate,** is determined by a forecasting parameter such as the length of a moving average or the size of one or more smoothing constants. It is possible to make this response rate parameter **dynamic** by making it a function of the forecast errors, that is, increasing the rate in proportion to the amount of deterioration in forecasts. It is also possible to combine the two adaptive approaches by increasing the response rate when an out-of-control signal occurs.

### 8.2.5 Corrective Actions

If tracking signals are being used, once an alarm is triggered, the intensity of the action taken to correct the faults will depend on the specific forecasting context. But in general, the possibilities include the following.

#### Single Forecast Schemes

- **change the model** We may wish to discard the previously fit model altogether and start over from scratch, repeating the entire model-fitting process on only the more recent data. This may require fitting a rather small history and will likely warrant closer than usual scrutiny (perhaps closer tolerances on the tracking signal) of the accuracy of the first few forecasts from the new model. Changing to an updating procedure, if possible, may be a wise choice at this time, since the series will have demonstrated instability.
- **reestimate model parameters** It may be sufficient to limit the action to reestimating some or all of the model parameters, discarding older data. Which parameters need to be refit will depend on the judgment of the analyst. Thus, for example, one may merely reestimate the slope and intercept of a linear trend model, leaving, say, the seasonal indexes unchanged.

#### Updating Schemes

- **reevaluate forecasting parameters** The scheme's response parameters may be adjusted to allow for developing dynamism in the series. Thus, one might shorten the length of a moving average or increase one or more of the smoothing constants in an exponential smoothing model.
- **reinitialize** New estimates of the model parameters, such as $\hat{\beta}_0(t)$, $\hat{\beta}_1(t)$, and $\hat{S}_i(t)$, may be substituted for the ones produced by the scheme, removing the influence of older observations. The new estimates can be fit to recent data, or subjectively estimated from inspection of the series or the forecast errors. Thus, we might examine the graph of a supposedly horizontal series, notice that a jump has occurred in the level, and subjectively estimate the new average level from the graph, allowing only observations *after* the apparent shift to influence the estimate.

## 8.3 Examples

### 8.3.1 Introduction

Deterioration in the quality of forecasts produced by a model that worked well initially is usually thought to have been caused by changes in the forces driving the series. Degradation can also occur quite quickly if, for one reason or another, the initial model fit was inaccurate. The following examples illustrate the kinds of problems which can occur.

### 8.3.2 Example 8.1—Consumer's Connection, Inc.

Consumer's Connection, Inc., is a large mail-marketing firm dealing in a line of consumer novelties that includes various household gadgets, specialty stationery, inexpensive electronics, small appliances, and so forth. For inventory control purposes, the firm forecasts biweekly demand for a large number of product items. Examination of the recent history of the demand for its small cordless pencil sharpener has led to choosing a single forecast, horizontal model with an estimated biweekly demand of $\hat{\beta}_0 = 410$ units. Figure 8.3 and Table 8.1 show how this forecasting model performed over the first seven months of use. From the graph of the series we can discern a tendency to underforecast for the last several periods of operation.

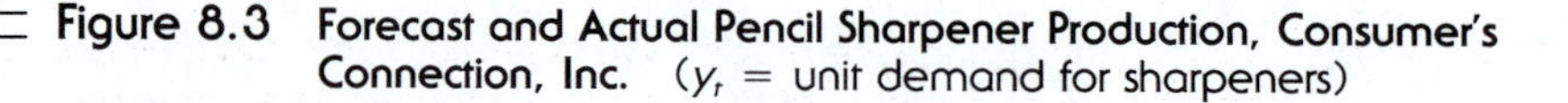

**Figure 8.3** **Forecast and Actual Pencil Sharpener Production, Consumer's Connection, Inc.** ($y_t$ = unit demand for sharpeners)

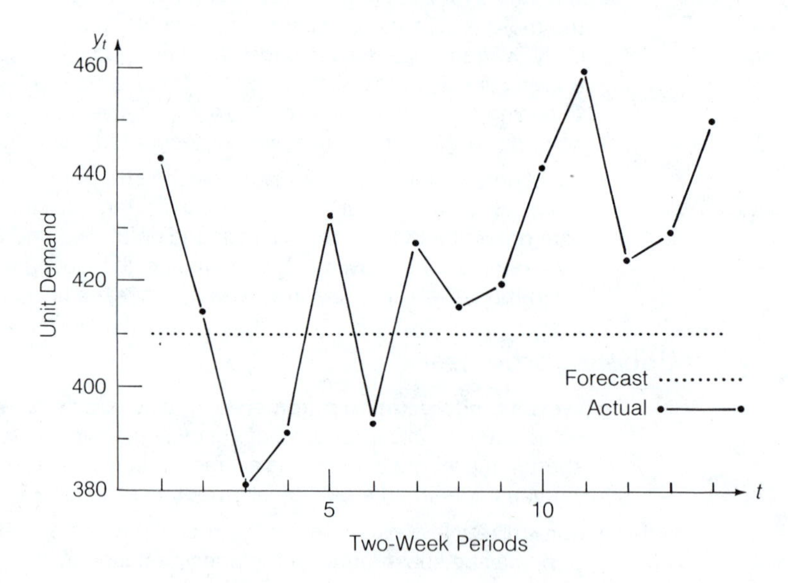


### 8.3.3 Example 8.2—Caston & Sons Fittings

Caston & Sons manufactures a special type of pipe-connecting device for which it holds the exclusive patent rights. To use the device, a buyer must submit precut lengths of steel pipe. Caston custom fits the connector to each end of these pieces of pipe.

**Table 8.1 Forecast and Actual Demand for Pencil Sharpeners, Consumer's Connection, Inc., March–August 1986** $y_t$ = unit demand

| Two-Week Period | $t$ | $y_t$ | $\hat{y}_t$ | $e_t$ |
|---|---|---|---|---|
| Mar 03 – Mar 11 | 1 | 443 | 410 | 33 |
| Mar 17 – Mar 28 | 2 | 414 | 410 | 4 |
| Mar 30 – Apr 11 | 3 | 381 | 410 | −29 |
| Apr 14 – Apr 25 | 4 | 391 | 410 | −19 |
| Apr 28 – May 02 | 5 | 432 | 410 | 22 |
| May 05 – May 16 | 6 | 393 | 410 | −17 |
| May 19 – May 30 | 7 | 427 | 410 | 17 |
| Jun 02 – Jun 13 | 8 | 415 | 410 | 5 |
| Jun 16 – Jun 27 | 9 | 419 | 410 | 9 |
| Jun 30 – Jul 11 | 10 | 441 | 410 | 31 |
| Jul 14 – Jul 25 | 11 | 459 | 410 | 49 |
| Jul 28 – Aug 08 | 12 | 424 | 410 | 14 |
| Aug 11 – Aug 22 | 13 | 429 | 410 | 19 |
| Aug 25 – Sep 05 | 14 | 450 | 410 | 40 |

The process requires special, expensive, and difficult-to-obtain equipment, and Caston is constantly revising its methods to improve production efficiency. To help in quoting shipment dates to customers, management decides to forecast weekly production figures. Records for several months prior to the implementation of the forecasting scheme indicate a fairly constant average production level of about 1299 units per hour. Because it is anticipated that improvement efforts will likely slowly increase this level, a simple exponential smoothing scheme with a 10% smoothing constant is adopted. Experience with the new scheme for the first 16 weeks is shown in Table 8.2 and

**Table 8.2 Forecast and Actual Pipe Fitting Production, Caston & Sons Fittings** $y_t$ = average hourly fittings produced

| Week | $t$ | $y_t$ | $\hat{y}_t$ | $e_t$ |
|---|---|---|---|---|
| Jan 06 | 1 | 1291 | 1299 | −8 |
| Jan 13 | 2 | 1311 | 1298 | 13 |
| Jan 20 | 3 | 1284 | 1299 | −15 |
| Jan 27 | 4 | 1315 | 1298 | 17 |
| Feb 03 | 5 | 1309 | 1300 | 9 |
| Feb 10 | 6 | 1295 | 1301 | −6 |
| Feb 17 | 7 | 1317 | 1300 | 17 |
| Feb 24 | 8 | 1311 | 1302 | 9 |
| Mar 03 | 9 | 1307 | 1303 | 4 |
| Mar 10 | 10 | 1337 | 1303 | 34 |
| Mar 17 | 11 | 1325 | 1306 | 19 |
| Mar 24 | 12 | 1330 | 1308 | 22 |
| Mar 31 | 13 | 1315 | 1310 | 5 |
| Apr 07 | 14 | 1327 | 1311 | 16 |
| Apr 14 | 15 | 1325 | 1313 | 12 |
| Apr 21 | 16 | 1322 | 1314 | 8 |

graphed in Figure 8.4. It appears that around the ninth or tenth week there was a notable increase in production capacity. The heavily smoothed ($\alpha = 10\%$) SES scheme is still lagging far below this increase, even at the end of the sixteenth week.

**Figure 8.4 Forecast and Actual Production for Caston & Sons Fittings** ($y_t$ = average hourly fittings produced)

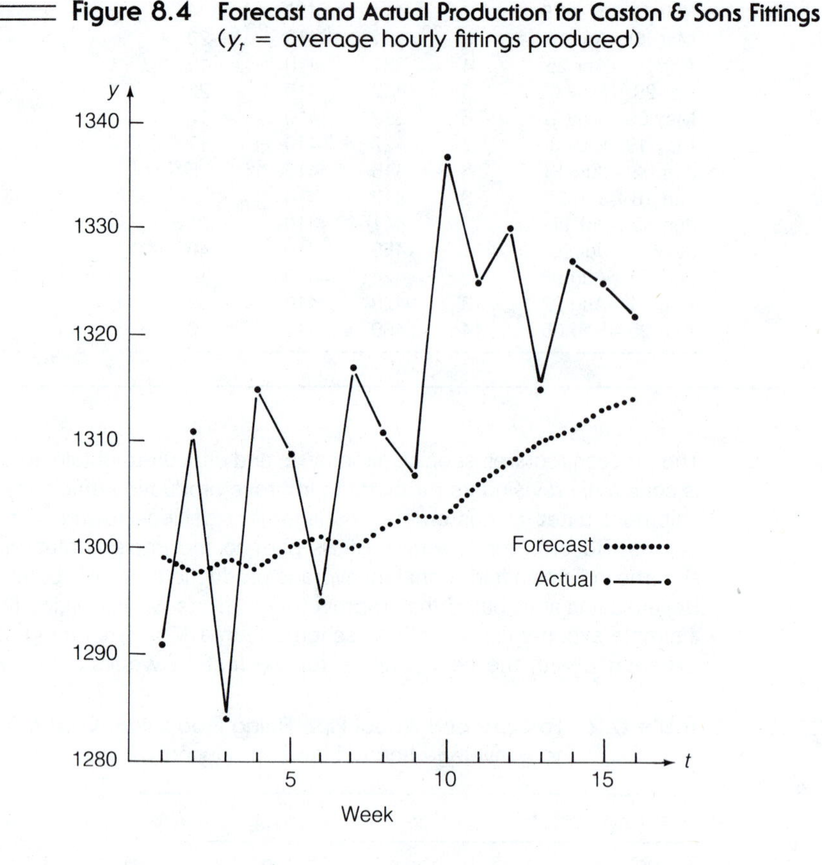


## 8.3.4 Example 8.3—Copy Machine Usage

To help maintain control of operating costs, a large company monitors and forecasts each month's copy machine usage. The records for the two years immediately prior to instituting its forecasting scheme indicated that the series was trended, so it was decided to use linear exponential smoothing. A parameter search with these past data indicated that $\alpha = .1$, $\gamma = .1$ would be reasonable smoothing constants. Backcasting was used to give initial estimates of the level, $\hat{T}_0(0) = 303.1$, and slope, $\hat{\beta}_1(0) = 5.0$, so the initial forecast for May 1985 was $\hat{y}_1 = 303.1 + 5.0 = 308.1$. Experience with the LES forecasting scheme for the first 20 months of use are given in Table 8.3 and graphed in Figure 8.5. It appears that there was a jump in the level of the series in May 1986, at which time the slope seemed to increase (estimates of the slope, delivered in the process of computing the LES forecasts, were increasing). Then, in November 1986 the level dropped and the slope appeared to return to its old, or perhaps a slightly

**Table 8.3** **Actual and Forecast Copier Usage, Example 8.3, Linear Exponential Smoothing ($\alpha$ = .1, $\gamma$ = .1)**
$y_t$ = average daily number of copies

| Month | $t$ | $y_t$ | $\hat{y}_t$ | $e_t$ |
|---|---|---|---|---|
| Sep 85 | 1 | 3001 | 3081 | −80 |
| Oct 85 | 2 | 3072 | 3122 | −50 |
| Nov 85 | 3 | 3256 | 3166 | 90 |
| Dec 85 | 4 | 3227 | 3225 | 2 |
| Jan 86 | 5 | 3159 | 3275 | −116 |
| Feb 86 | 6 | 3323 | 3312 | 11 |
| Mar 86 | 7 | 3461 | 3362 | 99 |
| Apr 86 | 8 | 3444 | 3422 | 22 |
| May 86 | 9 | 3864 | 3474 | 390 |
| Jun 86 | 10 | 4020 | 3567 | 453 |
| Jul 86 | 11 | 4065 | 3671 | 394 |
| Aug 86 | 12 | 4111 | 3773 | 338 |
| Sep 86 | 13 | 4151 | 3873 | 278 |
| Oct 86 | 14 | 4294 | 3970 | 324 |
| Nov 86 | 15 | 3867 | 4074 | −207 |
| Dec 86 | 16 | 3926 | 4123 | −197 |
| Jan 87 | 17 | 4007 | 4171 | −164 |
| Feb 87 | 18 | 4075 | 4221 | −146 |
| Mar 87 | 19 | 4195 | 4271 | −76 |
| Apr 87 | 20 | 4165 | 4327 | −162 |

**Figure 8.5** **Forecast and Actual Copier Usage**
($y_t$ = average number of copies per week)

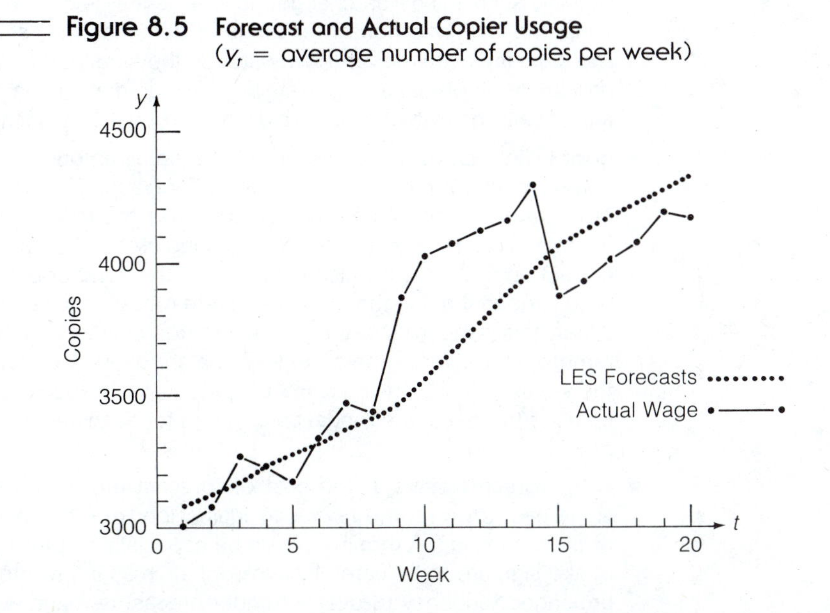


larger, value. During the period just after May 1986 and November 1986 a faster-tracking updating model would have improved the forecasts appreciably, though for other periods the more stable forecasts produced by the 10% smoothing constants would probably have been better. ●

# 8.4 The Decision to Monitor Forecasts

## 8.4.1 Introduction

Because of the nature and purpose of forecast monitors, decisions about their use are less statistical and more judgmental than in other aspects of modeling. A desire for adaptiveness reflects some degree of distrust of the series' stability or, perhaps, a lack of faith in the model that has been fit. These judgments may be based on the empirical record, but often they result from familiarity with the forces driving the series.

## 8.4.2 Conditions Requiring Monitors

Among the many conditions that may lead one to consider adaptive schemes are the following.

- **dynamic history** The history of the series may reveal changes where adaptiveness would have been desirable. These changes would tend to produce long runs of residuals of predominantly one sign (positive or negative). One way to check the data for such phenomena would be to run the proposed tracking signal through the history of the series to see if it would have been tripped at any point. As an alternative, one could retrospectively compare the forecasts that would have been produced by the proposed adaptive technique with those a similar, nonadaptive one would have produced.
- **anticipated change** The analyst, being familiar with the forces affecting the series, may feel that changes in one or more of these forces could occur, causing such things as changes in the series' level, slope, or growth rate. Thus, the level of a horizontal series of demands may shift upward, as large accounts are acquired, or downward, as they are lost. Changes in advertising policies or other marketing strategies may dramatically increase/decrease the rate of sales growth or cause rapid jumps in level, leaving the slope unchanged.
- **poor initial values** Cases where only a small amount of data is available to estimate model parameters or select forecasting parameters may produce inaccurate models. With adaptive schemes the initial performance of these estimates can be monitored as forecasting proceeds, helping to recognize and correct poor fits. It may also happen that the number of series to be forecast is so large that going through a complete model fit for each one is impractical. When this happens, very rough estimates or hastily determined model parameters for a very simple model often are used; e.g., one might forecast all the series with SES using $\alpha = 10\%$, $\hat{y}_{\text{init}} = y_1$. In such cases, adaptivity could be used to correct the model parameters for each series in the normal course of forecasting.
- **costly forecast errors** If the cost of forecast errors is quite high, a tracking signal can provide some insurance against long runs of moderate-to-large errors of the same sign. A monitor would be especially helpful if accumulated errors of like sign are to be carefully avoided, such as in inventory control, where a prolonged tendency to over- or underforecast demand would lead to greatly increased holding or stock-out costs, respectively.
- **need of an attention-drawing device** When a large number of series are being forecast, it is often impractical for the analyst to personally check and

adjust the model for each and every series. Under these conditions, a tracking signal can point out which models (if any) need attention and which do not. Sometimes trained analysts are not at hand to check how forecasts are developing. For untrained personnel, a tracking signal alarm can be used as a simple, easily explained indicator that attention from more knowledgeable sources is required.

# 8.5 Monitors and Adaptive Procedures

## 8.5.1 Introduction

The methods in common use for monitoring forecasting schemes all involve two quantities: (1) a *measure of performance* based on the forecast errors, $e_t = y_t - \hat{y}_t$, and (2) an *estimate of* how much *variability* to expect from the performance measure when the system is in control. The former is generally some measure of forecast bias, while the latter essentially involves estimating the standard deviation of the forecast errors. This latter estimate is needed so one can accurately differentiate between "in-control" and "out-of-control" conditions.

## 8.5.2 Measures of Performance

Performance measures are usually based on the supposition that, if the system is in control, forecasts should be unbiased, over- and underforecasting occurring about equally. The average error $\bar{e}$ should then be near zero. Out-of-control forecasts, on the other hand, will be biased, with either positive or negative errors predominating. The average (or sum) of the errors will then drift farther from zero. A number of different measures of forecast bias may be used, depending on the data-handling resources one is willing to allocate.

### Individual Errors, $e_t$

The simplest ongoing measure of performance at time $t$ is the single current forecast error $e_t$ itself. If we suppose that the $e_t$'s are approximately normally distributed with standard deviation $\sigma_e$, then, in a controlled system, approximately $(1 - \alpha') \times 100\%$ of the individual errors should fall within $z_{\alpha'/2}$ standard deviations of zero, where $z_{\alpha'/2}$ is the upper $(\alpha'/2) \times 100\%$ point of the standard normal distribution.

***Note:*** $\alpha'$ is used here to denote the *false-alarm* probability, i.e., the probability of an out-of-control value from a scheme that is in control. The prime is used to avoid confusion with the smoothing constant $\alpha$ in exponential smoothing schemes.

A value of $e_t$ could then be considered out of control if it were to fall outside of the interval $0 \pm z_{\alpha'/2} \cdot \sigma_e$, or equivalently, if

$$|e_t| > z_{\alpha'/2} \cdot \sigma_e \qquad (8.1)$$

We might consider it an alarm if any out-of-control error is encountered.

More conservatively, an alarm might be sounded only if two (or three, or however many) consecutive out-of-control errors occur.

Although the individual error tracking signal is very simple in concept and computation, it does not measure *aggregate* performance. As a result, it will tend to respond only to rather large disturbances in the series. More sensitive signals are usually based on an average of several errors.

### The Simple Cusum, $\text{SUM}_t$

As a measure of aggregate performance, it is natural to consider the sum of all past errors, referred to as the **cumulative sum** or the **simple cusum;** i.e.

$$\text{SUM}_t = \sum_1^t e_t \tag{8.2}$$

which can be written in updating form as

$$\text{SUM}_t = e_t + \text{SUM}_{t-1} \tag{8.3}$$

Notice that to update the simple cusum, only the previous cusum and the current error are needed; thus data handling is minimal.

Since errors are as likely to be positive as negative when the system is in control, the simple cusum should not drift too far from zero. If we are willing to assume that the errors are independent and approximately normally distributed, then the standard deviation of $\text{SUM}_t$ will be $\sigma_e\sqrt{t}$, and the simple cusum could be considered out of control when

$$|\text{SUM}_t| > z_{\alpha'/2} \cdot \sigma_e\sqrt{t} \tag{8.4}$$

or, if constant limits are preferred,

$$\frac{|\text{SUM}_t|}{\sqrt{t}} > z_{\alpha'/2} \cdot \sigma_e \tag{8.5}$$

### The Moving Total Error, $\text{MTE}_t$

A shortcoming of the simple cusum is that, if forecast errors are independent, the importance given to the current error diminishes with time. If the scheme has been operating in control for a long period, then shifts out of control, it will be some time (due to division by $\sqrt{t}$, to be discussed shortly) before the new, out-of-control errors gather enough weight to trip an alarm signal. The longer the system has been in control, the worse the problem becomes. In addition, when adaptive schemes are being used, we are generally more interested in how the system has been doing lately than in its ancient history. An aggregate measure that avoids this problem is the **moving cusum** or **moving total error** (of length $k$), which we will denote, at time $t$, as

$$\begin{aligned} \text{MTE}_t &= \text{sum of the most recent } k \text{ errors} \\ &= \sum_{t-k+1}^{t} e_t \end{aligned} \tag{8.6}$$

or, in updating form,

$$\text{MTE}_t = \text{MTE}_{t-1} + (e_t - e_{t-k}) \tag{8.7}$$

If we again assume that the $e_t$'s are uncorrelated and approximately normally distributed, the standard deviation of $\text{MTE}_t$ is $\sigma_e\sqrt{k}$ and control limits would be

$$|\text{MTE}_t| > z_{\alpha'/2} \cdot \sigma_e\sqrt{k} \tag{8.8}$$

In choosing the length of the moving average there is a trade-off between tracking and smoothing, much the same as when the moving average is used as a forecasting device. Longer length moving totals will be more sensitive to smaller disturbances in the series but will not respond as quickly as shorter length ones. The amount of information that must be carried from one period to the next (the most recent $k$ errors) is also a consideration. When large numbers of series are involved, this data storage may be the primary concern, in which case the smoothed error described next may be preferred.

### The Smoothed Error, $E_t$

Exponential smoothing may also be used to compute an average that discounts older errors. The (exponentially) **smoothed error,** $E_t$, is computed in the same fashion as for SES forecasts. That is, using smoothing constant $\eta$:

$$E_t = \eta \cdot e_t + (1 - \eta) \cdot E_{t-1} \tag{8.9}$$

or

$$E_t = E_{t-1} + \eta(e_t - E_{t-1}) \tag{8.10}$$

which results in exponentially diminishing weights $\eta$, $\eta(1 - \eta)$, $\eta(1 - \eta)^2$, $\eta(1 - \eta)^3$, ... for $e_t$, $e_{t-1}$, $e_{t-2}$, $e_{t-3}$, ... as with SES forecasts. If one assumes that $\eta$ is small, the $e_t$'s are relatively uncorrelated, and it may be shown that the standard deviation of $E_t$ is approximately

$$\sigma_e\sqrt{\frac{\eta}{2 - \eta}} \tag{8.11}$$

If normality is also assumed, approximate control limits (when $\eta$ is small) would be

$$|E_t| > z_{\alpha'/2} \cdot \sigma_e\sqrt{\frac{\eta}{2 - \eta}} \tag{8.12}$$

To start the smoothing an initial value $E_{\text{init}}$ is needed. It is reasonable (and common) practice to use $E_{\text{init}} = 0$.

## 8.5.3 Measures of Variability

Because the standard deviation of the forecast errors ($\sigma_e$) is rarely known, the expressions for control limits given in the previous discussions are useful only if an estimate, $\hat{\sigma}_e$, of $\sigma_e$ is available. By substituting $\hat{\sigma}_e$ for $\sigma_e$, approximate control limits can be obtained. There are a number of different ways to proceed, depending once more upon data management resources and availability.

### Historical Data, $s_e$

If a reasonably large stable history of the series (say, $n$ observations) is available, $\sigma_e$ may be estimated in the usual manner; that is,

$$\hat{\sigma}_e = s_e = \sqrt{\frac{\sum e^2}{n}} \tag{8.13}$$

If no history is available, the scheme will have to run for some time to obtain several observed forecast errors from which an estimate as above can be obtained at time $t$ as

$$\hat{\sigma}_e = s_e = \sqrt{\frac{\sum e^2}{t}} \tag{8.14}$$

This estimate may be updated for each new observation if so desired by letting

$$\text{SSE}_t = \sum_1^t e_t^2$$

Then, for updating,

$$\text{SSE}_t = \text{SSE}_{t-1} + e_t^2 \tag{8.15}$$

and

$$\hat{\sigma}_e = s_e = \sqrt{\frac{\text{SSE}_t}{t}} \tag{8.16}$$

### The Smoothed (Root) Mean Square Error, $\text{MSE}_t$ ($\text{RMSE}_t$)

It is also possible to exponentially smooth the squared errors, then take the square root to get an estimate of $\sigma_e$. Using smoothing constant $\eta$, we let

$$\text{MSE}_t = \eta \cdot e_t^2 + (1 - \eta) \cdot \text{MSE}_{t-1} \tag{8.17}$$

or, equivalently,

$$\text{MSE}_t = \text{MSE}_{t-1} + \eta(e_t^2 - \text{MSE}_{t-1}) \tag{8.18}$$

Then

$$\hat{\sigma}_e = \text{RMSE}_t = \sqrt{\text{MSE}_t} \tag{8.19}$$

This method also requires an initial value, $\text{MSE}_{\text{init}}$. It may be obtained by computing $s_e$ from a history of the series, as mentioned above, and letting $\text{MSE}_{\text{init}} = s_e^2$. Minimally, we can start smoothing at time $t = 2$, letting $\text{MSE}_1 = e_1^2$ and allowing the scheme to run for several periods (*burn in,* so to speak) before permitting an alarm signal.

### The Smoothed Absolute Error, $MAD_t$

Because it is conceptually simpler and somewhat easier to calculate, a smoothed form of the mean absolute deviation of the $e_t$'s called the **smoothed absolute error** is often used to estimate $\sigma_e$. It is computed using

$$MAD_t = \eta \cdot |e_t| + (1 - \eta) \cdot MAD_{t-1} \tag{8.20}$$

or

$$MAD_t = MAD_{t-1} + \eta(|e_t| - MAD_{t-1}) \tag{8.21}$$

For the normal and other common distributions, it has been shown that the mean absolute deviation is about 80% of the standard deviation, or, conversely, that the standard deviation is about 1.25 times the mean absolute deviation. We may, therefore, approximate $\sigma_e$ by using, in the expressions for control limits:

$$\hat{\sigma}_e = 1.25 \cdot MAD_t \tag{8.22}$$

For an initial value in the smoothing process some history from the forecast scheme is again needed. We then set

$$MAD_{init} = \frac{\sum_{\text{history}} |e|}{(\text{number of observations})} = \frac{\Sigma |e|}{n} \tag{8.23}$$

where $n$ is the number of historical errors. If no history is available, we can start smoothing at time $t = 2$ using $MAD_1 = |e_1|$ and let the monitor burn in for several periods before permitting alarms.

A desirable, and often overlooked, advantage of using the MAD to approximate $\sigma_e$ is that it responds less dramatically than the RMSE to the first few large forecast errors that occur when a process goes out of control (and the true mean error $\mu_e$ is no longer zero). Both the MAD and RMSE give distortedly large estimates of $\sigma_e$ under these circumstances, but the MAD less so. The net result is that monitors using the MAD will tend to respond somewhat more quickly to out-of-control conditions.

**Measures of Variability and Performance Used in Adaptive Schemes**

| | Performance Measure | Standard Deviation (Independent Errors) |
|---|---|---|
| **Individual Errors:** | $e_t$ | $\sigma_e$ |
| **Simple Cusum:** | $SUM_t = \sum_{1}^{t} e_t = e_t + SUM_{t-1}$ | $\sigma_{SUM} = \sigma_e\sqrt{t}$ |
| **Moving Total Error:** | $MTE_t = \sum_{t-k+1}^{t} e_t = MTE_{t-1} + (e_t - e_{t-k})$ | $\sigma_{MTE} = \sigma_e\sqrt{k}$ |
| **Smoothed Error:** | $E_t = \eta \cdot e_t + (1 - \eta) \cdot E_{t-1}$ | $\sigma_E = \sigma_e \cdot \sqrt{\frac{\eta}{2 - \eta}}$ |

**Estimates of $\sigma_e$**

**Historical Data:** $\hat{\sigma}_e = s_e = \sqrt{\dfrac{\Sigma e^2}{n}}$

**Forecasting History:** $\hat{\sigma}_e = s_e = \sqrt{\dfrac{\Sigma e^2}{t}}$

**Smoothed Squared Error:** $\text{MSE}_t = \eta \cdot e_t^2 + (1 - \eta) \cdot \text{MSE}_{t-1}$

$\hat{\sigma}_e = \text{RMSE}_t = \sqrt{\text{MSE}_t}$

**Smoothed Absolute Error:** $\text{MAD}_t = \eta \cdot |e_t| + (1 - \eta) \cdot \text{MAD}_{t-1}$

$\hat{\sigma}_e = 1.25 \cdot \text{MAD}_t$

## 8.5.4 Tracking Signals

Any performance measure may be paired with any variability measure to form a tracking signal, but some pairings are more natural and convenient (and popular) in given circumstances. For example, it is natural to pair measures derived from smoothing and use them to track smoothing schemes, or to use historical estimates with single forecast schemes. In addition, some signals have been studied in more depth than others, and more theoretical results are available for these signals. To illustrate all of the possible combinations would not only be prohibitive but confusing. Instead, we will concentrate on the more common and well-studied pairings. From these, the reader should gain an adequate idea of how adaptive methods are applied in different circumstances.

***Note:*** In the discussions throughout the remainder of this chapter, it will be convenient to use $\text{TS}_t$ as a generic notation for the value (at time $t$) of whatever tracking signal is being considered.

## 8.5.5 Single Forecasts

Although single forecast models are not often made adaptive, they make a reasonable starting point for study.

**EXAMPLE 8.4** *A Single Forecast Model. Consumer's Connection, Inc.*

To show how a tracking signal could have been incorporated into the Consumer's Connection single forecast model, we first need to look at the historical data used to obtain the model. These data are shown in Table 8.4. The mean ($\bar{y} = 410$) was used as the estimate of $\beta_0$ for the horizontal model ($y_t = \beta_0 + \epsilon_t$). The forecast for each subsequent period was predicated on the assumption that the level of demand would remain constant, at least over the period for which forecasts were to be generated. An estimate of the standard deviation of the errors is computed from the historical data, as shown in the table, giving $\hat{\sigma}_e = 19.6$. This value is used to compute the control limits for three of the four tracking signals shown. For these signals, control limits are computed using $z_{.025} \simeq 2.00$, that is, allowing a 5% false-alarm probability ($\alpha' = .05$)

**Table 8.4** Historical Data Used for Fitting Horizontal Single Forecast Model, Consumer's Connection, Inc. $n = 20$ observations

$y_t$ = unit biweekly demand for pencil sharpeners

| Period: | 1 | 2 | 3 | 4 | 5 | 6 | 7 | 8 | 9 | 10 |
|---|---|---|---|---|---|---|---|---|---|---|
| $y_t$: | 415 | 409 | 380 | 383 | 419 | 408 | 411 | 400 | 425 | 450 |
| **Period:** | 11 | 12 | 13 | 14 | 15 | 16 | 17 | 18 | 19 | 20 |
| $y_t$: | 424 | 419 | 401 | 443 | 422 | 396 | 391 | 427 | 372 | 400 |

$$\hat{\beta}_0 = \frac{\Sigma y}{n} = 410 \qquad \hat{\sigma}_e = \frac{\Sigma e^2}{n} = \frac{\Sigma (y - 410)^2}{20} = 19.6 \qquad \text{MAD} = \frac{\Sigma |y - 410|}{20} = 15.8$$

**Table 8.5** Tracking Signals for Consumer's Connection, Inc., Example 8.4

| Series | | Tracking Signal 1 | 2 | | 3 | 4 | | |
|---|---|---|---|---|---|---|---|---|
| $t$ | $y_t$ | $e_t$ | $\text{SUM}_t$ | $\text{SUM}_t/\sqrt{t}$ | $\text{MTE}_t$ | $E_t$ | $\text{MAD}_t$ | $E_t/\text{MAD}_t$ |
| 1 | 443 | 33 | 33 | 33.0 | — | 3.3 | 17.5 | .19 |
| 2 | 414 | 4 | 37 | 26.2 | — | 3.4 | 16.2 | .21 |
| 3 | 381 | −29 | 8 | 4.6 | — | 0.2 | 17.5 | .01 |
| 4 | 391 | −19 | −11 | −5.5 | −11 | −1.7 | 17.7 | −.10 |
| 5 | 432 | 22 | 11 | 4.5 | −22 | 0.7 | 18.1 | .04 |
| 6 | 393 | −17 | −6 | −2.4 | −43 | −1.1 | 18.0 | −.06 |
| 7 | 427 | 17 | 11 | 4.2 | 3 | 0.7 | 17.9 | .04 |
| 8 | 415 | 5 | 16 | 5.7 | 27 | 1.1 | 16.6 | .07 |
| 9 | 419 | 9 | 25 | 4.2 | 14 | 1.9 | 15.8 | .12 |
| 10 | 441 | 31 | 56 | 17.7 | 62 | 4.8 | 17.3 | .28 |
| 11 | 459 | **49*** | 105 | 31.7 | **94*** | 9.2 | 20.5 | .45 |
| 12 | 424 | 14** | 119 | 34.4 | | 9.7 | 19.9 | .49 |
| 13 | 429 | 19 | 138 | 38.2 | | 10.6 | 19.8 | .54 |
| 14 | 450 | 40 | 178 | **47.6*** | | 13.5 | 21.8 | **.62*** |

Control Limits: #1, 39.2; #2, 39.2; #3, 78.4; #4, .57

* Alarm ** Assuming no action was taken.

at any given time. All tracking signals, along with the data for each period, are shown in Table 8.5.

**Tracking Signal 1:** individual errors, $e_t$

**Control Limit:** $2\hat{\sigma}_e = 2 \times 19.6 = 39.2$ **(See Eq. 8.1)**

The errors are computed each time a new observation becomes available:

*At t = 1:* $\hat{y}_1 = 410 \qquad y_1 = 443 \qquad e_1 = y_1 - \hat{y}_1 = 443 - 410 = +33$

$\text{TS}_1 = +33$

*At t = 2:* $\hat{y}_2 = 410$ $y_2 = 414$ $e_2 = 414 - 410 = +4$

$TS_2 = +4$

*At t = 3:* $\hat{y}_3 = 410$ $y_3 = 381$ $e_3 = 381 - 410 = -29$

$TS_3 = -29$

and so forth. If it has been decided that any out-of-control observation constitutes an alarm, then an alarm occurs at $t = 11$, when $|TS_{11}| = |e_{11}| = 49$, which is greater than 39.2.

**Tracking Signal 2:** simple cusum, $SUM_t/\sqrt{t}$

**Control Limit:** $2\hat{\sigma}_e = 2 \times 19.6 = 39.2$ **(8.5)**

As each new observation becomes available, the error is computed and added into a running total, which is then divided by the square root of the current time period:

*At t = 1:* $y_1 = 443$ $e_1 = +33$ $SUM_1 = +33$

$$TS_1 = \frac{SUM_1}{\sqrt{1}} = +33.0$$

*At t = 2:* $y_2 = 414$ $e_2 = +4$ $SUM_2 = SUM_1 + e_2 = 33 + 4 = +37$

$$TS_2 = \frac{SUM_2}{\sqrt{2}} = +26.2$$

*At t = 3:* $y_3 = 381$ $e_3 = -29$ $SUM_3 = SUM_2 + e_3 = 37 - 29 = +6$

$$TS_3 = \frac{SUM_3}{\sqrt{3}} = +4.6$$

and so forth. An alarm occurs at time $t = 14$, when $|TS_{14}| = |SUM_{14}|/\sqrt{14} = 178/\sqrt{14} = 47.6$, which is greater than 39.2. At this point corrective action should be considered. If action is taken, the cusum should be zeroed out and $t$ set equal to 1 for the next forecast period.

**Tracking Signal 3:** moving total error ($k = 4$), $MTE_t$

**Control Limit:** $2\hat{\sigma}_e\sqrt{k} = 2 \times 19.6\sqrt{4} = 78.4$ **(See Eq. 8.8)**

Since $k = 4$, computation of the tracking signal cannot begin until $t = 4$. After that time, the moving total is updated by subtracting the four-period-old error from the current moving total and adding the new error (notice that four past errors must be carried from one time period to the next to allow the updating):

*At t = 4:* $e_1 = +33$ $e_2 = +4$ $e_3 = -29$ $e_4 = -19$

$TS_4 = MTE_4 = 33 + 4 - 29 - 19 = -11$

*At t = 5:* $y_5 = 432$ $e_5 = 432 - 410 = +22$

$e_2 = +4$ $e_3 = -29$ $e_4 = -19$ $e_5 = +22$

$TS_5 = MTE_5 = MTE_4 + e_5 - e_1 = -11 + (+22) - (+33) = -22$

*At t = 6:* $y_6 = 393$ $e_6 = -17$

$e_3 = -29$ $e_4 = -19$ $e_5 = +22$ $e_6 = -17$

$TS_6 = MTE_6 = MTE_5 + e_6 - e_2 = -22 + (-17) - (+4) = -43$

and so forth. An alarm occurs at $t = 11$, when $|TS_{11}| = |MTE_{11}| = 94$, which is greater than 78.4. Corrective action should be considered at that time. If action is taken, $t$ should be reset to 1 and tracking with the MTE ($k = 4$) cannot begin again until $t = 4$ (old $t = 15$). If we are willing to assume that $\sigma_e$ has not changed, we can continue to use 78.4 as the control limit. However, since there has apparently been a change in the level of the series, it is a good idea to compute a new estimate of $\sigma_e$ after several periods of new errors have been gathered, then recompute the control limit.

**Tracking Signal 4:** smoothed error ratio, $E_t/MAD_t$

**Control Limit:** for $\eta = .1$, $2.5\sqrt{\frac{\eta}{2-\eta}} = .57$ **(See Eqs. 8.12, 8.22)**

It is popular to use the ratio of the smoothed error to the smoothed mean absolute deviation as a tracking signal (often referred to as *Trigg's tracking signal*). Since $1.25 \cdot MAD_t$ is an estimate of $\sigma_e$, dividing by $MAD_t$ makes the control limits constant and, for $\alpha' = .05$, Equation 8.12 becomes

$$|E_t| > 2 \cdot 1.25 \cdot MAD_t \sqrt{\frac{\eta}{2-\eta}}$$

or

$$\frac{|E_t|}{MAD_t} > 2.5\sqrt{\frac{\eta}{2-\eta}} \tag{8.24}$$

To compute the tracking signal, then, as each error becomes known we (1) smooth the error, (2) smooth the absolute error, and (3) take the ratio. $E_{init}$ is taken to be zero, and $MAD_{init}$ computed from the historical data (see Table 8.4) is 15.8. (Notice, as a matter of interest here, that we could estimate the MAD of the historical data as $.8 \cdot \hat{\sigma}_e = .8 \times 19.6 = 15.7$, nearly the same result as computing the MAD directly.) The computation of the tracking signal would then proceed as follows.

*At $t = 1$:*

$$e_1 = +33$$

$$E_1 = \eta \cdot e_1 + (1-\eta) \cdot E_{init} = .1(+33) + .9(0) = 3.3$$

$$MAD_1 = \eta \cdot |e_1| + (1-\eta) \cdot MAD_{init} = .1 \times 33 + .9 \times 15.8 = 17.5$$

$$TS_1 = \frac{|E_1|}{MAD_1} = \frac{3.3}{17.5} = .19$$

*At $t = 2$:*

$$e_2 = +4$$

$$E_2 = .1(+4) + .9(3.3) = 3.4$$

$$MAD_2 = .1 \times 4 + .9 \times 17.5 = 16.2$$

$$TS_2 = \frac{|E_2|}{MAD_2} = \frac{3.4}{16.2} = .21$$

*At $t = 3$:*

$$e_3 = -29$$

$$E_3 = .1(-29) + .9(3.4) = 0.2$$

$$MAD_3 = .1 \times 29 + .9 \times 16.2 = 17.5$$

$$TS_3 = \frac{|E_3|}{MAD_3} = \frac{0.2}{17.5} = .01$$

and so forth. There is an alarm at $t = 14$, when $|TS_{14}| = |E_{14}|/MAD_{14} = .62$, which is greater than .57. Corrective action is taken, the smoothed error is reset to 0, and $t$ is reset to 1. The smoothed absolute error is not generally reset. ●

## 8.5.6 The Trigg and Brown Signals for Smoothing Schemes

Analysis of the distributional characteristics of tracking signals for updating schemes is greatly complicated because the forecast errors are not independent. Generally speaking, the more closely the scheme tracks, the greater will be the autocorrelation between errors. However, some results are known for the more widely used signals.

The two most popular signals for tracking exponential smoothing schemes are the **smoothed error ratio (Trigg's)** and the **simple cusum ratio (Brown's)** to the smoothed mean absolute error, i.e. $E_t/MAD_t$ and $SUM_t/MAD_t$, respectively (Brown, 1959; Trigg, 1964). When using these signals, the signal smoothing constant $\eta$ is usually chosen to be the same as the series smoothing constant $\alpha$. Several studies, both theoretical and by simulation, have been conducted with the purpose of determining control limits for these signals when tracking simple and double exponential smoothing models (see Chapters 2 and 3, respectively). The results of these studies vary somewhat, but the rules given in the table of approximate control limits should give approximations adequate for most practical purposes when $\eta = \alpha$ is less than about 0.40 (see Batty, 1969; Brown, 1967; Gardner, 1983; Montgomery and Johnson, 1976).

**Approximate Control Limits for Brown's and Trigg's Tracking Signals for SES and DES When $\eta = \alpha$ Is Small**

**Trigg's Signal, $SUM_t/MAD_t$:**

| Control Limits ($\alpha' = 5\%$) | | Control Limits ($\alpha' = 1\%$) | |
|---|---|---|---|
| SES | DES | SES | DES |
| $\pm 1.3\sqrt{\eta}$ | $\pm 1.2\sqrt{\eta}$ | $\pm 1.6\sqrt{\eta}$ | $\pm 1.4\sqrt{\eta}$ |

**Brown's Signal, $E_t/MAD_t$:**

Control Limits: $\pm z_{\alpha'/2} \cdot \sigma_{TS}$

$$\sigma_{TS} \simeq .88\sqrt{\frac{2-\eta}{1-(1-\eta)^{2m}}} \quad \text{where} \begin{cases} m = 1 \text{ (SES)} \\ m = 2 \text{ (DES)} \end{cases}$$

**EXAMPLE 8.5** *An SES Example, Caston & Sons Fittings*

By way of example, we now apply both the smoothed error and simple cusum ratios as they could have been used in tracking the SES scheme for the Caston & Sons series of Example 8.2. For both signals we use a 5% false alarm probability. For the simple cusum, since the forecasting scheme is SES the value of $m$ in the table is 1. The same smoothing constant will be used for forecasting and tracking, so $\alpha = \eta = .1$. The standard deviation of the signal is approximately

$$\sigma_{TS} \simeq .884\sqrt{\frac{2-\eta}{1-(1-\eta)^2}} = .884\sqrt{\frac{1.9}{1-.81}} = 2.80$$

and the critical value for $\alpha' = 5\%$ is $z_{\alpha'/2} \cdot \sigma = 2 \times 2.80 = 5.60$. For the smoothed error, SES, $\alpha' = 5\%$, and $\alpha = \eta = .1$ the critical value is $1.3\sqrt{\eta} = 1.3\sqrt{.1} = .41$.

Since past history at the time that SES forecasting was begun indicated that the average level of production was about 1299 per hour (see Example 8.2), that would have been our initial forecast; i.e. $\hat{y}_{init} = 1299$. Forecasting and tracking would have proceeded as shown in Table 8.6. Calculations for the initial few time periods follow.

**Table 8.6 Brown's and Trigg's Tracking Signals, Caston & Sons, Example 8.5**

| $t$ | $y_t$ | $\hat{y}_t$ | $e_t$ | $SUM_t$ | $E_t$ | $MAD_t$ | Brown's $TS_t$ | Trigg's $TS_t$ |
|---|---|---|---|---|---|---|---|---|
| 1 | 1291 | 1299 | −8 | −8 | −.8 | 8.0 | −1.00 | −.10 |
| 2 | 1311 | 1298 | 13 | 5 | .6 | 8.5 | .59 | .07 |
| 3 | 1284 | 1299 | −15 | −10 | −1.0 | 9.2 | −1.07 | −.11 |
| 4 | 1315 | 1298 | 17 | 7 | .8 | 10.0 | .70 | .08 |
| 5 | 1309 | 1300 | 9 | 16 | 1.6 | 9.9 | 1.62 | .16 |
| 6 | 1295 | 1301 | −6 | 10 | .8 | 9.5 | 1.01 | .08 |
| 7 | 1317 | 1300 | 17 | 27 | 2.4 | 10.3 | 2.62 | .23 |
| 8 | 1311 | 1302 | 9 | 36 | 3.1 | 10.2 | 3.53 | .30 |
| 9 | 1307 | 1303 | 4 | 40 | 3.2 | 9.6 | 4.17 | .33 |
| 10 | 1337 | 1303 | 34 | 74 | 6.3 | 12.0 | **6.17*** | **.53*** |
| 11 | 1325 | 1306 | 19 | 93 | 7.6 | 12.7 | 7.32** | .60** |
| 12 | 1330 | 1308 | 22 | 115 | 9.0 | 13.6 | 8.46 | .66 |
| 13 | 1315 | 1310 | 5 | 120 | 8.6 | 12.7 | 9.45 | .68 |
| 14 | 1327 | 1311 | 16 | 136 | 9.3 | 13.0 | 10.46 | .72 |
| 15 | 1325 | 1313 | 12 | 148 | 9.6 | 12.9 | 11.47 | .74 |
| 16 | 1322 | 1314 | 8 | 156 | 9.4 | 12.4 | 12.58 | .76 |

Control Limits: Brown's, 5.60; Trigg's, .41

* Alarm ** Assuming no action was taken.

**Tracking Signal 5:** the simple cusum ratio, $SUM_t/MAD_t$

**Control Limit:** 5.60 (as previously calculated)

***At t = 1:***

$\hat{y}_1 = 1299 \qquad y_1 = 1291 \qquad e_1 = 1291 - 1299 = -8$

$\hat{y}_2 = 1299 + .1(-8) = 1298$

$SUM_1 = -8 \qquad MAD_1 = |e_1| = 8$

$$TS_1 = \frac{SUM_1}{MAD_1} = \frac{-8}{8} = -1$$

***Note:*** Because the MAD is computed directly from the series, not from the history, we must allow for burn-in. Thus, we do not really pay attention to the TS value for several periods.

***At t = 2:***

$\hat{y}_2 = 1298 \qquad y_2 = 1311 \qquad e_2 = 13$

$\hat{y}_3 = 1298 + .1 \times 13 = 1299$

$SUM_2 = SUM_1 + e_2 = -8 + 13 = +5$

$$\text{MAD}_2 = \eta \cdot |e_2| + (1 - \eta) \cdot \text{MAD}_1 = .1 \times 13 + .9 \times 8 = 8.5$$

$$\text{TS}_2 = \frac{\text{SUM}_2}{\text{MAD}_2} = \frac{5}{8.5} = .59$$

*At $t = 3$:*

$$\hat{y}_3 = 1299 \qquad y_3 = 1284 \qquad e_4 = -15$$

$$\hat{y}_4 = 1299 + .1(-15) = 1298$$

$$\text{SUM}_3 = +5 + (-15) = -10$$

$$\text{MAD}_3 = .1 \times 15 + .9 \times 8.5 = 9.2$$

$$\text{TS}_3 = \frac{-10}{9.2} = -1.09$$

*At $t = 4$:*

$$\hat{y}_4 = 1298 \qquad y_4 = 1315 \qquad e_4 = 17$$

$$\hat{y}_5 = 1298 + .1 \times 17 = 1300$$

$$\text{SUM}_4 = -10 + 17 = 7$$

$$\text{MAD}_4 = .1 \times 7 + .9 \times 9.2 = 10.0$$

$$\text{TS}_4 = \frac{7}{10} = .70$$

and so forth. The first alarm occurs at $t = 10$, when $|\text{TS}_{10}| = 74/12 = 6.17$, which is greater than 5.60. At this time corrective action should have been considered, since a tendency to underforecast had developed. Had this corrective action been taken,the cusum would have been reset to zero and $t$ reset to 1 as usual. The MAD would not have been reset, and tracking would continue as before. Notice that the signal would have stayed out of control for the remaining six weeks, sounding a continuous alarm unless some action were taken to correct the bad forecasts.

**Tracking Signal 6:** the smoothed error ratio, $E_t/\text{MAD}_t$

**Control Limit:** .41 (as previously calculated)

*At $t = 1$:*

$$\hat{y}_1 = 1299 \qquad y_1 = 1291 \qquad e_1 = 1291 - 1299 = -8$$

$$\hat{y}_2 = 1299 + .1(-8) = 1298 \text{ (as before)}$$

$$E_1 = \eta \cdot e_1 + (1 - \eta) \cdot E_{\text{init}} = .1(-8) + .9 \times 0 = -.8$$

$$\text{MAD}_1 = |e_1| = 8 \text{ (as before)}$$

$$\text{TS}_1 = \frac{E_1}{\text{MAD}_1} = \frac{-.8}{8} = -.1$$

***Note:*** Again, the TS would be ignored for several periods of burn-in.

*At $t = 2$:*

$$\hat{y}_2 = 1298 \qquad y_2 = 1311 \qquad e_2 = 13$$

$$\hat{y}_3 = 1299 \text{ (as before)}$$

$$E_2 = \eta \cdot e_2 + (1 - \eta) \cdot E_2 = .1 \times 13 + .9(-.8) = .6$$

$$\text{MAD}_2 = 8.5 \text{ (as before)}$$

$$\text{TS}_2 = \frac{E_2}{\text{MAD}_2} = \frac{.6}{8.5} = .07$$

*At t = 3:*

$$\hat{y}_3 = 1299 \qquad y_3 = 1284 \qquad e_4 = -15$$
$$\hat{y}_4 = 1298$$
$$E_3 = .1(-15) + .9 \times .6 = -1.0$$
$$\text{MAD}_3 = 9.2$$
$$\text{TS}_3 = \frac{-1.0}{9.2} = -.11$$

*At t = 4:*

$$\hat{y}_4 = 1298 \qquad y_4 = 1315 \qquad e_4 = 17$$
$$\hat{y}_5 = 1300$$
$$E_4 = .1 \times 17 + .9(-1.0) = .8$$
$$\text{MAD}_4 = 10.0$$
$$\text{TS}_4 = \frac{.8}{10} = .08$$

and so forth. The first alarm again occurs at $t = 10$, when $|\text{TS}_{10}| = 6.3/12 = .52$, which is greater than .41. If corrective action were taken, the smoothed error and $t$, but not the MAD, would be reset. (Here again, if no action were taken, the signal would have stayed out for the remaining six weeks.) ●

In Example 8.5, both signals sounded alarms at the same time. Indeed, the two signals are similar for $\eta = .1$, since $E_t$ is then about equivalent to a long moving average (about length 19, see Section 2.10). Studies (Gardner, 1983; 1985) seem to indicate that the performance of the smoothed error ratio in terms of the time required to detect out-of-control conditions deteriorates badly for smoothing constants larger than .10. These studies support a general overall preference for the simple cusum.

On the other hand, an advantage of the smoothed error is that, under certain conditions, it will be less likely than the simple cusum to generate false alarms. If a rather large random movement is followed by several small ones, as sometimes happens in a stable series, the large error will tend to dominate the cusum. The smoothed MAD becomes smaller as the several small errors occur and the simple cusum ratio will tend to go out. The smoothed error, however, will be reduced right along with the MAD, so the smoothed error ratio will be less likely to go out. (Incidently, the shortcomings of the cusum mentioned earlier in this section do not apply for smoothing schemes, because the forecast errors are autocorrelated.)

It is difficult to handle tracking signals for schemes other than SES or DES with $\alpha = \eta$. One approach might be to compute control limits as though the forecast errors were independent. The independence assumption may be approximately true if the scheme tracks slowly, e.g., for small smoothing constants. Another approach is to use simulation; that is, artificially create series whose characteristics are similar to the one being forecast. One generates several long, stable series (with a computer) and uses the proposed scheme and tracking signal on these artificial series, maintaining an empirical frequency distribution (histogram) of the values of the tracking signal. The upper and lower small percentage points of this simulated distribution may be used as control limits of tracking signals for the actual series.

The problem may not be as difficult in a practical setting as it first seems. One can use very approximate limits (say, ones based on assuming independent errors) and keep track of the forecasting performance manually for a time. If the approximations seem to be generating too many false alarms or taking too long to detect out-of-control conditions, larger or smaller values, respectively, may be substituted until more reasonable ones are determined.

### 8.5.7 Dynamic Response Parameters

A series for which an adaptive scheme is being considered is very often already recognized as being only locally stable, i.e., stable for relatively short periods of time. Therefore, the schemes for which monitors are being designed are usually updating schemes, and very often exponential smoothing schemes. The main reason for adding the monitor is to somehow track the object series rapidly only when necessary. We want to avoid tracking random movement when the series is stable, and turn up the tracking power when instability in the series' average level, rather than randomness, becomes the predominant source of forecast error. We can then switch back to low tracking when the series has once more become stable. In this way we sort of get the best of both worlds (smoothing and tracking).

With tracking signals and intervention, this high/low tracking switch is achieved manually. With smoothing schemes, it can be done by actually changing the value of the smoothing constant(s). The change can be made automatic or semiautomatic, depending on the problem context and the forecaster's preference. With semiautomatic schemes a real time element of judgment can be interjected into the decision as to whether to crank up the tracking power, while automatic schemes will perform the change mechanically, according to rules preset by the analyst.

***Note:*** Be cautious with automatic schemes. The trick is to design the system to respond to deteriorating forecast quality, but not to overrespond. A properly functioning automatic response scheme can be likened to an experienced driver who has learned to moderate steering responses, subconsciously avoiding overcorrections that must subsequently be corrected and that lead to the constant back-and-forth movement so characteristic of beginners. In a poorly functioning scheme, overcorrections can lead to subsequent overcorrections for the previous overcorrections, and so on, resulting in complete loss of control. When this happens in forecasting, the adaptivity itself goes out of control, producing wildly varying (and mostly inaccurate) forecasts. So when implementing fully automatic schemes, remember that such schemes are automatic, *not* intelligent. It is best to keep at least a cursory eye on them regardless.

### 8.5.8 An Automatic HI/LO Tracking Scheme

A very straightforward way to provide alternate fast and slow (HI and LO) tracking is to use a small smoothing constant until a tracking signal alarm is tripped, switch to a larger constant until a predetermined number of periods have elapsed without an alarm, then switch back to the smaller constant again. While this procedure is reasonable, it has some disadvantages:

- Because of the tracking signal's response time, HI tracking generally does not start until some time after it is actually needed.
- During the HI tracking period, random error will be tracked as faithfully as any parameter changes. It will require several periods where no parameter changes occur before the series will revert to LO tracking. During this latter period the random-error tracking will increase the average forecast errors.

With exponential smoothing, a simple way to alleviate these problems is to maintain two streams of forecasts: (1) a LO tracking scheme that produces the actual, reported forecasts, and (2) a HI tracking scheme that provides a value for reinitializing the LO tracking scheme when an alarm is tripped. Maintaining the second (HI) stream does not require a great deal of data handling because the calculations are identical to those for the LO stream except for the value of the smoothing constant. Only one additional piece of information need be carried from one period to the next, the HI forecast.

Notationally let us suppose that

$\alpha$ = the LO stream smoothing constant

$\alpha^*$ = the HI stream smoothing constant $(\alpha^* > \alpha)$

$\hat{y}_t$ = the LO stream forecast (the reported forecast for $y_t$)

$\hat{y}_t^*$ = the HI stream forecast

$e_t = y_t - \hat{y}_t$ = the LO forecast error

$e_t^* = y_t - \hat{y}_t^*$ = the HI forecast error

$TS_t$ = the tracking signal being used

Then, in error correction form, the updating equations are

$$\hat{y}_{t+1}^* = \hat{y}_t^* + \alpha^* \cdot e_t^*$$

$$\hat{y}_{t+1} = \hat{y}_t + \alpha \cdot e_t \qquad \text{if } TS_t \text{ is in control}$$

$$= \hat{y}_{t+1}^* \qquad \text{if } TS_t \text{ is out of control}$$

**EXAMPLE 8.6** *HI/LO Tracking for Caston & Sons*

Let us suppose an automatic HI/LO scheme had been used for the Caston & Sons series. Let $\alpha = .1$ as with the nonadaptive smoothing scheme, and $\alpha^* = .5$ (about equivalent to a moving average of length 3, exponential weights: 50%, 25%, 12.5%, 6.25%, 3.13%, . . .) for fast tracking. In addition, suppose the smoothed error-ratio tracking signal is being used ($E_{\text{init}} = 0$, $\text{MAD}_1 = |e_1|$, SES, $\eta = \alpha = .1$, $\alpha' = 5\%$, control limit = .41 as in Example 8.5). The forecasting would proceed as follows:

**Automatic HI/LO Tracking for SES, $\alpha < \alpha^*$**

$$\hat{y}_{t+1}^* = \hat{y}_t^* + \alpha^* \cdot e_t^*$$

$$\hat{y}_{t+1} = \hat{y}_t + \alpha \cdot e_t \qquad \text{if } TS_t \le CL$$

$$= \hat{y}_{t+1}^* \qquad \text{if } TS_t > CL$$

where $e_t = y_t - \hat{y}_t$ $\quad e_t^* = y_t - \hat{y}_t^*$

$TS_t$ = a tracking signal with control limit CL

*At t = 1:* $\hat{y}_1 = \hat{y}_1^* = \hat{y}_{\text{init}} = 1299$ $\quad y_1 = 1291$ $\quad e_1 = e_1^* = -8$

$$\left.\begin{aligned} E_1 &= -.8 \\ \text{MAD}_1 &= 8 \\ \text{TS}_1 &= -.1 \end{aligned}\right\} \text{as in Example 8.5}$$

$$\hat{y}_2^* = \hat{y}_1^* + \alpha^* \cdot e_1 = 1299 + .5(-8) = 1295$$

$$\hat{y}_2 = \hat{y}_1 + \alpha \cdot e_1 = 1299 + .1(-8) = 1298 \text{ (TS in control)}$$

*At t = 2:* $y_2 = 1311$ $\quad e_2 = 1311 - 1298 = 13$ $\quad e_2^* = 1311 - 1295 = 16$

The actual forecast error $e_t$ is used for computing the tracking signal. Thus,

$$E_2 = .6 \quad \text{MAD}_2 = 8.5 \quad \text{TS}_2 = .07 \text{ (in control)}$$

$$\hat{y}_3^* = \hat{y}_2^* + \alpha^* \cdot e_2^* = 1295 + .5 \times 16 = 1303$$

$$\hat{y}_3 = \hat{y}_2 + \alpha \cdot e_2 = 1298 + .1 \times 13 = 1299 \text{ (TS in control)}$$

*At t = 3:* $y_3 = 1284$ $\quad e_3 = -15$ $\quad e_3^* = -19$

$$E_3 = -1.0 \quad \text{MAD}_3 = 9.2 \quad \text{TS}_3 = -.11 \text{ (in control)}$$

$$\hat{y}_4^* = 1303 + .5(-19) = 1294$$

$$\hat{y}_4 = 1299 + .1(-15) = 1298 \text{ (TS in control)}$$

and so on, the forecasts remaining the same as they were before until an alarm is tripped (see Table 8.7).

*At t = 10:* $\hat{y}_{10} = 1303$ $\quad \hat{y}_{10}^* = 1309$ $\quad y_{10} = 1337$

$$e_{10} = 34 \quad e_{10}^* = 28$$

**Table 8.7 Forecasts of Caston & Sons Production Using HI/LO Tracking with the Smoothed Error Ratio Signal** $\alpha = .1, \alpha^* = .5, \eta = .1$

| $t$ | $y_t$ | $\hat{y}_t$ | $\hat{y}_t^*$ | $e_t$ | $e_t^*$ | $E_t$ | $\text{MAD}_t$ | $\text{TS}_t$ |
|---|---|---|---|---|---|---|---|---|
| 1 | 1291 | 1299 | (1299) | −8 | (−8) | −.8 | 8.0 | −.10 |
| 2 | 1311 | 1298 | (1295) | 13 | (16) | .6 | 8.5 | .07 |
| 3 | 1284 | 1299 | (1303) | −15 | (−19) | −1.0 | 9.2 | −.11 |
| 4 | 1315 | 1298 | (1294) | 17 | (21) | .8 | 10.0 | .08 |
| 5 | 1309 | 1300 | (1305) | 9 | (4) | 1.6 | 9.9 | .16 |
| 6 | 1295 | 1301 | (1307) | −6 | (−12) | .8 | 9.5 | .08 |
| 7 | 1317 | 1300 | (1301) | 17 | (16) | 2.4 | 10.3 | .23 |
| 8 | 1311 | 1302 | (1309) | 9 | (2) | 3.1 | 10.2 | .30 |
| 9 | 1307 | 1303 | (1310) | 4 | (−3) | 3.2 | 9.6 | .33 |
| 10 | 1337 | 1303 | (1309) | 34 | (28) | 0.0** | 12.0 | **.53*** |
| 11 | 1325 | 1323 | (1323) | 2 | (2) | .2 | 11.0 | .02 |
| 12 | 1330 | 1323 | (1324) | 7 | (6) | .9 | 10.6 | .08 |
| 13 | 1315 | 1324 | (1327) | −9 | (−12) | −.1 | 10.4 | −.01 |
| 14 | 1327 | 1323 | (1321) | 4 | (6) | .3 | 9.8 | .03 |
| 15 | 1325 | 1323 | (1324) | 2 | (1) | .5 | 9.0 | .06 |
| 16 | 1322 | 1323 | (1325) | −1 | (−3) | .4 | 8.2 | .05 |
| 17 | ? | 1323 | (1324) | | | | | |

* Alarm! ** Reset to 0.0.

$$E_{10} = .1 \times 34 + .9 \times 3.2 = 6.3 \qquad \text{MAD}_{10} = 12.0$$

$$\text{TS}_{10} = \frac{6.3}{12} = .53 > .41 \text{ (alarm!)}$$

Now, $E_{10}$ is reset to 0, and the HI forecast will become the actual forecast for $t = 11$. The MAD is not affected.

$$\hat{y}^*_{11} = \hat{y}^*_{10} + \alpha^* \cdot e^*_{10} = 1309 + .5 \times 28 = 1323$$

$$\hat{y}_{11} = \hat{y}^*_{11} = 1323 \text{ (TS out of control)}$$

***At $t = 11$:***

$$y_{11} = 1325 \qquad e_{11} = e^*_{11} = 1325 - 1323 = 2$$

$$E_{11} = .1 \times 2 + .9 \times 0 = .2 \qquad \text{MAD}_{11} = .1 \times 2 + .9 \times 12.0 = 11.0$$

$$\text{TS}_{11} = \frac{.2}{11.0} = .02 < .41 \text{ (in control)}$$

$$\hat{y}^*_{11} = 1323 + .5 \times 2 = 1324$$

$$\hat{y}_{11} = 1323 + .1 \times 2 = 1323 \text{ (TS in control)}$$

***At $t = 12$:***

$$y_{12} = 1330 \qquad e_{12} = 7 \qquad e^*_{12} = 6$$

$$E_{12} = .1 \times 7 + .9 \times .2 = .9 \qquad \text{MAD}_{12} = .1 \times 7 + .9 \times 11 = 10.6$$

$$\text{TS}_{12} = \frac{.9}{10.6} = .08 < .41 \text{ (in control)}$$

$$\hat{y}^*_{13} = 1324 + .5 \times 6 = 1327$$

$$\hat{y}_{13} = 1323 + .1 \times 7 = 1324 \text{ (TS in control)}$$

and so on. Forecasting remains in control for the remainder of the 16 periods. The complete set of forecasts is given in Table 8.7, and the original SES forecasts along with the HI/LO forecasts are graphed in Figure 8.6 for comparison. Notice that the jump to the HI forecast at $t = 10$ resulted in much better forecasts than those produced by the nonadaptive SES scheme.

**Figure 8.6 SES and HI/LO Forecasts for Caston & Sons Series**
$y_t$ = average hourly fittings produced

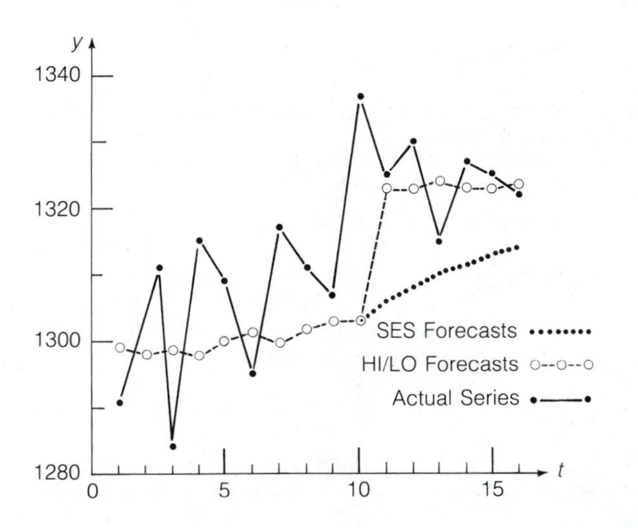

### 8.5.9 Adaptive Response Rate Exponential Smoothing (ARRES)

Another automatic adaptive scheme that is quite popular calculates the smoothing parameter for any time period as a function of the system's overall performance (Trigg and Leach, 1967). The smoothed error ratio is used as a measure of system performance. Because of the way this signal is computed, its absolute value must always lie somewhere between 0 and 1, exactly the range over which the smoothing constant may vary. The closer the signal is to 1.0, the worse the system's performance has been and the more tracking, it would seem, is needed. Setting $\alpha$ equal to the signal will provide this high tracking. Conversely, if the system has been performing well, no particular tracking is needed and the signal will be close to zero, again about where we would like $\alpha$ to be. We can then compute a value of $\alpha$ to use for each time period by letting $\alpha_t$ = "value $\alpha$ to be used in computing $\hat{y}_{t+1} = \text{TS}_t$." The updating equations are shown in the table nearby. There are two varieties of ARRES. In version I, the current value of the smoothing constant, $\alpha_t = |E_t|/\text{MAD}_t$, is used to generate $\hat{y}_{t+1}$. Some studies have reported that reacting so quickly can result in the type of instability to which automatic schemes are prone. These studies recommend version II, in which the reaction is delayed for one period by using $\alpha_{t-1}$ (rather than $\alpha_t$) to generate $y_{t+1}$ at time $t$ (Shone, 1967). The slight delay damps the response somewhat, and helps avoid the instability. It is also generally recommended that, in schemes with multiple smoothing constants (such as LES or HWS), only $\alpha$ be made adaptive. It should also be noted that some investigators have found that ARRES works poorly while the series is stable because of the randomness inherent in the $e_t$'s. Some even question whether the advantages gained from the overall increase in response to instability is worth the loss in accuracy for the stable portions of the series. These conclusions would certainly seem to point to the use of ARRES only in situations where frequent instability of the series is very likely (see summary for references).

**Adaptive Response Rate Exponential Smoothing, ARRES**
**The Trigg–Leach Method**

**Updating Equations:**

$$E_t = \eta \cdot e_t + (1 - \eta) \cdot E_{t-1}$$

$$\text{MAD}_t = \eta \cdot |e_t| + (1 - \eta) \cdot \text{MAD}_{t-1}$$

$$\alpha_t = \frac{|E_t|}{\text{MAD}_t}$$

**Forecasts:** For the various exponential smoothing models (SES, LES, etc.), with $\alpha$ = smoothing constant for the level, $\hat{T}_t(t)$:

**Version I:** $\alpha = \alpha_t$

**Version II:** $\alpha = \alpha_{t-1}$

**EXAMPLE 8.7** *An ARRES Example, Copier Usage*

The copier usage series of Example 8.3 is a reasonable one for which to try the ARRES method. It has undergone two seeming episodes of instability in its rather short forecasting history. The graph of forecast errors in Figure 8.5 shows that the LES scheme being used did not respond well to these episodes. Let us see how the two versions of ARRES would have compared, starting with ARRES I. The process will be initialized the same way as for the LES scheme, i.e., $\hat{T}_0(0) = 3031$ and $\hat{\beta}_1(0) = 50$. We will use $E_{init} = 0$ as usual and start with the same smoothing parameters, $\alpha_1 = \gamma = \eta = .10$, and adapt $\alpha$ thereafter.

*At t = 1:*

$$\hat{y}_1 = \hat{T}_1(0) = \hat{T}_0(0) + \hat{\beta}_1(0) = 3031 + 50 = 3081$$

$$y_1 = 3001 \qquad e_1 = -80$$

$$E_1 = \eta \cdot e_1 + (1 - \eta) \cdot E_{init} = .1(-80) + .9 \times 0 = -8$$

$$\text{MAD}_1 = |e_1| = 80$$

$\alpha_1 = .1$ (The initial value happens to be same as $|-8|/80$ only because $\eta = .1$ is being used.)

$$\hat{T}_1(1) = \hat{T}_1(0) + \alpha_1 \cdot e_1 = 3081 + .1(-80) = 3073$$

$$\hat{\beta}_1(1) = \hat{\beta}_1(0) + \alpha_1 \cdot \gamma \cdot e_1 = 50 + .1 \times .1(-80) = 49$$

$$\hat{y}_2(1) = \hat{T}_1(1) + \hat{\beta}_1(1) = 3073 + 49 = 3122$$

*At t = 2:*

$$y_2 = 3072 \qquad e_2 = 3072 - 3122 = -50$$

$$E_2 = .1(-50) + .9(-8) = -12$$

$$\text{MAD}_2 = .1 \times 50 + .9 \times 8 = 77$$

$$\alpha_2 = \frac{|-12|}{77} = .16$$

$$\hat{T}_2(2) = 3122 + .16(-50) = 3114$$

$$\hat{\beta}_1(2) = 49 + .16 \times .1(-50) = 48$$

$$\hat{y}_3(2) = \hat{T}_2(2) + \hat{\beta}_1(2) = 3114 + 48 = 3162$$

*At t = 3:*

$$y_3 = 3256 \qquad e_3 = 94$$

$$E_3 = .1 \times 94 + .9(-12) = -2$$

$$\text{MAD}_3 = .1 \times 94 + .9 \times 77 = 79$$

$$\alpha_3 = \frac{2}{79} = .02$$

$$\hat{T}_3(3) = 3162 + .02(94) = 3164$$

$$\hat{\beta}_1(3) = 48 + .02 \times .1 \times 94 = 48$$

$$\hat{y}_4(3) = 3164 + 48 = 3212$$

and so on. The value of $\alpha_t$ stays relatively small until about $t = 8$, then climbs rapidly, pulling the forecasts up toward the increased level of the series. The errors are reduced quite rapidly compared to those of the LES scheme. Unfortunately, the large errors that caused the increase in $\alpha_t$ are smoothed very heavily ($\eta = .1$) and so keep the tracking signal high for several periods after the errors have come back down again. When the shift downward in level takes place, around $t = 15$, $E_t$ cannot be pulled down quite rapidly enough to forestall a rather large overshoot and the consequent very large negative forecast errors for $t = 15$, 16, and 17. When the forecasting stops at $t = 20$, the tracking signal has again grown very large because of these large negative errors. If a shift up developed in the near future, an even worse undershoot might well occur. One can see how the stability of the adaptive structure can rapidly deteriorate and produce wildly variable and inaccurate forecasts if up/down shifts in the time series occur with the wrong frequency. If we now switch to ARRES II, the response is slowed somewhat, moderating the overshoot and producing overall better forecasts than ARRES I. The only difference, remember, is that the use of any given $\alpha_t$ value is delayed one time period. We will use $\alpha_0 = .1$ and $\alpha_1 = .1$ as initial values.

***At t = 1:***

$$\hat{y}_1 = 3081 \qquad y_1 = 3001 \qquad e_1 = -80$$

$$E_1 = -8 \qquad \text{MAD}_1 = |e_1| = 80$$

$$\alpha_0 = \alpha_1 = .1 \text{ (the initial values)}$$

$$\hat{T}_1(1) = \hat{T}_1(0) + \alpha_0 \cdot e_1 = 3081 + .1(-80) = 3073$$

$$\hat{\beta}_1(1) = \hat{\beta}_1(0) + \alpha_0 \cdot \gamma \cdot e_1 = 50 + .1 \times .1(-80) = 49$$

$$\hat{y}_2(1) = \hat{T}_1(1) + \hat{\beta}_1(1) = 3073 + 49 = 3122$$

***At t = 2:***

$$y_2 = 3072 \qquad e_2 = -50$$

$$E_2 = -12 \qquad \text{MAD}_2 = 77$$

$$\alpha_2 = \frac{|-12|}{77} = .16 \qquad (\alpha_1 = .1)$$

$$\hat{T}_2(2) = \hat{T}_2(1) + \alpha_1 \cdot e_2 = 3122 + .1(-50) = 3117$$

$$\hat{\beta}_1(2) = \hat{\beta}_1(1) + \alpha_1 \cdot \gamma \cdot e_2 = 49 + .1 \times .1(-50) = 49$$

$$\hat{y}_3(2) = \hat{T}_2(2) + \hat{\beta}_1(2) = 3117 + 49 = 3166$$

***At t = 3:***

$$y_3 = 3256 \qquad e_3 = 90$$

$$E_3 = .1 \times 90 + .9(-12) = -2$$

$$\text{MAD}_3 = .1 \times 90 + .9 \times 77 = 78$$

$$\alpha_3 = \frac{2}{78} = .03 \qquad (\alpha_2 = .16)$$

$$\hat{T}_3(3) = \hat{T}_3(2) + \alpha_2 \cdot e_3 = 3166 + .16(90) = 3180$$

$$\hat{\beta}_1(3) = \hat{\beta}_1(2) + \alpha_2 \cdot \gamma \cdot e_3 = 49 + .16 \times .1 \times 90 = 50$$

$$\hat{y}_4(3) = 3180 + 50 = 3230$$

and so on. The complete set of forecasts for ARRES I and II are given in Table 8.8. The errors for the nonadaptive LES scheme and the two adaptive schemes are given in Table 8.9 for comparison. The ARRES I and II forecasts for $t = 7$ through $t = 20$ are also graphed in Figure 8.7, where the effect of the delayed response can be seen in the reduced magnitude and duration of the overshoot.

**Table 8.8 ARRES I and II Forecasts for Copier Usage**
$y_t$ = Average Daily Number of Copies

| | | ARRES I | | | | | ARRES II | | | | |
|---|---|---|---|---|---|---|---|---|---|---|---|
| $t$ | $y_t$ | $\hat{y}_t$ | $e_t$ | $E_t$ | $MAD_t$ | $\alpha_t$ | $\hat{y}_t$ | $e_t$ | $E_t$ | $MAD_t$ | $\alpha_{t-1}$ |
| 1 | 3001 | 3081 | −80 | −8 | 8 | .10 | 3081 | −80 | −8 | 80 | .10 |
| 2 | 3071 | 3122 | −50 | −12 | 77 | .16 | 3122 | −50 | −12 | 77 | .10 |
| 3 | 3256 | 3162 | 94 | −2 | 79 | .02 | 3166 | 90 | −2 | 78 | .16 |
| 4 | 3227 | 3212 | 15 | 0 | 72 | .00 | 3230 | −3 | −2 | 71 | .03 |
| 5 | 3159 | 3260 | −101 | −10 | 75 | .13 | 3280 | −121 | −14 | 76 | .03 |
| 6 | 3323 | 3294 | 29 | −6 | 71 | .09 | 3326 | −3 | −13 | 69 | .18 |
| 7 | 3461 | 3344 | 117 | 6 | 75 | .08 | 3375 | 86 | −3 | 70 | .19 |
| 8 | 3444 | 3402 | 42 | 10 | 72 | .14 | 3443 | 1 | −3 | 63 | .04 |
| 9 | 3864 | 3457 | 407 | 49 | 105 | .47 | 3495 | 369 | 35 | 94 | .04 |
| 10 | 4020 | 3716 | 304 | 75 | 125 | .60 | 3564 | 456 | 77 | 130 | .37 |
| 11 | 4065 | 3984 | 81 | 76 | 121 | .63 | 3803 | 262 | 95 | 143 | .59 |
| 12 | 4111 | 4126 | −15 | 66 | 110 | .60 | 4043 | 68 | 93 | 136 | .66 |
| 13 | 4151 | 4207 | −56 | 54 | 105 | .52 | 4179 | −28 | 80 | 125 | .68 |
| 14 | 4294 | 4265 | 29 | 52 | 97 | .53 | 4249 | 45 | 77 | 117 | .64 |
| 15 | 3867 | 4369 | −502 | −4 | 138 | .03 | 4370 | −503 | 19 | 156 | .66 |
| 16 | 3926 | 4444 | −518 | −55 | 176 | .31 | 4098 | −172 | 0 | 157 | .12 |
| 17 | 4007 | 4354 | −347 | −84 | 193 | .44 | 4134 | −127 | −13 | 154 | .00 |
| 18 | 4075 | 4259 | −184 | −94 | 192 | .49 | 4191 | −116 | −23 | 150 | .08 |
| 19 | 4195 | 4217 | −22 | −87 | 175 | .50 | 4237 | −42 | −25 | 140 | .15 |
| 20 | 4165 | 4253 | −88 | −87 | 166 | .52 | 4286 | −121 | −35 | 138 | .18 |
| | | | MAD = 154 copies | | | | | MAD = 137 copies | | | |

**Table 8.9 LES and ARRES I, II Forecast Errors, Copier Usage**

| $t$ | LES $e_t$ | ARRES I $e_t$ | ARRES II $e_t$ |
|---|---|---|---|
| 1 | −80 | −80 | −80 |
| 2 | −50 | −50 | −50 |
| 3 | 90 | 94 | 90 |
| 4 | 2 | 15 | −3 |
| 5 | −116 | −101 | −121 |
| 6 | 11 | 29 | −3 |
| 7 | 99 | 117 | 86 |
| 8 | 22 | 42 | 1 |
| 9 | 390 | 407 | 369 |
| 10 | 453 | 304 | 456 |
| 11 | 394 | 81 | 262 |
| 12 | 338 | −15 | 68 |
| 13 | 278 | −56 | −28 |
| 14 | 324 | 29 | 45 |
| 15 | −207 | −502 | −503 |
| 16 | −197 | −518 | −172 |
| 17 | −164 | −347 | −127 |
| 18 | −146 | −184 | −116 |
| 19 | −76 | −22 | −42 |
| 20 | −162 | −88 | −121 |
| MAD: | 180 | 154 | 137 |

**Figure 8.7 ARRES Forecasts for Copier Usage Series**
$y_t$ = average number of copies per week

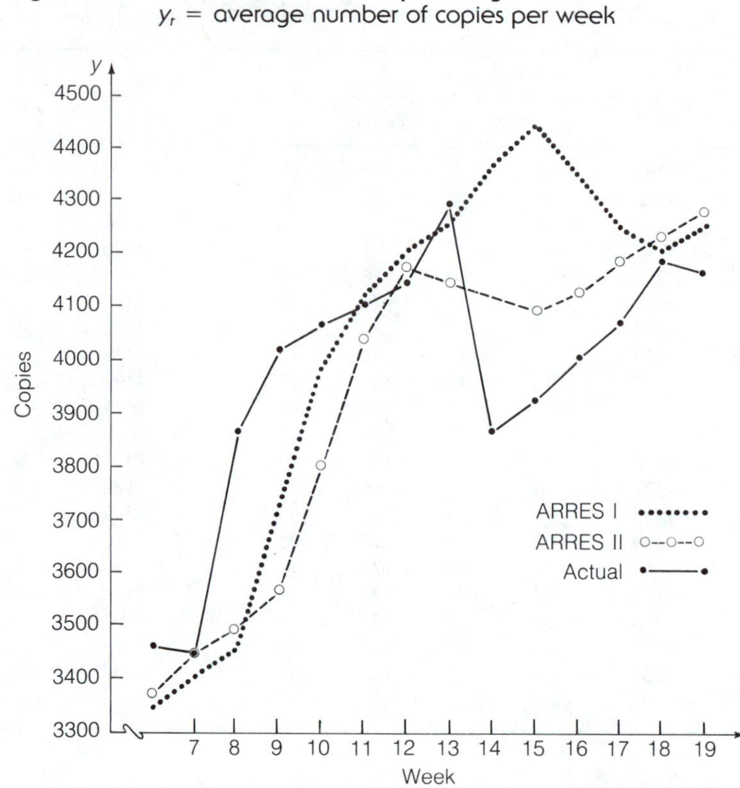


## 8.5.10 Summary

In an ongoing forecasting system, there is little doubt that monitoring forecast accuracy is a good idea. Monitoring methodology has been developed to apply principally to updating schemes, since they are the most likely to be ongoing. This monitoring can be done through the use of tracking signals that call the analyst's attention to developing problems and allow for corrective intervention. It is also possible to fashion schemes that require little or no outside attention. Such schemes either automatically reestimate a model's forecast parameters when a tracking signal fires, or continually readjust response parameters.

The tracking signals in use are virtually all based on the stream of forecast errors emanating from the ongoing process. The signals use performance measures that range from single current errors, to simple or moving cusum errors, to exponentially smoothed average errors. The series' inherent random variability is measured either by the historic or smoothed RMSE or MAD, with smoothed MAD probably being the most popular. The most widely used and studied signals are the simple cusum ratio (Brown's) and smoothed error ratio (Trigg's).

Research in this area is still underway, but recent performance studies (Gardner, 1983; 1985) indicate that the smoothed error (rather unexpectedly and for complicated reasons) is probably a poor choice, unless $\alpha$ is small (less than .2), and that the performance of the simple cusum (in terms of time required to detect shifts) is superior.

Automatic schemes are, themselves, difficult to control and so can easily result in highly unstable forecasts. Schemes such as ARRES cause accuracy to deteriorate under stable conditions, and there is some doubt that the improvement they offer for unstable conditions warrants their use (Gardner and Dannenbring, 1980; Gardner, 1983; Gardner, 1985; Makridakis and Hibon, 1979; Makridakis et al., 1982).

So in light of these results, and subject, of course, to individual circumstances, use of Brown's simple cusum ratio with intelligent intervention and appraisal is a reasonable, conservative overall choice.

## 8.6 Exercises

**8.1** What is a forecast monitor?

**8.2** In a general sense, what is meant by *adaptivity?*

**8.3** Under what conditions is a forecast monitor advisable?

**8.4** Why cannot simply using a fast tracking scheme take the place of forecast monitors?

**8.5** What are the performance measures used in constructing tracking signals? What is the basic rationale for these measures? Give the advantages and disadvantages of each.

**8.6** What is meant by *in control* and *out of control* when applied to forecasting schemes?

**8.7** What differentiates adaptive and nonadaptive single forecast schemes?

**8.8** What differentiates adaptive and nonadaptive updating forecast schemes?

**8.9** Why is it necessary to have a variability measure as well as a performance measure when constructing a forecast monitor?

**8.10** Contrast the different variability measures used in forecast monitors.

**8.11** What is meant by *dynamic response rate?*

**8.12** What is the main shortcoming of automatic forecasting schemes?

**8.13** Below are the weekly gross sales receipts ($1,000) for a retail discount book store during a seven-week period. Suppose this series had been

| $t$ | 1 | 2 | 3 | 4 | 5 | 6 | 7 |
|---|---|---|---|---|---|---|---|
| $y_t$ | 14.3 | 20.6 | 17.2 | 19.8 | 24.9 | 21.8 | 28.7 |

forecast with a no-trend model, $\hat{\beta}_0 = 18.0$. Calculate the following performance measures for times $t = 1$ to 7:

**a.** $e_t$

**b.** $\text{SUM}_t$

**c.** $\text{MTE}_t$, $k = 3$, $t = 3, \ldots, 7$

**d.** $E_t$ ($E_{\text{init}} = 0$)

**8.14** Suppose in Exercise 8.13 that the estimate $\hat{\beta}_0 = 18.0$ was obtained at time $t = 1$ based on the previous 16 weeks of historical data. The following quantities were also computed for these data:

$$\sum_{\text{history}} |y_t - 18| = 33.1 \qquad \sum_{\text{history}} (y_t - 18)^2 = 100$$

Compute the indicated variability measures for $t = 1, 2, \ldots, 7$:

**a.** $\hat{\sigma}_\epsilon$ = the historical root mean square error = $s_e'$

**b.** $\hat{\sigma}_\epsilon = \sqrt{\text{SSE}_t/t}$ (historical starting value)

**c.** $\hat{\sigma}_\epsilon = \sqrt{\text{SSE}_t/t}$ (no history)

**d.** $\hat{\sigma}_\epsilon = \sqrt{\text{MSE}_t}$ (historical starting value, $\eta = .1$)

**e.** $\hat{\sigma}_\epsilon = \sqrt{\text{MSE}_t}$ (no history, $\eta = .1$)

**f.** $\hat{\sigma}_\epsilon = 1.25 \cdot \text{MAD}_t$ (historical starting value, $\eta = .1$)

**g.** $\hat{\sigma}_\epsilon = 1.25 \cdot \text{MAD}_t$ (no history, $\eta = .1$)

**8.15** Calculate the control limits and values of the following tracking signals for the book store series of Exercise 8.13 ($\hat{\sigma}_\epsilon' = s_e'$):

**a.** $e_t$

**b.** $\text{SUM}_t/\sqrt{t}$

**c.** $\text{MTE}_t$, $k = 3$

**d.** $E_t/\text{MAD}_t$ ($\eta = .1$)

**8.16** Using least squares, fit a linear trend model to the data in Exercise 8.13 (weeks 1–7) and test for linear trend using a $t$-test. Using the resulting trend model, forecast sales for the next seven weeks (weeks 8–14). Using the indicated tracking signals, track the forecasts for the seven weeks assuming the actual observations turned out as below. Note any alarms that may occur ($\alpha' = .05$). (A computer printout from the regression analysis for weeks 1–7 is given below.)

| $t$ | 8 | 9 | 10 | 11 | 12 | 13 | 14 |
|---|---|---|---|---|---|---|---|
| $y_t$ | 17.5 | 17.0 | 22.9 | 16.9 | 22.1 | 21.2 | 16.0 |

**a.** $e_t$, $\hat{\sigma}_\epsilon$ = the standard error from the regression analysis

**b.** $\text{SUM}_t/\sqrt{t}$, $\hat{\sigma}_\epsilon = \sqrt{\text{MSE}_t}$, $\eta = .2$, $\text{MSE}_{\text{init}}$ = mean square error from the regression analysis

**c.** $\text{SUM}_t/\sqrt{t}$, $\hat{\sigma}_\epsilon = \sqrt{\text{MSE}_t}$, $\eta = .1$, $\text{MSE}_{\text{init}} = e_1^2$

**d.** $\text{MTE}_t$, $k = 3$, $\hat{\sigma}_\epsilon$ = standard error from the regression analysis

**e.** $E_t/\text{MAD}_t$, $\eta = .1$, $\text{MAD}_{\text{init}}$ = (the standard error from the regression analysis)/1.25

**Regression Printout for Weeks 1–7**

```
THE REGRESSION EQUATION IS
Y = 13.4 + 1.90 T

PREDICTOR        COEF          STDEV      T-RATIO
CONSTANT       13.429          2.223         6.04
T               1.9036         0.4971        3.83

S = 2.63        R-SQ = 74.6%      R-SQ(ADJ) = 69.5%

ANALYSIS OF VARIANCE

SOURCE        DF          SS          MS
REGRESSION     1      101.46      101.46
ERROR          5       34.6         6.92
TOTAL          6      136.06
```

**8.17** For a large apartment building, the following linear-trend, multiplicative seasonal model has been fit to $y_t$ = quarterly maintenance expenses ($1,000):

$\hat{T}_t = 2.4 + .3t$ Origin: IV, 1988
Units ($t$): Quarterly

$\hat{S}_1 = 1.21$ $\hat{S}_2 = .92$ $\hat{S}_3 = .84$ $\hat{S}_4 = 1.03$

Use this model to forecast 1989 and 1990, and then suppose the actual values were:

**Actual Maintenance Expenses (*Post* Model Fit)**

| | Quarter | | | |
|---|---|---|---|---|
| | I | II | III | IV |
| 1989 | 5.02 | 1.63 | 3.42 | 2.83 |
| 1990 | 3.36 | 4.10 | 5.09 | 5.82 |

Compute the indicated tracking signals. Are there any alarms after $t = 4$ ($\alpha' = .05$)?

**a.** $\text{SUM}_t/\sqrt{t}$, $\hat{\sigma}_\epsilon = \sqrt{\text{MSE}_t}$, $\eta = .1$

**b.** $E_t/\text{MAD}_t$, $\hat{\sigma}_\epsilon = 1.25 \cdot \text{MAD}_t$

**8.18** A wholesale paper supplier wishes to forecast monthly demand for its medium-weight house brand copy paper, which it sells in boxes of 10 reams each. Simple exponential smoothing with $\alpha = .2$ is to be used. Suppose that for the next 10 months $y_t$ = actual demand (boxes) is:

| $t$ | 1 | 2 | 3 | 4 | 5 | 6 | 7 | 8 | 9 | 10 |
|---|---|---|---|---|---|---|---|---|---|---|
| $y_t$ | 174 | 143 | 152 | 173 | 164 | 171 | 198 | 200 | 183 | 170 |

**a.** Suppose this series were forecast with the proposed scheme. What would be the forecast for $y_{11}$?

**b.** Suppose this series were forecast with the proposed scheme but adding Brown's tracking signal with $\alpha' = .05$, $\eta = \alpha$, and reinitializing to the current value of the series when an alarm sounded. What would be the forecast for $y_{11}$? (Sound no alarms until after $t = 3$.)

**c.** Repeat part (b) using Trigg's tracking signal.

**8.19** The time series below consists of quarterly values of a production efficiency index used by a certain manufacturer.

| $t$ | 1 | 2 | 3 | 4 | 5 | 6 | 7 | 8 | 9 | 10 | 11 | 12 |
|---|---|---|---|---|---|---|---|---|---|---|---|---|
| $y_t$ | 82 | 84 | 85 | 83 | 88 | 90 | 90 | 99 | 107 | 101 | 106 | 110 |

**a.** Suppose SES with $\alpha = .3$ and Brown's tracking signal with $\eta = .2$, $\alpha' = .01$ were used to forecast this time series, reinitializing $\hat{\beta}_0$ to the most current value of the series when an alarm is tripped. What would the forecast for $y_{13}$ be?

**b.** What would the forecast in part (a) be without using a tracking signal?

**8.20** A regional wholesale supplier of supermarket products recognizes that there is considerable volatility in demand among the several hundred lines of products the firm handles. Because of this volatility, and because of the large number of different series to be forecast, it is decided to use an automatic smoothing scheme, either HI/LO tracking with Brown's tracking signal ($\eta = \alpha = .1$, $\alpha^* = .6$, and $\alpha' = .05$), or Trigg-Leach ARRES (I or II) with $\eta = .2$. Suppose the series below, $y_t$ = biweekly orders for microwave popcorn (cases), were to be forecast.

**Biweekly Orders**

| $t$ | $y_t$ | $t$ | $y_t$ |
|---|---|---|---|
| 1 | 42 | 9 | 62 |
| 2 | 40 | 10 | 58 |
| 3 | 31 | 11 | 65 |
| 4 | 43 | 12 | 65 |
| 5 | 49 | 13 | 51 |
| 6 | 47 | 14 | 53 |
| 7 | 41 | 15 | 40 |
| 8 | 54 | 16 | 36 |

**a.** What would be the forecast for $y_{17}$, the RMSE, and Theil's $U$ using the HI/LO scheme?

**b.** What would be the forecast for $y_{17}$, the RMSE, and Theil's $U$ using the ARRES I scheme?

**c.** What would be the forecast for $y_{17}$, the RMSE, and Theil's $U$ using the ARRES II scheme?

**d.** What effect would using $\eta = .1$ have on the results of parts (b) and (c)?

**8.21** Forecast the paper demand data of Exercise 8.18 using HI/LO tracking with $\eta = \alpha = .1$, $\alpha^* = .6$, $\alpha' = .05$, and Trigg's tracking signal.

**8.22** Forecast the production efficiency series of Exercise 8.19 using ARRES I with $\eta = .1$

**8.23** Forecast the production efficiency series of Exercise 8.19 using ARRES II with $\eta = .1$

## References


Batty, M. (1969). "Monitoring an Exponential Smoothing Forecasting System." *Operational Research Quarterly* 20: 319–325.

Brown, R. G. (1967). *Decision Rules for Inventory Management.* New York: Holt, Rinehart and Winston.

Brown, R. G. (1959). *Statistical Forecasting for Inventory Control.* New York: McGraw-Hill.

Gardner, E. S. (1985). "Exponential Smoothing: The State of the Art." *J. Forecasting* 4: 1–28.

Gardner, E. S. (1983). "Automatic Monitoring of Forecast Errors." *J. Forecasting* 2: 1–21.

Gardner, E. S., and Dannenbring, D. G. (1980). "Forecasting with Exponential Smoothing: Some Guidelines for Model Selection." *Decision Sciences* 11: 370–383.

Gilchrist, W. (1976). *Statistical Forecasting.* New York: Wiley.

Makridakis, S., Andersen, A., Carbone, R., Fildes, R., Hibon, M., Lewandowski, R., Newton, J., Parzen, R., and Winkler, R. (1982). "The Accuracy of Extrapolation (Time Series) Methods: Results of a Forecasting Competition." *J. Forecasting* 1: 111–153.

Makridakis, S., and Hibon, M. (1979). "Accuracy of Forecasting: An Empirical Investigation," with discussion. *J. Roy. Stat. Soc.* A142: 97–145.

Montgomery, D. C., and Johnson, L. A. (1976). *Forecasting and Time Series Analysis.* New York: McGraw-Hill.

Shone, M. L. (1967). "Viewpoints." *Operational Research Quarterly,* 18: 318–319.

Trigg, D. W. (1964). "Monitoring a Forecasting System." *Operational Research Quarterly* 15: 271–274.

Trigg, D. W., and Leach, A. G. (1967). "Exponential Smoothing with an Adaptive Response Rate." *Operational Research Quarterly* 18: 53–59.

# APPENDIX AA

## Critical Values for $R$ in a Two-Tailed Runs Test for Stationarity ($\alpha = .10$)

$n$ = the number of observations in the time series

$m = n/2$ for $n$ even

$= (n - 1)/2$ for $n$ odd

$R$ = the number of runs above and below the median

Reject $H_o$ if $R \leq R_L$ or if $R \geq R_U$.

| $m$ | $R_L$ | | $R_U$ |
|---|---|---|---|
| 1–4 | — | not attainable | — |
| 5 | 2 | | 10 |
| 6 | 3 | | 11 |
| 7 | 3 | | 13 |
| 8 | 4 | | 14 |
| 9 | 5 | | 15 |
| 10 | 6 | | 16 |
| 11 | 7 | | 17 |
| 12 | 7 | | 17 |
| 13 | 8 | | 19 |
| 14 | 9 | | 20 |
| 15 | 10 | | 22 |
| 16 | 11 | | 23 |
| 17 | 11 | | 25 |
| 18 | 12 | | 26 |
| 19 | 13 | | 27 |
| 20 | 14 | | 28 |

*Source:* Adapted from F. S. Swed and C. Eisenhart (1943). "Tables for Testing Randomness of Grouping in a Sequence of Alternatives." *Annals of Mathematical Statistics* 14: 66–87. Used with permission of the Institute of Mathematical Statistics.

# APPENDIX BB

## Normal Curve Areas

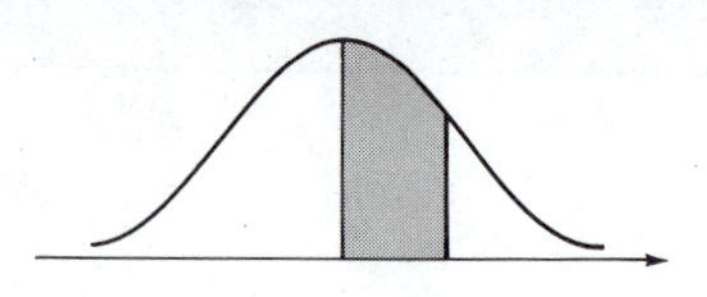

| $z$ | .00 | .01 | .02 | .03 | .04 | .05 | .06 | .07 | .08 | .09 |
|---|---|---|---|---|---|---|---|---|---|---|
| 0.0 | 0.0000 | 0.0040 | 0.0080 | 0.0120 | 0.0160 | 0.0199 | 0.0239 | 0.0279 | 0.0319 | 0.0359 |
| 0.1 | 0.0398 | 0.0438 | 0.0478 | 0.0517 | 0.0557 | 0.0596 | 0.0636 | 0.0675 | 0.0714 | 0.0753 |
| 0.2 | 0.0793 | 0.0832 | 0.0871 | 0.0910 | 0.0948 | 0.0987 | 0.1026 | 0.1064 | 0.1103 | 0.1141 |
| 0.3 | 0.1179 | 0.1217 | 0.1255 | 0.1293 | 0.1331 | 0.1368 | 0.1406 | 0.1443 | 0.1480 | 0.1517 |
| 0.4 | 0.1554 | 0.1591 | 0.1628 | 0.1664 | 0.1700 | 0.1736 | 0.1772 | 0.1808 | 0.1844 | 0.1879 |
| 0.5 | 0.1915 | 0.1950 | 0.1985 | 0.2019 | 0.2054 | 0.2088 | 0.2123 | 0.2157 | 0.2190 | 0.2224 |
| 0.6 | 0.2257 | 0.2291 | 0.2324 | 0.2357 | 0.2389 | 0.2422 | 0.2454 | 0.2486 | 0.2517 | 0.2549 |
| 0.7 | 0.2580 | 0.2611 | 0.2642 | 0.2673 | 0.2704 | 0.2734 | 0.2764 | 0.2794 | 0.2823 | 0.2852 |
| 0.8 | 0.2881 | 0.2910 | 0.2939 | 0.2967 | 0.2995 | 0.3023 | 0.3051 | 0.3078 | 0.3106 | 0.3133 |
| 0.9 | 0.3159 | 0.3186 | 0.3212 | 0.3238 | 0.3264 | 0.3289 | 0.3315 | 0.3340 | 0.3365 | 0.3389 |
| 1.0 | 0.3413 | 0.3438 | 0.3461 | 0.3485 | 0.3508 | 0.3531 | 0.3554 | 0.3577 | 0.3599 | 0.3621 |
| 1.1 | 0.3643 | 0.3665 | 0.3686 | 0.3708 | 0.3729 | 0.3749 | 0.3770 | 0.3790 | 0.3810 | 0.3830 |
| 1.2 | 0.3849 | 0.3869 | 0.3888 | 0.3907 | 0.3925 | 0.3944 | 0.3962 | 0.3980 | 0.3997 | 0.4015 |
| 1.3 | 0.4032 | 0.4049 | 0.4066 | 0.4082 | 0.4099 | 0.4115 | 0.4131 | 0.4147 | 0.4162 | 0.4177 |
| 1.4 | 0.4192 | 0.4207 | 0.4222 | 0.4236 | 0.4251 | 0.4265 | 0.4279 | 0.4292 | 0.4306 | 0.4319 |
| 1.5 | 0.4332 | 0.4345 | 0.4357 | 0.4370 | 0.4382 | 0.4394 | 0.4406 | 0.4418 | 0.4429 | 0.4441 |
| 1.6 | 0.4452 | 0.4463 | 0.4474 | 0.4484 | 0.4495 | 0.4505 | 0.4515 | 0.4525 | 0.4535 | 0.4545 |
| 1.7 | 0.4554 | 0.4564 | 0.4573 | 0.4582 | 0.4591 | 0.4599 | 0.4608 | 0.4616 | 0.4625 | 0.4633 |
| 1.8 | 0.4641 | 0.4649 | 0.4656 | 0.4664 | 0.4671 | 0.4678 | 0.4686 | 0.4693 | 0.4699 | 0.4706 |
| 1.9 | 0.4713 | 0.4719 | 0.4726 | 0.4732 | 0.4738 | 0.4744 | 0.4750 | 0.4756 | 0.4761 | 0.4767 |
| 2.0 | 0.4773 | 0.4778 | 0.4783 | 0.4788 | 0.4793 | 0.4798 | 0.4803 | 0.4808 | 0.4812 | 0.4817 |
| 2.1 | 0.4821 | 0.4826 | 0.4830 | 0.4834 | 0.4838 | 0.4842 | 0.4846 | 0.4850 | 0.4854 | 0.4857 |
| 2.2 | 0.4861 | 0.4864 | 0.4868 | 0.4871 | 0.4875 | 0.4878 | 0.4881 | 0.4884 | 0.4887 | 0.4890 |
| 2.3 | 0.4893 | 0.4896 | 0.4898 | 0.4901 | 0.4904 | 0.4906 | 0.4909 | 0.4911 | 0.4913 | 0.4916 |
| 2.4 | 0.4918 | 0.4920 | 0.4922 | 0.4925 | 0.4927 | 0.4929 | 0.4931 | 0.4932 | 0.4934 | 0.4936 |
| 2.5 | 0.4938 | 0.4940 | 0.4941 | 0.4943 | 0.4945 | 0.4946 | 0.4948 | 0.4949 | 0.4951 | 0.4952 |
| 2.6 | 0.4953 | 0.4955 | 0.4956 | 0.4957 | 0.4959 | 0.4960 | 0.4961 | 0.4962 | 0.4963 | 0.4964 |
| 2.7 | 0.4965 | 0.4966 | 0.4967 | 0.4968 | 0.4969 | 0.4970 | 0.4971 | 0.4972 | 0.4973 | 0.4974 |
| 2.8 | 0.4974 | 0.4975 | 0.4976 | 0.4977 | 0.4977 | 0.4978 | 0.4979 | 0.4979 | 0.4980 | 0.4981 |
| 2.9 | 0.4981 | 0.4982 | 0.4982 | 0.4983 | 0.4984 | 0.4984 | 0.4985 | 0.4985 | 0.4986 | 0.4986 |
| 3.0 | 0.4987 | 0.4987 | 0.4987 | 0.4988 | 0.4988 | 0.4989 | 0.4989 | 0.4989 | 0.4990 | 0.4990 |

**Percentage Points:**

| $\alpha$ | .100 | .050 | .025 | .010 | .005 |
|---|---|---|---|---|---|
| $z_\alpha$ | 1.282 | 1.645 | 1.960 | 2.326 | 2.576 |

*Source:* Computed using Minitab software.

# APPENDIX CC

## Critical Values of Kendall's tau

| | α (one-tailed test) or α/2 (two-tailed test) | | | | |
|---|---|---|---|---|---|
| $n$ | .100 | .050 | .025 | .010 | .005 |
| 4 | .67 | .67 | — | — | — |
| 5 | .60 | .60 | .80 | .80 | — |
| 6 | .47 | .60 | .73 | .73 | .87 |
| 7 | .43 | .52 | .62 | .71 | .81 |
| 8 | .36 | .50 | .57 | .64 | .71 |
| 9 | .33 | .44 | .50 | .61 | .67 |
| 10 | .33 | .42 | .47 | .56 | .60 |

- Reject $H_o$ when $\tau$ *exceeds* the tabled entry. That is, for $n = 9$ and a 5% level of significance:
  Reject $H_o$ if $|\tau| > .50$ (two-tailed test)
  Reject $H_o$ if $\tau > .44$ (upper one-tailed test)
- For $n > 10$, under $H_o$, $\tau$ is approximately normal,

$$\mu_\tau = 0 \quad \text{and} \quad \sigma_\tau = \sqrt{\frac{2(2n + 5)}{9n(n - 1)}}$$

*Source:* Adapted from E. S. Pearson and H. O. Hartley (1966). *Biometrika Tables for Statisticians,* 3rd ed., Vol. 1. Cambridge, England: Cambridge University Press, Table 45. Used with permission of the *Biometrika* Trustees.

# APPENDIX DD

## Critical Values (Approximate) of Spearman's rho

$n$ = sample size

$$r_s = 1 - \frac{6 \cdot \Sigma d_r^2}{n(n^2 - 1)}$$

| One-Tailed α: / Two-Tailed α: | .001 / .002 | .005 / .010 | .010 / .020 | .025 / .050 | .050 / .100 | .100 / .200 | $n$ |
|---|---|---|---|---|---|---|---|
| | — | — | — | — | .8000 | .8000 | 4 |
| | — | — | .9000 | .9000 | .8000 | .7000 | 5 |
| | — | .9429 | .8857 | .8286 | .7714 | .6000 | 6 |
| | .9643 | .8929 | .8571 | .7450 | .6786 | .5357 | 7 |
| | .9286 | .8571 | .8095 | .7143 | .6190 | .5000 | 8 |
| | .9000 | .8167 | .7667 | .6833 | .5833 | .4667 | 9 |
| | .8667 | .7818 | .7333 | .6364 | .5515 | .4424 | 10 |
| | .8364 | .7545 | .7000 | .6091 | .5273 | .4182 | 11 |
| | .8182 | .7273 | .6713 | .5804 | .4965 | .3986 | 12 |
| | .7912 | .6978 | .6429 | .5549 | .4780 | .3791 | 13 |
| | .7670 | .6747 | .6220 | .5341 | .4593 | .3626 | 14 |
| | .7464 | .6536 | .6000 | .5179 | .4429 | .3500 | 15 |
| | .7265 | .6324 | .5824 | .5000 | .4265 | .3382 | 16 |
| | .7083 | .6152 | .5637 | .4853 | .4118 | .3260 | 17 |
| | .6904 | .5975 | .5480 | .4716 | .3994 | .3148 | 18 |
| | .6737 | .5825 | .5333 | .4579 | .3895 | .3070 | 19 |
| | .6586 | .5684 | .5203 | .4451 | .3789 | .2977 | 20 |
| | .6455 | .5545 | .5078 | .4351 | .3688 | .2909 | 21 |
| | .6318 | .5426 | .4963 | .4241 | .3597 | .2829 | 22 |
| | .6186 | .5306 | .4852 | .4150 | .3518 | .2767 | 23 |
| | .6070 | .5200 | .4748 | .4061 | .3435 | .2704 | 24 |
| | .5962 | .5100 | .4654 | .3977 | .3362 | .2646 | 25 |
| | .5856 | .5002 | .4564 | .3894 | .3299 | .2588 | 26 |
| | .5757 | .4915 | .4481 | .3822 | .3236 | .2540 | 27 |
| | .5660 | .4828 | .4401 | .3749 | .3175 | .2490 | 28 |
| | .5567 | .4744 | .4320 | .3685 | .3113 | .2443 | 29 |
| | .5479 | .4665 | .4251 | .3620 | .3059 | .2400 | 30 |

*Source:* G. J. Glasser and R. F. Winter (1961). "Values of the Coefficient of Rank Correlation for Testing the Hypothesis of Independence." *Biometrika* 48:444–448. Reprinted by permission of the *Biometrika* trustees. Includes corrections of W. J. Conover (1971). *Practical Nonparametric Statistics,* New York: Wiley.

# APPENDIX EE

## Critical Values of von Neumann's Ratio (Mean Square Successive Difference Test)

$$M = \frac{SS_{\Delta y}}{SS_{yy}} = \frac{\sum_{2}^{n} (\Delta y)^2}{\sum_{1}^{n} (y - \bar{y})^2}$$

Lower critical value: $M_{1-p}$ in table

Upper critical value: $M_p = 4 - M_{1-p}$

One-tailed test: $p = \alpha$ Two-tailed test: $p = \alpha/2$

| | $p$ | | | | $p$ | | | | $p$ | | |
|---|---|---|---|---|---|---|---|---|---|---|---|
| $n$ | .10 | .05 | .01 | $n$ | .10 | .05 | .01 | $n$ | .10 | .05 | .01 |
| 10 | 1.251 | 1.062 | 0.752 | 30 | 1.543 | 1.418 | 1.195 | 100 | 1.745 | 1.674 | 1.542 |
| 11 | 1.280 | 1.096 | 0.792 | 32 | 1.557 | 1.436 | 1.218 | 110 | 1.757 | 1.689 | 1.563 |
| 12 | 1.306 | 1.128 | 0.828 | 34 | 1.569 | 1.451 | 1.239 | 120 | 1.767 | 1.702 | 1.581 |
| 13 | 1.329 | 1.156 | 0.862 | 36 | 1.581 | 1.466 | 1.259 | 130 | 1.776 | 1.714 | 1.597 |
| 14 | 1.351 | 1.182 | 0.893 | 38 | 1.592 | 1.480 | 1.277 | 140 | 1.784 | 1.724 | 1.611 |
| 15 | 1.370 | 1.205 | 0.922 | 40 | 1.602 | 1.492 | 1.293 | 150 | 1.792 | 1.733 | 1.624 |
| 16 | 1.388 | 1.227 | 0.949 | 42 | 1.611 | 1.504 | 1.309 | 160 | 1.798 | 1.741 | 1.636 |
| 17 | 1.405 | 1.247 | 0.974 | 44 | 1.620 | 1.515 | 1.324 | 170 | 1.804 | 1.749 | 1.647 |
| 18 | 1.420 | 1.266 | 0.998 | 46 | 1.628 | 1.525 | 1.338 | 180 | 1.810 | 1 756 | 1.656 |
| 19 | 1.434 | 1.283 | 1.020 | 48 | 1.635 | 1.534 | 1.351 | 190 | 1.815 | 1.763 | 1.665 |
| 20 | 1.447 | 1.300 | 1.041 | 50 | 1.642 | 1.544 | 1.363 | 200 | 1.819 | 1.768 | 1.674 |
| 21 | 1.460 | 1.315 | 1.060 | 55 | 1.659 | 1.564 | 1.391 | 250 | 1.838 | 1.793 | 1.708 |
| 22 | 1.471 | 1.329 | 1.078 | 60 | 1.673 | 1.582 | 1.415 | 300 | 1.852 | 1.811 | 1.733 |
| 23 | 1.482 | 1.342 | 1.096 | 65 | 1.685 | 1.598 | 1.437 | 350 | 1.863 | 1.825 | 1.752 |
| 24 | 1.492 | 1.355 | 1.112 | 70 | 1.697 | 1.612 | 1.457 | 400 | 1.872 | 1.836 | 1.768 |
| 25 | 1.502 | 1.367 | 1.128 | 75 | 1.707 | 1.625 | 1.474 | 450 | 1.879 | 1.845 | 1.781 |
| 26 | 1.511 | 1.378 | 1.143 | 80 | 1.716 | 1.636 | 1.490 | 500 | 1.886 | 1.853 | 1.793 |
| 27 | 1.520 | 1.389 | 1.157 | 85 | 1.724 | 1.647 | 1.505 | 600 | 1.895 | 1.866 | 1.811 |
| 28 | 1.528 | 1.399 | 1.170 | 90 | 1.732 | 1.657 | 1.518 | 800 | 1.909 | 1.884 | 1.836 |
| 29 | 1.535 | 1.409 | 1.183 | 95 | 1.739 | 1.666 | 1.531 | 1000 | 1.919 | 1.896 | 1.853 |

*Source:* Reproduced from L. S. Nelson (1980). "The Mean Square Successive Difference Test." *Journal of Quality Technology* 12(3): 175. Used by permission of the Editor, *Journal of Quality Technology.*

# APPENDIX FF

## Critical Values of chi-Square

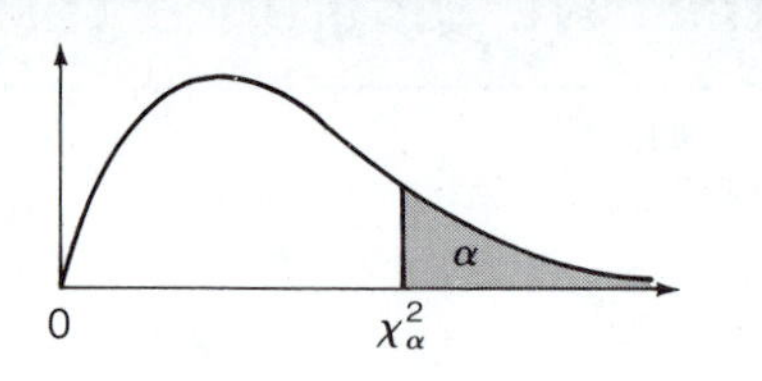


| df | $\chi^2_{.100}$ | $\chi^2_{.050}$ | $\chi^2_{.025}$ | $\chi^2_{.010}$ | $\chi^2_{.005}$ |
|---|---|---|---|---|---|
| 1 | 2.706 | 3.841 | 5.024 | 6.635 | 7.879 |
| 2 | 4.605 | 5.991 | 7.378 | 9.210 | 10.597 |
| 3 | 6.251 | 7.815 | 9.348 | 11.345 | 12.838 |
| 4 | 7.779 | 9.488 | 11.143 | 13.277 | 14.860 |
| 5 | 9.236 | 11.070 | 12.832 | 15.086 | 16.750 |
| 6 | 10.645 | 12.592 | 14.449 | 16.812 | 18.547 |
| 7 | 12.017 | 14.067 | 16.013 | 18.475 | 20.278 |
| 8 | 13.362 | 15.507 | 17.535 | 20.090 | 21.955 |
| 9 | 14.684 | 16.919 | 19.023 | 21.666 | 23.589 |
| 10 | 15.987 | 18.307 | 20.483 | 23.209 | 25.188 |
| 11 | 17.275 | 19.675 | 21.920 | 24.725 | 26.757 |
| 12 | 18.549 | 21.026 | 23.337 | 26.217 | 28.299 |
| 13 | 19.812 | 22.362 | 24.736 | 27.688 | 29.819 |
| 14 | 21.064 | 23.685 | 26.119 | 29.141 | 31.319 |
| 15 | 22.307 | 24.996 | 27.488 | 30.578 | 32.801 |
| 16 | 23.542 | 26.296 | 28.845 | 31.999 | 34.266 |
| 17 | 24.769 | 27.587 | 30.191 | 33.408 | 35.718 |
| 18 | 25.989 | 28.869 | 31.526 | 34.805 | 37.156 |
| 19 | 27.204 | 30.143 | 32.852 | 36.191 | 38.582 |
| 20 | 28.412 | 31.410 | 34.169 | 37.566 | 39.996 |
| 21 | 29.615 | 32.670 | 35.479 | 38.932 | 41.400 |
| 22 | 30.813 | 33.924 | 36.781 | 40.289 | 42.795 |
| 23 | 32.007 | 35.172 | 38.076 | 41.638 | 44.181 |
| 24 | 33.196 | 36.415 | 39.363 | 42.978 | 45.556 |
| 25 | 34.381 | 37.652 | 40.646 | 44.313 | 46.926 |
| 26 | 35.563 | 38.885 | 41.923 | 45.641 | 48.289 |
| 27 | 36.741 | 40.113 | 43.194 | 46.962 | 49.644 |
| 28 | 37.916 | 41.337 | 44.460 | 48.278 | 50.992 |
| 29 | 39.087 | 42.557 | 45.722 | 49.586 | 52.333 |
| 30 | 40.256 | 43.773 | 46.979 | 50.891 | 53.670 |
| 40 | 51.805 | 55.758 | 59.341 | 63.690 | 66.764 |
| 50 | 63.167 | 67.505 | 71.420 | 76.154 | 79.490 |
| 60 | 74.397 | 79.081 | 83.297 | 88.377 | 91.947 |
| 70 | 85.527 | 90.531 | 95.023 | 100.424 | 104.213 |
| 80 | 96.578 | 101.879 | 106.628 | 112.328 | 116.320 |
| 90 | 107.564 | 113.144 | 118.133 | 124.110 | 128.287 |
| 100 | 118.497 | 124.341 | 129.559 | 135.801 | 140.158 |

*Source:* Computed using Minitab software.

# APPENDIX GG

## Critical Values of Student's $t$

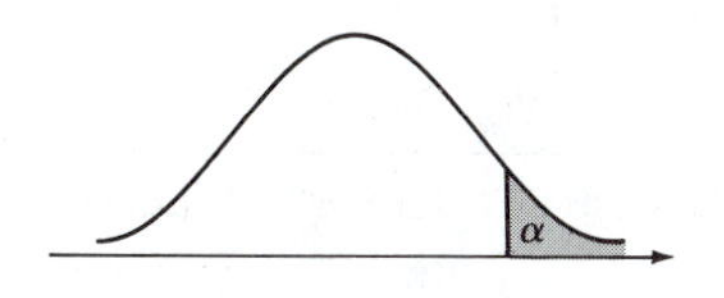


| df | $t_{.100}$ | $t_{.050}$ | $t_{.025}$ | $t_{.010}$ | $t_{.005}$ |
|---|---|---|---|---|---|
| 1 | 3.078 | 6.314 | 12.706 | 31.820 | 63.655 |
| 2 | 1.886 | 2.920 | 4.303 | 6.965 | 9.925 |
| 3 | 1.638 | 2.353 | 3.182 | 4.541 | 5.841 |
| 4 | 1.533 | 2.132 | 2.776 | 3.747 | 4.604 |
| 5 | 1.476 | 2.015 | 2.571 | 3.365 | 4.032 |
| 6 | 1.440 | 1.943 | 2.447 | 3.143 | 3.707 |
| 7 | 1.415 | 1.895 | 2.365 | 2.998 | 3.499 |
| 8 | 1.397 | 1.860 | 2.306 | 2.896 | 3.355 |
| 9 | 1.383 | 1.833 | 2.262 | 2.821 | 3.250 |
| 10 | 1.372 | 1.812 | 2.228 | 2.764 | 3.169 |
| 11 | 1.363 | 1.796 | 2.201 | 2.718 | 3.106 |
| 12 | 1.356 | 1.782 | 2.179 | 2.681 | 3.055 |
| 13 | 1.350 | 1.771 | 2.160 | 2.650 | 3.012 |
| 14 | 1.345 | 1.761 | 2.145 | 2.624 | 2.977 |
| 15 | 1.341 | 1.753 | 2.131 | 2.602 | 2.947 |
| 16 | 1.337 | 1.746 | 2.120 | 2.583 | 2.921 |
| 17 | 1.333 | 1.740 | 2.110 | 2.567 | 2.898 |
| 18 | 1.330 | 1.734 | 2.101 | 2.552 | 2.878 |
| 19 | 1.328 | 1.729 | 2.093 | 2.539 | 2.861 |
| 20 | 1.325 | 1.725 | 2.086 | 2.528 | 2.845 |
| 21 | 1.323 | 1.721 | 2.080 | 2.518 | 2.831 |
| 22 | 1.321 | 1.717 | 2.074 | 2.508 | 2.819 |
| 23 | 1.319 | 1.714 | 2.069 | 2.500 | 2.807 |
| 24 | 1.318 | 1.711 | 2.064 | 2.492 | 2.797 |
| 25 | 1.316 | 1.708 | 2.060 | 2.485 | 2.787 |
| 26 | 1.315 | 1.706 | 2.056 | 2.479 | 2.779 |
| 27 | 1.314 | 1.703 | 2.052 | 2.473 | 2.771 |
| 28 | 1.313 | 1.701 | 2.048 | 2.467 | 2.763 |
| 29 | 1.311 | 1.699 | 2.045 | 2.462 | 2.756 |
| 30 | 1.310 | 1.697 | 2.042 | 2.457 | 2.750 |
| 40 | 1.303 | 1.684 | 2.021 | 2.423 | 2.704 |
| 60 | 1.296 | 1.671 | 2.000 | 2.390 | 2.660 |
| 120 | 1.289 | 1.658 | 1.980 | 2.358 | 2.617 |
| $\infty$ | 1.282 | 1.645 | 1.960 | 2.326 | 2.576 |

*Source:* Computed using Minitab software.

# APPENDIX HH

## Critical Values of the *F*-Distribution ($\alpha = .10$)

| $\nu_2$ \ $\nu_1$ | NUMERATOR DEGREES OF FREEDOM 1 | 2 | 3 | 4 | 5 | 6 | 7 | 8 | 9 |
|---|---|---|---|---|---|---|---|---|---|
| DENOMINATOR DEGREES OF FREEDOM 1 | 39.86 | 49.50 | 53.59 | 55.83 | 57.24 | 58.20 | 58.91 | 59.44 | 59.86 |
| 2 | 8.53 | 9.00 | 9.16 | 9.24 | 9.29 | 9.33 | 9.35 | 9.37 | 9.38 |
| 3 | 5.54 | 5.46 | 5.39 | 5.34 | 5.31 | 5.28 | 5.27 | 5.25 | 5.24 |
| 4 | 4.54 | 4.32 | 4.19 | 4.11 | 4.05 | 4.01 | 3.98 | 3.95 | 3.94 |
| 5 | 4.06 | 3.78 | 3.62 | 3.52 | 3.45 | 3.40 | 3.37 | 3.34 | 3.32 |
| 6 | 3.78 | 3.46 | 3.29 | 3.18 | 3.11 | 3.05 | 3.01 | 2.98 | 2.96 |
| 7 | 3.59 | 3.26 | 3.07 | 2.96 | 2.88 | 2.83 | 2.78 | 2.75 | 2.72 |
| 8 | 3.46 | 3.11 | 2.92 | 2.81 | 2.73 | 2.67 | 2.62 | 2.59 | 2.56 |
| 9 | 3.36 | 3.01 | 2.81 | 2.69 | 2.61 | 2.55 | 2.51 | 2.47 | 2.44 |
| 10 | 3.29 | 2.92 | 2.73 | 2.61 | 2.52 | 2.46 | 2.41 | 2.38 | 2.35 |
| 11 | 3.23 | 2.86 | 2.66 | 2.54 | 2.45 | 2.39 | 2.34 | 2.30 | 2.27 |
| 12 | 3.18 | 2.81 | 2.61 | 2.48 | 2.39 | 2.33 | 2.28 | 2.24 | 2.21 |
| 13 | 3.14 | 2.76 | 2.56 | 2.43 | 2.35 | 2.28 | 2.23 | 2.20 | 2.16 |
| 14 | 3.10 | 2.73 | 2.52 | 2.39 | 2.31 | 2.24 | 2.19 | 2.15 | 2.12 |
| 15 | 3.07 | 2.70 | 2.49 | 2.36 | 2.27 | 2.21 | 2.16 | 2.12 | 2.09 |
| 16 | 3.05 | 2.67 | 2.46 | 2.33 | 2.24 | 2.18 | 2.13 | 2.09 | 2.06 |
| 17 | 3.03 | 2.64 | 2.44 | 2.31 | 2.22 | 2.15 | 2.10 | 2.06 | 2.03 |
| 18 | 3.01 | 2.62 | 2.42 | 2.29 | 2.20 | 2.13 | 2.08 | 2.04 | 2.00 |
| 19 | 2.99 | 2.61 | 2.40 | 2.27 | 2.18 | 2.11 | 2.06 | 2.02 | 1.98 |
| 20 | 2.97 | 2.59 | 2.38 | 2.25 | 2.16 | 2.09 | 2.04 | 2.00 | 1.96 |
| 21 | 2.96 | 2.57 | 2.36 | 2.23 | 2.14 | 2.08 | 2.02 | 1.98 | 1.95 |
| 22 | 2.95 | 2.56 | 2.35 | 2.22 | 2.13 | 2.06 | 2.01 | 1.97 | 1.93 |
| 23 | 2.94 | 2.55 | 2.34 | 2.21 | 2.11 | 2.05 | 1.99 | 1.95 | 1.92 |
| 24 | 2.93 | 2.54 | 2.33 | 2.19 | 2.10 | 2.04 | 1.98 | 1.94 | 1.91 |
| 25 | 2.92 | 2.53 | 2.32 | 2.18 | 2.09 | 2.02 | 1.97 | 1.93 | 1.89 |
| 26 | 2.91 | 2.52 | 2.31 | 2.17 | 2.08 | 2.01 | 1.96 | 1.92 | 1.88 |
| 27 | 2.90 | 2.51 | 2.30 | 2.17 | 2.07 | 2.00 | 1.95 | 1.91 | 1.87 |
| 28 | 2.89 | 2.50 | 2.29 | 2.16 | 2.06 | 2.00 | 1.94 | 1.90 | 1.87 |
| 29 | 2.89 | 2.50 | 2.28 | 2.15 | 2.06 | 1.99 | 1.93 | 1.89 | 1.86 |
| 30 | 2.88 | 2.49 | 2.28 | 2.14 | 2.05 | 1.98 | 1.93 | 1.88 | 1.85 |
| 40 | 2.84 | 2.44 | 2.23 | 2.09 | 2.00 | 1.93 | 1.87 | 1.83 | 1.79 |
| 60 | 2.79 | 2.39 | 2.18 | 2.04 | 1.95 | 1.87 | 1.82 | 1.77 | 1.74 |
| 120 | 2.75 | 2.35 | 2.13 | 1.99 | 1.90 | 1.82 | 1.77 | 1.72 | 1.68 |
| $\infty$ | 2.71 | 2.30 | 2.08 | 1.94 | 1.85 | 1.77 | 1.72 | 1.67 | 1.63 |

*Source:* Adapted from M. Merrington and C. M. Thompson (1943). "Tables of Percentage Points of the Inverted Beta (*F*)-Distribution." *Biometrika* 33:73–88. Used by permission of the *Biometrika* Trustees.

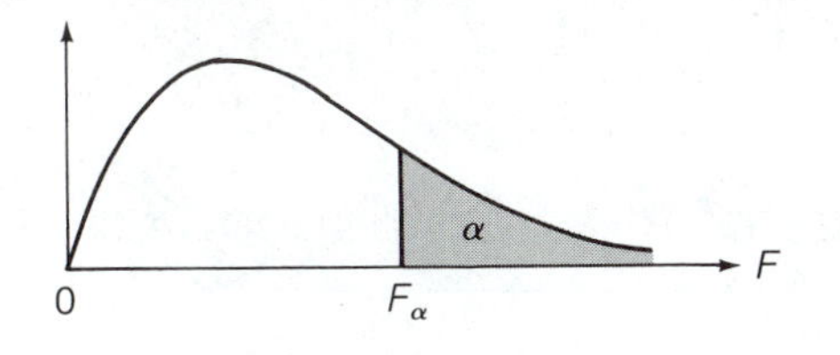


| $\nu_2$ \ $\nu_1$ | NUMERATOR DEGREES OF FREEDOM 10 | 12 | 15 | 20 | 24 | 30 | 40 | 60 | 120 | ∞ |
|---|---|---|---|---|---|---|---|---|---|---|
| 1 | 60.19 | 60.71 | 61.22 | 61.74 | 62.00 | 62.26 | 62.53 | 62.79 | 63.06 | 63.33 |
| 2 | 9.39 | 9.41 | 9.42 | 9.44 | 9.45 | 9.46 | 9.47 | 9.47 | 9.48 | 9.49 |
| 3 | 5.23 | 5.22 | 5.20 | 5.18 | 5.18 | 5.17 | 5.16 | 5.15 | 5.14 | 5.13 |
| 4 | 3.92 | 3.90 | 3.87 | 3.84 | 3.83 | 3.82 | 3.80 | 3.79 | 3.78 | 3.76 |
| 5 | 3.30 | 3.27 | 3.24 | 3.21 | 3.19 | 3.17 | 3.16 | 3.14 | 3.12 | 3.10 |
| 6 | 2.94 | 2.90 | 2.87 | 2.84 | 2.82 | 2.80 | 2.78 | 2.76 | 2.74 | 2.72 |
| 7 | 2.70 | 2.67 | 2.63 | 2.59 | 2.58 | 2.56 | 2.54 | 2.51 | 2.49 | 2.47 |
| 8 | 2.54 | 2.50 | 2.46 | 2.42 | 2.40 | 2.38 | 2.36 | 2.34 | 2.32 | 2.29 |
| 9 | 2.42 | 2.38 | 2.34 | 2.30 | 2.28 | 2.25 | 2.23 | 2.21 | 2.18 | 2.16 |
| 10 | 2.32 | 2.28 | 2.24 | 2.20 | 2.18 | 2.16 | 2.13 | 2.11 | 2.08 | 2.06 |
| 11 | 2.25 | 2.21 | 2.17 | 2.12 | 2.10 | 2.08 | 2.05 | 2.03 | 2.00 | 1.97 |
| 12 | 2.19 | 2.15 | 2.10 | 2.06 | 2.04 | 2.01 | 1.99 | 1.96 | 1.93 | 1.90 |
| 13 | 2.14 | 2.10 | 2.05 | 2.01 | 1.98 | 1.96 | 1.93 | 1.90 | 1.88 | 1.85 |
| 14 | 2.10 | 2.05 | 2.01 | 1.96 | 1.94 | 1.91 | 1.89 | 1.86 | 1.83 | 1.80 |
| 15 | 2.06 | 2.02 | 1.97 | 1.92 | 1.90 | 1.87 | 1.85 | 1.82 | 1.79 | 1.76 |
| 16 | 2.03 | 1.99 | 1.94 | 1.89 | 1.87 | 1.84 | 1.81 | 1.78 | 1.75 | 1.72 |
| 17 | 2.00 | 1.96 | 1.91 | 1.86 | 1.84 | 1.81 | 1.78 | 1.75 | 1.72 | 1.69 |
| 18 | 1.98 | 1.93 | 1.89 | 1.84 | 1.81 | 1.78 | 1.75 | 1.72 | 1.69 | 1.66 |
| 19 | 1.96 | 1.91 | 1.86 | 1.81 | 1.79 | 1.76 | 1.73 | 1.70 | 1.67 | 1.63 |
| 20 | 1.94 | 1.89 | 1.84 | 1.79 | 1.77 | 1.74 | 1.71 | 1.68 | 1.64 | 1.61 |
| 21 | 1.92 | 1.87 | 1.83 | 1.78 | 1.75 | 1.72 | 1.69 | 1.66 | 1.62 | 1.59 |
| 22 | 1.90 | 1.86 | 1.81 | 1.76 | 1.73 | 1.70 | 1.67 | 1.64 | 1.60 | 1.57 |
| 23 | 1.89 | 1.84 | 1.80 | 1.74 | 1.72 | 1.69 | 1.66 | 1.62 | 1.59 | 1.55 |
| 24 | 1.88 | 1.83 | 1.78 | 1.73 | 1.70 | 1.67 | 1.64 | 1.61 | 1.57 | 1.53 |
| 25 | 1.87 | 1.82 | 1.77 | 1.72 | 1.69 | 1.66 | 1.63 | 1.59 | 1.56 | 1.52 |
| 26 | 1.86 | 1.81 | 1.76 | 1.71 | 1.68 | 1.65 | 1.61 | 1.58 | 1.54 | 1.50 |
| 27 | 1.85 | 1.80 | 1.75 | 1.70 | 1.67 | 1.64 | 1.60 | 1.57 | 1.53 | 1.49 |
| 28 | 1.84 | 1.79 | 1.74 | 1.69 | 1.66 | 1.63 | 1.59 | 1.56 | 1.52 | 1.48 |
| 29 | 1.83 | 1.78 | 1.73 | 1.68 | 1.65 | 1.62 | 1.58 | 1.55 | 1.51 | 1.47 |
| 30 | 1.82 | 1.77 | 1.72 | 1.67 | 1.64 | 1.61 | 1.57 | 1.54 | 1.50 | 1.46 |
| 40 | 1.76 | 1.71 | 1.66 | 1.61 | 1.57 | 1.54 | 1.51 | 1.47 | 1.42 | 1.38 |
| 60 | 1.71 | 1.66 | 1.60 | 1.54 | 1.51 | 1.48 | 1.44 | 1.40 | 1.35 | 1.29 |
| 120 | 1.65 | 1.60 | 1.55 | 1.48 | 1.45 | 1.41 | 1.37 | 1.32 | 1.26 | 1.19 |
| ∞ | 1.60 | 1.55 | 1.49 | 1.42 | 1.38 | 1.34 | 1.30 | 1.24 | 1.17 | 1.00 |

(Row axis label: DENOMINATOR DEGREES OF FREEDOM)

## Critical Values of the *F*-Distribution ($\alpha$ = .05)

| $\nu_2$ \ $\nu_1$ | NUMERATOR DEGREES OF FREEDOM 1 | 2 | 3 | 4 | 5 | 6 | 7 | 8 | 9 |
|---|---|---|---|---|---|---|---|---|---|
| DENOMINATOR DEGREES OF FREEDOM 1 | 161.4 | 199.5 | 215.7 | 224.6 | 230.2 | 234.0 | 236.8 | 238.9 | 240.5 |
| 2 | 18.51 | 19.00 | 19.16 | 19.25 | 19.30 | 19.33 | 19.35 | 19.37 | 19.38 |
| 3 | 10.13 | 9.55 | 9.28 | 9.12 | 9.01 | 8.94 | 8.89 | 8.85 | 8.81 |
| 4 | 7.71 | 6.94 | 6.59 | 6.39 | 6.26 | 6.16 | 6.09 | 6.04 | 6.00 |
| 5 | 6.61 | 5.79 | 5.41 | 5.19 | 5.05 | 4.95 | 4.88 | 4.82 | 4.77 |
| 6 | 5.99 | 5.14 | 4.76 | 4.53 | 4.39 | 4.28 | 4.21 | 4.15 | 4.10 |
| 7 | 5.59 | 4.74 | 4.35 | 4.12 | 3.97 | 3.87 | 3.79 | 3.73 | 3.68 |
| 8 | 5.32 | 4.46 | 4.07 | 3.84 | 3.69 | 3.58 | 3.50 | 3.44 | 3.39 |
| 9 | 5.12 | 4.26 | 3.86 | 3.63 | 3.48 | 3.37 | 3.29 | 3.23 | 3.18 |
| 10 | 4.96 | 4.10 | 3.71 | 3.48 | 3.33 | 3.22 | 3.14 | 3.07 | 3.02 |
| 11 | 4.84 | 3.98 | 3.59 | 3.36 | 3.20 | 3.09 | 3.01 | 2.95 | 2.90 |
| 12 | 4.75 | 3.89 | 3.49 | 3.26 | 3.11 | 3.00 | 2.91 | 2.85 | 2.80 |
| 13 | 4.67 | 3.81 | 3.41 | 3.18 | 3.03 | 2.92 | 2.83 | 2.77 | 2.71 |
| 14 | 4.60 | 3.74 | 3.34 | 3.11 | 2.96 | 2.85 | 2.76 | 2.70 | 2.65 |
| 15 | 4.54 | 3.68 | 3.29 | 3.06 | 2.90 | 2.79 | 2.71 | 2.64 | 2.59 |
| 16 | 4.49 | 3.63 | 3.24 | 3.01 | 2.85 | 2.74 | 2.66 | 2.59 | 2.54 |
| 17 | 4.45 | 3.59 | 3.20 | 2.96 | 2.81 | 2.70 | 2.61 | 2.55 | 2.49 |
| 18 | 4.41 | 3.55 | 3.16 | 2.93 | 2.77 | 2.66 | 2.58 | 2.51 | 2.46 |
| 19 | 4.38 | 3.52 | 3.13 | 2.90 | 2.74 | 2.63 | 2.54 | 2.48 | 2.42 |
| 20 | 4.35 | 3.49 | 3.10 | 2.87 | 2.71 | 2.60 | 2.51 | 2.45 | 2.39 |
| 21 | 4.32 | 3.47 | 3.07 | 2.84 | 2.68 | 2.57 | 2.49 | 2.42 | 2.37 |
| 22 | 4.30 | 3.44 | 3.05 | 2.82 | 2.66 | 2.55 | 2.46 | 2.40 | 2.34 |
| 23 | 4.28 | 3.42 | 3.03 | 2.80 | 2.64 | 2.53 | 2.44 | 2.37 | 2.32 |
| 24 | 4.26 | 3.40 | 3.01 | 2.78 | 2.62 | 2.51 | 2.42 | 2.36 | 2.30 |
| 25 | 4.24 | 3.39 | 2.99 | 2.76 | 2.60 | 2.49 | 2.40 | 2.34 | 2.28 |
| 26 | 4.23 | 3.37 | 2.98 | 2.74 | 2.59 | 2.47 | 2.39 | 2.32 | 2.27 |
| 27 | 4.21 | 3.35 | 2.96 | 2.73 | 2.57 | 2.46 | 2.37 | 2.31 | 2.25 |
| 28 | 4.20 | 3.34 | 2.95 | 2.71 | 2.56 | 2.45 | 2.36 | 2.29 | 2.24 |
| 29 | 4.18 | 3.33 | 2.93 | 2.70 | 2.55 | 2.43 | 2.35 | 2.28 | 2.22 |
| 30 | 4.17 | 3.32 | 2.92 | 2.69 | 2.53 | 2.42 | 2.33 | 2.27 | 2.21 |
| 40 | 4.08 | 3.23 | 2.84 | 2.61 | 2.45 | 2.34 | 2.25 | 2.18 | 2.12 |
| 60 | 4.00 | 3.15 | 2.76 | 2.53 | 2.37 | 2.25 | 2.17 | 2.10 | 2.04 |
| 120 | 3.92 | 3.07 | 2.68 | 2.45 | 2.29 | 2.17 | 2.09 | 2.02 | 1.96 |
| $\infty$ | 3.84 | 3.00 | 2.60 | 2.37 | 2.21 | 2.10 | 2.01 | 1.94 | 1.88 |

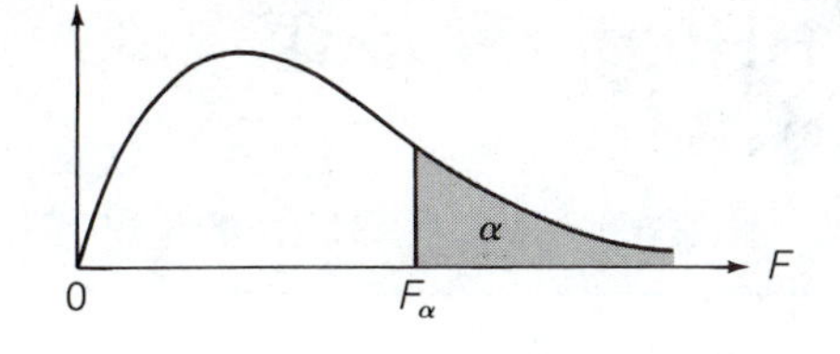


| $\nu_2$ \ $\nu_1$ | NUMERATOR DEGREES OF FREEDOM | | | | | | | | | |
|---|---|---|---|---|---|---|---|---|---|---|
| | 10 | 12 | 15 | 20 | 24 | 30 | 40 | 60 | 120 | ∞ |
| 1 | 241.9 | 243.9 | 245.9 | 248.0 | 249.1 | 250.1 | 251.1 | 252.2 | 253.3 | 254.3 |
| 2 | 19.40 | 19.41 | 19.43 | 19.45 | 19.45 | 19.46 | 19.47 | 19.48 | 19.49 | 19.50 |
| 3 | 8.79 | 8.74 | 8.70 | 8.66 | 8.64 | 8.62 | 8.59 | 8.57 | 8.55 | 8.53 |
| 4 | 5.96 | 5.91 | 5.86 | 5.80 | 5.77 | 5.75 | 5.72 | 5.69 | 5.66 | 5.63 |
| 5 | 4.74 | 4.68 | 4.62 | 4.56 | 4.53 | 4.50 | 4.46 | 4.43 | 4.40 | 4.36 |
| 6 | 4.06 | 4.00 | 3.94 | 3.87 | 3.84 | 3.81 | 3.77 | 3.74 | 3.70 | 3.67 |
| 7 | 3.64 | 3.57 | 3.51 | 3.44 | 3.41 | 3.38 | 3.34 | 3.30 | 3.27 | 3.23 |
| 8 | 3.35 | 3.28 | 3.22 | 3.15 | 3.12 | 3.08 | 3.04 | 3.01 | 2.97 | 2.93 |
| 9 | 3.14 | 3.07 | 3.01 | 2.94 | 2.90 | 2.86 | 2.83 | 2.79 | 2.75 | 2.71 |
| 10 | 2.98 | 2.91 | 2.85 | 2.77 | 2.74 | 2.70 | 2.66 | 2.62 | 2.58 | 2.54 |
| 11 | 2.85 | 2.79 | 2.72 | 2.65 | 2.61 | 2.57 | 2.53 | 2.49 | 2.45 | 2.40 |
| 12 | 2.75 | 2.69 | 2.62 | 2.54 | 2.51 | 2.47 | 2.43 | 2.38 | 2.34 | 2.30 |
| 13 | 2.67 | 2.60 | 2.53 | 2.46 | 2.42 | 2.38 | 2.34 | 2.30 | 2.25 | 2.21 |
| 14 | 2.60 | 2.53 | 2.46 | 2.39 | 2.35 | 2.31 | 2.27 | 2.22 | 2.18 | 2.13 |
| 15 | 2.54 | 2.48 | 2.40 | 2.33 | 2.29 | 2.25 | 2.20 | 2.16 | 2.11 | 2.07 |
| 16 | 2.49 | 2.42 | 2.35 | 2.28 | 2.24 | 2.19 | 2.15 | 2.11 | 2.06 | 2.01 |
| 17 | 2.45 | 2.38 | 2.31 | 2.23 | 2.19 | 2.15 | 2.10 | 2.06 | 2.01 | 1.96 |
| 18 | 2.41 | 2.34 | 2.27 | 2.19 | 2.15 | 2.11 | 2.06 | 2.02 | 1.97 | 1.92 |
| 19 | 2.38 | 2.31 | 2.23 | 2.16 | 2.11 | 2.07 | 2.03 | 1.98 | 1.93 | 1.88 |
| 20 | 2.35 | 2.28 | 2.20 | 2.12 | 2.08 | 2.04 | 1.99 | 1.95 | 1.90 | 1.84 |
| 21 | 2.32 | 2.25 | 2.18 | 2.10 | 2.05 | 2.01 | 1.96 | 1.92 | 1.87 | 1.81 |
| 22 | 2.30 | 2.23 | 2.15 | 2.07 | 2.03 | 1.98 | 1.94 | 1.89 | 1.84 | 1.78 |
| 23 | 2.27 | 2.20 | 2.13 | 2.05 | 2.01 | 1.96 | 1.91 | 1.86 | 1.81 | 1.76 |
| 24 | 2.25 | 2.18 | 2.11 | 2.03 | 1.98 | 1.94 | 1.89 | 1.84 | 1.79 | 1.73 |
| 25 | 2.24 | 2.16 | 2.09 | 2.01 | 1.96 | 1.92 | 1.87 | 1.82 | 1.77 | 1.71 |
| 26 | 2.22 | 2.15 | 2.07 | 1.99 | 1.95 | 1.90 | 1.85 | 1.80 | 1.75 | 1.69 |
| 27 | 2.20 | 2.13 | 2.06 | 1.97 | 1.93 | 1.88 | 1.84 | 1.79 | 1.73 | 1.67 |
| 28 | 2.19 | 2.12 | 2.04 | 1.96 | 1.91 | 1.87 | 1.82 | 1.77 | 1.71 | 1.65 |
| 29 | 2.18 | 2.10 | 2.03 | 1.94 | 1.90 | 1.85 | 1.81 | 1.75 | 1.70 | 1.64 |
| 30 | 2.16 | 2.09 | 2.01 | 1.93 | 1.89 | 1.84 | 1.79 | 1.74 | 1.68 | 1.62 |
| 40 | 2.08 | 2.00 | 1.92 | 1.84 | 1.79 | 1.74 | 1.69 | 1.64 | 1.58 | 1.51 |
| 60 | 1.99 | 1.92 | 1.84 | 1.75 | 1.70 | 1.65 | 1.59 | 1.53 | 1.47 | 1.39 |
| 120 | 1.91 | 1.83 | 1.75 | 1.66 | 1.61 | 1.55 | 1.50 | 1.43 | 1.35 | 1.25 |
| ∞ | 1.83 | 1.75 | 1.67 | 1.57 | 1.52 | 1.46 | 1.39 | 1.32 | 1.22 | 1.00 |

DENOMINATOR DEGREES OF FREEDOM

## Critical Values of the *F*-Distribution ($\alpha = .01$)

| $\nu_2$ \ $\nu_1$ | NUMERATOR DEGREES OF FREEDOM 1 | 2 | 3 | 4 | 5 | 6 | 7 | 8 | 9 |
|---|---|---|---|---|---|---|---|---|---|
| 1 | 4,052 | 4,999.5 | 5,403 | 5,625 | 5,764 | 5,859 | 5,928 | 5,982 | 6,022 |
| 2 | 98.50 | 99.00 | 99.17 | 99.25 | 99.30 | 99.33 | 99.36 | 99.37 | 99.39 |
| 3 | 34.12 | 30.82 | 29.46 | 28.71 | 28.24 | 27.91 | 27.67 | 27.49 | 27.35 |
| 4 | 21.20 | 18.00 | 16.69 | 15.98 | 15.52 | 15.21 | 14.98 | 14.80 | 14.66 |
| 5 | 16.26 | 13.27 | 12.06 | 11.39 | 10.97 | 10.67 | 10.46 | 10.29 | 10.16 |
| 6 | 13.75 | 10.92 | 9.78 | 9.15 | 8.75 | 8.47 | 8.26 | 8.10 | 7.98 |
| 7 | 12.25 | 9.55 | 8.45 | 7.85 | 7.46 | 7.19 | 6.99 | 6.84 | 6.72 |
| 8 | 11.26 | 8.65 | 7.59 | 7.01 | 6.63 | 6.37 | 6.18 | 6.03 | 5.91 |
| 9 | 10.56 | 8.02 | 6.99 | 6.42 | 6.06 | 5.80 | 5.61 | 5.47 | 5.35 |
| 10 | 10.04 | 7.56 | 6.55 | 5.99 | 5.64 | 5.39 | 5.20 | 5.06 | 4.94 |
| 11 | 9.65 | 7.21 | 6.22 | 5.67 | 5.32 | 5.07 | 4.89 | 4.74 | 4.63 |
| 12 | 9.33 | 6.93 | 5.95 | 5.41 | 5.06 | 4.82 | 4.64 | 4.50 | 4.39 |
| 13 | 9.07 | 6.70 | 5.74 | 5.21 | 4.86 | 4.62 | 4.44 | 4.30 | 4.19 |
| 14 | 8.86 | 6.51 | 5.56 | 5.04 | 4.69 | 4.46 | 4.28 | 4.14 | 4.03 |
| 15 | 8.68 | 6.36 | 5.42 | 4.89 | 4.56 | 4.32 | 4.14 | 4.00 | 3.89 |
| 16 | 8.53 | 6.23 | 5.29 | 4.77 | 4.44 | 4.20 | 4.03 | 3.89 | 3.78 |
| 17 | 8.40 | 6.11 | 5.18 | 4.67 | 4.34 | 4.10 | 3.93 | 3.79 | 3.68 |
| 18 | 8.29 | 6.01 | 5.09 | 4.58 | 4.25 | 4.01 | 3.84 | 3.71 | 3.60 |
| 19 | 8.18 | 5.93 | 5.01 | 4.50 | 4.17 | 3.94 | 3.77 | 3.63 | 3.52 |
| 20 | 8.10 | 5.85 | 4.94 | 4.43 | 4.10 | 3.87 | 3.70 | 3.56 | 3.46 |
| 21 | 8.02 | 5.78 | 4.87 | 4.37 | 4.04 | 3.81 | 3.64 | 3.51 | 3.40 |
| 22 | 7.95 | 5.72 | 4.82 | 4.31 | 3.99 | 3.76 | 3.59 | 3.45 | 3.35 |
| 23 | 7.88 | 5.66 | 4.76 | 4.26 | 3.94 | 3.71 | 3.54 | 3.41 | 3.30 |
| 24 | 7.82 | 5.61 | 4.72 | 4.22 | 3.90 | 3.67 | 3.50 | 3.36 | 3.26 |
| 25 | 7.77 | 5.57 | 4.68 | 4.18 | 3.85 | 3.63 | 3.46 | 3.32 | 3.22 |
| 26 | 7.72 | 5.53 | 4.64 | 4.14 | 3.82 | 3.59 | 3.42 | 3.29 | 3.18 |
| 27 | 7.68 | 5.49 | 4.60 | 4.11 | 3.78 | 3.56 | 3.39 | 3.26 | 3.15 |
| 28 | 7.64 | 5.45 | 4.57 | 4.07 | 3.75 | 3.53 | 3.36 | 3.23 | 3.12 |
| 29 | 7.60 | 5.42 | 4.54 | 4.04 | 3.73 | 3.50 | 3.33 | 3.20 | 3.09 |
| 30 | 7.56 | 5.39 | 4.51 | 4.02 | 3.70 | 3.47 | 3.30 | 3.17 | 3.07 |
| 40 | 7.31 | 5.18 | 4.31 | 3.83 | 3.51 | 3.29 | 3.12 | 2.99 | 2.89 |
| 60 | 7.08 | 4.98 | 4.13 | 3.65 | 3.34 | 3.12 | 2.95 | 2.82 | 2.72 |
| 120 | 6.85 | 4.79 | 3.95 | 3.48 | 3.17 | 2.96 | 2.79 | 2.66 | 2.56 |
| ∞ | 6.63 | 4.61 | 3.78 | 3.32 | 3.02 | 2.80 | 2.64 | 2.51 | 2.41 |

DENOMINATOR DEGREES OF FREEDOM

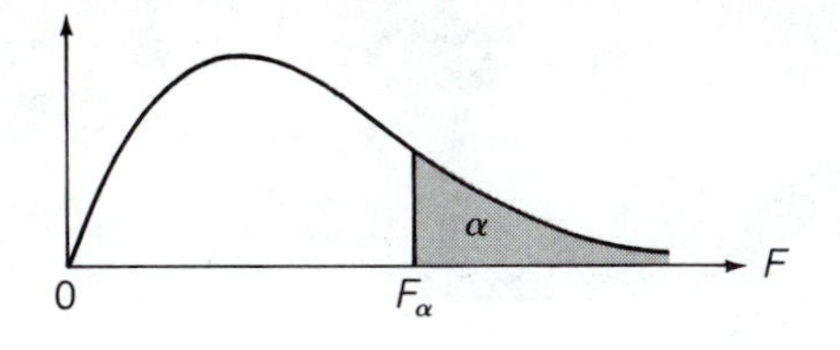


| $\nu_2$ \ $\nu_1$ | NUMERATOR DEGREES OF FREEDOM 10 | 12 | 15 | 20 | 24 | 30 | 40 | 60 | 120 | ∞ |
|---|---|---|---|---|---|---|---|---|---|---|
| DENOMINATOR DEGREES OF FREEDOM 1 | 6,056 | 6,106 | 6,157 | 6,209 | 6,235 | 6,261 | 6,287 | 6,313 | 6,339 | 6,366 |
| 2 | 99.40 | 99.42 | 99.43 | 99.45 | 99.46 | 99.47 | 99.47 | 99.48 | 99.49 | 99.50 |
| 3 | 27.23 | 27.05 | 26.87 | 26.69 | 26.60 | 26.50 | 26.41 | 26.32 | 26.22 | 26.13 |
| 4 | 14.55 | 14.37 | 14.20 | 14.02 | 13.93 | 13.84 | 13.75 | 13.65 | 13.56 | 13.46 |
| 5 | 10.05 | 9.89 | 9.72 | 9.55 | 9.47 | 9.38 | 9.29 | 9.20 | 9.11 | 9.02 |
| 6 | 7.87 | 7.72 | 7.56 | 7.40 | 7.31 | 7.23 | 7.14 | 7.06 | 6.97 | 6.88 |
| 7 | 6.62 | 6.47 | 6.31 | 6.16 | 6.07 | 5.99 | 5.91 | 5.82 | 5.74 | 5.65 |
| 8 | 5.81 | 5.67 | 5.52 | 5.36 | 5.28 | 5.20 | 5.12 | 5.03 | 4.95 | 4.86 |
| 9 | 5.26 | 5.11 | 4.96 | 4.81 | 4.73 | 4.65 | 4.57 | 4.48 | 4.40 | 4.31 |
| 10 | 4.85 | 4.71 | 4.56 | 4.41 | 4.33 | 4.25 | 4.17 | 4.08 | 4.00 | 3.91 |
| 11 | 4.54 | 4.40 | 4.25 | 4.10 | 4.02 | 3.94 | 3.86 | 3.78 | 3.69 | 3.60 |
| 12 | 4.30 | 4.16 | 4.01 | 3.86 | 3.78 | 3.70 | 3.62 | 3.54 | 3.45 | 3.36 |
| 13 | 4.10 | 3.96 | 3.82 | 3.66 | 3.59 | 3.51 | 3.43 | 3.34 | 3.25 | 3.17 |
| 14 | 3.94 | 3.80 | 3.66 | 3.51 | 3.43 | 3.35 | 3.27 | 3.18 | 3.09 | 3.00 |
| 15 | 3.80 | 3.67 | 3.52 | 3.37 | 3.29 | 3.21 | 3.13 | 3.05 | 2.96 | 2.87 |
| 16 | 3.69 | 3.55 | 3.41 | 3.26 | 3.18 | 3.10 | 3.02 | 2.93 | 2.84 | 2.75 |
| 17 | 3.59 | 3.46 | 3.31 | 3.16 | 3.08 | 3.00 | 2.92 | 2.83 | 2.75 | 2.65 |
| 18 | 3.51 | 3.37 | 3.23 | 3.08 | 3.00 | 2.92 | 2.84 | 2.75 | 2.66 | 2.57 |
| 19 | 3.43 | 3.30 | 3.15 | 3.00 | 2.92 | 2.84 | 2.76 | 2.67 | 2.58 | 2.49 |
| 20 | 3.37 | 3.23 | 3.09 | 2.94 | 2.86 | 2.78 | 2.69 | 2.61 | 2.52 | 2.42 |
| 21 | 3.31 | 3.17 | 3.03 | 2.88 | 2.80 | 2.72 | 2.64 | 2.55 | 2.46 | 2.36 |
| 22 | 3.26 | 3.12 | 2.98 | 2.83 | 2.75 | 2.67 | 2.58 | 2.50 | 2.40 | 2.31 |
| 23 | 3.21 | 3.07 | 2.93 | 2.78 | 2.70 | 2.62 | 2.54 | 2.45 | 2.35 | 2.26 |
| 24 | 3.17 | 3.03 | 2.89 | 2.74 | 2.66 | 2.58 | 2.49 | 2.40 | 2.31 | 2.21 |
| 25 | 3.13 | 2.99 | 2.85 | 2.70 | 2.62 | 2.54 | 2.45 | 2.36 | 2.27 | 2.17 |
| 26 | 3.09 | 2.96 | 2.81 | 2.66 | 2.58 | 2.50 | 2.42 | 2.33 | 2.23 | 2.13 |
| 27 | 3.06 | 2.93 | 2.78 | 2.63 | 2.55 | 2.47 | 2.38 | 2.29 | 2.20 | 2.10 |
| 28 | 3.03 | 2.90 | 2.75 | 2.60 | 2.52 | 2.44 | 2.35 | 2.26 | 2.17 | 2.06 |
| 29 | 3.00 | 2.87 | 2.73 | 2.57 | 2.49 | 2.41 | 2.33 | 2.23 | 2.14 | 2.03 |
| 30 | 2.98 | 2.84 | 2.70 | 2.55 | 2.47 | 2.39 | 2.30 | 2.21 | 2.11 | 2.01 |
| 40 | 2.80 | 2.66 | 2.52 | 2.37 | 2.29 | 2.20 | 2.11 | 2.02 | 1.92 | 1.80 |
| 60 | 2.63 | 2.50 | 2.35 | 2.20 | 2.12 | 2.03 | 1.94 | 1.84 | 1.73 | 1.60 |
| 120 | 2.47 | 2.34 | 2.19 | 2.03 | 1.95 | 1.86 | 1.76 | 1.66 | 1.53 | 1.38 |
| ∞ | 2.32 | 2.18 | 2.04 | 1.88 | 1.79 | 1.70 | 1.59 | 1.47 | 1.32 | 1.00 |

# APPENDIX II

## Durbin–Watson Test ($\alpha = .05$)

| $n$ | $k = 1$ | | $k = 2$ | | $k = 3$ | | $k = 4$ | | $k = 5$ | |
|---|---|---|---|---|---|---|---|---|---|---|
| | $d_L$ | $d_U$ | $d_L$ | $d_U$ | $d_L$ | $d_U$ | $d_L$ | $d_U$ | $d_L$ | $d_U$ |
| 15 | 1.08 | 1.36 | 0.95 | 1.54 | 0.82 | 1.75 | 0.69 | 1.97 | 0.56 | 2.21 |
| 16 | 1.10 | 1.37 | 0.98 | 1.54 | 0.86 | 1.73 | 0.74 | 1.93 | 0.62 | 2.15 |
| 17 | 1.13 | 1.38 | 1.02 | 1.54 | 0.90 | 1.71 | 0.78 | 1.90 | 0.67 | 2.10 |
| 18 | 1.16 | 1.39 | 1.05 | 1.53 | 0.93 | 1.69 | 0.82 | 1.87 | 0.71 | 2.06 |
| 19 | 1.18 | 1.40 | 1.08 | 1.53 | 0.97 | 1.68 | 0.86 | 1.85 | 0.75 | 2.02 |
| 20 | 1.20 | 1.41 | 1.10 | 1.54 | 1.00 | 1.68 | 0.90 | 1.83 | 0.79 | 1.99 |
| 21 | 1.22 | 1.42 | 1.13 | 1.54 | 1.03 | 1.67 | 0.93 | 1.81 | 0.83 | 1.96 |
| 22 | 1.24 | 1.43 | 1.15 | 1.54 | 1.05 | 1.66 | 0.96 | 1.80 | 0.86 | 1.94 |
| 23 | 1.26 | 1.44 | 1.17 | 1.54 | 1.08 | 1.66 | 0.99 | 1.79 | 0.90 | 1.92 |
| 24 | 1.27 | 1.45 | 1.19 | 1.55 | 1.10 | 1.66 | 1.01 | 1.78 | 0.93 | 1.90 |
| 25 | 1.29 | 1.45 | 1.21 | 1.55 | 1.12 | 1.66 | 1.04 | 1.77 | 0.95 | 1.89 |
| 26 | 1.30 | 1.46 | 1.22 | 1.55 | 1.14 | 1.65 | 1.06 | 1.76 | 0.98 | 1.88 |
| 27 | 1.32 | 1.47 | 1.24 | 1.56 | 1.16 | 1.65 | 1.08 | 1.76 | 1.01 | 1.86 |
| 28 | 1.33 | 1.48 | 1.26 | 1.56 | 1.18 | 1.65 | 1.10 | 1.75 | 1.03 | 1.85 |
| 29 | 1.34 | 1.48 | 1.27 | 1.56 | 1.20 | 1.65 | 1.12 | 1.74 | 1.05 | 1.84 |
| 30 | 1.35 | 1.49 | 1.28 | 1.57 | 1.21 | 1.65 | 1.14 | 1.74 | 1.07 | 1.83 |
| 31 | 1.36 | 1.50 | 1.30 | 1.57 | 1.23 | 1.65 | 1.16 | 1.74 | 1.09 | 1.83 |
| 32 | 1.37 | 1.50 | 1.31 | 1.57 | 1.24 | 1.65 | 1.18 | 1.73 | 1.11 | 1.82 |
| 33 | 1.38 | 1.51 | 1.32 | 1.58 | 1.26 | 1.65 | 1.19 | 1.73 | 1.13 | 1.81 |
| 34 | 1.39 | 1.51 | 1.33 | 1.58 | 1.27 | 1.65 | 1.21 | 1.73 | 1.15 | 1.81 |
| 35 | 1.40 | 1.52 | 1.34 | 1.58 | 1.28 | 1.65 | 1.22 | 1.73 | 1.16 | 1.80 |
| 36 | 1.41 | 1.52 | 1.35 | 1.59 | 1.29 | 1.65 | 1.24 | 1.73 | 1.18 | 1.80 |
| 37 | 1.42 | 1.53 | 1.36 | 1.59 | 1.31 | 1.66 | 1.25 | 1.72 | 1.19 | 1.80 |
| 38 | 1.43 | 1.54 | 1.37 | 1.59 | 1.32 | 1.66 | 1.26 | 1.72 | 1.21 | 1.79 |
| 39 | 1.43 | 1.54 | 1.38 | 1.60 | 1.33 | 1.66 | 1.27 | 1.72 | 1.22 | 1.79 |
| 40 | 1.44 | 1.54 | 1.39 | 1.60 | 1.34 | 1.66 | 1.29 | 1.72 | 1.23 | 1.79 |
| 45 | 1.48 | 1.57 | 1.43 | 1.62 | 1.38 | 1.67 | 1.34 | 1.72 | 1.29 | 1.78 |
| 50 | 1.50 | 1.59 | 1.46 | 1.63 | 1.42 | 1.67 | 1.38 | 1.72 | 1.34 | 1.77 |
| 55 | 1.53 | 1.60 | 1.49 | 1.64 | 1.45 | 1.68 | 1.41 | 1.72 | 1.38 | 1.77 |
| 60 | 1.55 | 1.62 | 1.51 | 1.65 | 1.48 | 1.69 | 1.44 | 1.73 | 1.41 | 1.77 |
| 65 | 1.57 | 1.63 | 1.54 | 1.66 | 1.50 | 1.70 | 1.47 | 1.73 | 1.44 | 1.77 |
| 70 | 1.58 | 1.64 | 1.55 | 1.67 | 1.52 | 1.70 | 1.49 | 1.74 | 1.46 | 1.77 |
| 75 | 1.60 | 1.65 | 1.57 | 1.68 | 1.54 | 1.71 | 1.51 | 1.74 | 1.49 | 1.77 |
| 80 | 1.61 | 1.66 | 1.59 | 1.69 | 1.56 | 1.72 | 1.53 | 1.74 | 1.51 | 1.77 |
| 85 | 1.62 | 1.67 | 1.60 | 1.70 | 1.57 | 1.72 | 1.55 | 1.75 | 1.52 | 1.77 |
| 90 | 1.63 | 1.68 | 1.61 | 1.70 | 1.59 | 1.73 | 1.57 | 1.75 | 1.54 | 1.78 |
| 95 | 1.64 | 1.69 | 1.62 | 1.71 | 1.60 | 1.73 | 1.58 | 1.75 | 1.56 | 1.78 |
| 100 | 1.65 | 1.69 | 1.63 | 1.72 | 1.61 | 1.74 | 1.59 | 1.76 | 1.57 | 1.78 |

*Source:* From J. Durbin and G. S. Watson (1951). "Testing for Serial Correlation in Least Squares Regression, II." *Biometrika* 38:159–178. Used by permission of the *Biometrika* trustees.

# Index